SAXON® GEOMETRY

Reteaching Masters

SAXON®

An Imprint of HMH Supplemental Publishers

www.SaxonPublishers.com
1-800-531-5015

ISBN 13: 978-1-6027-7560-2

ISBN 10: 1-6027-7560-5

Printed in the United States of America

6 2026
4500930133

Contents

Reteaching Masters

Name ______________________ Date __________ Class __________

Reteaching 1

Points, Lines, and Planes

Points, lines, and planes are the building blocks for all other geometric figures. Both lines and planes extend through space forever.

Use the diagram on the right to answer the following questions.

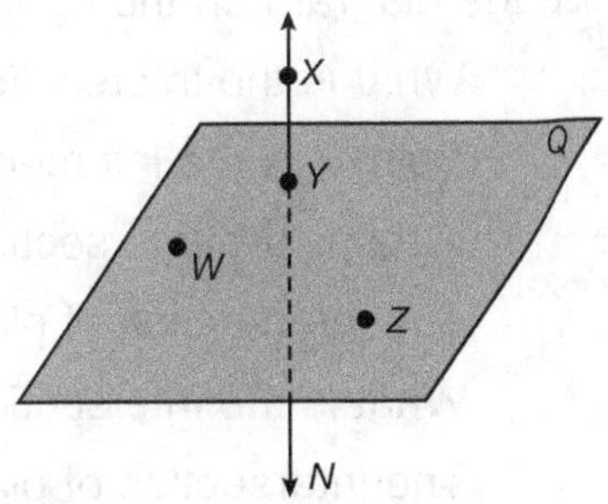

a. Give two different names for the line.

Two possible names for the line are line *n* and $\overleftrightarrow{XY}$.

b. Name two collinear points and three noncollinear points.

Points *X* and *Y* are collinear, and points *X*, *Y*, and *Z* are noncollinear.

c. Give two different names for the plane.

Two possible names for the plane are plane *WYZ* and plane *Q*.

Practice

Use the diagram on the right to complete questions 1–5.

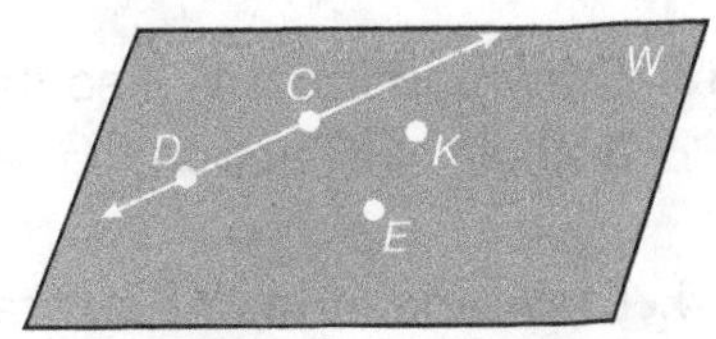

1. Label the plane as plane *W*.
2. Draw and label coplanar points *C*, *D*, and *E* on the plane.
3. Draw line *CD* on the plane.
4. Place point *K* on the plane so that it is collinear with point *C* but not on $\overleftrightarrow{CD}$.
5. What is another name for plane *W*?

____

Use the diagram on the right to complete questions 6–10.

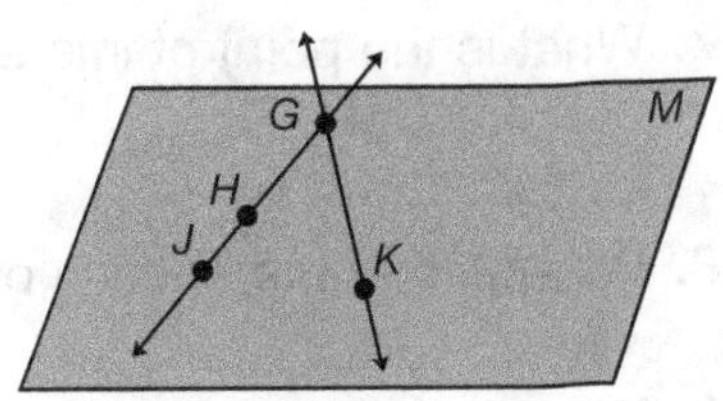

6. Give another name for plane *M*. ______________
7. Name three collinear points. ______________
8. Give three points that are noncollinear. ______________
9. Are any of the points noncoplanar? ___
10. Name two coplanar lines. ______________

Two lines intersect at exactly one point. When two planes intersect, their intersection is an infinite number of points and creates a line. When a plane and a line intersect, their intersection may be just a point or the entire line.

Use the diagram on the right to answer the following questions.

a. What is the intersection of $\overleftrightarrow{MN}$ and $\overleftrightarrow{PQ}$

Point L is the intersection of $\overleftrightarrow{MN}$ and $\overleftrightarrow{PQ}$.

b. What is the intersection of planes S and T?

The intersection of planes S and T is $\overleftrightarrow{MN}$.

c. What is the intersection of plane T and $\overleftrightarrow{PQ}$?

The intersection of plane T and is $\overleftrightarrow{PQ}$.

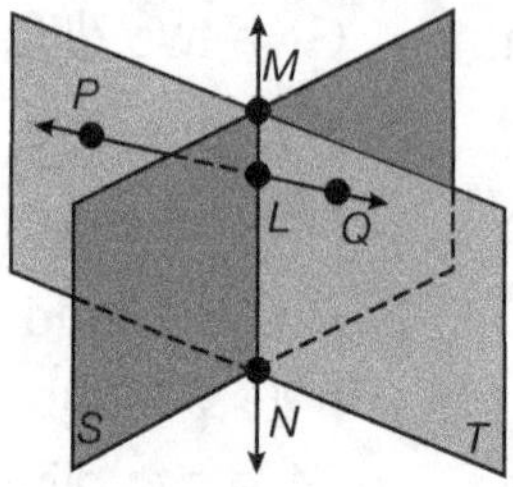

Practice

Use the diagram on the right to complete questions 15–19.

11. Label the planes A and B.

12. Draw and label the intersection of planes A and B as $\overleftrightarrow{GH}$.

13. Draw and label $\overleftrightarrow{WX}$ so that it intersects plane A at every point and plane B at point G.

14. Draw and label $\overleftrightarrow{YZ}$ so that it intersects $\overleftrightarrow{WX}$ at point G. The two lines are not on the same plane.

15. Draw and label point T so that it is coplanar with points Y and Z.

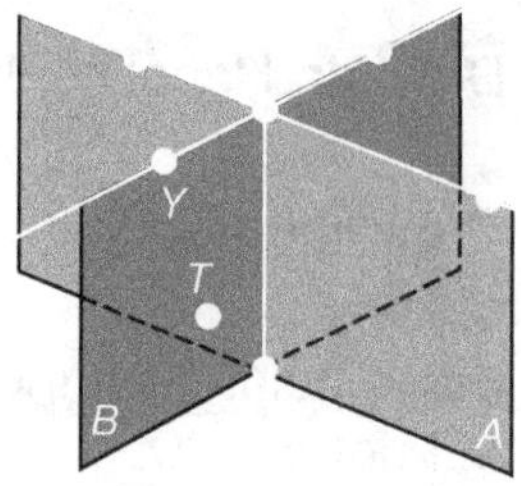

Identify each of the following from the diagram.

16. What is the intersection of plane J and $\overleftrightarrow{QR}$? ________

17. What is the point of intersection of line P and $\overleftrightarrow{MN}$?

____

18. What is the intersection of planes J and K? ______

19. What is the intersection of $\overleftrightarrow{QR}$ and $\overleftrightarrow{QS}$? ________

20. What is the intersection of plane K and $\overleftrightarrow{MN}$?

____

21. What is the intersection of plane J and $\overleftrightarrow{MN}$? ___

22. What is the intersection of plane K and $\overleftrightarrow{QR}$? ___

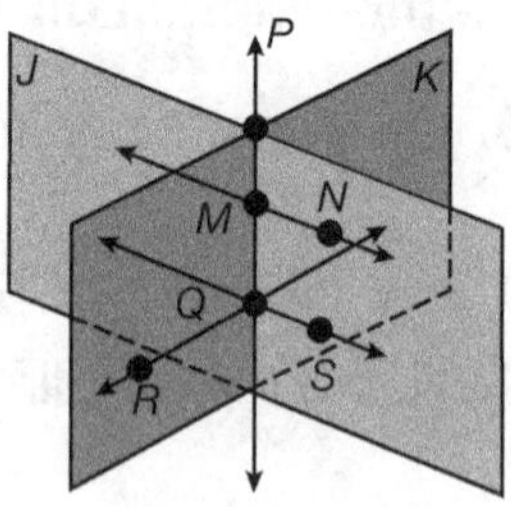

Name ______________________ Date __________ Class __________

Reteaching 2
Segments

You have learned that a line is a straight path that extends forever. Now you will work with line segments.

Find each distance.

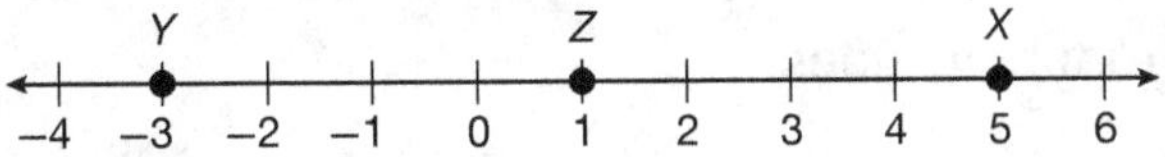

a. *XY*

Point $X = 5$ and point $Y = -3$.

So $|\text{Point } X - \text{Point } Y| = |5 - (-3)| = |8| = 8$.

b. *YZ*

Point $Y = -3$, and point $Z = 1$.

So $|\text{Point } Y - \text{Point } Z| = |(-3) - 1| = |-4| = 4$.

Practice

Complete the steps to find each distance.

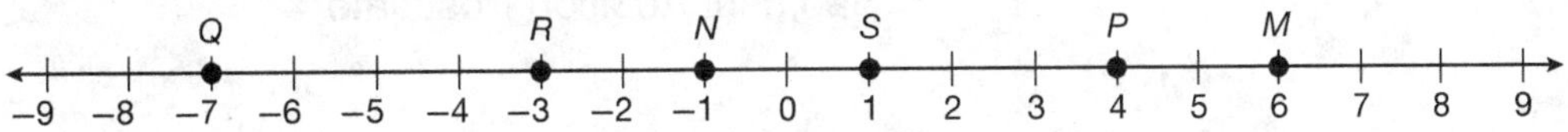

1. Find *PQ*.

$|\text{Point } P - \text{Point } Q| = |4 - \_\_\_\_| = |\_\_| \_\_$

2. Find *NS*.

$|\text{Point } N - \text{Point } S| = |\_\_\_\_ -1| = |\_\_\_| = \_$

3. Find *MR*.

$|\text{Point } M - \text{Point } R| = |\_ - (-3)| = |\_| = \_$

Find each distance.

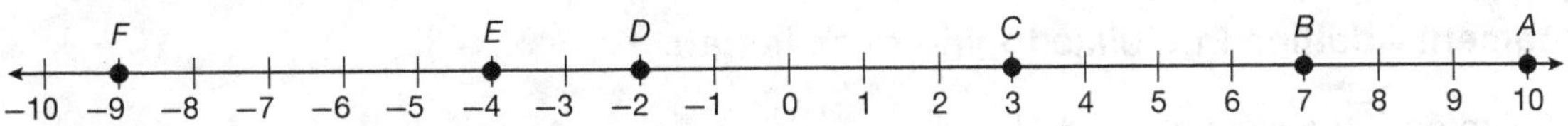

4. *AB* _ **5.** *BD* _ **6.** *DE* _

7. *DC* _ **8.** *FD* _ **9.** *AF* __

Use the Segment Addition Postulate to find each length.

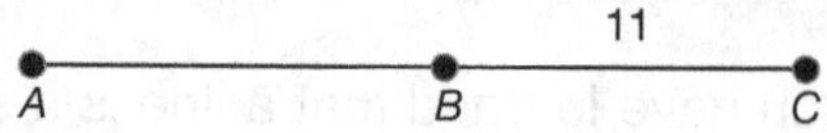

a. Find AB if $AC = 28$ and $BC = 11$.

$AB + BC = AC$	Segment Addition Postulate
$AB + 11 = 28$	Substitute.
$AB + 11 - 11 = 28 - 11$	Subtract 11 from both sides.
$AB = 17$	Simplify.

b. Find PR in terms of x.

$PR = PQ + QR$	Segment Addition Postulate
$PR = (3x + 4) + (4x - 1)$	Substitute.
$PR = 7x + 3$	Simplify.

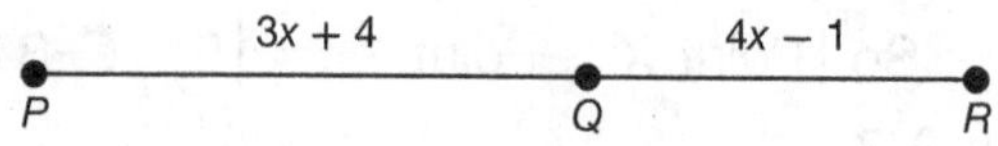

Practice

Complete the steps to find each length.

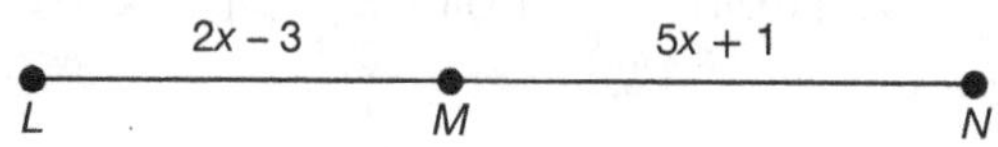

10. Find LN in terms of x.

$LN =$ ___ + ___ Segment Addition Postulate

$LN = (2x - 3) + ($______$)$ __________

$LN =$ ______ Simplify.

11. Point G lies on $\overline{FH}$ between F and H.

Find GH if $FG = 15$ and $FH = 34$.

$FH = FG + GH$ __________________

$34 =$ ___ $+ GH$ Substitute.

$34 -$ ___ $= 15 + GH -$ ___ __________________

___ $= GH$ Simplify.

Use the Segment Addition Postulate to find each length.

12. Point M lies on $\overline{LN}$ between L and N.

Find MN if $LN = 32$ and $LM = 14$. ___

13. Find WY in terms of n. __________

14. Point D lies on $\overline{CF}$ between C and F.

Find CD if $DF = 22$ and $CF = 45$. ___

Name ______________________ Date ____________ Class ____________

Reteaching 3
Angles

You have learned about lines and line segments. Now you will learn about rays and angles.

An **acute angle** measures greater than 0° and less than 90°.

An **obtuse angle** measures greater than 90° and less than 180°.

A **right angle** measures exactly 90°.

A **straight angle** measures exactly 180°.

Classify each angle and use a protractor to find its measure.

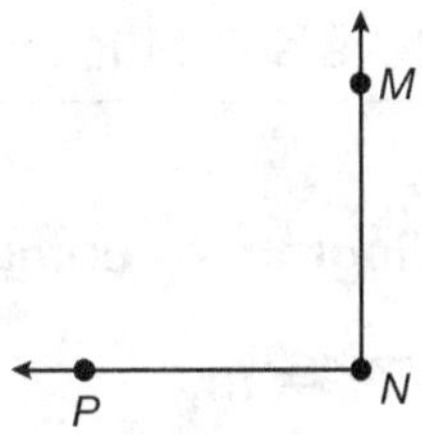

a. ∠*MNP*

Angle *MNP* is a right angle with a measure of 90°.

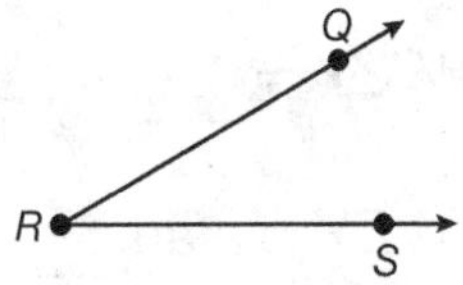

b. ∠*QRS*

Angle *QRS* is an acute angle with a measure of 30°.

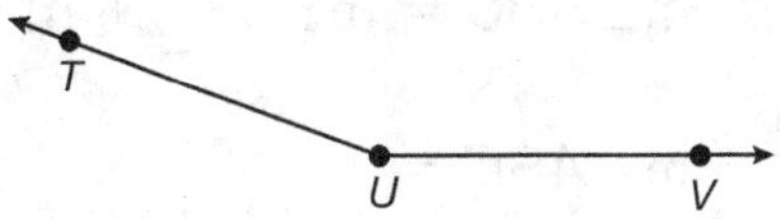

c. ∠*TUV*

Angle *TUV* is an obtuse angle with a measure of 160°.

Practice

Measure each angle and complete each statement.

1. Angle *QRT* is a/an ______ angle with a measure of 57°.

2. Angle *TRS* is an obtuse angle with a measure of ______.

3. Angle *QRS* is a/an ______ angle with a measure of 180°.

Give two names for each angle. Classify the angle and use a protractor to find its measure.

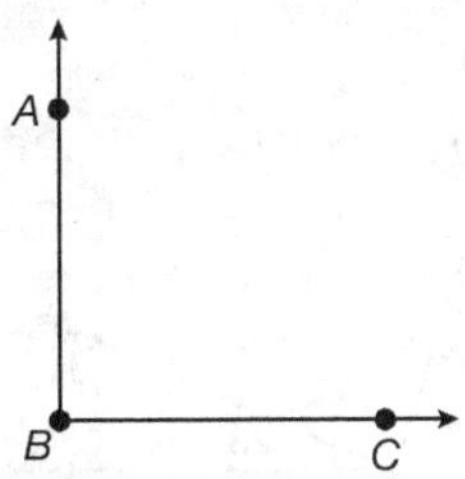

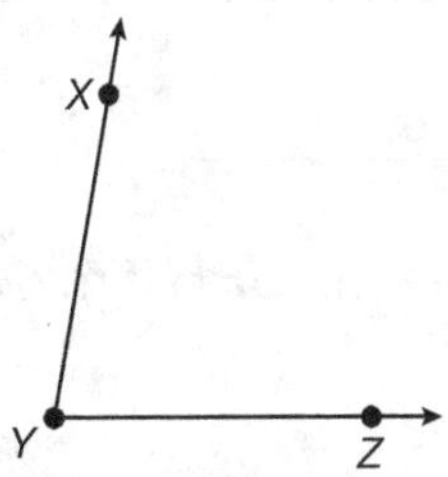

4. ______________________

5. ______________________

The measure of $\angle KLN$ is 42° and the measure of $\angle NLJ$ is 83°.
Find m$\angle KLJ$ and then classify the angle.

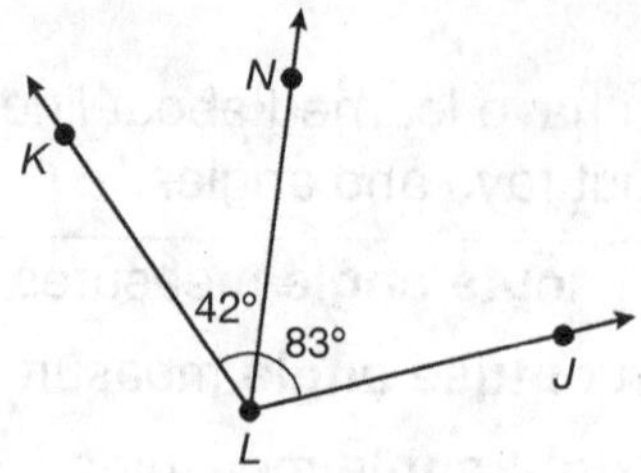

m$\angle KLJ$ = m$\angle KLN$ + m$\angle NLJ$ — Angle Addition Postulate

m$\angle KLJ$ = 42° + 83° — Substitute.

m$\angle KLJ$ = 125° — Simplify.

Angle *KLJ* is an obtuse angle.

Practice

Use the diagram to complete each statement.

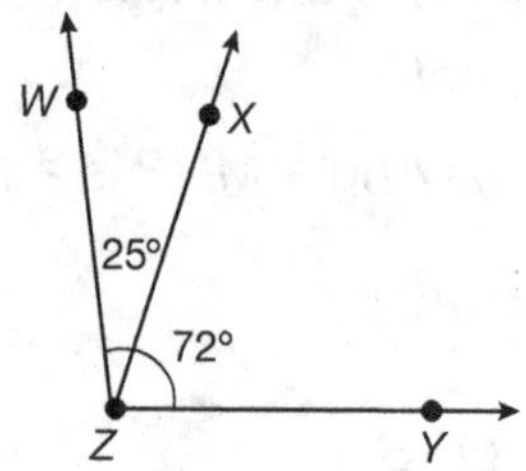

6. m$\angle WZY$ = m$\angle$____ + m$\angle$____ — Angle Addition Postulate

m$\angle WZY$ = ___ + 72° — ____________

m$\angle WZY$ = ___ — Simplify.

m$\angle WZY$ is a/an ______ angle.

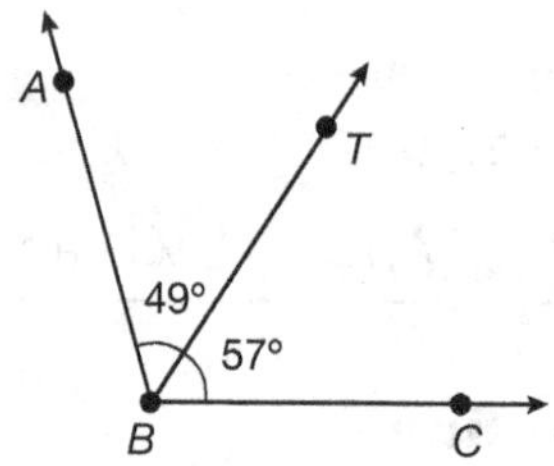

7. m$\angle ABC$ = m$\angle$____ + m$\angle$____ — Angle Addition Postulate

m$\angle ABC$ = ___ + ___ — Substitute.

m$\angle ABC$ = ____ — __________

$\angle ABC$ is a/an ______ angle.

Use the diagram to classify and find the measure of each angle.

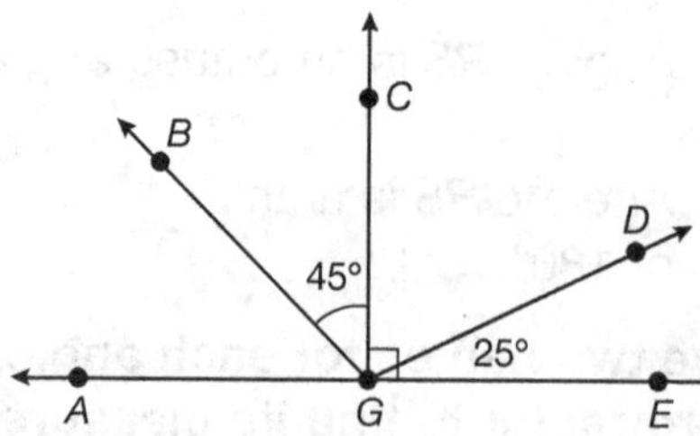

8. $\angle AGC$ __________

9. $\angle AGB$ __________

10. $\angle CGD$ __________

11. $\angle BGD$ __________

12. $\angle AGD$ __________

13. Determine the measure m$\angle JFH$ if m$\angle KFG$ is 125°.

m$\angle JFH$ = ___

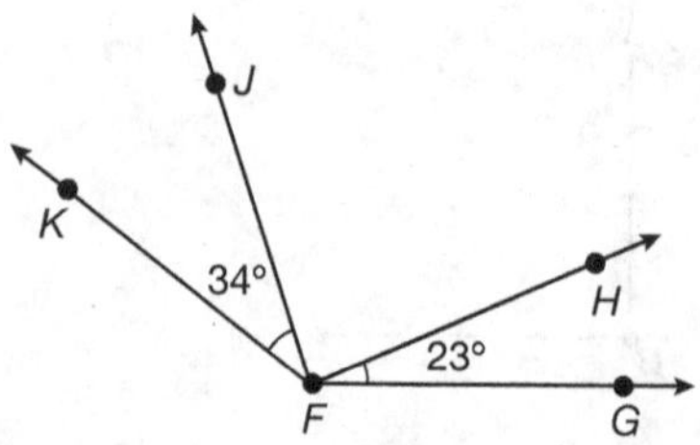

Name ______________________________ Date __________ Class __________

Reteaching 4

Postulates and Theorems About Points, Lines, and Planes

You have learned about points, lines, and planes. Now you will learn about the postulates and theorems that explain the relationships between and among points, lines, and planes.

Name the following.

a. five points
   V, W, X, Y, Z

b. two planes
   planes *G* and *H*

c. two lines
   $\overleftrightarrow{XY}$ and $\overleftrightarrow{WZ}$

d. four coplanar points
   W, *X*, *Y*, and *Z*

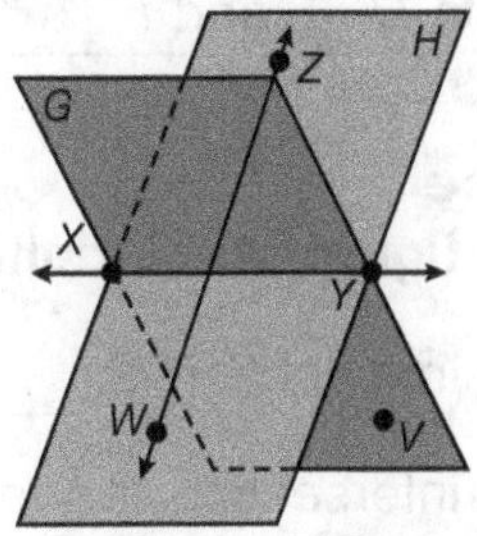

Practice

Complete the following statements.

1. Through any ____ points there is exactly one line.

2. Through any _____ noncollinear points there exists exactly one _____.

3. Give three conditions for defining a plane. Draw a figure to display each condition.

a. ______________________

b. ______________________________

c. ____________________

a. Identify the intersection of planes P and Q.
   The intersection of planes P and Q is $\overleftrightarrow{EG}$.
b. Identify a point of intersection of plane P and $\overleftrightarrow{CD}$.
   Point F
c. Identify all points of intersection of lines on plane Q.
   Points D, E, and F

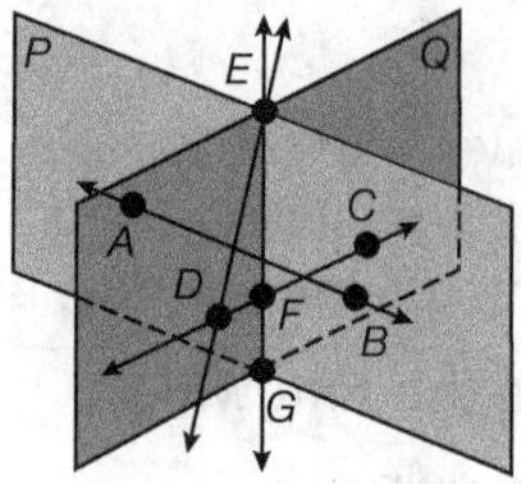

Practice

Use the figure at the right to complete problems 4–8.

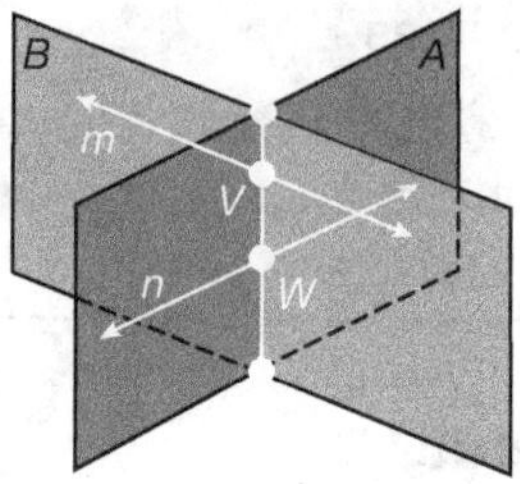

4. The intersection of ________ and ________ is $\overleftrightarrow{XY}$.

5. The intersection of line __ and line __ is point V.

6. The intersection of line __ and line __ is point W.

7. Plane A and line m intersect at ________.

8. Plane __ and line __ intersect at point W.

9. Plane A and line n intersect at ______.

10. Line m and plane B intersect at ______.

Complete the following statements.

11. If two planes intersect, then their intersection is ______.

12. ____ ______ define a line, _____ ______ points define a plane, and ____ noncoplanar ______ define space.

Use the figure to answer problems 13–19

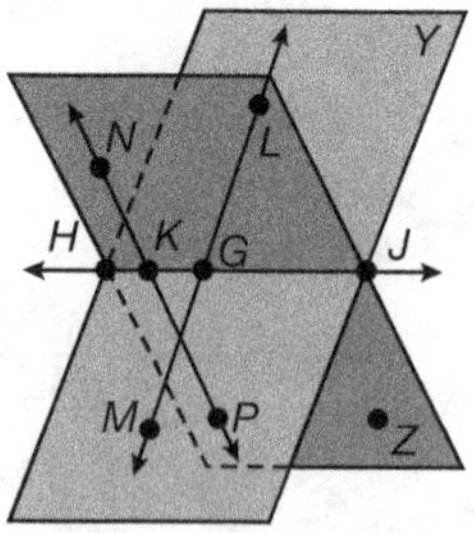

13. What is the intersection of planes Y and Z? ________

14. Identify a line on plane Z. ___

15. Identify a line that intersects $\overleftrightarrow{HJ}$. What is the intersection?

____

16. Are points P and L coplanar? __

17. Identify three points that are coplanar but not collinear.

____

18. What is the intersection of plane Z and $\overleftrightarrow{LM}$? ________

19. What is the intersection of plane Z and $\overleftrightarrow{NP}$? ___

Name ______________________ Date __________ Class __________

Reteaching 5

More Theorems About Lines and Planes

You will learn about special relationships of lines and planes.

Both planes and lines can be **parallel** or **perpendicular**. Lines can have one additional relationship, which is called **skew**.

parallel	lines that lie in the same plane and do not intersect	$\overleftrightarrow{LK} \parallel \overleftrightarrow{MN}$
perpendicular	lines that form 90° angles	$\overleftrightarrow{LK} \perp \overleftrightarrow{JK}$
skew	lines that do not lie in the same plane and do not intersect	$\overleftrightarrow{JK}$ and $\overleftrightarrow{MN}$

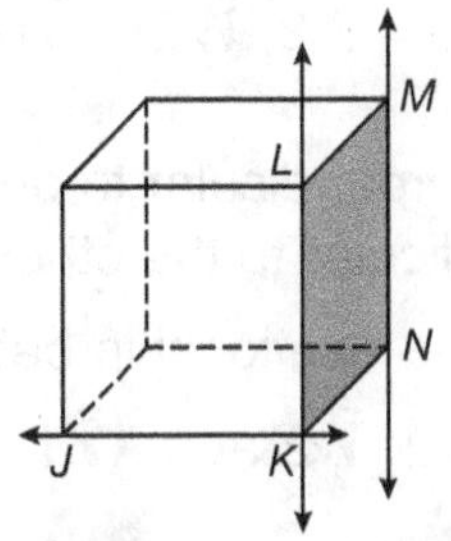

If two parallel planes are cut by a third plane, then the lines of intersection are parallel.

Planes *GHI* and *KLM* are parallel.

Name two pairs of parallel lines.

$\overleftrightarrow{GH} \parallel \overleftrightarrow{KL}$ and $\overleftrightarrow{IJ} \parallel \overleftrightarrow{MN}$

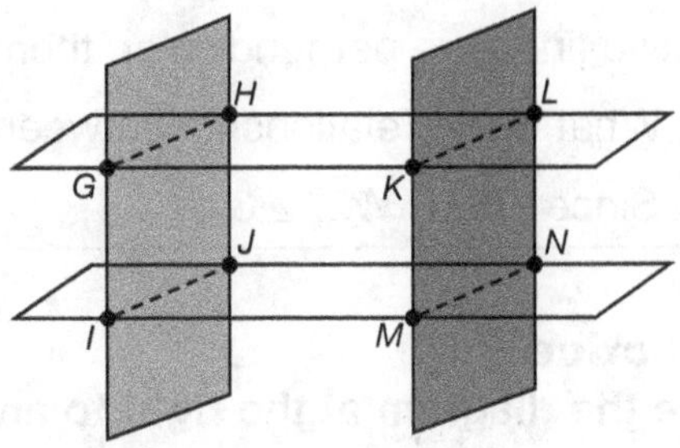

Practice

Use the diagram at the right to complete 1–7.

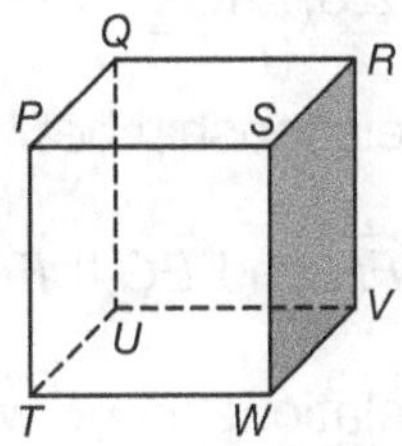

1. $\overleftrightarrow{PQ} \parallel \overleftrightarrow{TU}$ and $\overleftrightarrow{PQ} \parallel$ ___
2. $\overleftrightarrow{PS} \parallel$ ___ and $\overleftrightarrow{PS} \parallel$ ___
3. $\overleftrightarrow{PT} \perp$ ___ and $\overleftrightarrow{PT} \perp$ ___ on plane *PQR*
4. $\overleftrightarrow{RV} \perp$ ___ and $\overleftrightarrow{RV} \perp$ ___ on plane *TUV*
5. Name a line that is skew to $\overleftrightarrow{SW}$. ______________
6. Name a line that is parallel to $\overleftrightarrow{UV}$. ______________
7. Name a line that is perpendicular to $\overleftrightarrow{UV}$. ______________

Use the diagram at the right to answer each question.

Given: Plane *ABC* is parallel to plane *EFG*

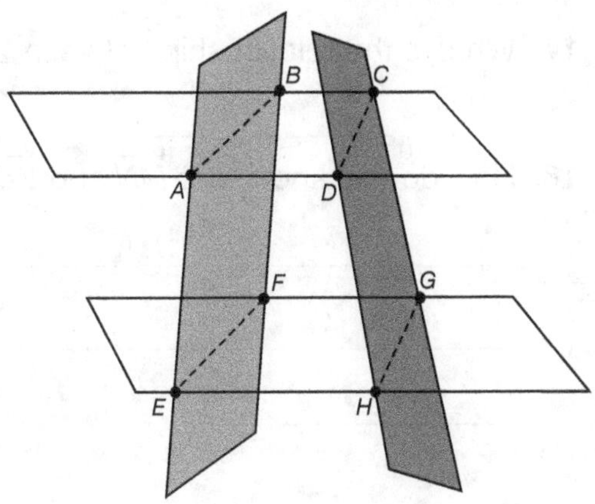

8. Name two pairs of parallel lines. ______________
9. Which line will intersect $\overleftrightarrow{CD}$? __________
10. Which line will intersect $\overleftrightarrow{EF}$? __________
11. Which line is skew to $\overleftrightarrow{AB}$? __________
12. Which line is skew to $\overleftrightarrow{EF}$? __________

In the figure, $\angle 1 \cong \angle 2$, $\overleftrightarrow{XY} \parallel \overleftrightarrow{WZ}$, and $\overleftrightarrow{RS} \perp \overleftrightarrow{WZ}$.

If two lines form congruent adjacent angles, then they are perpendicular.

a. What is the relationship between $\overleftrightarrow{WZ}$ and $\overleftrightarrow{TU}$?

Since $\angle 1 \cong \angle 2$, $\overleftrightarrow{WZ} \perp \overleftrightarrow{TU}$.

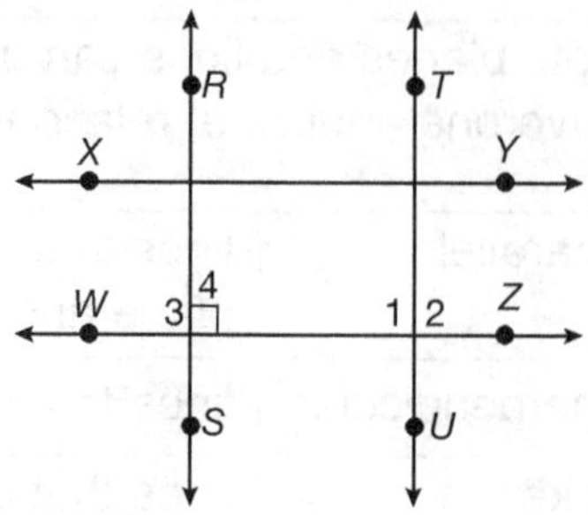

If a line is perpendicular to one of two parallel lines, then it is perpendicular to the other one.

b. What is the relationship between XY and RS?

Since $\overleftrightarrow{RS} \perp \overleftrightarrow{WZ}$ and $\overleftrightarrow{XY} \parallel \overleftrightarrow{WZ}$, $\overleftrightarrow{RS} \perp \overleftrightarrow{XY}$.

If two lines are perpendicular, then they form congruent adjacent angles.

c. What is the relationship between $\angle 3$ and $\angle 4$?

Since $\overleftrightarrow{RS} \perp \overleftrightarrow{WZ}$, $\angle 3 \cong \angle 4$.

Practice

Use the diagram at the right to answer 13–15.

Given: $\overleftrightarrow{EG} \parallel \overleftrightarrow{FH}$ and $\angle 2 \cong \angle 3$

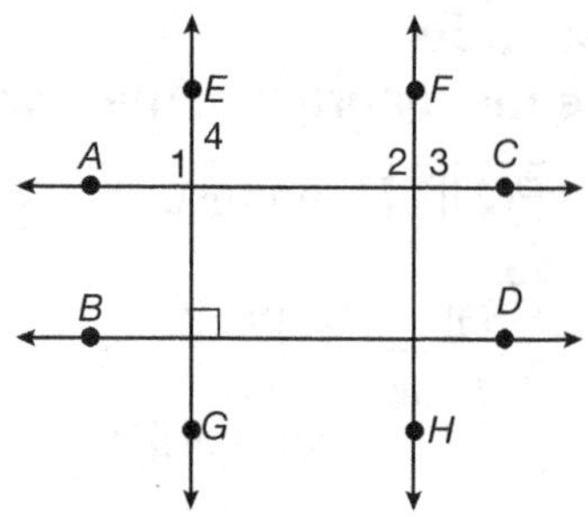

13. What is the relationship between $\overleftrightarrow{AC}$ and $\overleftrightarrow{FH}$?

Since $\angle 2 \cong \angle 3$, $\overleftrightarrow{AC}$ __ $\overleftrightarrow{FH}$

14. What is the relationship between $\overleftrightarrow{AC}$ and $\overleftrightarrow{EG}$?

Since $\overleftrightarrow{AC} \perp \overleftrightarrow{FH}$ and $\overleftrightarrow{EG} \parallel \overleftrightarrow{FH}$, then ________.

15. What is the relationship between $\angle 1$ and $\angle 4$?

Since $\overleftrightarrow{EG} \perp \overleftrightarrow{AC}$, ________.

Use the diagram at the right to answer 16–18.

Given: $\overleftrightarrow{MN} \perp \overleftrightarrow{RV}$ and $\overleftrightarrow{WX} \parallel \overleftrightarrow{RV}$

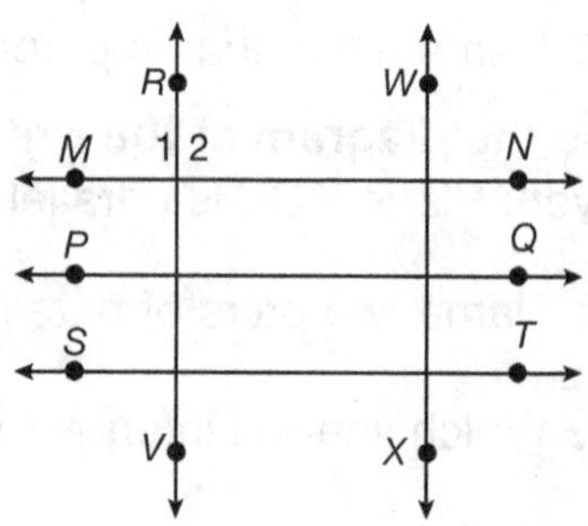

16. Name each line parallel to $\overleftrightarrow{MN}$ ____________

17. What is the relationship between $\angle 1$ and $\angle 2$?

____

18. How do you know that $\overleftrightarrow{MN}$ and $\overleftrightarrow{WX}$ are perpendicular?

____

____

____

Name ______________________ Date __________ Class __________

Reteaching 6

Identifying Pairs of Angles

You have solved problems involving pairs of lines. Now you will solve problems involving pairs of angles.

Complementary and Supplementary Angles

Two angles are **complementary angles** if their combined measures total 90°.

Two angles are **supplementary angles** if their combined measures total 180°.

Find the complement of ∠ABC.

Step 1: $m\angle ABC = 37°$

Let x = complement of $\angle ABC$.

Step 2: $m\angle ABC$ + complement of $\angle ABC = 90°$

$$37° + x = 90°$$

$$37° + x - 37° = 90° - 37°$$

$$x = 53°$$

The measure of the complement of $\angle ABC = 53°$

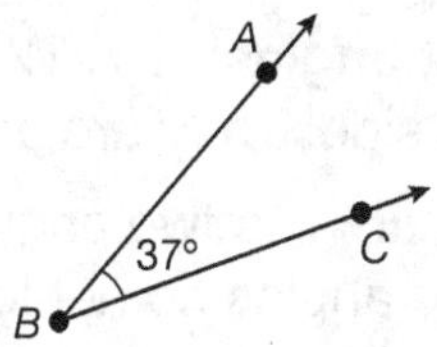

Practice

Complete the steps to find the supplement of ∠*LMN*. Let *x* represent the measure of the complement of ∠*LMN*.

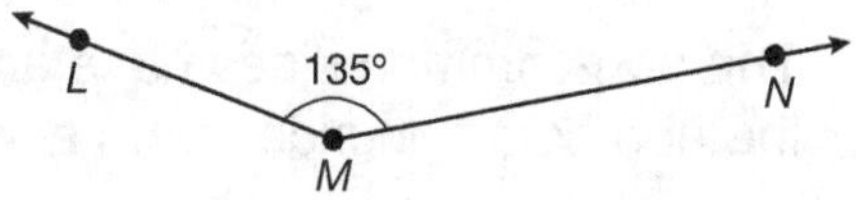

1. $135° + x =$ ____

$135° + x -$ ____ $= 180° -$ ____

____ $=$ ____

The measure of the supplement of ∠*LMN* is ____

Find the measure of each of the following angles.

2. complement of ∠*RST* ____

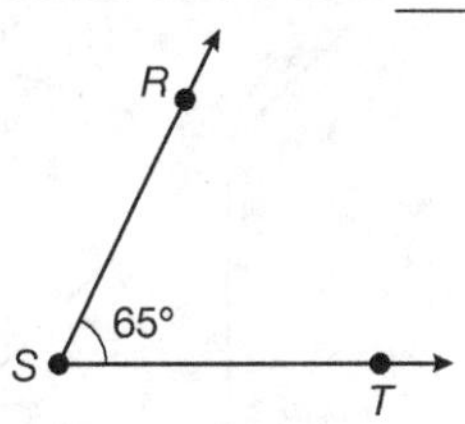

3. supplement of ∠*DEF* ____

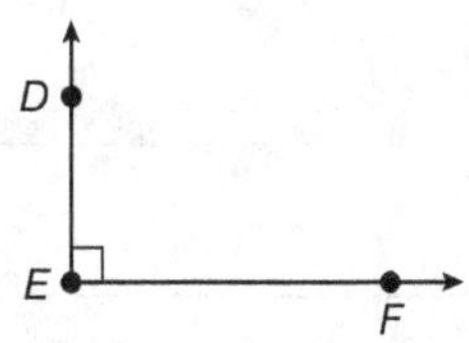

4. supplement of ∠*RST* ____

5. complement of ∠*DEF* ____

Pairs of Angles

Adjacent angles have the same vertex and share a common side. In the figure, $\angle CDE$ is adjacent to $\angle EDG$.

A **linear pair** is formed by two adjacent angles whose non-common sides are opposite rays. The sum of the measures of a linear pair is 180°. $\angle FDC$ and $\angle CDG$ form a linear pair.

Vertical angles are nonadjacent angles formed by two intersecting lines. $\angle FDC$ and $\angle EDG$ are vertical angles.

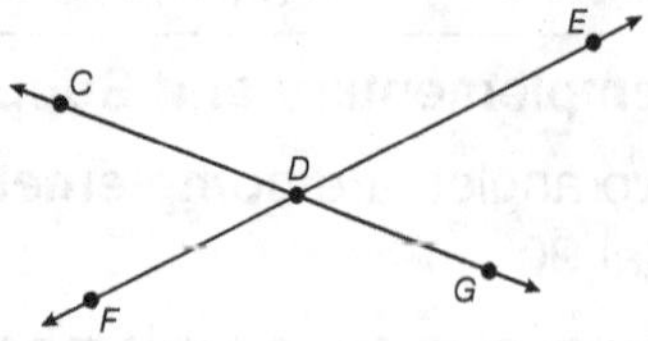

Tell whether $\angle YZW$ and $\angle WZX$ are adjacent angles, form a linear pair, or are vertical angles.

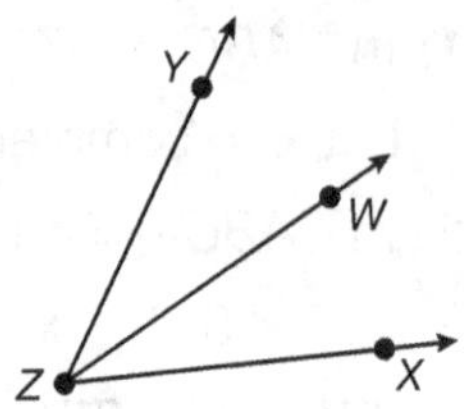

Adjacent angles: $\angle YZW$ and $\angle WZX$ have the same vertex and a common side. They are adjacent angles.

Linear pair: The two angles together do not make an angle that is 180°. The angles do not form a linear pair.

Vertical angles: The two angles are not vertical angles, because they are adjacent angles.

Practice

Complete the steps to show that two angles form a linear pair.

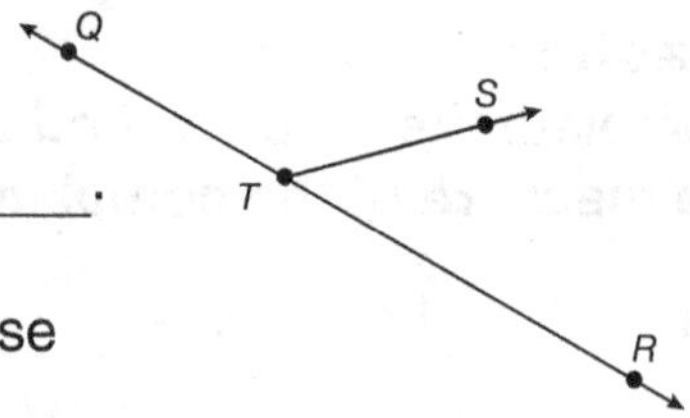

6. $\angle QTS$ and $\angle STR$ have the same ______ and a common ________.

   The non-common sides have an angle of measure _____ because the non-common sides form a straight _____.

   $\angle QTS$ and $\angle STR$ form a __________.

Tell whether the pair of angles are adjacent angles, form a linear pair, or are vertical angles.

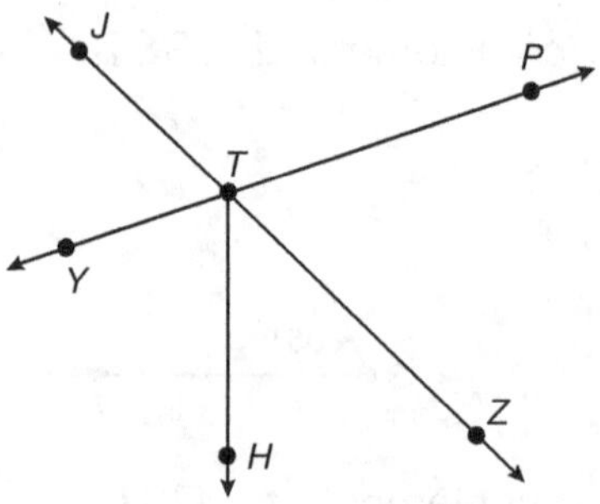

7. $\angle JTY$ and $\angle YTH$ ____________________

8. $\angle YTZ$ and $\angle PTZ$ ____________________

9. $\angle JTP$ and $\angle YTZ$ ____________________

10. $\angle PTZ$ and $\angle HTZ$ ____________________

Name ______________________ Date __________ Class ___________

Reteaching 7

Using Inductive Reasoning to Make Conjectures

You have solved problems involving pairs of angles. Now you will use inductive reasoning to make conjectures.

Making Conjectures

When you make a general rule or conclusion based on a pattern, you are using **inductive reasoning.** A conclusion based on a pattern is a called a **conjecture.**

Find the next two terms in the pattern.

−8, −3, 2, 7, . . .

Step 1: Study the pattern and try to find a mathematical relationship between the numbers. Test your conjecture on the given numbers.

Step 2: The correct conjecture is that each term is 5 more than the previous term.

$-8 + 5 = -3$

$-3 + 5 = 2$

$2 + 5 = 7$

Step 3: Find the next term by adding 5 to the last term: $7 + 5 = 12$.

The next term is $12 + 5 = 17$. The next two terms in the pattern are 12 and 17.

Practice

Complete the steps to find the next two items in the pattern.

1.

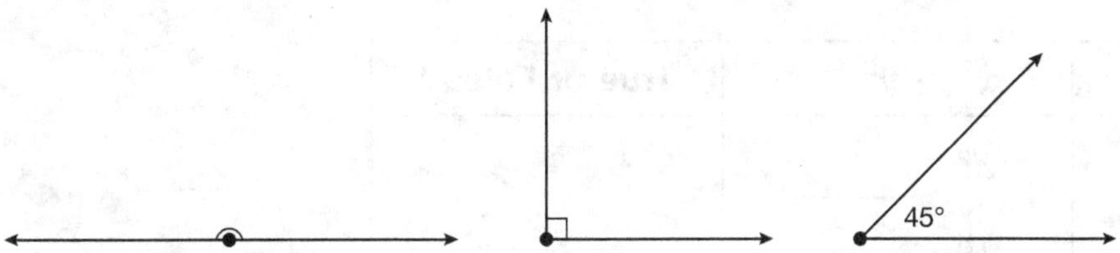

The first angle has measure ________.

The second angle has measure ________. It is ________ of 180°.

The measure of the third angle is ________. It is ________ of 90°.

The measure of the fourth angle is ________ of 45°, or ________.

The measure of the fifth angle is ________ of 22.5°, or ________.

Find the next two items in each pattern.

2.

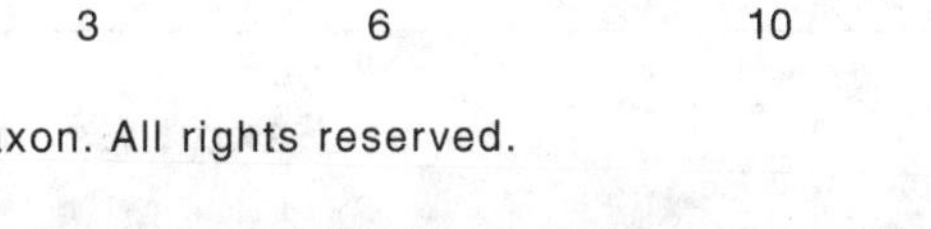

Proving a Conjecture Is False

Since a conjecture is an educated guess, it may be true, or it may be false. It takes only one example to prove that a conjecture is false.

Show that the conjecture is false.

Conjecture: For any integer n, $n \leq 4n$.

Step 1: Make a table of sample values of n.

Step 2: Substitute each value into the inequality $n \leq 4n$ and determine whether that value makes the inequality true or false.

n	$n \leq 4n$	True or False?
3	$3 \leq 4(3)$ $3 \leq 12$	true
0	$0 \leq 4(0)$ $0 \leq 0$	true
−2	$-2 \leq 4(-2)$ $-2 \leq -8$	false

$n = -2$ makes the inequality false, so the conjecture is false.

Practice

Complete the table to show that the conjecture is false.

3. Conjecture: For any real numbers x and y, if $x > y$, then $x^2 > y^2$.

x	y	$x^2 > y^2$	True or False?
4	3	$4^2 >$ ____ $16 >$ ____	
5	2	$5^2 >$ ____ $25 >$ ____	
−4	−5	$(-4)^2 >$ ____ $16 >$ ____	

Show that each conjecture is false by finding one case that makes the conjecture false.

4. For any number n, $2n < n^2$.

____

5. For any integer n, $n > -n$.

____

6. For any integer a, $a > \frac{a}{2}$.

____

7. For any integer n, $\frac{n}{2n} = \frac{1}{2}$.

____

Name ____________________ Date __________ Class __________

Reteaching 8

Using Formulas in Geometry

You have worked with conjectures. Now you will use formulas in geometry.

Perimeter and Area of Rectangles and Triangles

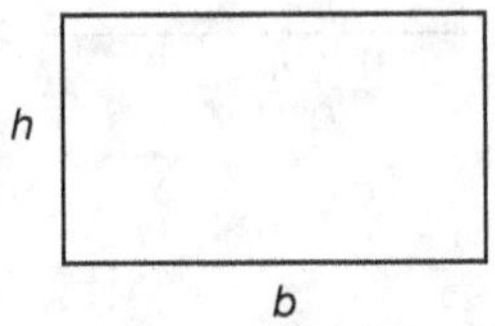

$P = 2h + 2b$

$A = bh$

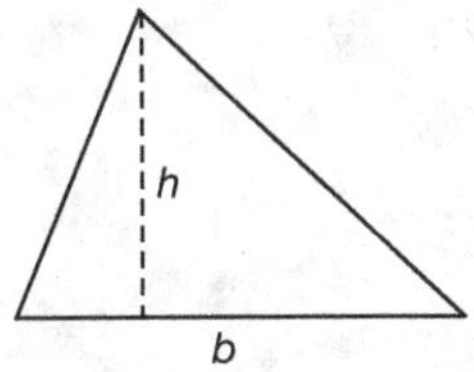

$P =$ sum of the side lengths

$A = \frac{1}{2}bh$

Find the perimeter of the rectangle.

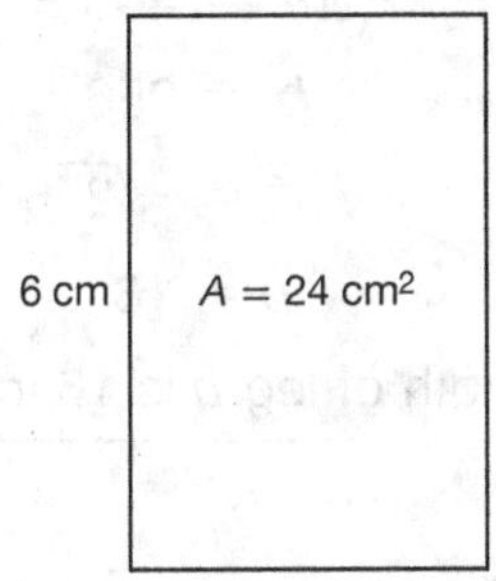

Step 1: Determine the base. Substitute the given information into the formula and solve for *h*.

$A = bh$

$24 = (6)h$

$\frac{24}{6} = \frac{(6)h}{6}$

$4 = h$

Step 2: Substitute $b = 4$ and $h = 6$ into the perimeter formula and simplify: $P = 2b + 2h = 2(4) + 2(6) = 20$ cm.

The perimeter of the rectangle is 20 centimeters.

Practice

Complete the steps to find the area of the triangle.

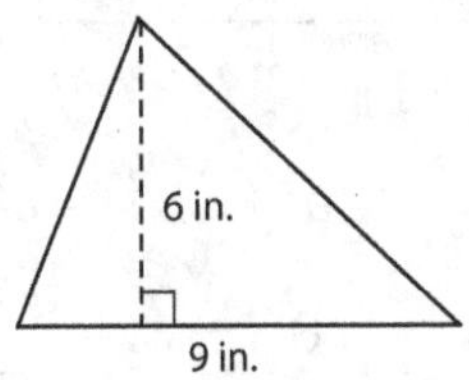

1. $A = \frac{1}{2}bh$

$A = \frac{1}{2}(9)(\_\_)$

$A = \frac{1}{2}(\_\_)$

$A =$ ___ in.^{2}

Find each measurement.

2. area of the rectangle

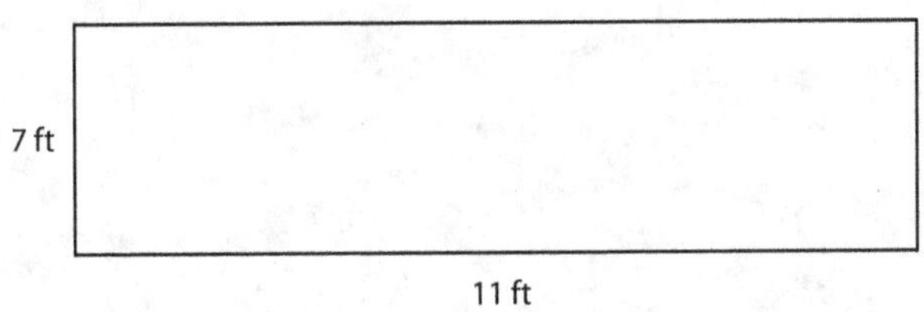

______

3. perimeter of the triangle

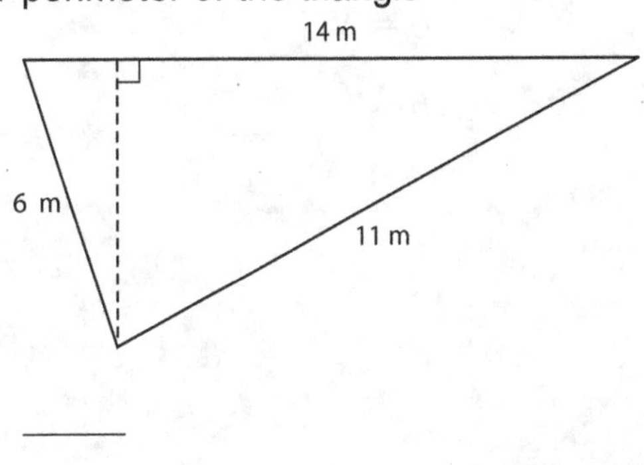

______

The Pythagorean Theorem

In a right triangle, the sum of the square of the legs, *a* and *b,* is equal to the square of the hypotenuse, *c*: $a^2 + b^2 = c^2$

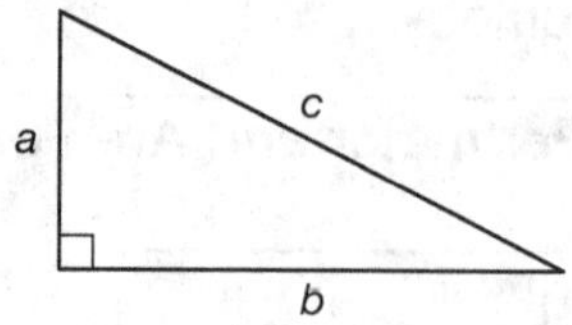

Use the Pythagorean theorem to solve for the length of *b*.

Step 1: Substitute $a = 12$ and $c = 20$.

$a^2 + b^2 = c^2$

$12^2 + b^2 = 20^2$

Step 2: Simplify.

$144 + b^2 = 400$

$144 + b^2 - 144 = 400 - 144$

$b^2 = 256$

$\sqrt{b^2} = \sqrt{256}$

$b = 16$

The length of leg *b* is 16 inches.

Practice

Complete the steps in problems 4 and 5.

4. Find the length of side *b*.

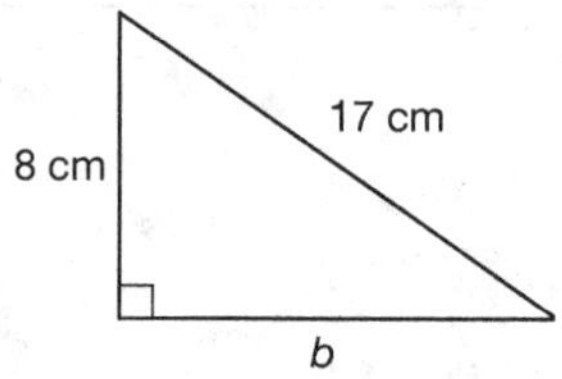

$a^2 + b^2 = c^2$

$8^2 + b^2 =$ ____

____ $+ b^2 =$ ____

$64 + b^2 -$ ____ $= 289 -$ ____

$b^2 =$ ____

$b = \sqrt{}$ ____

$b =$ ____

5. Find the hypotenuse.

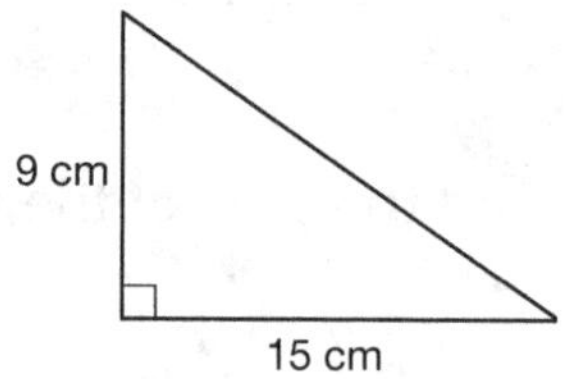

$a^2 + b^2 = c^2$

$9^2 + 15^2 =$ ____

____ $+ 225 =$ ____

____ $=$ ____

____ $= c^2$

____ $= c$

Name ______________________ Date __________ Class __________

Reteaching 9

Finding Length: Distance Formula

You have worked with geometric formulas. Now you will use the distance formula.

Distance Between Two Points on a Line

To find the distance between two points on a number line, take the absolute value of the difference between the points' coordinates.

$$d = |a_2 - a_1|$$

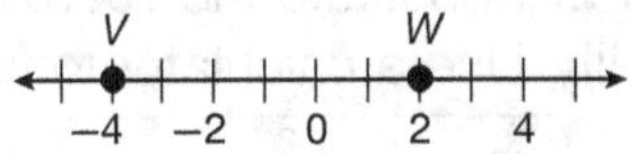

Find the distance between the points on a number line.

Step 1: Choose a point to be a_1. The other point will be a_2.

$a_1 = -4$ $a_2 = 2$

Step 2: Substitute the values into the formula and simplify.

$d = |a_2 - a_1|$

$= |2 - (-4)|$

$= |2 + 4|$

$= |6| = 6$

The distance is 6 units.

Practice

Complete the steps to find the distance between the points on the number line.

1. $d = |a_2 - a_1|$

$= \left|-7 - (\_)\right|$

$= \left|-7 - \_\right|$

$= |\_\_\_| = \_$

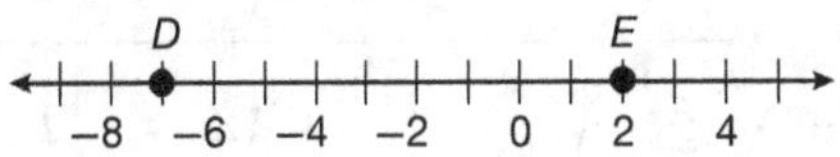

Find the distance between each pair of points.

2.

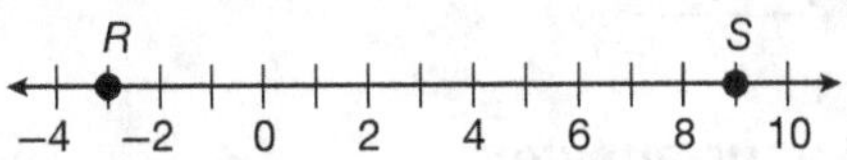

___

3.

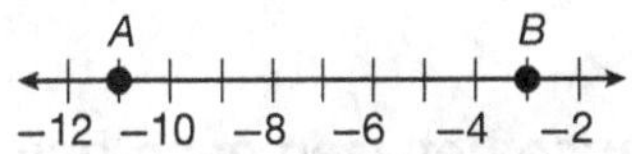

_

4.

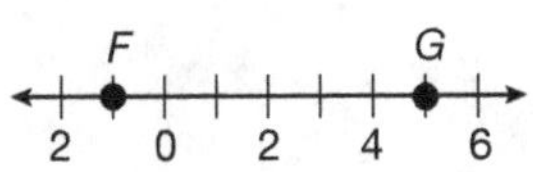

_

5.

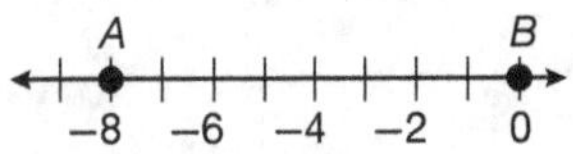

_

Distance on a Coordinate Plane

The **Distance Formula** can be used to find the distance between two points, (x_1, y_1) and (x_2, y_2), in a coordinate plane: $d = \sqrt{(x_2 - x_1)^2 + (y_2 - y_1)^2}$.

Find the distance between the two points. Round your answer to the nearest tenth.

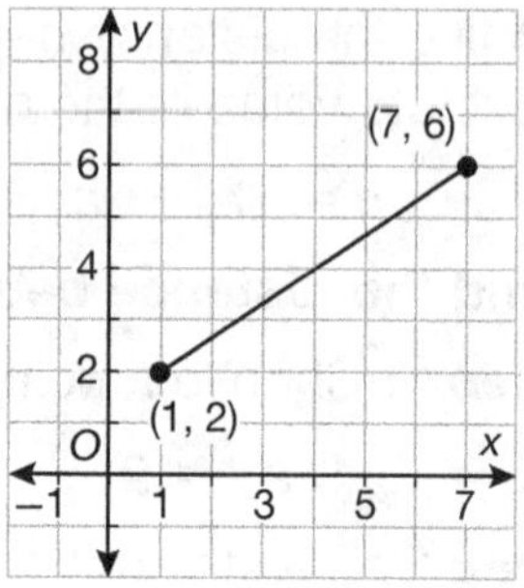

Step 1: Let $(1, 2) = (x_1, y_1)$ and $(7, 6) = (x_2, y_2)$.

Step 2: Substitute the coordinates into the distance formula and simplify. Use a calculator to find the square root.

$$
\begin{aligned}
d &= \sqrt{(x_2 - x_1)^2 + (y_2 - y_1)^2} \\
&= \sqrt{(7 - 1)^2 + (6 - 2)^2} \\
&= \sqrt{(6)^2 + (4)^2} \\
&= \sqrt{36 + 16} \\
&= \sqrt{52} \\
&\approx 7.2
\end{aligned}
$$

Practice

Complete the steps to find the distance between each pair of points. Round your answer to the nearest tenth.

6. (2, 4) and (−3, 9)

$$
\begin{aligned}
d &= \sqrt{(x_2 - x_1)^2 + (y_2 - y_1)^2} \\
&= \sqrt{(2 - (\underline{\quad}))^2 + (4 - \underline{\quad})^2} \\
&= \sqrt{(\underline{\quad})^2 + (-5)^2} \\
&= \sqrt{\underline{\quad} + 25} \\
&= \sqrt{\underline{\quad}} \\
&\approx 7.1
\end{aligned}
$$

7. (−7, −2) and (4, 1)

$$
\begin{aligned}
d &= \sqrt{(x_2 - x_1)^2 + (y_2 - y_1)^2} \\
&= \sqrt{(\underline{\quad} - 4)^2 + (\underline{\quad} - 1)^2} \\
&= \sqrt{(\underline{\quad})^2 + (\underline{\quad})^2} \\
&= \sqrt{\underline{\quad} + 9} \\
&= \sqrt{\underline{\quad}} \\
&\approx 11.4
\end{aligned}
$$

Find the distance between each pair of points. Round your answer to the nearest tenth.

8. (6, 2) and (−1, −5)

___

9. (0, 4) and (−8, 0)

___

10. (8, −3) and (5, 5)

___

11. (4, −2) (−7, 1)

___

Name ______________________ Date __________ Class __________

Reteaching 10

Using Conditional Statements

You have found the distance between two points. Now you will work with conditional statements.

Hypothesis and Conclusion

A **conditional statement** is a statement in the form "If *p,* then *q,*" where *p* is the **hypothesis** and *q* is the **conclusion.**

For example:

If two lines are parallel, then they do not intersect.

The hypothesis comes after the word *if.*

The conclusion comes after the word *then.*

Identify the hypothesis and conclusion of the conditional statement.

If a figure is a quadrilateral, then it has four sides.

Step 1: The hypothesis is the phrase that follows the word *if*: A figure is a quadrilateral.

Step 2: The conclusion is the phrase that follows the word *then*: It has four sides.

Practice

Complete the steps to identify the hypothesis and conclusion of the statements.

1. If x is an even number, then x is divisible by 2.

   Hypothesis: ______________

   Conclusion: ______________

2. If two angles are supplementary, then they form a linear pair.

   Hypothesis: ______________

   Conclusion: ______________

For each conditional statement, underline the hypothesis and double-underline the conclusion.

3. If two angles are not adjacent, then they cannot be a linear pair.

4. If the weather is rainy, then the football team will not practice after school.

5. If $3x - 4 = 11$, then $x = 5$.

Truth Value of a Conditional Statement

Some conditional statements are true, while others are false. This is called the **truth value** of a conditional statement. A statement is false only when the hypothesis is true and the conclusion is false.

Determine whether the conditional statement is true or false. If it is false, explain your reasoning.

If an acute angle measures 140°, then it is called a straight angle.

Step 1: Determine whether the hypothesis is true or false. The hypothesis is that an acute angle measures 140°. The hypothesis is false because the measure of an acute angle is less than 90°.

Step 2: Determine whether the conditional statement is true or false. This statement has a false hypothesis. When the hypothesis is false, the conditional statement as a whole has a truth value of "true." The statement cannot have a truth value of "false" unless a situation exists in which the hypothesis is true.

Practice

Circle the correct answers for each conditional statement.

6. If a square has a side length of 5 centimeters, then its area is 20 square centimeters.

The hypothesis is true /false.

The conclusion is true/ false .

The conditional situation is true/ false .

7. If two angles are right angles, then they are congruent.

The hypothesis is true /false.

The conclusion is true /false.

The conditional situation is true /false.

Determine whether the conditional statement is true or false. If it is false, explain your reasoning.

8. If an angle is obtuse, then it is not a right angle. ____

9. If two right angles are complementary, then they are not congruent. ____

Name ______________________________ Date ___________ Class ___________

Reteaching INV 1

Transversals and Angle Relationships

Transversals

A transversal is a line that intersects two or more coplanar lines at different points. Line a is the transversal in the picture to the right.

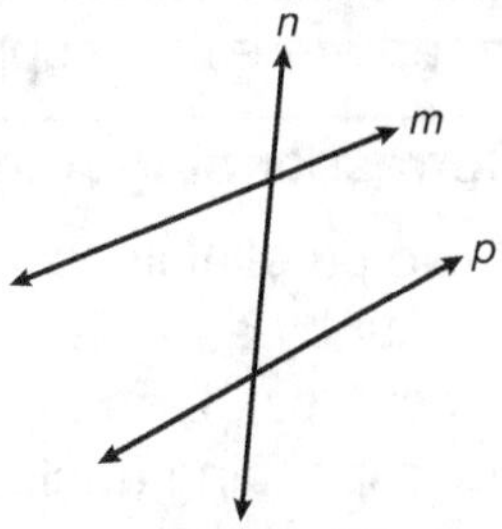

When two lines are intersected by a transversal, the angle pairs are classified by type.

Classification	Example	Classification	Example
A pair of **corresponding angles** are two angles that lie on the same side of the transversal and on the same sides of the other two lines.	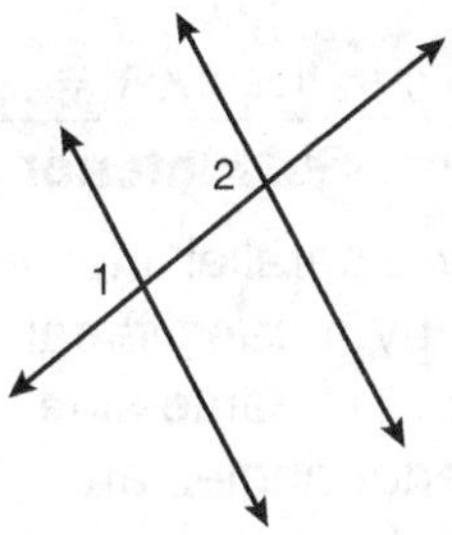	A pair of **alternate exterior angles** are two angles that lie on opposite sides of the transversal and outside the other two lines.	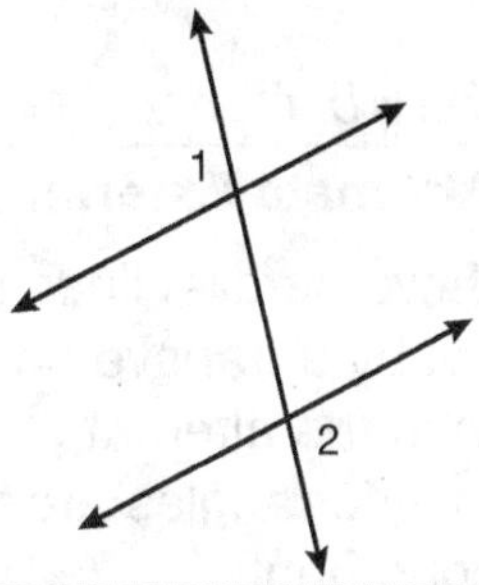
A pair of **alternate interior angles** are two nonadjacent angles that lie on opposite sides of the transversal and between the other two lines.	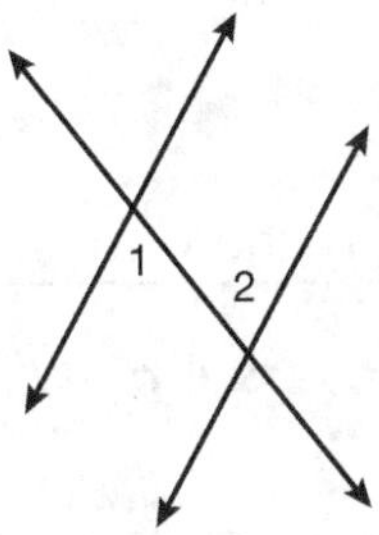	A pair of **same-side interior angles** are two angles that lie on the same side of the transversal and between the other two lines; also called **consecutive interior angles**.	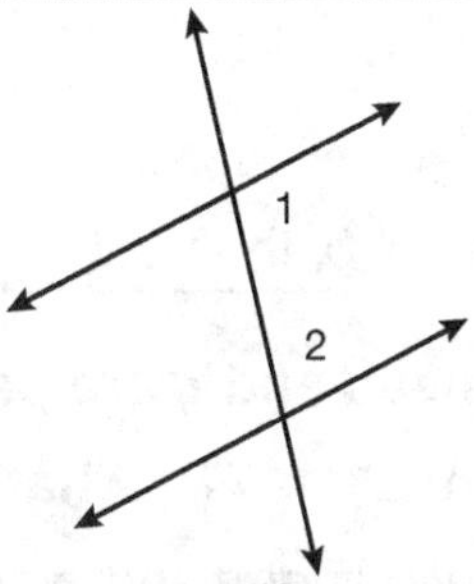

Give an example of a pair of alternate exterior angles.

One pair of alternate exterior angles is given by ∠2 and ∠7.

Another pair of alternate exterior angles is given by ∠1 and ∠8.

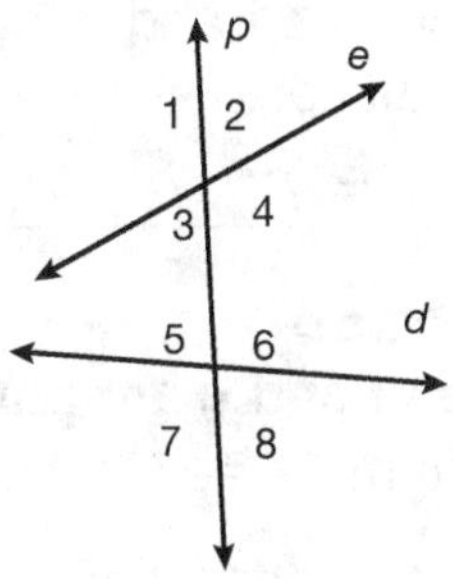

Practice

Complete each statement with the correct term.

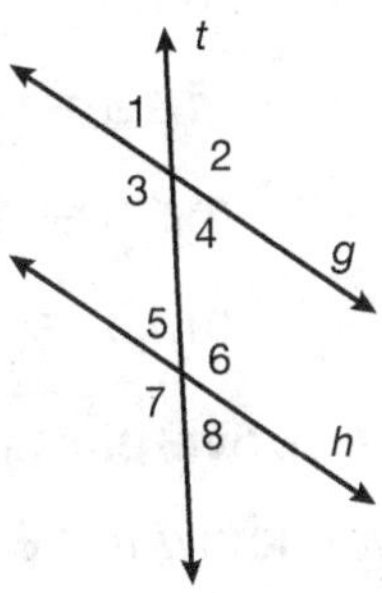

1. Line *t* is the __________ of lines *g* and *h*.
2. ∠3 and ∠5 are a pair of ____________________.
3. ∠2 and ∠6 are a pair of ____________________.
4. ∠1 and ∠8 are a pair of ____________________.
5. ∠4 and ∠5 are a pair of ____________________.

Transversals and Parallel Lines

When a transversal intersects parallel lines, the angle pairs that are formed are either supplementary or congruent.

Corresponding Angles Postulate

If two parallel lines are cut by a transversal, then the corresponding angles are congruent.

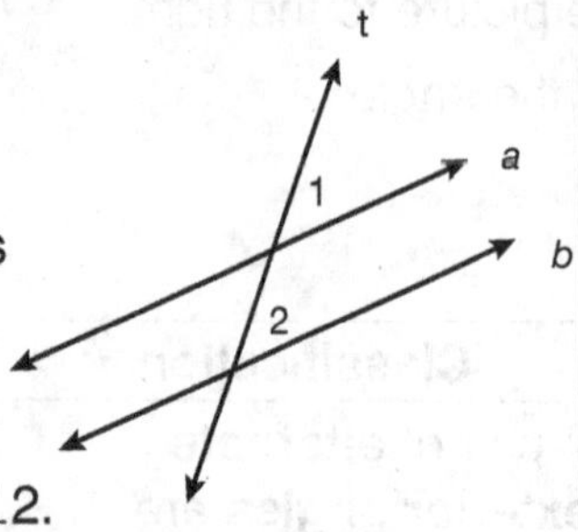

If $a \parallel b$, then $\angle 1 \cong \angle 2$.

Alternate Interior Angles Theorem

If two parallel lines are cut by a transversal, then the alternate interior angles are congruent.

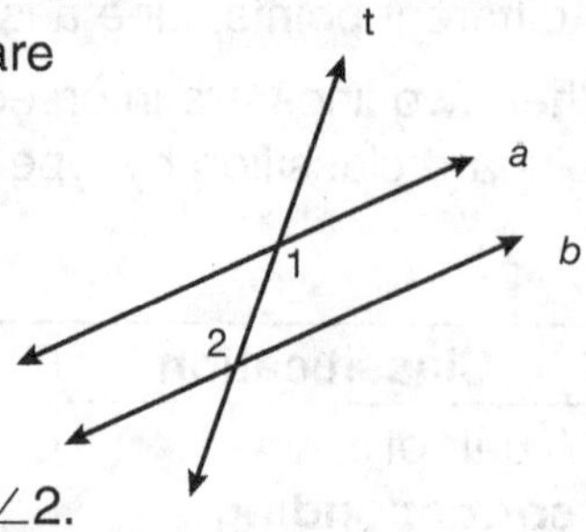

If $a \parallel b$, then $\angle 1 \cong \angle 2$.

Alternate Exterior Angles Theorem

If two parallel lines are cut by a transversal, then the alternate exterior angles are congruent.

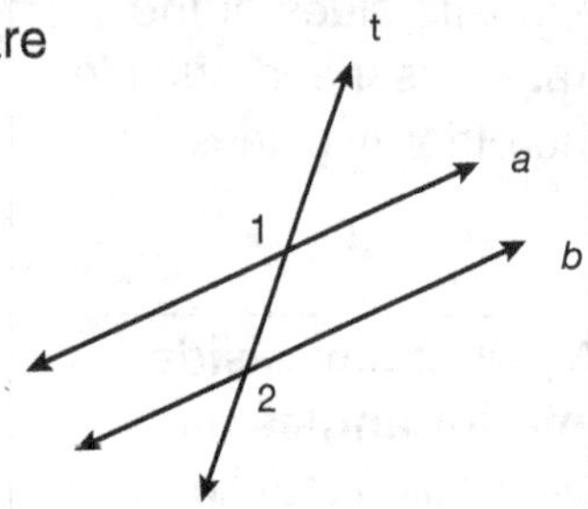

If $a \parallel b$, then $\angle 1 \cong \angle 2$.

Same-Side Interior Angles Theorem

If two parallel lines are cut by a transversal, then the same-side interior angles are supplementary.

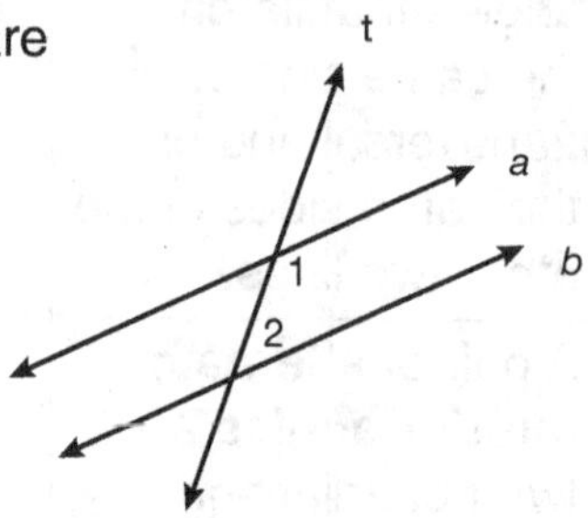

If $a \parallel b$, then $\angle 1 + \angle 2 = 180°$.

Lines *f* and *g* are parallel lines intersected by transversal *c*.

If $m\angle 1 = 77°$, what is $m\angle 7$?

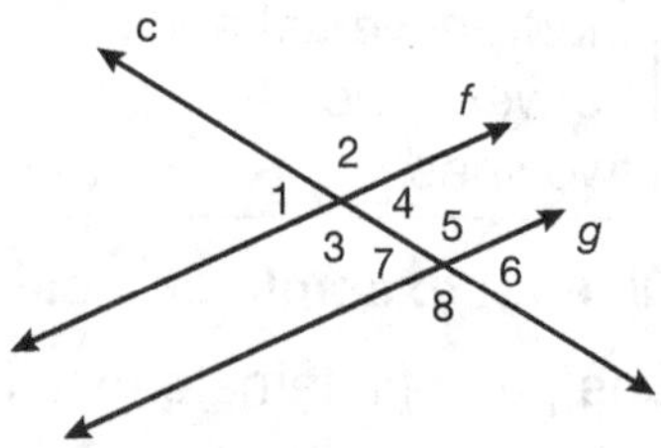

Since lines *f* and *g* are parallel and $\angle 1$ and $\angle 7$ are corresponding angles, $\angle 1$ and $\angle 7$ are congruent by the Corresponding Angles Postulate.

$m\angle 7 = 77°$

Practice

Lines *f* and *g* are parallel. Complete the steps.

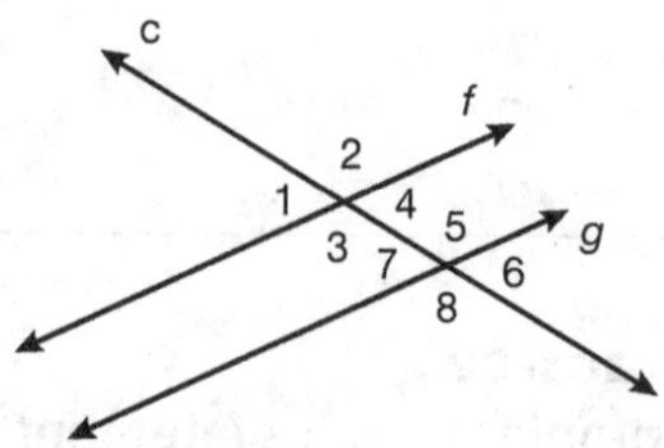

6. If $m\angle 7 = 77°$, find $m\angle 3$.

   $\angle 7$ and $\angle 3$ are ____________ interior angles.

   $m\angle 7 + m\angle 3 =$ ____

   ____ $+ m\angle 3 =$ ____

   $m\angle 3 = 180° -$ ____

   $m\angle 3 =$ ____

Use the above picture to answer each question.

7. If you know $m\angle 8$, is it possible to know $m\angle 2$? ______________________________

   ____

Name ______________________ Date __________ Class __________

Reteaching 11

Finding Midpoints

You know that a midpoint divides a segment into two congruent parts.

Now you will determine the midpoint for a line segment.

> The midpoint of a segment is found by taking the average of the two coordinates: $c = \frac{a + b}{2}$.
>
> **What is the midpoint of *A* and *B* on the number line?**
>
> What is the coordinate of *A*? 1
>
> What is the coordinate of *B*? 9
>
> **Substitute the coordinates into the formula and simplify.**
>
> $c = \frac{a + b}{2}$
>
> $c = \frac{1 + 9}{2}$
>
> $c = \frac{10}{2}$
>
> $c = 5$
>
> The midpoint of *A* and *B* is 5.
>
> 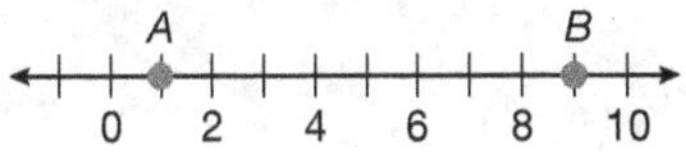
>

Practice

Complete the steps to determine the midpoint of *C* and *D* on the number line.

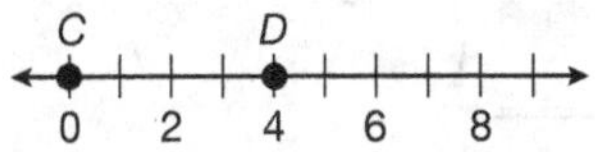

1. What is the coordinate of *C*? ___

   What is the coordinate of *D*? ___

   $c = \frac{a + b}{2}$

   $c = \frac{0 + \_\_\_}{2}$

   $c = \frac{\_\_\_}{2}$

   $c =$ ___

   The midpoint of *C* and *D* is ___.

Determine the midpoint of each segment.

2. ___

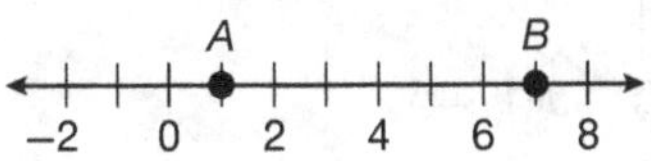

3. ___

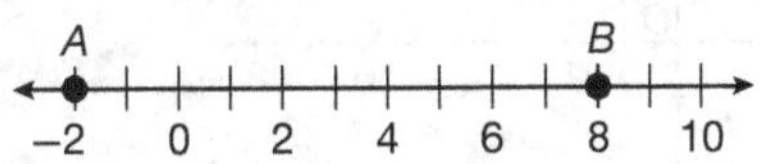

4. ___

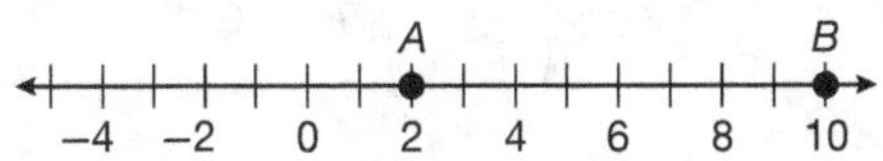

The midpoint of a line segment in a coordinate plane can be found by using the Midpoint Formula.

$M = \left(\frac{x_1 + x_2}{2}, \frac{y_1 + y_2}{2}\right)$

Determine the midpoint of line segment $\overline{GH}$ connecting (1, 2) and (7, 6).

Determine the *x* and *y* coordinates for each point.
Substitute the coordinates into the formula and simplify.

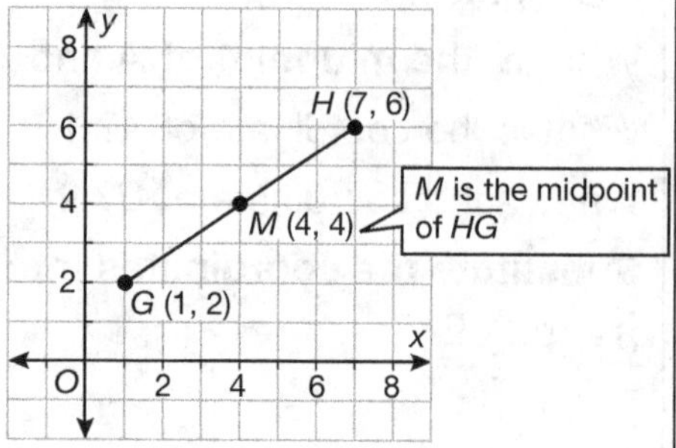

$M = \left(\frac{x_1 + x_2}{2}, \frac{y_1 + y_2}{2}\right)$

$M = \left(\frac{1 + 7}{2}, \frac{2 + 6}{2}\right)$

$M = \left(\frac{8}{2}, \frac{8}{2}\right)$

$M = (4, 4)$

The midpoint is (4, 4).

Practice

Complete the steps to determine the midpoint of the given segment.

5. $M = \left(\frac{x_1 + x_2}{2}, \frac{y_1 + y_2}{2}\right)$

$M = \left(\frac{-3 + \_\_}{2}, \frac{2 + \_\_\_\_\_\_}{2}\right)$

$M = \left(\frac{-2}{2}, \frac{\_\_\_}{2}\right)$

$M =$ **(−1, _____)**

The midpoint is ________.

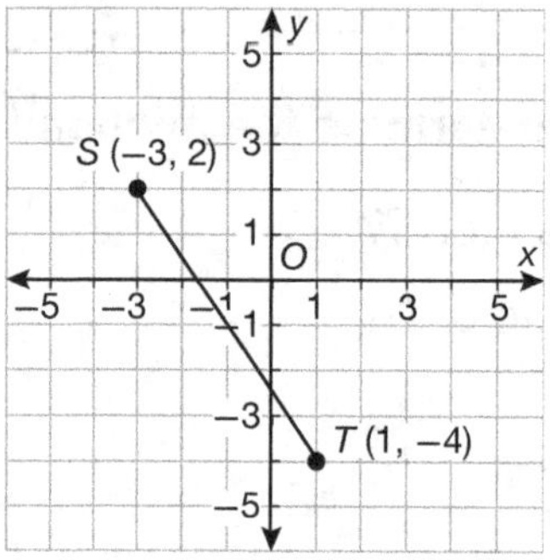

Determine the midpoint.

6. _____

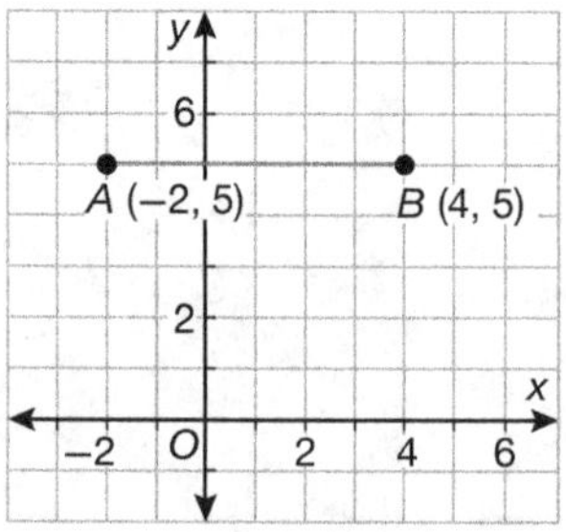

7. _____

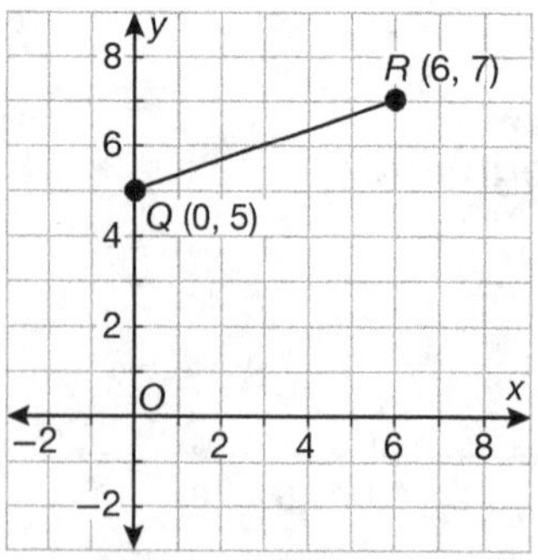

8. _____

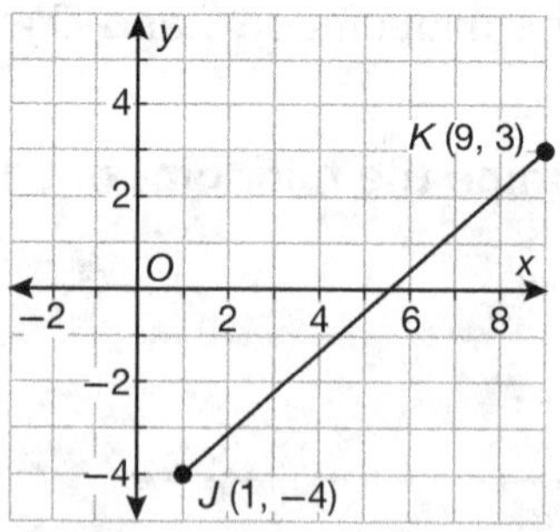

Name ______________________ Date ____________ Class ____________

Reteaching 12

Proving Lines Parallel

You have worked with parallel lines. Now, you will prove that lines are parallel using the converse of theorems.

Converse of the Corresponding Angles Postulate: If two lines are cut by a transversal and the corresponding angles are congruent, then the lines are parallel.

Converse of the Alternate Interior Angles Theorem: If two lines are cut by a transversal and the alternate interior angles are congruent, then the lines are parallel.

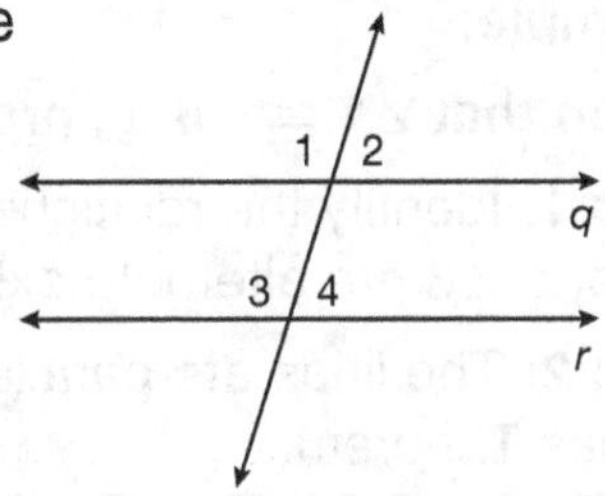

Example:

Given that $\angle 1 \cong \angle 3$, prove that lines *q* and *r* are parallel.

Step 1: Identify the relationship between the two angles.

$\angle 1$ and $\angle 3$ are corresponding angles.

Step 2: The lines are parallel by the Converse of the Corresponding Angles Postulate.

Practice

Complete the steps to determine whether the lines are parallel.

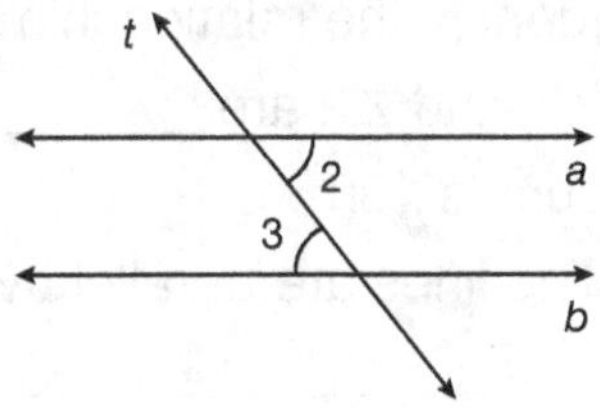

1. Given that $\angle 2 \cong \angle 3$, prove that lines *a* and *b* are parallel.

Identify the relationship between $\angle 2$ and $\angle 3$.

$\angle 2$ and $\angle 3$ are ______________________.

The lines are parallel by the ______________

______________________ Theorem.

Given the information in each exercise, state the reason why lines *j* and *k* are parallel.

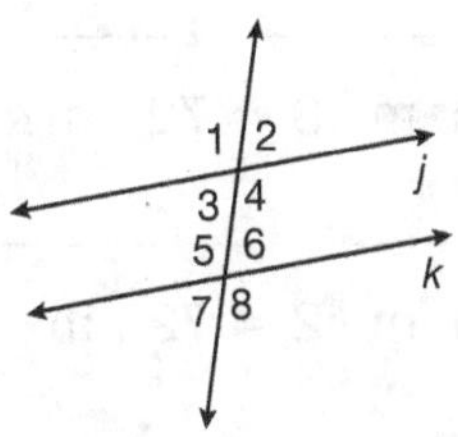

2. Given: $\angle 2 \cong \angle 6$

____

3. Given: $\angle 3 = \angle 62°$, $\angle 6 = \angle 62°$

____

4. Given: $\angle 1 \cong \angle 5$

____

5. Given: $\angle 4 \cong \angle 5$

____

Converse of the Alternate Exterior Angles Theorem: If two lines are cut by a transversal and the alternate exterior angles are congruent, then the lines are parallel.

Converse of the Same-Side Interior Angles Theorem: If two lines are cut by a transversal and the same-side interior angles are supplementary, then the lines are parallel.

Example:

Given that $\angle 1 \cong \angle 8$ ⊥, prove that lines *j* and *k* are parallel.

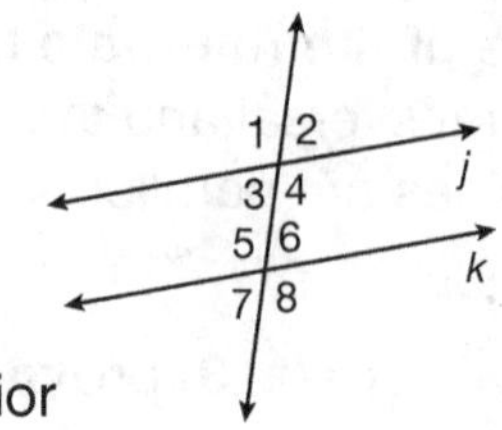

Step 1: Identify the relationship between the two angles.
$\angle 1$ and $\angle 8$ are alternate exterior angles.

Step 2: The lines are parallel by the Converse of the Alternate Exterior Angles Theorem.

Practice

Complete the steps to determine whether the lines are parallel.

6. Given that $m\angle 1 + m\angle 2 = +180°$ prove that lines *s* and *t* are parallel.

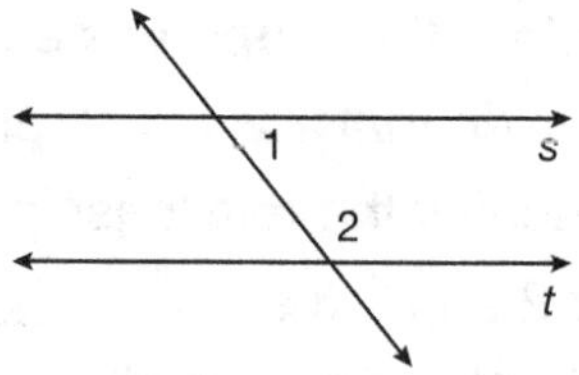

Identify the relationship between $\angle 1$ and $\angle 2$.

$\angle 1$ and $\angle 2$ are ____________________________ whose sum is 180°.

The lines are parallel by the ____________________ ______________________________ Theorem.

Given the information in each exercise, state the reason why lines *j* and *k* are parallel.

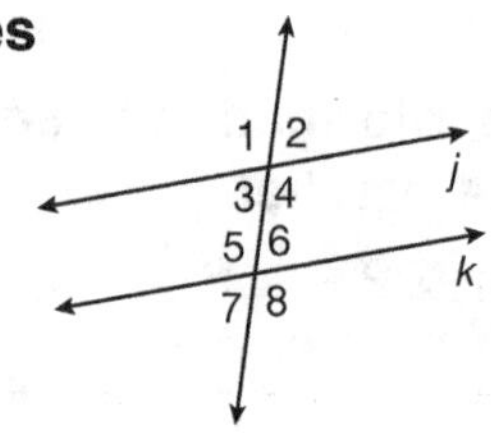

7. Given: $\angle 2 \cong \angle 7$

____

8. Given: $m\angle 3 = 72°$, $m\angle 5 = 108°$

____

9. Given: $m\angle 2 = 72°$, $m\angle 7 = 72°$

____

10. Given: $m\angle 4 + m\angle 6 = 108°$

____

Name ______________________________ Date __________ Class __________

Reteaching 13

Introduction to Triangles

You know that a triangle is a three-sided polygon. Now you will classify triangles by their sides and angles.

You can classify triangles by their angle measures.

Acute Triangle	Right Triangle	Obtuse Triangle
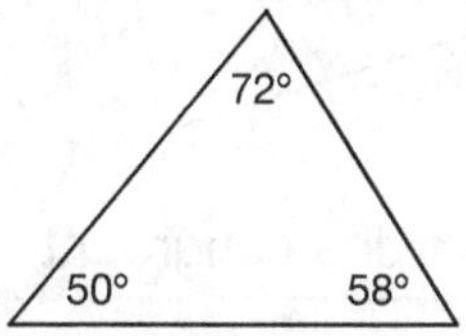	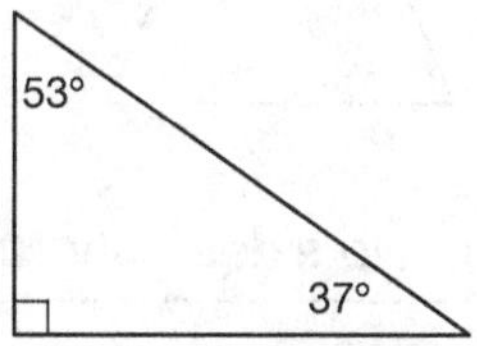	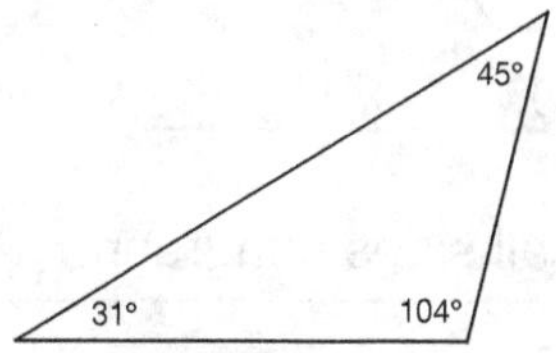
all acute angles	one right angle	one obtuse angle

Use angle measures to classify the triangle. Identify the measures of each angle.

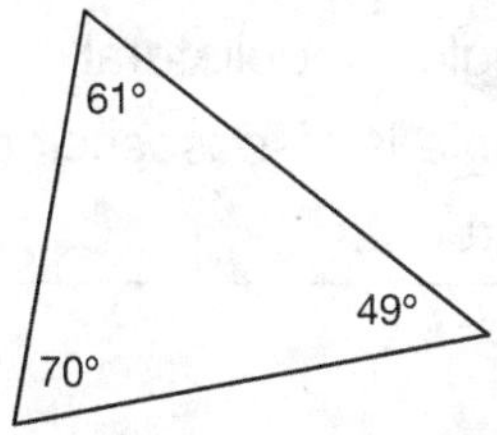

61° = acute   70° = acute   49° = acute

All three angles are acute.

The triangle is acute.

Practice

Complete the steps to classify each triangle by its angle measures.

1. 36° = _____
   54° = _____
   90° = _____

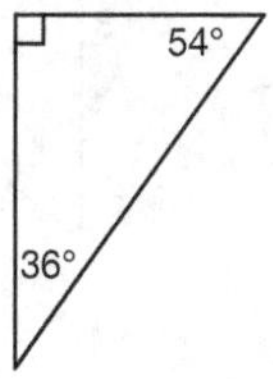

   Triangle is right.

2. 47° = acute
   103° = _____
   30° = _____

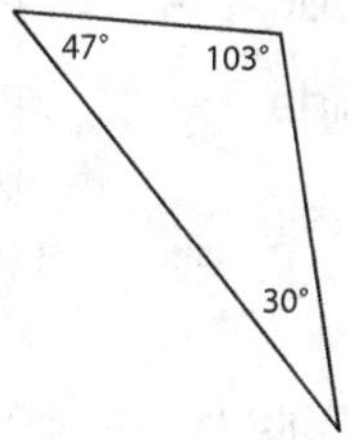

   Triangle is _____.

Classify each triangle by its angle measures.

3. _____

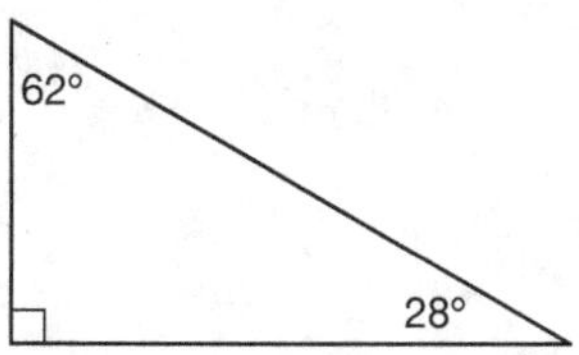

4. _____

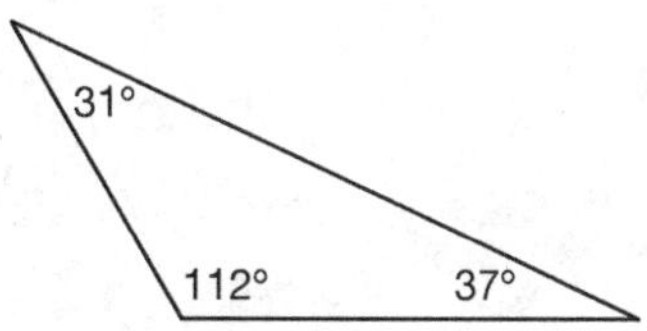

5. _____

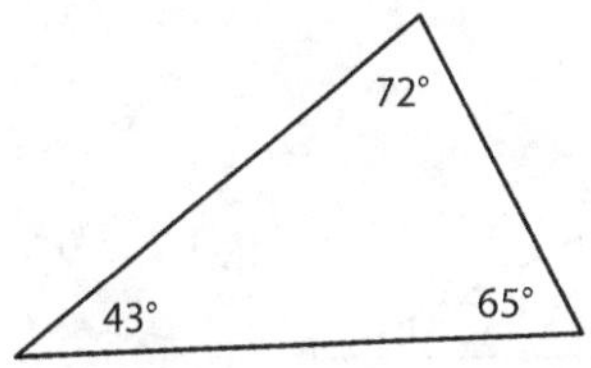

You can also classify triangles by their side lengths.

Equilateral Triangle	Isosceles Triangle	Scalene Triangle
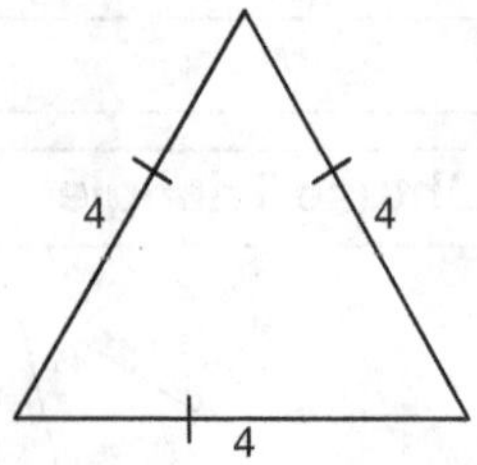 all sides congruent	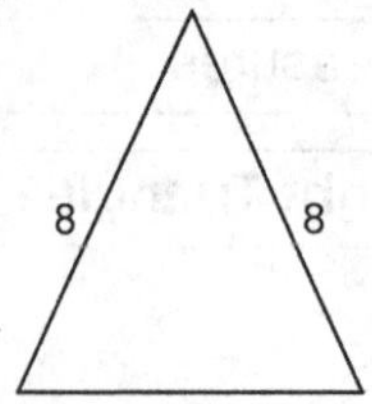 at least two sides congruent	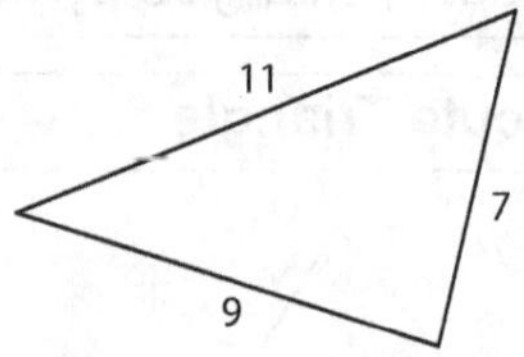 no sides congruent

Classify the triangle by its side lengths.

Three sides are the same length.

The triangle is equilateral.

The triangle is also isosceles because at least two sides are congruent.

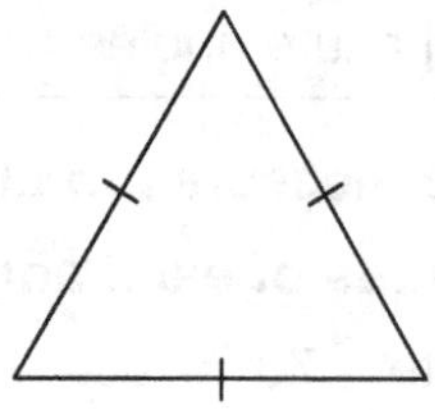

Practice

Complete the steps to classify each triangle by its side lengths.

6. one side = _

one side = 4

one side = _

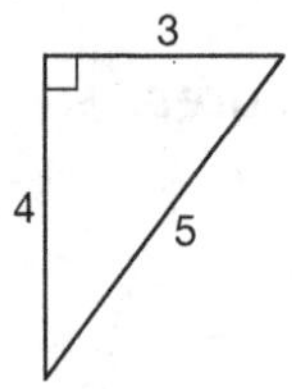

Triangle is scalene.

7. one side = _

one side = _

one side = 7

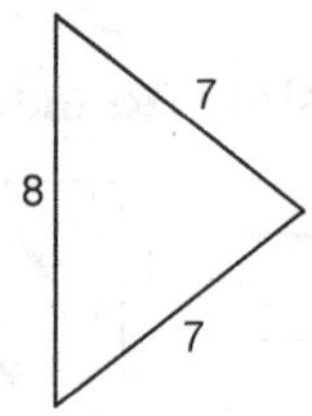

Triangle is ________.

Classify each triangle by its side lengths.

8. ________

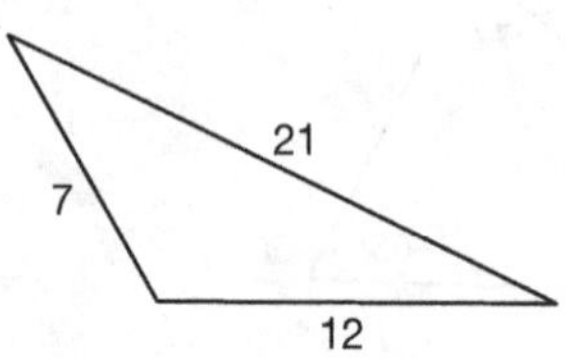

9. ________

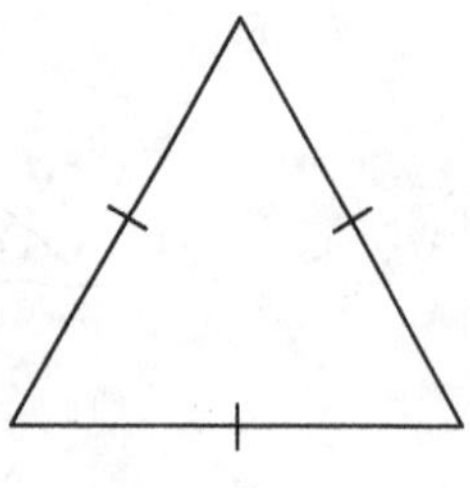

10. ________

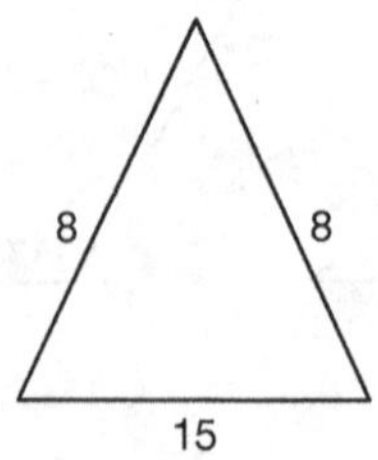

Name ______________________________ Date ____________ Class ______________

Reteaching 14

Disproving Conjectures with Counterexamples

You know that a statement that is believed to be true but has not been proved is a conjecture. Now, you will disprove conjectures with counterexamples.

Geometric Conjectures

A counterexample is an example that proves a conjecture or statement is false.

Use the conjecture to answer a and b.

If $\angle A$ is an acute angle, then $\angle A = 45°$.

a. What is the hypothesis and conclusion of the conjecture?

Hypothesis: $\angle A$ is an acute angle.

Conclusion: $\angle A = 45°$

b. Find a counterexample to the conjecture.

A counterexample would be an example of an angle for which the hypothesis is true but the conclusion is false.

An acute angle has any measure between 0° and 90°.
Counterexample: An angle of 55° is an acute angle, but it is not 45°.

Practice

Complete the steps to find a counterexample to the conjecture.

1. If two angles are congruent, then they are vertical angles.

Hypothesis: ______________________________

Conclusion: ______________________________

Counterexample: Two angles can be congruent in measure but not be vertical angles.

Determine the hypothesis and conclusion and find a counterexample to the conjecture.

2. If a shape is a quadrilateral, then it is a parallelogram.

Hypothesis: ______________________________

Conclusion: ________________________

Counterexample: ____

____

Algebraic Conjectures

Find a counterexample to the conjectures.

a. Conjecture: The difference of two integers is a smaller number than either of the original numbers.

Counterexample: The equation $-2 - (-3) = 1$ shows that the difference of two integers can be a larger number than either of the two original numbers.

b. Conjecture: If x is an integer, then $-x < 0$.
Counterexample: If $x = -4$, then $-(-4)$ is not less than 0.

Practice

Complete the steps to find a counterexample to the algebraic conjecture.

3. If x is an even number, then $x + 2$ is divisible by 4.

Counterexample: The expression _______ is not divisible by 4.

4. If x and y are two different integers, then $x - y > y - x$.

Counterexample: ____

____

Find a counterexample to each algebraic conjecture.

5. If $x^2 = 16$, then $x = 4$.

Counterexample: _______

6. If $x < 0$, then $x^2 < 1$.

Counterexample: ________________________

7. If a number is a perfect square, then its square root is even.

Counterexample: ____

____

8. If $(x + 7)(x - 1) = 0$, then $x = 1$.

Counterexample: _______

9. If $x + y = 24$, then $x = 15$ and $y = 9$.

Counterexample: ____

Name ______________________ Date __________ Class __________

Reteaching 15

Introduction to Polygons

You have worked with congruent line segments and angles. Now you will work with polygons.

Polygons are named for the number of their sides. Some common names are given in the table.

A polygon is **equiangular** if all the angles are congruent.

A polygon is **equilateral** if all the sides are congruent.

A polygon that is both equiangular and equilateral is called a **regular** polygon.

A polygon that is not equiangular and not equilateral is called an **irregular** polygon.

Number of Sides	Polygon
3	Triangle
4	Quadrilateral
5	Pentagon
6	Hexagon
7	Heptagon
8	Octagon
9	Nonagon
10	Decagon

Name the polygon. Determine whether it is equiangular, equilateral, regular, irregular, or more than one of these.

The polygon has 6 sides.

The sides and angles are all congruent.

It is a regular hexagon.

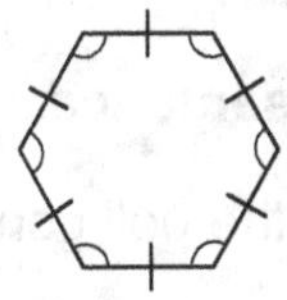

Practice

Name the polygon. Determine whether it is equiangular, equilateral, regular, irregular, or more than one of these.

1.

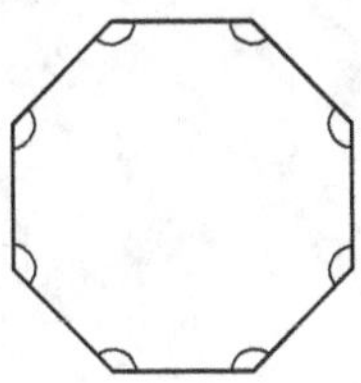

The polygon has _ sides.

The angles are congruent, but the sides are not.

The polygon is an ________.

It is equiangular and ________.

2.

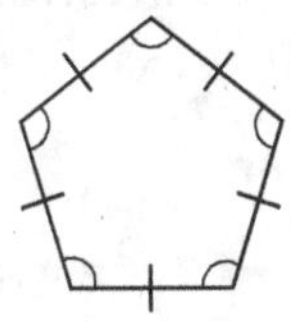

The polygon has _ sides.

The sides and angles are all congruent.

The polygon is a ________________.

Interior and Exterior Angles

An **interior angle** is an angle that is inside a shape.

An **exterior angle** is any angle that is between any side of a shape and a line extended from the adjacent side.

Determine whether each angle is interior or exterior.

∠1 is formed by the side of the shape and a line extended from the adjacent side. It is an exterior angle.

∠2 is inside of the polygon. It is an interior angle.

∠3 is inside the polygon. It is an interior angle.

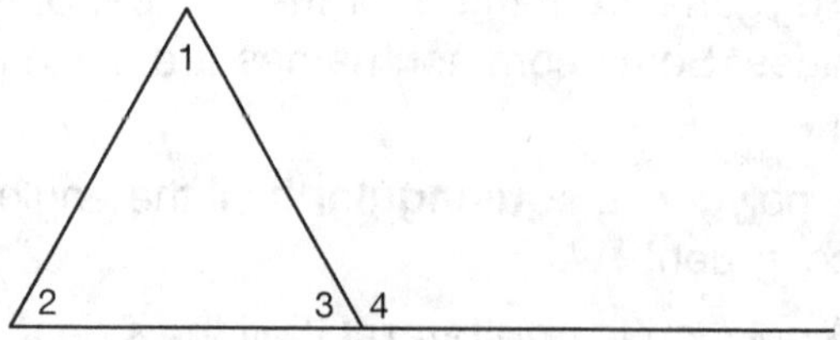

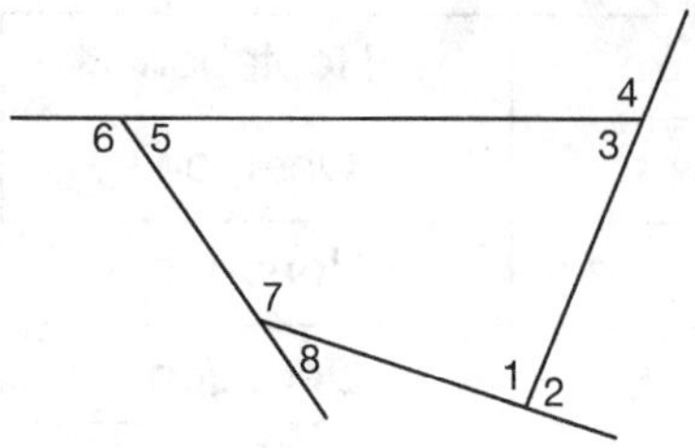

Practice

Complete each sentence.

3. ∠1 is inside the polygon, so it is an ________ angle.

4. ∠4 is formed by the side of the shape and a line extending outside the shape, so it is an ________ angle.

Determine whether each angle is an interior angle or an exterior angle.

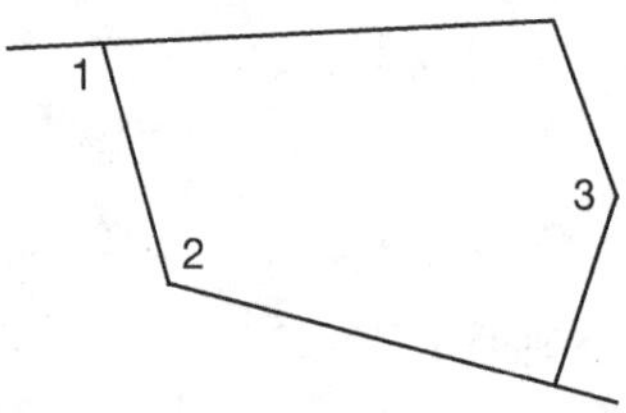

5. ∠1 ________ angle

6. ∠8 ________ angle

7. ∠3 ________ angle

8. ∠2 ________ angle

9. ∠6 ________ angle

10. ∠5 ________ angle

Name ______________________________ Date ___________ Class ___________

Reteaching 16

Finding Slopes and Equations of Lines

You have worked with ordered pairs. Now you will find the slope and equation of the line between two ordered pairs.

Slope

The **slope** of a line describes how steep the line is. You can find the slope by writing the ratio of the **rise** to the **run.**

slope $= \frac{\text{rise}}{\text{run}} = \frac{3}{6} = \frac{1}{2}$

You can use a formula to calculate the slope m of the line through points (x_1, y_1) and (x_2, y_2).

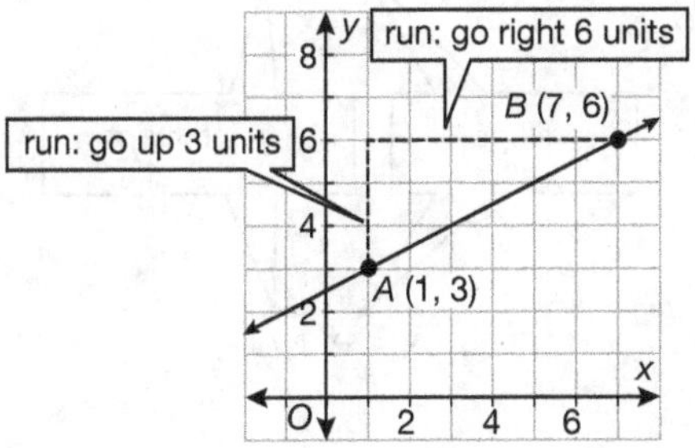

Find the slope m of $\overleftrightarrow{AB}$ using the formula.
Substitute (1, 3) for (x_1, y_1) and (7, 6) for (x_2, y_2).

$m = \frac{y_2 - y_1}{x_2 - x_1}$

$m = \frac{6 - 3}{7 - 1}$ Substitute.

$m = \frac{3}{6}$ Simplify.

$m = \frac{1}{2}$ Simplify.

Practice

Complete the steps to find the slope of each line.

1. $m = \frac{y_2 - y_1}{x_2 - x_1}$

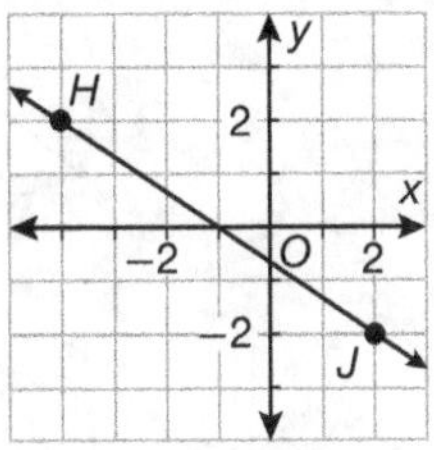

$m = \frac{-2 - \_\_\_}{2 - \_\_\_}$

$m = \frac{\_\_\_}{6}$

$m = -\frac{\_\_\_}{3}$

2. $m = \frac{y_2 - y_1}{x_2 - x_1}$

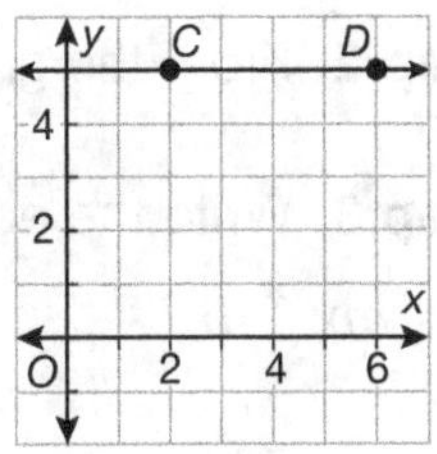

$m = \frac{\_\_\_ - \_\_\_}{\_\_\_ - 2}$

$m = \frac{\_\_\_}{4}$

$m = \_\_\_$

Use the slope formula to determine the slope of each line.

3. (0, −2)(1, 6) $m =$ ___

4. (3, 1)(6 , −3) $m =$ ___

Equations of Lines

The slope-intercept form of a line is one way of writing a linear equation using the slope *m* and the *y*-intercept *b* of the line.

Slope-Intercept Form	Example
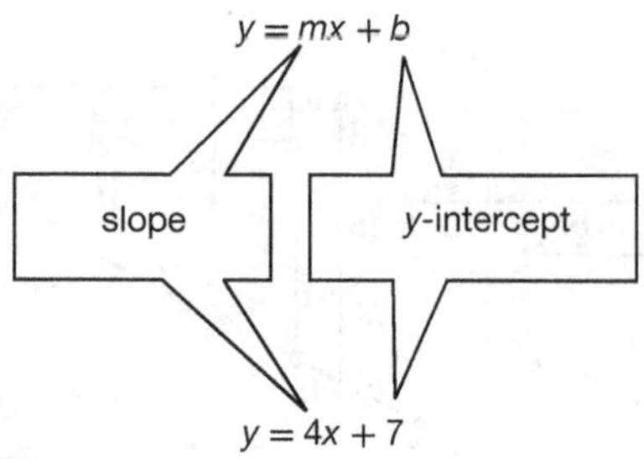	**Write the equation of the line through (0, 1) and (2, 7) in slope-intercept form.** **Step 1:** Find the slope. $m = \frac{y_2 - y_1}{x_2 - x_1}\ \frac{7 - 1}{2 - 0} = \frac{6}{2} = 3$ **Step 2:** The *y*-intercept is (0, 1), so 1 is the value of *b*. **Step 3:** Write the equation. $y = mx + b$ $y = 3x + 1$ Substitute 3 for *m* and 1 for *b*

Practice

Complete the steps to write the equation of the line in slope-intercept form.

5. **Step 1:** Determine two points on the line to find the slope.

   Use points (4, 2) and (–4, –4).

   $m = \frac{y_2 - y_1}{x_2 - x_1}\ \frac{-4 - \_\_\_}{-4 - \_\_\_} = \frac{\_\_\_}{-8} = \frac{3}{\_\_\_}$

   Step 2: From the graph the *y*-intercept is (0, –1).

   Step 3: Write the equation.

   $y = mx + b$

   $y = \_\_\_ x - 1$

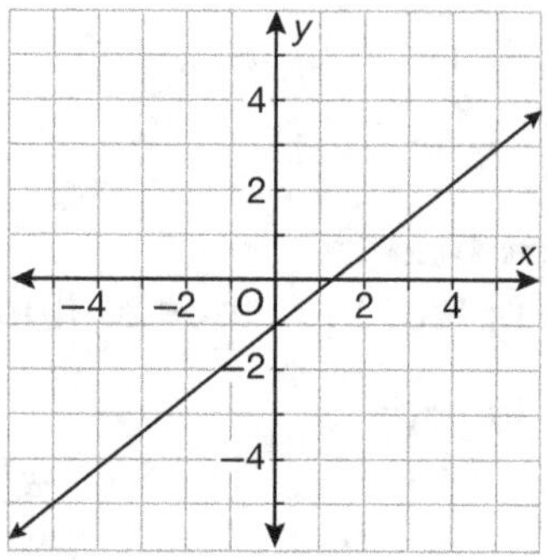

Write the equation of each line.

6. the line through (0, 2) and (5, 8)

   ____

7. the line through (0, 5) and (–2, 6)

   ____

Name ______________________ Date __________ Class __________

Reteaching 17

More Conditional Statements

Now you are going to look at the converse of a statement which results from switching the hypothesis and conclusion.

Given the conditional statement below, state the converse.

If x is an even number, then x is divisible by 2.

Hypothesis	…x is an even number….
Conclusion	…x is divisible by 2….
Converse	If x is divisible by 2, then x is an even number.

Is the converse a true statement?

The converse is a true statement. We know that if a number is divisible by 2, then it is an even number.

Practice

Complete the statements for the hypothesis, conclusion, and converse.

1. If a line containing points *J*, *K*, and *L* lies in a plane, then *J*, *K*, and *L* are coplanar.

   Hypothesis: A line containing points *J*, *K*, and *L* lies in a plane.

   Conclusion: ______________________

   Converse: If *J*, *K*, and *L* are coplanar, then ______________________ ______________________

2. If it is Tuesday, then play practice is at 6:00.

   Hypothesis: ____________

   Conclusion: Play practice is at 6:00.

   Converse: ______________________, then it is Tuesday.

Identify the hypothesis and conclusion for each statement. Then, state the converse.

3. If you buy this cell phone, then you will receive ten free ringtones.

   Hypothesis: ______________________

   Conclusion: ______________________

   Converse: ______________________ ______________________

Two other conditional statements can be formed from the hypothesis and conclusion.

Inverse: This is formed when the hypothesis and conclusion are negated.

Contrapositive: This is formed by both exchanging and negating the hypothesis and conclusion.

Statement	Example
Conditional	If a figure is a square, then it has four right angles. (Hypothesis: a figure is a square; Conclusion: it has four right angles)
Converse	**Switch the hypothesis and conclusion.** If a figure has four right angles, then it is a square.
Inverse	**Negate the hypothesis and conclusion.** If a figure is not a square, then it does not have four right angles.
Contrapositive	**Switch and negate the hypothesis and conclusion.** If a figure does not have four right angles, then it is not a square.

Practice

Complete the statements of the converse, inverse, and contrapositive.

4. If an animal is an armadillo, then it is nocturnal.

Converse: If an animal is nocturnal, then ____________________

Inverse: If an animal is not an armadillo, then ____________________

Contrapositive: If an animal is not nocturnal, then ____________________

Identify the hypothesis and conclusion of each statement. Then, state the converse, inverse, and contrapositive.

5. If an angle has a measure less than 908, then it is acute.

Converse: ____________________

Inverse: ____________________

Contrapositive: ____________________

6. If $y = 1$, then $y^2 = 1$.

Converse: ____________________

Inverse: ____________________

Contrapositive: ____________________

Name ______________________ Date __________ Class __________

Reteaching 18

Triangle Theorems

You have worked with different angle measures and classified angles in triangles. Now you will work with special angle relationships in triangles.

According to the **Triangle Angle Sum Theorem,** the sum of the angle measures of a triangle is 180°.

Find the measure of $\angle L$.

Step 1: Write the equation. $m\angle J + m\angle K + m\angle L = 180°$

Step 2: Substitute. $62° + 73° + m\angle L = 180°$

Step 3: Solve for $m\angle L$. $135° + m\angle L = 180°$

$m\angle L = 45°$

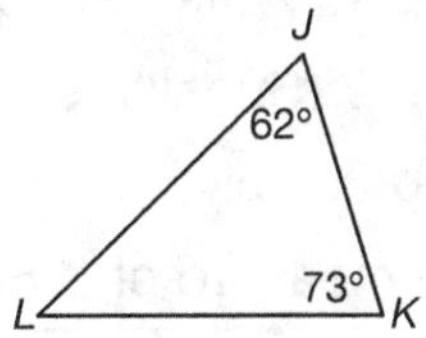

The measure of $\angle L$ is 45°.

Practice

Complete the steps to determine the measure of the missing angle.

1.

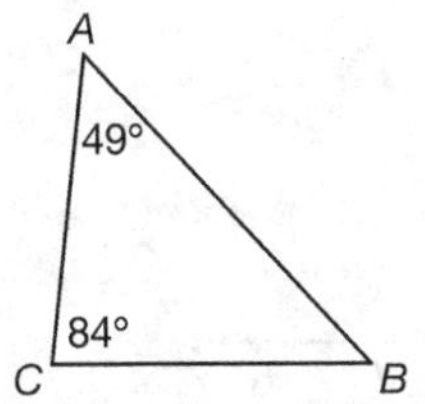

$m\angle A + m\angle B + m\angle C =$ ____

____ $+ m\angle B +$ ____ $=$ ____

____ $+ m\angle B =$ ____

$m\angle B =$ ____

2.

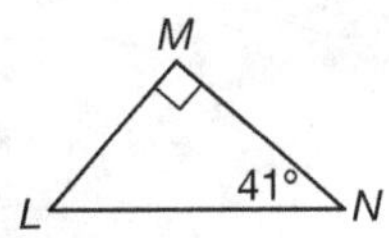

$m\angle L + m\angle M + m\angle N =$ ____

$m\angle L + 90° +$ ____ $=$ ____

$m\angle L +$ ____ $=$ ____

$m\angle L =$ ____

Find the measure of the missing angle.

3.

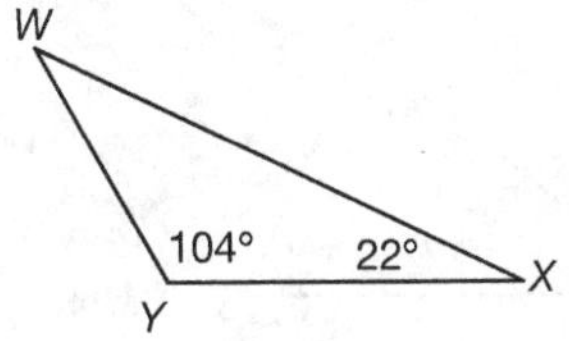

____

4.

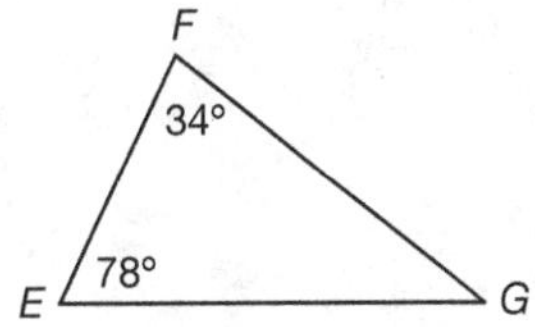

____

An **exterior angle** of a triangle is formed by one side of the triangle and the extension of an adjacent side.

The **Exterior Angle Theorem** states that the measure of an exterior angle of a triangle is equal to the sum of the measures of its remote interior angles.

Find the measure of ∠*FHJ*.

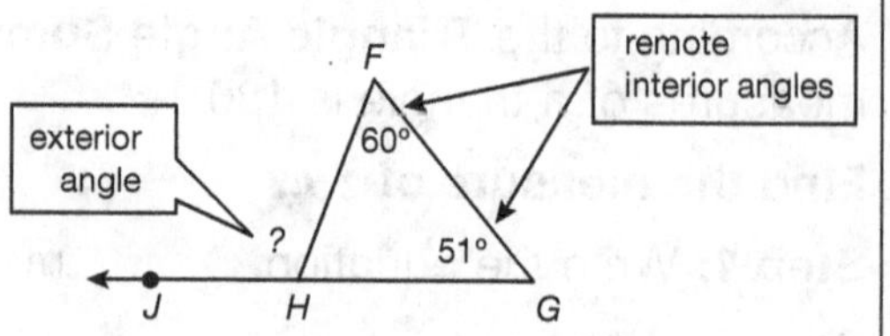

Step 1: Write the equation. $m\angle F + m\angle G = m\angle FHJ$

Step 2: Substitute. $60° + 51° = m\angle FHJ$

Step 3: Solve. $111° = m\angle FHJ$

The measure of ∠*FHJ* is 111°.

Practice

Complete the steps to determine the measure of the angle.

5. ∠*ABD*

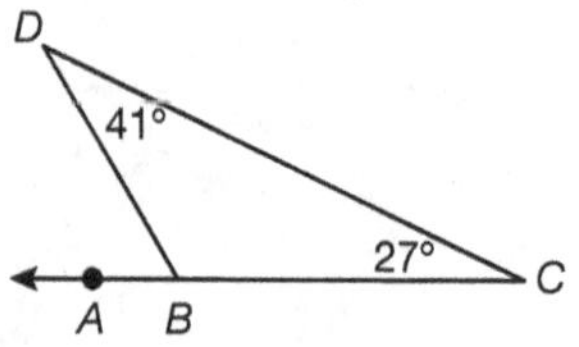

$m\angle D +$ ______ $= m\angle ABD$

$41° +$ ___ $= m\angle ABD$

___ $= m\angle ABD$

6. ∠*HJK*

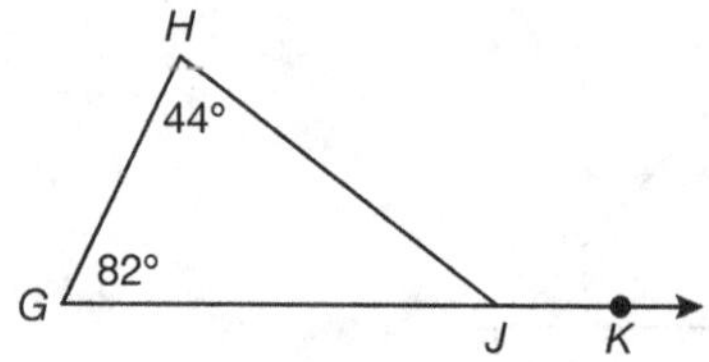

______ $+ m\angle H = m\angle HJK$

___ $+ 44° = m\angle HJK$

_____ $= m\angle HJK$

Find the measure of the angle.

7. ∠*MNP*

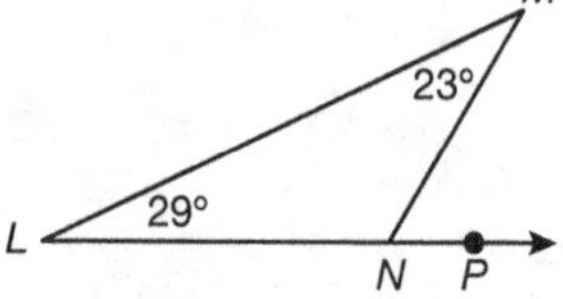

___

8. ∠*QRS*

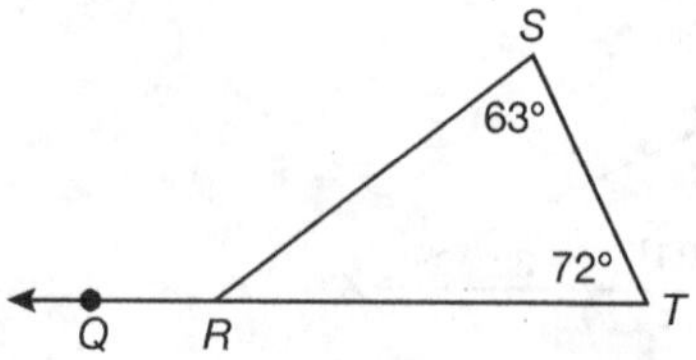

_____

Name ______________________ Date __________ Class __________

Reteaching 19

Introduction to Quadrilaterals

A quadrilateral is a polygon with four sides. Specific properties of figures are listed in the table below.

Figure	Properties
Parallelogram	Both pairs of opposite sides are parallel.
Kite	It has exactly two pairs of congruent consecutive sides.
Trapezoid	Exactly one pair of opposite sides is parallel.
Trapezium	No sides are parallel.
Rectangle	It is a parallelogram with four right angles.
Rhombus	It is a parallelogram with four congruent sides.
Square	It is a parallelogram with four right angles and four congruent sides.

Classify the quadrilateral. Give multiple names if possible.

Quadrilateral *EFGH*:

Sides $\overline{EF}$ and $\overline{HG}$ are parallel.

Sides $\overline{HE}$ and $\overline{GF}$ are parallel.

A figure with opposite sides parallel is a parallelogram.

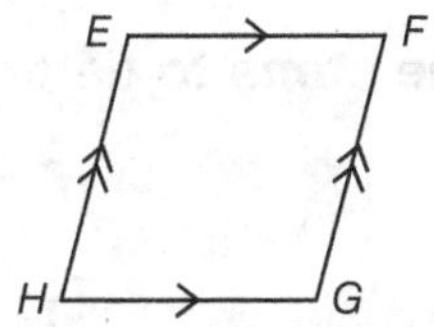

Practice

Complete the steps to classify the quadrilateral. Give multiple names if possible.

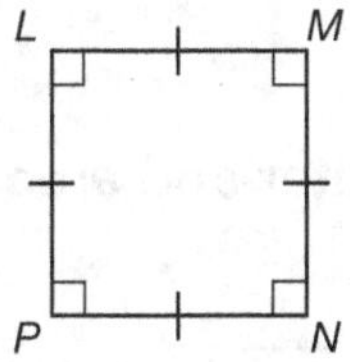

1. $\angle L$, $\angle M$, $\angle N$, and $\angle P$ are ____ angles.

$\overline{LM}$, ____ , ____ , and ____ are congruent sides.

The figure is a ______.

It is also a parallelogram, __________, and a rectangle.

Classify the quadrilaterals. Give multiple names if possible.

2.

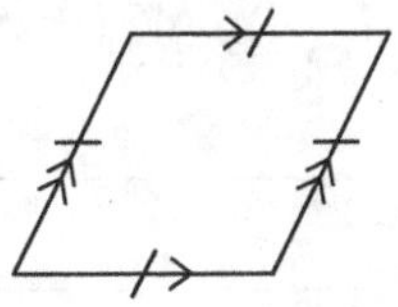

____

3.

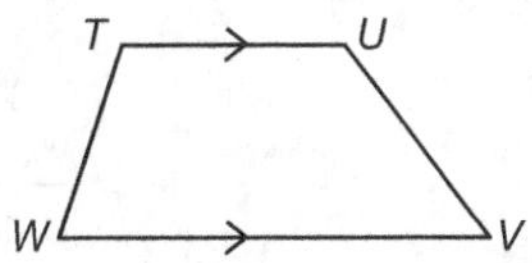

____

Determine the perimeter, area, length and width of this rectangle.

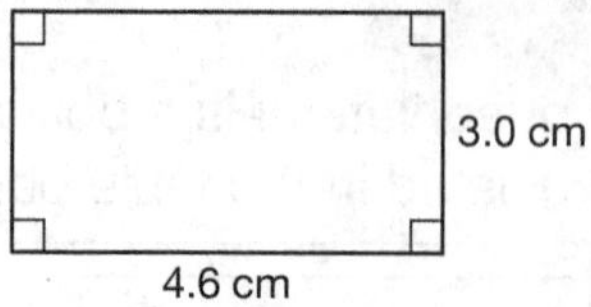

The length is 4.6 centimeters, and the width is 3.0 centimeters.

The perimeter is the sum of the side lengths.

$P = 3.0 + 4.6 + 3.0 + 4.6$

$P = 2(3.0) + 2(4.6)$

$P = 15.2$

The perimeter of the rectangle is 15.2 centimeters.

The area is the side length times the side width.

$A = lw$

$A = (4.6)(3.0)$

$A = 13.8$

The area of the rectangle is 13.8 cm^2.

Practice

Complete the steps to determine the perimeter and area of the figure.

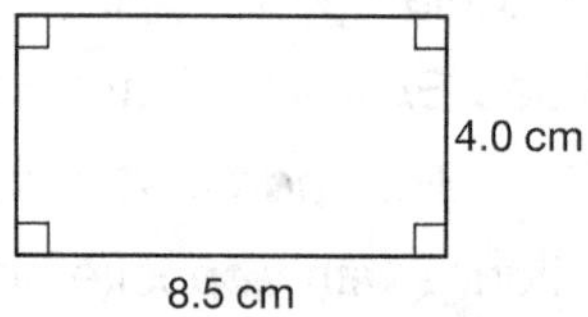

4. Perimeter

$P =$ ___ + ___ + 4.0 + 8.5

$P = 2$ ___ + 2 _____

$P =$ ___ cm

Area

$A = lw$

$A =$ _____ (4)

$A =$ ___ cm^2

Find the perimeter and area of each figure.

5.

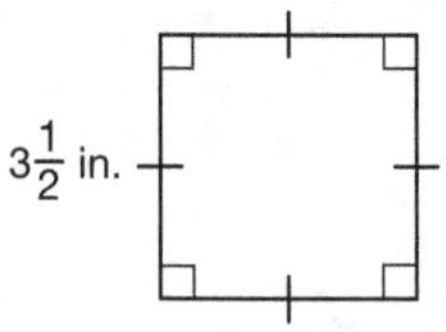

____

____

6.

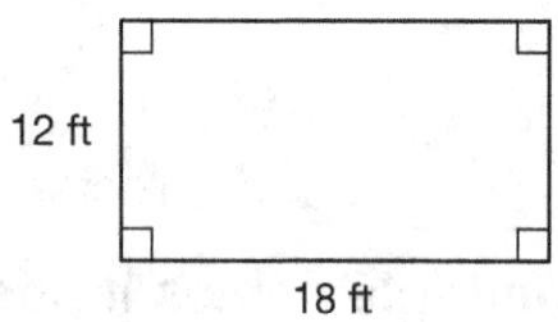

____

____

7.

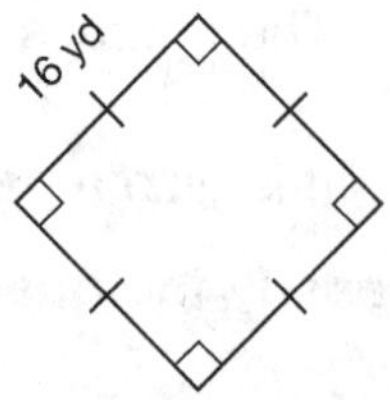

____

____

Name ______________________ Date __________ Class __________

Reteaching 20

Interpreting Truth Tables

You have worked with conditional statements. Now you will work with biconditional statements and truth tables.

> A **biconditional statement** combines a conditional statement (if *p*, then *q*) with its converse (if *q*, then *p*).
>
> **Conditional:** *p* ... *q*
> If the sides of a triangle are congruent, then the angles are congruent.
>
> **Converse:** *q* ... *p*
> If the angles of a triangle are congruent, then the sides are congruent.
>
> **Biconditional:** *p* ... *q*
> The sides of a triangle are congruent if and only if the angles are congruent.

Practice

Complete the statements for the converse and biconditional.

1. If you can download six songs for $5.94, then each song costs $0.99.

   Converse: If each song costs $0.99, then ______________ ______________.

   Biconditional: You can download six songs for $5.94 if and only if ______________.

2. If Lindsay works on the yearbook, then she does not play soccer.

   Converse: ______________ then she works on the yearbook.

   Biconditional: ______________ if and only if she does not play soccer.

For each conditional, write the converse and a biconditional statement.

3. If a figure has ten sides, then it is a decagon.

   Converse: ______________

   Biconditional: ______________

4. An angle is obtuse if it measures between 90 and 180 degrees.

   Converse: ______________ ______________

   Biconditional: ______________ ______________

A compound statement combines two statements using *and* or *or.*
A compound statement that uses *and* is called a **conjunction.** A compound statement that uses *or* is called a **disjunction.**

The table below shows when a conjunction or disjunction is true or false.

p	*q*	Conjunction: *p* and *q*	Disjunction: *p* or *q*
T	T	T	T
T	F	F	T
F	T	F	T
F	F	F	F

Example: Write a conjunction using the two statements and determine whether the conjunction is true or false. All squares are rectangles.
A foot is 12 inches.

Conjunction: All squares are rectangles, and a foot is 12 inches.

The conjunction is **true** since both statements are true.

Example: Write a disjunction using the two statements and determine whether the disjunction is true or false. Pine trees are evergreens. Giraffes are blue.

Disjunction: Pine trees are evergreens, or giraffes are blue.

The disjunction is **true** since one statement is true.

Practice

Complete the statements for the conjunction and disjunction and determine whether the statement is true or false.

5. A triangle has three sides. An octagon has three sides.

Conjunction: A triangle has three sides, and ______________________.

Disjunction: ______________________, or an octagon has three sides.

The conjunction is _____.

The disjunction is true since one of the statements is true.

Write the conjunction and disjunction and determine whether the statement is true or false.

6. A parallelogram has opposite parallel sides. A square has four congruent sides.

Conjunction: ____

____

Disjunction: ____

____

The conjunction is _____. The disjunction is _____.

Name ______________________ Date __________ Class __________

Reteaching

Proving the Pythagorean Theorem

INV 2

Right Triangles

A **right triangle** is a triangle with one right angle. The side of the right triangle that is opposite the **right angle** is called the hypotenuse. The other two sides of the triangle are called the **legs**.

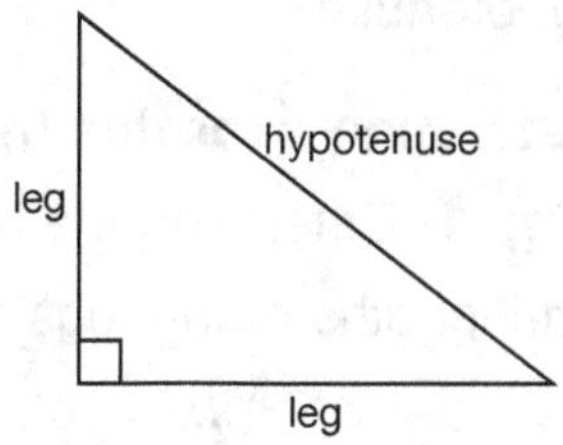

The **Pythagorean theorem** states that, for a right triangle, the sum of the squares of the lengths of the legs is equal to the square of the length of the hypotenuse.

Explain how the right triangle satisfies the Pythagorean theorem.

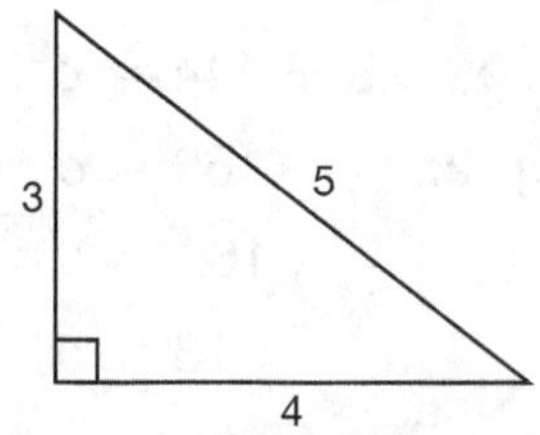

The square of the length of one leg is 3^2, or 9. The square of the length of the other leg is 4^2, or 16. The square of the length of the hypotenuse is 5^2, or 25. The sum of the squares of the lengths of the legs is $9 + 16 = 25$, which is equal to the square of the hypotenuse's length.

Practice

1. Draw a right triangle with side lengths of 9 centimeters, 12 centimeters, and 15 centimeters.
   Use a protractor to ensure that one angle is 90°.
   Draw and cut out three squares: one with side length of 9 centimeters, one with side length of 12 centimeters, and one with side length of 15 centimeters.
   Place each square next to the corresponding side of the triangle.
2. The length of the hypotenuse is ______.
3. The lengths of the legs are _____ and ______.
4. The area of the square on the hypotenuse is ________.
5. The area of the squares on the legs are _______ and ________.
6. Calculate the area of each of the squares that you constructed.
   Show that the sum of the areas of the two smaller squares equals the area of the largest square.
   ____
7. How does the length of each side of the triangle relate to the area of the corresponding square?
   ____
   ____

The Pythagorean Theorem

The Pythagorean theorem states that $a^2 + b^2 = c^2$, where *a* and *b* represent the lengths of the legs and *c* represents the length of the hypotenuse.

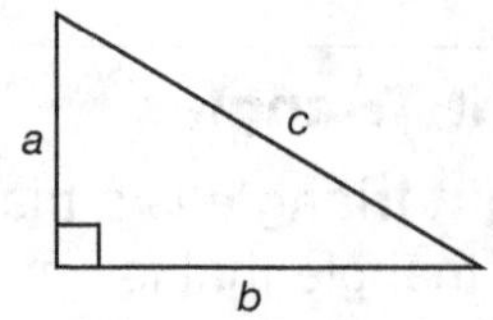

Determine *c* for this triangle.

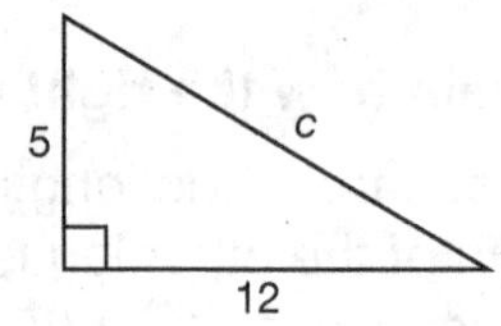

Step 1: Determine the lengths of the legs of the right triangle.

The lengths of the legs are 5 and 12.

$a^2 + b^2 = c^2$

$5^2 + 12^2 = c^2$ Substitute values.

$25 + 144 = c^2$ Simplify.

Step 2: $169 = c^2$ Simplify.

$\sqrt{169} = \sqrt{c^2}$ Take square roots of both sides.

$13 = c$

The value of *c* is 13.

Practice

Complete the steps to find the length of the indicated side.

8. A right triangle has a hypotenuse that measures 26 inches and one leg that measures 10 inches. What is the length of the other leg?

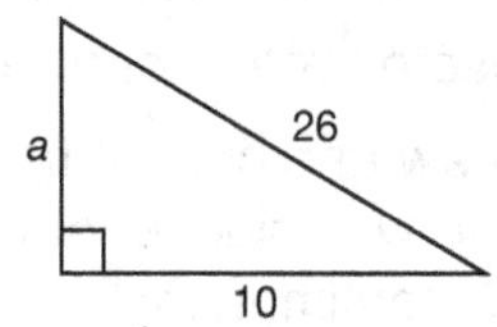

Represent the length of the missing side by the letter ___.

$a^2 + b^2 =$ ___

$a^2 + 10^2 =$ ___

___ $+ 100 =$ ___

$a^2 + 100 -$ ___ $=$ ___ $- 100$

$a^2 =$ ___

$a =$ ___

The length of the leg is ___ inches.

Find the length of the indicated side.

9. A right triangle has a hypotenuse that measures 17 centimeters and one leg that measures 8 centimeters. What is the length the other leg? ______

10. A right triangle has one leg that measures 20 meters and one leg that measures 15 meters. What is the length of the hypotenuse? _____

11. A right triangle has a hypotenuse that measures 10 inches and one leg that measures 6 inches. What is the length of the other leg? ____.

12. A right triangle has one leg that measures 9 yards and a hypotenuse that measures 15 yards. What is the length of the other leg? ______

Name ______________________ Date __________ Class __________

Reteaching 21

Laws of Detachment and Syllogism

With deductive, reasoning you will use facts, definitions, and properties to draw conclusions and prove whether those conclusions are valid.

The **Law of Detachment** allows you to draw conclusions from true facts.

Given	If you have \$2, then you can buy a snack. You have \$2.	If you have \$2, then you can buy a snack. You can buy a snack.
Therefore	You can buy a snack.	You have \$2.
Valid Conclusion?	Yes, the conditional and the hypothesis are both true.	No, the hypothesis may or may not be true. For example, if you borrowed \$2 from someone, you could also buy a snack.

Practice

Complete a valid concluding statement using the Law of Detachment.

1. If you ride the Titan roller coaster in Arlington, Texas, then you will drop 255 feet.

Michael rode the Titan roller coaster.

Therefore, Michael dropped ____ feet.

2. A segment that is a diameter of a circle has endpoints on the circle.

$\overline{GH}$ is the diameter of a circle.

Therefore, $\overline{GH}$ has endpoints __________.

For the following statements, use the Law of Detachment to write a valid concluding statement.

3. Opposite rays are two rays that have a common endpoint and form a line.

Therefore, ___ and ___ are opposite rays.

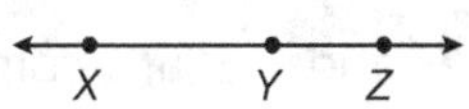

4. If a person sees penguins, then he or she is in Antarctica.

Catherine sees penguins.

Therefore, ______________________.

5. If the sum of the digits of a number can be divided by 3, then the number can be divided by 3.

The number 171 has a sum of 9.

Therefore, ______________________.

Another valid form of deductive reasoning is the **Law of Syllogism**; it is similar to the Transitive Property of Equality.

Transitive Property of Equality	Law of Syllogism
If $y = 10x$ and $10x = 20$, then $y = 20$.	**Given:** If you have a horse, then you have to feed it. If you have to feed a horse, then you have to get up early every morning. **Conclusion:** If you have a horse, then you have to get up early every morning.

Practice
Determine whether each conclusion is valid by the Law of Syllogism.

6. **Given:** If you buy a car, then you can drive to school. If you can drive to school, then you will not ride the bus.

   Conclusion: If you buy a car, then you will not ride the bus. _____

7. **Given:** If $\angle K$ is obtuse, then it does not have a measure of 90°. If an angle does not have measure of 90°, then it is not a right angle.

   Conclusion: If $\angle K$ is obtuse, then it is not a right angle. _____

8. **Given:** If two segments are congruent, then they have the same measure. If two segments have the same measure of 6.5 centimeters, then they are congruent.

   Conclusion: If two segments are congruent, then they each have a measure of 6.5 centimeters. _____

Draw a conclusion from the given information.

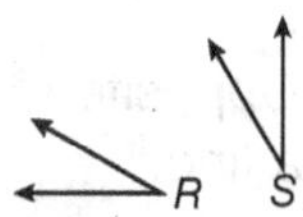

9. If two angles are alternate interior angles, then they are congruent. If two angles are congruent, then they have the same measure. $\angle 1$ and $\angle 2$ are alternate interior angles.

   ____

10. If $\angle R$ and $\angle S$ are complementary to the same angle, then the two angles are congruent. If two angles are congruent, then they have the same angle measure. $\angle R$ and $\angle S$ are complementary to the same angle.

   ____

Name ______________________ Date __________ Class ____________

Reteaching 22

Finding Areas of Quadrilaterals

You have found the area of a triangle. Now you will find the area of different quadrilaterals.

Area of a Trapezoid

A trapezoid with bases b^1 and b^2 and height h has area $A = \frac{1}{2}(b_1 + b_2)h.$

Find the area of the trapezoid.

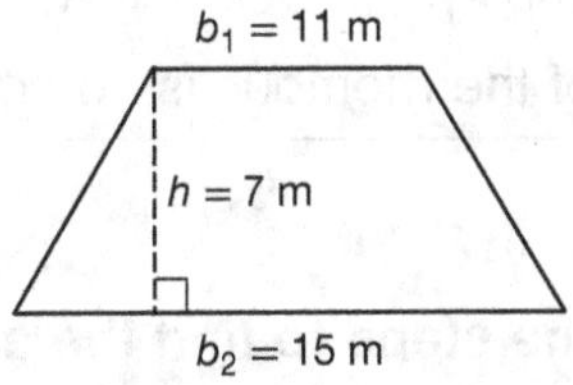

Step 1: Write the formula. $A = \frac{1}{2}(b_1 + b_2)h$

Step 2: Substitute. $A = \frac{1}{2}(11\text{ m} + 15\text{ m})(7\text{ m})$

Step 3: Simplify. $A = \frac{1}{2}(26\text{ m})(7\text{ m})$

$A = 91\text{ m}^2$

The area of the trapezoid is 91 m^2.

Practice

Complete the steps to find the area of each trapezoid.

1.

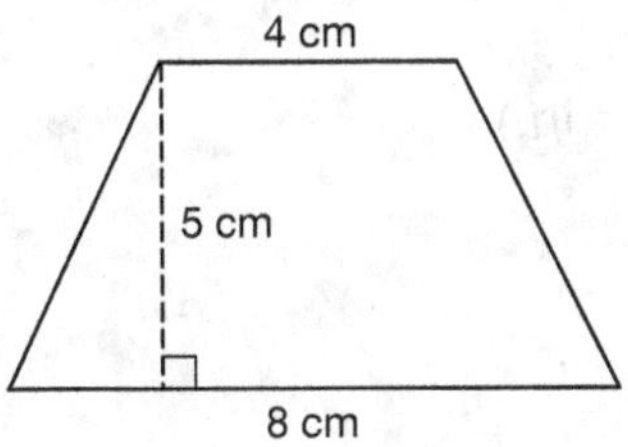

$A = \frac{1}{2}(b_1 + b_2)h$

$A = \frac{1}{2}(4\text{ cm} + \_\text{ cm})(\_\text{ cm})$

$A = \frac{1}{2}(\_\_\text{ cm})(\_\text{ cm})$

$A = \_\_\text{ cm}^2$

2.

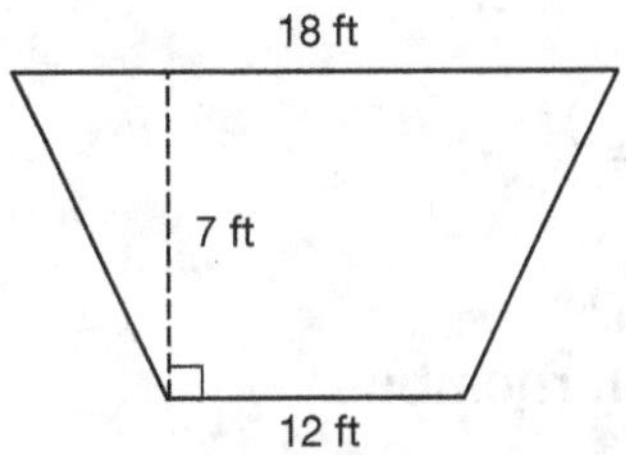

$A = \frac{1}{2}(b_1 + b_2)h$

$A = \frac{1}{2}(18\text{ ft} + \_\_\text{ ft})(\_\text{ ft})$

$A = \frac{1}{2}(\_\_\text{ ft})(\_\text{ ft})$

$A = \_\_\text{ ft}^2$

Find the area of each trapezoid.

3. ______

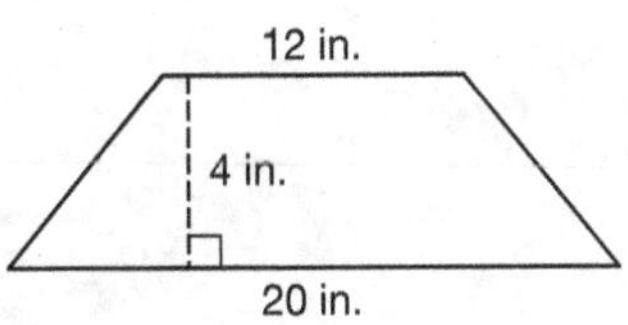

4. ______

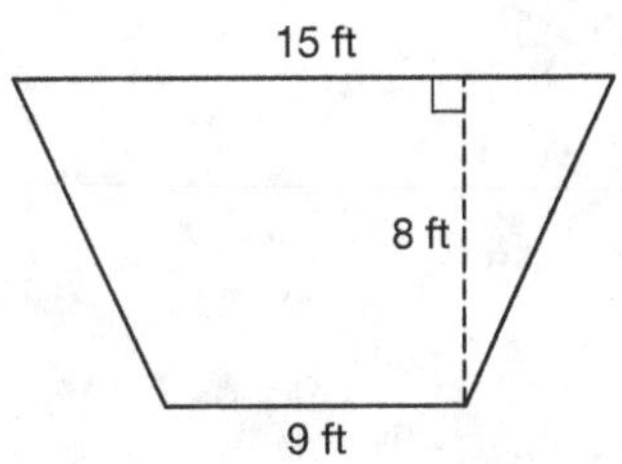

Area of a Rhombus

A rhombus with diagonals d^1 and d^2 has area $A = \frac{1}{2}d_1d_2$.

Find the area of the rhombus.

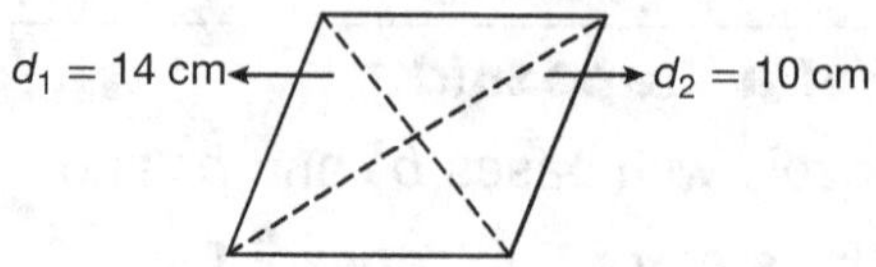

Step 1: Write the formula. $A = \frac{1}{2}d_1d_2$

Step 2: Substitute. $A = \frac{1}{2}(14 \text{ cm})(10 \text{ cm})$

Step 3: Simplify. $A = 70 \text{ cm}^2$

The area of the rhombus is 70 cm^2.

Practice

Complete the steps to find the area of each rhombus.

5.

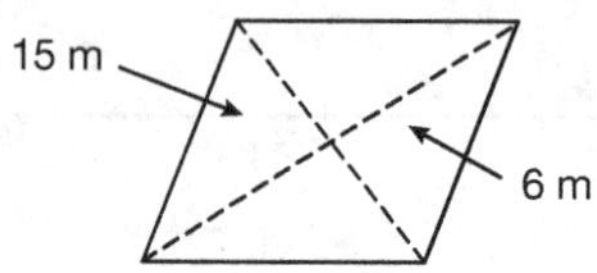

$A = \frac{1}{2}d_1d_2$

$A = \frac{1}{2}(15 \text{ m})(\_\_\_\_)$

$A = \_\_\_ \text{ m}^2$

6.

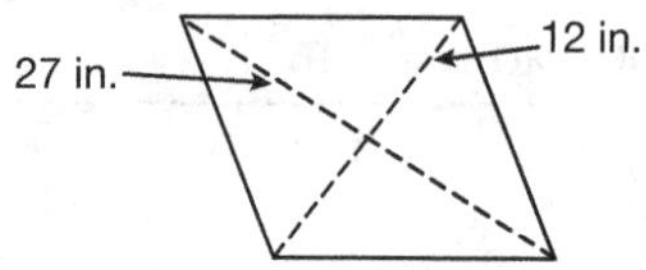

$A = \frac{1}{2}d_1d_2$

$A = \frac{1}{2}(\_\_ \text{ in.})(\_\_ \text{ in.})$

$A = \_\_\_\_ \text{ in}^2$

Find the area of each rhombus.

7. ________

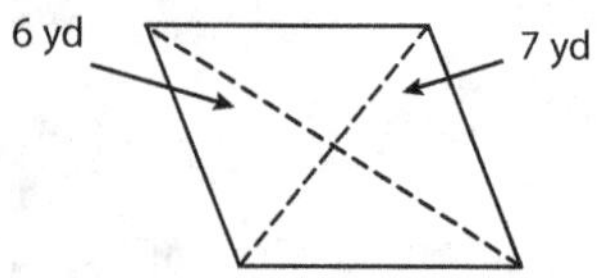

8. ________

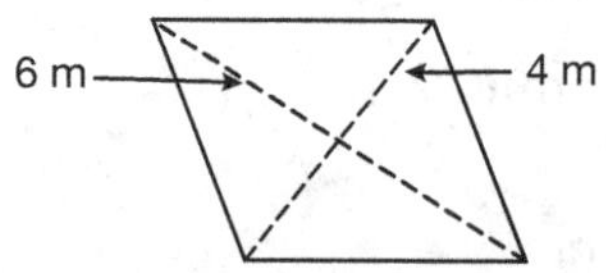

9. Find the area of a rhombus in which $d_1 = 14$ in. and $d_2 = 48$ in.

____

10. Find the d_2 of a rhombus in which $d_1 = 15$ m and A $= 180 \text{ m}^2$.

____

Name ______________________ Date __________ Class __________

Reteaching 23

Introduction to Circles

Now you will find the circumference and area of circles

Circumference of Circles

A circle with radius r has circumference $C = 2\pi r$.

Find the circumference of the circle to the nearest hundredth inch. Use 3.14 for π.

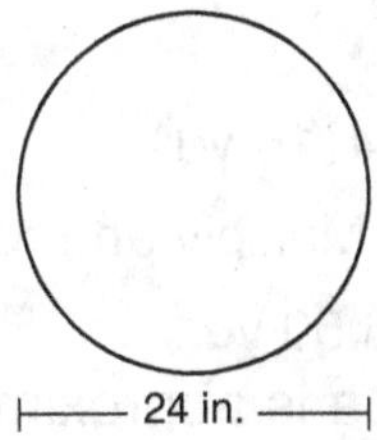

Step 1: Divide the diameter by 2 to find the radius of the circle.

$24 \text{ in.} \div 2 = 12 \text{ in.}$

The radius r is 12 in.

Step 2: Substitute $r = 12$ in. and $\pi = 3.14$ into the formula.

$C = 2\pi r$

$C \approx 2(3.14)(12 \text{ in.})$

Step 3: Multiply and round to the nearest hundredth inch.

$C \approx 75.36$ in.

The circumference is approximately 75.36 inches.

Practice

Complete the steps to find the circumference. Use 3.14 for π.

1.

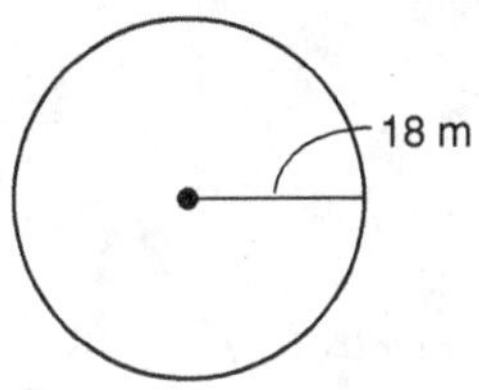

$r =$ ______

$C = 2\pi r$

$C \approx 2($______$)($______$)$

$C \approx$ __________

2.

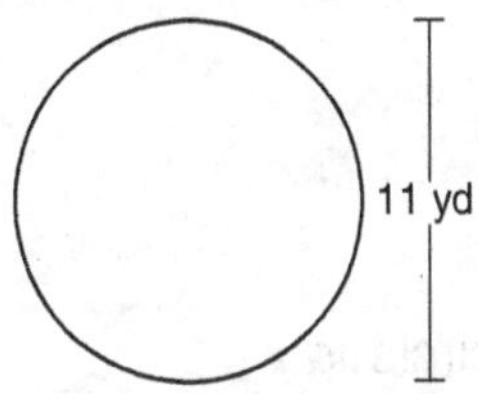

$r =$ ______

$C = 2\pi r$

$C \approx 2($______$)($______$)$

$C \approx$ __________

Find the circumference of each circle. Use 3.14 for π.

3.

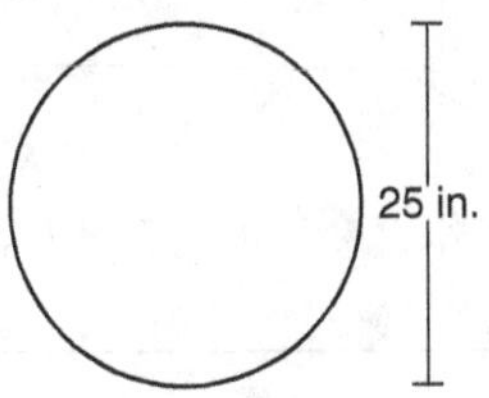

____

4.

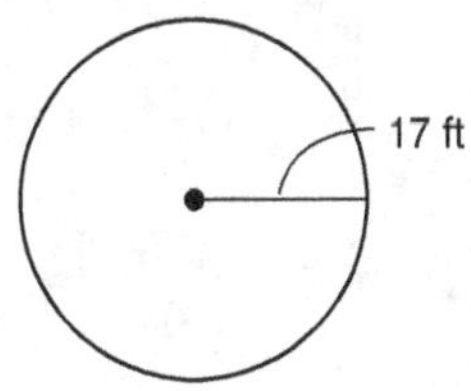

____

Area of Circles

A circle with radius r has area $A = \pi r^2$.

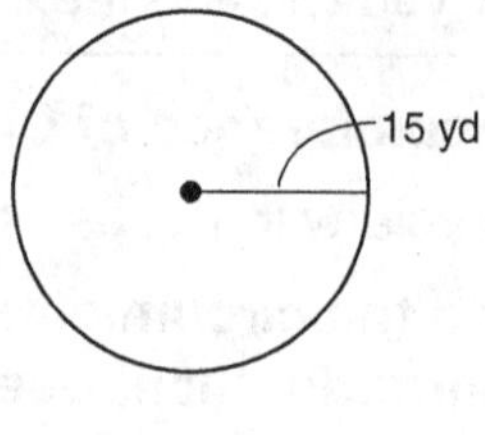

Find the area of the circle to the nearest hundredth square yard. Use 3.14 for π.

Step 1: Find the radius of the circle. The radius r is 15 yd.

Step 2: Substitute r = 15 yd and π = 3.14 into the formula for the area of a circle.

$A = \pi r^2$

$A \approx 3.14(15 \text{ yd})^2$

Step 3: Multiply and round to the nearest hundredth yard.

$A \approx 706.50 \text{ yd}^2$

The area is approximately 706.50 yd^2.

Practice

Complete the steps to find the area of the given circle to the nearest hundredth square unit. Then, complete the sentence. Use 3.14 for π.

5.

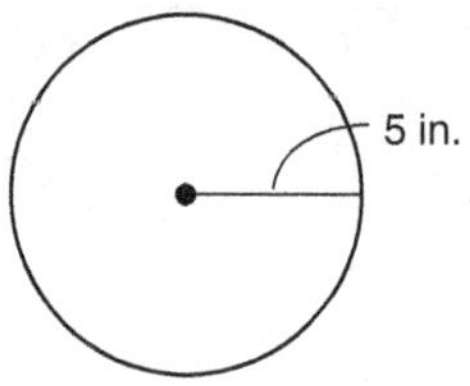

The radius of the circle is r = ______

$A = \pi r^2$

$A \approx$ ______(______)^2

$A \approx$ ____________

6.

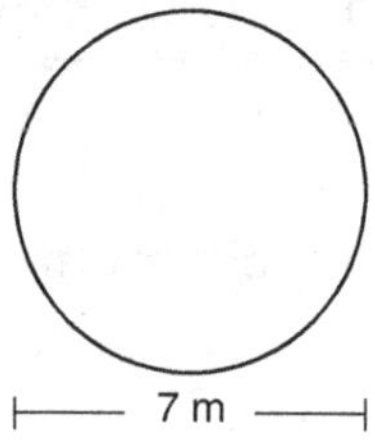

The radius of the circle is r = ______.

$A = \pi r^2$

$A \approx$ ______(______)^2

$A \approx$ ____________

Find the area of the given circle to the nearest hundredth. Use 3.14 for π.

7.

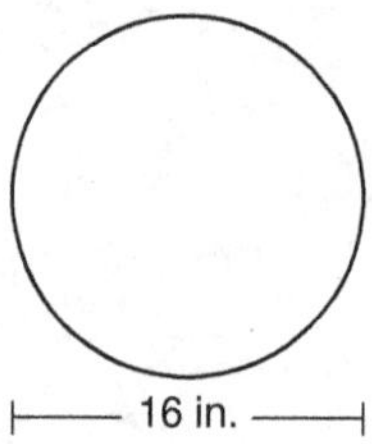

____

8.

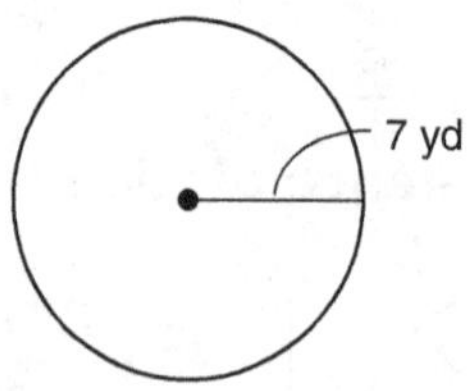

____

Name ______________________ Date __________ Class __________

Reteaching 24

Algebraic Proofs

You have learned about deductive reasoning. Now you will work with algebraic proofs.

Algebraic Proofs

An **algebraic proof** uses algebraic properties—including the Distributive Property and the properties of equality—to show how a problem is solved step-by-step. The original equation is the **Given** statement.

Solve the equation $5(x - 2) = x + 6$ Provide a justification for each step.

$5(x - 2) = x + 6$	**Given**
$5x - 10 = x + 6$	Distributive Property
$5x - 10 + 10 = x + 6 + 10$	Addition Property of Equality
$5x = x + 16$	Simplify
$5x - x = x + 16 - x$	Subtraction Property of Equality
$4x = 16$	Simplify
$\frac{4x}{4} = \frac{16}{4}$	Division Property of Equality
$x = 4$	Simplify

Practice

Complete the proof.

$2x + 6 = x - 1$	Given
$2x + 6 - 6 = x - 1 - 6$	______________________
1. ___ $= x - 7$	________
$2x -$ ___ $= x - 7 - x$	Subtraction Property of Equality
$x =$ ___	________
2. $\frac{x + 2}{4} = 7$	______
$4\left(\frac{x + 2}{4}\right) = 4(7)$	______________________
$x + 2 =$ ___	________
$x + 2 - 2 =$ ___ $- 2$	Subtraction Property of Equality
$x =$ ___	________

The steps of the algebraic proof for solving the equation $\frac{8x-6}{3} = 3$ are given below in the correct order. However, the justifications for each step are out of order.

Rewrite the incorrect justifications.

	Incorrect Order	**Correct Order**
$\frac{8x-6}{3} = 6$	Addition Property of Equality	Given
$3\left(\frac{8x-6}{3}\right) = 3(6)$	Division Property of Equality	Multiplication Property of Equality
$8x - 6 = 18$	Simplify	
$8x - 6 + 6 = 18 + 6$	Given	Addition Property of Equality
$8x = 24$	Simplify	
$\frac{8x}{8} = \frac{24}{8}$	Simplify	Division Property of Equality
$x = 3$	Simplify	

Practice

The steps for the proof below are in the correct order, but some of the justifications are not. Rewrite the incorrect justifications on the lines.

3.

$4x + 8 = 2(3x - 1)$	Given	
$4x + 8 = 6x - 2$	Addition Property of Equality	________________
$4x + 8 - 8 = 6x - 2 - 8$	Division Property of Equality	________________ ________
$4x = 6x - 10$	Simplify	
$4x - 6x = 6x - 10 - 6x$	Distributive Property	________________ ________
$-2x = -10$	Subtraction Property of Equality	______
$\frac{-2x}{-2} = \frac{-10}{-2}$	Simplify	______________ ________
$x = 5$	Simplify	

Name ______________________________ Date ____________ Class ______________

Reteaching 25
Congruent Triangles

The **corresponding sides** and the **corresponding angles** of a figure are the sides and angles that are in the same position.

Identify the corresponding angles and sides for ΔABC and ΔMNP.

The names of the triangles show the following:

Corresponding Angles

$\angle A$ corresponds to $\angle M$.

$\angle B$ corresponds to $\angle N$.

$\angle C$ corresponds to $\angle P$.

Corresponding Sides

$\Delta\overline{AB}$ corresponds to $\Delta\overline{MN}$.

$\Delta\overline{BC}$ corresponds to $\Delta\overline{NP}$.

$\Delta\overline{AC}$ corresponds to $\Delta\overline{MP}$.

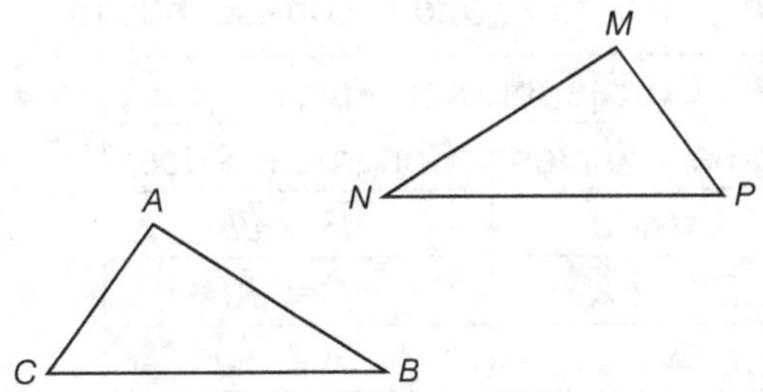

Practice

Complete each statement to identify the corresponding angles and sides for ΔDEF and ΔRST.

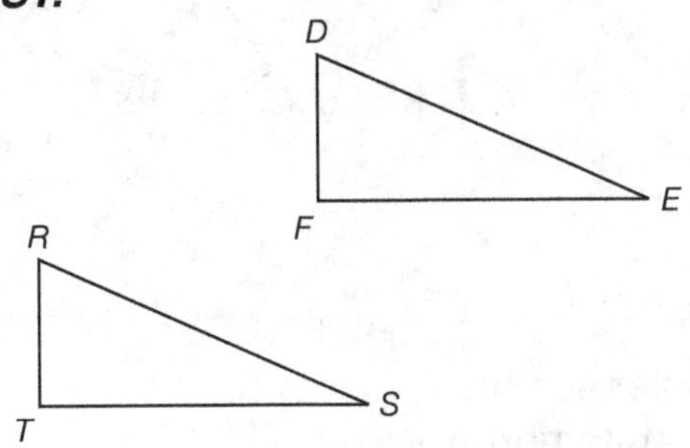

1. $\angle D$ corresponds to _____.

   $\angle E$ corresponds to _____.

   _____ corresponds to $\angle T$.

2. $\overline{DE}$ corresponds to _____.

   $\overline{DF}$ corresponds to _____.

   _____ corresponds to $\overline{ST}$.

Identify the corresponding angles and sides for ΔXYZ and ΔKLM.

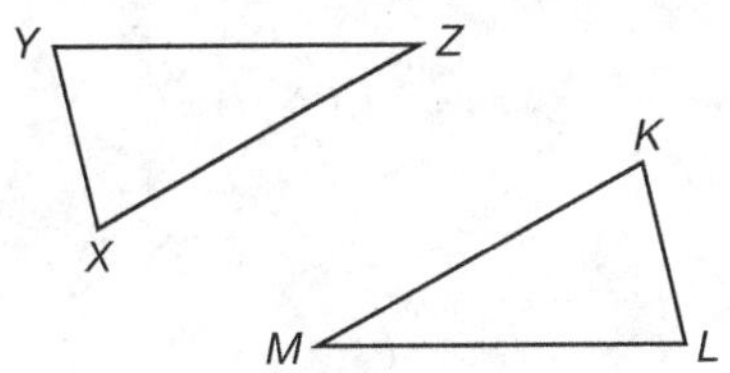

3. ______________________________.

   ______________________________.

   ______________________________.

4. ______________________________.

   ______________________________.

   ______________________________.

Triangles are **congruent triangles** when all of their corresponding sides and corresponding angles are congruent.

ΔABC and ΔJKL are congruent triangles.

Write six congruence statements for the corresponding sides and angles.

In these congruent triangles, A corresponds to J, B corresponds to K, and C corresponds to L.

Corresponding Parts	
Congruent Angles	Congruent Sides
$\angle A \cong \angle J$	$\overline{AB} \cong \overline{JK}$
$\angle B \cong \angle K$	$\overline{BC} \cong \overline{KL}$
$\angle C \cong \angle L$	$\overline{CA} \cong \overline{LJ}$

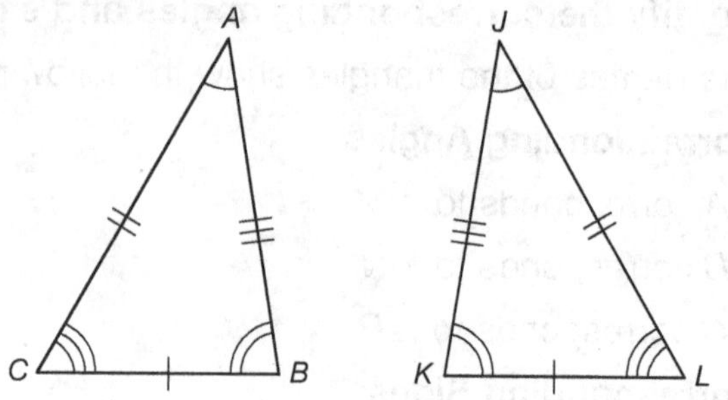

$\Delta ABC \cong \Delta JKL$

Practice

Complete the six congruence statements for the corresponding sides and angles.

5. Given: $\Delta XYZ \cong \Delta NPQ$.

$\angle X \cong$ ____ $\qquad \overline{XY} \cong$ ____

____ $\cong \angle P$ $\qquad \overline{YZ} \cong$ ____

$\angle Z \cong$ ____ $\qquad$ ____ $\cong \angle NQ$

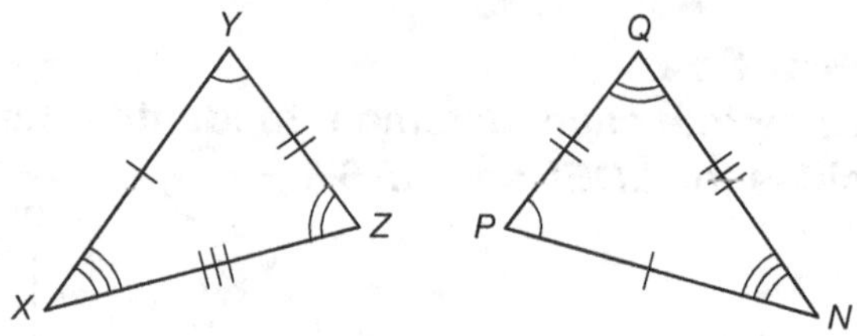

Identify the congruent sides and angles for the two triangles. Write six congruence statements.

6. $\angle A \cong$ ____ $\qquad \overline{AB} \cong$ ____

____ $\cong$ ____ $\qquad$ ____ $\cong$ ____

____ $\cong$ ____ $\qquad$ ____ $\cong$ ____

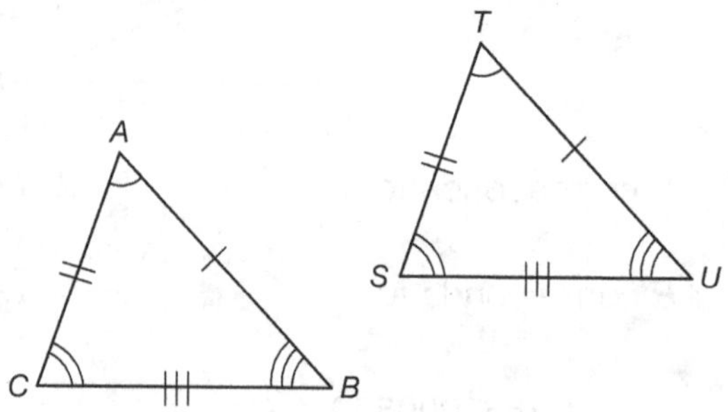

7. ____ $\cong$ ____ $\qquad$ ____ $\cong$ ____

____ $\cong$ ____ $\qquad$ ____ $\cong$ ____

____ $\cong$ ____ $\qquad$ ____ $\cong$ ____

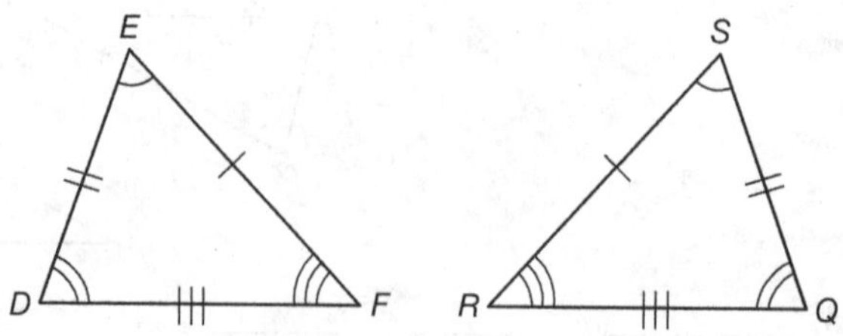

Name ______________________ Date ____________ Class ____________

Reteaching 26

Central Angles and Arc Measure

You have solved problems involving congruent triangles. Now you will solve problems involving measure.

Congruent Arcs

Two arcs that are in the same circle or in congruent circles and that have the same measure are called congruent arcs.

The measure of $\overset{\frown}{AB}$ is given by the expression $4x - 12$. The measure of $\overset{\frown}{ST}$ is given by the expression $5x + 30$. It is given that $\overset{\frown}{AB} \cong \overset{\frown}{ST}$

Determine the value of *x* and the measure of each arc.

Step 1: Since the arcs are congruent, set the expressions for their measures equal.

$4x - 12 = 5x + 30$

Step 2: Solve for *x*.

$$4x - 12 = 5x - 30$$
$$4x - 12 + 12 = 5x - 30 + 12$$
$$4x = 5x - 18$$
$$4x - 5x = 5x - 18 - 5x$$
$$-x = -18$$
$$x = 18$$

Step 3: The value of *x* is 18. Therefore, $m\overset{\frown}{AB} = 4(18) - 12 = 72 - 12 = 60°$.

Since the arcs are congruent, $m\overset{\frown}{ST} = 60°$.

Practice

Complete the steps to determine the value of *x* and the measure of each arc. Each pair of arcs is congruent.

1. $m\overset{\frown}{FG}$ is given by $5x - 14$.

$m\overset{\frown}{JK}$ is given by $3x + 20$.

$5x - 14 =$ ________

$5x -$ ____ $+$ ____ $= 3x + 20 +$ ____

$5x = 3x +$ ____

$5x -$ ____ $= 3x + 34 -$ ____

____ $= 34$

$x =$ ____

$m\overset{\frown}{FG} = 5($____$) - 14 =$ ____

$m\overset{\frown}{JK} =$ ____

2. $m\overset{\frown}{AB}$ is given by $3x + 25$.

$m\overset{\frown}{RS}$ is given by $6x - 11$.

________ $=$ **$6x - 11$**

$3x +$ ____ $-$ ____ $= 6x - 11 -$ ____

$3x = 6x -$ ____

$3x -$ ____ $= 6x - 36 -$ ____

________ $= -36$

$x =$ ____

$m\overset{\frown}{AB} = 3($____$) + 25 =$ ____

$m\overset{\frown}{RS} =$ ____

The Arc Addition Postulate

The measure of an arc formed by two adjacent arcs is the sum of the measures of the two arcs.

Write an expression that represents the measure of the arc from *A* to *C*. Find the measure of the arc.

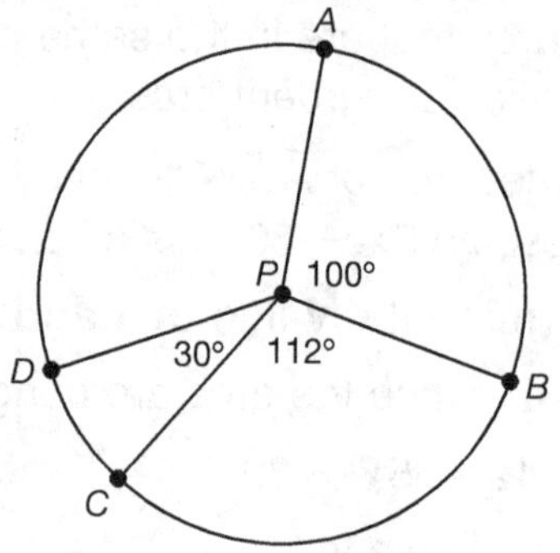

Step 1: Since measure of the arc *A* to *C* is greater than 180°, write the arc as $\overset{\frown}{ABC}$. The adjacent arcs that make up $\overset{\frown}{ABC}$ have measures less than 180°. Thus, write the arcs as $\overset{\frown}{AB}$ and $\overset{\frown}{BC}$. Therefore, $m\overset{\frown}{ABC} + m\overset{\frown}{AB} + m\overset{\frown}{BC}$.

Step 2: $m\overset{\frown}{ABC} = m\overset{\frown}{AB} + m\overset{\frown}{BC}$

$= 100° + 112°$

$= 212°$

Therefore, $m\overset{\frown}{ABC} = 212°$.

Practice

Use the Arc Addition Postulate to complete the steps.

3.

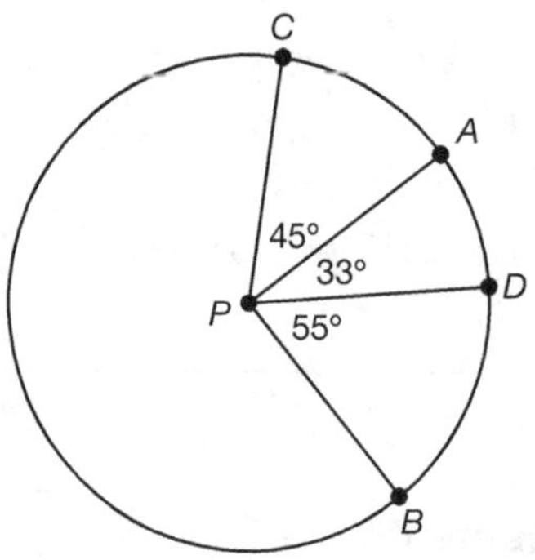

$m\overset{\frown}{CD} =$ _____ $+ m\overset{\frown}{AD}$

$= 45° +$ _____

$=$ _____

4.

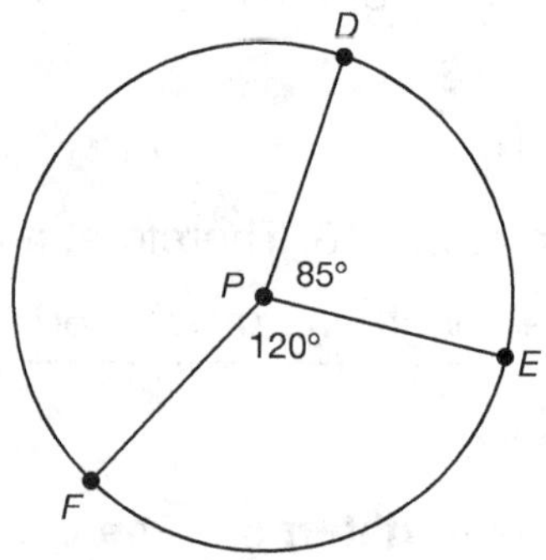

$m\overset{\frown}{DEF} =$ _____ $+ m\overset{\frown}{EF}$

$= 85° +$ _____

$=$ _____

Use the Arc Addition Postulate to find the measure of the given arc.

5. $m\overset{\frown}{ABC} =$ _____

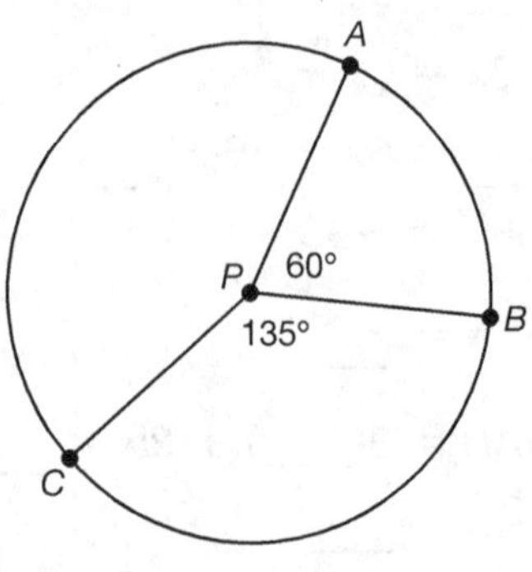

6. $m\overset{\frown}{BD} =$ _____

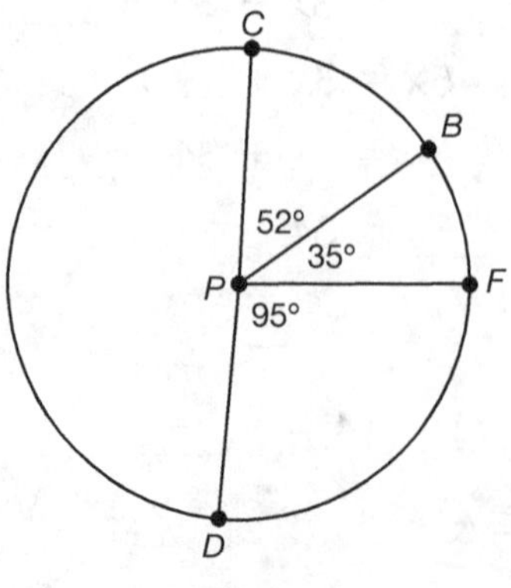

Name ______________________ Date __________ Class __________

Reteaching 27
Two-Column Proofs

You have worked with algebraic proofs. Now you will work with geometric two-column proofs.

Fill in the justifying statements to support the following.

Given: $\overrightarrow{BD}$ is the angle bisector of $\angle ABC$, and $\angle ABC \cong \angle 1$.

Prove: $\angle DBC \cong \angle 1$

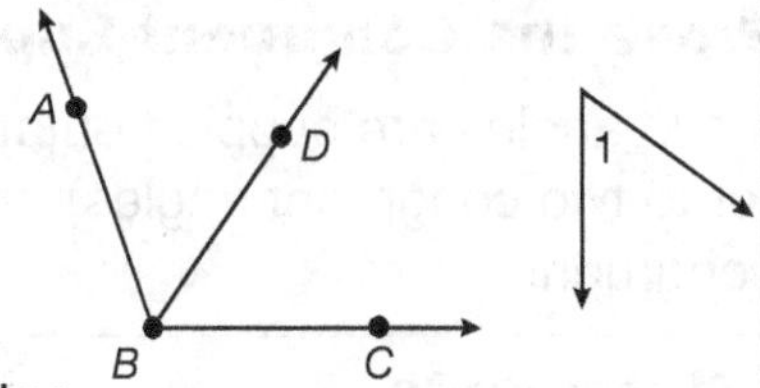

1. $\overrightarrow{BD}$ is the angle bisector of $\angle ABC$. — 1. Given
2. $\angle ABD \cong \angle DBC$ — 2. Definition of angle bisector
3. $\angle ABD \cong \angle 1$ — 3. Given
4. $\angle DBC \cong \angle 1$ — 4. Transitive Property of $\cong$

Practice

Write the letter of the correct justification next to each step. Use one justification twice.

Given: $\overrightarrow{JL}$ is the bisector of $\angle KJM$, and $\angle 1 \cong \angle 3$.

Prove: $\angle 2 \cong \angle 3$.

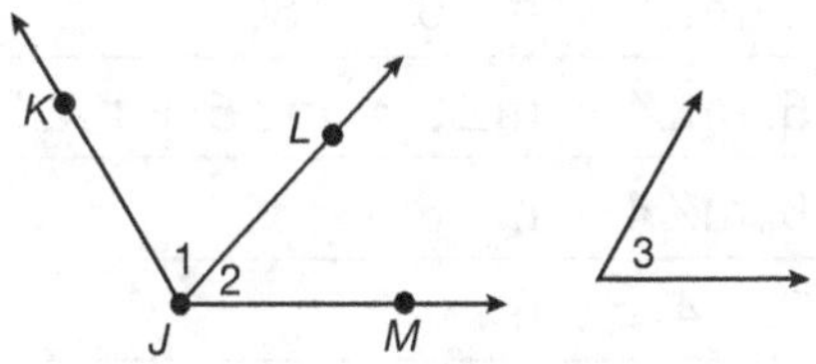

1. $\overrightarrow{JL}$ is the bisector of $\angle KJM$. _
2. $\angle 2 \cong \angle 1$. _
3. $\angle 1 \cong \angle 3$. _
4. $\angle 2 \cong \angle 3$. _

A. Definition of $\angle$bisector

B. Given

C. Transitive Property of $\cong$

Write a justification for each step to complete the proof.

Given: N is the midpoint of $\overline{MP}$, Q is the midpoint of $\overline{RP}$, and $\overline{PQ} \cong \overline{NM}$.

Prove: $\overline{PN} \cong \overline{QR}$

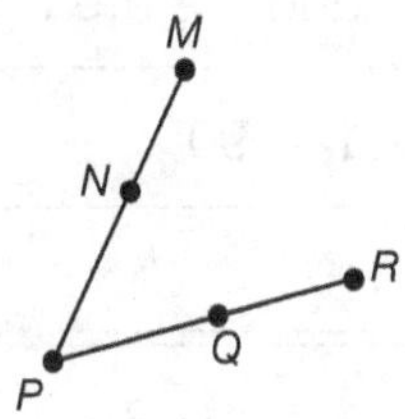

1. N is the midpoint of $\overline{MP}$. — 1. ____________
2. Q is the midpoint of $\overline{RP}$. — 2. ____________
3. $\overline{PN} \cong \overline{NM}$ — 3. ____________
4. $\overline{PQ} \cong \overline{NM}$ — 4. ____________
5. $\overline{PN} \cong \overline{PQ}$ — 5. ____________
6. $\overline{PQ} \cong \overline{QR}$ — 6. ____________
7. $\overline{PN} \cong \overline{QR}$ — 7. ____________

Two-column proofs have a format composed of five parts: given statements(s), prove statement, diagram, statements, and reasons.

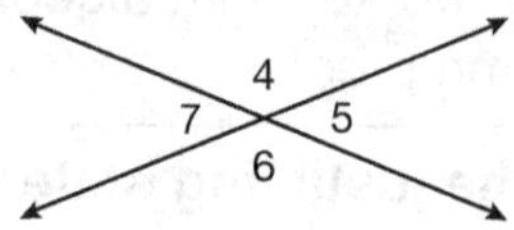

Prove the Congruent Supplements Theorem:

If two angles are supplementary to the same angle (or to two congruent angles), then the two angles are congruent.

Statements	**Reasons**
1. ∠4 and ∠5 are supplementary.	1. Given
2. ∠5 and ∠6 are supplementary.	2. Given
3. $m\angle 5 + m\angle 6 = 180°$	3. Definition of supplementary angles
4. $m\angle 5 + m\angle 6 = 180°$	4. Definition of supplementary angles
5. $m\angle 4 + m\angle 5 = m\angle 5 + m\angle 6$	5. Substitution Property of Equality
6. $m\angle 4 = m\angle 6$	6. Subtraction Property of Equality
7. $\angle 4 \cong \angle 6$	7. Definition of congruent angles

Practice

Fill in the blanks to complete the two-column proof of the Right Angle Congruence Theorem.

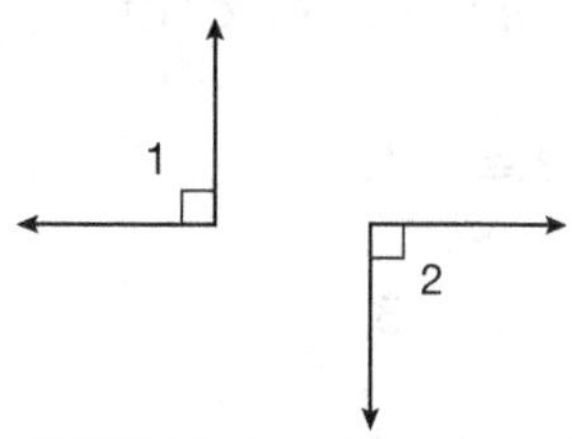

Given: ∠1 and ∠2 are right angles.

Prove: $\angle 1 \cong \angle 2$

Statements	**Reasons**
1. ∠1 and ∠2 are right angles.	1. ______________
2. $m\angle 1 = 90°$	2. ______________
3. ______________	3. Definition of right angle
4. $m\angle 1 = m\angle 2$	4. ______________
5. ______________	5. Definition of congruent angles

Name ______________________ Date __________ Class __________

Reteaching 28

Triangle Congruence: SAS

You have worked with congruent figures. Now you will determine whether two triangles are congruent.

Determine whether the pair of triangles is congruent by the SAS Postulate.
If two sides and the included angle of one triangle are congruent to two sides and the included angle of another triangle, then the triangles are congruent by side-angle-side congruence.

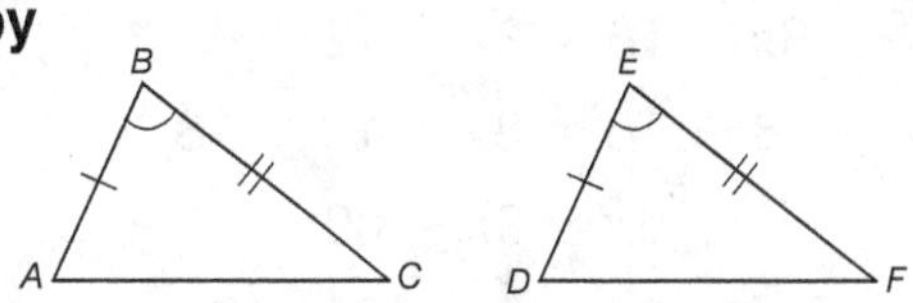

Step 1: Identify the congruent sides: $\overline{AB} \cong \overline{DE}$; $\overline{BC} \cong \overline{EF}$.

Step 2: Identify the congruent angle: $\angle B \cong \angle E$.

Step 3: Is the angle between the congruent sides in both figures? yes

Step 4: The two triangles are congruent by SAS.

Practice

Complete the steps to determine whether the pair of triangles is congruent by the SAS Postulate.

1.

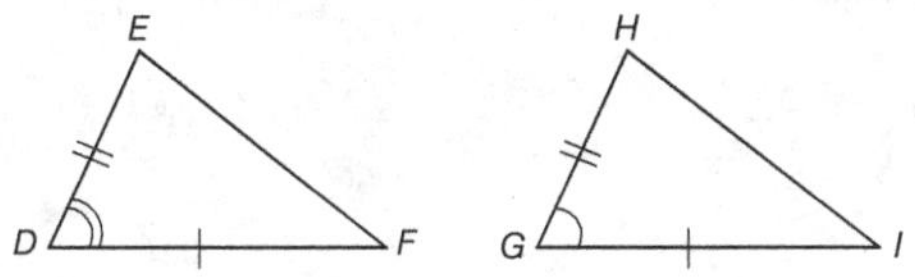

$\overline{DF} \cong$ ____; $\overline{DE} \cong$ ____

$\angle D \cong$ ____

The angle ___ between the two congruent sides.

Are the two triangles congruent? _____

2.

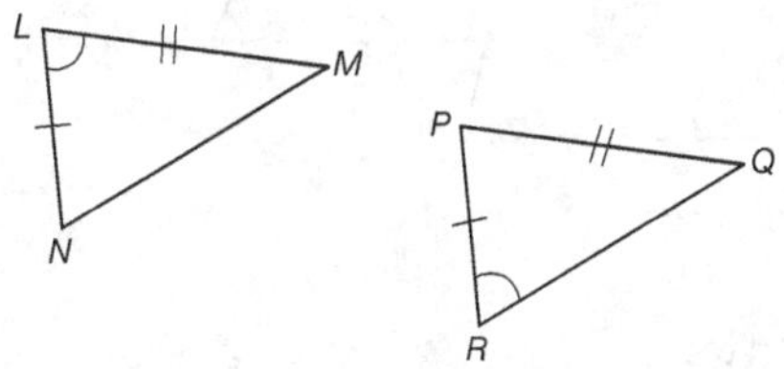

$\overline{NL} \cong$ ____; $\overline{LM} \cong$ ____

$\angle L \cong$ ____

The angle _______ between the two congruent sides.

Are the two triangles congruent? _____

Determine whether the pair of triangles is congruent by the SAS Postulate.

3. _____

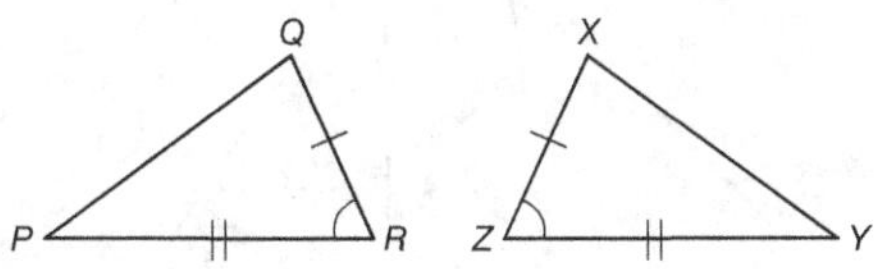

4. _____

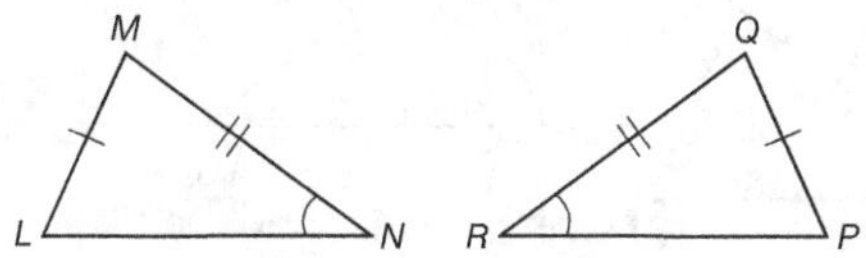

Find the value of *x* that makes the triangles congruent.

For the two triangles to be congruent, the measures of the included angles must be equal.

Set the equations equal to each other and solve.

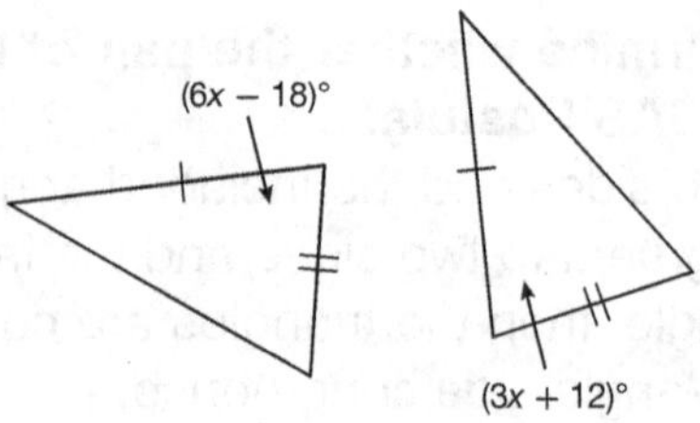

$$6x - 18 = 3x + 12$$
$$6x - 18 + 18 = 3x + 12 + 18$$
$$6x = 3x + 30$$
$$6x - 3x = 3x + 30 - 3x$$
$$3x = 30$$
$$\frac{3x}{3} = \frac{30}{3}$$
$$x = 10$$

Practice

Complete the steps to find the value of *x* that makes the triangles congruent.

5.

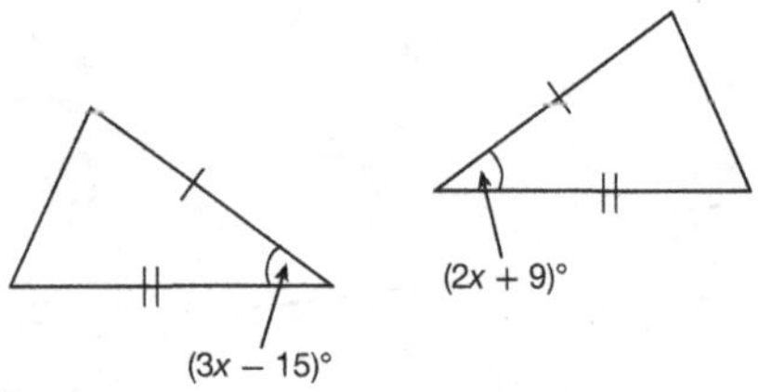

$3x -$ ____ $=$ ____ $+ 9$

$3x -$ ____ $+$ ____ $= 2x + 9 +$ ____

$3x = 2x +$ ____

$3x -$ ____ $= 2x + 24 -$ ____

$x =$ ____

6.

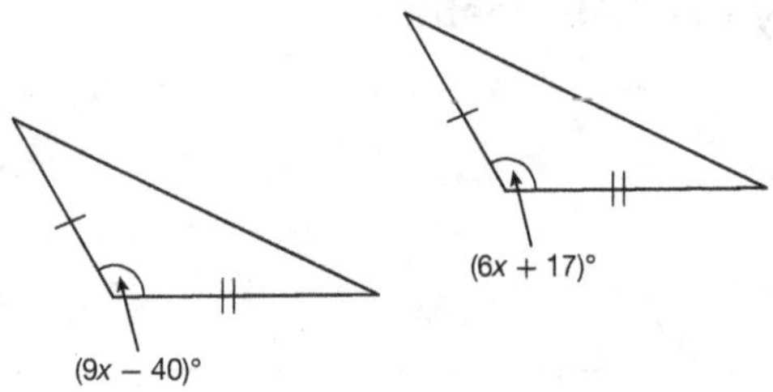

$9x - 40 = 6x +$ ____

$9x - 40 +$ ____ $= 6x +$ ____ $+$ ____

$9x = 6x +$ ____

$9x -$ ____ $= 6x +$ ____ $-$ ____

$3x =$ ____

$\frac{3x}{\_\_\_\_} = \frac{\_\_\_\_}{\_\_\_\_}$

$x =$ ____

Find the value of *x* that makes the triangles congruent.

7. ______

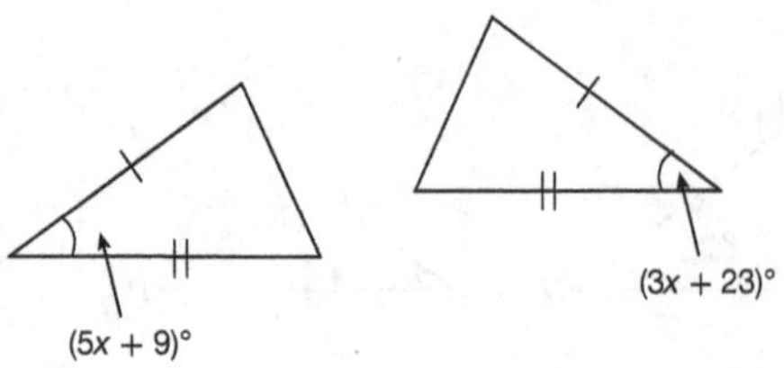

8. ______

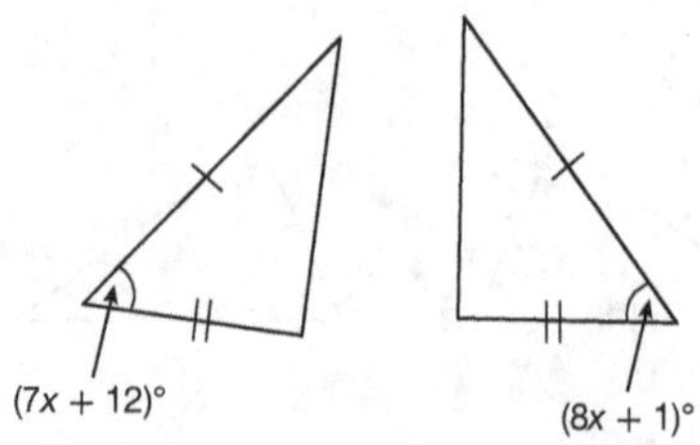

Name ______________________ Date __________ Class __________

Reteaching 29
Using the Pythagorean Theorem

You have solved problems involving triangle congruence with SAS.
Now you will solve problems using the Pythagorean theorem.

Pythagorean Triples

A set of three nonzero whole numbers a, b, and c such that $a^2 + b^2 = c^2$ is a **Pythagorean Triple.**

Find the unknown length in the triangle.

Do the side lengths form a Pythagorean Triple?

Step 1: Since the length of the hypotenuse is the unknown, $c = x$.

The lengths of the legs, which are represented by a and b, are 7 feet and 24 feet. It does not matter which value is a and which is b.

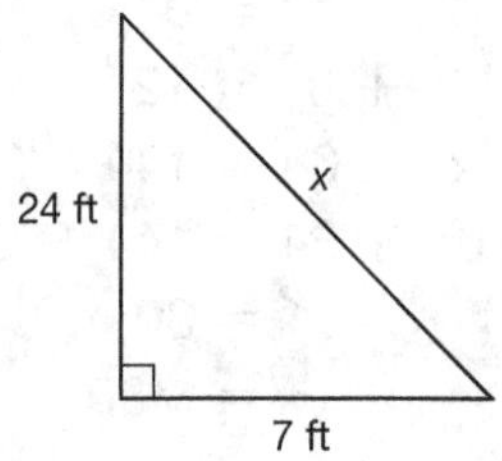

Step 2: Substitute $a = 7$, $b = 24$, and $c = x$ into the equation.

Solve for x.

$$a^2 + b^2 = c^2$$
$$7^2 + 24^2 = x^2$$
$$49 + 576 = x^2$$
$$625 = x^2$$
$$25 = x$$

The length of the hypotenuse is 25 feet.

The set (7, 24, 25) is a **Pythagorean Triple** because $7^2 + 24^2 = 25^2$.

Practice

Complete the steps to find the unknown length in the triangle.
Determine whether the side lengths form a Pythagorean Triple.

1.

$$a^2 + b^2 = c^2$$
$$\_\_\_^2 + x^2 = \_\_\_^2$$
$$\_\_\_ + x^2 = \_\_\_\_$$
$$\_\_\_ + x^2 - 64 = 289 - \_\_\_$$
$$x^2 = \_\_\_\_$$
$$x = \_\_\_$$

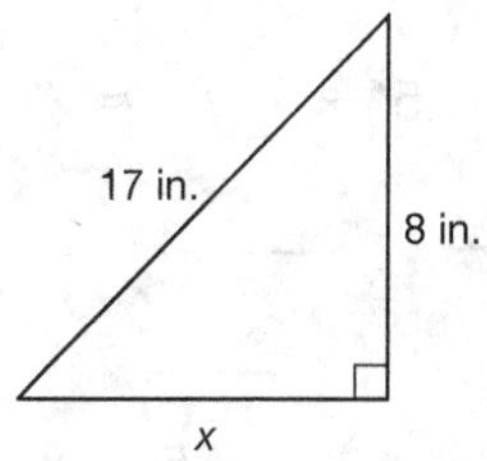

The length of the unknown leg is __________

Do the side lengths form a Pythagorean Triple? __________

Find the unknown length in the triangle. Do the side lengths form a Pythagorean Triple?

2. ______________

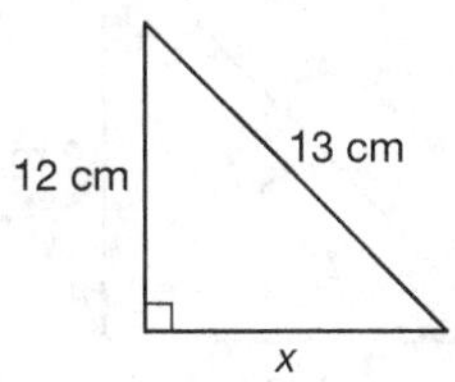

3. ______________

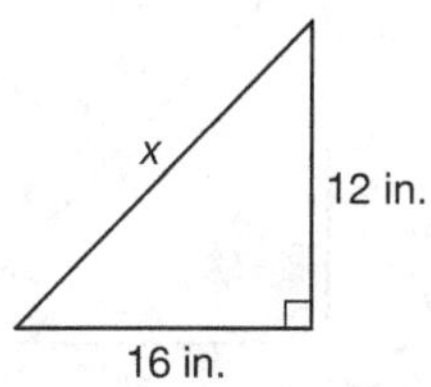

Simplifying Radicals

One or more side lengths of a right triangle may be written as a **radical expression.** A radical expression is any expression that contains a root. Radical expressions are usually reduced by simplifying the root.

Find the value of *x*. Give your answer in simplified radical form.

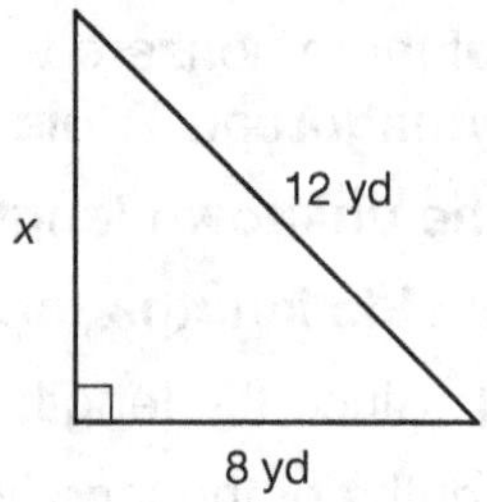

Step 1: The length of the hypotenuse is 12 yards. One of the legs is 8 yards long. Since the length of other leg is unknown, it is represented by *x*.

Step 2: Substitute *x*, 8, and 12 into the equation $a^2 + b^2 = c^2$ and solve for *x*.

$$a^2 + b^2 = c^2$$
$$x^2 + 8^2 = 12^2$$
$$x^2 + 64 = 144$$
$$x^2 + 64 - 64 = 144 - 64$$
$$x^2 = 80$$
$$x = \sqrt{80}$$

Step 3: Simplify the radical. Factor out all perfect square factors of the number under the radical sign. The largest perfect square that is a factor of 80 is 16, so write 80 as 16×5.

$x = \sqrt{16 \times 5} = 4\sqrt{5}$

Therefore, the length of the unknown leg is $4\sqrt{5}$ yards.

Practice

Complete the steps to find the value of x and write the answer in simplified radical form.

4.

$$a^2 + b^2 = c^2$$
$$5^2 + \_\_\_^2 = x^2$$
$$\_\_\_\_ + \_\_\_\_ = x^2$$
$$\_\_\_\_ = x^2$$
$$\sqrt{\_\_\_ \times 5} = x$$
$$\_\_\_\sqrt{5} = x$$

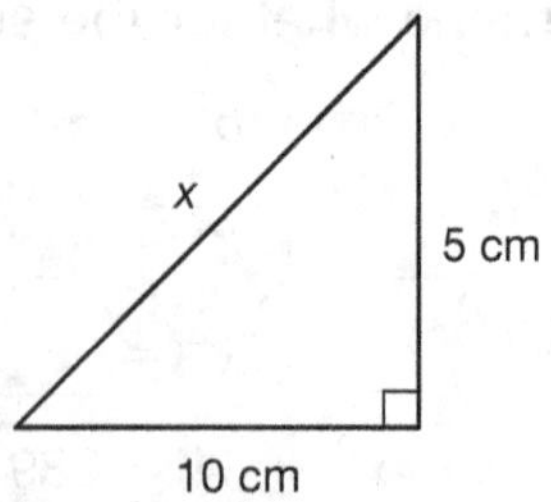

The length of the unknown side is ____________________.

Find the value of *x*. Give your answer in simplified radical form.

5. ____________

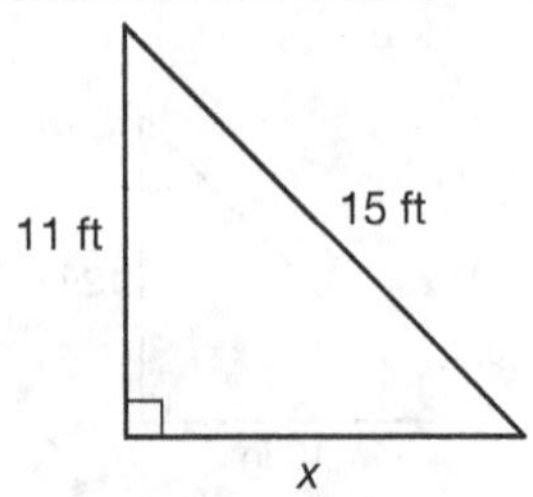

6. ____________

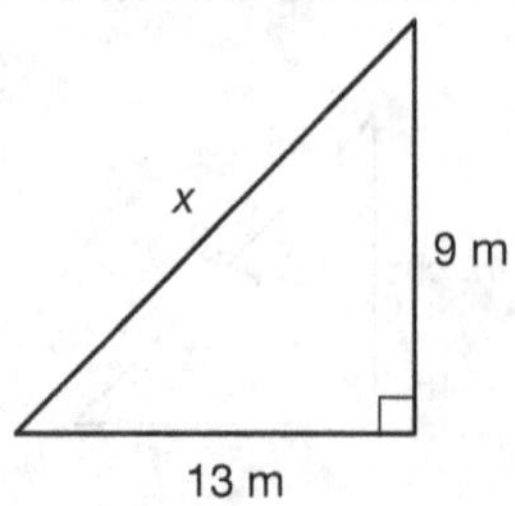

Name ______________________________ Date __________ Class ____________

Reteaching 30

Triangle Congruence Theorems: ASA and AAS

You have worked with triangle congruence using Side-Angle-Side. Now you will determine triangle congruence by two other methods.

Determine whether triangles ΔKLM and ΔNPQ are congruent by the Angle-Side-Angle (ASA) Postulate.

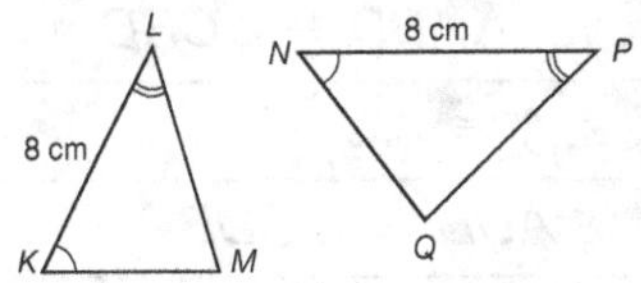

If two angles and the included side of one triangle are congruent to two angles and the included side of another triangle, then the triangles are congruent.

Step 1: Identify the congruent side: $\overline{KL} \cong \overline{NP}$.

Step 2: Identify the congruent angles: $\angle K \cong \angle N$, and $\angle L \cong \angle P$.

Step 3: Is the congruent side between the congruent angles in both figures? yes

Step 4: The two triangles are congruent by ASA.

Practice

Fill in the blanks to determine whether each pair of triangles is congruent by ASA.

1.

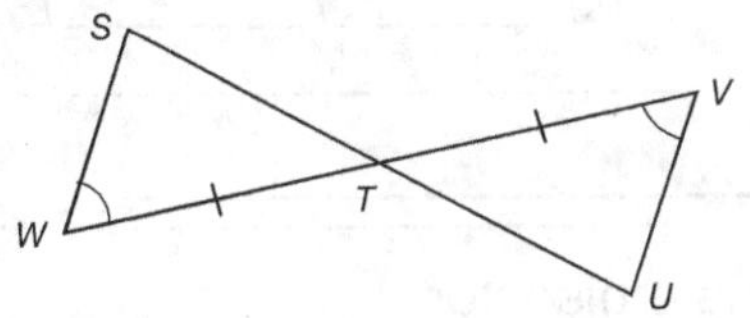

$\overline{TW} \cong$ ___, $\angle W \cong \angle$___

$\angle STW \cong \angle UTV$ by Vertical Angles Theorem.

The congruent side _____ between the two congruent angles.

Are the two triangles congruent by ASA?

______

2.

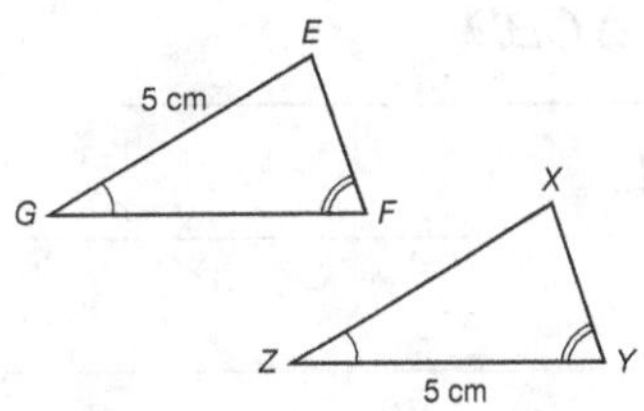

$\angle G \cong \angle$___, $\angle F \cong \angle$___

The congruent side _______ between the the two congruent angles.

Are the two triangles congruent by ASA?

_____

Determine whether each pair of triangles is congruent by ASA.

3. ______

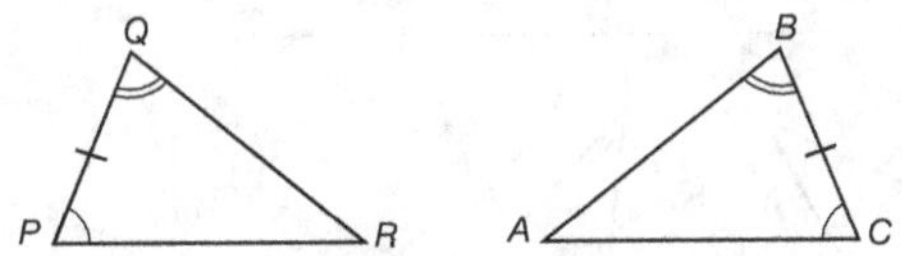

4. _____

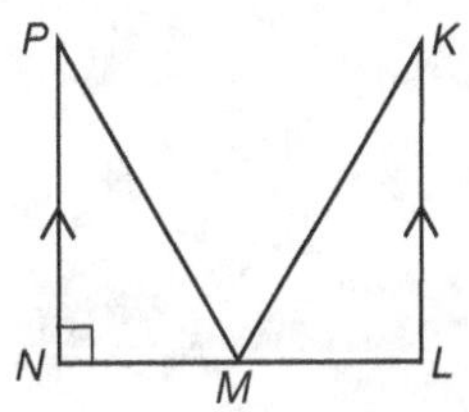

Angle-Angle-Side (AAS) Congruence Postulate: If two angles and the non-included side of one triangle are congruent to two angles and the corresponding non-included side of another triangle, then the triangles are congruent.

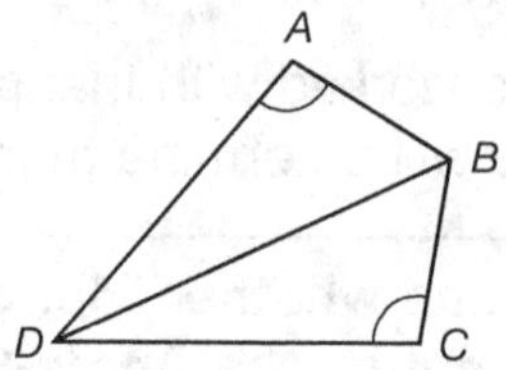

Given: $\overline{BD}$ is the angle bisector of $\angle ADC$.

Prove: $\Delta ABD \cong \Delta CBD$

1. $\angle A \cong \angle C$	1. Given
2. $\angle ADB \cong \angle CDB$	2. Definition of angle bisector
3. $\overline{BD} \cong \overline{BD}$	3. Reflexive Property of $\cong$
4. $\Delta ABD \cong \Delta CBD$	4. ASA Congruence Postulate

Practice

Complete the steps to prove the triangles are congruent by AAS.

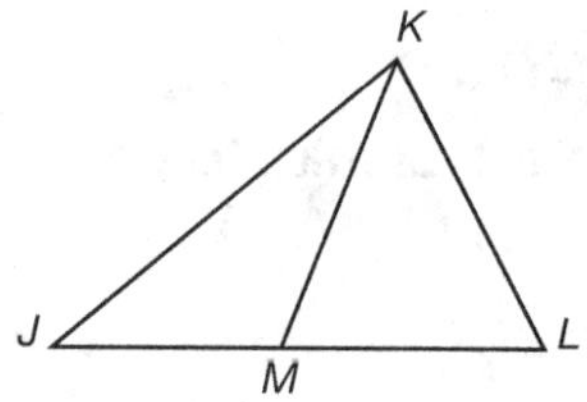

5. Given: $\overline{KM}$ bisects $\angle JKL$, $m\angle J = x$, $m\angle L = x$.

Prove: $\Delta ABD \cong \Delta CBD$

1. $\overline{KM}$ bisects $\angle JKL$.	1. ________________
2. $\angle JKM \cong$ ________	2. Definition of angle bisector
3. $m\angle J = x$, $m\angle L =$ ____	3. Given
4. $m\angle J =$ ________	4. Transitive Property of Equality
5. $\angle J = \angle L$	5. Definition of____________
6. $\overline{KM} \cong \overline{KM}$	6. Reflexive Property of congruence
7. $\Delta ABD \cong \Delta CBD$	7. ____________

Determine whether you can use the AAS Congruence Postulate to prove the triangles are congruent. If not, tell what else you need to know.

7. ______

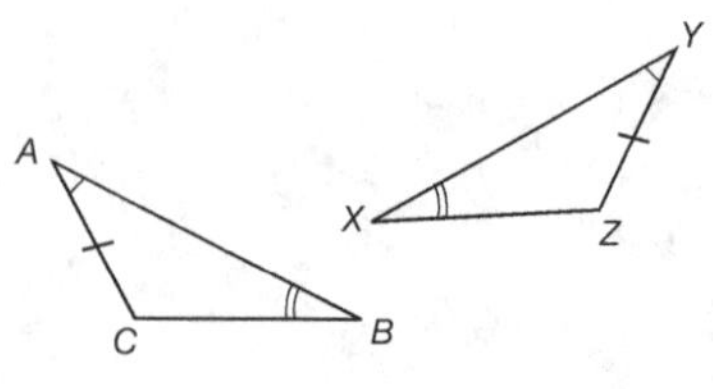

8. no; need to know

____

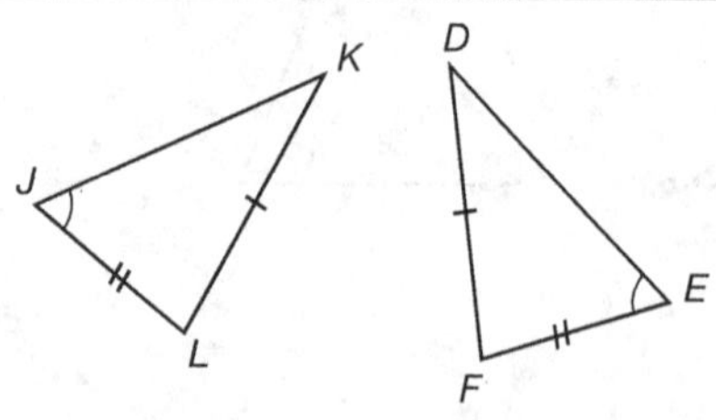

Name ______________________ Date __________ Class __________

Reteaching

INV 3

Exploring Angles of Polygons

You have learned to identify interior and exterior angles in polygons.
Now you will determine angle measures in regular polygons.

Interior Angles

Sum of the Interior Angles of a Polygon	**Interior Angle Measure of a Regular Polygon**
$(n - 2)180°$, where n is the number of sides of the polygon.	$\frac{(n-2)180°}{n}$, where n is the number of sides of the polygon.
Example: **Determine the sum of the interior angles of an octagon.** **Step 1:** How many sides are in an octagon? 8 **Step 2:** Substitute and solve the formula. $(n - 2)180° = (8 - 2)180°$ $= 6 \cdot 180°$ $= 1080°$	**Example:** **Determine the interior angle measure of a regular octagon.** **Step 1:** How many sides are in an octagon? 8 **Step 2:** Substitute and solve the formula. $\frac{(n-2)180°}{n} = \frac{(8-2)180°}{8}$ $= \frac{6 \cdot 180°}{8}$ $= 135°$

Practice

Complete the statements to solve each problem.

1. What is the sum of the interior angles of a regular 15-gon?

   How many sides are in a 15-gon? ___

   Substitute and simplify the formula.

   $(n - 2)180° = (\_\_ - 2)180°$

   $= \_\_ \cdot 180°$

   $= \_\_\_\_$

2. What is the interior angle measure of a regular 12-gon?

   How many sides are in a 12-gon? ___

   Substitute and simplify the formula.

   $\frac{(n-2)180°}{n} = \frac{(\_\_ - 2)180°}{12}$

   $= \frac{\_\_ \cdot 180°}{12}$

   $= \_\_\_\_$

3. The sum of the measures of the interior angles of a regular 14-gon is ______.

4. The measure of an interior angle of a regular 30-gon is ____.

5. The sum of the measures of the interior angles of a regular 18-gon is ______.

Exterior and Central Angles

Exterior Angle Measure of a Regular Polygon	**Central Angle Measure of a Regular Polygon**
The measure of each exterior angle $= \frac{360°}{n}$, where n is the number of sides in the polygon.	The measure of each central angle $= \frac{360°}{n}$, where n is the number of sides in the polygon.
Example: **Determine the measure of each exterior angle of a regular 12-gon.** **Step 1:** How many sides are in a 12-gon? 12 **Step 2:** Substitute and solve the formula. $\frac{360°}{n} = \frac{360°}{12}$ $= 30°$	**Example:** **Determine the central angle of a regular 24-gon.** **Step 1:** How many sides are in a 24-gon? 24 **Step 2:** Substitute and solve the formula. $\frac{360°}{n} = \frac{360°}{24}$ $= 15°$

Practice

Complete the statements to solve each problem.

6. **Determine the measure of each exterior angle of a regular 15-gon. How many sides are in a 15-gon?** ___

   $\frac{360°}{n} = \frac{360°}{\_\_\_\_}$

   $= \_\_\_\_$

7. Determine the central angle of a regular 30-gon.
   How many sides are in a 30-gon? ___

   $\frac{360°}{n} = \frac{360°}{\_\_\_\_}$

   $= \_\_\_\_$

8. Determine the measure of each exterior angle for a regular 20-gon. _____
9. Determine the central angle of a regular 36-gon. _____
10. Determine the measure of each exterior angle of a regular 40-gon. _____
11. Determine the central angle of a regular 72-gon. _____

Name ____________________ Date __________ Class __________

Reteaching 31

Flowchart and Paragraph Proofs

You have worked with triangle congruence (ASA and AAS). Now you will write flowchart and paragraph proofs.

Flowchart Proofs

You can use the flowchart proof to prove conjectures are true.

Flowchart Proofs use boxes and arrows.

Steps go from top to bottom, as shown by arrows. The justification or reason for each step is written below the box.

Given: $\angle 1$ and $\angle 2$ are right angles.

Prove: $\angle 1 \cong \angle 2$

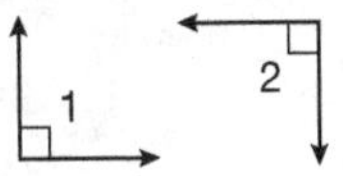

Flowchart Proof

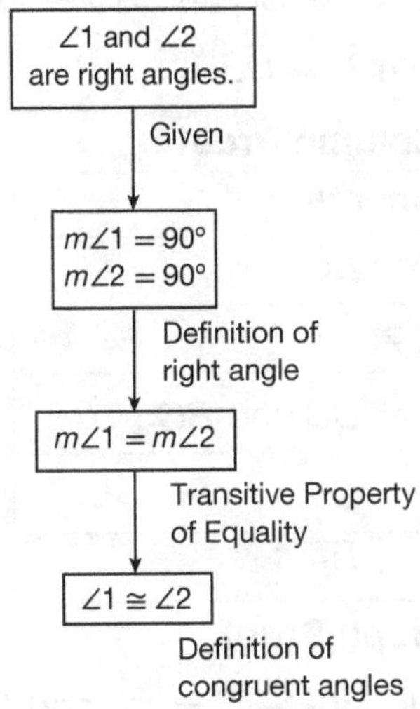

Practice

1. Complete the flowchart proof for the given two-column proof.

   Given: V is the midpoint of $\overline{SW}$, and W is the midpoint of $\overline{VT}$.

   Prove: $\overline{SV}$ is congruent to $\overline{WT}$.

Two-Column Proof

Statements	Reasons
1. V is the midpoint of $\overline{SW}$.	1. Given
2. W is the midpoint of $\overline{VT}$.	2. Given
3. $\overline{SV}$ is congruent to $\overline{VW}$. $\overline{VW}$ is congruent to $\overline{WT}$.	3. Definition of midpoint
4. $\overline{SV}$ is congruent to $\overline{WT}$.	4. Transitive Property of Equality

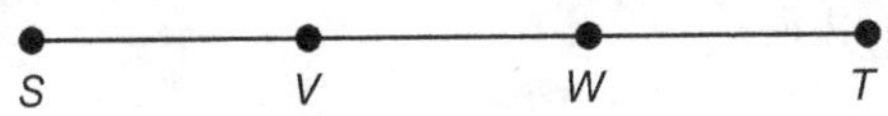

Flowchart Proof

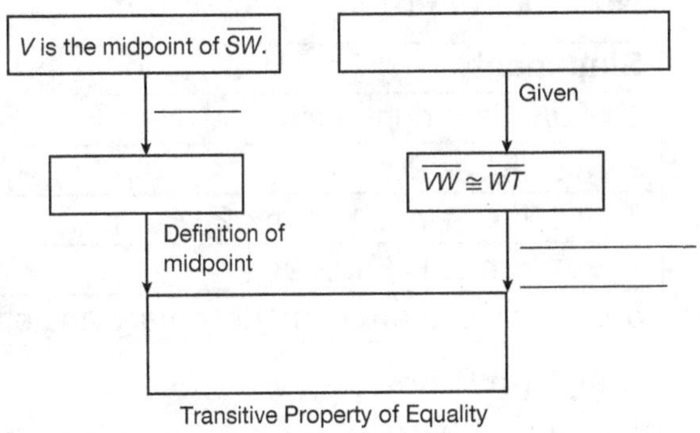

Paragraph Proofs

To write a paragraph proof, use sentences to write a paragraph that presents the statements and reasons.

Use the two-column proof to write a paragraph proof.

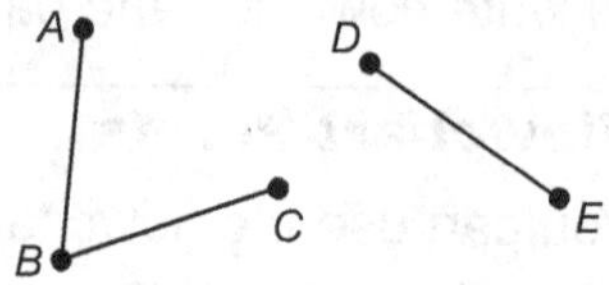

Given: $\overline{AB} \cong \overline{BC}$ and $\overline{BC} \cong \overline{DE}$.

Prove: $\overline{AB} \cong \overline{DE}$

Two-Column Proof

Statements	Reasons
1. $\overline{AB} \cong \overline{BC}$	1. Given
2. $\overline{BC} \cong \overline{DE}$	2. Given
3. $AB = BC$ and $BC = DE$	3. Definition of congruent segments
4. $AB = DE$	4. Transitive Property of Equality
5. $\overline{AB} \cong \overline{DE}$	5. Definition of congruent segments

Paragraph Proof

It is given that $\overline{AB} \cong \overline{BC}$ and $\overline{BC} \cong \overline{DE}$, so $AB = BC$ and $BC = DE$ by the definition of congruent segments. $AB = DE$ by the Transitive Property of Equality. Thus, $\overline{AB} \cong \overline{DE}$ by the definition of congruent segments.

Practice

2. Use the given two-column proof to complete the paragraph proof.

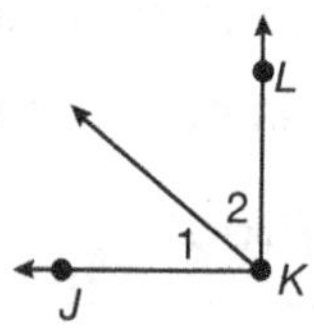

Given: $\angle JKL$ is a right angle.

Prove: $\angle 1$ and $\angle 2$ are complementary angles.

Two-Column Proof:

Statements	Reasons
1. $\angle JKL$ is a right angle.	1. Given
2. $m\angle JKL = 90°$	2. Definition of right angle
3. $m\angle JKL = m\angle 1 + m\angle 2$	3. Angle Addition Postulate
4. $90° = m\angle 1 + m\angle 2$	4. Substitution
5. $\angle 1$ and $\angle 2$ are complementary angles.	5. Definition of complementary angles

Paragraph Proof

Since it is given that ______ is a right angle, $m\angle JKL =$ ____ by the definition of a ____________. By the Angle Addition Postulate, ______________________________. Using ______________, $90° =$ ________________. Thus, by the definition of ____________________ ________, $\angle 1$ and ____ are complementary angles.

Name ______________________________ Date __________ Class __________

Reteaching 32

Altitudes and Medians of Triangles

You have written flowchart and paragraph proofs. Now you will solve problems involving altitudes and medians of triangles.

Medians of Triangles

In the picture, $\overline{AH}$, $\overline{BJ}$, and $\overline{CG}$ are **medians of a triangle.** They each join a vertex and the midpoint of the opposite side. The point of intersection of the medians is called the **centroid** of ΔABC.

Centroid Theorem

The centroid of a triangle is located $\frac{2}{3}$ of the distance from each vertex to the midpoint of the opposite side.

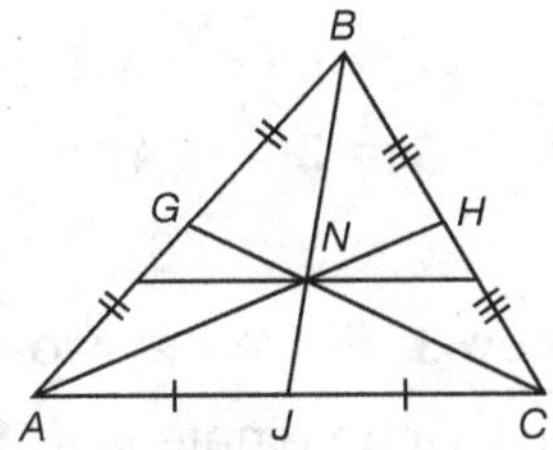

Example:

Given: $\overline{AH}$, $\overline{BJ}$, and $\overline{CG}$ are medians of ΔABC.

Conclusion: $AN = \frac{2}{3}AH,\ CN = \frac{2}{3}CG,\ BN = \frac{2}{3}BJ$

In ΔABC, suppose $AH = 18$ and $BN = 10$.

Use the Centroid Theorem to find *AN* and *BJ*.

$AN = \frac{2}{3}AH$	Centroid Theorem	$BN = \frac{2}{3}BJ$	Centroid Theorem
$AN = \frac{2}{3}(18)$	Substitute 18 for *AH*.	$10 = \frac{2}{3}BJ$	Substitute 10 for *BN*.
$AN = 12$	Simplify.	$15 = BJ$	Simplify.

Practice

In ΔQRS, $RX = 48$, and $QW = 30$. Fill in the steps to find each length.

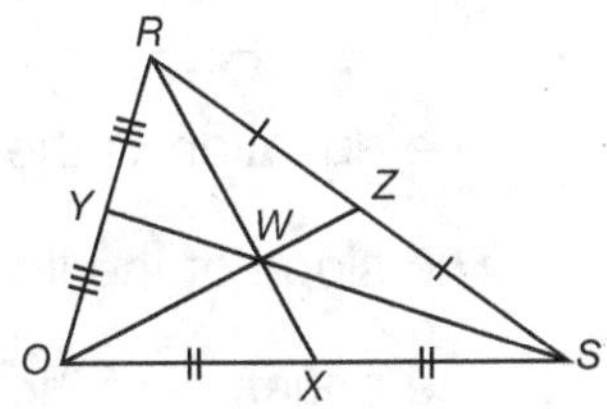

1. *QZ*

_____ $= \frac{2}{3}QZ$

_____ $= \frac{2}{3}QZ$

_____ $= QZ$

2. *RW*

$RW = \frac{2}{3}$_____

$RW = \frac{2}{3}$_____

$RW =$ _____

In ΔHJK, $HD = 21$, and $BK = 18$. Find each length.

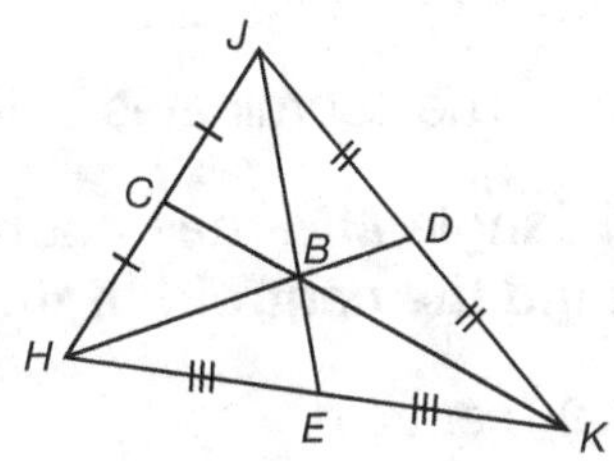

3. *HB*

_____

4. *BD*

_____

5. *CK*

_____

6. *CB*

_____

Finding a Centroid on the Coordinate Plane

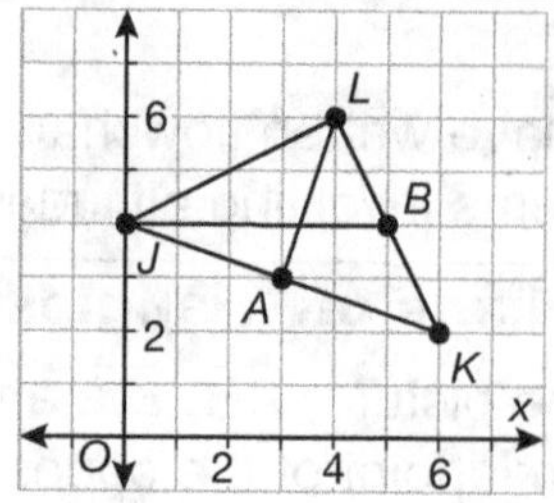

Find the centroid of ΔJKL with vertices at $J(0, 4)$, $K(6, 2)$ and $L(4, 6)$

Step 1: Find the midpoint of two of the sides of the triangle. Since the three medians meet at the same point, the intersection of any two will give the location of the centroid. The midpoint of $\overline{JK}$ is (3, 3). The midpoint of is $\overline{LK}$ is (5, 4).

Step 2: Graph the two medians and find the equation of each median. $\overline{JB}$ is a horizontal line, so the equation is y = 4. To find the equation of $\overline{LA}$, use the slope and the coordinates of one endpoint. The slope of is $\overline{LA}$ is 3.

$y - y_1 = m(x - x_1)$ Point-slope Formula

$y - 6 = 3(x - 4)$ Substitute.

$y = 3x - 6$ Simplify.

Step 3: Solve the two equations as a system: $y = 4$ and $y = 3x - 6$.

The y-coordinate is 4. Solve for the x-coordinate:

$4 = 3x - 6$

$3.3 = x$

The x-coorinate is 3.3. The centroid is located at (3.3, 4).

Practice

Triangle *DEF* has coordinates *D*(−1, 7), *E*(0, 3), and *F*(−5, 5). Complete the steps to find the centroid of the triangle.

7. The midpoint of $\overline{DE}$ is (_____, _____), and the midpoint of $\overline{DF}$ is (_____, _____).

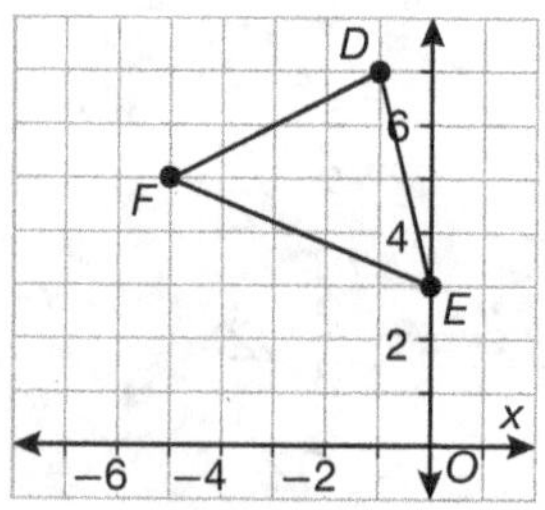

The equation of the median from (−0.5, 5) to *F* is $y =$ ___.

The slope of the line from (−3, 6) to *E* is ____.

The equation of the line from (−3, 6) to *E* is ____________.

To find the coordinates of the centroid, solve the system of equations: ______, and ____________.

The coordinates of the centroid are (___, ___).

Triangle *MNP* has coordinates *M*(−3, 3), *N*(−1, 6), and *P*(1, −1). Find the centroid of the triangle.

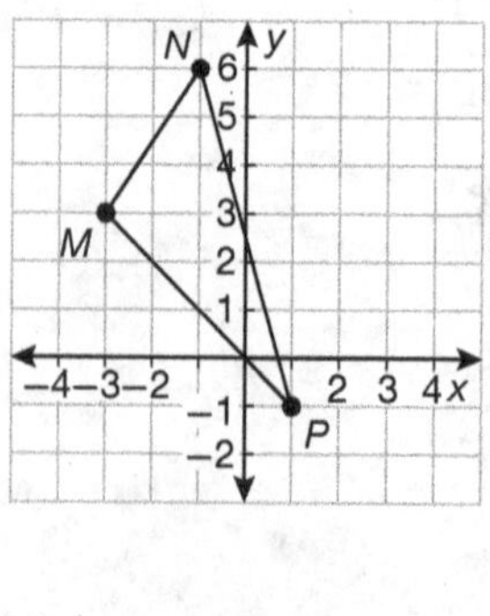

8. centroid __________

Name _______________________ Date __________ Class __________

Reteaching 33

Converse of the Pythagorean Theorem

You have worked with medians of triangles. Now you will solve problems that use the Pythagorean theorem.

The Pythagorean Theorem

The **Pythagorean theorem** states the following relationship among the lengths of the legs, *a* and *b*, and the length of the hypotenuse, *c*, of any right triangle.

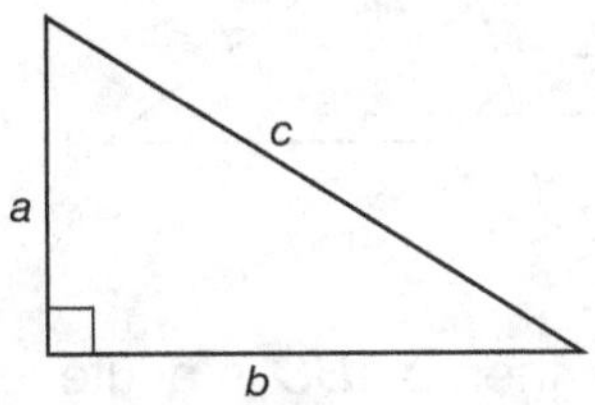

$$a^2 + b^2 = c^2$$

Use the Pythagorean theorem to find the value of *x* in the triangle.

Substitute 6 for a, *x* for *b*, and 12 for *c* in the equation $a^2 + b^2 = c^2$ and solve for *x*.

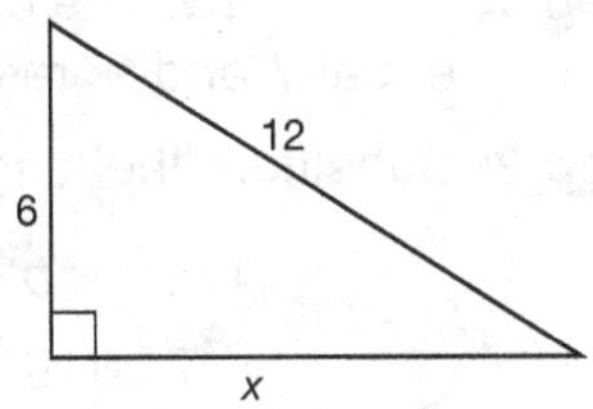

$a^2 + b^2 = c^2$	Pythagorean theorem
$6^2 + x^2 = 12^2$	Substitute
$36 + x^2 = 144$	Simplify
$x^2 = 108$	Subtract
$x = \sqrt{108}$	Square root
$x = \sqrt{36 \times 3} = 6\sqrt{3}$	Simplify radical

The length of the unknown side is $6\sqrt{3}$.

Practice

Complete the steps to find the value of *x*.

1. $\_ + x^2 = \_^2$

$\_\_ + x^2 = \_\_$

$\_\_ + x^2 - 25 = 64 - \_\_$

$x^2 = \_\_$

$x = \_\_\_$

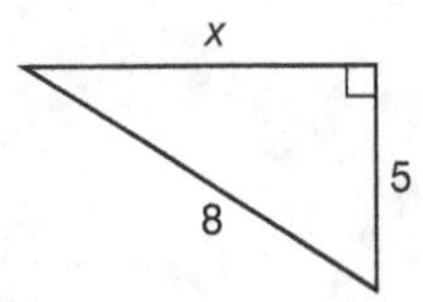

Find the value of x. Give your answer in simplest radical form.

2.

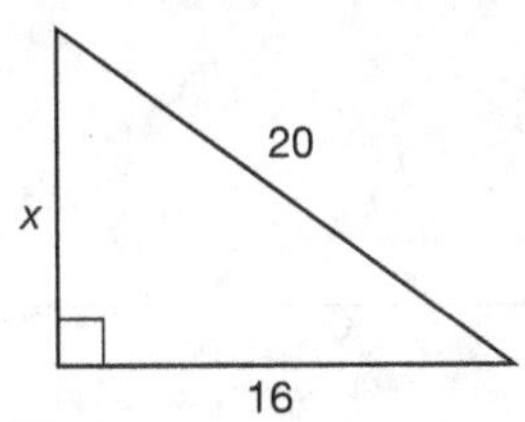

____

3.

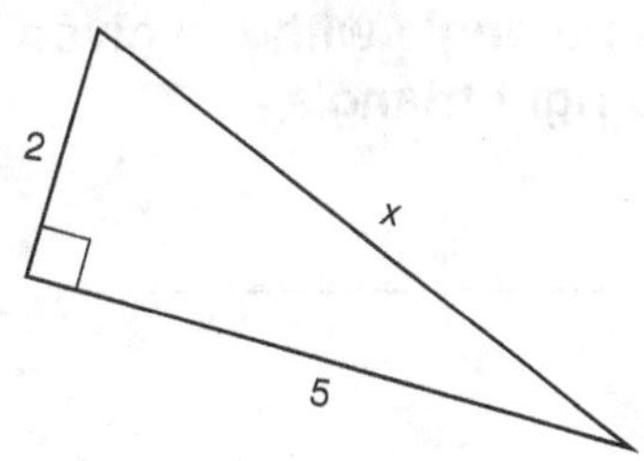

____

Pythagorean Inequality Theorem

You can use the following theorem to classify triangles by their angles if you know their side lengths. Always use the length of the longest side for c.

Pythagorean Inequality Theorem	
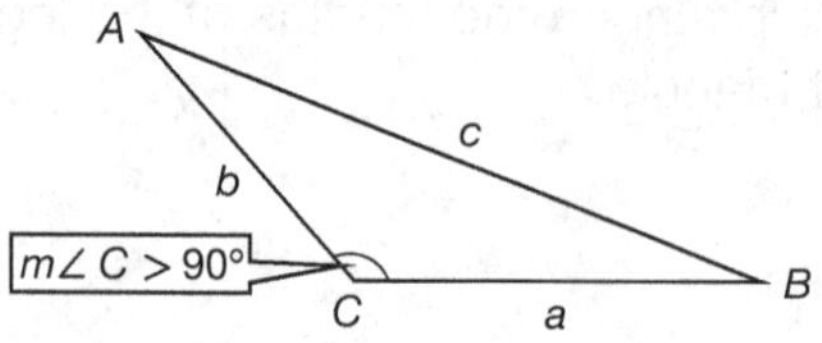 If $a^2 + b^2 < c^2$, then ΔABC is obtuse.	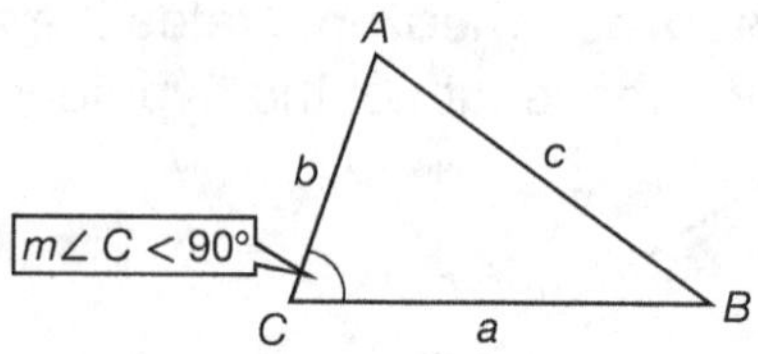 If $a^2 + b^2 > c^2$, then ΔABC is acute.

Determine if ΔABC is an obtuse, acute, or right triangle.

Step 1: Let $c = 13$. The other two lengths, a and b, will equal 7 and 9, respectively.

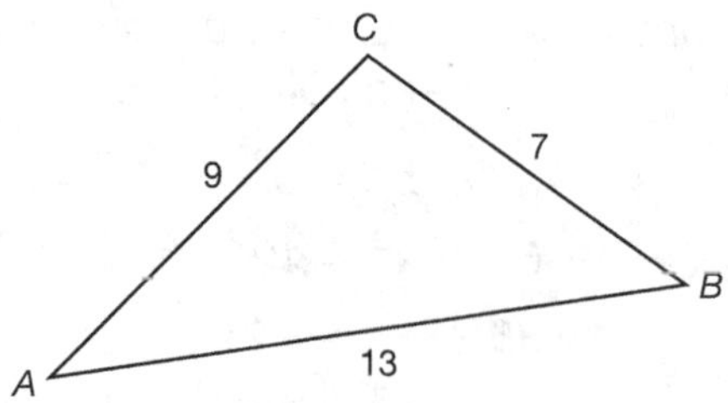

Step 2: Substitute the lengths into $a^2 + b^2 \overset{?}{=} c^2$.

$$a^2 + b^2 \overset{?}{=} c^2$$
$$7^2 + 9^2 \overset{?}{=} 13^2$$
$$49 + 81 \overset{?}{=} 169$$
$$130 < 169$$

Since $a^2 + b^2 < c^2$, the triangle is an obtuse triangle.

Practice

Complete the steps to determine if the given side lengths form an acute, obtuse, or right triangle.

4. 10, 11, 12

___ $+ 11^2 \overset{?}{=}$ ___

$100 + 121 \overset{?}{=}$ ___

___ > 144

The triangle is a/an __________ triangle.

5. 3, 4, 5

$3^2 +$ ___ $\overset{?}{=}$ ___

$9 + 16 \overset{?}{=}$ ___

___ $= 25$

The triangle is a/an __________ triangle.

Determine whether the triangle with the given side lengths is an obtuse, an acute, or a right triangle.

6. 8, 13, 16

____

7. 4, 7, 8

____

8. 8, 8, 11

____

9. 9, 12, 15

____

Name ______________________ Date __________ Class __________

Reteaching 34

Properties of Parallelograms

All parallelograms have the following properties.

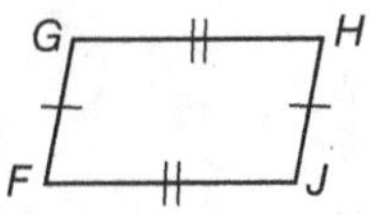

$\overline{FG} \cong \overline{JH}$
$\overline{GH} \cong \overline{FJ}$

Opposite sides are congruent.

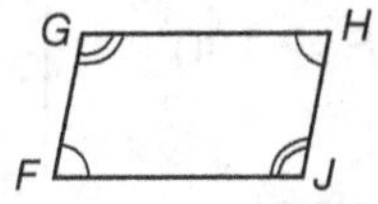

$\angle F \cong \angle H$
$\angle G \cong \angle J$

Opposite angles are congruent.

Prove opposite sides of a parallelogram are congruent.

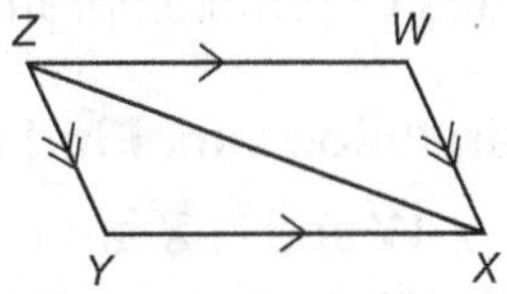

Given: *WXYZ* is a parallelogram.

Prove: $\overline{WZ} \cong \overline{XY}$, $\overline{WX} \cong \overline{ZY}$

Solution

Statements	Reasons
1. *WXYZ* is a parallelogram.	1. Given
2. $\overline{WZ} \parallel \overline{XY}$, $\overline{WZ} \parallel \overline{XY}$	2. Definition of parallelogram
3. $\overline{ZX} \cong \overline{ZX}$	3. Reflexive property
4. $\angle WXZ \cong \angle XZY$, $\angle ZXY \cong \angle WZX$	4. Corresponding Angles Postulate
5. $\Delta WZX \cong \Delta YXZ$	5. ASA triangle congruence.
6. $\overline{WZ} \cong \overline{XY}$, $\overline{WX} \cong \overline{ZY}$	6. Corresponding Parts of Congruent Triangles are Congruent.

Practice

Complete the paragraph proof to prove that opposite angles of a parallelogram are congruent.

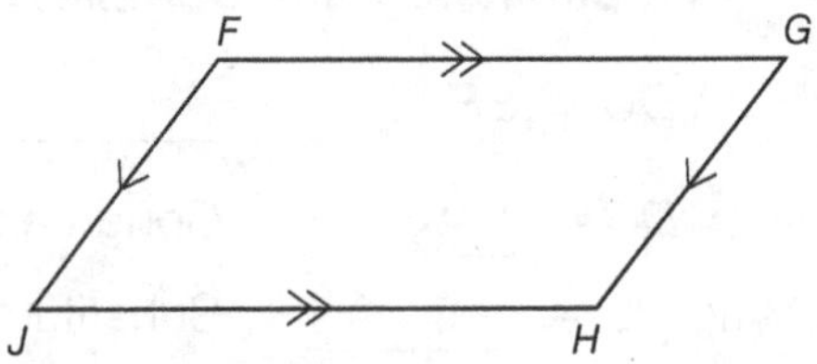

7. Given: *FGHJ* is a parallelogram.

Prove: $\angle G \cong \angle J$, $\angle F \cong \angle H$

It is given that _____ is a parallelogram. By the definition of a parallelogram, $\overline{GH} \parallel$ __, and $\overline{FG} \parallel$ __ . Since same-side interior angles are supplementary, $m\angle G + m\angle H = 180°$, and _____________ $= 180°$. For the same reason $m\angle G + m\angle F =$ ____ and _____________ $= 180°$. Using substitution, $m\angle H + m\angle J =$ _____ $+ m\angle H$. By the Subtraction Property of Equality, $m\angle H + m\angle J - m\angle H = m\angle G + m\angle H -$ _____. Simplifying, $m\angle J =$ _____. Using __________, $m\angle F + m\angle J = m\angle H + m\angle J$. By the Subtraction Property of ________, $m\angle F + m\angle J - m\angle J = m\angle H + m\angle J -$ _____. Simplifyig, ____________. By the definition of ______________, _________ and $\angle F \cong \angle H$.

More Properties of Parallelograms

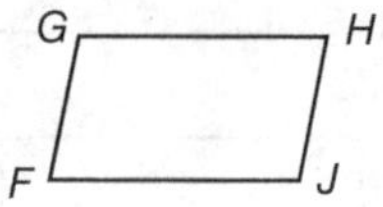

$m\angle F + m\angle G = 180°$
$m\angle G + m\angle H = 180°$
$m\angle H + m\angle J = 180°$
$m\angle J + m\angle F = 180°$

Consecutive angles are supplementary.

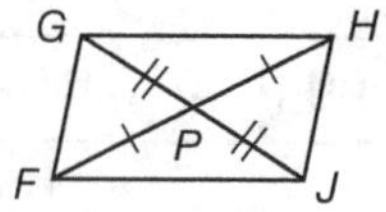

$\overline{FP} \cong \overline{HP}$
$\overline{GP} \cong \overline{JP}$

The diagonals bisect each other.

You can use the properties of parallelograms to find the measures of angles and sides of parallelograms.

***WXYZ* is a parallelogram. Find $m\angle X$.**

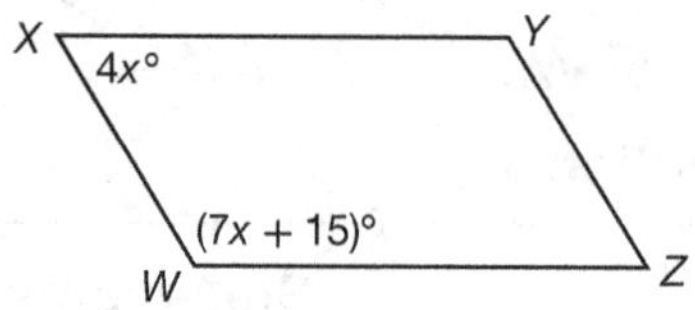

Step 1: Since $\angle W$ and $\angle X$ are consecutive angles, they are supplementary. Thus, $m\angle W + m\angle X = 180°$.

Step 2: Substitute $m\angle W = (7x + 15)°$ and $m\angle X = (4x)°$ into $m\angle W + m\angle X = 180°$ and solve for x.

$m\angle W + m\angle X = 180°$	Consecutive angles are supplementary.
$(7x + 15) + 4x = 180$	Substitute the given values.
$11x + 15 = 180$	Combine like terms.
$11x = 165$	Subtract 15 from both sides.
$x = 15$	Divide both sides by 11.

Step 3: Substitute $x = 15$ into the expression for $m\angle X$.

$m\angle X = (4x)° = [4(15)]° = 60°$

Practice

***CDEF* is a parallelogram. Complete the steps to find the length of $\overline{DC}$.**

8. $\overline{DC} \cong \overline{EF}$ ________ sides are congruent

$DC =$ ___ Congruent sides have the same ______.

$4w + 8 =$ ________ Substitute.

$8 =$ ___ $+ 1$ Subtract $4w$ from both sides.

___ $= w$ Subtract 1 from both sides.

$CD = 4w + 8 = 4($ ___ $) + 8 =$ ___

Find each measure for the parallelogram *CDEF*.

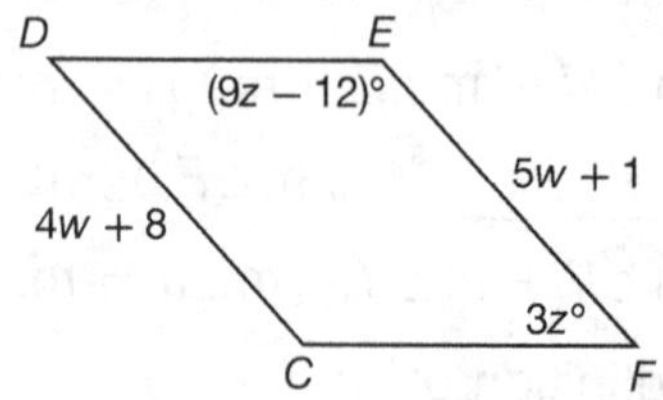

9. EF

___

10. $m\angle F$

___

Name ______________________________ Date __________ Class __________

Reteaching 35

Finding Areas of Sectors and Arc Length

You have worked with properties of parallelograms. Now you will find arc lengths and the area of sectors.

Arc Length

Arc length is the distance along a part of the circumference of a circle measured in linear units. The arc length of a circle is given by the formula $L = 2\pi r\left(\frac{m^\circ}{360^\circ}\right)$. The variable m stands for the degree measure of the arc.

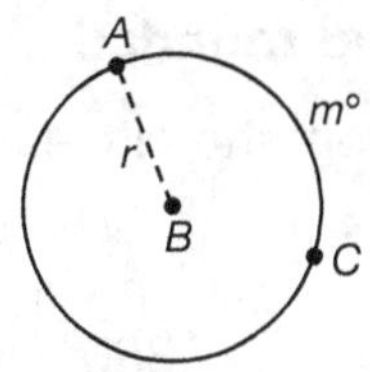

Find the arc length of $\overset{\frown}{JK}$. Give the answer in terms of π and rounded to the nearest hundredth.

From the circle, $r = 9$ cm, and $m^\circ = 84^\circ$. Substitute the values into the formula.

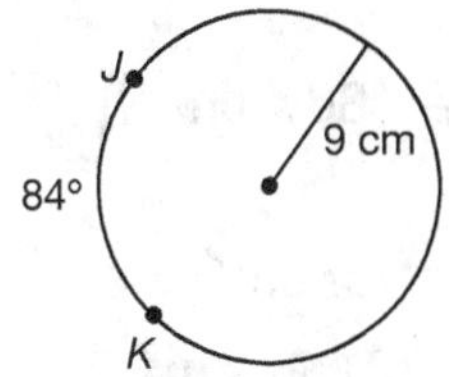

$L = 2\pi r\left(\frac{m^\circ}{360^\circ}\right)$	Formula for arc length.
$= 2\pi(9\text{ cm})\left(\frac{84^\circ}{360^\circ}\right)$	Substitute 9 cm and 84°.
$= \frac{21}{5}\pi$ cm	Solve for the answer in terms of π.
≈ 13.19 cm	Round to the nearest hundredth.

Practice

Complete the steps to find the arc length. Give your answer in terms of π and rounded to the nearest hundredth.

1. $\overset{\frown}{AB}$

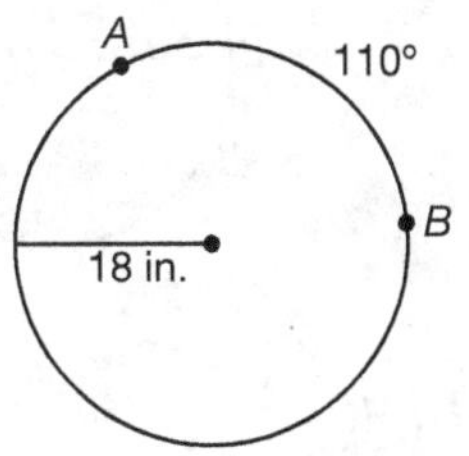

$L = 2\pi r\left(\frac{m^\circ}{360^\circ}\right)$

$= 2\pi(\_\_\_\_)\left(\frac{\_\_\_\_}{360^\circ}\right)$

$=$ ____ in.

$\approx$ ______ in.

2. $\overset{\frown}{WX}$

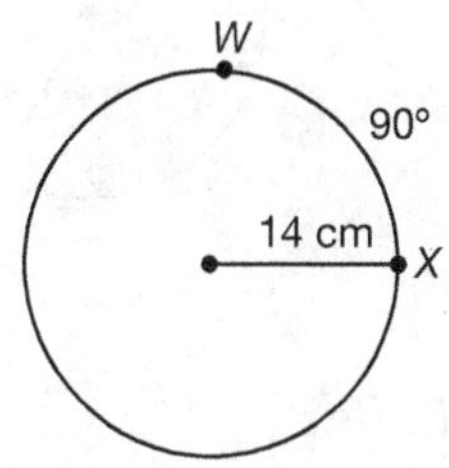

$L = 2\pi r\left(\frac{m^\circ}{360^\circ}\right)$

$= 2\pi(\_\_\_\_)\left(\frac{\_\_\_\_}{360^\circ}\right)$

$=$ ____ cm

$\approx$ ______ cm

Sector Area

A **sector of a circle** is a region bounded by two radii of the circle and their intercepted arc.

The area of a sector of a circle is given by the formula $A = \pi r^2\left(\frac{m^\circ}{360^\circ}\right)$.

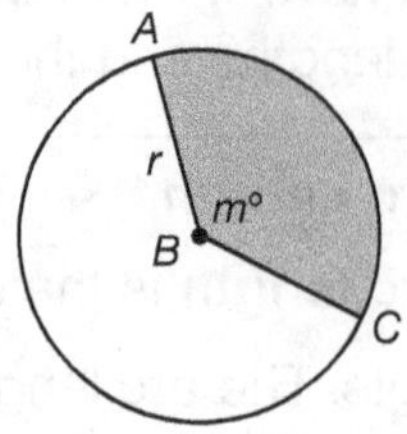

Find the area of the sector *CDE*. Give the answer in terms of π and rounded to the nearest hundredth.

From the circle, $r = 6$ cm, and $m^\circ = 70^\circ$. Substitute the values into the formula.

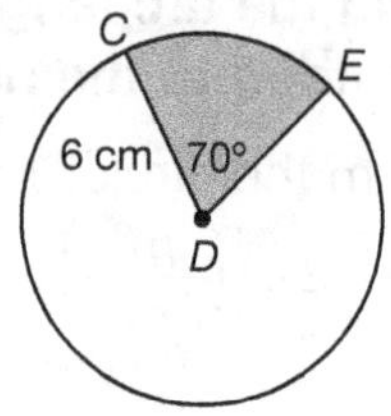

$A = \pi r^2\left(\frac{m^\circ}{360^\circ}\right)$	Formula for sector area.
$= \pi(6 \text{ cm}^2)\left(\frac{70^\circ}{360^\circ}\right)$	Substitute 6 cm and 70°.
$= (36\pi \text{ cm}^2)\left(\frac{70^\circ}{360^\circ}\right)$	Simplify.
$= 7\pi \text{ cm}^2$	Solve for the answer in terms of π.
$\approx 21.99 \text{ cm}^2$	Round to the nearest hundredth.

Practice

Complete the steps to find the area of each sector. Give your answer in terms of π and rounded to the nearest hundredth.

5. sector *QRS*

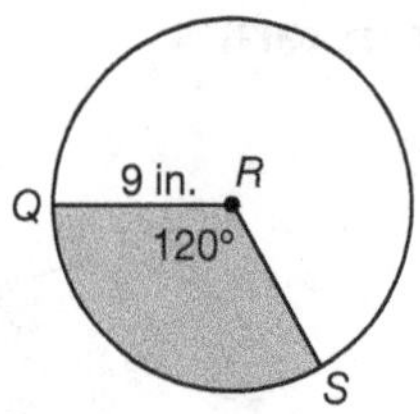

$A = \pi r^2\left(\frac{m^\circ}{360^\circ}\right)$

$= \pi\left(\_\_\_\_\right)^2\left(\frac{\_\_\_\_}{360^\circ}\right)$

$= \left(\_\_\_\_\_\_\right)\left(\frac{120^\circ}{360^\circ}\right)$

$= \_\_\_\_ \text{ in}^2$

$\approx \_\_\_\_ \text{ in}^2$

6. sector *JHG*

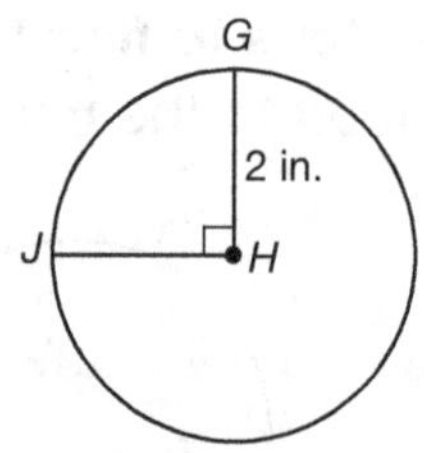

$A = \pi r^2\left(\frac{m^\circ}{360^\circ}\right)$

$= \pi\left(\_\_\_\_\right)^2\left(\frac{\_\_\_\_}{360^\circ}\right)$

$= \left(\_\_\_\_\_\_\right)\left(\frac{90^\circ}{360^\circ}\right)$

$= \_\_ \text{ in}^2$

$\approx \_\_\_\_ \text{ in}^2$

Name ______________________________ Date ____________ Class ____________

Reteaching 36

Right Triangle Congruence Theorems

Hypotenuse-Angle Congruence Theorem: If the hypotenuse and an acute angle of one right triangle are congruent to the hypotenuse and an acute angle of another right triangle, then the triangles are congruent.

Use the Hypotenuse-Angle Congruence Theorem to prove that $\triangle ABC \cong \triangle JKL$.

Statement	Reason
1. $\triangle ABC$ and $\triangle JKL$ are right triangles.	1. Given
2. $\angle A \cong \angle J$	2. Given
3. $\overline{AB} \cong \overline{JK}$	3. Given
4. $\triangle ABC \cong \triangle JKL$	4. HA Congruence Theorem

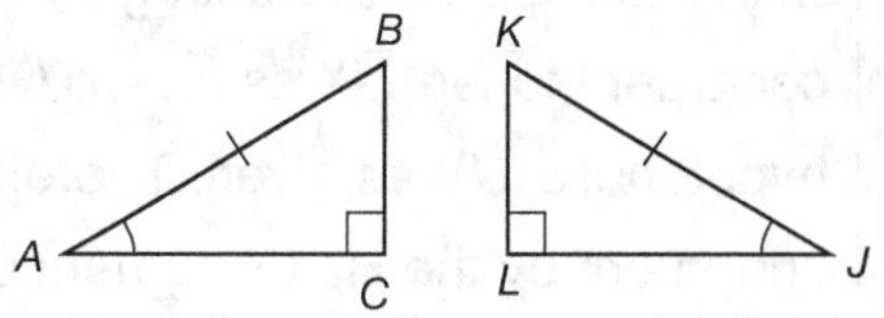

Practice

1. Use the Hypotenuse- Angle Congruence Theorem to show that $\triangle GHI \cong \triangle MNP$.

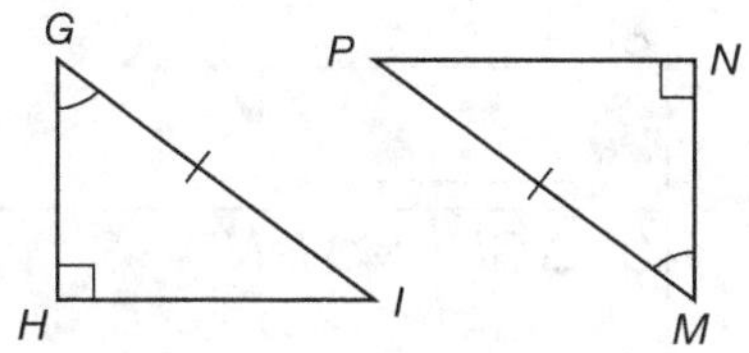

Statement	Reason
1. $\triangle GHI$ and $\triangle MNP$ are right triangles.	1. ______
2. __________	2. Given
3. ____________	3. Given
4. $\triangle GHI \cong \triangle MNP$	4. ________________

2. Use the Hypotenuse-Angle Congruence Theorem to prove that $\triangle HJK > \triangle XYZ$.

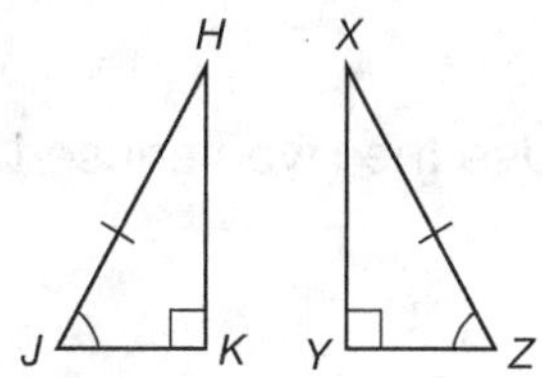

Hypotenuse-Leg Congruence Theorem: If the hypotenuse and a leg of one right triangle are congruent to the hypotenuse and a leg of another right triangle, then the triangles are congruent.

Use the Hypotenuse-Leg Congruence Theorem to prove that $\Delta RST \cong \Delta UVX$.

ΔRST and ΔUVX are both given as right triangles. Leg $\overline{RT}$ is congruent to leg $\overline{UX}$ as given. Also, hypotenuse $\overline{RS}$ is congruent to hypotenuse $\overline{UV}$ as given. Therefore, $\Delta RST \cong \Delta UVX$ are congruent by the HL Congruence Theorem.

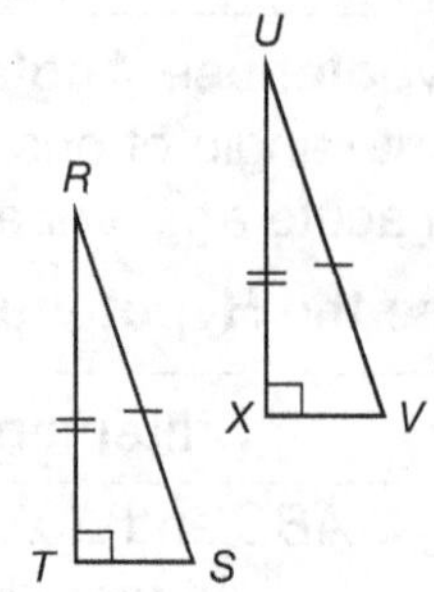

Practice

3. Use the Hypotenuse-Leg Congruence Theorem to prove that $\Delta PQR \cong \Delta CDE$.

Statement	Reason
1. ΔPQR and ΔCDE are right triangles.	1. ______
2. ______________ ___________	2. Given
3. $\Delta GHI \cong \Delta MNP$	4. ______________

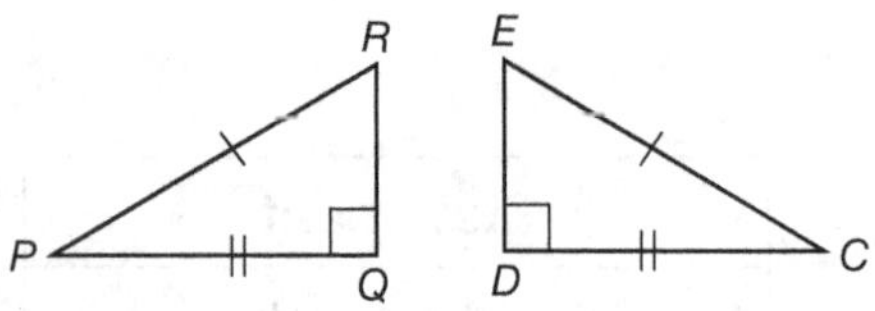

4. Use the Hypotenuse-Leg Congruence Theorem to prove that $\Delta ABC \cong \Delta VWX$.

5. Use the Hypotenuse-Leg Congruence Theorem to prove that $\Delta EFG \cong \Delta MNP$.

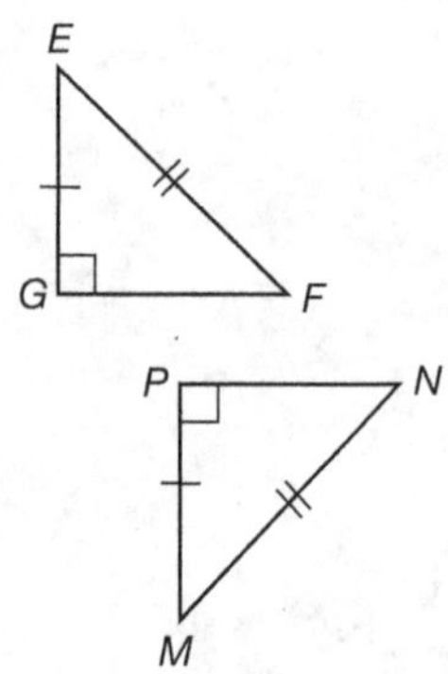

Name ______________________ Date __________ Class __________

Reteaching 37

Writing Equations of Parallel and Perpendicular Lines

Find the slopes of lines parallel and perpendicular to line *k*.

First, find the slope of line *k*.

$m = \frac{y_1 - y_2}{x_1 - x_2}$

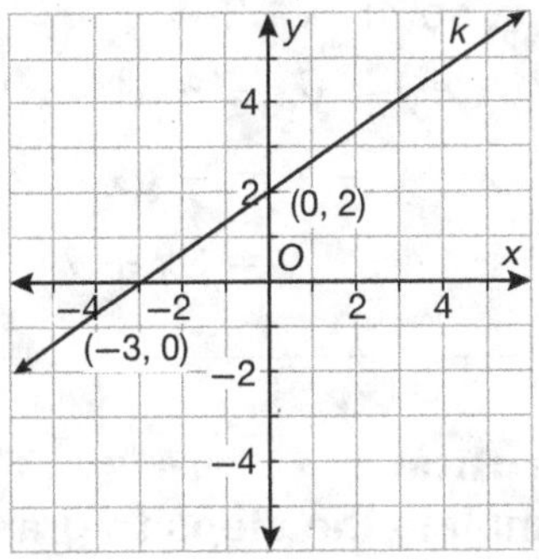

Choose two points on line *k* to substitute into the equation.

$(x_1, y_1) = (0, 2)$ and $(x_2, y_2) = (-3, 0)$

$m = \frac{2 - 0}{0 - (-3)} = \frac{2}{3}$

Since parallel lines have the same slope, the slope of a line parallel to line *k* is $\frac{2}{3}$.

Since the slopes of two perpendicular lines are opposite reciprocals, the slope of a line perpendicular to line *k* is $-\frac{3}{2}$.

Practice

Complete the steps to find the slopes of lines parallel and perpendicular to each line.

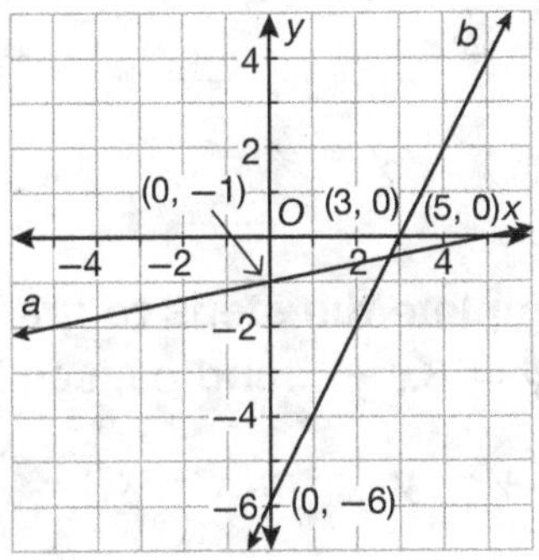

1. line *a*: $m_a = \frac{(-1) - 0}{\_\_\_\_ - \_\_\_\_} = \frac{\_\_\_\_}{\_\_\_\_}$

Slope of line parallel to line *a*: ____

Slope of line perpendicular to line *a*: ____

2. line *b*: $m_b = \frac{\_\_\_\_ - \_\_\_\_}{3 - 0} = \_\_\_\_$

Slope of line parallel to line *b*: ____

Slope of line perpendicular to line *b*: ____

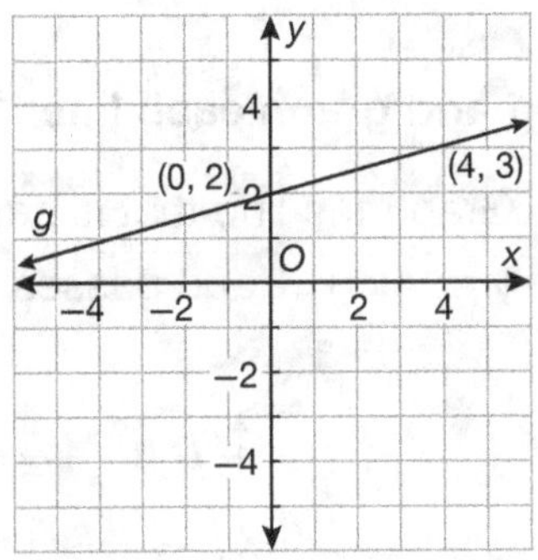

Find the slopes of the lines parallel and perpendicular to each line.

3. line *g*

Slope of line parallel to line *g:* ____

Slope of line perpendicular to line *g:* ____

4. line *h*

Slope of line parallel to line *h*: ________

Slope of line perpendicular to line *h*: ____

Graph the line that is perpendicular to the line $y = \frac{1}{3}x + 4$ and passes through (2, 1).

The slope of the perpendicular line is the opposite reciprocal of $\frac{1}{3}$, so $m = -3$.

$$y - y_1 = m(x - x_1) \quad \text{Point-slope formula}$$
$$y - 1 = -3(x - 2) \quad \text{Substitute.}$$
$$y = 3x + 7 \quad \text{Solve.}$$

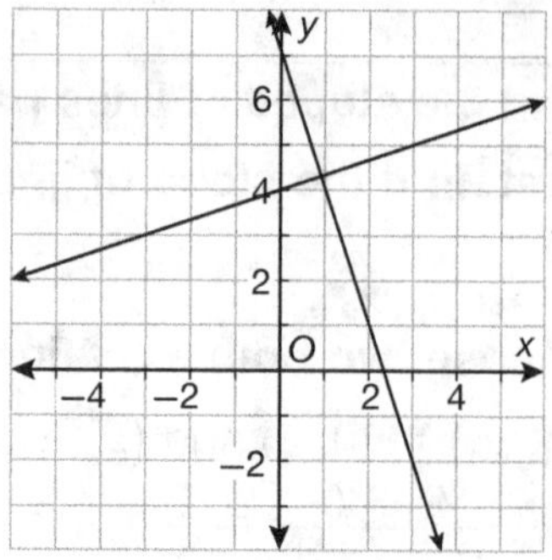

Practice

Complete the steps to graph the line that is perpendicular to $y = 2x - 1$ and passes through (2, 3).

5. $y - y_1 = m(x - x_1)$

$y - \_\_\_\_ = \_\_\_\_ (x - \_\_\_\_)$

$y = \_\_\_\_ (x + \_\_\_\_)$

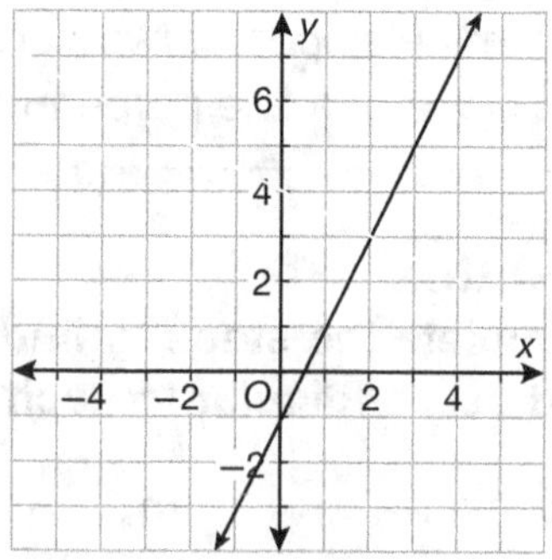

Complete the steps to graph the line that is parallel to $y = 3x + 2$ and passes through (1, –2).

6. $y - y_1 = m(x - x_1)$

$y + \_\_\_\_ = \_\_\_\_ (x - \_\_\_\_)$

$y = \_\_\_\_ x - \_\_\_\_$

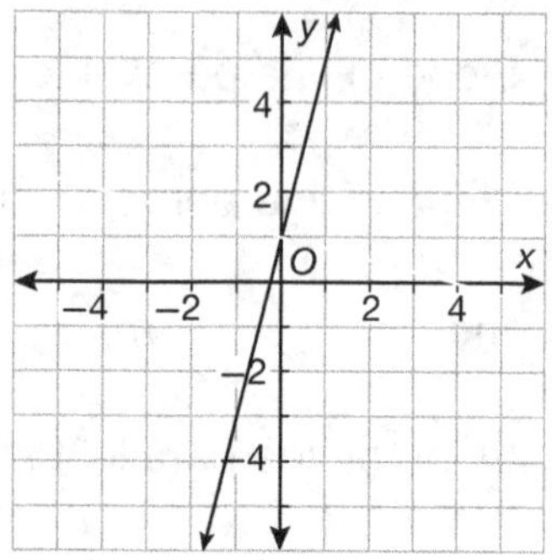

Find and graph each line.

7. Graph the line that is perpendicular to $y = 4x + 1$ and passes through (−4, 5).

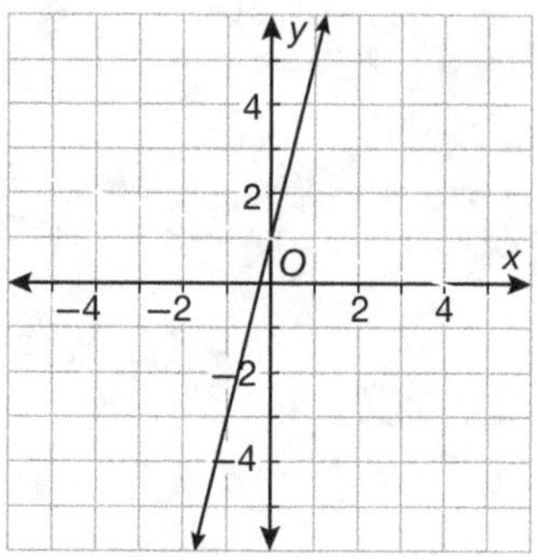

8. Graph the line that is parallel to $y = -\frac{1}{2}x - 3$ and passes through (–4, 1).

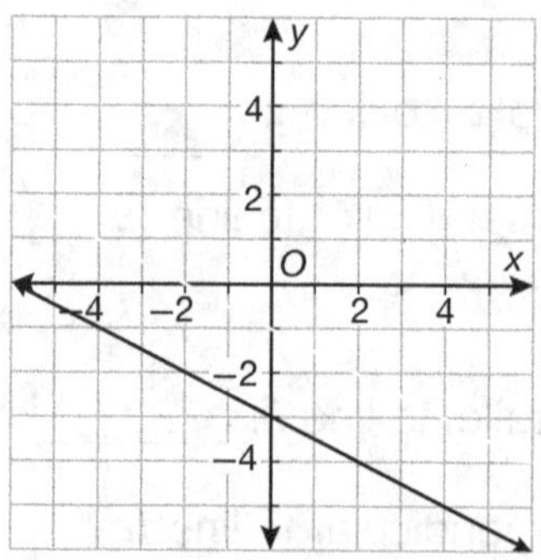

Name ______________________ Date __________ Class __________

Reteaching 38

Perpendicular and Angle Bisectors of Triangles

In the diagram, $\overline{XZ}$ bisects $\angle X$. According to the **Triangle Angle Bisector Theorem,** $\overline{WZ}$ and $\overline{ZY}$ are proportional to $\overline{WX}$ and $\overline{XY}$. The proportion is $\frac{\overline{WX}}{\overline{WZ}} = \frac{\overline{XY}}{\overline{ZY}}$.

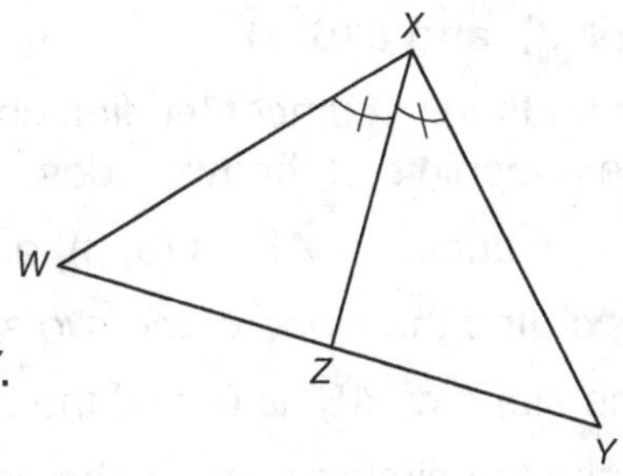

Find the length of $\overline{WX}$ if $\overline{WZ} = 9$, $\overline{XY} = 24$, and $\overline{ZY} = 10$.

$\frac{\overline{WX}}{\overline{WZ}} = \frac{\overline{XY}}{\overline{ZY}}$ Triangle Angle Bisector Theorem

$\frac{n}{9} = \frac{24}{10}$ Substitute the values for $\overline{WZ}$, $\overline{XY}$, and $\overline{ZY}$.

$10n = 216$ Cross multiply.

$n = 21.6$ Solve.

So, $\overline{WX} = 21.6$.

Practice

1. Find the length of $\overline{TV}$ if $\overline{RS} = 18$, $\overline{ST} = 8$ and $\overline{RV} = 20$.

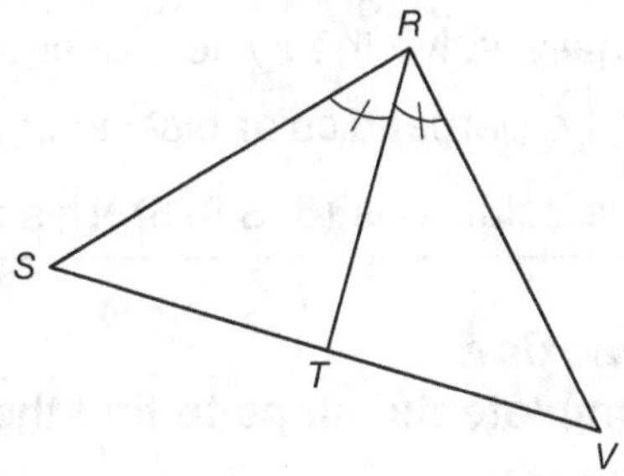

$\frac{18}{\_\_\_} = \frac{\_\_\_}{n}$

___ n = ___

n = ___

2. Find the length of $\overline{RV}$ if $\overline{RS} = 27$, $\overline{ST} = 12$, and $\overline{TV} = 11$.

$\frac{\_\_\_}{\_\_\_} = \frac{n}{\_\_\_}$

___ n = ___

n = ___

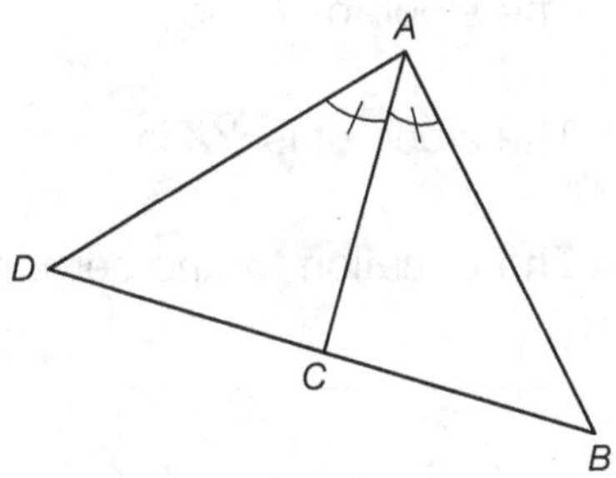

3. Find the length of $\overline{BC}$ if $\overline{AD} = 15$, $\overline{AB} = 14$, and $\overline{DC} = 9$. ___

4. Find the length of $\overline{AD}$ if $\overline{AB} = 30$, $\overline{BC} = 12$, and $\overline{DC} = 15$. ___

5. Find the length of $\overline{DC}$ if $\overline{AD} = 21$, $\overline{AB} = 25$, and $\overline{BC} = 15$. ___

6. Find the length of $\overline{AB}$ if $\overline{AD} = 28$, $\overline{DC} = 8$, and $\overline{BC} = 6$. ___

The circumcenter of a triangle is the point at which all the perpendicular bisectors of the sides intersect. The circumcenter is equidistant from all the vertices of the triangle.

Find the circumcenter of the triangle with vertices at $A(3, 3)$, $B(9, 3)$, and $C(6, 7)$.

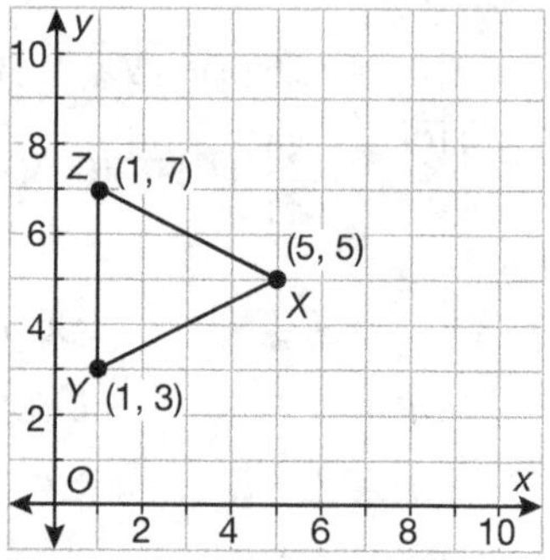

To find the circumcenter, first choose two sides to use. Then, find the midpoints of the two sides.

The midpoint of $\overline{AB}$ is (6, 3), and the midpoint of $\overline{AC}$ is (4.5, 5).

Next, find the slope of the two sides.

The slope of $\overline{AB}$ is 0, and the slope of $\overline{AC}$ is $\frac{4}{3}$.

Since the circumcenter is the point at which the perpendicular bisectors intersect, find the equations of the lines perpendicular to $\overline{AB}$ and $\overline{AC}$.

The equation of the line perpendicular to $\overline{AB}$ through (6, 3) is $x = 6$, and the equation of the line perpendicular to $\overline{AC}$ through (4.5, 5) is $y = -\frac{3}{4}x + \frac{67}{8}$.

Finally, solve the system of equations to find the point of intersection of the perpendicular bisectors: $x = 6$ and $y = -\frac{3}{4}x + \frac{67}{8}$.

The solution is (6, 3.875). This is the circumcenter of triangle *ABC*.

Practice

Complete the steps to find the circumcenter of the triangle.

7. ΔXYZ with vertices at $X(5, 5)$, $Y(1, 3)$, and $Z(1, 7)$.

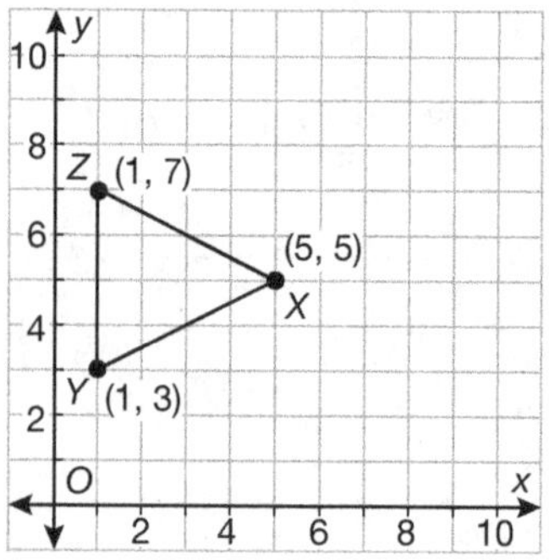

   The midpoint of $\overline{XY}$ is ______.

   The midpoint of $\overline{YZ}$ is ______.

   The slope of $\overline{XY}$ is ______.

   The slope of is $\overline{YZ}$ is __________.

   The equation for the perpendicular bisector of $\overline{XY}$ is

   ________________.

   The equation for the perpendicular bisector of $\overline{YZ}$ is __________.

   The circumcenter for ΔXYZ is ______.

8. Find the circumcenter of the triangle with vertices at $A(2, 4)$, B(8, 4), and C(6, 10). ________

Name ______________________________ Date __________ Class __________

Reteaching 39

Inequalities in a Trianglee

You learned that the sum of the angles inside a triangle is 180°.
Now you will relate the angle measures to the side lengths of a triangle.

A triangle's largest side is always opposite its largest angle, and its smallest side is always opposite its smallest angle.

Order the angle measures of ΔABC from least to greatest.

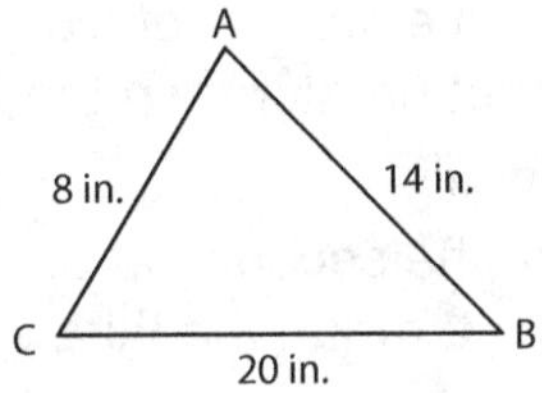

The shortest side length is $\overline{AC}$, so the smallest angle in the triangle is $\angle B$.

The longest side is $\overline{CB}$, so $\angle A$ is the largest angle.

Finally, $\overline{AB}$ is the middle side length, so $\angle C$ is the middle angle measure.

Therefore, the order of the angle measures from least to greatest is $\angle B$, $\angle C$, $\angle A$.

Practice

Fill in the blanks to put the angle measures in order from least to greatest for ΔFGH.

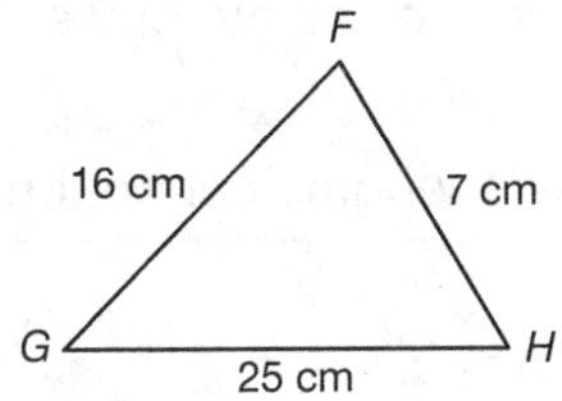

1. The shortest side is ___, so the smallest angle is ___.

The longest side is ___, so the largest angle is ___.

The remaining angle, ___, is the middle angle measure.

Therefore, the order of the angles from least to greatest measure is ____________.

Fill in the blanks to put the angle measures in order from least to greatest for ΔXYZ.

2. The measure of $\angle Y$ is ____.

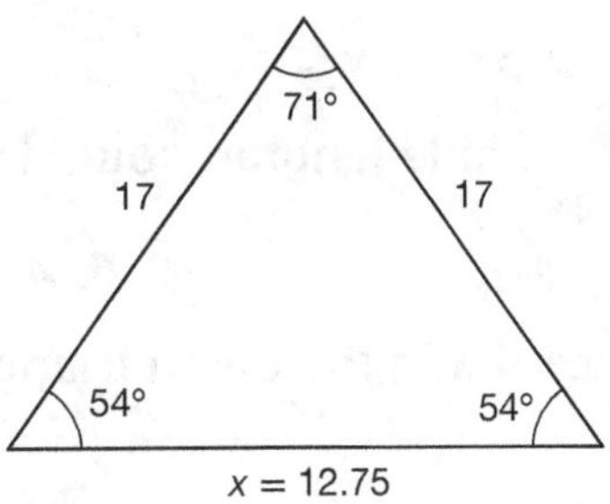

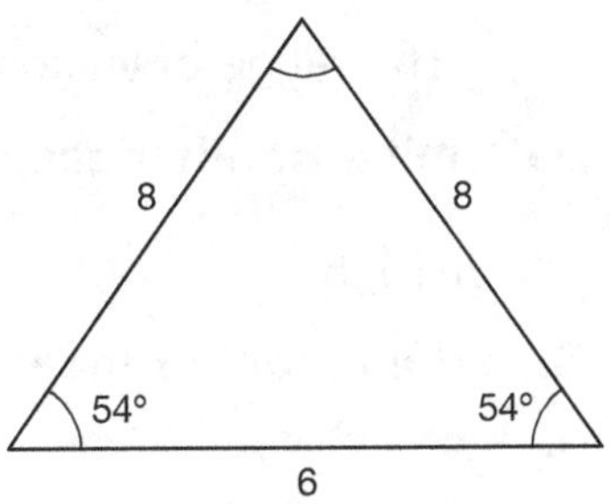

Therefore, ___ is the longest side. The angle with the smallest measure is ___, so the shortest side length is ___.

So the middle side length is ___.

Therefore, the order of the side lengths from least to greatest is __________.

Triangle Inequality Theorem: The sum of the lengths of any two sides of a triangle must be greater than the length of the third side.

Determine whether each set of side lengths could form a valid triangle.

(4, 5, 9) and (9, 11, 12)

For the first set of side lengths, the sum of 4 and 5 is equal to 9. This is not greater than the third side, so these side lengths cannot be a triangle.

For the second set of side lengths, the sum of any pair of sides is greater than the third side, so these lengths do form a triangle.

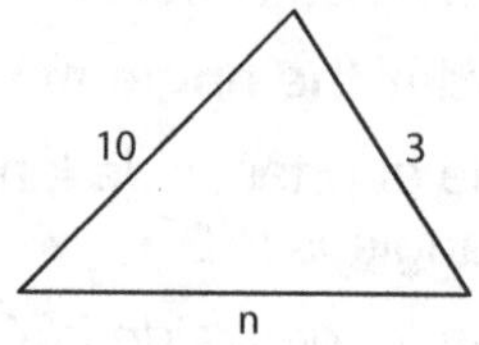

What is the range for the values of n in the given triangle?

Using the **Triangle Inequality Theorem**, the sum of any two sides must be greater than the third side.

First, write inequalities for the sum of each pair of sides.

$10 + 3 > n \quad 10 + n > 3 \quad 3 + n > 10$

$13 > n \quad n > -7 \quad n > 7$

Then, eliminate any side lengths that are not possible and find the solution.

Side lengths cannot be negative, so the solution is $7 < n < 13$.

Practice

Fill in the blanks to find the range for the values of *x*.

3. ___ + ___ $> x$ ___ $+ x >$ ___

___ $> x$ $x >$ ___

___ $+ x >$ ___

$x >$ ___

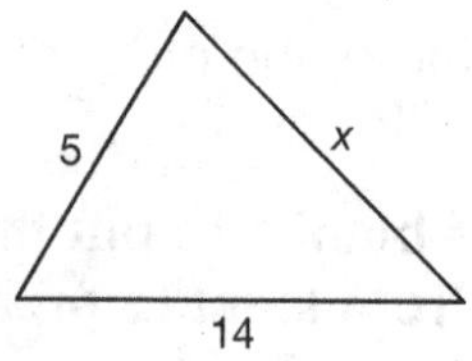

The range of values for x is ___ $< x <$ ___.

Determine whether each set of side lengths could form a triangle.

4. 6, 10, 5 ___

5. 4, 8, 15 ___

Give the range for the values of *x* in the given triangles.

6.

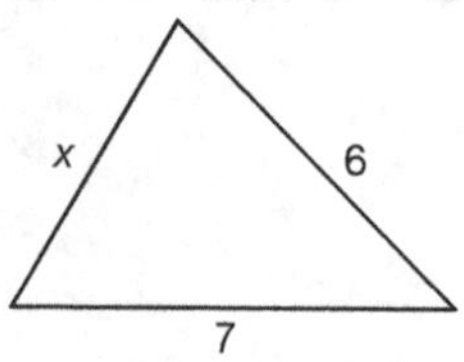

___

7.

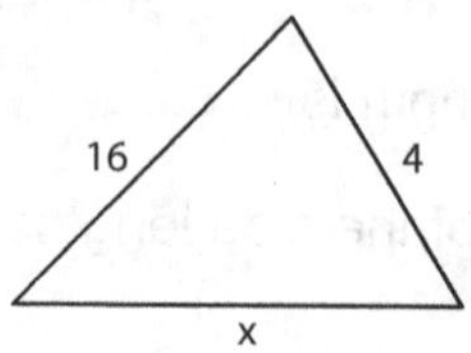

___

Name ______________________ Date __________ Class __________

Reteaching 40

Finding Perimeters and Areas of Composite Figures

You have learned how to find the perimeter and area of polygons and circles. Now you will find the perimeter and area of composite figures.

Area of Composite Figures

Find the area of this composite figure.

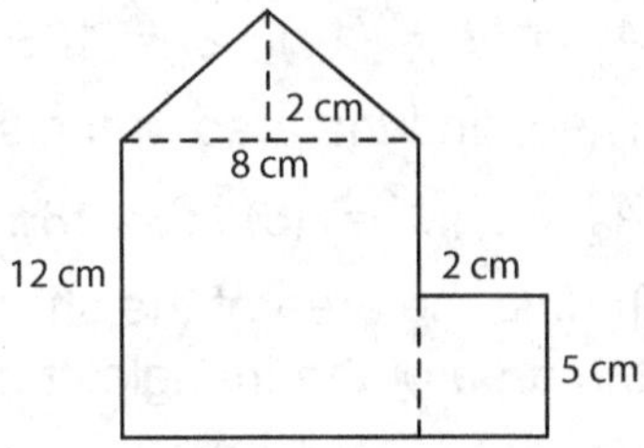

The figure is made up of a triangle and two rectangles.
First, find the area of each shape in the figure.

The area of the triangle is $A_1 = \frac{1}{2}bh = \frac{1}{2}(8)(2) = 8 \text{ cm}^2$.

The area of the large rectangle is $A_2 = bh = (8)(12) = 96 \text{ cm}^2$.

The area of the small rectangle is $A_3 = bh = (2)(5) = 10 \text{ cm}^2$.

Add all the areas. By the **Angle Addition Postulate**, the area of the composite figure is $A = 8 + 96 + 10 = 114 \text{ cm}^2$.

Practice

Complete the steps to find the area of each figure.

1. $A_1 = bh = (\underline{\quad})(\underline{\quad}) = \underline{\quad}$ in^2

   $A_2 = \frac{1}{2}bh = \frac{1}{2}(\underline{\quad})(\underline{\quad}) = \underline{\quad}$ in^2

   $A = \underline{\quad} + \underline{\quad} = \underline{\quad}$ in^2

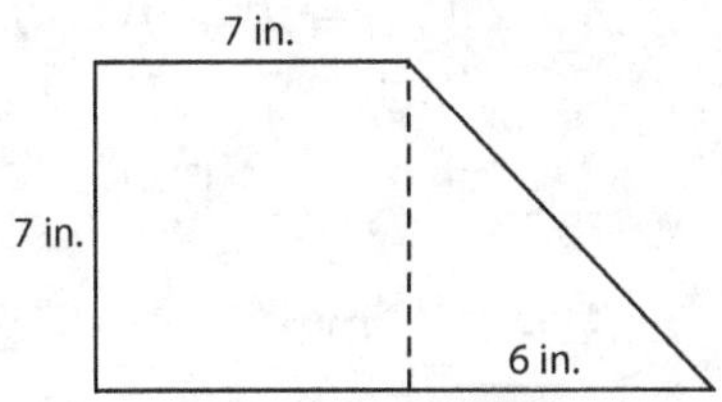

2. $A_2 = bh = (\underline{\quad})(\underline{\quad}) = \underline{\quad}$ m^2

   $A_2 = bh = (\underline{\quad})(\underline{\quad}) = \underline{\quad}$ m^2

   $A = \underline{\quad} + \underline{\quad} = \underline{\quad}$ m^2

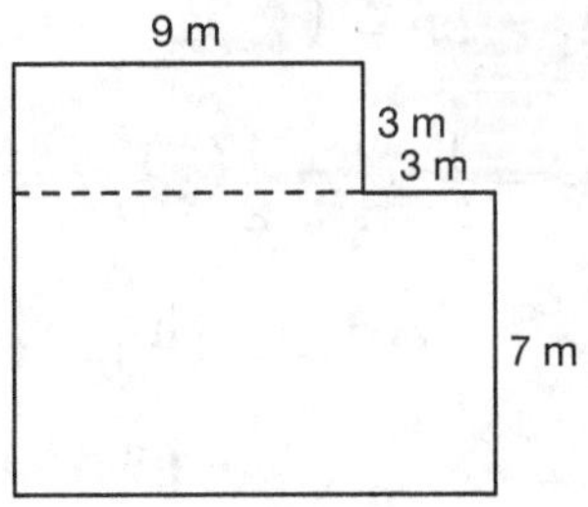

Find the area of each composite figure. When necessary, write your answer in terms of π.

3.

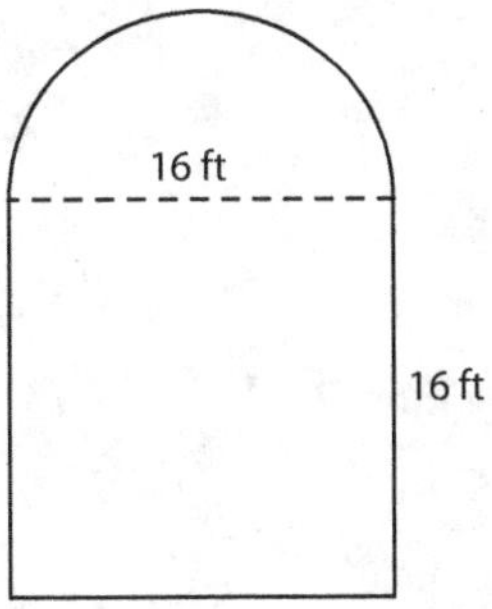

____

4.

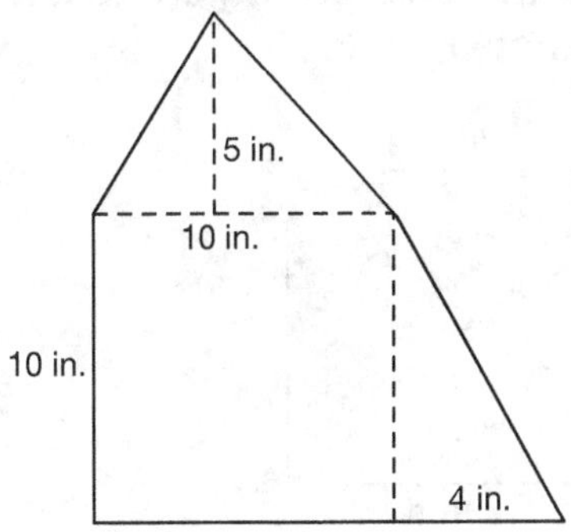

____

Area of Shaded Regions

Find the area of the shaded region. Leave your answer in terms of π.

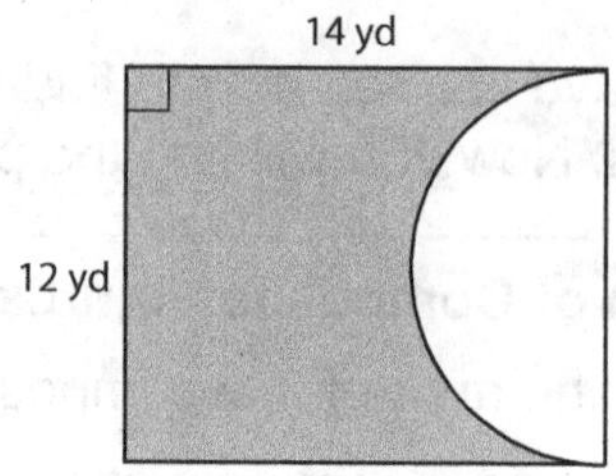

The figure is made up of a semicircle inside a rectangle.

First, find the area of the whole rectangle.

$A_1 = bh = (14)(12) = 168 \text{ yd}^2$

Then, find the area of the semicircle.

$A_2 = \frac{1}{2}\pi r^2 = (6)^2\pi = 18\pi \text{ yd}^2$

To find the area of the shaded region, we need to know the area of the rectangle without the area of the triangle. If A is the area of the shaded region, then $A_1 = A + A_2$.

So, $168 = A + 18\pi$, and $A = 168 - 18\pi \text{ yd}^2$.

Practice

Fill in the blanks to find the area of the shaded region for each figure.

5. $A_1 = bh = (\_\_)(\_\_) = \_\_\_ \text{ mm}^2$

   $A_2 = \frac{1}{2}bh = \frac{1}{2}(\_\_)(\_\_) = \_\_ \text{ mm}^2$

   $\_\_\_ = A + \_\_ \text{ mm}^2$

   $A = \_\_\_ \text{ mm}^2$

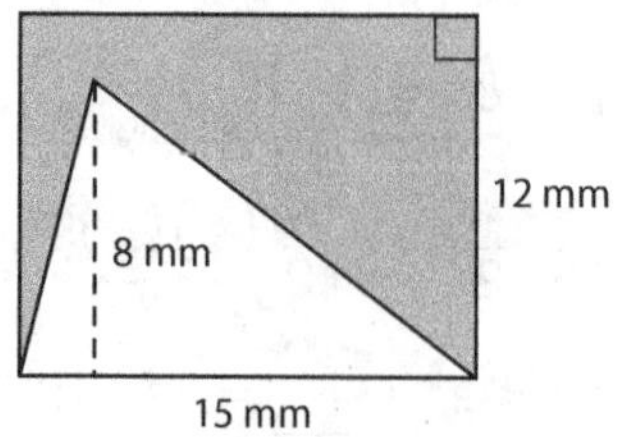

6. $A_1 = bh = (\_\_)(\_\_) = \_\_\_ \text{ ft}^2$

   $A_2 = \frac{1}{2}\pi r^2 = \frac{1}{2}(\_)^2\pi = \_\ \pi \text{ ft}^2$

   $\_\_\_ = A + \_\_\_ \text{ ft}^2$

   $A = \_\_\_\_\_\_ \text{ ft}^2$

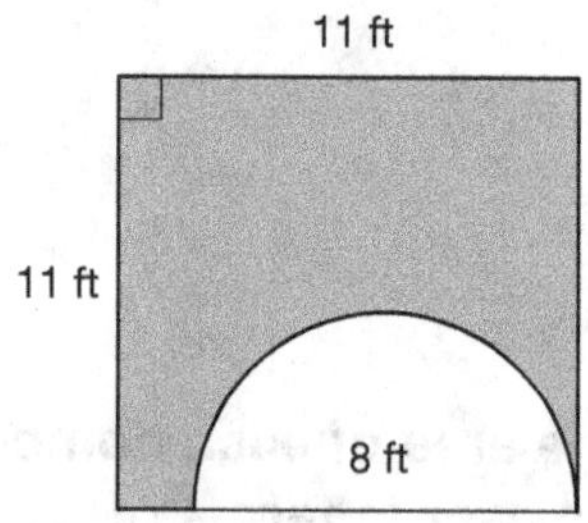

Find the area of the shaded region. When necessary, write your answer in terms of π.

7.

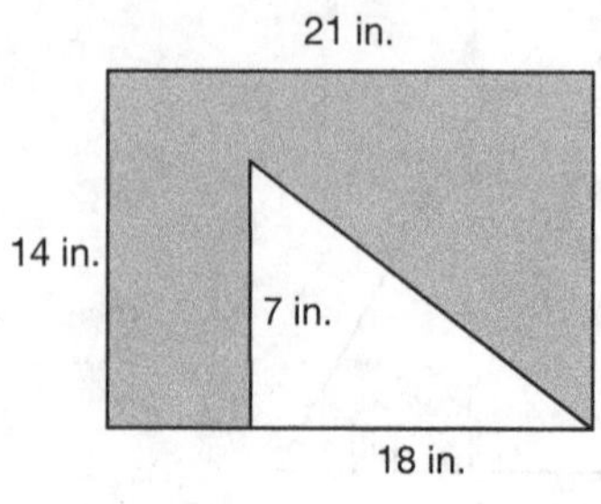

8.

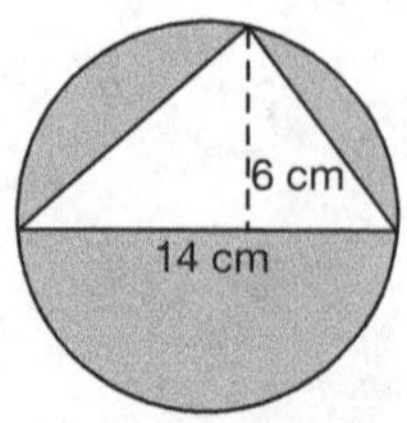

Diameter = 14 cm

________ ________

Name ______________________________ Date __________ Class ____________

Reteaching
Inequalities in Two Triangles

INV 4

You have worked with segments and angles in triangles. Now you will explore inequalities with triangles.

Hinge Theorem

If two sides of one triangle are congruent to those of another triangle, and if the included angle of the first triangle is greater than that of the second triangle, then the third side of the first triangle is longer than that of the second triangle.

Example:

Compare the measures of *BC* and *EF*.

Step 1: Compare the angle measures of the sides opposite *BC* and *EF*.

The opposite angle of $BC = 42°$.

The opposite angle of $EF = 39°$.

Step 2: Make a comparison.

The side opposite the greater angle is longer.

$BC > EF$

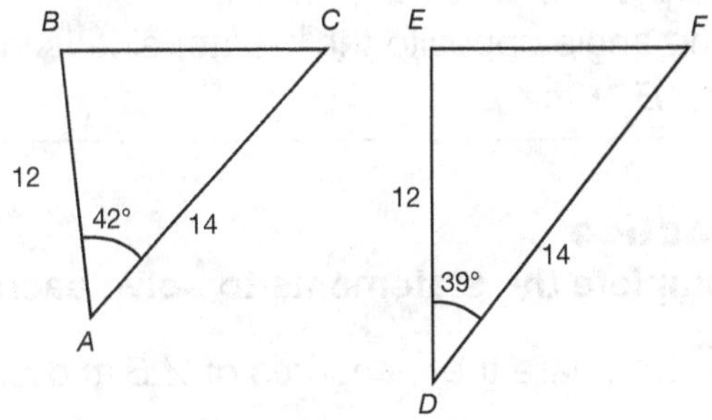

Practice

Complete the statements to solve each problem.

1. Compare the measures of *AB* and *DE*.

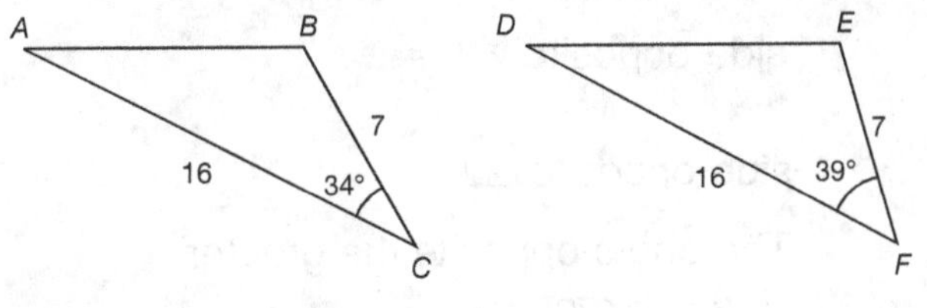

AB opposite angle = ___

DE opposite angle = ___

___ $<$ *DE*

2. Compare the lengths of *KL* and *NP*.

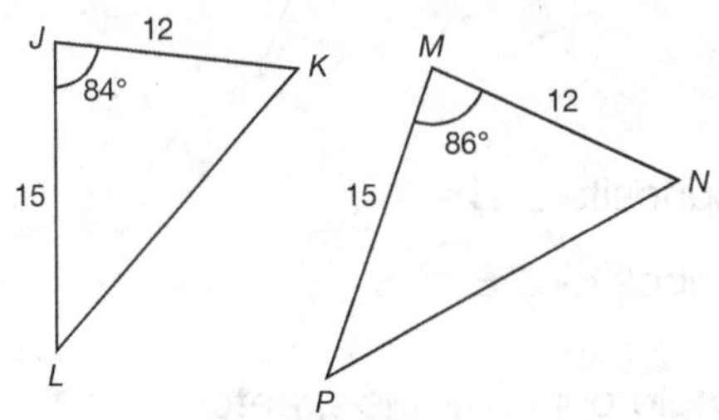

KL opposite angle = ___

NP opposite angle = ___

___ $<$ ___

Compare the measures.

3. Compare the measures of *VW* and *YZ*.

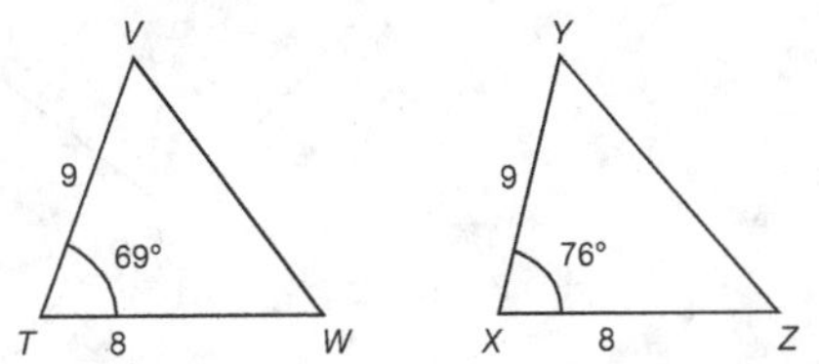

____

4. Compare the measures of *JL* and *MP*.

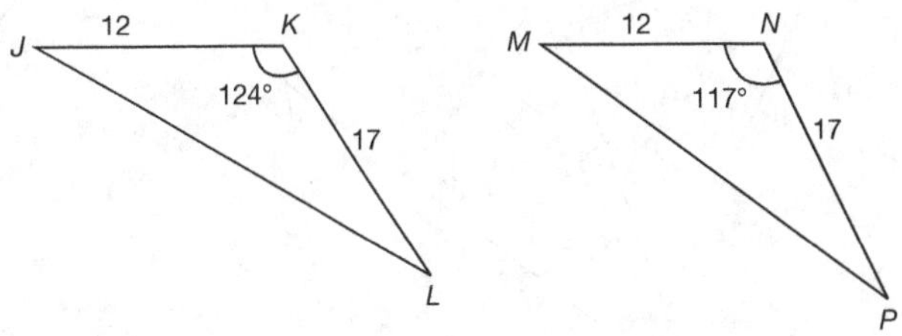

____

Converse of the Hinge Theorem

If two sides of one triangle are congruent to those of another triangle, and if the third side of the first triangle is longer than that of the second triangle, then the measure of the angle opposite the third side of the first triangle is greater than that of the second triangle.

Example:

Compare the measures of ∠B and ∠E.

Step 1: Compare the measures of the lengths of the sides opposite the angles.

The side opposite ∠B = 4.

The side opposite ∠E = 4.5.

Step 2: Make a comparison.

The angle opposite the greater side is larger.

m∠E > m∠B

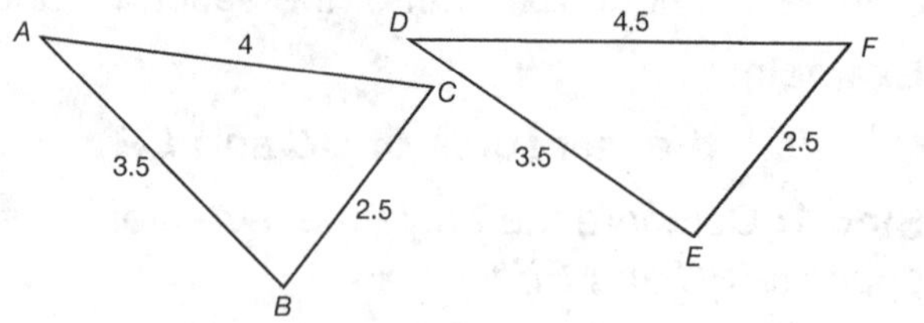

Practice

Complete the statements to solve each problem.

5. Compare the measures of ∠B and∠E.

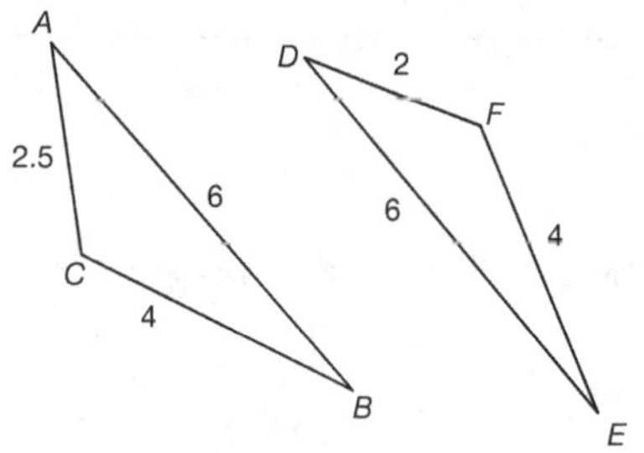

side opposite ∠B = ___

side opposite ∠E = _

The angle opposite the greater side is larger.

m∠B __ m∠ E

6. Compare the measures of ∠B and ∠E.

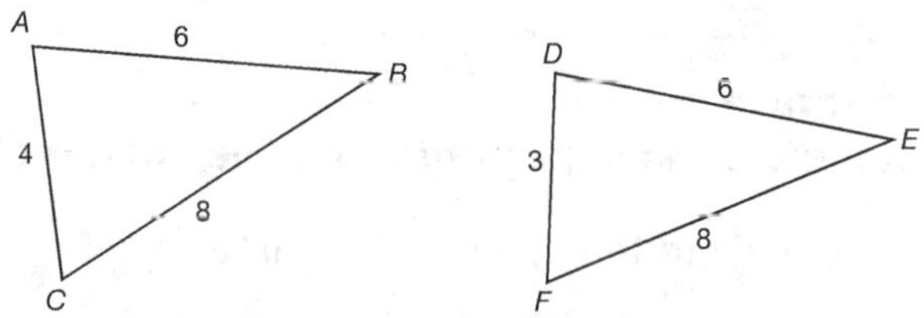

side opposite ∠B = _

side opposite ∠E = _

The angle opposite the greater side is larger.

m∠ _ > m∠ _

Compare the measures.

7. Compare the measures of ∠C and ∠F.

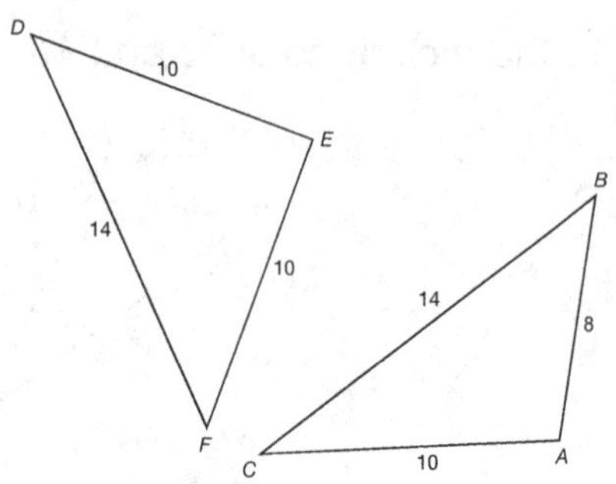

____

8. Compare the measures of ∠C and ∠F.

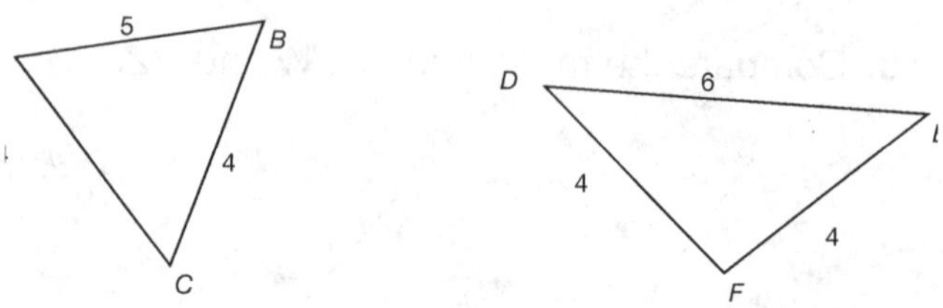

____

Name ______________________ Date __________ Class __________

Reteaching 41

Ratios, Proportions, and Similarity

You have worked with corresponding sides and angles. Now you will work with ratios and proportions.

A proportion is an equation stating that two ratios are equal. In every proportion, the product of the extremes equals the product of the means.

Cross Products Property	In a proportion, if $\frac{a}{b} = \frac{c}{d}$ and b and $d \neq 0$, then $ad = bc$. 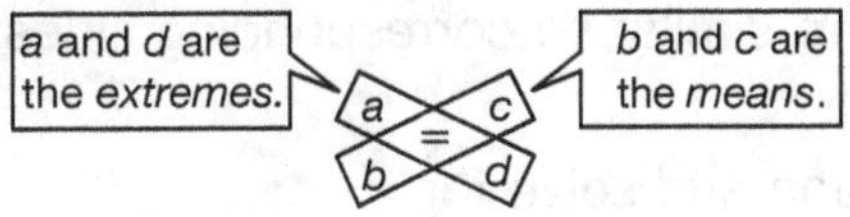

You can solve a proportion like $\frac{x}{8} = \frac{35}{56}$ by finding the cross products.

$\frac{x}{8} = \frac{35}{56}$

$x(56) = 8(35)$ Cross Products Property

$56x = 280$ Simplify.

$x = 5$ Divide both sides by 56.

Practice

Complete the statements to solve each proportion.

1. $\frac{9}{w} = \frac{36}{28}$

$w($__$) = 9($__$)$ Cross Products Property

__ w = ____ Simplify.

w = __ Divide both sides by 36.

2. $\frac{x}{32} = \frac{15}{16}$

$x($__$)$ = __(35) Cross Products Property

____ = ____ Simplify.

x = __ Divide both sides by 16.

Solve each proportion.

3. $\frac{24}{42} = \frac{y}{7}$ ______

4. $\frac{40}{49} = \frac{x}{98}$ ______

5. $\frac{20}{25} = \frac{y}{45}$ ______

6. $\frac{27}{63} = \frac{24}{x}$ ______

In **similar polygons,** the corresponding angles are congruent, and the corresponding sides are proportional.

A **similarity ratio** is the ratio of two corresponding linear measurements in a pair of similar figures.

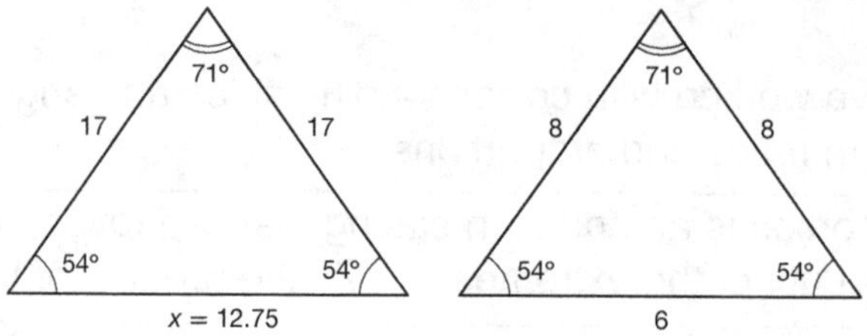

Find the unknown side length in the two similar triangles.

The triangles are similar so corresponding sides are in proportion.

Write a proportion and solve for x.

$\frac{17}{x} = \frac{8}{6}$

$(8)x = 17(6)$	Cross Products Property
$8x = 102$	Simplify.
$x = 12.75$	Divide both sides by 8.

The length of the missing side is 12.75.

Practice

Complete the statement to solve for x.

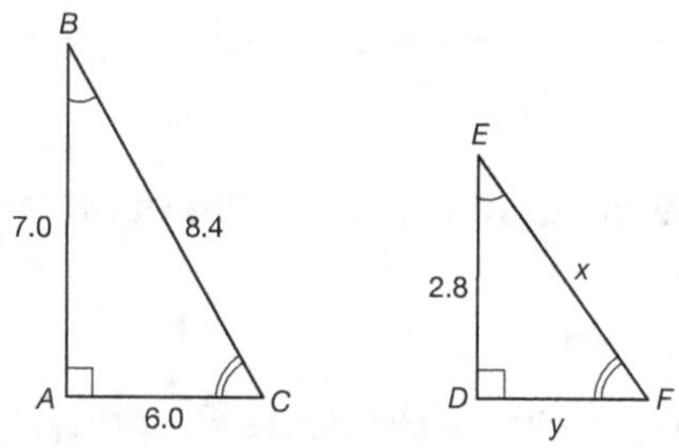

7. $\frac{7.0}{8.4} = \frac{2.8}{x}$

$7(x) =$ ____ (8.4)	Cross Products Property
$7x =$ ______	Simplify.
$x =$ ______	Divide

Solve for the unknown side lengths in the similar figures.

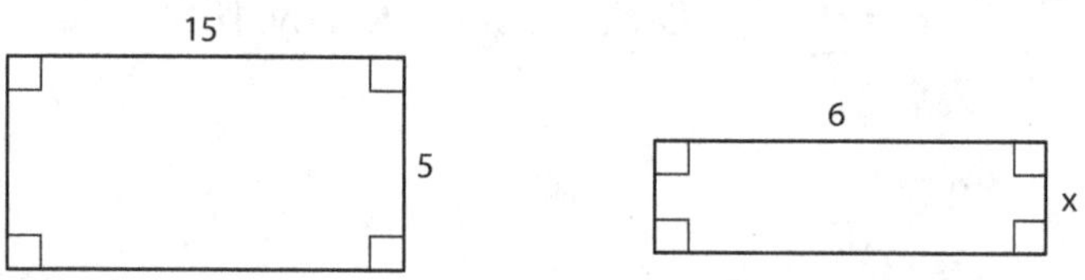

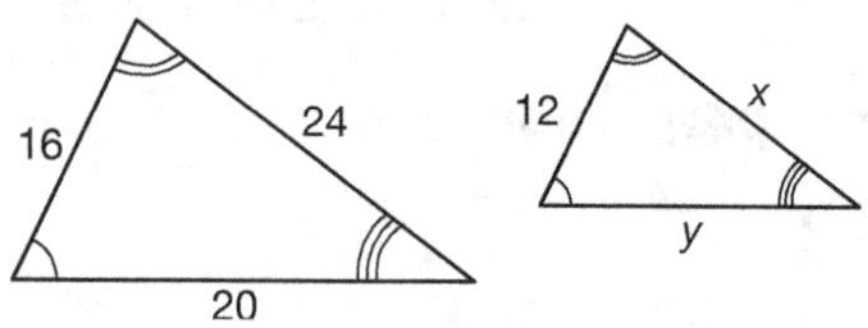

8. ______

9. ____________

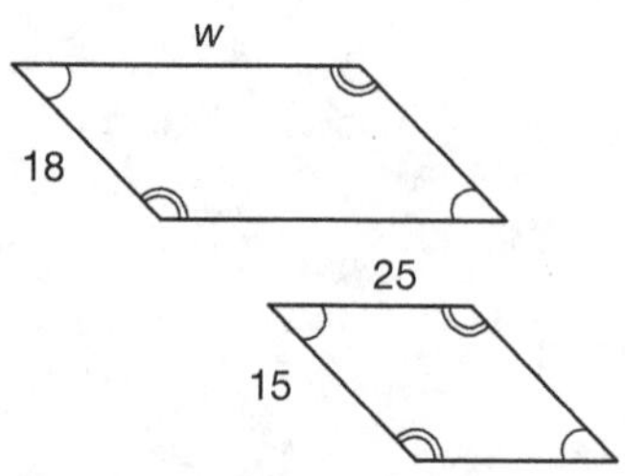

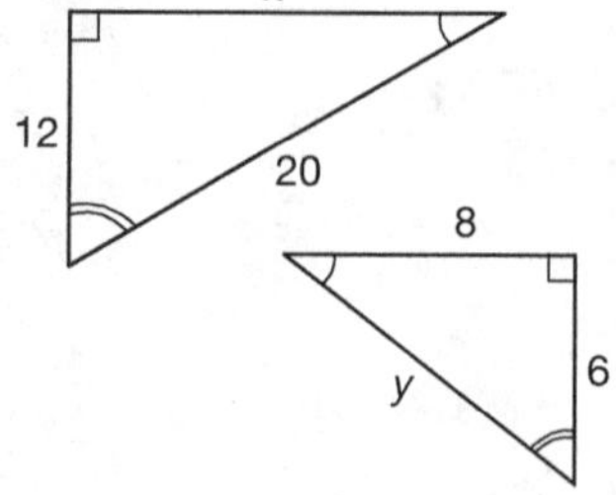

10. ____________

11. ____________

Name ______________________ Date __________ Class __________

Reteaching 42

Finding Distance from a Point to a Line

You have worked with finding distance on a coordinate plane. Now you will find the shortest distance between a point and a line.

Finding Distance

The shortest distance between a point and a line is always a perpendicular segment.

Find the distance from the point (1, 3) to the line $y = 1$.	Find the distance from the point (4, 2) to the line $x = 1$.
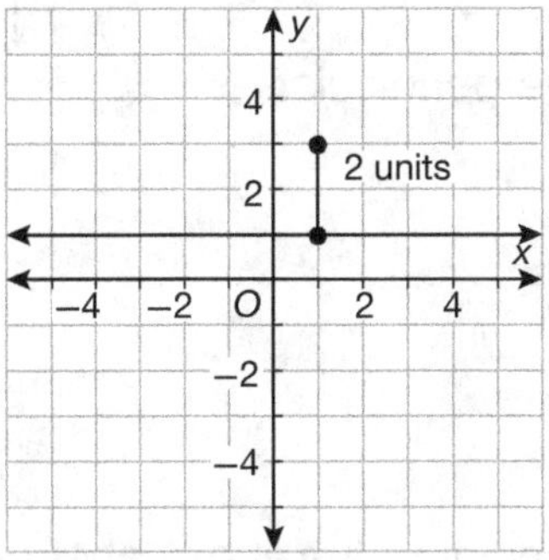	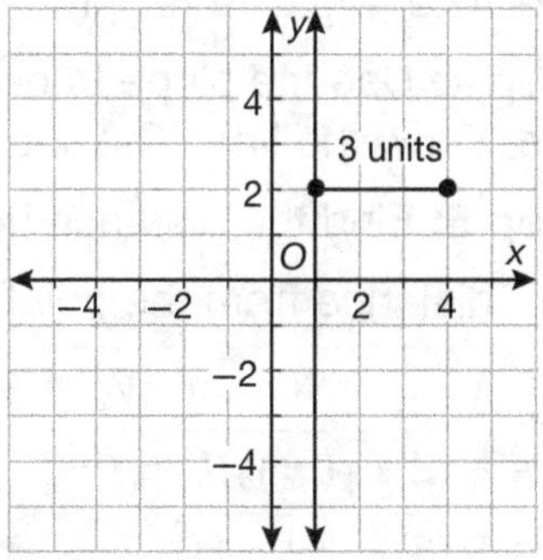
Step 1: Draw a point on the line $y = 1$, perpendicular from the given point. **Step 2:** Count the number of units between the two points. **Step 3:** The distance between the points is 2 units.	**Step 1:** Draw a point on the line $x = 1$, perpendicular from the given point. **Step 2:** Count the number of units between the two points. **Step 3:** The distance between the points is 3 units.

Practice

Complete the statements to find the distance between the point and the line.

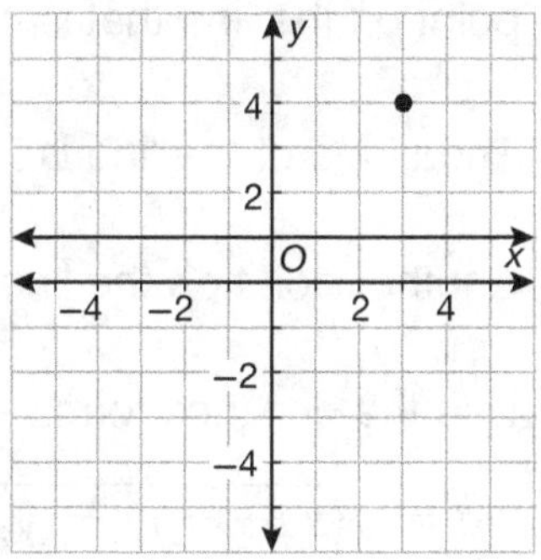

1. Find the distance between the point (3, 4) and the line $y = 1$.

   The point on the line perpendicular to the given point is (3, __).

   Count the number of units between the two points: ________.

   The distance is __ units.

2. Find the distance between the point (2, 3) and the line $y = 5$.

   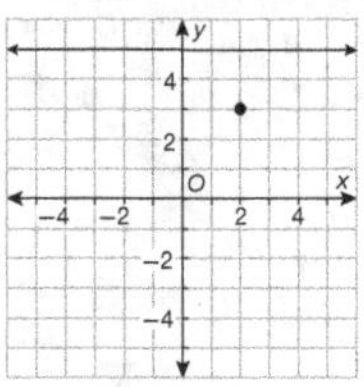

   __ units

3. Find the distance between the line $x = 4$ and the point (–3, 1).

   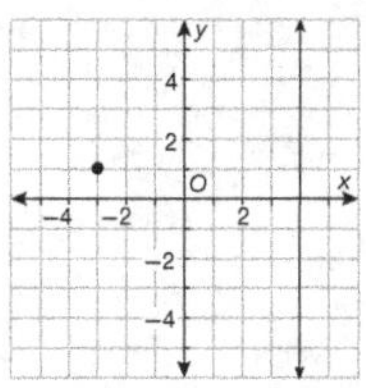

   __ units

Finding the Closest Point

Given the equation $y = 3x + 1$ and the point $W(2, -3)$, find the point on the line that is closest to W. Find the distance from W to the line.

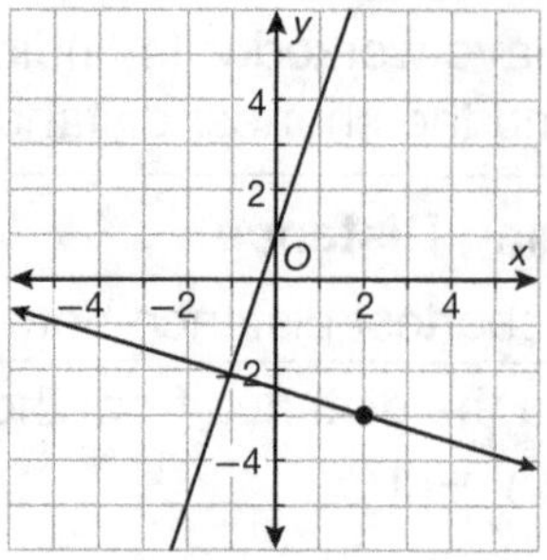

Step 1: Draw the line and point on the coordinate grid.

Step 2: Determine the slope of the line. The slope of the line is 3.

Step 3: Determine the slope of the line perpendicular to the given line. The negative reciprocal of 3 is $-\frac{1}{3}$.

Step 4: Use the slope to determine more points on the perpendicular line. Draw the line. The lines intersect at (–1, –2).

Step 5: Find the distance between (2, –3) and (–1, –2).

The distance from the point to the line is about 3.16 units.

$$\begin{aligned} d &= \sqrt{(x_2 - x_1)^2 + (y_2 - y_1)^2} \\ &= \sqrt{(2 - (-1))^2 + (-3 - (-2))^2} \\ &= \sqrt{(3)^2 + (-1)^2} \\ &= \sqrt{10} \\ &\approx 3.16 \end{aligned}$$

The distance from the point to the line is about 3.16 units.

Practice

Complete the statements to find the distance.

4. Given the equation $y = 2x - 1$ and the point (2, –2), find the point on the line that is closest to the point.

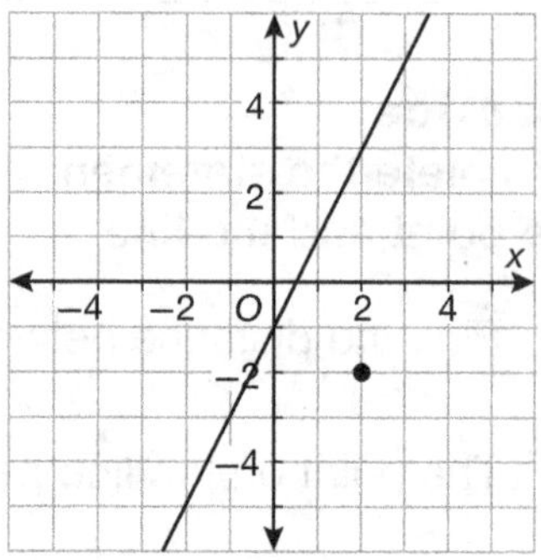

The slope of the line is __. The negative reciprocal is ___.

Another point on the line is (0, ___).

Use the distance formula.

$$\begin{aligned} d &= \sqrt{(x_2 - x_1)^2 + (y_2 - y_1)^2} \\ &= \sqrt{(\_ - 0)^2 + (\_\_ - (-1))^2} \\ &= \sqrt{(\_)^2 + (-1)^2} \\ &= \sqrt{\_} \\ &\approx \_\_\_ \end{aligned}$$

The distance is about ____ units.

Name ________________________________ Date __________ Class __________

Reteaching 43

Chords, Secants, and Tangents

You have working with radii and diameters of a circle. Now, you will work with other types of lines and segments in circles.

Lines that Intersect Circles

A chord is a segment whose endpoints lie on a circle. A diameter is a chord.

A **secant** is a line that intersects a circle at two points.

A **tangent** is a line in the same plane as a that of a circle that intersects the circle at exactly one point, called the **point of tangency.**

Radii and diameters also intersect circles.

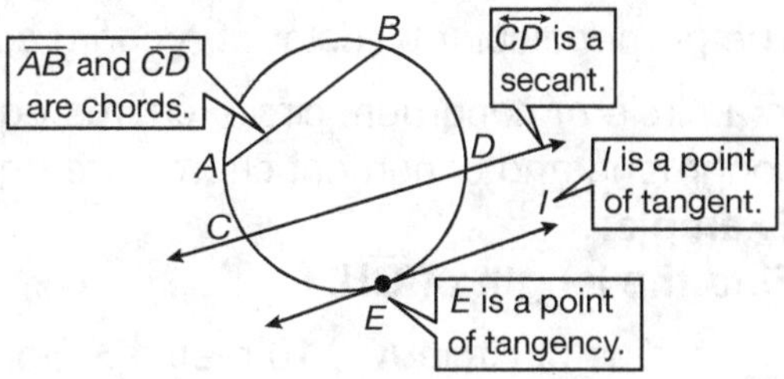

Practice

Complete the statements to identify each line or segment that intersects the circle.

1. Chord: $\overline{FG}$

Secant: _____

Tangent: ______

Diameter: $\overline{FG}$

Radii: $\overline{HF}$ and ___

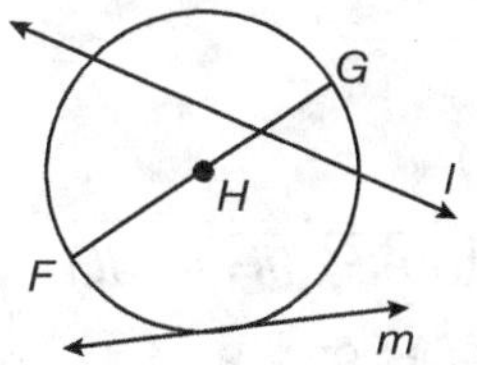

Identify each line or segment that intersects each circle.

2.

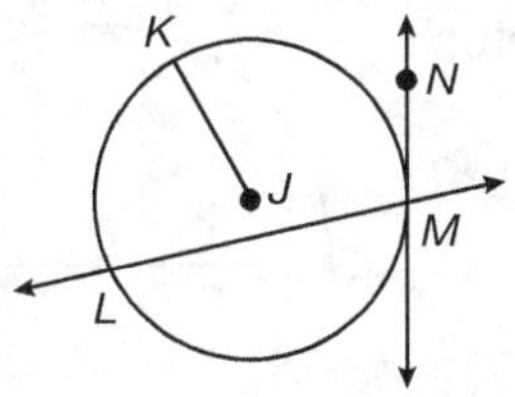

Chord: ___

Secant: ___

Tangent: ___

Diameter: _____

Radii: ___

Point of tangency: __

3.

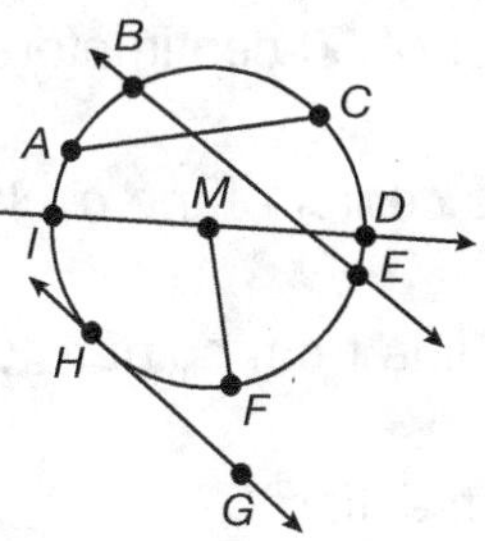

Chord: __________

Secant: ___

Tangent: ___

Diameter: __

Radii: ____________

Point of tangency: __

Chord–Diameter Relationships

If a diameter is perpendicular to a chord, then it bisects the chord and the chord's arcs.

If a diameter bisects a chord other than another diameter, then it is perpendicular to the chord.

The perpendicular bisector of a chord contains the center of the circle.

In a circle or congruent circle, chords equidistant from the center are congruent, and congruent chords are equidistant from the center of the circle.

Example:

Find the length of $\overline{GH}$.

The radius of circle K is 10 meters, and KJ is 6 meters.

First, draw a radius $\overline{KH}$.

Then, use the Pythagorean theorem to find $\overline{JH}$.

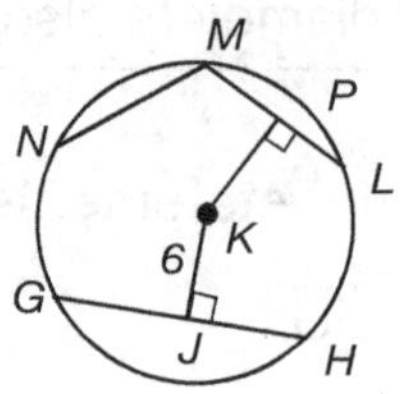

$KJ^2 + JH^2 = KH^2$

$6^2 + b^2 = 10^2$

$36 + b^2 = 100$ $\qquad JH = 8\text{m}$

$b^2 = 64$

$b = 8$

Since $\overline{KJ}$ is perpendicular to $\overline{GH}$, $\overline{JH} = \overline{JG}$.

Therefore, the length of $GH = 8 + 8 = 16$, or 16 meters.

Practice

Complete the statements to find the missing distance.

4. A circle has a diameter of 50 centimeters. Chord JK is 24 centimeters.

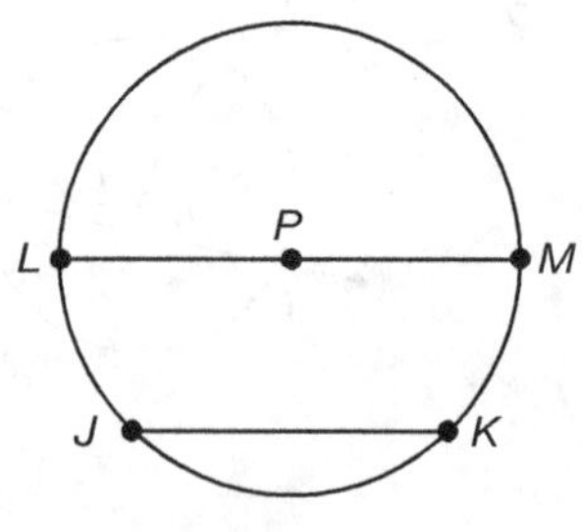

How far is JK from the center of the circle to the nearest hundredth?

Draw a perpendicular line from point P to chord JK.

Label the point of intersection C.

$PC^2 + CK^2 = PK^2$

$PK =$ __ cm; $CK = 12$ cm

$PC^2 +$ __$^2 = 25^2$

$PC^2 +$ ___ $=$ ___

$PC^2 =$ ___

$PC \approx$ ___

Name ______________________ Date __________ Class __________

Reteaching 44
Applying Similiarity

You have worked with proportions to find unknown sides in similar triangles. Now you will work with other types of similar polygons.

Using Similarity to Find Unknown Measures
You can use properties of similar polygons to solve proportions.

The figures in the diagram are similar. Find the values of *x* and *y*.

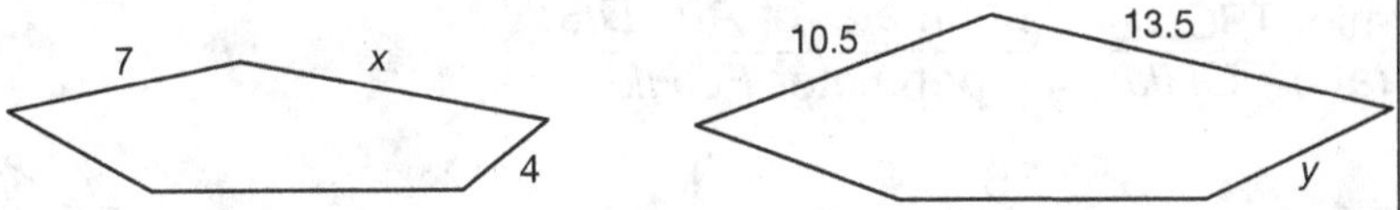

Step 1: Write a similarity ratio. $\frac{7}{10.5}$

Step 2: Write and solve a proportion to solve for *x* and *y*.

$\frac{7}{10.5} = \frac{x}{13.5}$ $\qquad$ $\frac{7}{10.5} = \frac{4}{y}$

$10.5x = 7(13.5)$ $\qquad$ $7y = 4(10.5)$

$10.5x = 94.5$ $\qquad$ $7y = 42$

$x = 9$ $\qquad$ $y = 6$

Practice

The figures in the diagram are similar. Complete the statements to find the missing values.

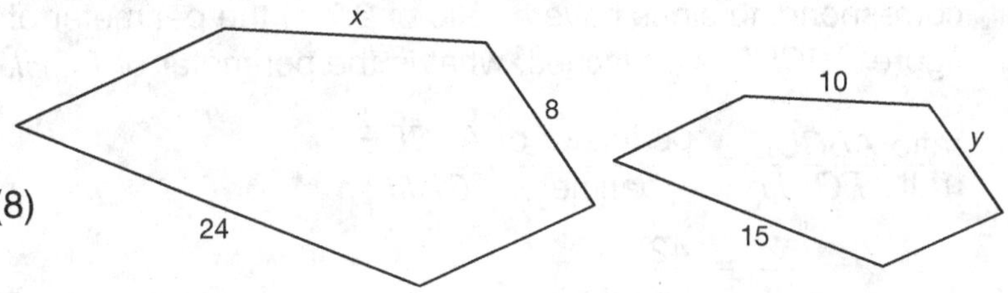

1. $\frac{24}{15} = \frac{x}{\_\_}$ $\qquad$ $\frac{24}{15} = \frac{\_\_}{y}$

$15x = \_\_(10)$ $\qquad$ $24y = \_\_(8)$

$15x = \_\_\_$ $\qquad$ $24y = \_\_\_$

$x = \_\_$ $\qquad$ $y = \_$

The figures in the diagram are similar. Solve for the missing lengths.

2. 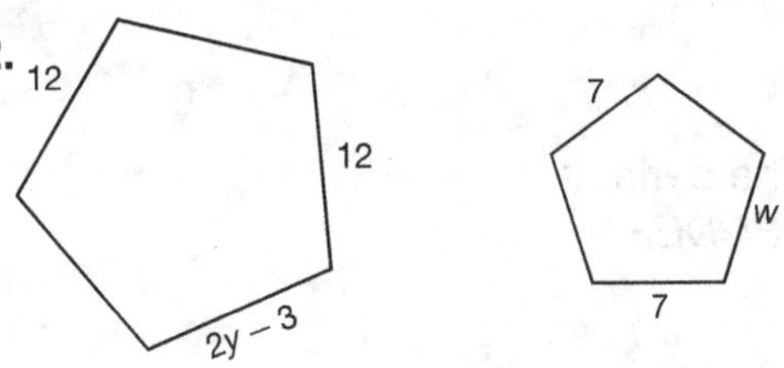

3. The similarity ratio is 3:2.

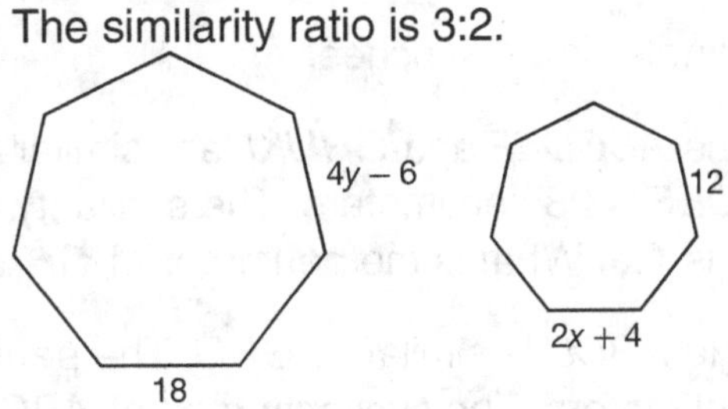

______________________ $\qquad$ ______________________

Applying Similarity to Perimeter

Figures *ABCDE* and *FGHIJ* are similar polygons. Their corresponding sides have a ratio of 3:5. If the perimeter of figure *ABCDE* is 36 inches, what is the perimeter of *FGHIJ?*

Step 1: The figures are similar polygons. The ratio of the perimeters is 3:5.

Step 2: Set up a proportion using the ratio 3:5.

$$\frac{\text{ratio } ABCDE}{\text{ratio } FGHIJ} = \frac{\text{perimeter of } ABCDE}{\text{perimeter } FGHIJ}$$

$$\frac{3}{5} = \frac{36}{x}$$

$$3x = 36(5)$$

$$3x = 180$$

$$x = 60$$

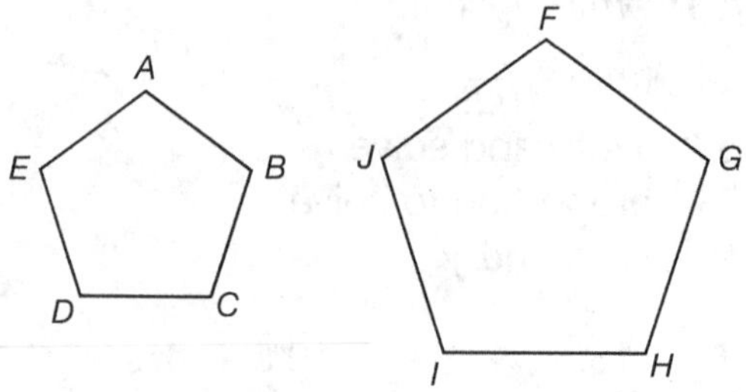

The perimeter of *FGHIJ* is 60 inches.

Practice

Complete the statements to find the perimeter of the similar figure.

4. Figures *ABCDE* and *FGHIJ* are similar polygons. Their corresponding sides have a ratio of 2:7. If the perimeter of figure *ABCDE* is 42 inches, what is the perimeter of *FGHIJ?*

$$\frac{\text{ratio } ABCDE}{\text{ratio } FGHIJ} = \frac{\text{perimeter of } ABCDE}{\text{perimeter } FGHIJ}$$

$$\frac{2}{\_\_} = \frac{42}{\_\_}$$

$$2x = \_\_\ 42)$$

$$2x = \_\_\_$$

$$x = \_\_\_$$

The perimeter is ____ inches.

5. Hexagons *ABCDEF* and *GHIJKL* are similar figures. The perimeter of *ABCDEF* is 28 centimeters. The similarity ratio of *ABCDEF* to *GHIJKL* is 4:9. What is the perimeter of *GHIJKL?* ____________

6. Rectangle *ABCD* is similar to *GHIJ*. The perimeter of *ABCD* is 56 centimeters. The similarity ratio of *ABCD* to *GHIJ* is 1:16. What is the perimeter of *GHIJ?* ____________

7. Figures *LMNP* and *QRST* are similar figures. The ratio of their corresponding sides is 3:8. If the perimeter of *LMNP* is 18 centimeters what is the perimeter of *QRST?* ____________

Name ______________________ Date __________ Class __________

Reteaching 45

Introduction to Coordinate Proofs

You have worked with a coordinate plane and proofs. Now you will work with proofs on a coordinate plane.

Positioning a Figure in the Coordinate Plane

A **coordinate proof** is a proof that uses coordinate geometry and algebra. In a coordinate proof, the first step is to position a figure in a plane. There are several ways you can do this to make your proof easier.

Keep the figure in Quadrant 1 by using the origin as a vertex.	Center a side of the figure at the origin.	Use one or both axes as sides of the figure.
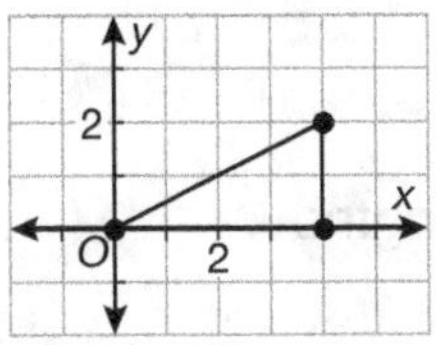	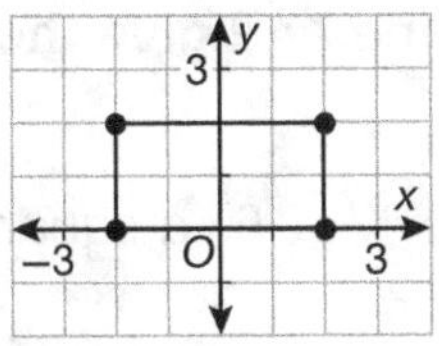	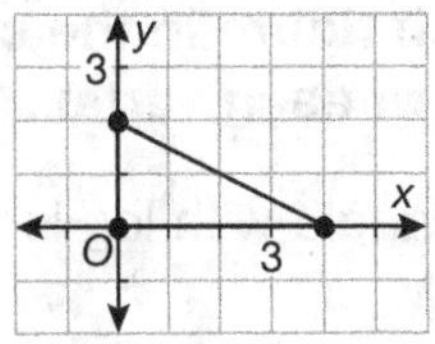

Practice

Position each figure in the coordinate plane and give the coordinates of each vertex.

1. a square with side length 4 units

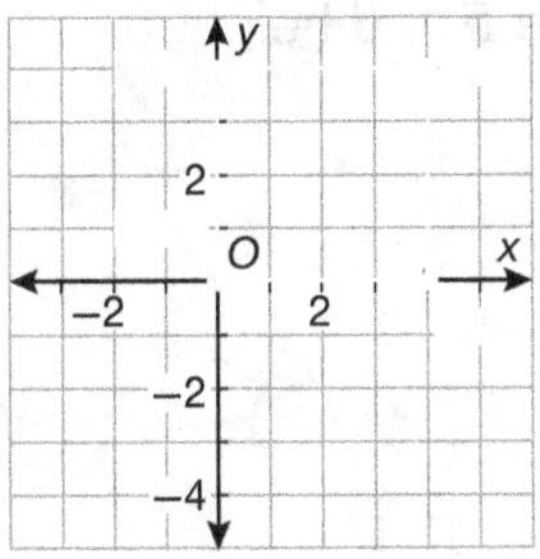

2. a right triangle with leg lengths 3 and 4 units

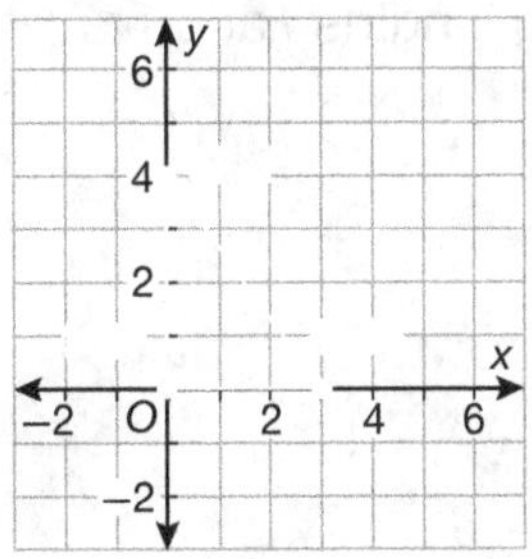

3. a triangle with base 8 units and height 2 units

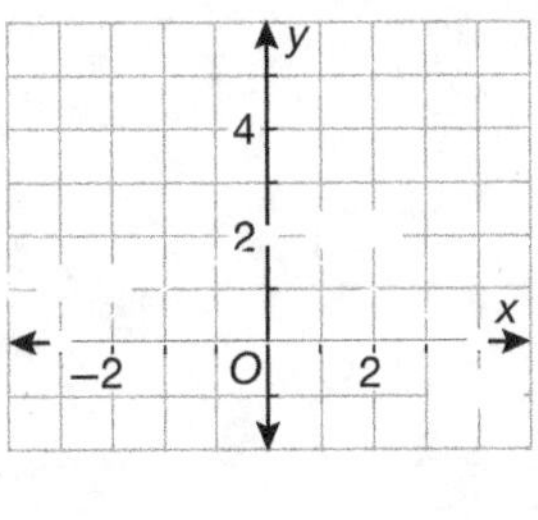

4. a rectangle with length 6 units and width 3 units

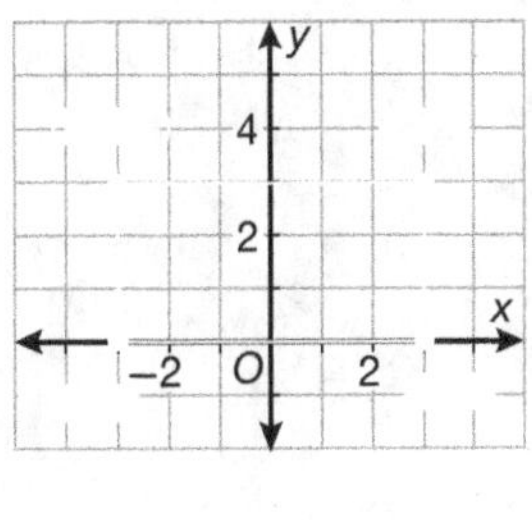

You can prove that a statement about a figure is true without knowing the side lengths. To do this, assign variables as the coordinates of the vertices.

Position a right triangle with leg lengths *c* and *d* in the coordinate plane and give the coordinates of each vertex.

Step 1: Place one vertex at the origin and label (0, 0).

Step 2: The vertex on the *x*-axis is *c* units away. Label (*c*, 0).

Step 3: The vertex on the *y*-axis is *d* units up. Label (0, *d*).

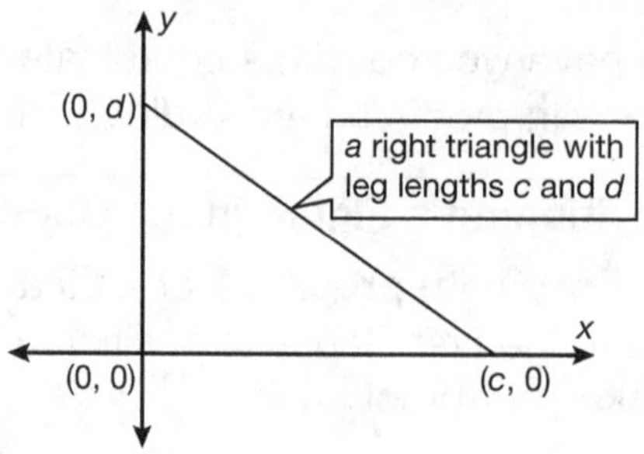

Practice

Position each figure in the coordinate plane and give the coordinates of each vertex.

5. a right triangle with leg lengths *s* and *t*

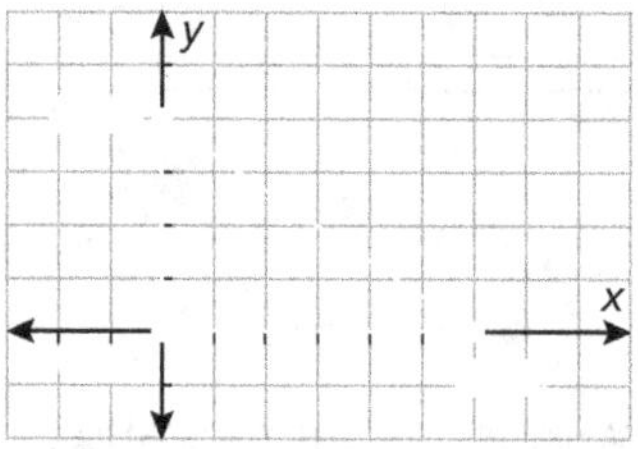

6. a square with side lengths *k*

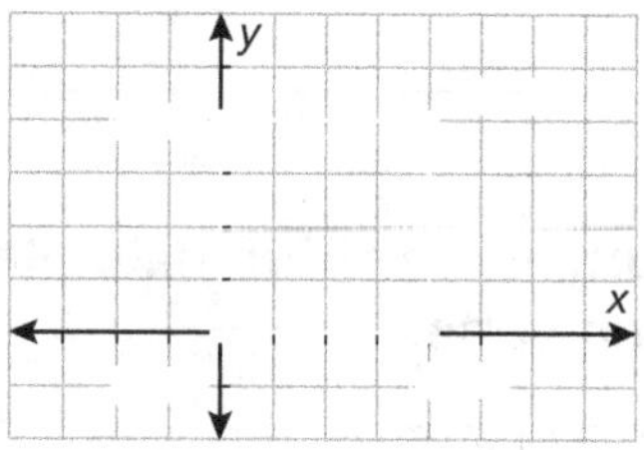

7. a rectangle with leg lengths *l* and *w*

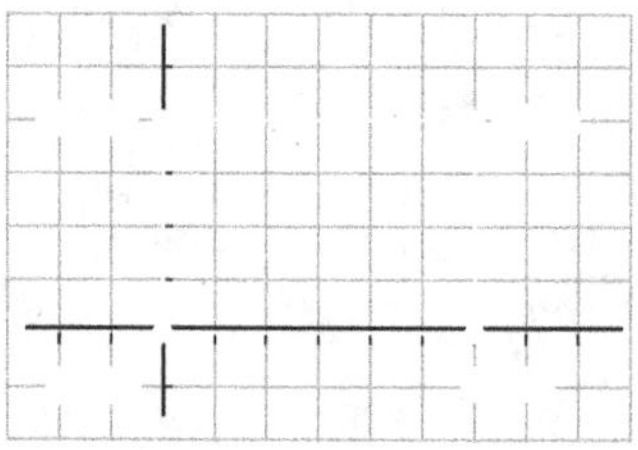

8. a triangle with base *b* and height *h*

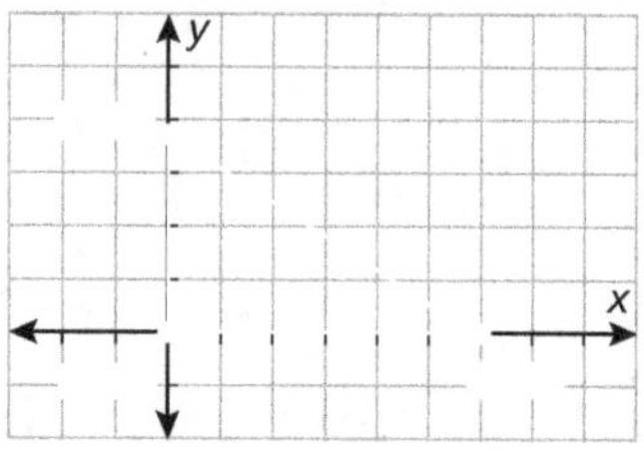

Name ______________________ Date __________ Class __________

Reteaching 46

Triangle Similarity: AA, SAS, SSS

You have worked with coordinate proofs. Now you will work with triangle similarity.

Triangle Similarity: AA, SSS

Angle-Angle (AA) Similarity

If two angles of one triangle are congruent to two angles of another triangle, then the triangles are similar.

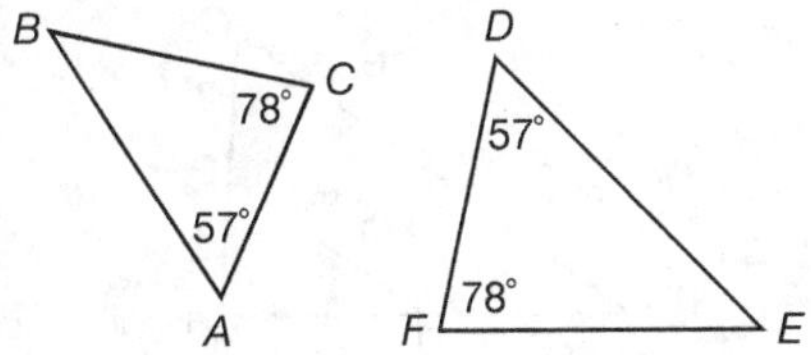

$\Delta ABC \sim \Delta DEF$

Side-Side-Side (SSS) Similarity

If the three sides of one triangle are proportional to the three corresponding sides of another triangle, then the triangles are similar.

$$\frac{TU}{XY} = \frac{UV}{YZ} = \frac{VT}{ZX} = \frac{3}{1}$$

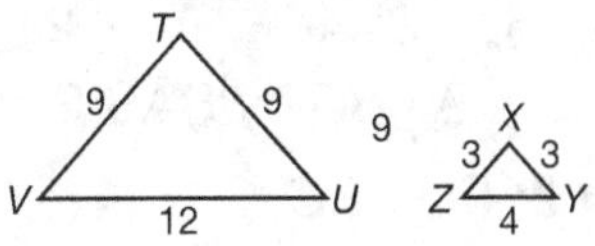

$\Delta TUV \sim \Delta XYZ$

Given the values of the two triangles as shown, show that the triangles are similar triangles.

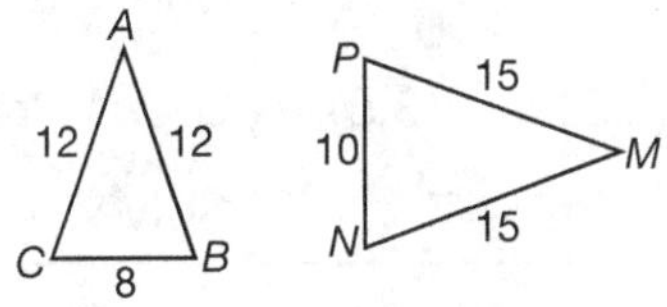

Step 1: Write similarity ratios for pairs of sides. Let the lengths of ΔABC be the top number of the proportions.

$\overline{AC} : \overline{MP}$ $\quad$ $\overline{AB} : \overline{MN}$ $\quad$ $\overline{BC} : \overline{NP}$

$\frac{AC}{MP} = \frac{12}{15} = \frac{4}{5}$ $\quad$ $\frac{AB}{MN} = \frac{12}{15} = \frac{4}{5}$ $\quad$ $\frac{BC}{NP} = \frac{8}{10} = \frac{4}{5}$

Step 2: Since the ratio of each pair of sides reduces to the same fraction, all of the side pairs of the two triangles are proportional. Therefore, $\Delta ABC \sim \Delta MNP$ by SSS Similarity.

Practice

Complete the steps to show that the two triangles are similar.

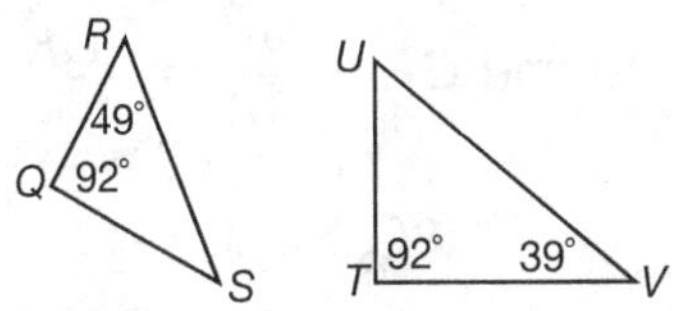

1. $\angle Q \cong$ _____ by the definition of __________ angles.

   $m\angle S =$ _____ by the Triangle Angle Sum Theorem.

   $\angle S \cong$ _____ by the definition of __________ angles.

   Therefore, $\Delta QRS \cong \Delta TUV$ by ______ Similarity.

Explain how the triangles are similar and write a similarity statement.

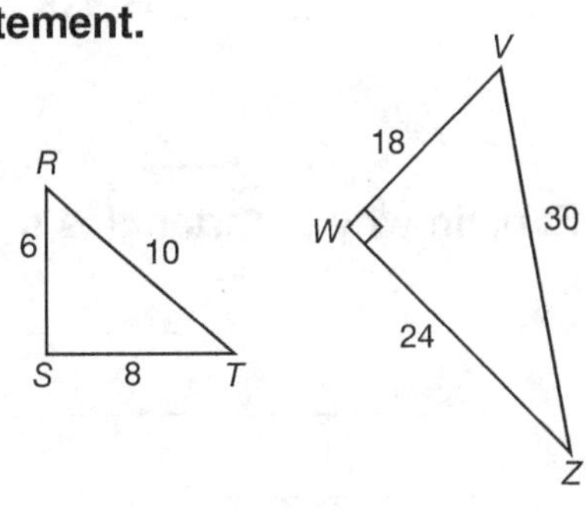

2. ______________________________

   ____

Triangle Similarity: SAS

Side-Angle-Side (SAS) Similarity	
If two sides of one triangle are proportional to two sides of another triangle and their included angles are congruent, then the triangles are similar.	$\frac{AB}{DE} = \frac{AC}{DF} = \frac{5}{6}$ $\angle A = \angle D$ $\triangle ABC \sim \triangle DEF$

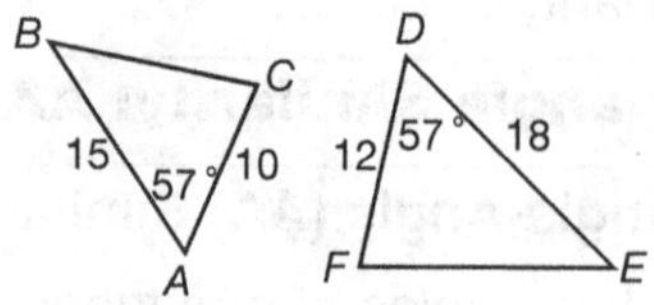

Show that $\triangle ADE \sim \triangle ABC$ Then, find BC.

Step 1: $\angle A \cong \angle A$ by the Reflexive Property.

$\frac{AD}{AB} = \frac{3}{6} = \frac{1}{2}, \frac{AE}{AC} = \frac{2}{4} = \frac{1}{2}$

$\triangle ADE \sim \triangle ABC$ by SAS Similarity.

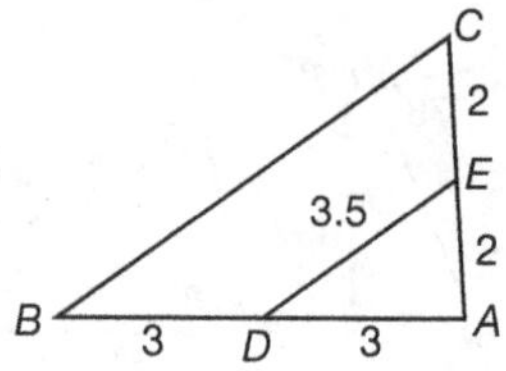

Step 2: Find BC.

$\frac{AD}{AB} = \frac{DE}{BC}$	Proportional Corresponding Sides
$\frac{3}{6} = \frac{3.5}{BC}$	Substitute.
$3(BC) = 6(3.5)$	Cross Products Property
$BC = 7$	Solve.

Practice

Complete the steps to show that the triangles are similar. Then, find GK.

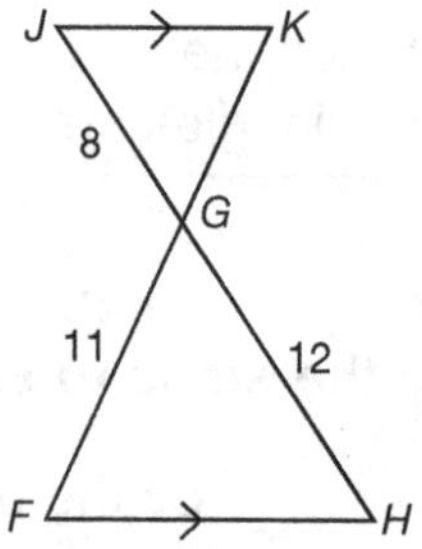

3. $\overline{JK} \parallel \overline{FH}$, so $\angle J \cong$ _____ and $\angle K \cong$ _____ by the Alternate _______________ Theorem.

   $\triangle$_________ $\sim \triangle$_________ by _____ Similarity.

   To find GK: $\frac{8}{12} = \frac{GK}{\_\_\_\_}$

   8(_____) = _____(_____)

   _____ = 12(GK)

   ______ = GK

   $\frac{}{\_\_\_\_\_}$ = GK

Explain why the triangles are similar and then find US.

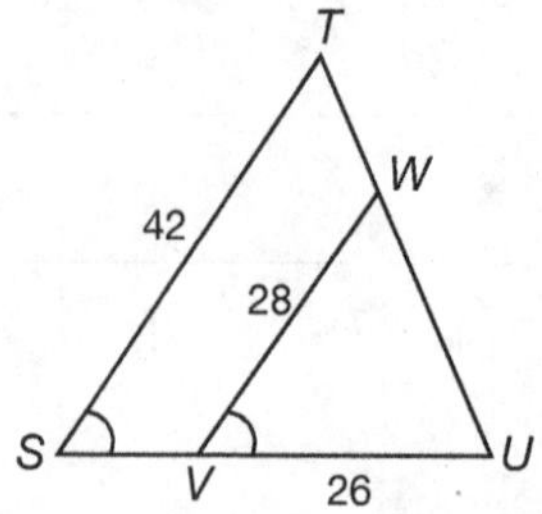

4. ____

____

____

Name ______________________ Date __________ Class __________

Reteaching 47

Circles and Inscribed Angles

Inscribed Angles

The measure of an inscribed angle is half the measure of its intercepted arc. In the picture, $\angle ABC$ is an inscribed angle and $\overset{\frown}{AC}$ is its intercepted arc on the circle.

$m\angle ABC = \frac{1}{2} m\overset{\frown}{AC}$

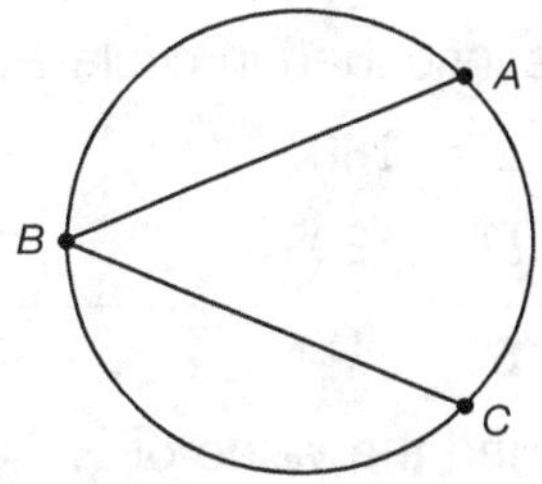

If an inscribed angle intercepts a semicircle, then it is a right angle. $\angle ACB$ intersects the semicircle, so $m\angle ACB = 90°$.

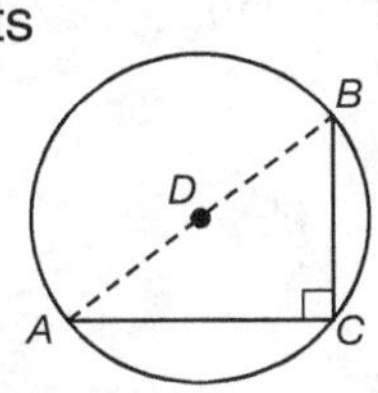

If inscribed angles intercept the same arc, then they are congruent. $\angle ABC$ and $\angle ADC$ intercept $\overset{\frown}{AC}$, so $\angle ABC \cong \angle ADC$.

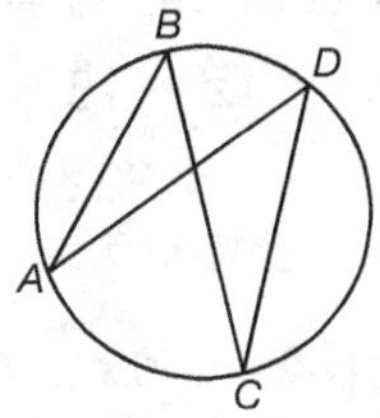

Find $m\angle GFJ$ and $m\overset{\frown}{HF}$.

Step 1: Since $m\overset{\frown}{GJ} = 110°$, $m\angle GFJ = \frac{1}{2} m\overset{\frown}{GJ} = \frac{1}{2}(110°) = 55°$.

Step 2: Since $m\angle FJH = 36°$, then $\frac{1}{2} m\overset{\frown}{HF} = m\angle FJH = 36°$.

Solve for $m\overset{\frown}{HF}$: $\frac{1}{2} m\overset{\frown}{HF} = 36°$

$m\overset{\frown}{HF} = 2(36°)$

$m\overset{\frown}{HF} = 72°$

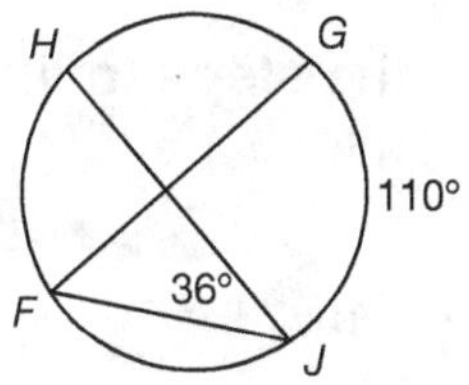

Practice

Complete the steps to solve for *x*.

1. Since $\angle RQS$ intercepts a __________, $m\angle RQS$ = ___.

$(5x + 8)° =$ ___

$5x + 8 = 90$

$5x =$ ___

$x =$ ____

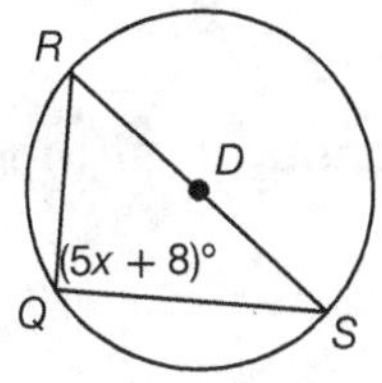

Find each value.

2. z

___

3. $m\angle FJH$

___

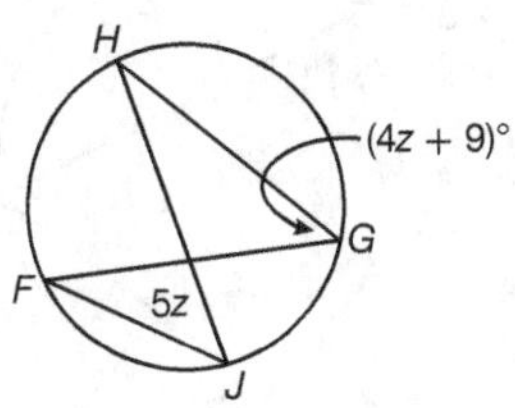

Inscribed Quadrilaterals

If a quadrilateral is inscribed in a circle, then it has supplementary opposite angles.

ABCD is inscribed in circle *E*.

$\angle A + \angle C = 180°$

$\angle B + \angle D = 180°$

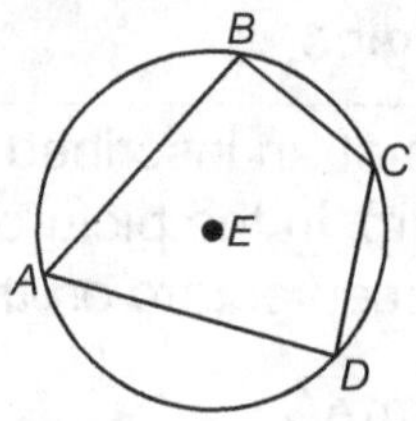

Find m∠G.

Step 1: Find the value of *z*.

$m\angle E + m\angle G = 180°$	*EFGH* is inscribed in a circle.
$8z + 14z - 18 = 180$	Substitute.
$22z = 198$	Simplify.
$z = 9$	

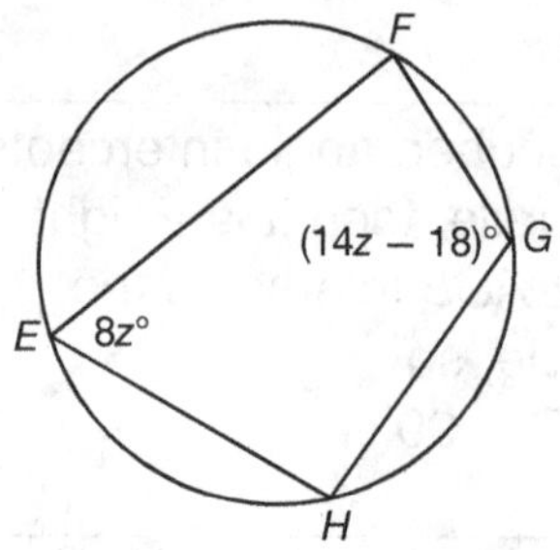

Step 2: Find m∠G.

$m\angle G = 14z - 18 = 14(9) - 18 = 108°$

Practice

Complete the steps to find the measure of the given angles.

4. $\angle R$

$m\angle R + m\angle T =$ ____

$7y +$ ____ $= 180$

____ $= 180$

$y =$ ___

$m\angle R = 7y = 7($__$) =$ ___

5. $\angle T$

$m\angle T = 11y$

$m\angle T = 11($__$)$

$m\angle T =$ ____

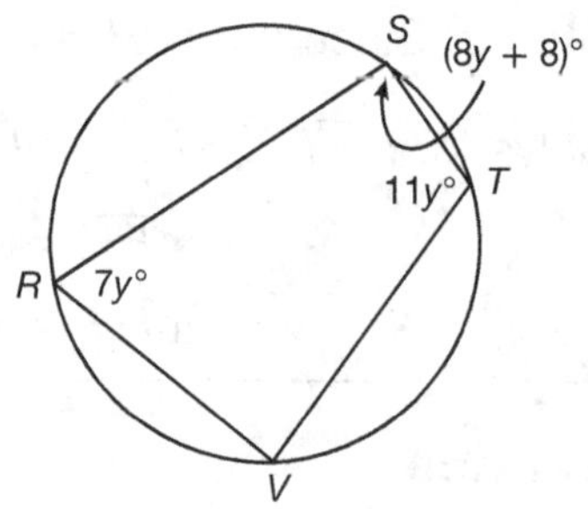

Find the measures of the given angles in RSTV.

6. $\angle S$ ___

7. $\angle V$ ___

Find the angle measures of each quadrilateral.

8. *ABCD*

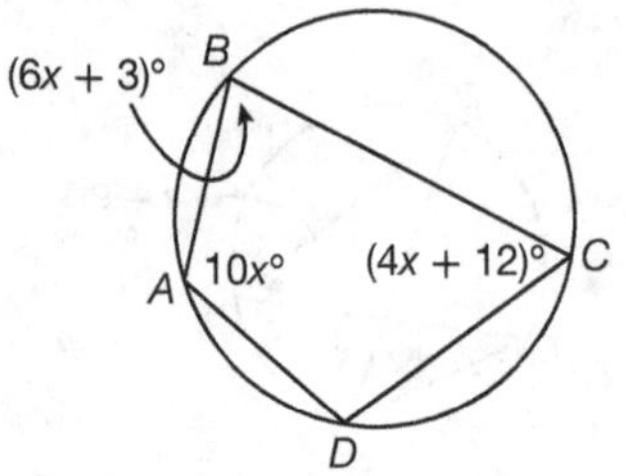

____

9. *JKLM*

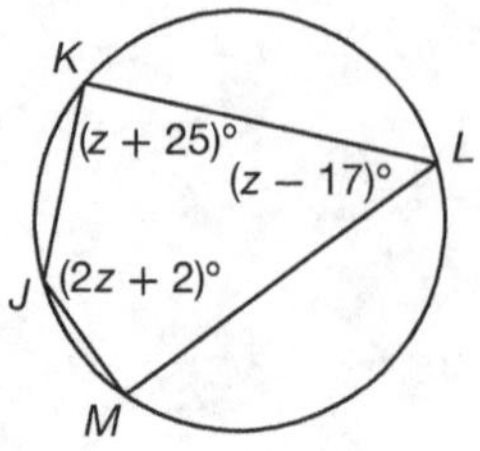

____

Name ______________________ Date __________ Class __________

Reteaching 48

Indirect Proofs

You have worked with inscribed angles in circles. Now you will work with indirect proofs.

Indirect Proofs

In **direct reasoning,** you begin with a true hypothesis and prove that a conclusion is true. In an **indirect proof**, you begin by assuming that the conclusion is false. You then show that this assumption leads to a contradiction.

To write an indirect proof, first identify the given information and the conclusion to be proved. Next, assume either that the conclusion is false or that the opposite of the conclusion is true.

Identify the conclusion to be proved and the opposite of the conclusion to be assumed for the following statement.

If two angles are not congruent, then they are not both right angles.

Step 1: Identify the conclusion to be proved.

Prove: ∠1 and ∠2 are not both right angles.

Step 2: When writing an indirect proof, assume that the opposite of the conclusion is true.

Assume: ∠1 and ∠2 are both right angles.

Practice

Complete the steps to write the conclusion and the opposite of the conclusion for each statement.

1. If $a^2 \neq b^2$, then $a \neq b$, for any real numbers, a and b.

   Conclusion: _____

   Assume: _____

2. If the sum of the measures of two angles is not equal to 180°, then the two angles are not supplementary.

   Conclusion: ∠1 and ∠2 are not supplementary.

   Assume: ______________________

Write the conjecture and the opposite of the conclusion for each statement.

3. If ∠1 is not congruent to ∠1 ∠2, then line a intersects line b.

   Conclusion: ______________________

   Assume: Line _ is parallel to line b.

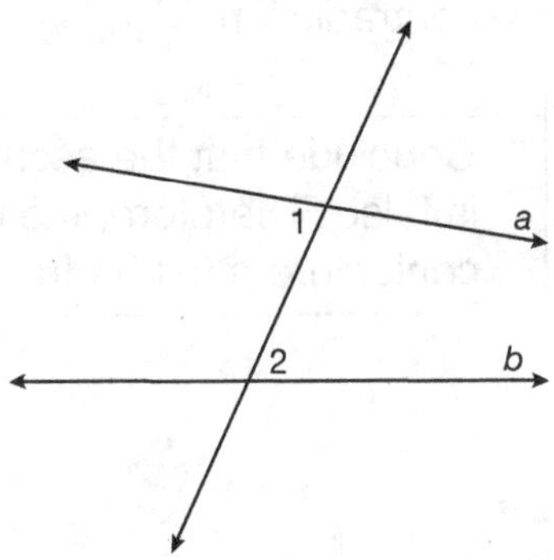

4. If ΔRST is acute, then it does not have a right angle.

   Conclusion: ______________________

   Assume: ______________________

Writing an Indirect Proof

Consider the statement:

Two acute angles with a common side do not form a linear pair.

Here are the steps to write an indirect proof of the statement.

Steps	Example
1. Identify the conjecture to be proved.	Given: $\angle 1$ and $\angle 2$ are acute angles with a common side. Prove: $\angle 1$ and $\angle 2$ do not form a linear pair.
2. Assume the opposite of the conclusion is true.	Assume: $\angle 1$ and $\angle 2$ form a linear pair.
3. Use direct reasoning to show that the assumption leads to a contradiction.	$m\angle 1 + m\angle 2 = 180°$ by the definition of a linear pair. Since $m\angle 1 < 90°$ and $m\angle 2 < 90°$, $m\angle 1 + m\angle 2 < 180°$. This is a contradiction.
4. Conclude that the assumption is false. Therefore, the original conjecture must be true.	The assumption that $\angle 1$ and $\angle 2$ form a linear pair is false. Therefore, $\angle 1$ and $\angle 2$ do not form a linear pair.

Practice

Complete the indirect proof by following the directions in each step.

5. An obtuse triangle cannot have a right angle.

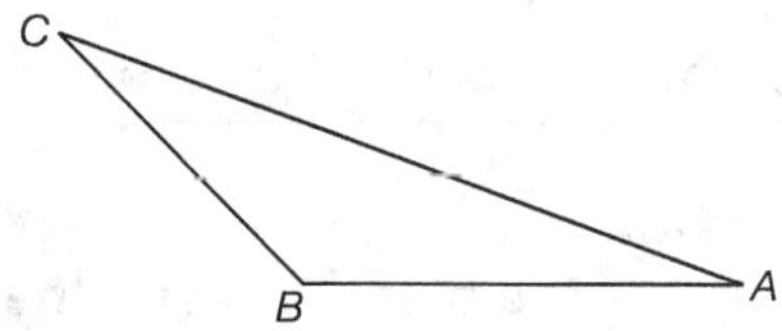

∠B is an obtuse angle.

Steps	Proof
1. Identify the conjecture to be proved.	Given: ΔABC is an obtuse triangle. $\angle B$ is an obtuse angle. Prove: ΔABC ______________________
2. Assume the opposite of the conclusion is true. Write the assumption.	Assume: ______________________ Let $\angle A$ be a right angle.
3. Use direct reasoning to show that the assumption leads to a contradiction.	If $\angle A$ is a right angle, then ______________________ ______________________
4. Conclude that the assumption is false. Therefore, the original conjecture must be true.	The assumption that ______________________ is false. Therefore, ______________________

Name ____________________ Date __________ Class __________

Reteaching 49

Introduction to Solids

You have worked with indirect proofs. Now you will work with solid geometry.

Three-dimensional figures, or **solids**, can have flat or curved surfaces.

Prisms and pyramids are named by the shape of their **bases**.

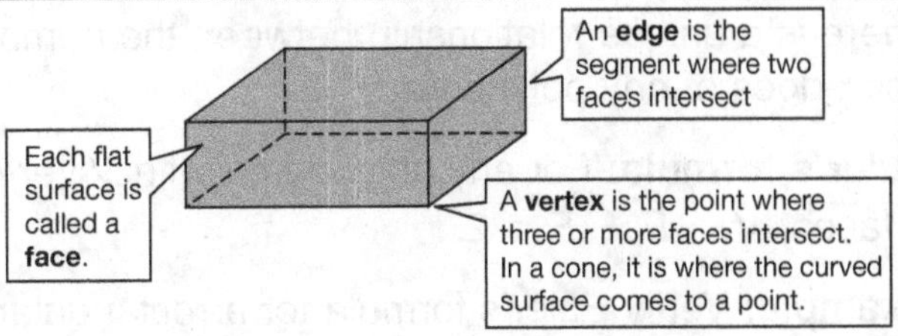

Prisms		Pyramids		Cylinders	Cones
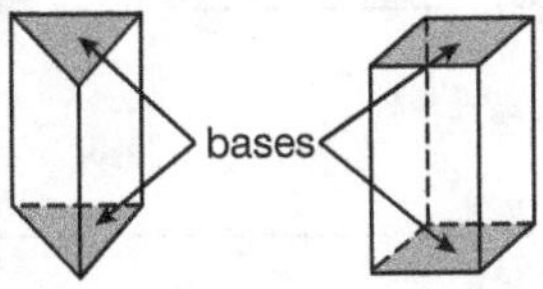		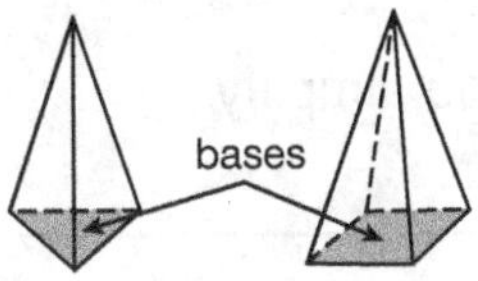		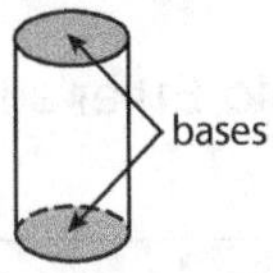	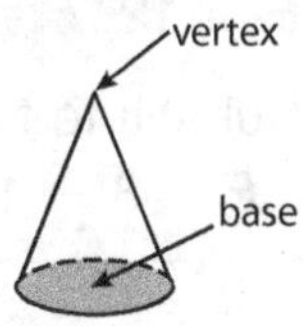
triangular prism	rectangular prism	triangular pyramid	rectangular pyramid	Neither cylinders nor cones have edges.	

Identify the figure. Name the vertices, edges, and bases.

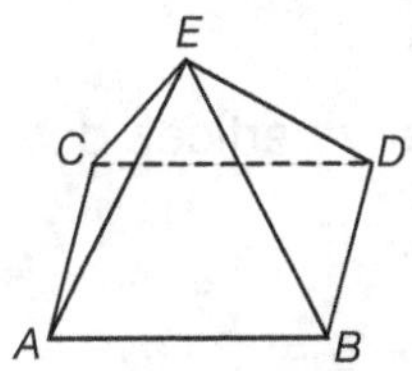

Step 1: The figure is a rectangular pyramid.

Step 2: The vertices are *A, B, C, D,* and *E.* The edges are $\overline{AB}$, $\overline{BD}$, $\overline{CD}$, $\overline{AC}$, $\overline{AE}$, $\overline{BE}$, $\overline{CE}$, and $\overline{DE}$. The base is quadrilateral *ABDC.*

Practice

Complete the steps to identify the figure and name the vertices, edges, and bases.

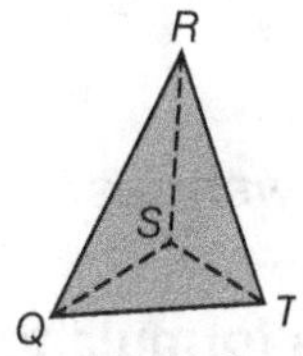

1. This is a ____________________.

The vertices are *Q, R,* _, and _.

The edges are $\overline{RQ}$, ___, ___, ___, ___, and ___.

The base is Δ ____.

Identify each figure. Name the vertices, edges, and bases.

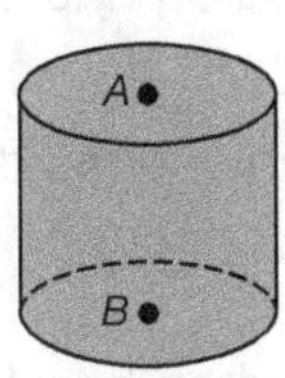

2. ____

____

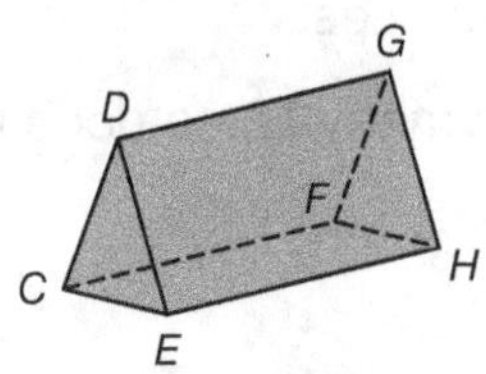

3. ____

____

____

Euler's Formula

A closed three-dimensional figure formed by four or more polygons that intersect only at their edges is called a **polyhedron**; for example, prisms and pyramids.

There is a unique relationship between the number of faces, vertices, and edges of any polyhedron.

Euler's formula: For any polyhedron with V vertices, E edges, and F faces, $V - E + F = 2$.

Example: Verify Euler's formula for a rectangular prism.

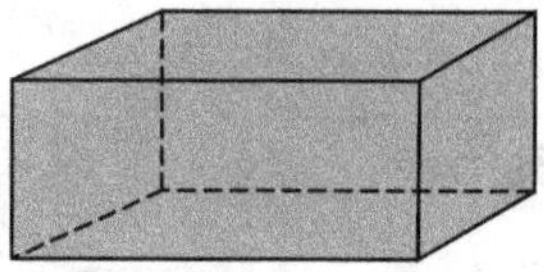

Step 1: Find the number of vertices, edges, and faces.
$V = 8$, $E = 12$, $F = 6$

Step 2: Substitute the values into Euler's formula and simplify.
$V - E + F = 8 - 12 + 6 = 2$

Practice

Complete the steps to find the missing number.

4. A triangular prism has 5 faces and 9 edges.

How many vertices does it have?

$V - E + F = \_$

$V - \_ + 5 = 2$

$V - \_ = \_$

$V = \_$

It has _ vertices.

5. A rectangular pyramid has 5 faces and 5 vertices.

How many edges does it have?

$V - E + F = \_$

$\_ - E + \_ = 2$

$\_\_ - E = \_$

$-E = \_\_$

$E = 8$

It has _ edges

Use Euler's formula to solve each exercise.

6. A polyhedron has 6 vertices and 4 faces.

How many edges does it have?

_

7. A polyhedron has 12 faces and 18 vertices.

How many edges does it have?

__

8. A polyhedron has 16 edges and 10 vertices.

How many faces does it have?

_

9. A polyhedron has 9 vertices and 12 edges.

How many faces does it have?

_

Name ______________________ Date ____________ Class ____________

Reteaching 50
Geometric Mean

You have worked with solids. Now you will work with geometric mean.

Geometric Mean

The geometric mean for two positive numbers, *a* and *b*, is the positive number *x* such that $\frac{a}{x} = \frac{x}{b}$.

Find the geometric mean of 4 and 12.

Let x represent the geometric mean of 4 and 12. Write the algebraic expression for finding the geometric mean and solve for x:

$\frac{a}{x} = \frac{x}{b}$	
$\frac{4}{x} = \frac{x}{12}$	Substitude 4 and 12.
$(x) = 4(12)$	Find the cross product.
$x^2 = 48$	
$x = \sqrt{48}$	Find the square root.
$x \approx 6.9$	Round to the nearest tenth.

Practice

Complete the steps to find the geometric mean for each pair of numbers. Round your answer to the nearest tenth.

1. 9 and $\frac{2}{3}$

$\frac{9}{x} = \frac{x}{\_\_\_}$

$x(\_\_\_) = \_\left(\frac{2}{3}\right)$

$x^2 = \_$

$x = \_\_\_$

$x \approx \_\_\_$

2. 8 and 15

$\frac{\_\_\_}{x} = \frac{x}{\_\_\_}$

$x(\_\_\_) = \_(15)$

$x^2 = \_\_\_$

$x = \_\_\_$

$x \approx \_\_\_$

3. 10 and 6

$\frac{10}{x} = \frac{x}{\_\_\_}$

$x(\_\_\_) = 10(\_)$

$x^2 = \_\_\_$

$x = \_\_\_$

$x \approx \_\_\_$

Find the geometric mean for each pair of numbers. Round your answer to the nearest tenth.

4. 7 and $\frac{1}{3}$ _____

5. 9 and 4 _____

6. 14 and 5 _____

7. 25 and $\frac{3}{5}$ _____

8. 11 and 6 _____

9. 4 and 16 _____

When an altitude is drawn from the vertex of the 90° angle of a right triangle to the hypotenuse of the right triangle, some useful relationships are formed.

If the altitude is drawn to the hypotenuse of a right triangle, then the two triangles formed are similar both to each other and to the original triangle.

$\Delta JLM \sim \Delta MLK \sim \Delta JMK$

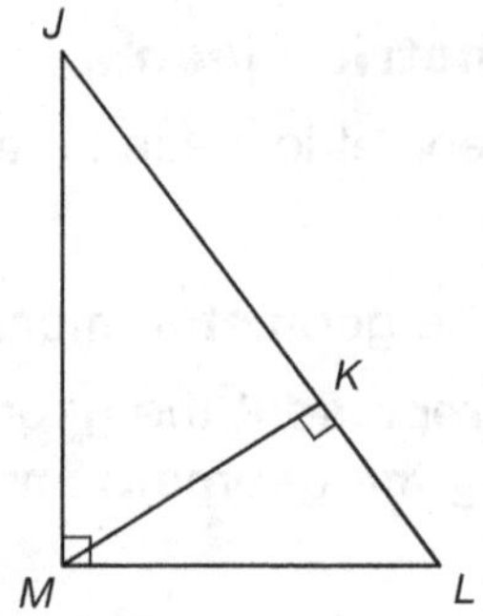

The length of the altitude is the geometric mean between the segments of the hypotenuse.	The length of a leg is the geometric mean between the hypotenuse and the segment of the hypotenuse that is closer to that leg.

Example:

Given ΔXYZ, find the missing value, t.

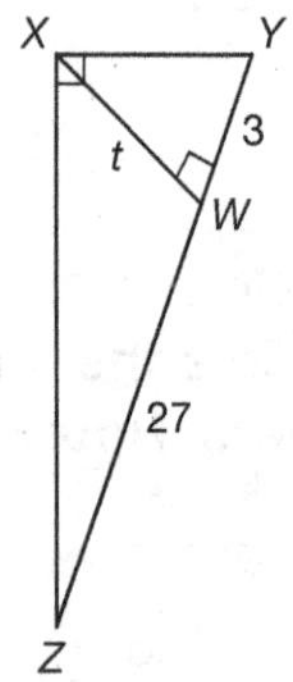

Step 1: Since $\overline{XW}$ is an altitude of a right triangle, t is the geometric mean of the segments of the hypotenuse. The segments measure 27 and 3.

Since t is the geometric mean of 27 and 3, write $\frac{27}{t} = \frac{t}{3}$.

Step 2: Solve for t: $\frac{27}{t} = \frac{t}{3}$.

$t^2 = 81$

$t = 9$

Practice

10. Find the value of a and b.

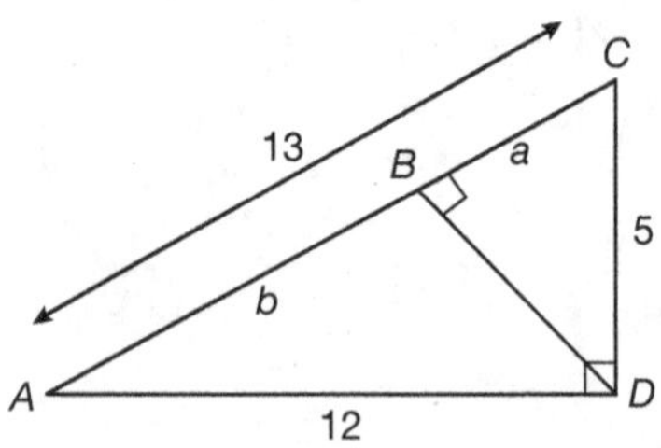

Since a and b are the legs of the triangle, each one is a geometric mean of the hypotenuse and the segment closer to that leg.

$\frac{13}{5} = \frac{5}{\_\_\_}$ $\qquad$ $\frac{13}{12} = \frac{\_\_\_}{\_\_\_}$

$\_\_\_(a) = (5)\_\_\_$ $\qquad$ $\_\_\_(b) = (12)\_\_\_$

$\_ = \frac{25}{\_\_\_}$ $\qquad$ $b = \frac{\_\_\_}{\_\_\_}$

$a = \_\_\_$ $\qquad$ $b = \_\_\_$

Name ______________________ Date __________ Class __________

Reteaching INV 5
Nets

You have worked with two and three-dimensional figures before. Now you'll work with nets, which are 2-D representations of 3-D figures.

Making a 3-D Figure from a Net

A net is a two-dimensional representation of a 3-D figure. Each shape in the net represents one face of the figure. By folding along the edges of the figures, the net can be made into the 3-D figure it represents.

Example: Figure 1 shows a net of a cube. Each square represents one face of the cube. In figure 2 the squares are being folded upwards so that it becomes the cube in figure 3.

Figure 1

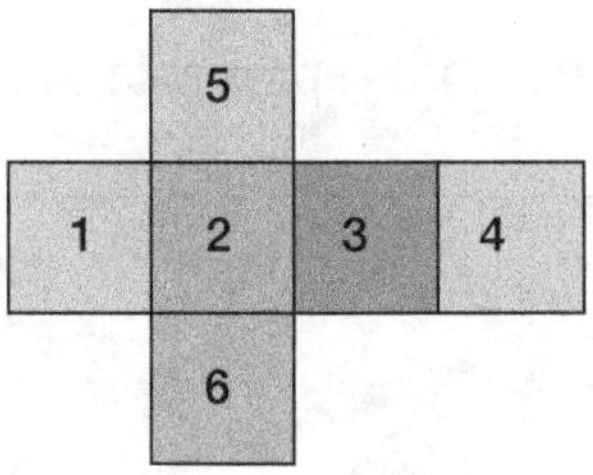

Figure 2

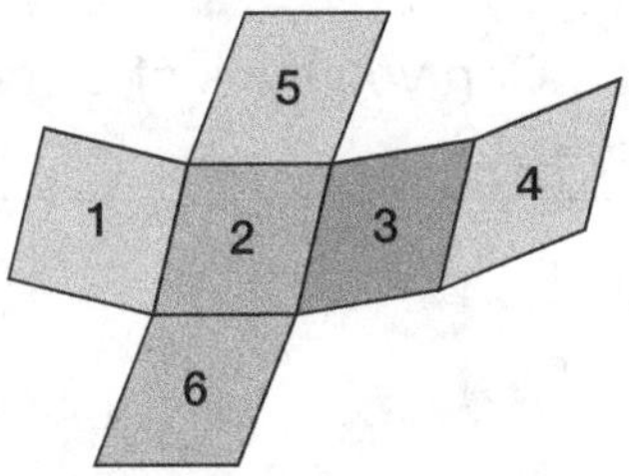

Figure 3

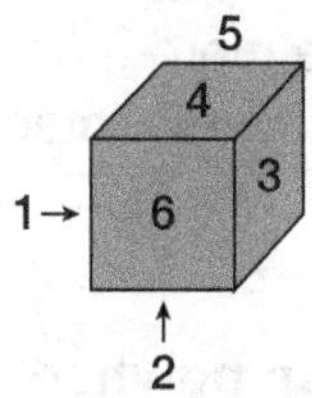

Practice

Copy the nets given in problems 1–3. Form the 3-D figures they represent. State the name of each figure.

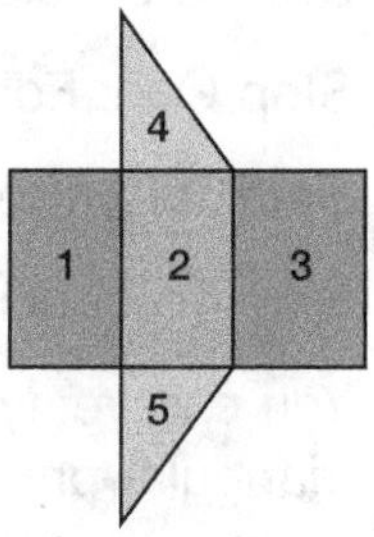

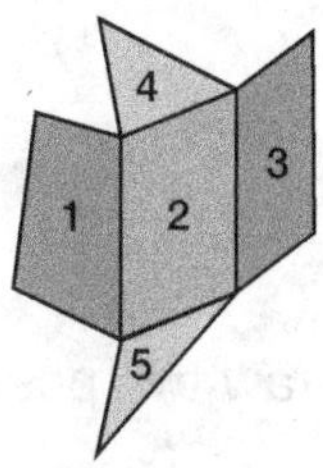

1. Step One: Fold the triangles up along the edges of the rectangle.

   Step Two: The figure has two triangular bases and three rectangular faces. The figure is a ______.

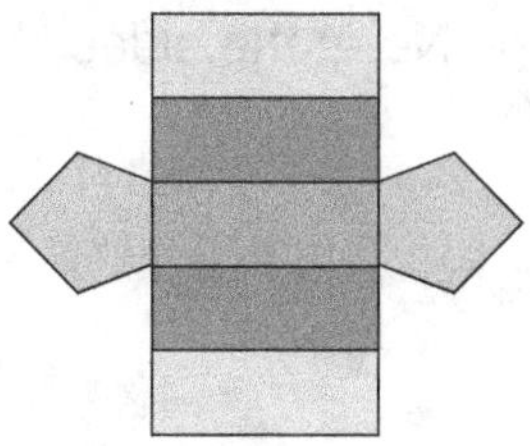

2. ________________

3. Can you form a rectangular prism with the net in figure 5? If so, form it. If not, explain why.

   ____

   ____

   ____

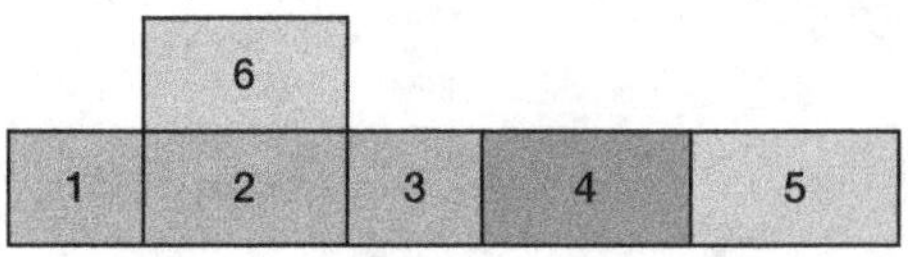

Making Nets of a 3-D Figure

A 3-D figure with flat faces is called a polyhedron. To construct the net of a figure, imagine unfolding it along its edges so that it is a flat shape.

Example: The cube in figure 1 unfolds in figure 2 to become the net in figure 3.

Figure 1 Figure 2 Figure 3

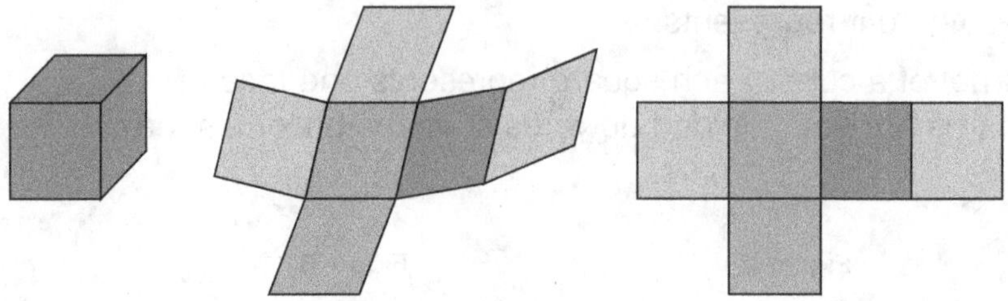

Most polyhedrons have more than one net. Can you think of a second net that represents the same cube as the one represented by figure 1?

Practice

Draw a net for the given polyhedron.

4. Draw a possible net for this triangular prism.

Step One: Fold the triangles out.

Step Two: Fold the rectangles out.

Step Three: Draw the net.

You can verify your answer by re-creating the triangular prism using the net you've made.

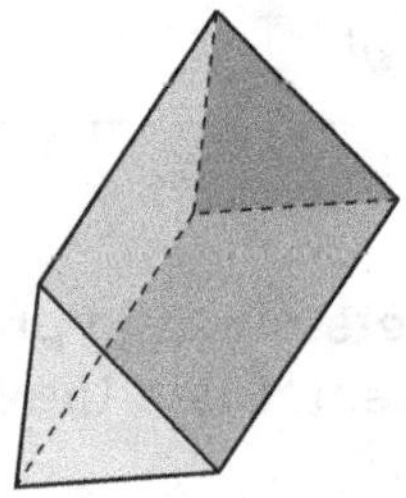

5. Draw two possible nets for this truncated pyramid.

(Note: the sides are trapezoids and the top and bottom are squares)

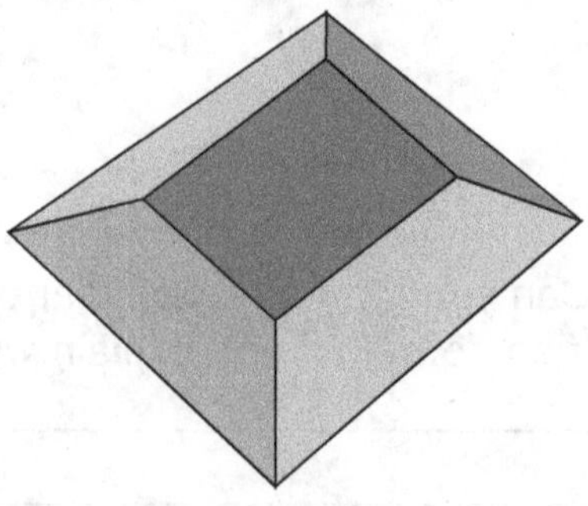

Name ______________________ Date ____________ Class ____________

Reteaching 51

Properties of Isosceles and Equilateral Triangles

You have worked with different types of triangles. Now you will work specifically with isosceles and equilateral triangles.

Isosceles Triangles

Isosceles Triangle Theorem

If two sides of a triangle are congruent, then the angles opposite the sides are congruent.

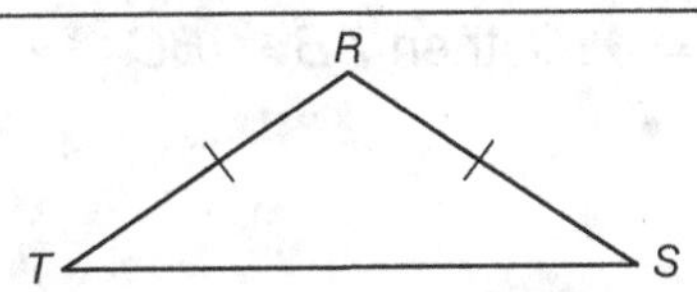

If $\overline{RT} \cong \overline{RS}$, then $\angle T \cong \angle S$.

Converse of Isosceles Triangle Theorem

It two angles of a triangle are congruent, then the sides opposite those angles are congruent.

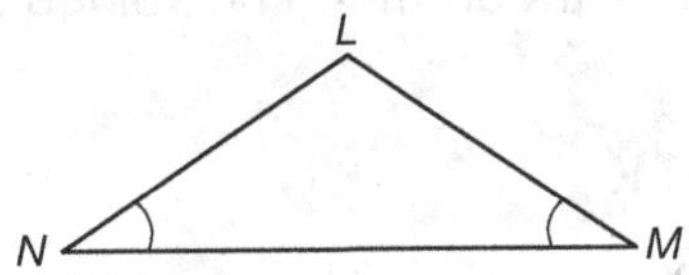

If $\angle N \cong \angle M$, then $\overline{LN} \cong \overline{LM}$.

You can use these theorems to find angle measures in isosceles triangles.

Find **m$\angle E$ in ΔDEF.**

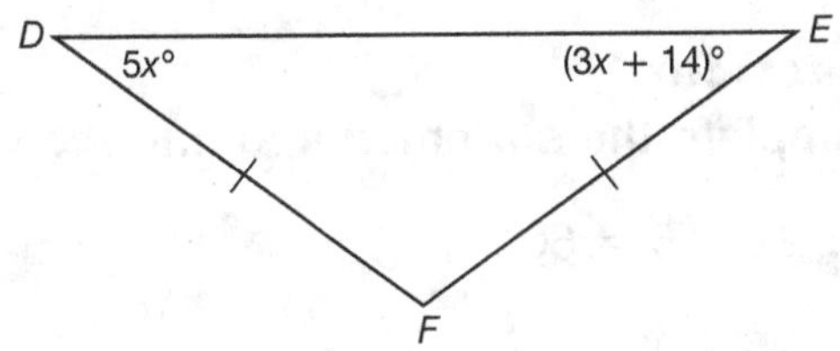

$m\angle D = m\angle E$	Isosceles Triangle Theorem
$5x = 3x + 14$	Substitute.
$2x = 14$	Subtract $3x$ from both sides.
$x = 7$	Divide both sides by 2.
$m\angle E = 3(7) + 14 = 35°$	

Practice

Complete the statements to find m$\angle A$ and m$\angle C$.

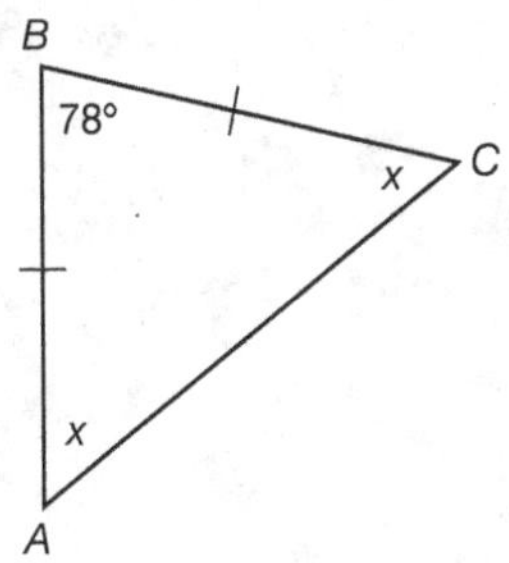

1. $m\angle A + m\angle B + m\angle C = 180°$

   ______ + ______ + $x° = 180°$

   ______ $+ 78° = 180°$

   $2x =$ ______

   $x =$ ______

Find each angle measure.

2. $m\angle Q$ ______

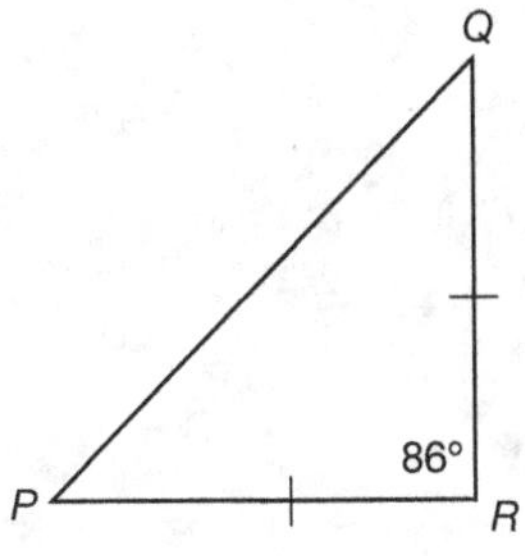

3. $m\angle H$ ______

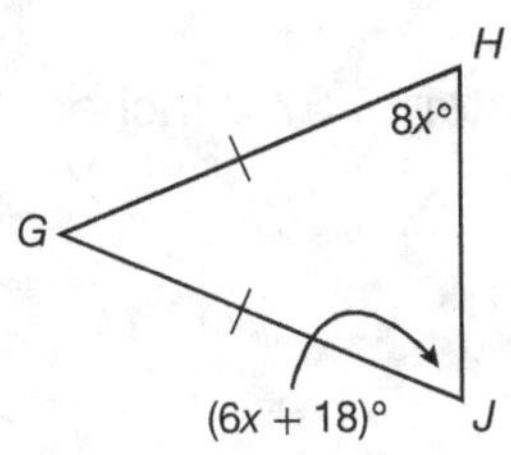

4. $m\angle M$ ______

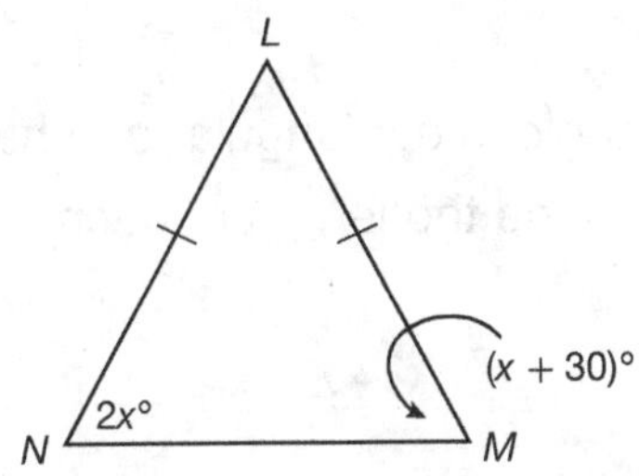

Equilateral Triangles

Equilateral Triangle Corollary

If a triangle is equiangular, then it is equilateral.

If $\angle A \cong \angle B \cong \angle C$, then $\overline{AB} \cong \overline{BC} \cong \overline{CA}$.

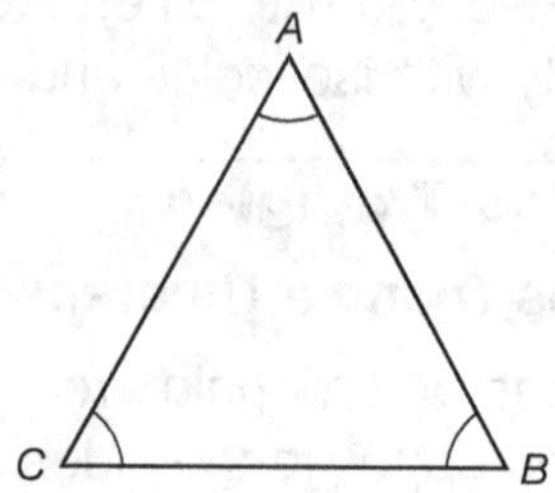

You can use this theorem to find values in equilateral triangles.

Find *x* in ΔSTV.

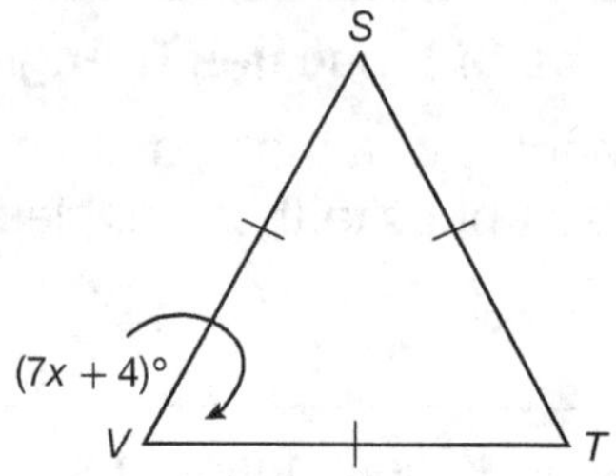

$7x + 4 = 608$

$7x = 56°$ Subtract 4 from both sides.

$x = 8$ Divide both sides by 7.

Practice

Complete the statements to find the value.

5. $2x - 6 = 60$

______ $= 66$

$x =$ ______

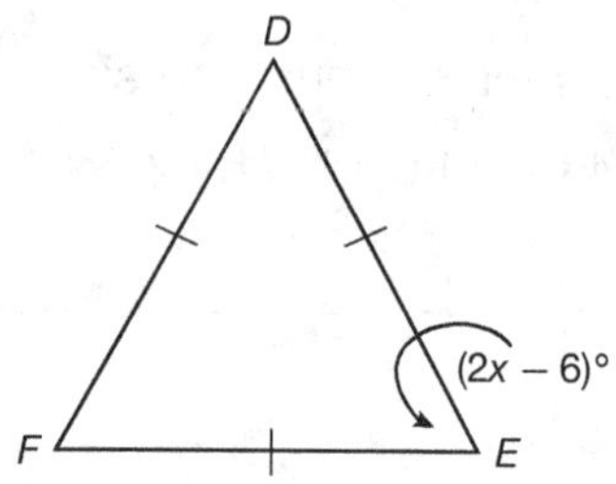

Find each value.

6. $n =$ ______

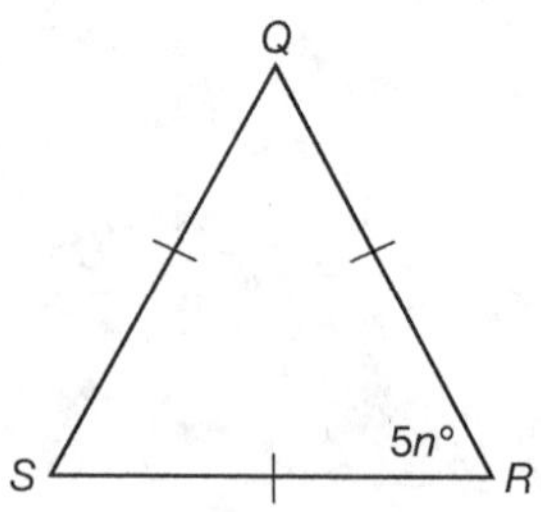

7. $VT =$ ______

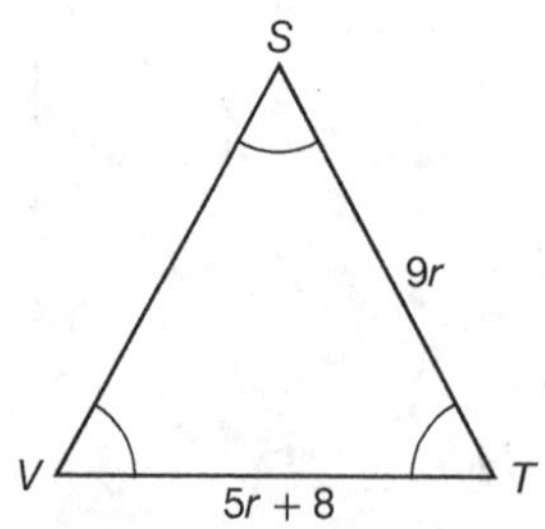

8. $MN =$ ______

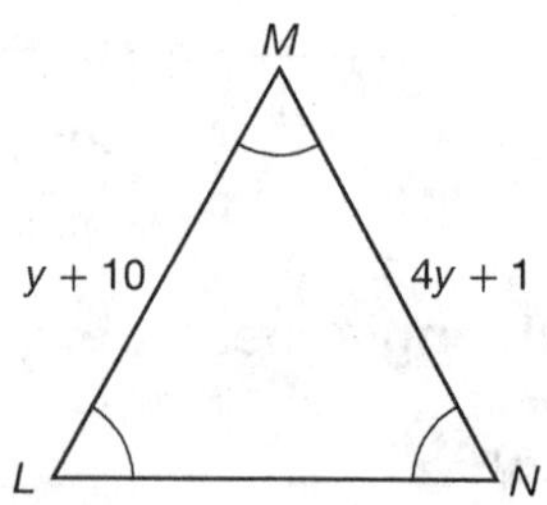

9. A triangle is equiangular and has a perimeter of 37.2 inches. Determine the lenght of each side. ____________

Name ______________________ Date __________ Class __________

Reteaching 52

Properties of Rectangles, Rhombuses, and Squares

You have worked with special triangles. Now you will work with properties of parallelograms.

Properties of Special Parallelograms

Properties of Rectangles	**Properties of a Rhombus**
The diagonals of a rectangle are congruent. 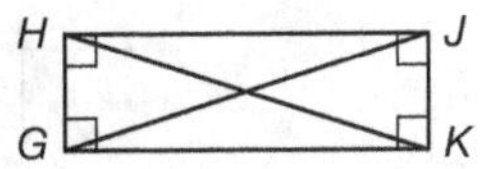 $\overline{GJ} \cong \overline{HK}$	The diagonals of a rhombus are perpendicular. Each diagonal of a rhombus bisects opposite angles. 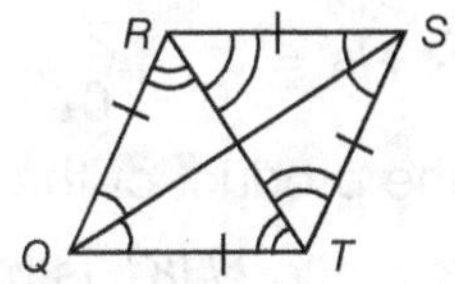 $\overline{QS} \perp \overline{RT}$ $\angle RQS \cong \angle SQT$

ABCD is a rectangle. Find the length of $\overline{CD}$ and of $\overline{BD}$.

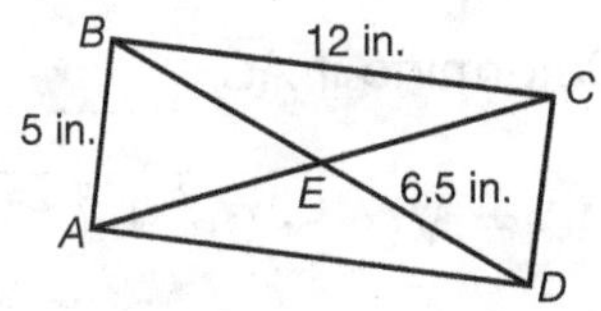

$CD = 5$ Opposite sides of a rectangle are congruent.

$\overline{BE} \cong \overline{ED}$ Diagonals of a parallelogram bisect each other.

It is given that $ED = 6.5$. Therefore, $BE = 6.5$ also.

$BD = BE + ED$ Segment Addition Postulate

$BD = 6.5 + 6.5 + 13$ inches Substitution

Practice

Complete the statements.

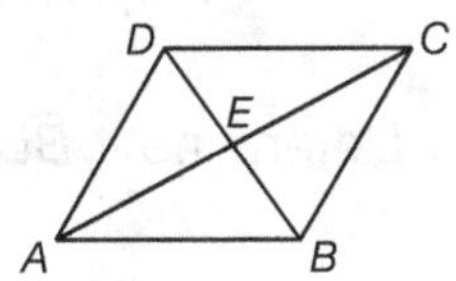

1. If $m\angle ABC = 128$, then $m\angle ABE =$ __. Diagonals of a rhombus bisect opposite angles.

   Use the sum of the angles in a quadrilateral to find and $m\angle BCD$.

   $128° + 128° + x + x = 360°$

   ____ $+ 2x =$ ____

   $2x =$ ____

   $x =$ ___

   $m\angle BCD =$ ___

   $m\angle BCE =$ ___

***KLMN* is a rhombus. Find each measure.**

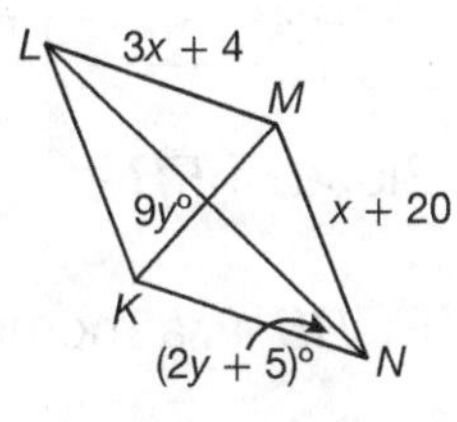

2. $KL =$ __

3. $m\angle MNK =$ ___

Using Properties of Parallelograms

HJKL is a parallelogram. Decide what type of parallelogram it is by using the properties of rectangles and rhombuses.

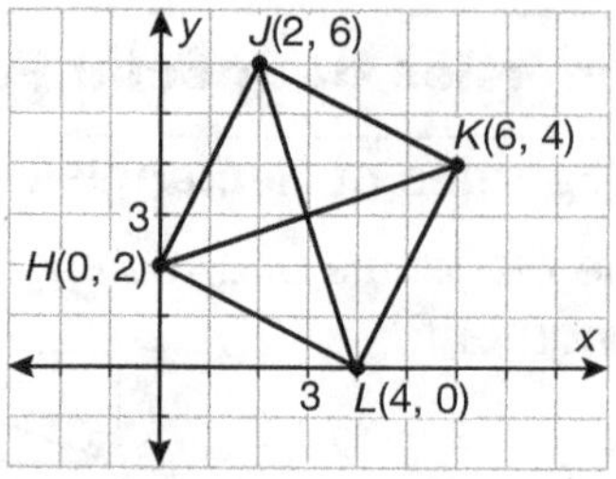

Step 1: Determine if $\overline{HK} \cong \overline{JL}$. Use the distance formula.

$HK = \sqrt{(6-0)^2 + (4-2)^2} = 2\sqrt{10}$

$JK = \sqrt{(4-2)^2 + (0-6)^2} = 2\sqrt{10}$

$HK = JL = 2\sqrt{10}$, so $\overline{HK} \cong \overline{JL}$, and *HJKL* is a rectangle.

Step 2: Determine if $\overline{HK} \perp \overline{JL}$ Find the slope.

slope of $\overline{HK} = \frac{4-2}{6-0} = \frac{1}{3}$ slope of $\overline{JL} = \frac{0-6}{4-2} = -3$

Since the product of the slopes is -1, $\overline{HK} \perp \overline{JL}$, and *HJKL* is a rhombus.

Step 3: Since *HJKL* is both a rectangle and a rhombus, it is a square.

Practice

4. Determine if $\overline{AC} \cong$ ___.

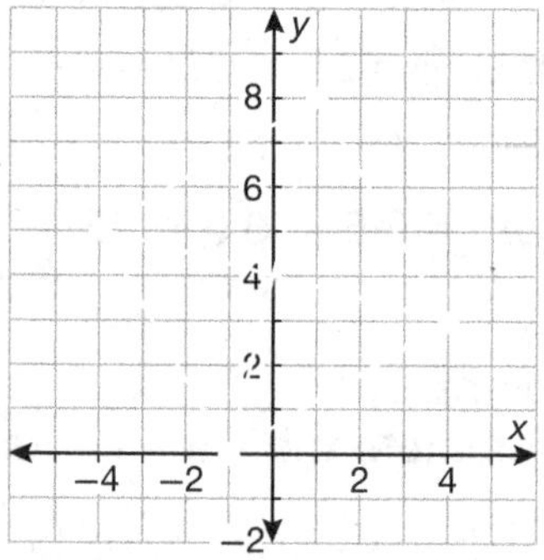

$AC = \sqrt{(\_ - 0)^2 + (\_ - (-1))^2}$

$= \sqrt{\_\_ + 4} = \sqrt{\_\_}$ $AC =$ ___, so $\overline{AC} \cong \overline{BD}$.

$\_\_ = \sqrt{(\_ - 3)^2 + (-4 - 4)^2}$ *ABCD* is a ________.

$= \sqrt{\_ + 64} = \sqrt{\_\_}$

5. Determine if $\overline{BD} \perp$ ___. Find the slope.

slope of $\overline{BD} = \frac{\_ - 3}{\_\_ - 4} = \frac{\_}{-8} = -\frac{1}{4}$ slope of $\overline{AC} = \frac{\_ - 0}{1 - (-1)} = \frac{\_}{2} = \_$

The product of the slopes is –1, so $\overline{BD} \perp$ ___.

ABCD is a ________. Since ABCD is both a ________ and a ________, it is a _______.

6. Find the lengths and the slopes of the diagonals. Determine what type of parallelogram *PQRS* is.

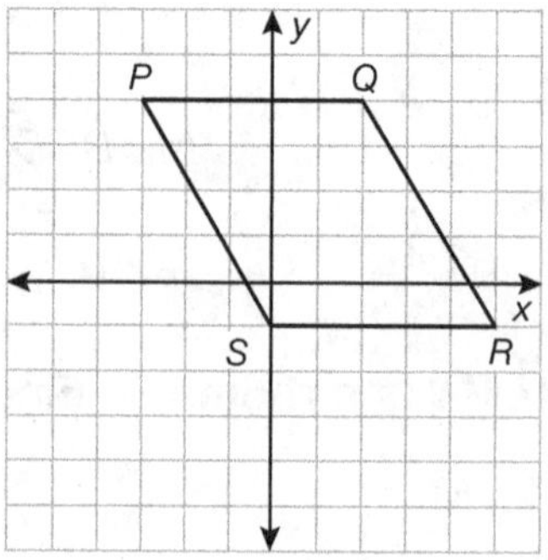

PR = _____, and *QS* = _____.

PQRS _____ a rectangle.

Slope of *PR* = ____, and slope of *QS* = _.

PQRS _ a rhombus.

Name ______________________ Date __________ Class __________

Reteaching 53

45°-45°-90° Right Triangles

You have worked with right triangles. Now you will work with a particular kind of right triangle.

Special Right Triangle

Side Lengths in 45°-45°-90° Triangles	Example
In a 45°-45°-90° right triangle, both legs are congruent, and the length of the hypotenuse is the length of a leg multiplied by $\sqrt{2}$.	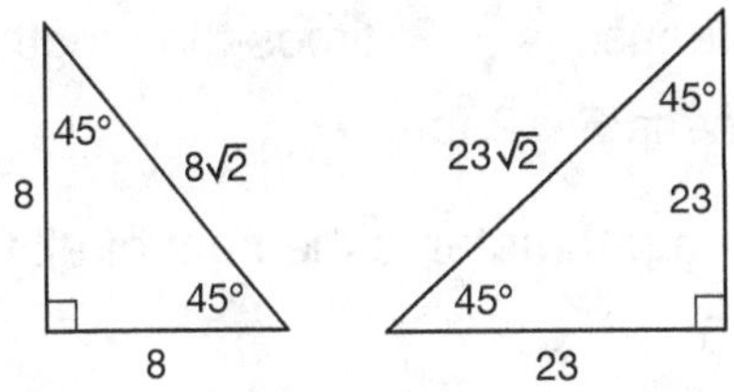

Find the value of *x* in ΔEFG

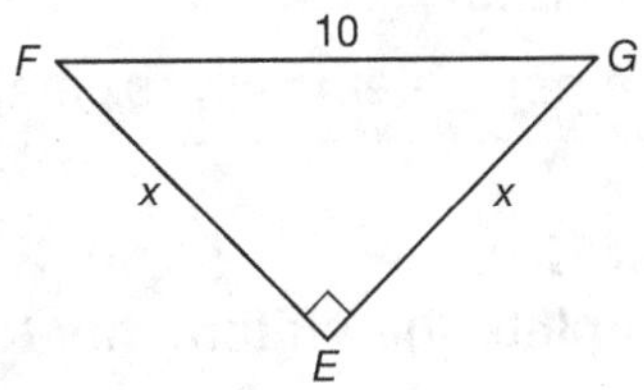

The hypotenuse has length 10.

$10 = x\sqrt{2}$	Hypotenuse is $\sqrt{2}$ times the length of a leg.
$\dfrac{10}{\sqrt{2}} = \dfrac{x\sqrt{2}}{\sqrt{2}}$	Divide both sides by $\sqrt{2}$.
$5\sqrt{2} = x$	Rationalize the denominator.

Practice

Complete the statements to find the value of *x*.

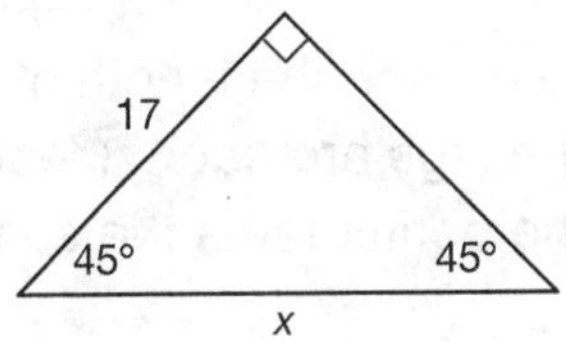

1. The leg is given. It is ___.

   The hypotenuse is the length of the leg times $\sqrt{2}$.

   The hypotenuse is ______, so $x =$ ______

Find the value of *x*.

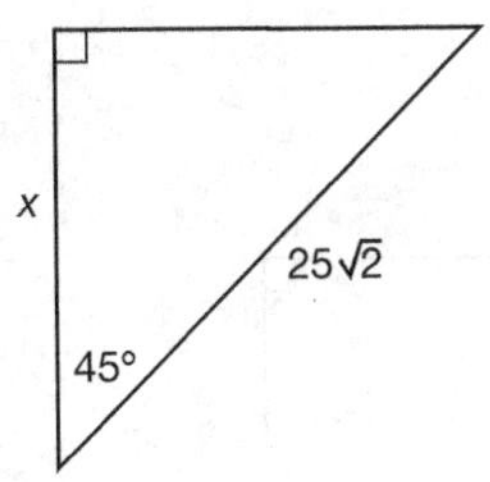

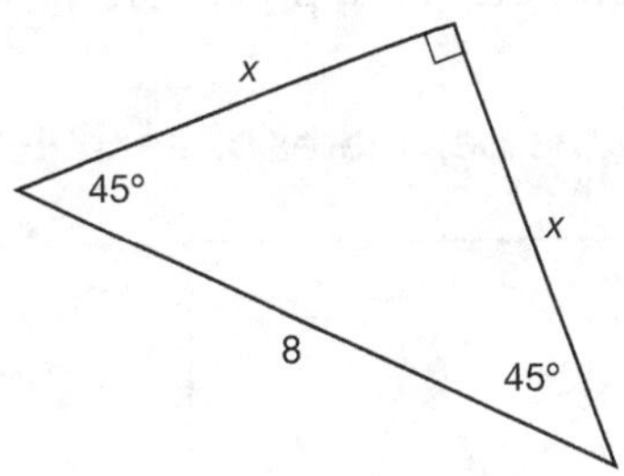

2. $x =$ ___

3. $x =$ ______

Finding the Perimeter of a 45°-45°-90° Right Triangle

Find the perimeter of the triangle.

Step 1: Determine the length of each leg.

The legs are each 5 feet long.

Step 2: Determine the length of the hypotenuse.

The hypotenuse is $\sqrt{2}$ times the length of a leg, so the hypotenuse is $5\sqrt{2}$ feet.

Step 3: The perimeter is the sum of all the sides.

$P = 5 + 5 + 5\sqrt{2}$

$P = 10 + 5\sqrt{2}$

$P \approx 17.07$

The perimeter is approximately 17 feet.

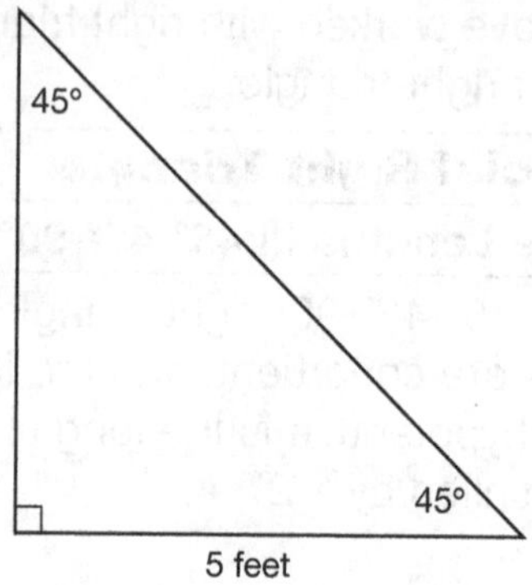

Practice

Complete the statements to determine the perimeter of the triangle.

4. Determine the length of the hypotenuse.

The hypotenuse is ______.

Determine the length of each leg.
The legs are each 3 feet long.
The perimeter is the sum of all the sides.

$P =$ _ $+ 3 +$ ______

$P =$ _ $+ 3\sqrt{2}$

$P \approx$ ______

The perimeter is approximately __ feet.

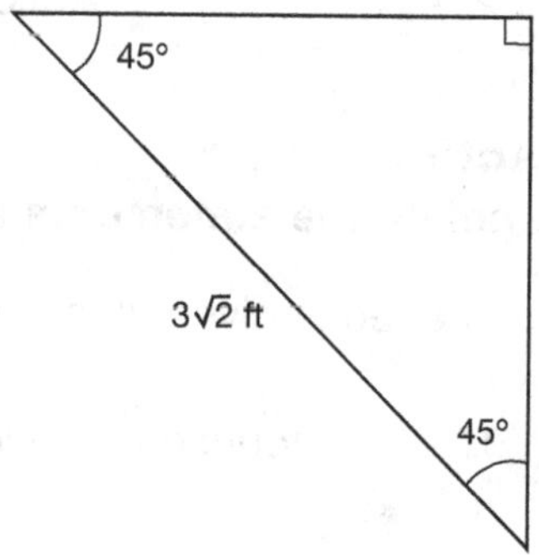

Determine the perimeter of each 45°-45°-90° right triangle.

5.

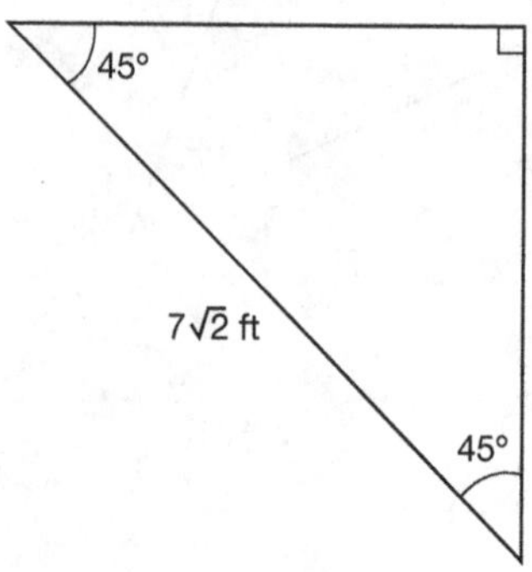

____

6.

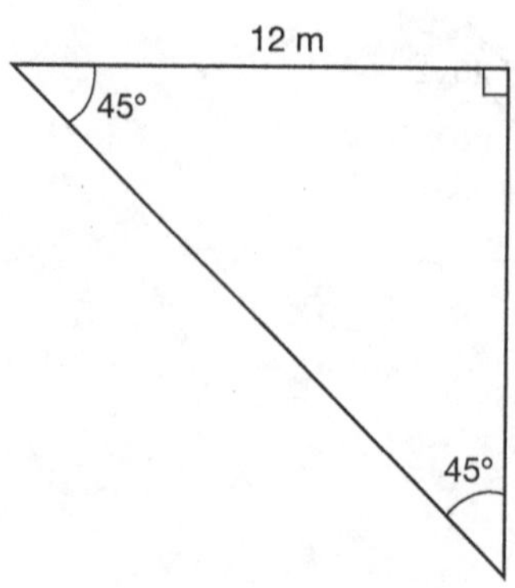

____

Name ______________________ Date __________ Class __________

Reteaching 54

Representing Solids

You have worked with different solids. Now you will work on drawing different perspectives of solids.

Drawing in One-Point Perspective

In a one-point perspective drawing, non-vertical lines are drawn so that they meet at a vanishing point.

Example:

Draw a triangular prism in one-point perspective.

Step 1:

Step 1 Draw a horizontal line and a vanishing point on the line. Draw a triangle below the line.

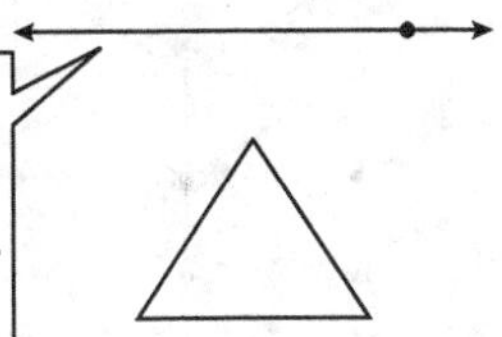

Step 2:

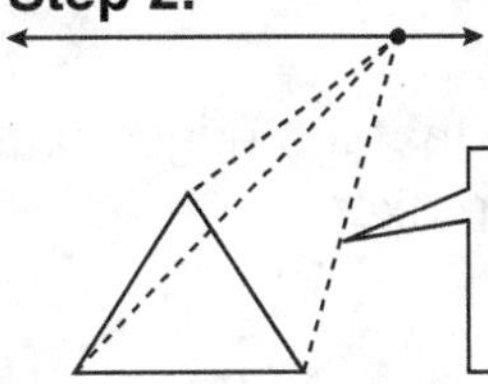

Step 2 From each vertex of the triangle, draw dashed segments to the vanishing point.

Step 3:

Step 3 Draw a smaller triangle with vertices on the dashed segments.

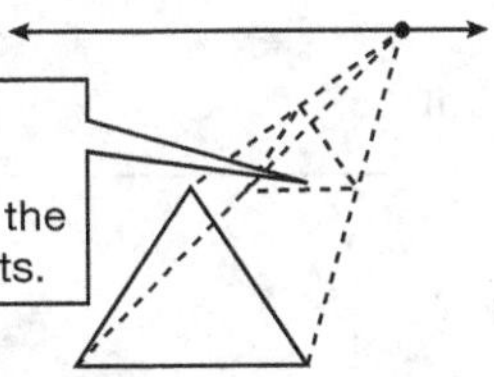

Step 4:

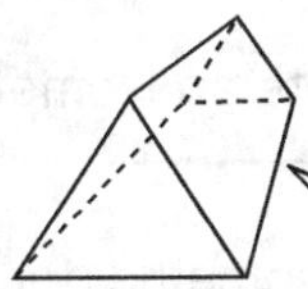

Step 4 Draw the edges of the prism. Use dashed lines for hidden edges. Erase segments that are not part of the prism.

Practice

1. Draw a rectangular prism in one-point perspective that has the given rectangle as its front face and the given point as its vanishing point.

Step 1: The rectangle is the front face of the solid. The given point is the vanishing point.

Step 2: Draw a dashed line from the vanishing point to each vertex of the rectangle.

Step 3: Draw a smaller rectangle with vertices on the dashed segments.

Step 4: Draw the edges of the prism. Use dashed lines for hidden edges. Erase segments that are not part of the prism.

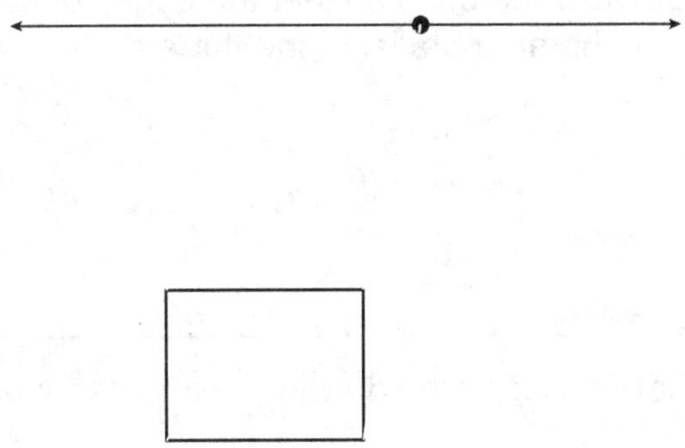

Creating Isometric Drawings

An isometric drawing is a way of drawing a solid figure by using isometric dot paper. The drawings can be made by using three axes that intersect to form 120° angles.

Example:

Create an isometric drawing of a triangular prism.

Step 1: Draw the three axes on isometric dot paper.

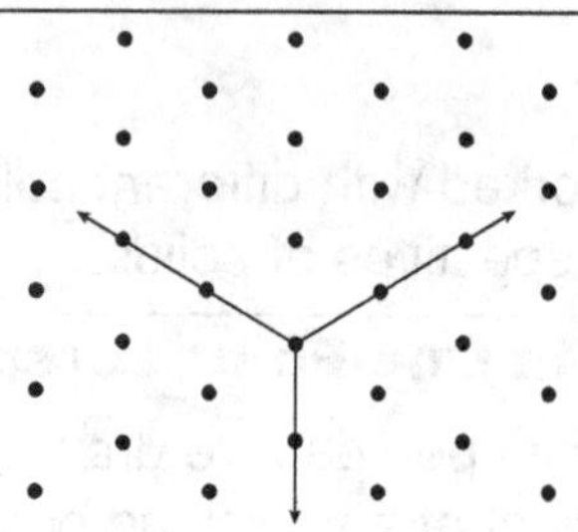

Step 2: Use the intersection of the three axes as the bottom vertex of the prism.

Step 3: Draw the prism so that the edges of the prism are parallel to the three axes.

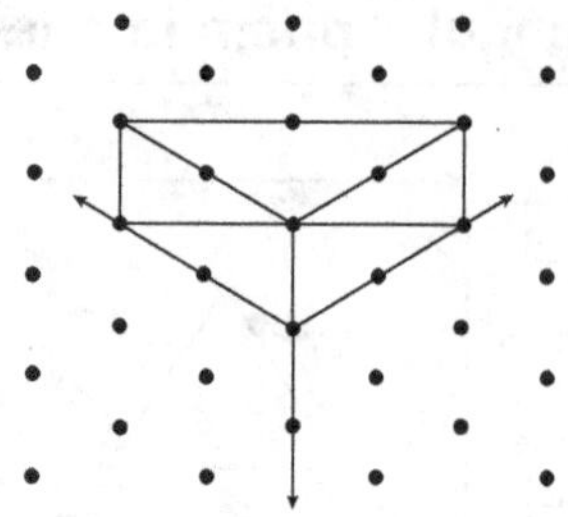

Practice

2. Follow the steps below to create an isometric drawing of a cube.

Step 1: Draw the three axes on isometric dot paper.

Step 2: Use the intersection of the three axes as the bottom vertex of the cube.

Step 3: Draw the cube so that the edges of the cube are parallel to the three axes.

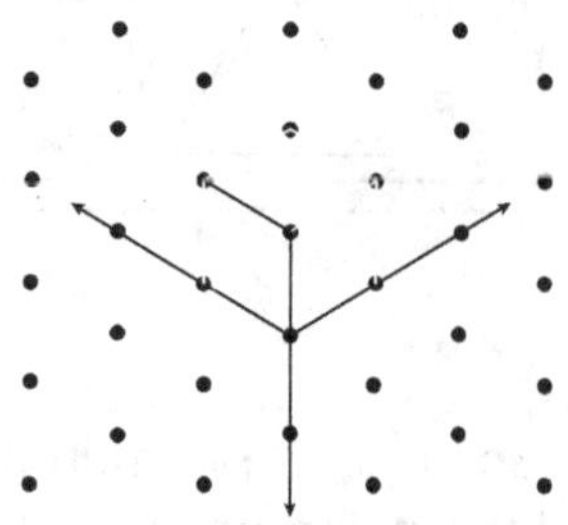

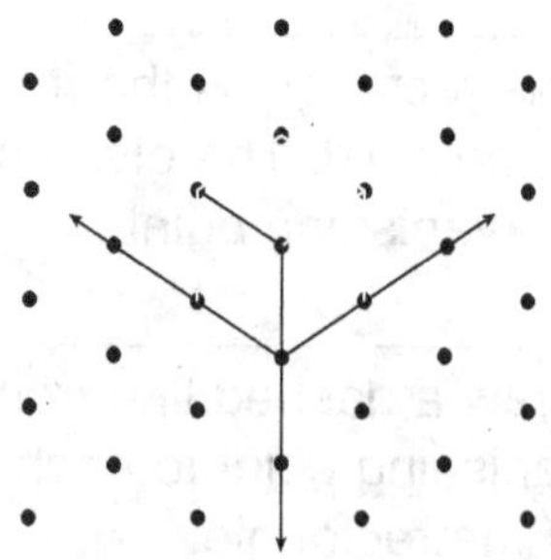

3. Create an isometric drawing of a rectangular prism.

Name ______________________ Date __________ Class __________

Reteaching 55

Triangle Midsegment Theorem

You have worked with midpoints. Now you will work with midsegments of triangles.

Triangle Midsegment Theorem

The segment joining the midpoints of two sides of a triangle is parallel to, and half the length of, the third side.

$\overline{PQ}$ is the midsegment of ΔLMN.

$\overline{PQ} \parallel \overline{LN}$, $PQ = \frac{1}{2}LN$

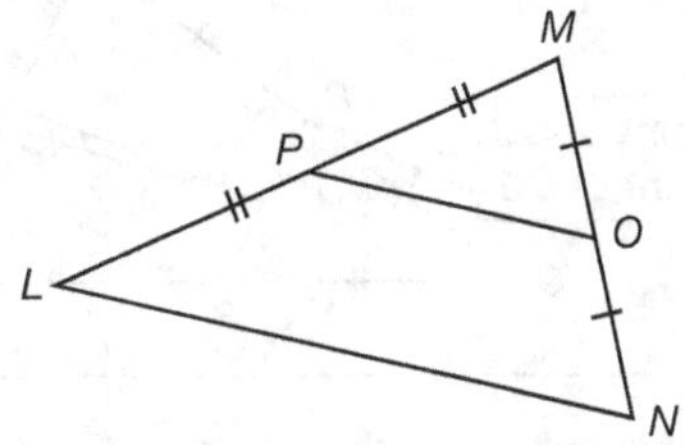

Example:

Find the following measures in ΔABC.

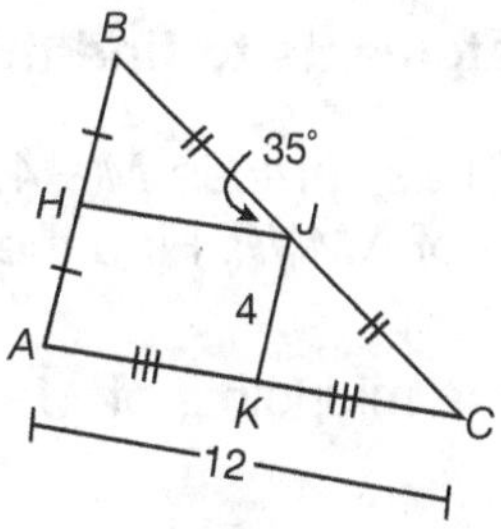

HJ

$HJ = \frac{1}{2}AC$ Δ Midsegement Theorem

$HJ = \frac{1}{2}(12)$ Substitute 12 for *AC*.

$HJ = 6$ Solve.

AB

$JK = \frac{1}{2}AB$ Δ Midsegement Theorem

$4 = \frac{1}{2}AB$ Substitute 4 for *JK*.

$8 = AB$ Solve.

m∠*BCA*

$HJ \parallel AC$ Δ Midsegment Theorem

$m\angle BCA = m\angle BJH$ Corresponding Angle Theroem

$m\angle BCA = 35°$ Substitute 35° for $m\angle BJH$.

Practice

Complete the statements to find the measure.

1. *VX*

$VX = \frac{1}{2}GH$

$VX = \frac{1}{2}(\_\_)$

$VX = \_\_$

2. *HJ*

$\_\_ = \frac{1}{2}HJ$

$\_\_ = \frac{1}{2}HJ$

$\_\_ = HJ$

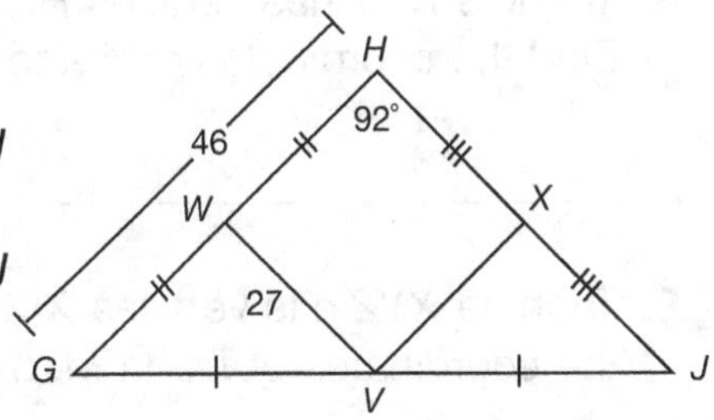

Find each measure.

3. *ST* ___

4. *DE* ___

5. m∠*DES* ___

6. m∠*RCD* ___

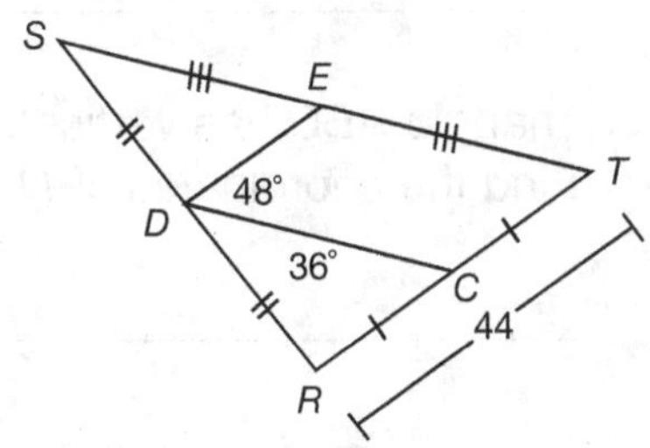

Finding Midpoints of Sides of Triangles

A midsegment of a triangle joins the midpoints of two sides of the triangle. Every triangle has three midsegments. We can use the Midpoint Formula to find the coordinates of the endpoints of a midsegment.

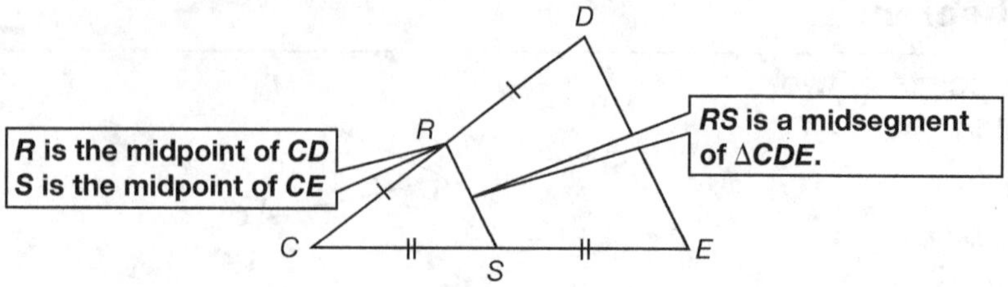

Practice

Complete the statements to find the coordinates of the midsegment.

7. Triangle *MNP* has vertices $M(-4, 7)$, $N(4, 5)$, and $P(2, -1)$. $\overline{UV}$ is a midsegment of ΔMNP. Find the coordinates of U and V.

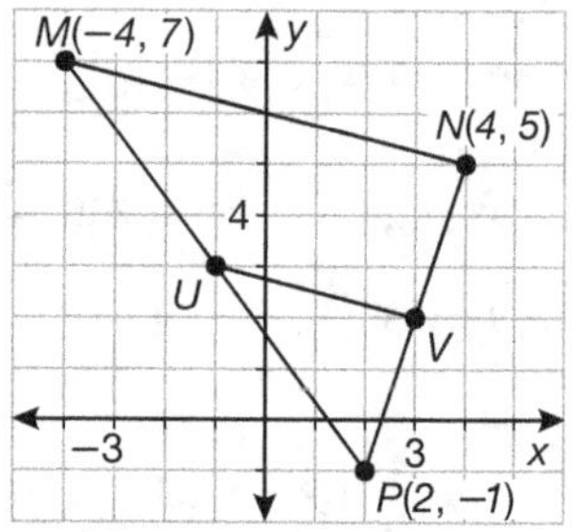

U and *V* are the midpoints of $\overline{MP}$ and $\overline{PN}$.

$U = \left(\frac{\_\_\_ + 2}{2}, \frac{\_\_\_ - 1}{2}\right)$ $\qquad$ $V = \left(\frac{\_\_\_ + 2}{2}, \frac{\_\_\_ - 1}{2}\right)$

$= \left(\frac{\_\_\_}{2}, \frac{\_\_\_}{2}\right)$ $\qquad$ $= \left(\frac{\_\_\_}{2}, \frac{\_\_\_}{2}\right)$

$= (\_\_\_, 3)$ $\qquad$ $= (\_\_, 2)$

Solve.

8. Triangle *RST* has vertices $R(0, -3)$, $S(2, 3)$, and $T(6, -1)$. $\overline{AB}$ is the midsegment of ΔRST. Find the coordinates of *A* and *B*. *A* and *B* are the midpoints of $\overline{RS}$ and $\overline{ST}$.

____

9. Triangle *XYZ* has vertices $X(3, 0)$, $Y(5, 6)$, and $Z(9, 2)$. $\overline{LM}$ is the midsegment of ΔXYZ. Find the coordinates of *L* and *M*. *L* and *M* are the midpoints of $\overline{XY}$ and $\overline{YZ}$.

____

10. Triangle *ABC* has vertices $A(1, 12)$, $B(9, 10)$, and $C(7, 4)$. $\overline{DE}$ is the midsegment of ΔABC. Find the coordinates of *D* and *E*. *D* and *E* are the midpoints of $\overline{BC}$ and $\overline{AC}$.

____

Name ____________________ Date __________ Class __________

Reteaching 56

30°-60°-90° Right Triangles

You have worked with 45°-45°-90° right triangles. Now you will work with 30°-60°-90° right triangles.

Special Right Triangle

30°-60°-90° Triangle	Example
In a 30°-60°-90° right triangle, the length of the hypotenuse is twice the length of the short leg, and the length of the longer leg is the length of the shorter leg times $\sqrt{3}$.	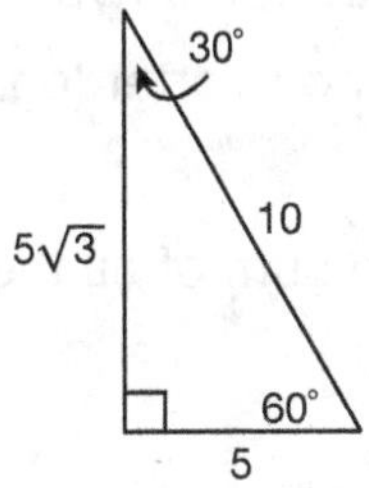 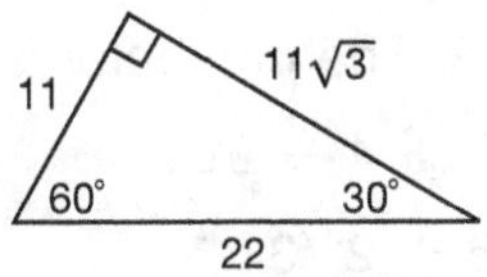

Example

Find the values of x and y in ΔHJK.

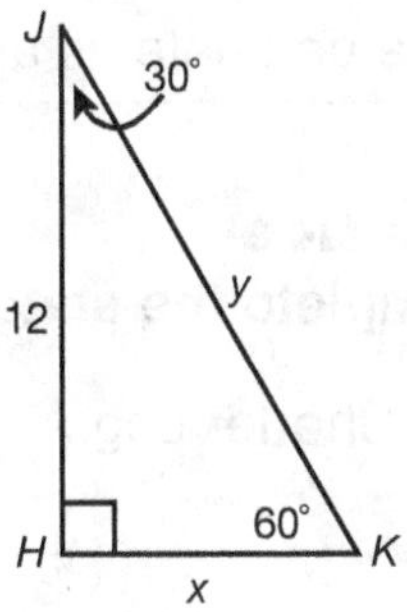

$12 = x\sqrt{3}$	Longer leg = shorter leg times $\sqrt{3}$.
$\frac{12}{\sqrt{3}} = \frac{x\sqrt{3}}{\sqrt{3}}$	Divide both sides by $\sqrt{3}$.
$4\sqrt{3} = x$	Rationalize the denominator.
$y = 2x$	Hypotenuse equals two times the shorter leg.
$y = 2(4\sqrt{3})$	Substitute $4\sqrt{3}$ for x.
$y = 8\sqrt{3}$	Simplify.

Practice

Complete the statements to find the value of x and y.

1. $18 = 2x$

__ $= x$

2. $y = x\sqrt{3}$

$y =$ __ $\sqrt{3}$

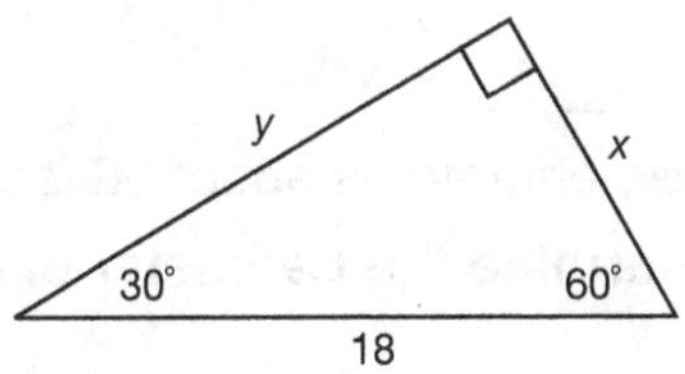

Find the value of x and y.

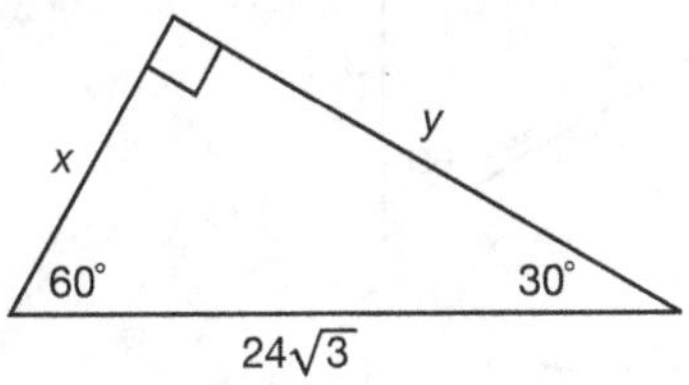

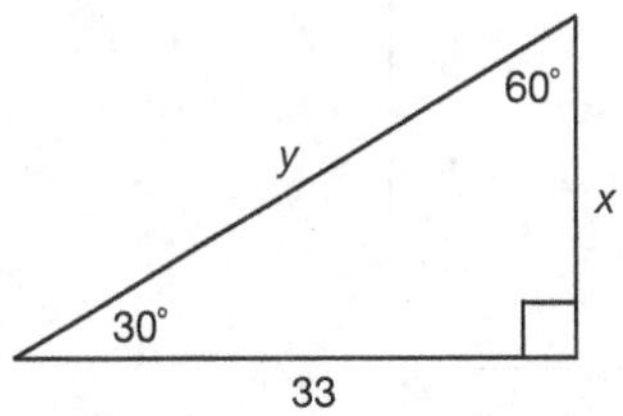

3. ____________________

4. ____________________

Finding the Perimeter of a 30°-60°-90° Right Triangle

Example:

Find the perimeter of the triangle.

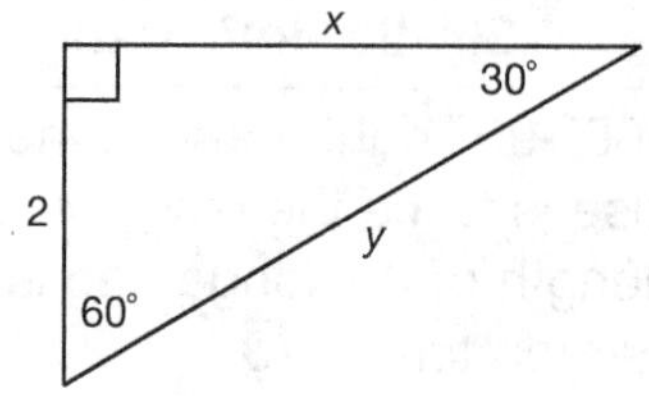

Step 1: Determine the length of each leg.

The short leg is 2 ft.

The longer leg is $2\sqrt{3}$ ft.

Step 2: Determine the length of the hypotenuse.

The hypotenuse is 2 times the shorter leg.

$2(2) = 4$ft

Step 3: The perimeter is the sum of all the sides.

$P = 2 + 4 + 2\sqrt{3}$

$P = 6 + 2\sqrt{3}$

$P \approx 9.46$

The perimeter is approximately 9 feet.

Practice

Complete the statements to determine the perimeter of the triangle.

5. Shorter Leg: Hypotenuse: Perimeter:

Shorter Leg	Hypotenuse	Perimeter
$\_\_ = x\sqrt{3}$	$y = 2x$	$P = \_\_\_\_ + \_\_\_\_ + 24$
$\dfrac{\_\_\_\_}{\sqrt{3}} = \dfrac{x\sqrt{3}}{\sqrt{3}}$	$y = 2(\_\_\_\_)$	$= \_\_\_\_ + 24$
$\_\_\_\_ = x$	$y = \_\_\_\_$	$\approx \_\_\_\_$

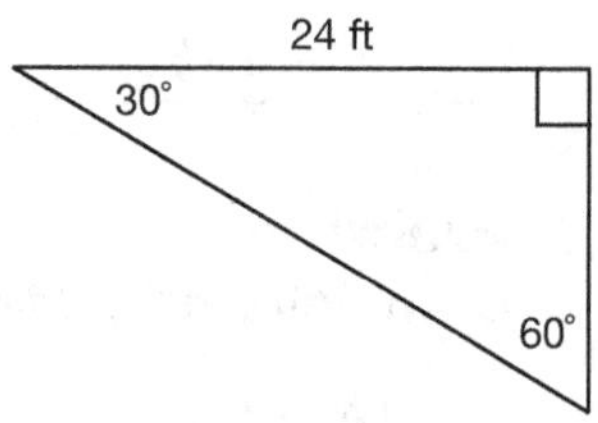

The perimeter is about 66 feet.

Determine the perimeter of each 30°-60°-90° triangle.

6.

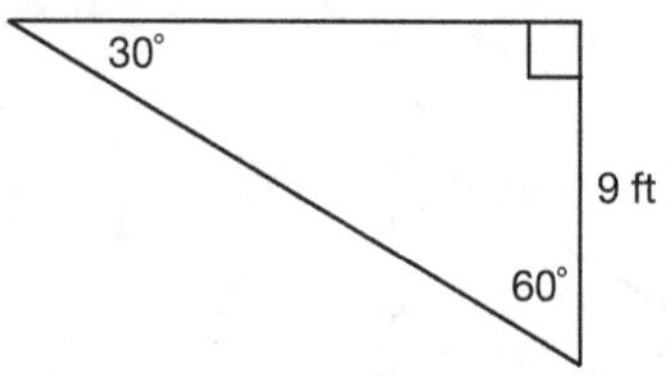

____

7.

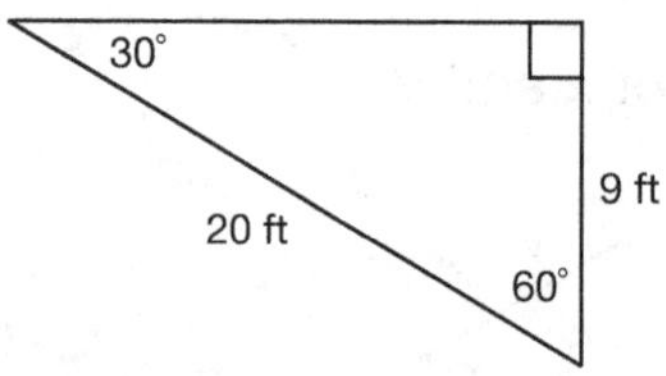

____

Name ______________________________ Date ____________ Class ____________

Reteaching 57
Finding Perimeter and Area with Coordinates

You have worked with special triangles. Now you will find perimeter and area with coordinates.

Finding Perimeter with Coordinates

Find the perimeter of rectangle *RSTV* with coordinates *R* (−5, 1), *S* (−3, 5), *T* (3, 2), and *V* (1, −2).

Step 1: Plot the points on a coordinate plane and draw *RSTV*.

Step 2: Use the distance formula to calculate the length of each side. Find the length of $\overline{RS}$ first.

$$d_{RS} = \sqrt{(x_2 - x_1)^2 + (y_2 - y_1)^2}$$
$$= \sqrt{(-3-(-5))^2 + (5 - 1)^2}$$
$$= \sqrt{(2)^2 + (4)^2}$$
$$= \sqrt{20}$$
$$= 2\sqrt{5}$$

Because the figure is a rectangle, $d_{TV} = 2\sqrt{5}$.

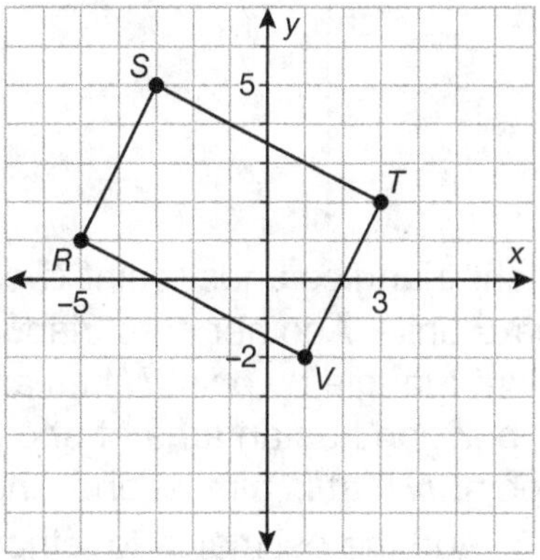

Step 3: Find the length of one of the longer sides.

$$d_{RV} = \sqrt{(1 - (-5))^2 + (-2 - 1)^2} = 3\sqrt{5}$$

Step 4: The length of the opposite side, *ST*, is $3\sqrt{5}$.

Step 5: Perimeter: $P = RS + ST + TV + RV$

$$P = 2\sqrt{5} + 3\sqrt{5} + 2\sqrt{5} + 3\sqrt{5}$$
$$P = 10\sqrt{5} \text{ or approximately } 22.36$$

Practice

Complete the steps to find the perimeter of the triangle.

1. $d_{CD} = \sqrt{(3 - (\_\_\_))^2 + (2 - (\_\_\_))^2}$

$= \sqrt{\_\_\_ + \_\_\_}$

$= \sqrt{\_\_\_}$

≈ 5.66

$d_{DB} = \sqrt{(5 - (\_\_\_))^2 + (0 - (\_\_\_))^2}$

$= \sqrt{\_\_\_}$

≈ 6.32

$d_{BC} = \sqrt{(5 - \_)^2 + (0 - \_)^2}$

$= \sqrt{\_ + 4}$

$= \sqrt{\_}$

≈ 2.83

$P \approx 5.66 + \_\_\_\_ + \_\_\_\_ \approx \_\_\_\_\_$

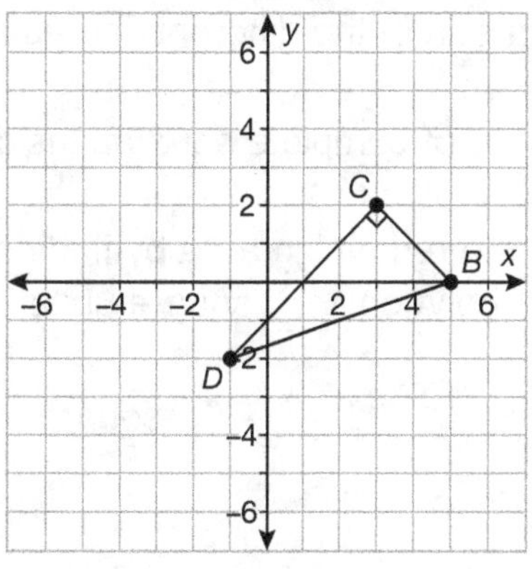

Estimating Area with Coordinates

Find the area of the polygon with vertices *N* (−4, −1), P(−1, 3), *Q* (4, 3), and *R* (2, −2).

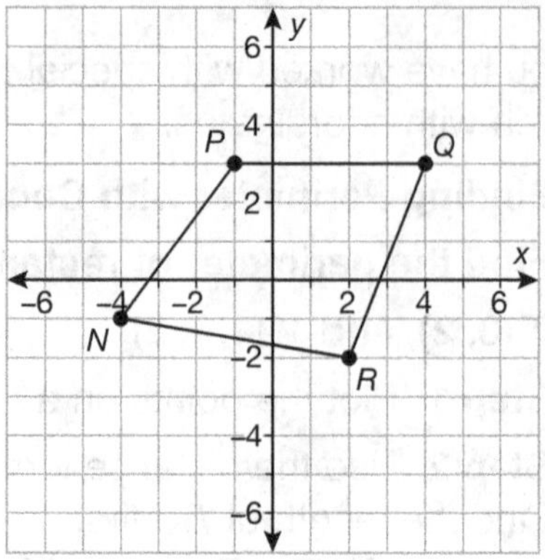

Step 1: Count all the squares completely inside the polygon: 17.

Step 2: Estimate the area of the remaining space.
Look for triangles.

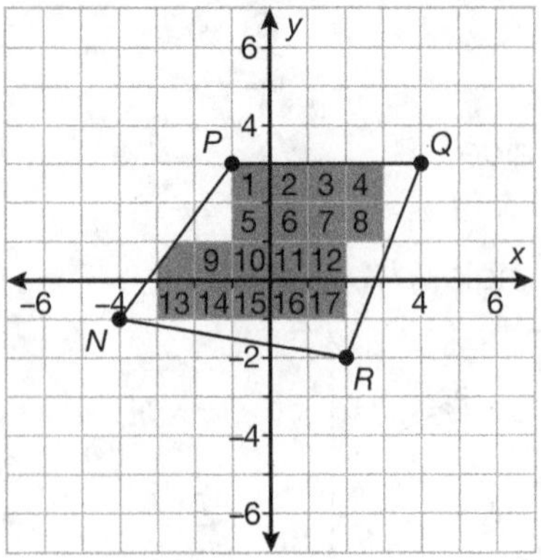

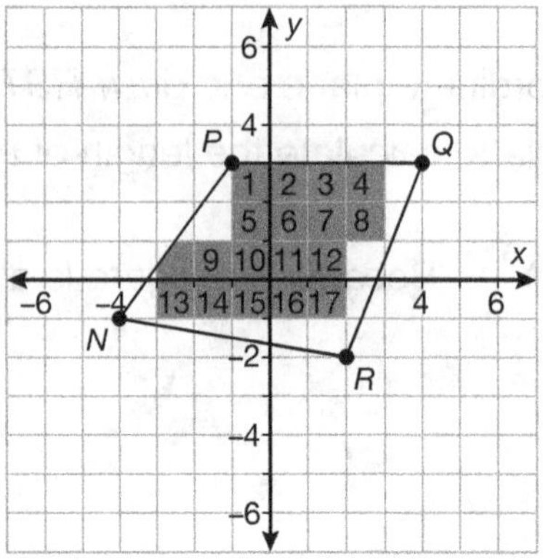

Two right triangles under point *Q* each have an area about 1.5 square units, for a total of 3 square units. Another right triangle by point *P* has an area of about 2 square units. The right triangle by point *N* has an area of about 1 square unit. The right triangle at the bottom of the polygon has an area of about 3 square units. There is one remaining area near point *N* that is almost one whole square unit.

Step 3: Add the estimates together to estimate the area of polygon *NPQR*.

A ≈ 17 + 3 + 2 + 1 + 3 + 1 ≈ 27 square units

Practice

Complete the steps to estimate the area of the polygon.

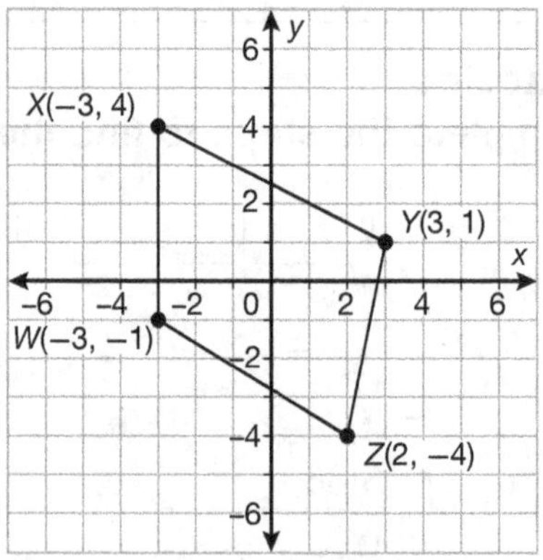

2. Find the areas of the different figures in the polygon.

The number of complete squares inside the polygon is ___ .

Estimate the remaining area by finding areas of right triangles inside the polygon: _ square units.

A ≈ 20 + ___ = ____ square units

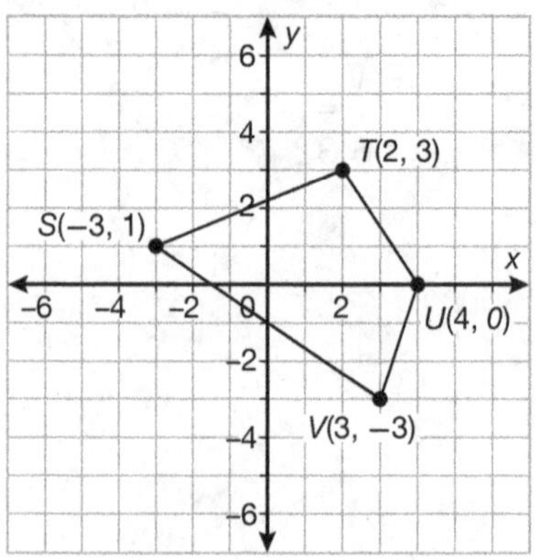

3. Find the areas of the different figures in the polygon.

The number of complete squares inside the polygon is ___.

Estimate the remaining area by finding areas of right triangles inside the polygon: ___ square units.

A ≈ 10 + ____ = ____ square units

Name ________________________ Date __________ Class __________

Reteaching 58

Tangents and Circles, Part 1

Tangent Lines and Angle Measures

A tangent line is a line that is in the same plane as a circle and intersects the circle at exactly one point.

Theorems		
If a line is tangent to a circle, then the line is perpendicular to a radius drawn to the point of tangency. $\overline{AB} \perp \overline{DC}$	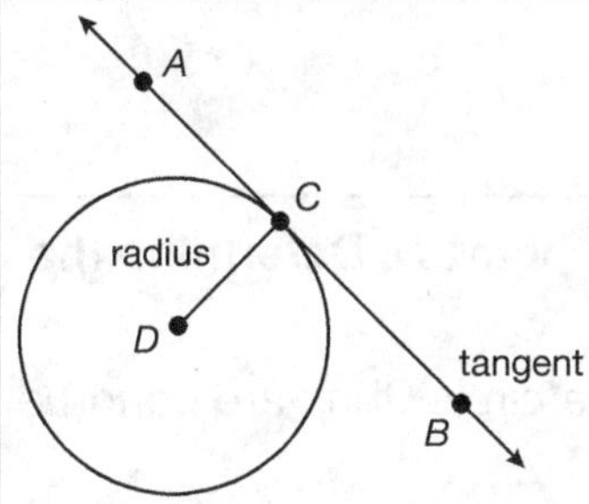	If a line in the plane of a circle is perpendicular to a radius at its endpoint on the circle, then the line is tangent to the circle. $\overline{AB}$ is tangent to $\odot D$.

Which line is tangent to $\odot A$ at point *E*? Line *b* is tangent to $\odot A$ at point *E*.

If $m\angle ABE = 45°$, find $m\angle BAE$.

Since $\overline{AE}$ is the radius of the circle, $\overline{AE} \perp b$ and $m\angle AEB = 90°$.

Therefore, ΔBAE is a right triangle.

$$m\angle ABE + m\angle BAE = 90°$$
$$45° + m\angle BAE = 90°$$
$$m\angle BAE = 90° - 45°$$
$$m\angle BAE = 45°$$

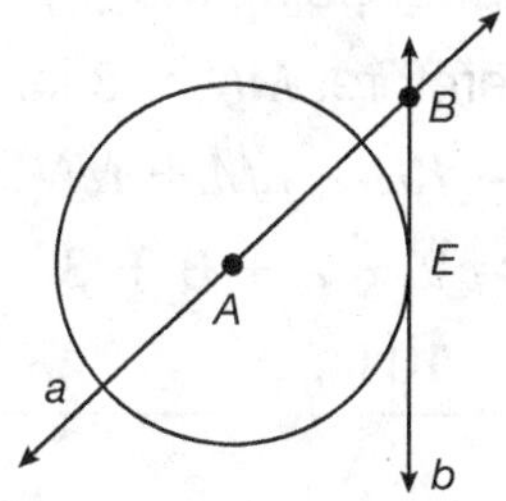

Practice

If line *m* is tangent to $\odot S$ at point *R* and $m\angle STR = 28°$, find $m\angle SRT$ and $m\angle RST$.

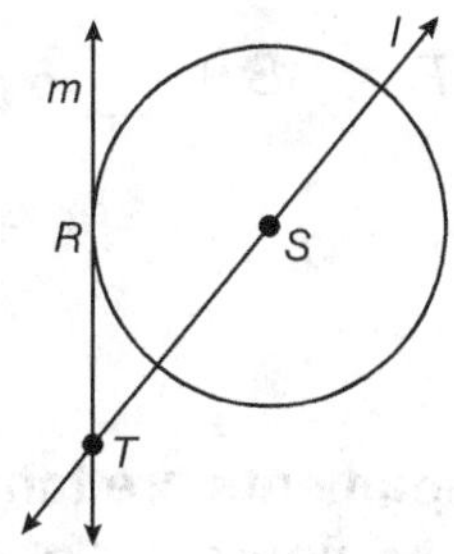

1. Since ___ is the radius of the circle and is perpendicular to

_______, $m\angle SRT =$ ___.

$m\angle STR + m\angle RST =$ ______

___ $+ m\angle RST = 90°$

$m\angle RST = 90° -$ ___

$m\angle RST =$ ___

Line *b* is tangent to $\odot J$ at point *N*, and $m\angle NJM = 32°$.

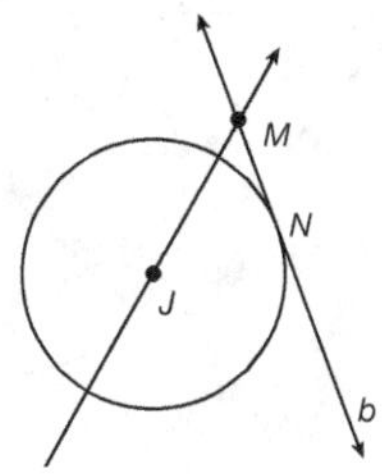

2. Find $m\angle JNM$.

___

3. Find $m\angle JMN$.

___

Applying Relationships of Tangents from an Exterior Point

Theorem

If two segments are tangent to a circle from the same exterior point, then they are congruent.

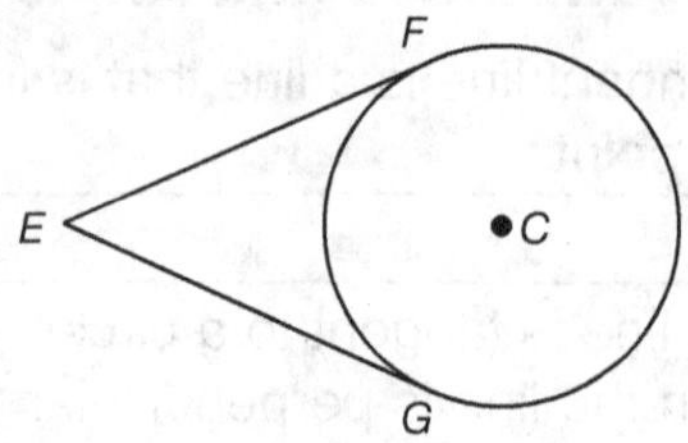

$\overline{EF} \cong \overline{EG}$

***$\overline{KN}$* and *$\overline{MN}$* are tangent to ⊙*L* from point *N*. Determine the perimeter of quadrilateral *KLMN*.**

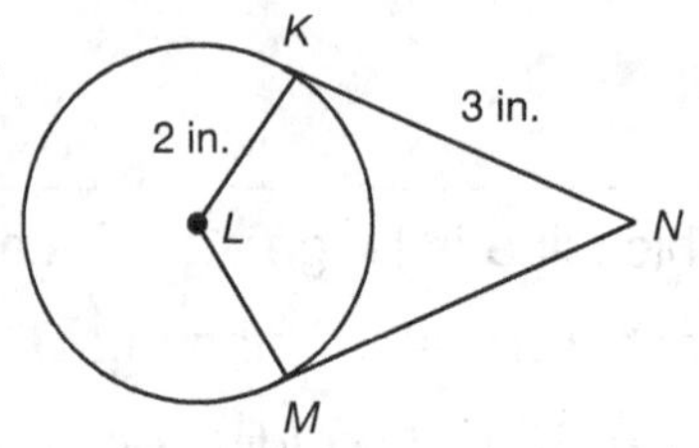

Since $\overline{LM}$ and $\overline{LK}$ are radii of the same circle, they are congruent. Therefore, $LM = 2$ in.

Since $\overline{KN}$ and $\overline{MN}$ are tangents to the same circle from the same exterior point, they are congruent.

Therefore, $MN = 3$ in.

$P = KL + LM + MN + KN$

$= 2 + 2 + 3 + 3$

$= 10$ in.

Practice

The segments are tangent to the circle from the same exterior point. Complete the steps to find the perimeter of the quadrilateral.

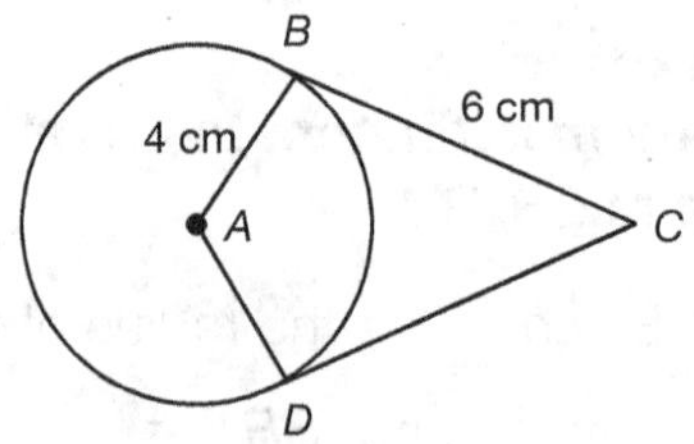

4. $AB =$ _____; $AD =$ _____

$BC =$ _____; $CD =$ _____

$P = AB +$ ___ $+ BC +$ ___

$P = 4 +$ _ $+ 6 +$ _

$P =$ __ cm

The segments are tangent to the circle from the same exterior points. Find the perimeter of each quadrilateral.

5.

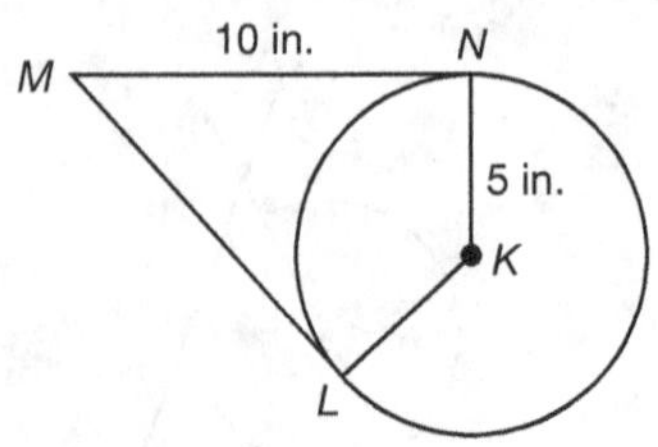

_____

6.

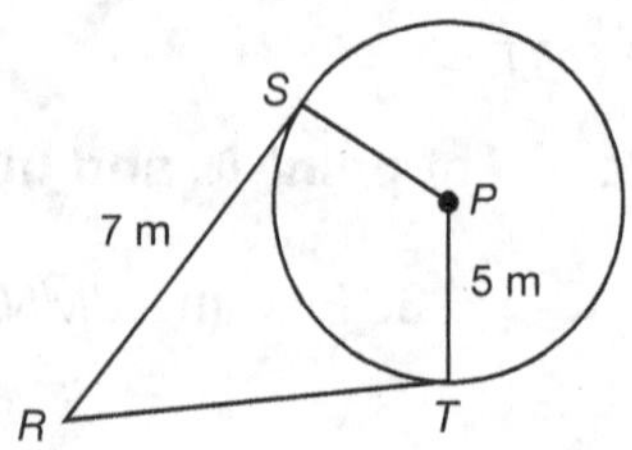

_____

Name ______________________ Date __________ Class __________

Reteaching 59

Finding Surface Area and Volumes of Prisms

You have worked with tangents and circles. Now you will find surface area and volume of prisms.

Surface Area of Prisms

The **lateral area** of a prism is the sum of the areas of all the lateral faces. A lateral face is not a base. The **surface area** is the total area of all faces.

The **lateral area** of a prism with base perimeter p and height h is $L = ph$.

The **surface area** of a prism with lateral area L and base area B is $S = L + 2B$.

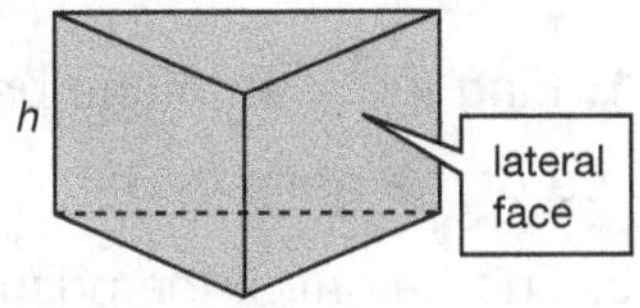

Find the lateral area and the surface area of the rectangular prism.

Step 1: The perimeter is $p = 2(5) + 2(6) = 22$ in.

Therefore, $L = ph = (22)(6) = 132$ in^2.

Step 2: Find the area of the base.

The area of the base is $(6)(5) = 30$ in^2.

Step 3: Find the total surface area.

$S = L + 2B = 132 + 2(30) = 192$ in^2.

The surface area of the prism is 192 square inches.

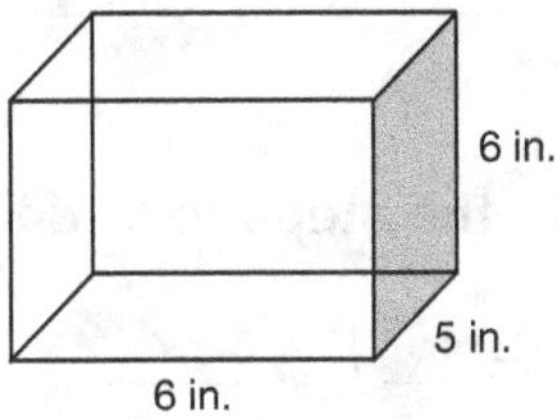

Practice

Complete the steps to find the lateral area and surface area of the prism.

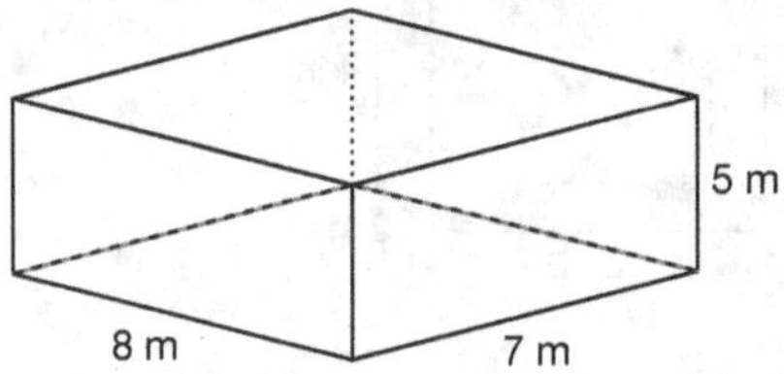

1. lateral area:

   $p = 2(8) + 2(\_) =$ ____

   surface area: area of each base = $(8)(\_) =$ _____

   $S = L + 2B =$ ____ $+ 2($____$) =$ ______

Find the lateral area and surface area of each prism.

2. 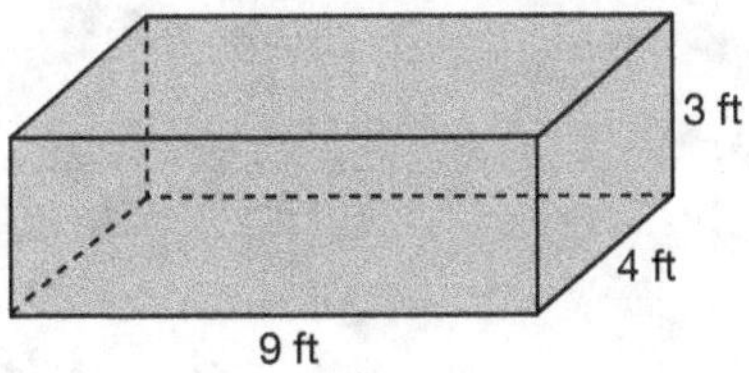

3. 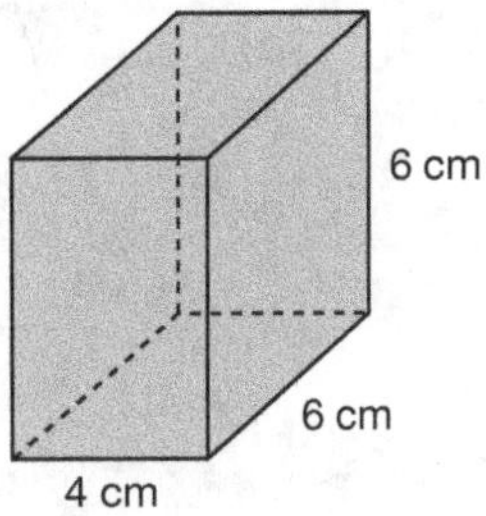

____________________ ____________________

Volume of Prisms

The volume of a prism with base area ***B*** and height ***h*** is $V = Bh$.

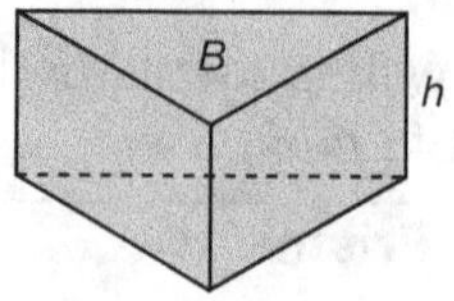

Find the volume of the right prism.

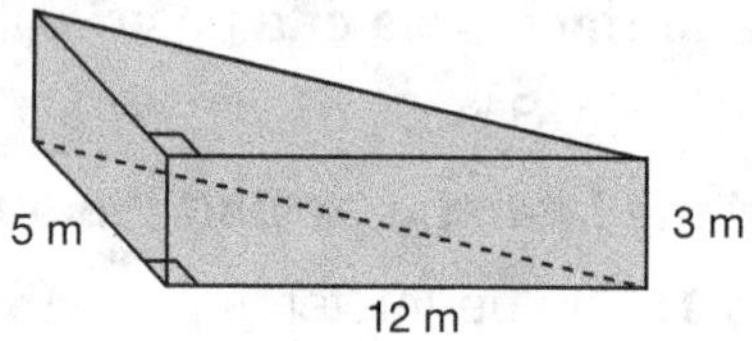

Step 1: Find the area of the base. The base is a right triangle.
The area of the base is $B = \frac{1}{2}(5)(12) = 30 \text{ m}^2$.
Step 2: The height of the prism is $h = 3$ m.
The volume of the prism is $V = Bh = (30)(3) = 90 \text{ m}^3$
The volume of the right prism is 90 cubic meters.

Practice

Complete the steps to find the volume of the prism.

4. $B = (2)(\_) =$ ______

The height of the prism is _____

$V = Bh = (\_\_)(4) =$ ______

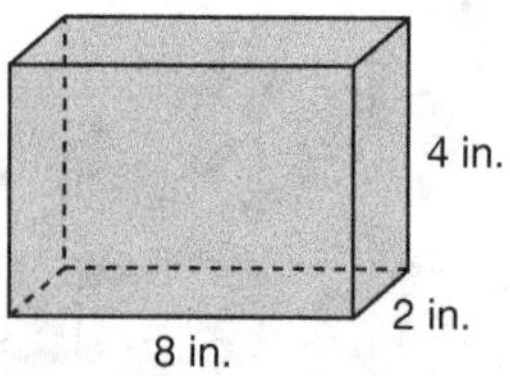

5. $B = \frac{1}{2}(4)(\_) =$ _____

The height of the prism is _____.

$V = Bh = (\_)(4) =$ ______

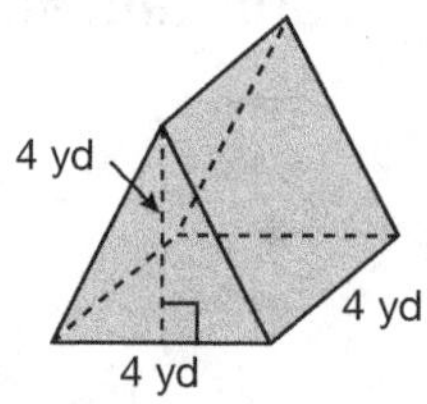

Find the volume of each prism.

6.

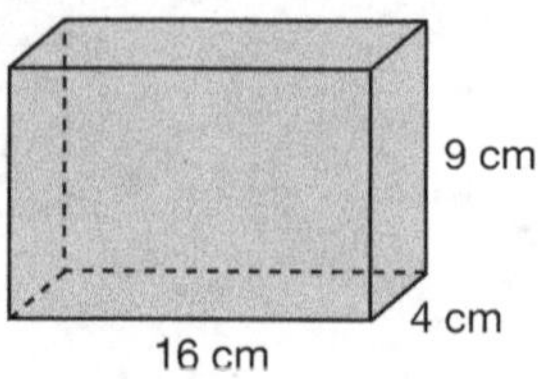

____

7.

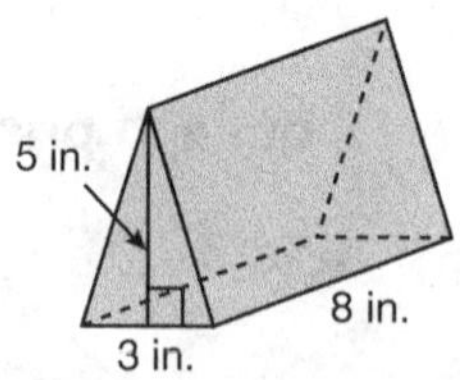

______

Name ______________________ Date __________ Class __________

Reteaching 60

Proportionality Theorems

You have worked with proportional relationships that exist within triangles. Now you will work with any line that intersects a triangle and that is parallel to a base.

Triangle Proportionality Theorem

If a line parallel to one side of a triangle intersects the other two sides, then it divides those sides proportionally.

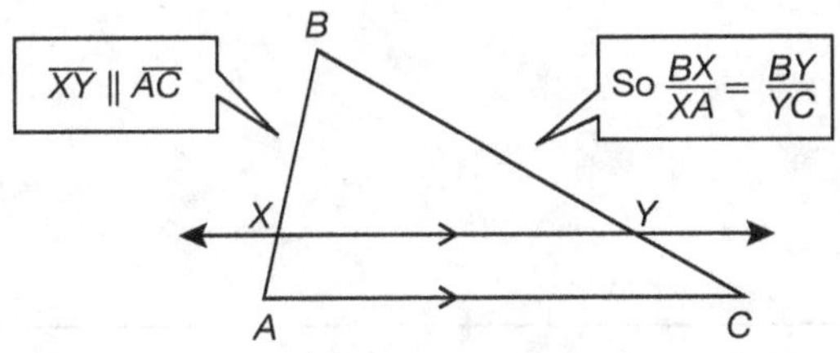

You can use the **Triangle Proportionality** Theorem to find lengths of segments in triangles.

Example:

Find the length of $\overline{EG}$.

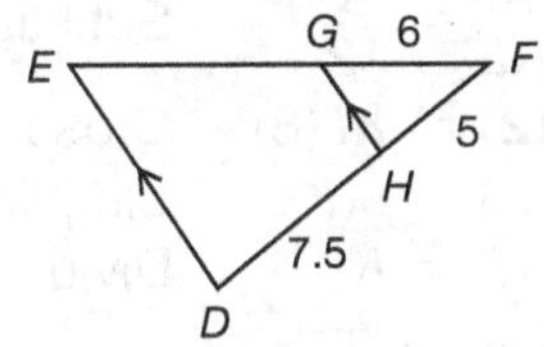

$\frac{EG}{GF} = \frac{DH}{HF}$	Triangle Proportionality Theorem
$\frac{EG}{6} = \frac{7.5}{5}$	Substitute the known values.
$EG(5) = 6(7.5)$	Cross Products Property
$5(EG) = 45$	Simplify.
$EG = 9$	Divide both sides by 5.

Practice

Complete the statements to find the length of $\overline{RQ}$.

1. $\frac{NR}{\_\_\_} = \frac{\_\_\_}{SQ}$

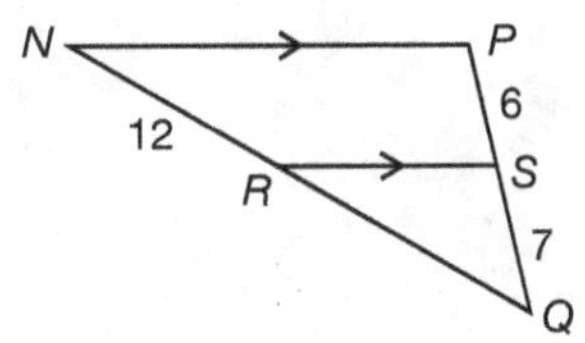

$\frac{\_\_\_}{RQ} = \frac{\_\_\_}{7}$

$\_\_(7) = \_(RQ)$

$\_\_ = 6(RQ)$

$\_\_ = RQ$

Find the length of each segment.

2. $\overline{JN}$ ____

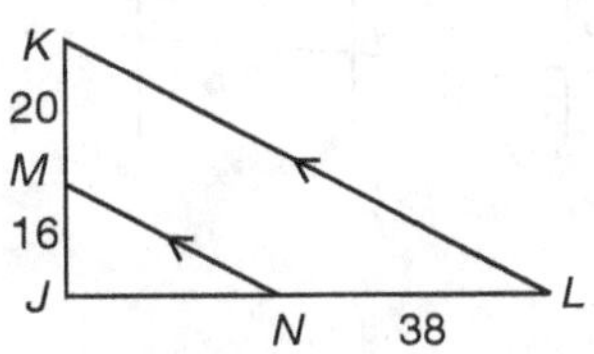

3. $\overline{BH}$ ____

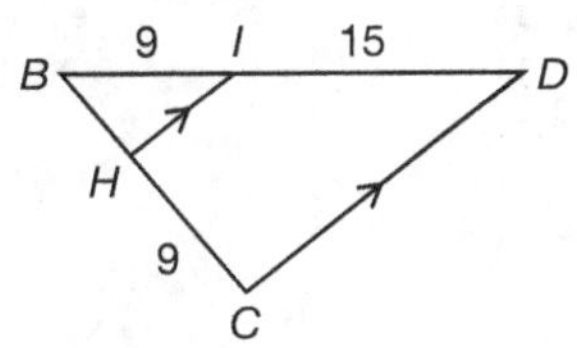

Parallel Lines and Transversals

If parallel lines intersect transversals, then they divide the transversals proportionally. $\frac{AC}{CE} = \frac{BD}{DF}$	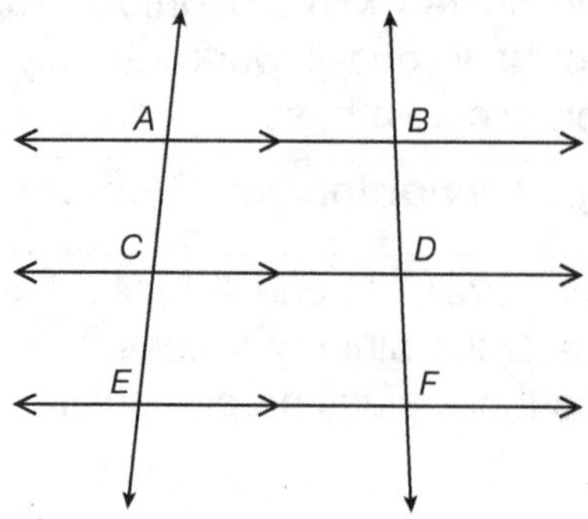

You can use the Triangle Proportionality Theorem to find lengths of segments.

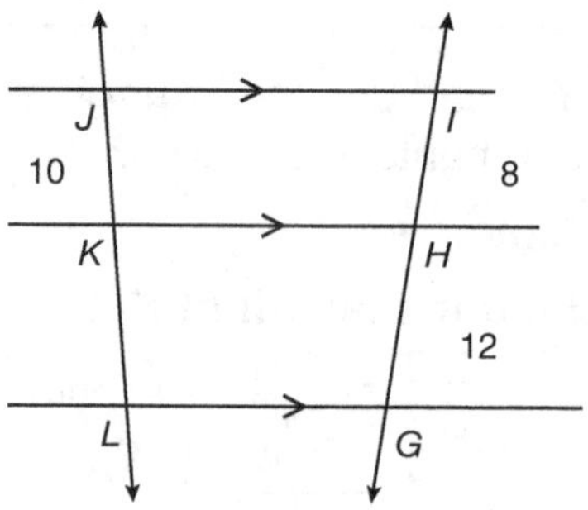

Find the length of $\overline{KL}$.

$\frac{JK}{KL} = \frac{IH}{HG}$	Triangle Proportionality Theorem
$\frac{10}{KL} = \frac{8}{12}$	Substitute the known values.
$10(12) = KL(8)$	Cross Products Property
$120 = 8(KL)$	Simplify.
$15 = KL$	Divide both sides by 8.

The length of $\overline{KL}$ is 15.

Practice

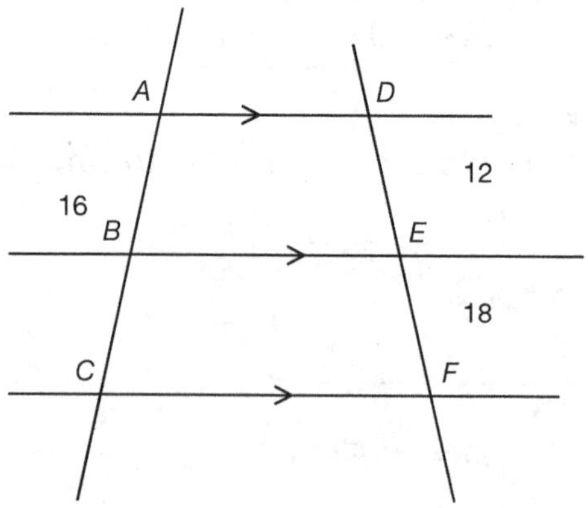

4. $\frac{\_\_\_}{BC} = \frac{\_\_\_}{EF}$

$\frac{\_\_\_}{BC} = \frac{\_\_\_}{18}$

$16(\quad) = BC(\quad)$

$\_\_\_ = 12(BC)$

$\_\_\_ = BC$

5. Find the length of segment LM _.

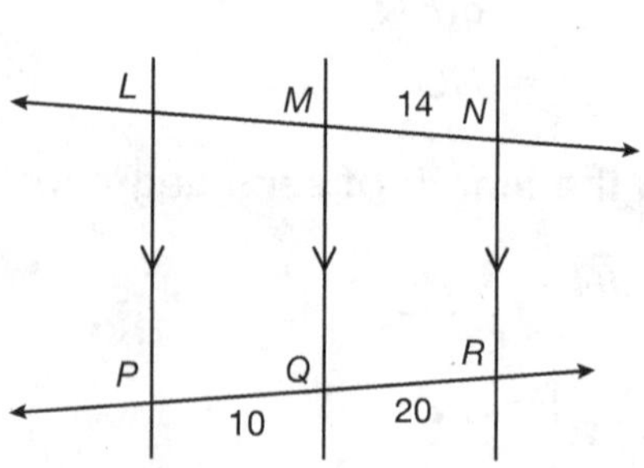

Name ______________________ Date __________ Class __________

Reteaching

Geometric Probability

INV 6

You have calculated probabilities of events that occur when coins are tossed and number cubes are rolled. Now you will learn about **geometric probability.**

Geometric Probability

You can find the probability of an event occurring by dividing the number of ways the event can occur by the total number of outcomes. To know the probability of something falling into a part of a whole area or region, you can use geometric probability, which is the ratio of the size of the partial region to that of the whole region. Depending on the situation, size can be measured in degrees, length, area, or volume.

$$P(\text{partial region}) = \frac{\text{size of partial region}}{\text{size of whole region}}$$

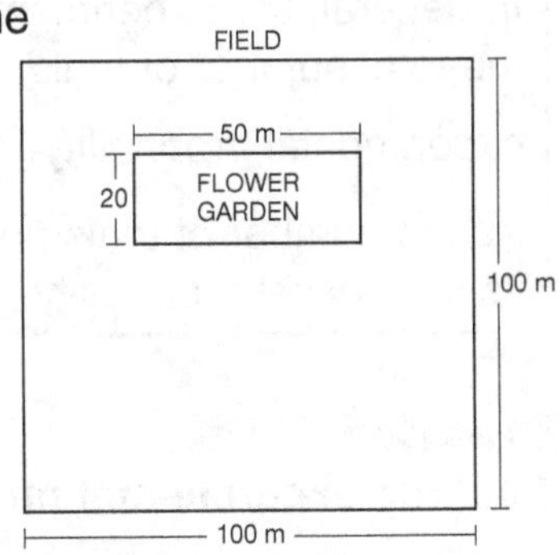

A small meteor lands in the field to the right. What is the probability that it has landed in the flower garden?

$$\text{area of the flower garden} = (20\ m)(50\ m) = 1000\ m^2$$

$$\text{area of the field} = (100\ m)(100\ m) = 10{,}000\ m^2$$

$$P(\text{meteor lands in flower garden}) = \frac{1000\ m^2}{10{,}000\ m^2} = 0.1 = 10\%$$

Practice

Compete the steps to find the probability.

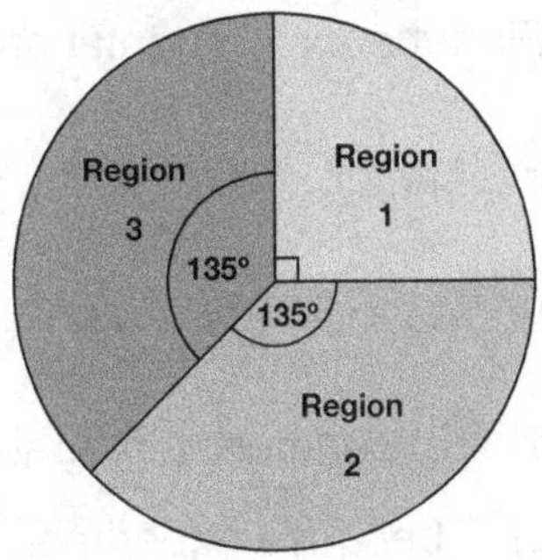

1. Find the probability that a dart thrown at the dartboard lands in region 3.
   The degree measure of region 3 is ______.
   The dartboard is in the shape of a circle.
   The number of degrees in a circle is ______.
   The probability of a dart hitting region 3 is $\frac{\text{region 3}}{\text{total degrees}}$ $\frac{\_\_\_}{\_\_\_}$ = ______, or ______%

Find the probabilities.

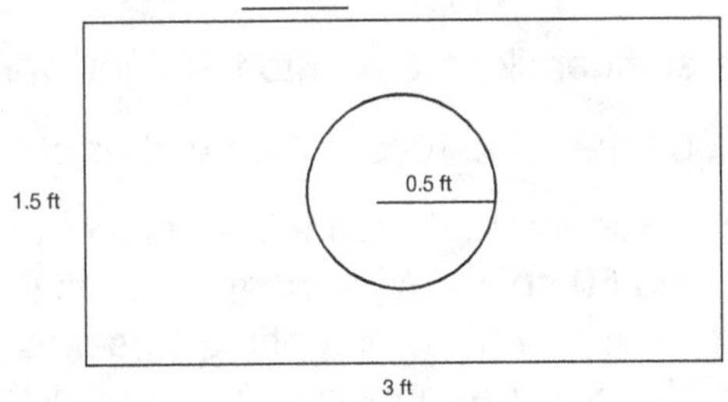

2. At the local pizza shop, if you drop a coin in the little bowl in the fish tank, you get a free slice of pizza.
   What is the probability of getting the coin in the fish tank if you have to drop it in while blindfolded?
   approximately ______%

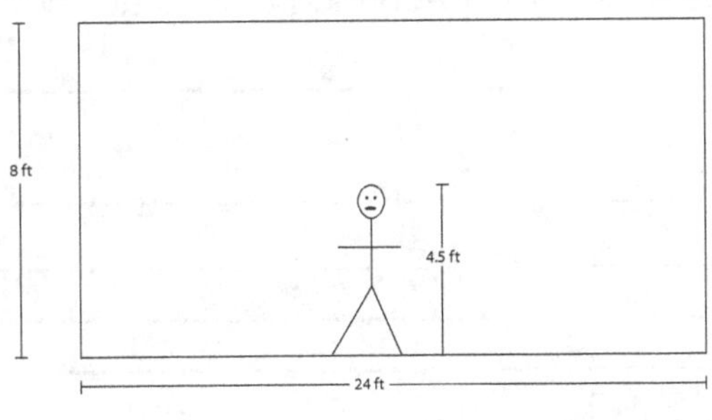

3. Find the probability that a soccer ball kicked on goal goes over the goalie's head. ______%

Experimental Probability

When you say that the probability of flipping a coin and getting heads is $\frac{1}{2}$, you are talking about the **theoretical probability** of flipping heads. If we flipped a coin 50 times, the **experimental probability** of heads would be the actual percentage of flips that resulted in heads.

If a coin is tossed 50 times and heads occurs 26 times, what is the experimental probability of heads?

$$P(\text{heads}) = \frac{\text{number of heads}}{\text{number of tosses}} = \frac{26 \text{ heads}}{50 \text{ tosses}} = 0.52 = 52\%$$

In general, the experimental probability of an event is the number of times it occurs divided by the total number of trials:

$$\text{experimental probability} = \frac{\text{\# of times the event occurred}}{\text{\# of trials}}$$

As the number of trials goes up, an event's experimental probability will get closer and closer to its theoretical probability.

Practice

Find the experimental probabilities.

4. Kyle rolled a number cube 50 times and got a one 9 times.

Find the experimental probability of a one.

$$\text{experimental probability of a one} = \frac{\text{number of ones}}{\text{number of trials}} = \frac{\_\_}{\_\_} = \_\_\_\_ = \_\_\%$$

What fraction will the experimental probability get closer to if the cube is tossed more?

The experimental probability will get closer to the ____________ probability of $\frac{\_}{\_}$ because experimental probability approaches theoretical probability as the number of trials ____________.

5. Make a spinner like the one to the right.

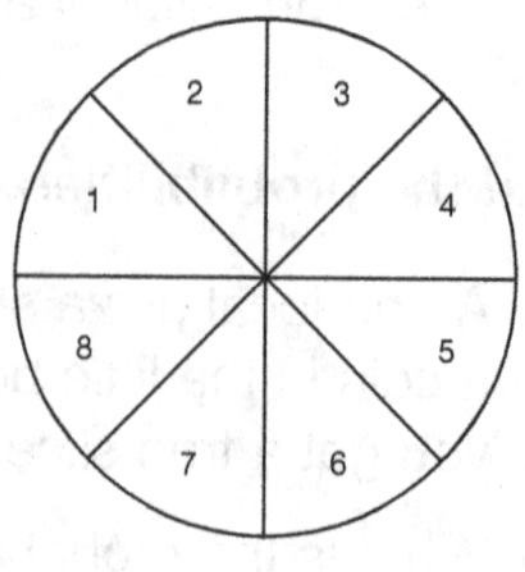

Spin it 50 times and record the number of times it lands on three.

Find the experimental probability of landing on three after 10, 25, and 50 spins. What number does it look like the experimental probability is getting closer to? Would it get closer to it if you performed more trials? Why?

____

____

____

____

____

Name ______________________ Date __________ Class __________

Reteaching 61

Quadrilateral Is a Parallelogram

You have worked with quadrilaterals. Now you will determine whether a quadrilateral is a parallelogram.

Identifying Parallelograms

A parallelogram is a quadrilateral with two pairs of parallel sides.

All parallelograms have the following properties:

Opposite sides are congruent.	Opposite angles are congruent.
$\overline{GF} \cong \overline{HJ}$ $\overline{GH} \cong \overline{FJ}$ 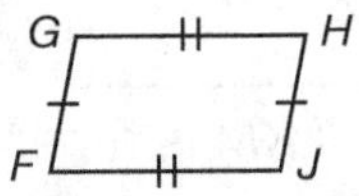	$\angle F \cong \angle H$ $\angle G \cong \angle J$ 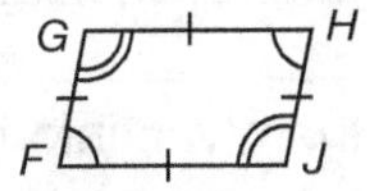

Example: In quadrilateral *JKLM*, $\overline{JK} \cong \overline{ML}$. Is *JKLM* a parallelogram?

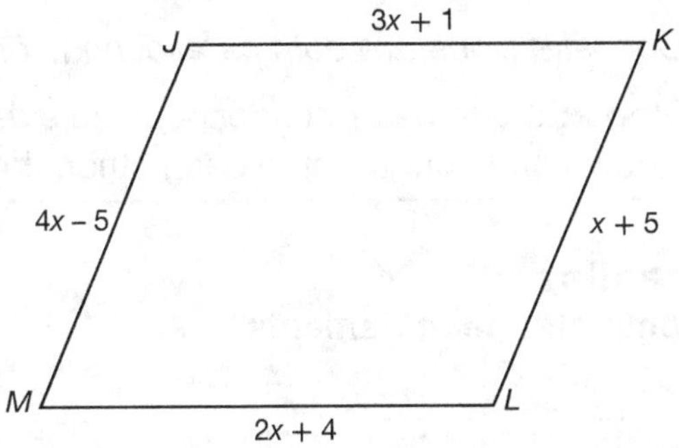

Step 1: Solve for *x*.

$JK = ML$	Given
$3x + 1 = 2x + 4$	Substitute.
$x = 3$	Solve.

Step 2: Substitute $x = 3$ to find *JM* and *KL*.

$JM = 4(3) - 5 = 7$

$KL = 3 + 4 = 7$

Since $\overline{JK} \cong \overline{ML}$ and $\overline{JM} \cong \overline{KL}$, *JKLM* is a parallelogram.

Practice

Complete the statements.

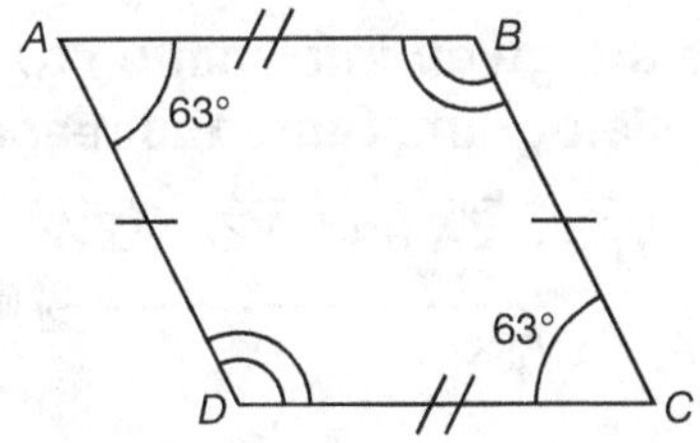

1. $\overline{AB} \cong$ ___; $\overline{AD} \cong$ ___ $\angle A \cong$ ___; $\angle B \cong$ ___

The figure is a parallelogram since _________ angles and sides are ___________.

Use the given information to determine whether the figures are parallelograms. State the reason.

2. Given: $\overline{AB} \cong \overline{DC}$

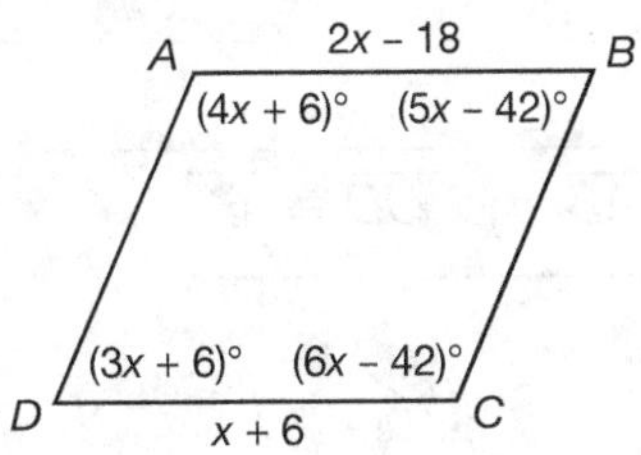

____

____

3. Given: $\overline{PR} \cong \overline{QS}$

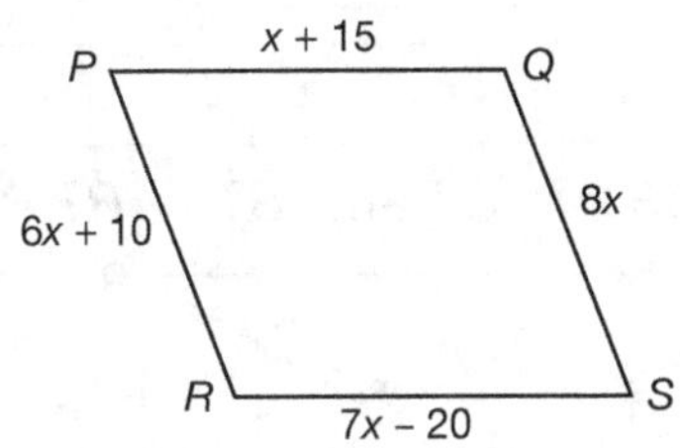

____

____

More on Identifying Parallelograms

One pair of opposite sides of a quadrilateral are both parallel and congruent to each other.	The diagonals bisect each other.
$\overline{AB} \parallel \overline{CD}$ $\overline{AB} \cong \overline{CD}$ 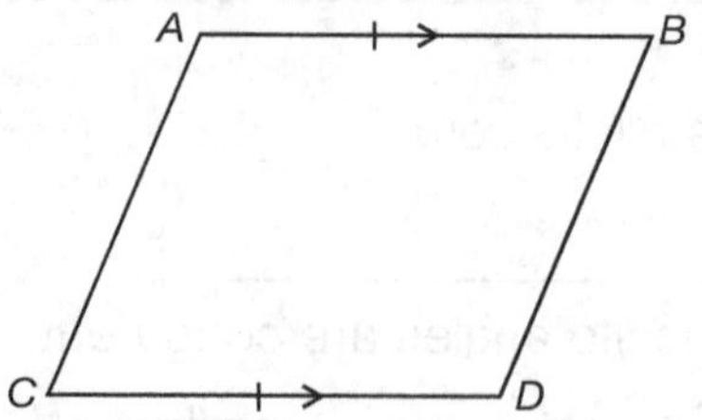	$\overline{FP} \cong \overline{PH}$ $\overline{GP} \cong \overline{PJ}$ 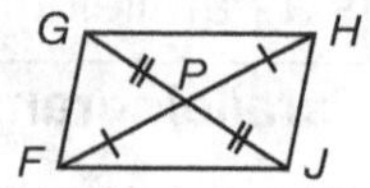

Example: Use the given information to decide whether *JLKM* is a parallelogram.

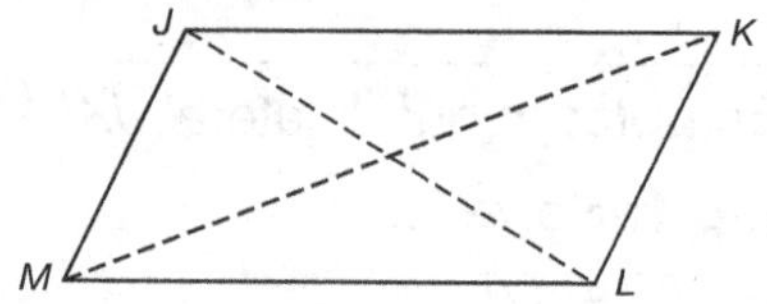

Given: $\overline{JK} \parallel \overline{LM}$ and $\overline{JK} \cong \overline{LM}$

Step 1: Is there a relationship between the sides?

Opposite sides are both parallel and congruent to each other.

Step 2: If one pair of corresponding sides of a quadrilateral are both parallel and congruent to each other, the figure is a parallelogram.

Practice

Complete the statements.

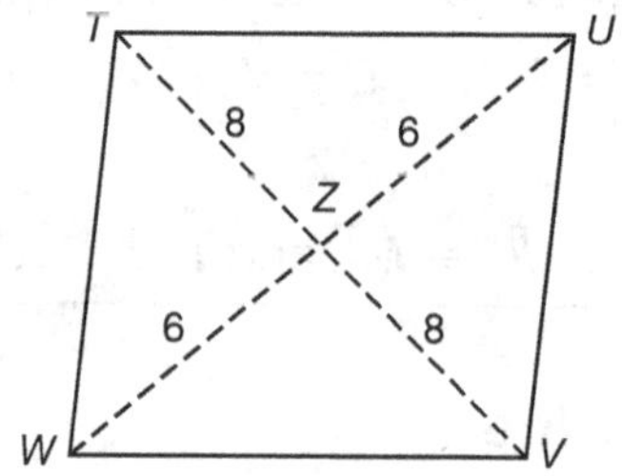

4. $\overline{TZ} \cong$ ___; $WZ \cong$ ___

The diagonals of the quadrilateral _______ each other.

The figure ___ a parallelogram.

Use the given information to determine whether the figure is a parallelogram. State the reason.

5. $\overline{VZ} \cong \overline{ZX}$ and $\overline{YZ} \cong \overline{ZW}$

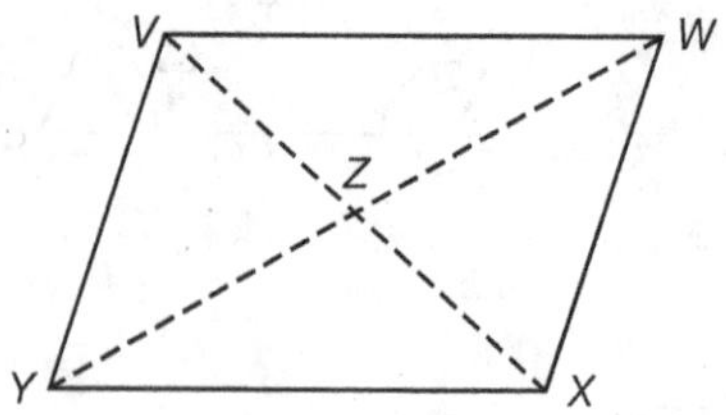

____

____

6. $\overline{AB} \cong \overline{DC}$ and $\overline{AB} \parallel \overline{DC}$

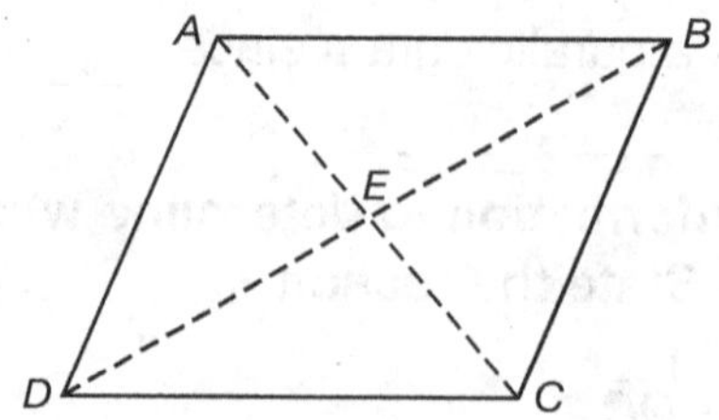

____

____

7. $\overline{PT} \cong \overline{QT}$ and $\overline{ST} \cong \overline{RT}$

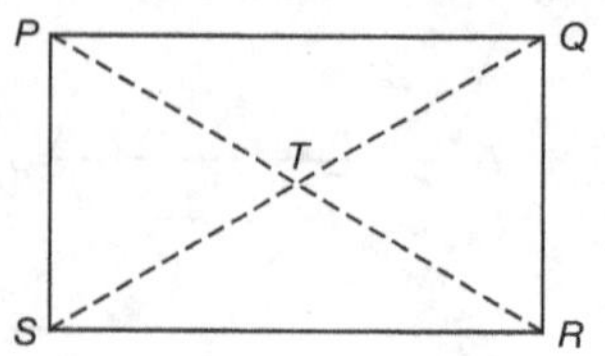

____

8. $\overline{AD} \cong \overline{AB}$ and $\overline{CD} \cong \overline{CB}$

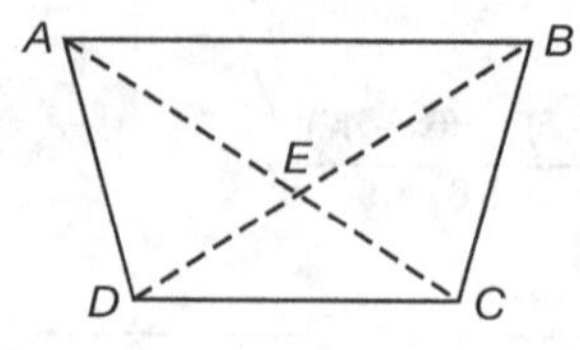

____

Name ______________________ Date __________ Class __________

Reteaching 62

Finding Surface Area and Volume of Cylinders

You have learned how to find the surface area and the volume of prisms.
Now you will find the surface area and the volume of cylinders.

To find the surface area of a cylinder, use the formula $S = 2\pi r^2 + 2\pi rh$ to combine the lateral area and the area of the two bases.

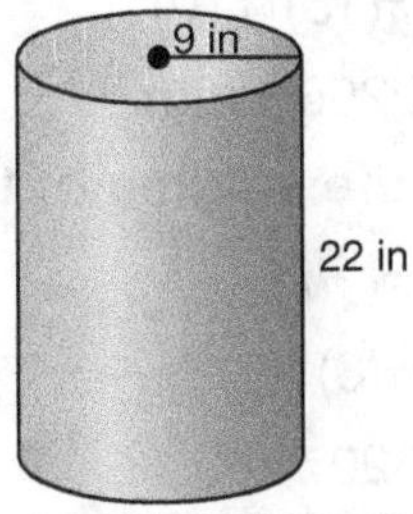

Find the area of the cylinder in terms of π.

$S = 2\pi r^2 + 2\pi rh$

$S = 2\pi(9)^2 + 2\pi(9)(22)$

$S = 162\pi + 396\pi$

$S = 558\pi \text{ in}^2$

Practice

Complete the steps to find the surface area of the cylinder in terms of π.

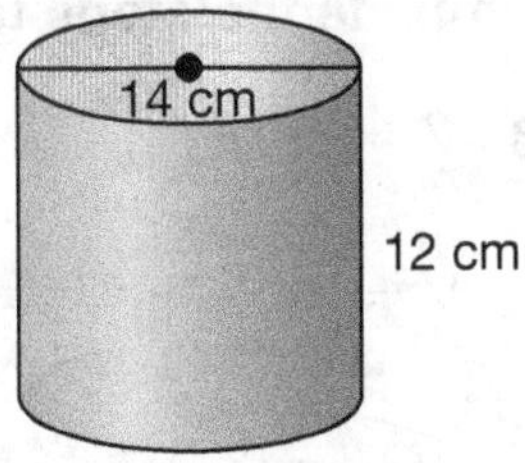

1. $S = 2\pi(\_)^2 + 2\pi(\_)(\_\_)$

   $S = \_\_\pi + \_\_\_\pi$

   $S = \_\_\_\pi \text{ cm}^2$

Find the surface area of each cylinder in terms of π.

2. 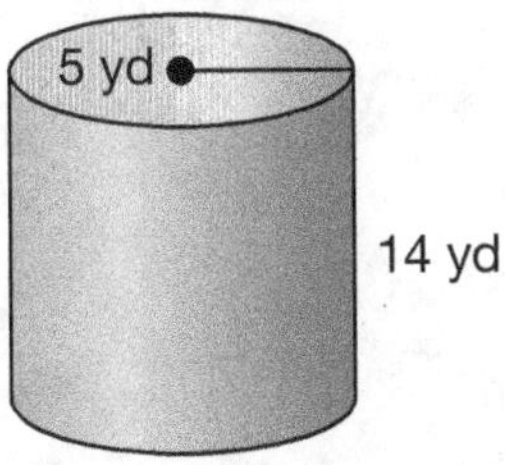

   ____

3. 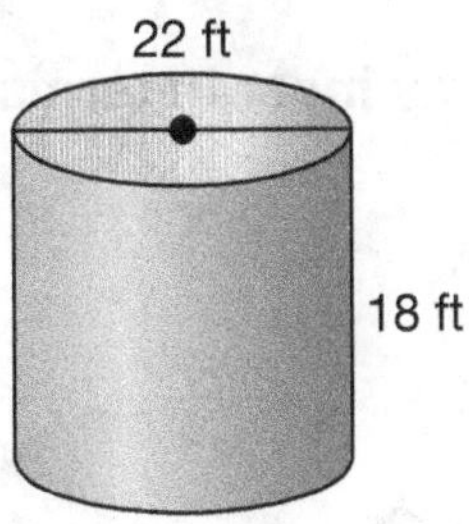

   ____

4. 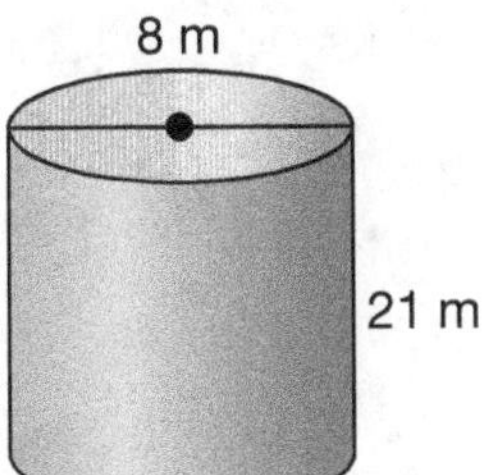

   ____

5. 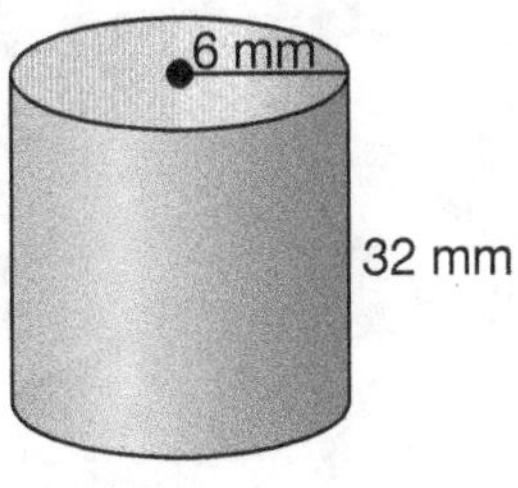

   ____

Both right cylinders and oblique cylinders each use the same formula, $V = \pi r^2 h$, for volume.

Find the volume of the right cylinder in terms of π.

$V = \pi r^2 h$

$V = \pi(15)^2(10)$

$V = 2250\pi \text{ mm}^3$

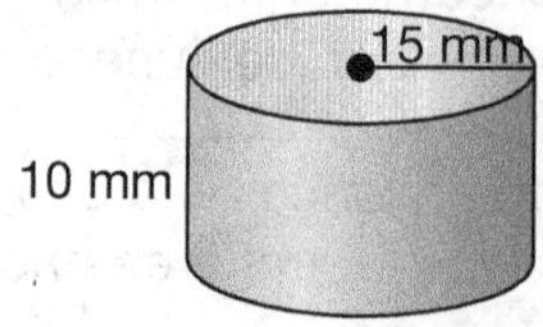

Find the volume of the oblique cylinder in terms of π.

$V = \pi r^2 h$

$V = \pi(8)^2(5)$

$V = 320\pi \text{ ft}^3$

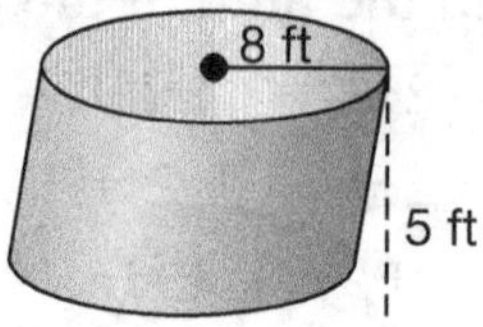

Practice

Complete the steps to find the volume of each cylinder in terms of π.

6. $V = \pi(\_)^2(\_)$

$V =$ ____ $\pi \text{ in}^3$

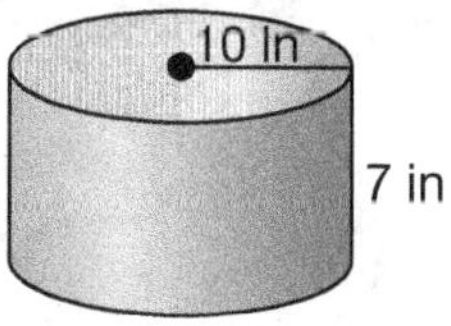

7. $V = \pi(\_\_)^2(\_\_)$

$V =$ _____ $\pi \text{ yd}^3$

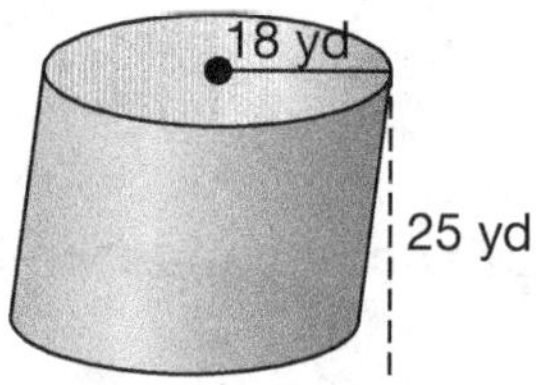

Find the volume of each cylinder in terms of π.

8.

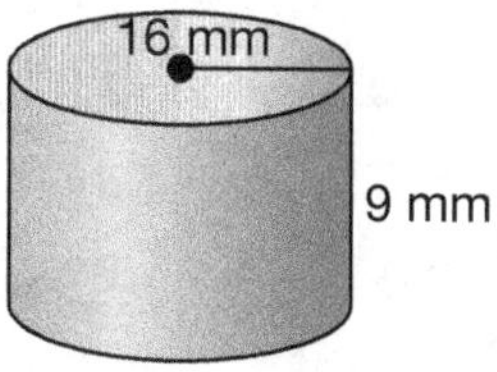

____

9.

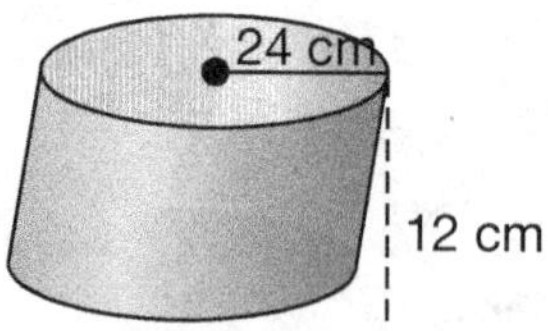

____

10.

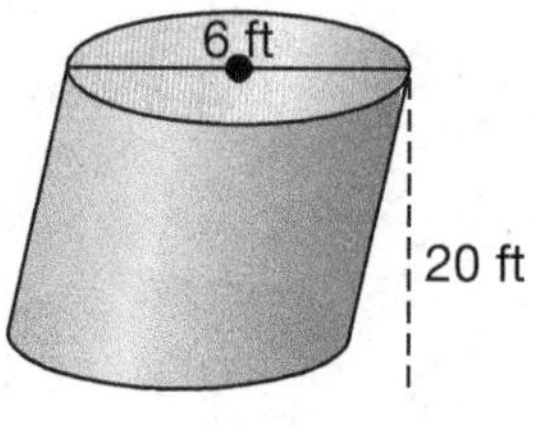

____

11.

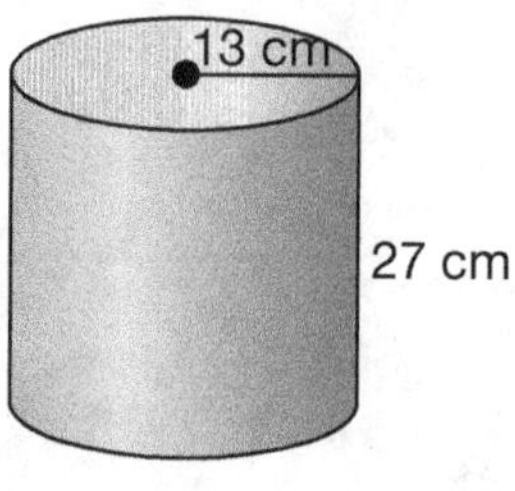

____

Name ______________________ Date ____________ Class ____________

Reteaching 63

Introduction to Vectors

You have learned how to find points on a coordinate plane. Now you will determine vectors and their magnitudes.

Find $|\vec{r}|$.

Step 1: Place M on the origin.

Step 2: N will be 4 units to the right and 5 units up, so the coordinates of the terminal point N are (4, 5).

Step 3: Use the distance formula to find the distance between $M(0, 0)$ and $N(4, 5)$.

$d = \sqrt{(x_2 - x_1)^2 + (y_2 - y_1)^2}$

$d = \sqrt{(4 - 0)^2 + (5 - 0)^2}$

$d = \sqrt{4^2 + 5^2} = \sqrt{41}$

Therefore, $|\vec{r}|$ *is* $\sqrt{41}$.

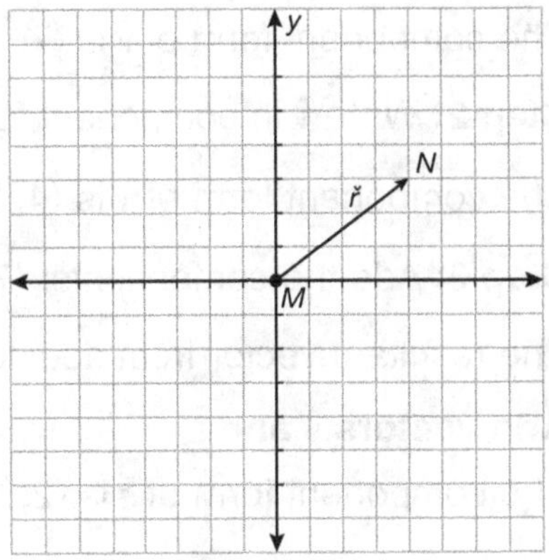

Practice

Complete the steps to find the magnitude of the vector.

1. Find $|\vec{p}|$.

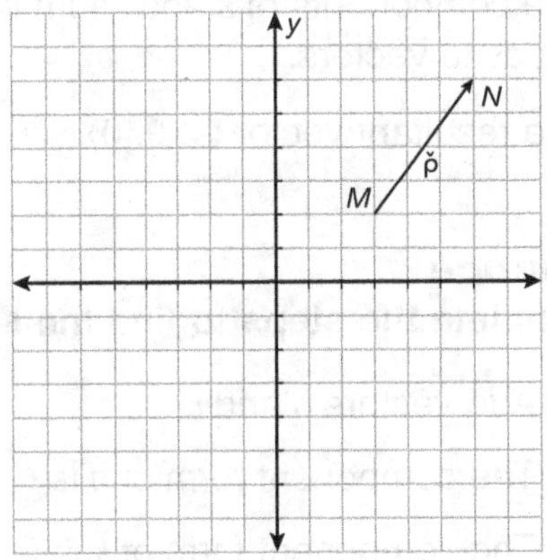

Place M on the origin. N is now at (3, _).

$d = \sqrt{(\_ - \_)^2 + (\_ - \_)^2}$

$d = \sqrt{(\_)^2 + (\_)^2} = 5$

So, $|\vec{p}| = 5$.

Find the magnitude of each vector in simplified radical form.

2. $|\vec{m}|$ ______

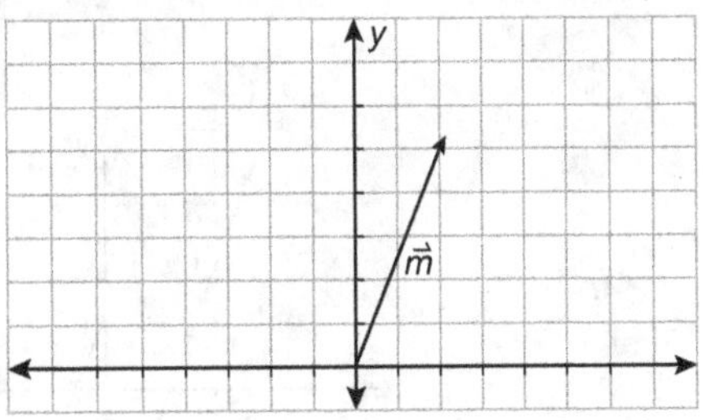

3. $|\vec{d}|$ ______

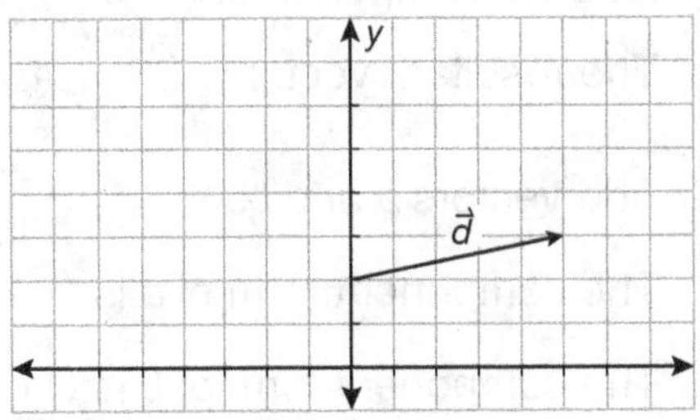

4. $|\vec{e}|$ ______

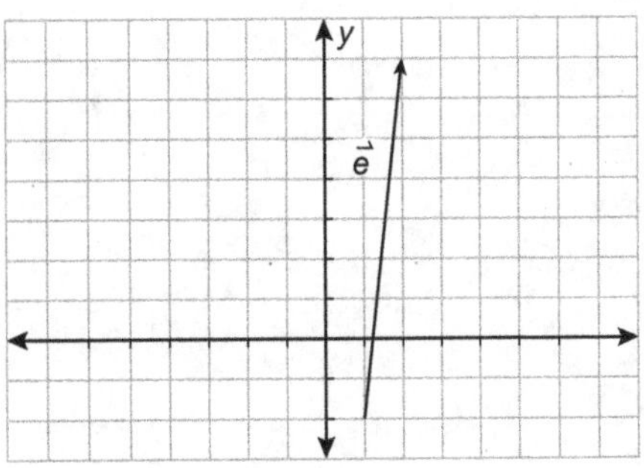

5. $|\vec{g}|$ ______

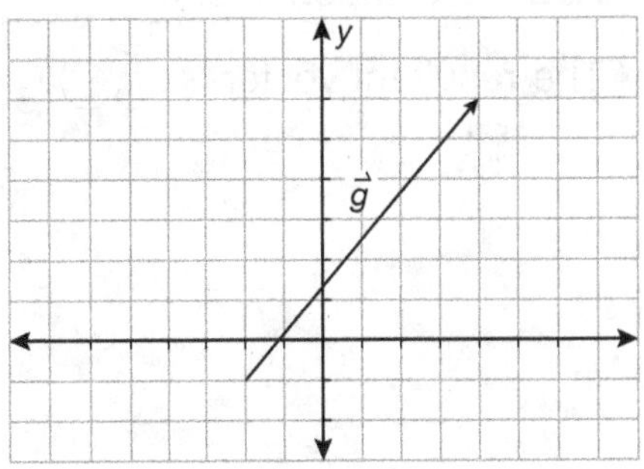

Add vectors $\vec{v}$ and $\vec{w}$.

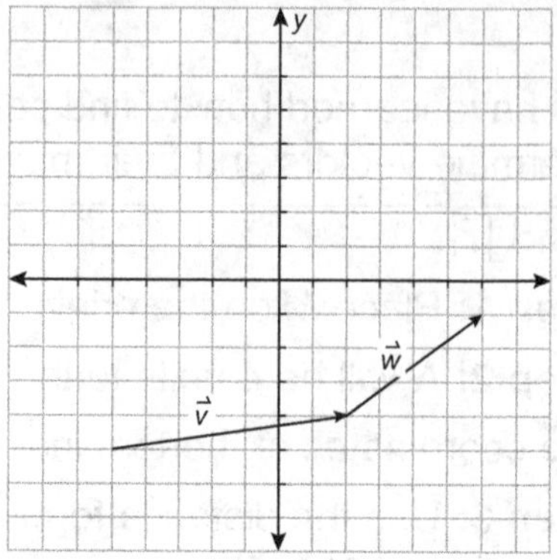

Step 1: Write $\vec{v}$ in component form.

The horizontal change is 7. The vertical change is 1.

The component form of $\vec{v}$ is $\langle 7, 1 \rangle$.

Step 2: Write $\vec{w}$ in component form.

The component form of $\vec{w}$ is $\langle 4, 3 \rangle$.

Step 3: Add the components: $\langle 7 + 4, 1 + 3 \rangle = \langle 11, 4 \rangle$.

The resultant vector from adding $\vec{v}$ and $\vec{w}$ is $\langle 11, 4 \rangle$.

Add vectors $\vec{s}$ and $\vec{t}$.

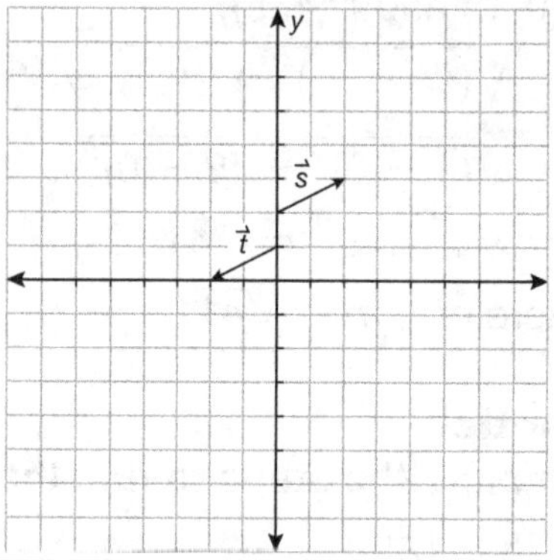

The component form of $\vec{s}$ is $\langle 2, 1 \rangle$.

The component form of $\vec{t}$ is $\langle -2, -1 \rangle$.

The components of $\vec{s}$ and $\vec{t}$ sum of 0 because they are opposite vectors.

The resultant vector is $\langle 0, 0 \rangle$.

Practice

Complete the steps to find the sum of the vectors.

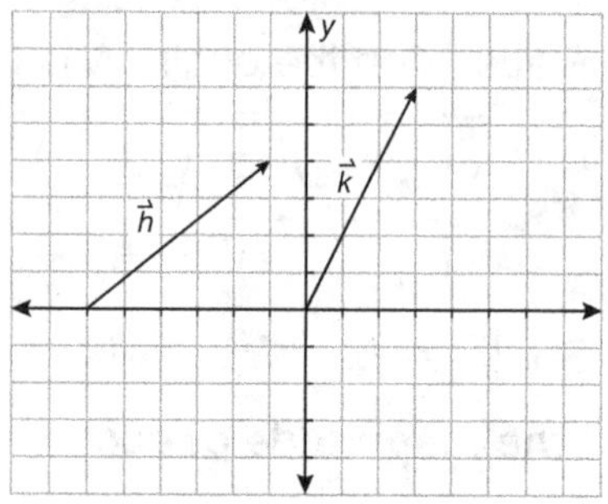

6. Add vectors $\vec{h}$ and $\vec{k}$.

The component form of $\vec{h}$ is ⟨_, _⟩.

The component form of $\vec{k}$ is ⟨_, _⟩.

Add the components: ⟨_ + _, _ + _⟩.

The resultant vector is ⟨_, ___⟩.

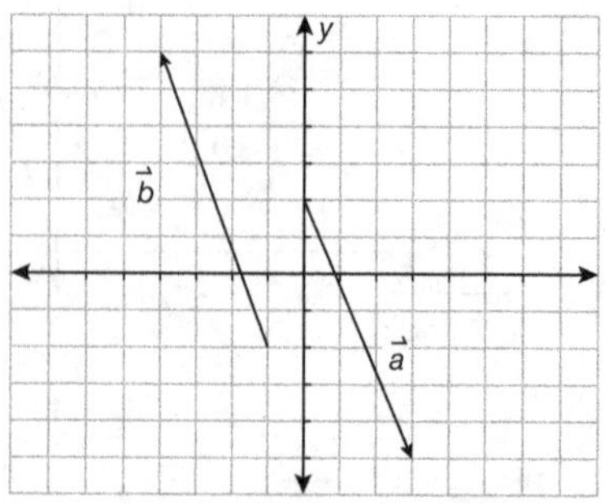

7. Add vectors $\vec{a}$ and $\vec{b}$.

The component form of $\vec{a}$ is ⟨_, ___⟩.

The component form of $\vec{b}$ is ⟨___, _⟩.

Add the components: ⟨_ + ___, ___ + _⟩.

The resultant vector is ⟨_, _⟩.

Name ______________________ Date __________ Class __________

Reteaching 64

Angles Interior to Circles

You have learned how to construct a tangent line on a circle. Now you will determine the measure of an angle formed by a chord and a line tangent to a circle.

The measure of an angle formed by a chord and a tangent is equal to half the measure of the arc that subtends it.

$m\angle ABC = \frac{1}{2} m\overset{\frown}{BEC}$

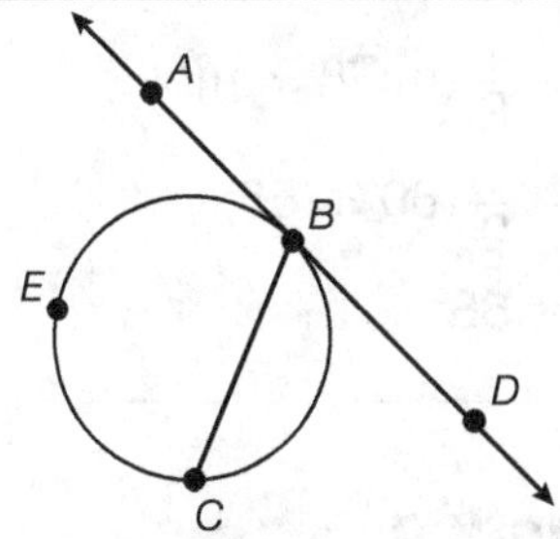

Given: $\overline{YZ}$ and $\overline{HI}$ are tangent lines.

Find the indicated measure.

a. $m\angle XYZ$

$m\angle XYZ = \frac{1}{2} m\overset{\frown}{XY}$

$m\angle XYZ = \frac{1}{2}(170°)$

$m\angle XYZ = 85°$

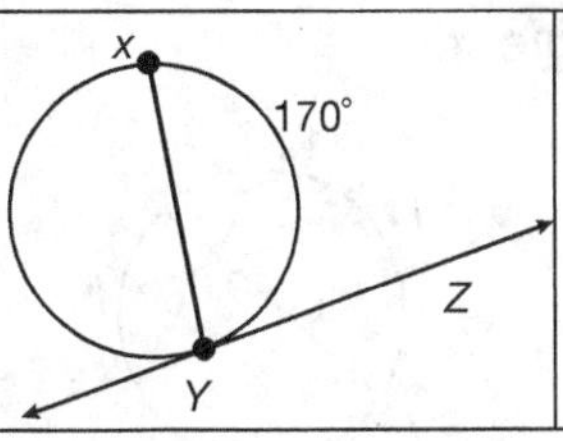

b. $m\overset{\frown}{GH}$

$m\angle GHI = \frac{1}{2} m\overset{\frown}{GH}$

$75° = \frac{1}{2} m\overset{\frown}{GH}$

$150° = m\overset{\frown}{GH}$

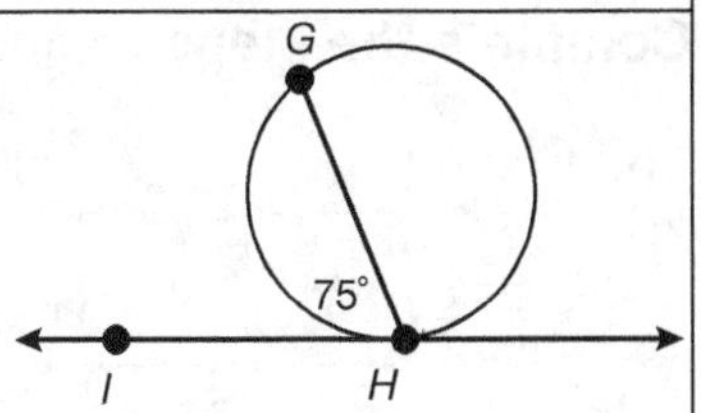

Practice

Complete the steps to find each measure.

1. $m\angle ABC$

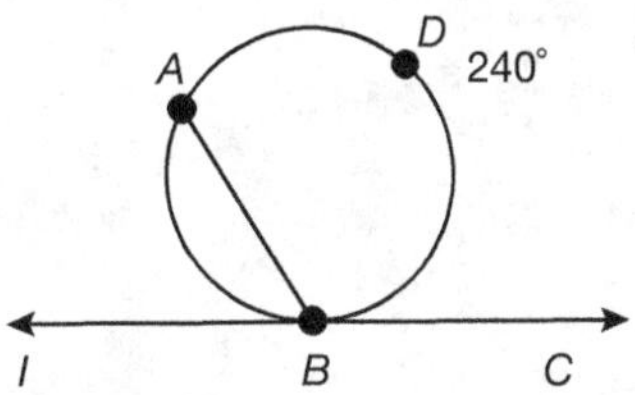

$m\angle ABC = \frac{1}{2} \overset{\frown}{ADB}$

$m\angle ABC = \frac{1}{2}(\_\_\_\_)$

$m\angle ABC = \_\_\_\_$

2. $m\overset{\frown}{PQR}$

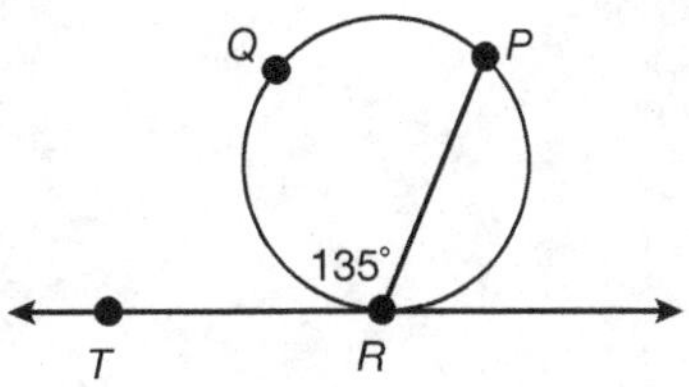

$m\angle PRT = \frac{1}{2} \_\_\_\_\_$

$\_\_\_\_ = \frac{1}{2} \_\_\_\_\_$

$\_\_\_\_ = m\overset{\frown}{PQR}$

Find the value of *x* in each figure.

3. ________

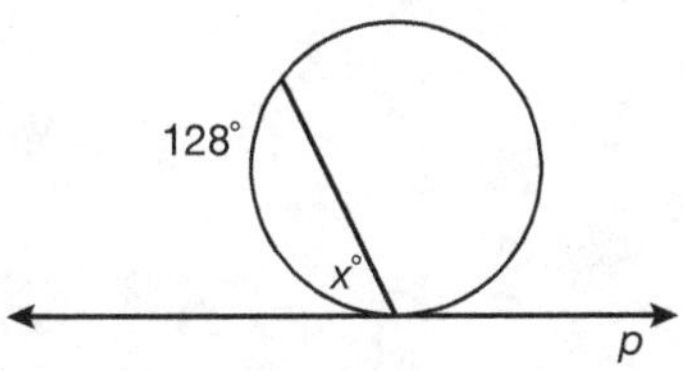

4. ________

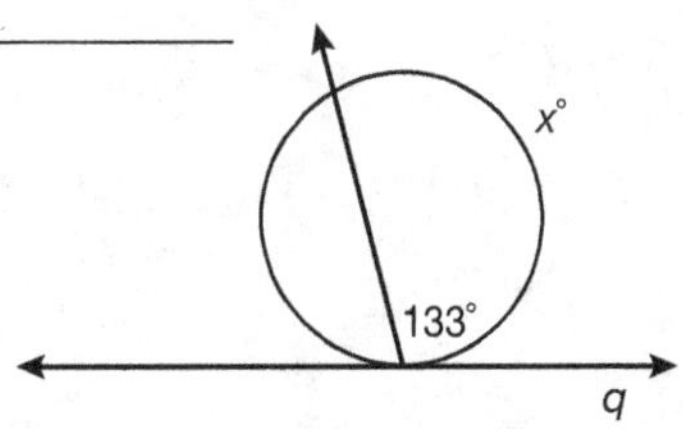

The measure of an angle formed by two chords intersecting in a circle is equal to half the sum of the intersected arcs.

$m\angle 1 = \frac{1}{2}(m\overset{\frown}{AD} + m\overset{\frown}{BC})$

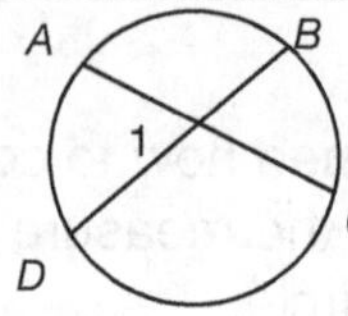

Find the measure of angle *x*.

$x = \frac{1}{2}(m\overset{\frown}{CF} + m\overset{\frown}{DE})$

$x = \frac{1}{2}(60 + 56)$

$x = 58°$

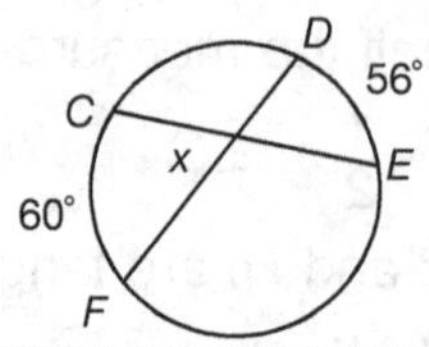

Practice

Complete the steps to find the measure of angle *x*.

5. $x = \frac{1}{2}(\_\_\_\_ + \_\_\_\_)$

$x = \frac{1}{2}(\_\_\_ + \_\_\_)$

$x = \_\_\_\_$

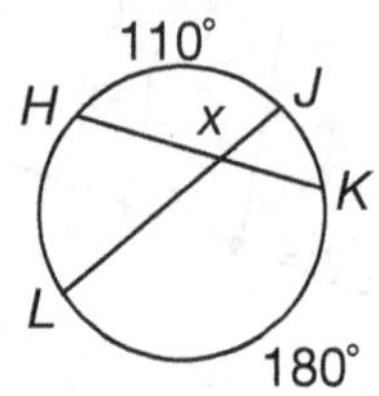

6. $x = \frac{1}{2}(\_\_\_\_ + \_\_\_\_)$

$x = \frac{1}{2}(\_\_ + \_\_)$

$x = \_\_\_$

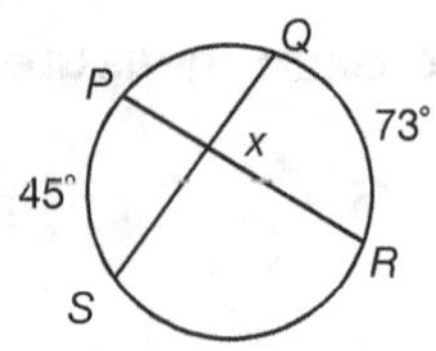

Find the measure of angle *x* for each circle.

7. ______

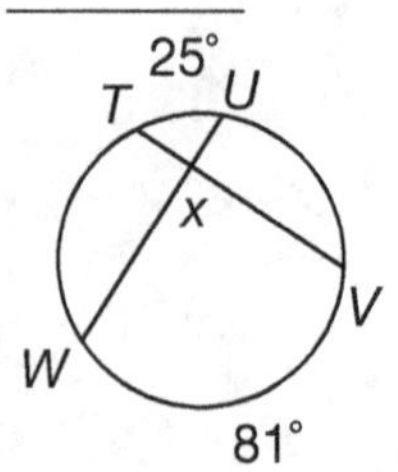

8. ______

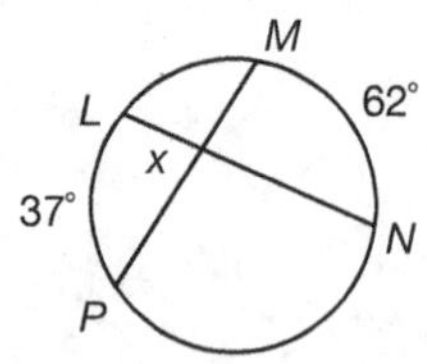

9. ______

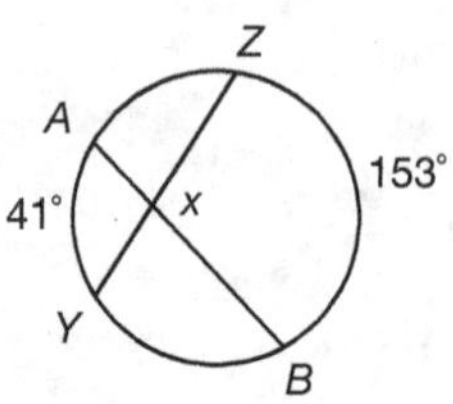

10. ______

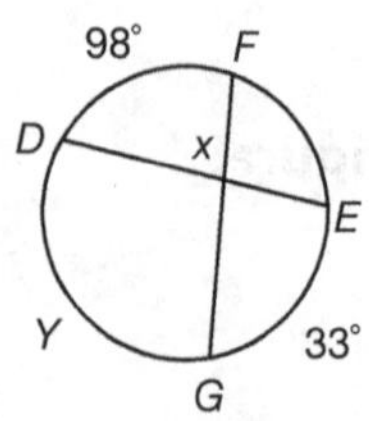

Name ______________________ Date __________ Class __________

Reteaching 65

Distinguishing Types of Parallelograms

A parallelogram is a rectangle if it has one 90° angle.

Example:

Is this parallelogram a rectangle?

Since the consecutive angles of a parallelogram are supplementary, we know that the measure of the two given expressions must add up to 180°.

Solve for x.

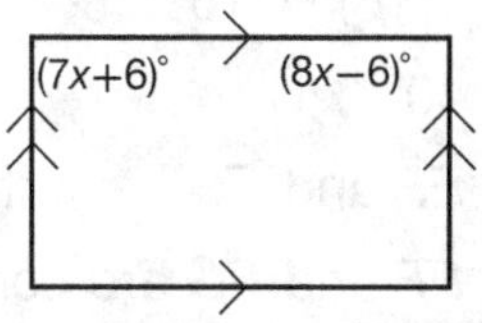

$(8x - 6) + (7x + 6) = 180$

$15x = 180$

$x = 12$

Substitute the value of x into one of the expressions to find the measure of the angle.

$8(12) - 6 = 90$

One of the angle measures of the parallelogram is 90°, so the parallelogram is a rectangle.

Practice

Complete the statements to distinguish the parallelogram.

1. Is parallelogram *MNOP* a rectangle?

   For a parallelogram to be a rectangle, it must have one ___ ______.

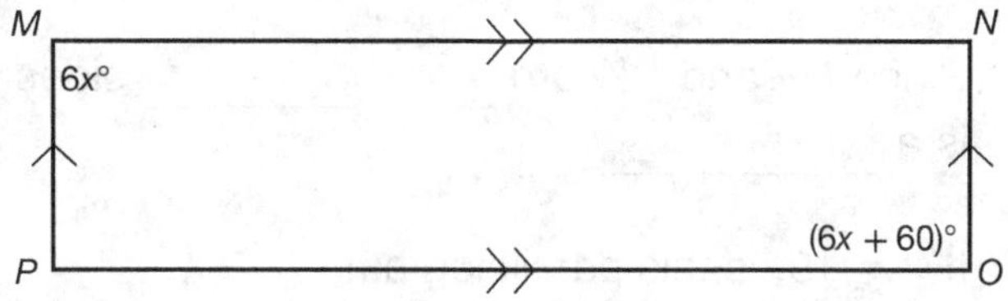

   __ = ________

   __ = __

   _ = __

   If $x = 15$, $\angle M = 6x = 6(\_\_) = \_\_$, and *MNOP* _ a rectangle.

2. Is this parallelogram a rectangle? __

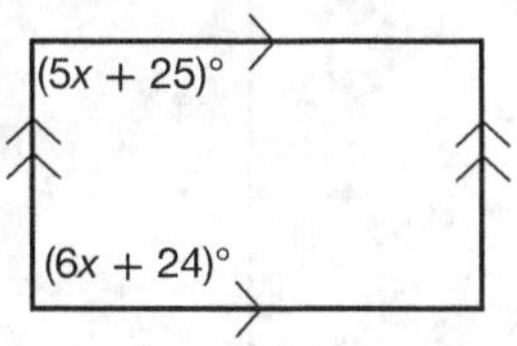

3. Is this parallelogram a rectangle? ___

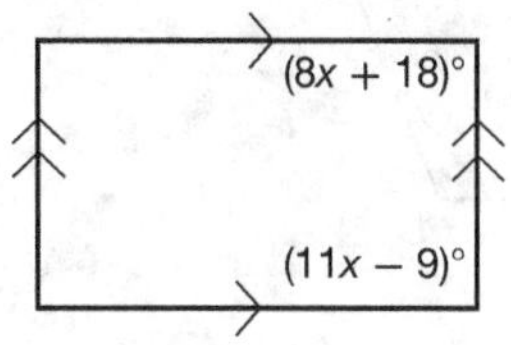

4. If $y = 10$, is this parallelogram a rectangle?

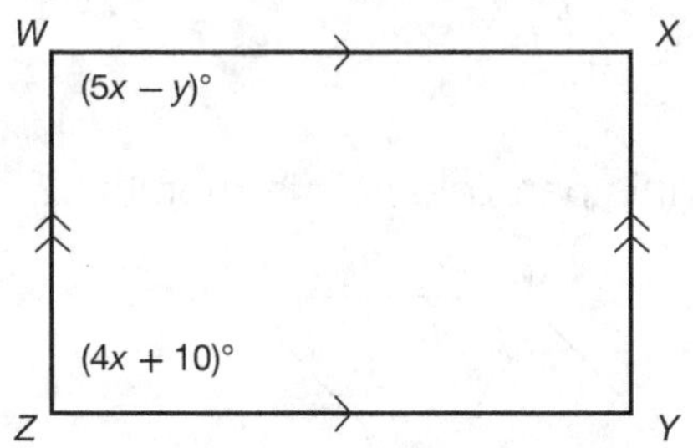

   ______

5. If $y = 3$, is this parallelogram a rectangle?

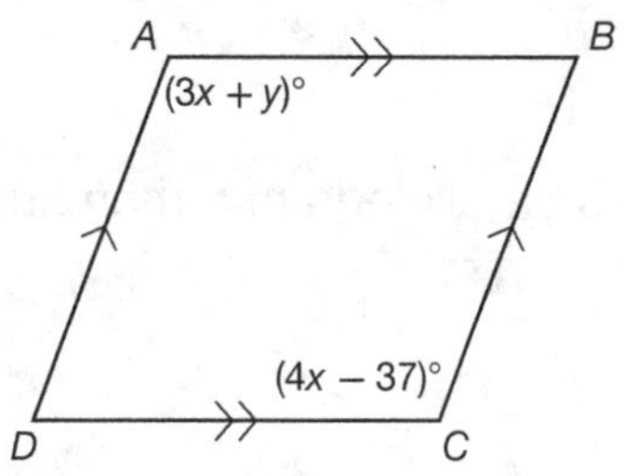

   _____

A parallelogram is a rhombus if it has two consecutive sides that are congruent.

Example:

If $x = 14$, is parallelogram $EFGH$ a rhombus?

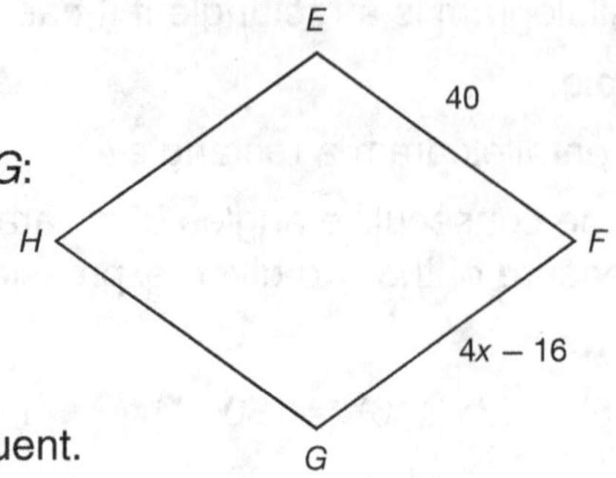

Since we know $x = 14$, we can find the length of side FG:

$FG = 4x - 16$

$FG = 4(16) - 16$

$FG = 40$

Since EF and FG have the same length, they are congruent.

Since EF and FG are consecutive sides of equal length, parallelogram $EFGH$ is a rhombus.

Practice

6. If $x = 2$, is the parallelogram a rhombus?

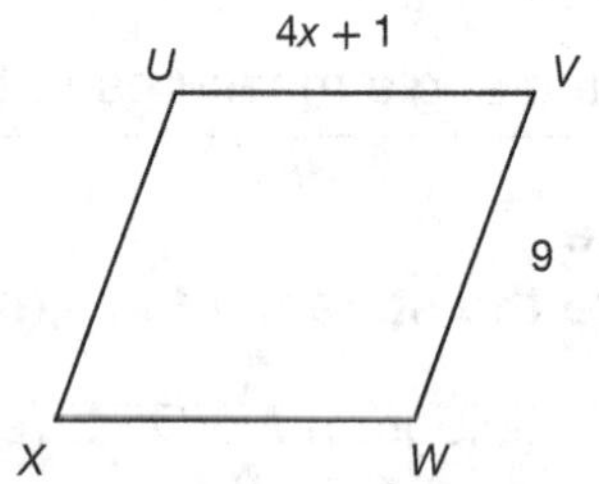

For a parallelogram to be a rhombus, ______________ sides must have equal length.

_______ = _

_________ = _

Since UV and VW are ______________ sides that have ________ lengths, the parallelogram is a ___________.

7. If $x = 19$, is this parallelogram a rhombus?

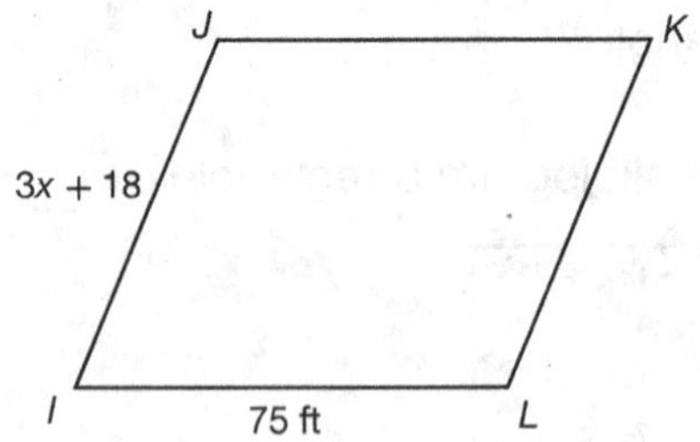

____

8. If $a = 3$, $b = 5$, $c = 4$, and $d = 4$, is this parallelogram a rhombus?

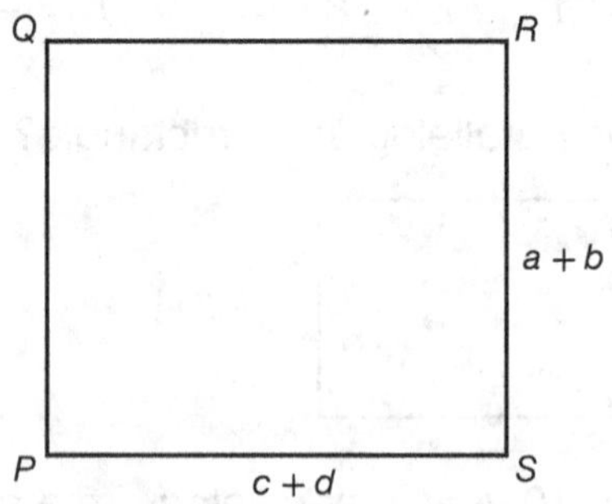

____

9. If $x \neq y$, is this parallelogram a rhombus? __

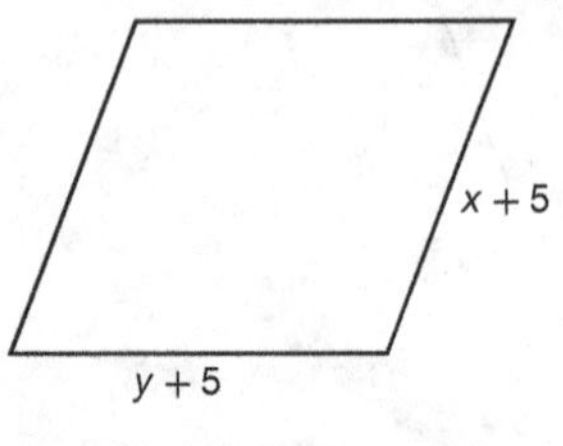

10. Is this parallelogram a rhombus? ___

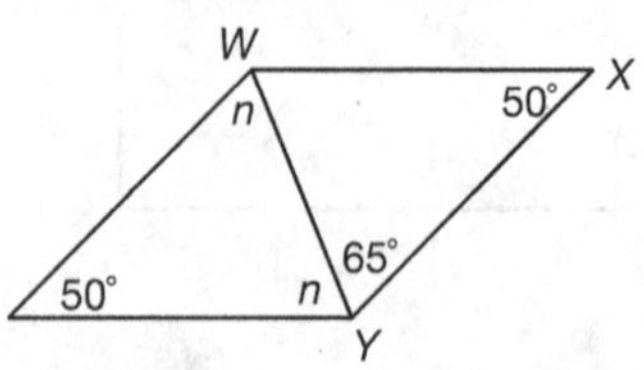

Name ______________________ Date __________ Class __________

Reteaching 66

Finding Perimeters and Areas of Regular Polygons

You have learned how to find perimeter and area of many types of figures. Now you will use formulas to find perimeters and areas of regular polygons.

The perimeter of a regular polygon can be found by multiplying the length of one of its sides by the number of sides it has.

Find the perimeter of the regular octagon.

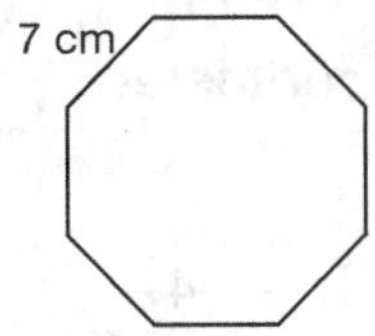

Because this is a regular octagon, the length of each side is the same as that of the labeled side, 7 centimeters. An octagon has 8 sides, so we can find the area by multiplying 8 by 7.

$P = ns$

$P = (8)(7 \text{ cm})$

$P = 56 \text{ cm}$

The perimeter of the regular octagon is 56 centimeters.

Practice

Complete the steps to find the perimeter of the regular polygon.

1. $P = (\_)(\_)$

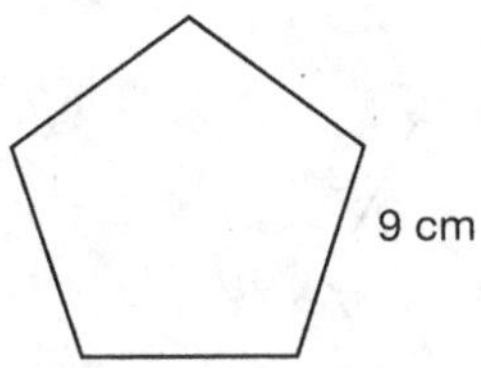

$P =$ ___ cm

2. $P = (\_)(\_\_\_)$

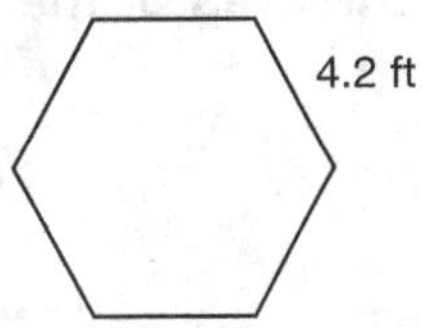

$P =$ _____ ft

Find the perimeter of each regular polygon.

3. _____

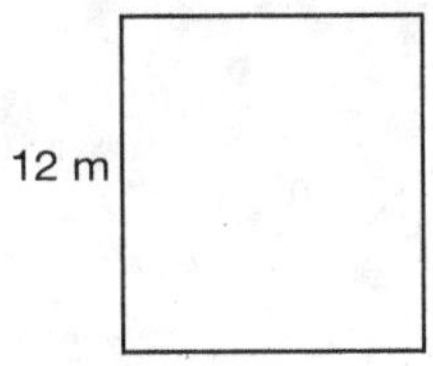

4. __________

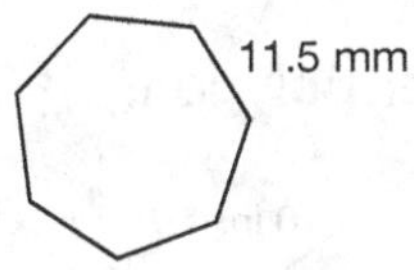

5. ________

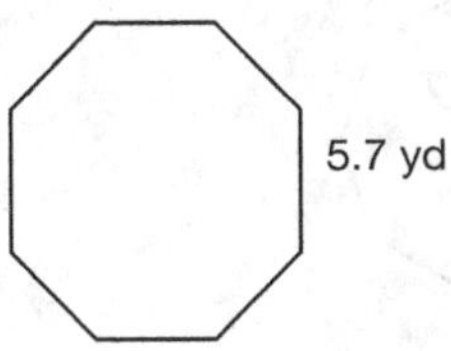

6. ________

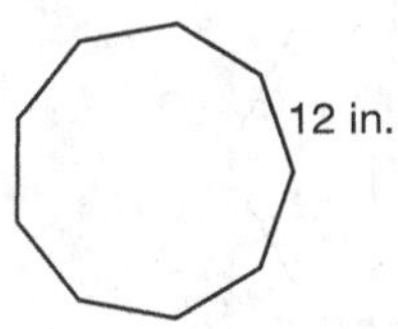

The area of a regular polygon is equal to its perimeter times the length of the apothem, divided by 2 ($A = \frac{1}{2}aP$).
Here we are not given the length of the apothem, so we have to find it. When the figure is a regular hexagon, we can form an equilateral triangle with the center of the hexagon as the top vertex. The apothem forms a 30°-60°-90° with the side and hypotenuse.

Use the Pythagorean Theorem to find that the length of the apothem is $7\sqrt{3}$ millimeters.

Find the perimeter:

$P = ns$

$P = (6)(14) = 84$

Find the area:

$A = \frac{1}{2}aP$

$A = \frac{1}{2}(7\sqrt{3})(84) = 294\sqrt{3}$

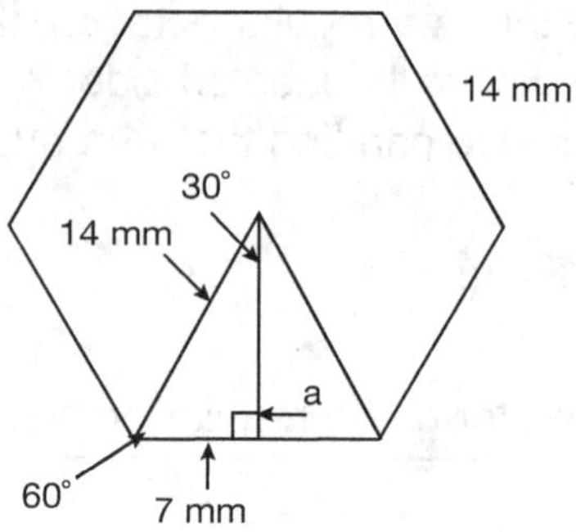

The area of this regular hexagon is $294\sqrt{3}$ mm^2.

Practice

Complete the steps to find the area of the regular polygon.

7. apothem = $4\sqrt{3}$ ft

$P = (\_)(\_)$

$A = \frac{1}{2}(\_\_\_\_)(\_\_) = \_\_\_\_\_\_$ ft^2

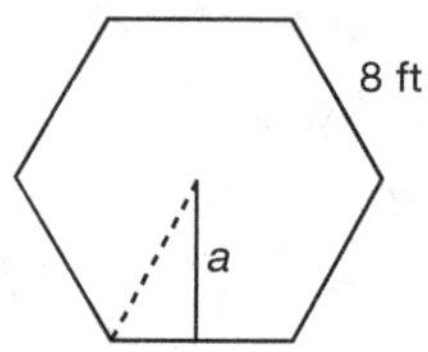

8. apothem = 11.5 cm

$P = (\_)(\_\_)$

$A = \frac{1}{2}(\_\_\_\_$ cm$)(__$ cm$) = ______$ cm^2

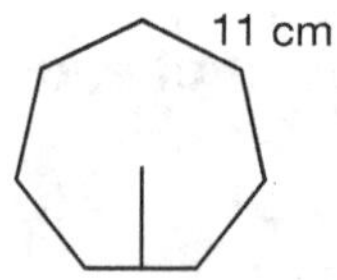

Find the area of each regular polygon.

9. ________

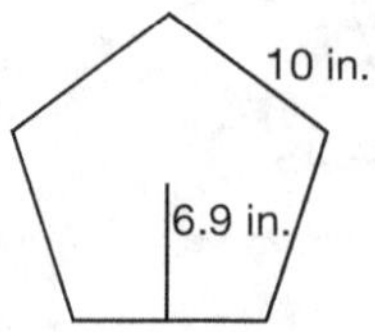

10. ________

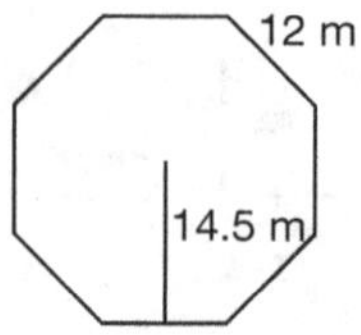

11. ________

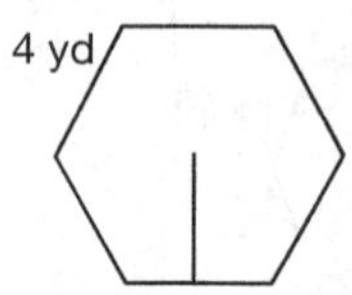

12. ________

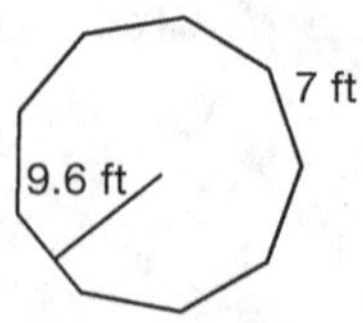

Name ______________________ Date __________ Class __________

Reteaching 67

Introduction to Transformations

You have worked with different figures in the coordinate plane. Now you will work with moving those figures around.

Transformations

An **isometry** is a transformation that does not change the size or the shape of a figure. There are different types of isometries.

A **translation** or a **slide** is a transformation in which all the points of a figure are moved the same distance in the same direction.

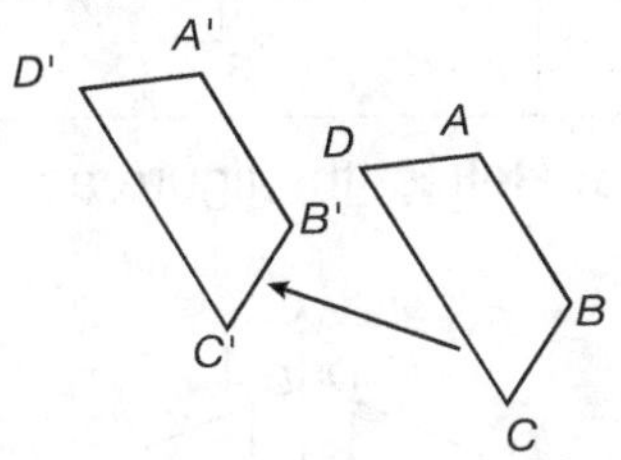

A **reflection** or a **flip** is a transformation that flips a figure across a line.

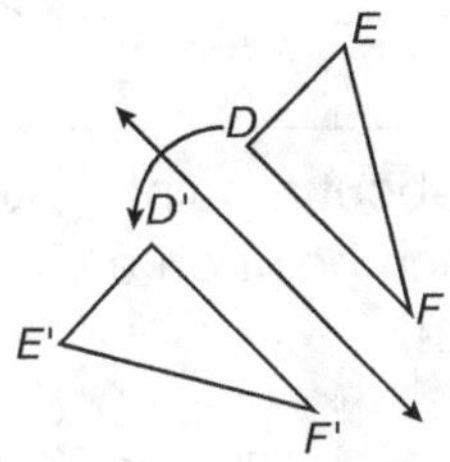

A **rotation** is a transformation that turns a figure around a fixed point, which is called the center of rotation.

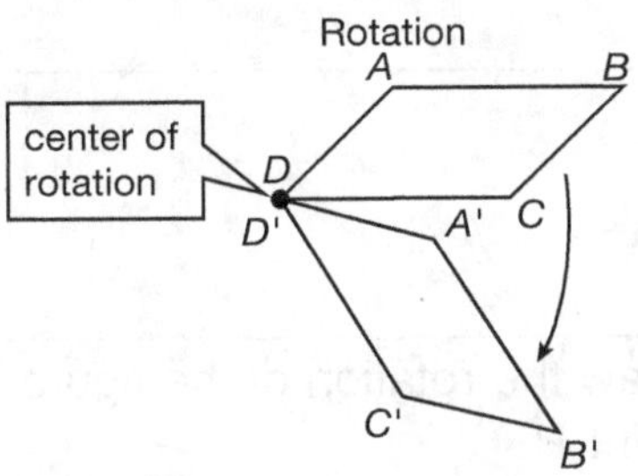

Example: Identify the type of transformation.

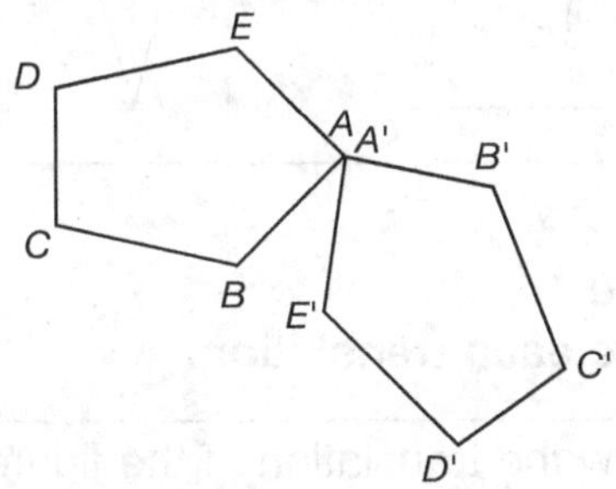

The figure is rotated about a point.

Practice

Complete the statement to determine the type of transformation.

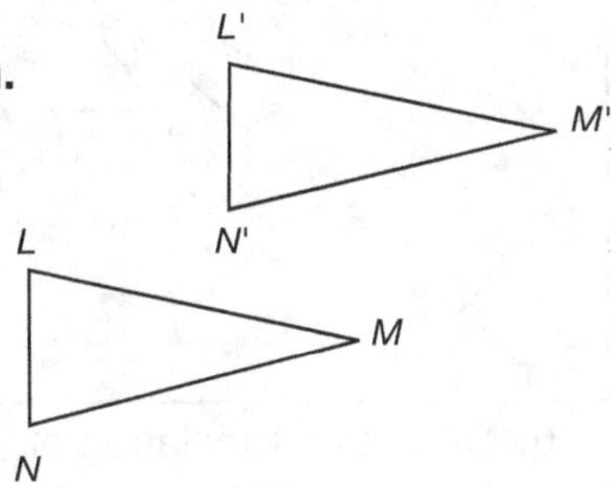

1. All the points of the figure are ______ the same distance in the same ______.

   The transformation is a __________ or a slide.

Determine the type of transformation.

2. __________

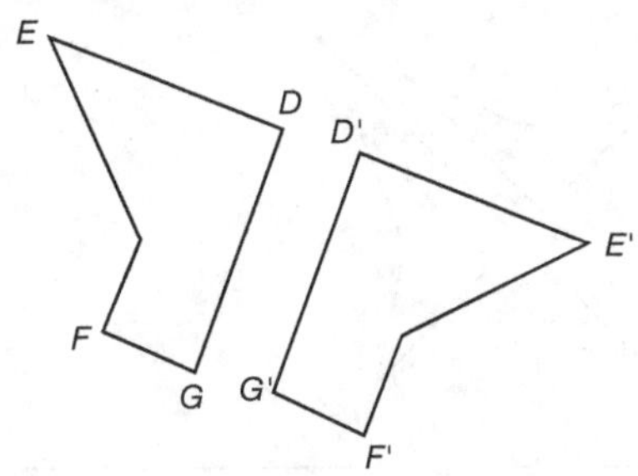

3. __________

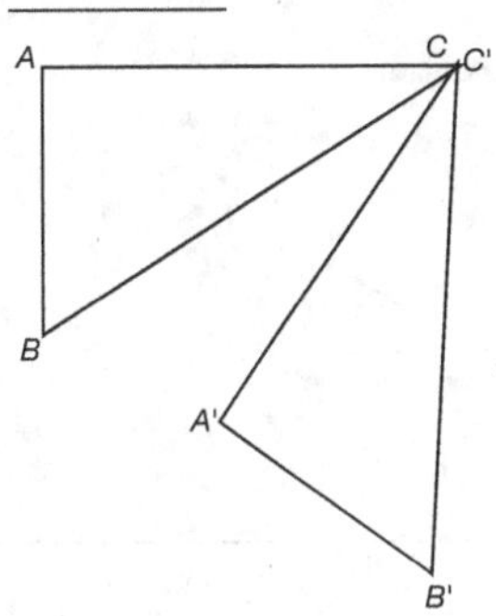

Performing Transformations

For a **translation** or a **slide**, move each point of the preimage the indicated distance and direction.

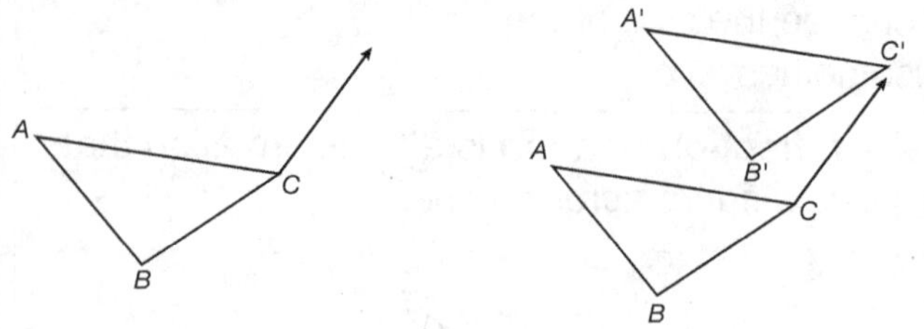

For a **reflection** or a **flip**, move each point across the line of reflection so that the point and its image are the same distance from the line.

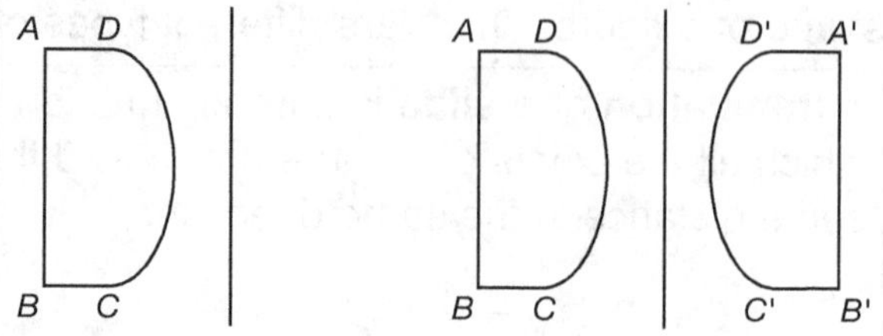

To **rotate** a figure about a point A, keep point A fixed and turn each point on a circular path around A, as shown.

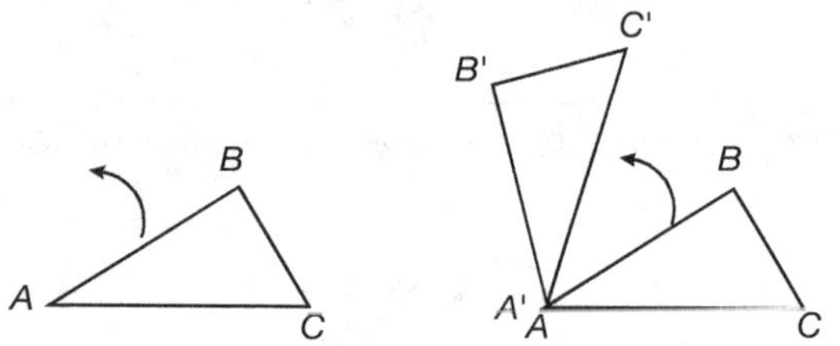

Example: Reflect the figure across the line.

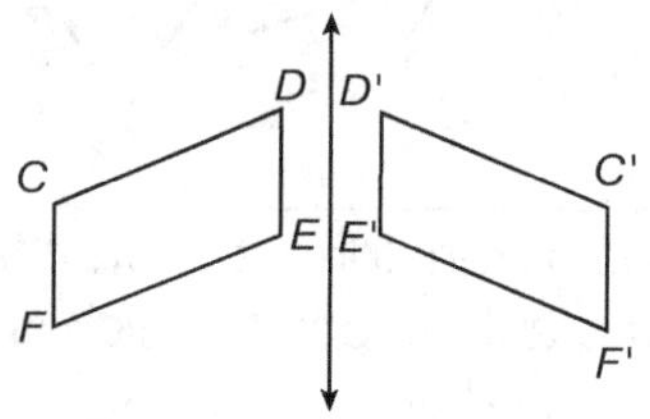

Practice

Complete each translation.

4. Draw the translation of the figure along the given path.

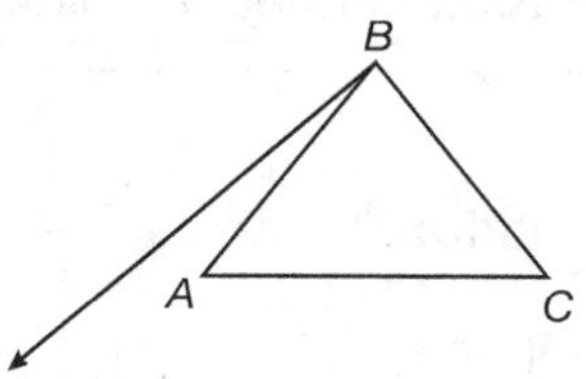

5. Draw the rotation of the figure about point P.

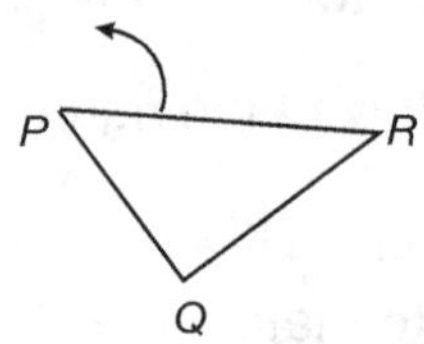

6. Draw the translation of the figure along the given path.

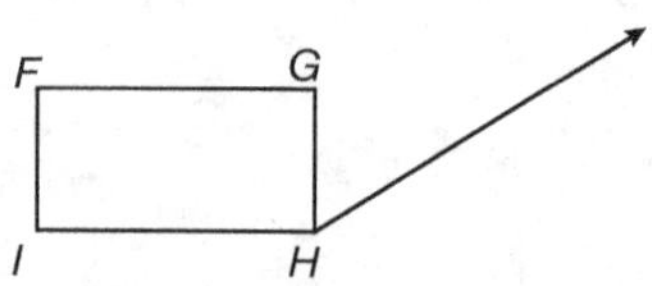

7. Draw the rotation of the figure across the line.

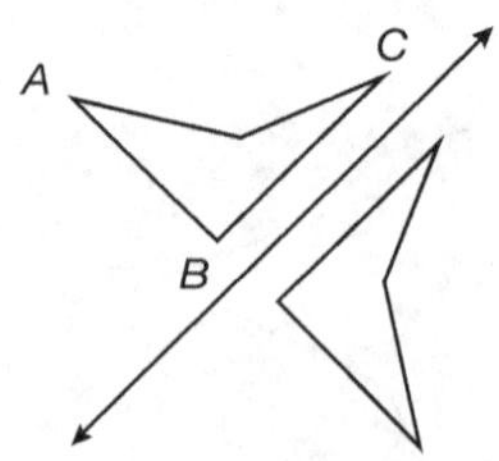

Name ______________________ Date __________ Class __________

Reteaching 68

Introduction to Trigonometric Functions

You have worked with right triangles. Now you will work with special relationships between the angles and sides in a right triangle.

Trigonometric Ratios

$\sin \angle = \dfrac{\text{leg opposite } \angle}{\text{hypotenuse}}$ $\cos \angle = \dfrac{\text{leg adjacent to } \angle}{\text{hypotenuse}}$ $\tan \angle = \dfrac{\text{leg opposite } \angle}{\text{leg adjacent to } \angle}$

Example:

Write each trigonometric ratio as a fraction.

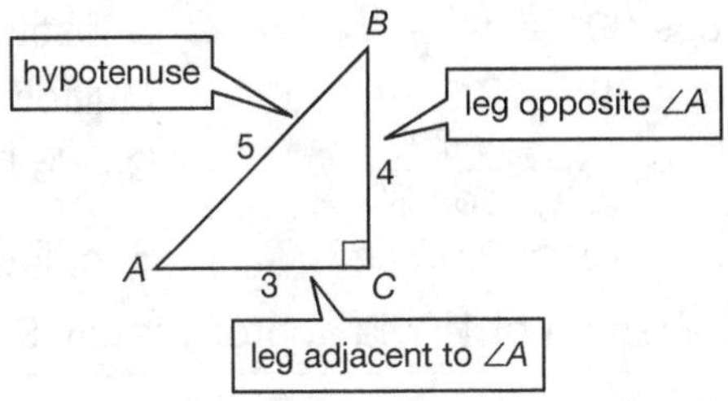

$\sin A = \dfrac{\text{leg opposite } \angle A}{\text{hypotenuse}} = \dfrac{4}{5}$

$\cos A = \dfrac{\text{leg adjacent to } \angle A}{\text{hypotenuse}} = \dfrac{3}{5}$

$\tan A = \dfrac{\text{leg opposite } \angle A}{\text{leg adjacent to } \angle A} = \dfrac{4}{3}$

Practice

Complete the statement to determine the given trigonometric ratio.

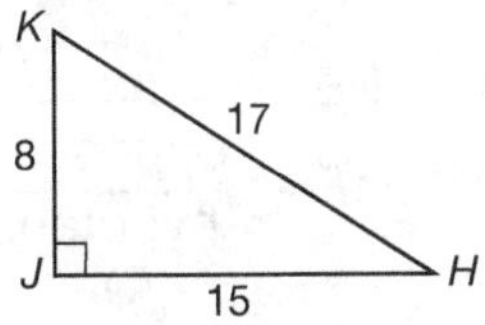

1. $\sin K = \dfrac{\text{leg opposite } \angle K}{\underline{\qquad\qquad}} = \dfrac{\underline{\quad}}{17}$

$\cos K = \dfrac{\text{leg adjacent to } \angle \underline{\quad}}{\text{hypotenuse}} = \dfrac{\underline{\quad}}{\underline{\quad}}$

$\tan K = \dfrac{\text{leg } \underline{\qquad\qquad} \angle K}{\text{leg } \underline{\qquad\qquad} \text{ to } \angle K} = \dfrac{\underline{\quad}}{8}$

Use the diagram from Exercise 1 to write each trigonometric ratio as a fraction.

2. $\sin H =$ ——

3. $\cos H =$ ——

4. $\tan H =$ ——

Write each trigonometric ratio as a fraction.

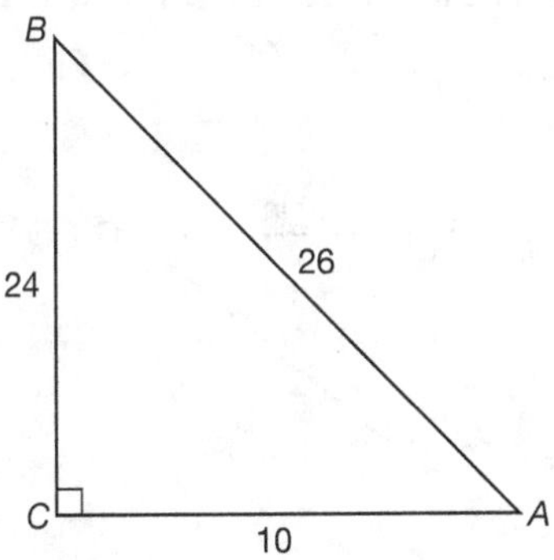

5. $\sin A =$ ——

6. $\cos A =$ ——

7. $\tan A =$ ——

8. $\sin B =$ ——

9. $\tan B =$ ——

10. $\cos B =$ ——

Trigonometric Ratios (continued)

You can use a calculator to find the value of trigonometric ratios:

$\cos 38^\circ \approx 0.7880107536$, or about 0.79.

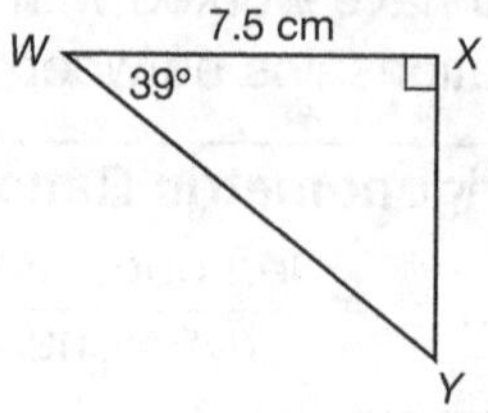

Example:

Find *WY*.

$\cos W = \dfrac{\text{adjacent leg}}{\text{hypotenuse}}$	Write a trigonometric ratio that involves *WY*.
$\cos 39^\circ = \dfrac{7.5 \text{ cm}}{x}$	Substitute the given values.
$x \cos 39^\circ = 7.5$	Multiply both sides by *x*.
$x = \dfrac{7.5}{\cos 39^\circ}$	Divide both sides by os 39°
$x \approx 9.65$ cm	Simplify.

The length of *WY* is approximately 9.65 centimeters.

Practice

Complete the steps to find *DE*.

11. $DE \approx$ _____

$\sin E = \dfrac{\text{opposite leg}}{\text{hypotenuse}}$

$\sin \_\_\_ = \dfrac{\_\_\_}{x}$

$x \sin 27^\circ =$ _____

$x = \dfrac{\_\_\_}{\sin 27^\circ}$

$x \approx$ _____ m

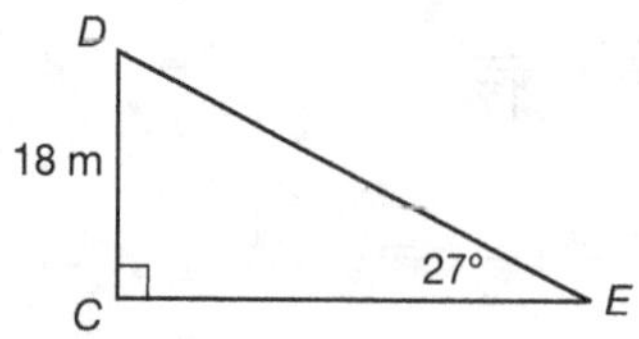

Use your calculator to find each trigonometric ratio. Round to the nearest hundredth.

12. $\sin 42^\circ$ ____ **13.** $\cos 89^\circ$ ____ **14.** $\tan 55^\circ$ ____

Find each length. Round to the nearest hundredth.

15. *FH* ______ **16.** *JK* ______ **17.** *US* ______

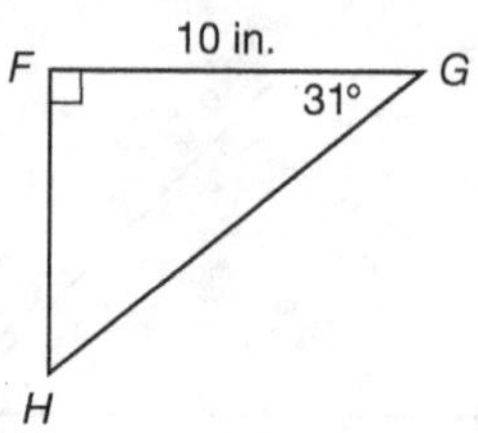

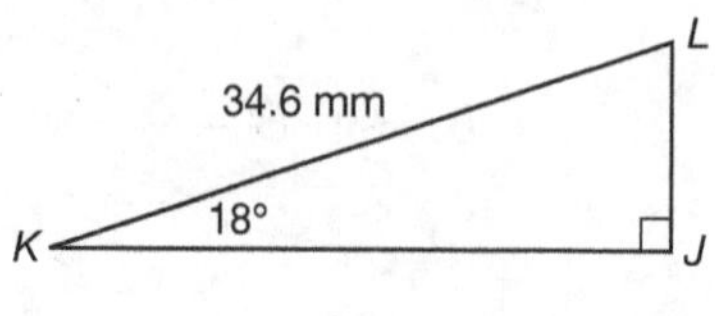

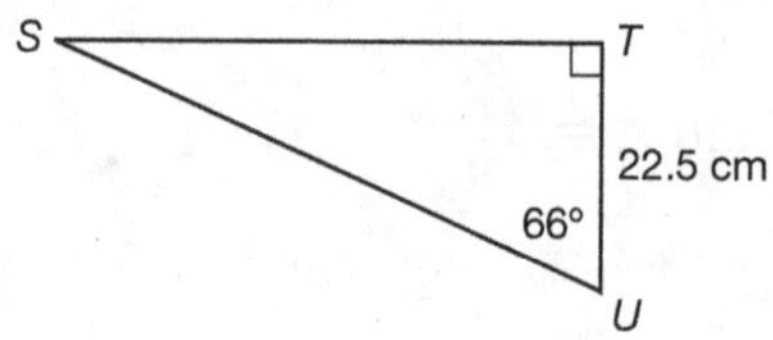

Name ______________________ Date __________ Class __________

Reteaching 69

Properties of Trapezoids and Kites

You have worked with triangles. Now you will find the lengths of sides of trapezoids and kites.

Trapezoid Midsegment Theorem

A **trapezoid** is a quadrilateral with one pair of parallel sides. The **midsegment** of a trapezoid is the segment whose endpoints are the midpoints of the legs.

The midsegment of a trapezoid is parallel to each base.

$\overline{AB} \parallel \overline{MN}$ and $\overline{AB} \parallel \overline{LP}$

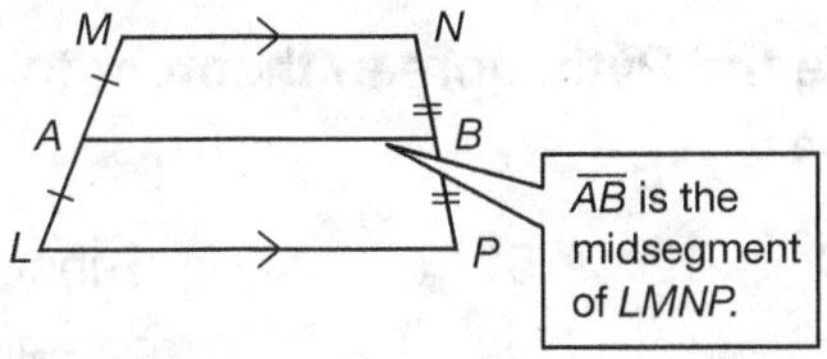

The length of the midsegment is one-half of the sum of the length of the bases.

$AB = \frac{1}{2}(MN + LP)$

Use the formula to find the length of the midsegment $\overline{KL}$.

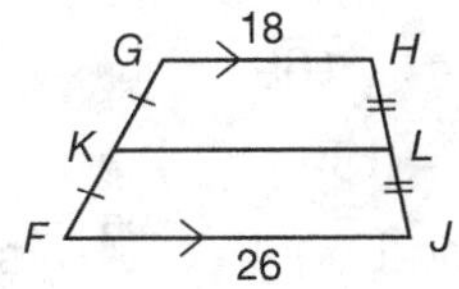

$KL = \frac{1}{2}(GH + FJ)$

$KL = \frac{1}{2}(18 + 26)$ Substitute.

$KL = \frac{1}{2}(44)$ Simplify.

$KL = 22$ Simplify.

Practice

Complete the steps to find each length.

1. *MN*

$MN = \frac{1}{2}(AB + CD)$

$MN = \frac{1}{2}(\_\_ + 17)$

$MN = \frac{1}{2}(\_\_)$

$MN = \_\_$

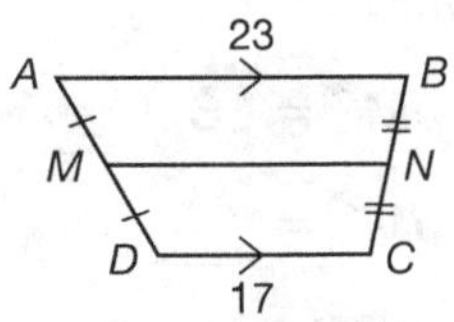

2. *FJ*

$XY = \frac{1}{2}(FJ + GH)$

$\_\_ = \frac{1}{2}(FJ + 48)$

$\_\_ = FJ + 48$

$FJ = \_\_$

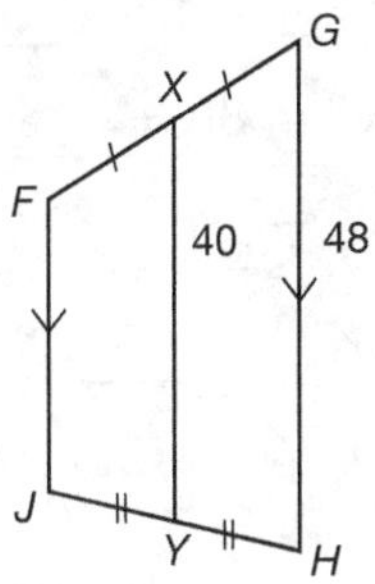

Find each length.

3. *KL*

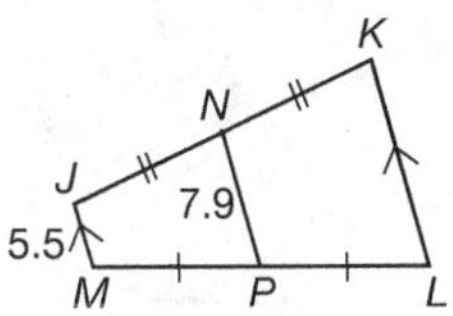

______

4. *TV*

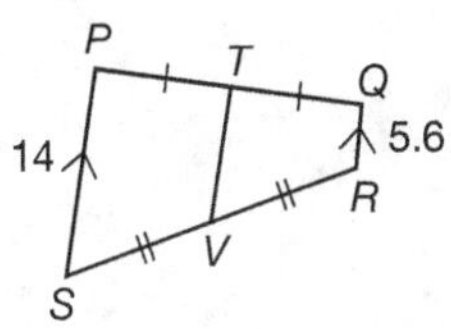

______

Diagonals of Kites

A **kite** is a quadrilateral with two pairs of congruent adjacent sides. The **diagonals** of a kite are perpendicular to each other.

$\overline{AC} \perp \overline{BD}$

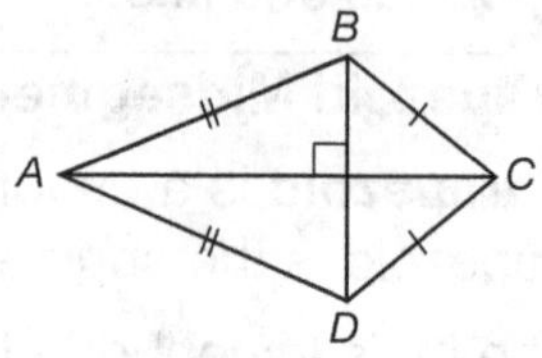

Because the diagonals of a kite are perpendicular to each other, you can use the Pythagorean theorem to find the length of each side.

Find the length of $\overline{AD}$.

Use the Pythagorean theorem to find the length of the side.

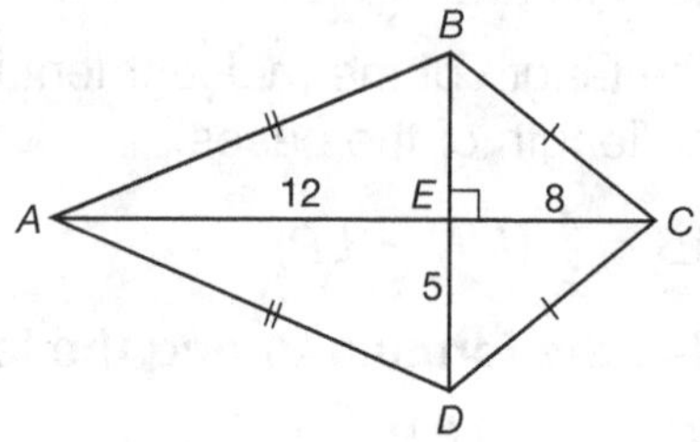

$AD^2 = AE^2 + ED^2$

$AD^2 = 12^2 + 5^2$ Substitute.

$AD^2 = 144 + 25$ Simplify.

$AD^2 = 169$ Simplify.

$AD = \sqrt{169}$ Solve.

$AD = 13$

So, the Length of $\overline{AD}$ is 13.

Practice

Complete the steps to find the lengths of $\overline{JK}$ and $\overline{JM}$. Round each length to the nearest tenth.

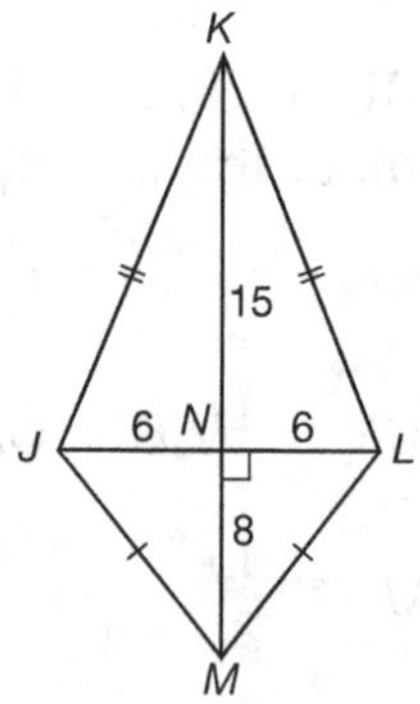

5. Use the Pythagorean theorem to find the length of each side.

$JK^2 = JN^2 + NK^2$ $JM^2 = JN^2 + NM^2$

$JK^2 = 6^2 + \underline{\quad}$ $JM^2 = 6^2 + \underline{\quad}$

$JK^2 = 36 + \underline{\quad}$ $JM^2 = 36 + \underline{\quad}$

$JK^2 = \underline{\quad}$ $JM^2 = \underline{\quad}$

$JK = \underline{\quad}$ $JM = \underline{\quad}$

$JK = \underline{\quad}$ $JM = \underline{\quad}$

Find each length. Round to the nearest tenth.

6. Find the length of $\overline{AD}$.

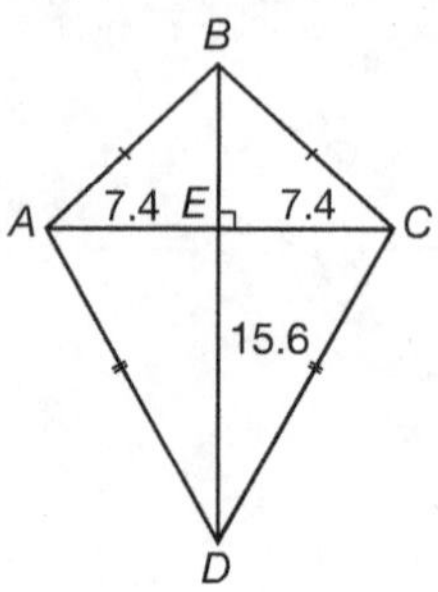

______

7. Find the length of $\overline{ED}$.

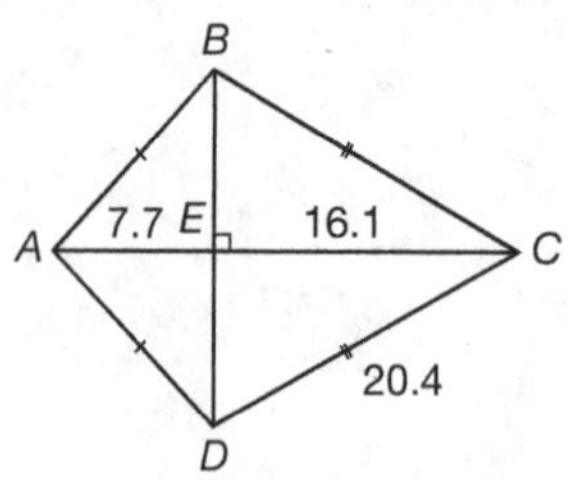

______

Name ______________________ Date __________ Class __________

Reteaching 70

Finding Surface Areas and Volumes of Pyramids

You have worked with surface area and volume of prisms. Now you will find the lateral area, the surface area, and the volume of pyramids.

Lateral and Surface Area of a Regular Pyramid

The **lateral area** of a regular pyramid with base perimeter P and **slant height** ℓ is $L = \frac{1}{2}P\ell$.

The **surface area** of a regular pyramid with lateral area L and base area B is $S = L + B$ or $S = \frac{1}{2}P\ell + B$.

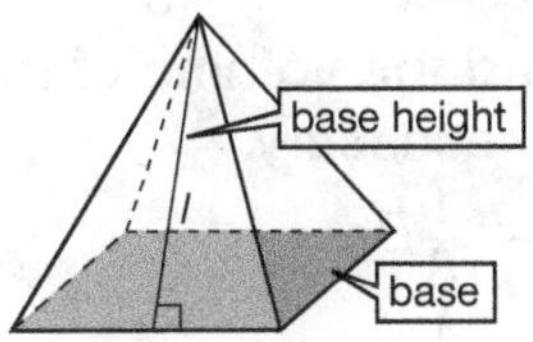

Find the lateral area and the surface area of the regular pyramid.

$L = \frac{1}{2}P\ell$	Lateral area formula
$L = \frac{1}{2}(32)(12)$	Substitute.
$L = 192 \text{ m}^2$	Simplify.
$S = L + B$	Surface area formula
$S = 192 + 6$	Substitute.
$S = 256 \text{ m}^2$	Simplify.

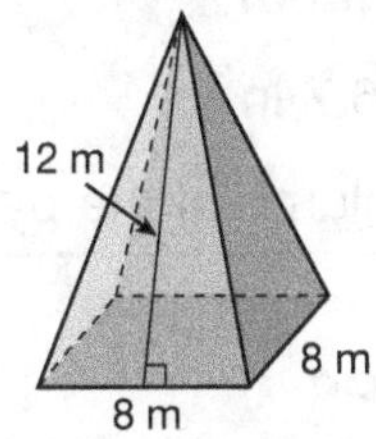

The lateral area is 192 square meters. The surface area is 256 square meters.

Practice

Complete the steps to find the lateral area and the surface area of the regular pyramid. Round your answer to the nearest tenth.

1. $L = \frac{1}{2}P\ell = \frac{1}{2}(\_\_\_)(12) = \_\_\_\_ \text{ ft}^2$

   $B = \frac{1}{2}\left(\frac{s\sqrt{3}}{2}\right)P = \frac{1}{2}\left(\frac{3\sqrt{3}}{2}\right)\_\_\_ = \_\_\_\_ \text{ ft}^2$

   $S = L + B = \_\_\_ + \_\_\_\_ = \_\_\_\_\_ \text{ ft}^2$

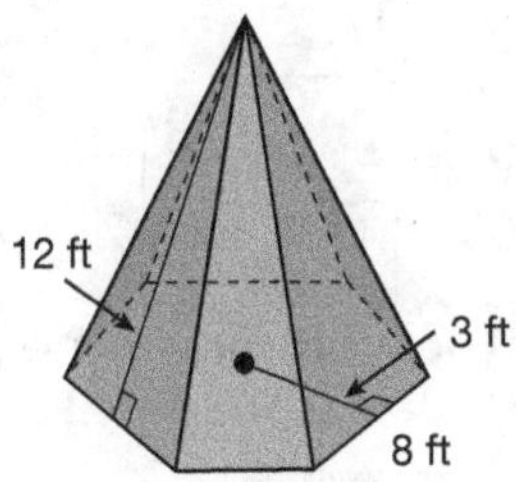

Find the lateral area and the surface area of the regular pyramid. Round to the nearest tenth.

2. $L =$ ______ in^2

   $S =$ ______ in^2

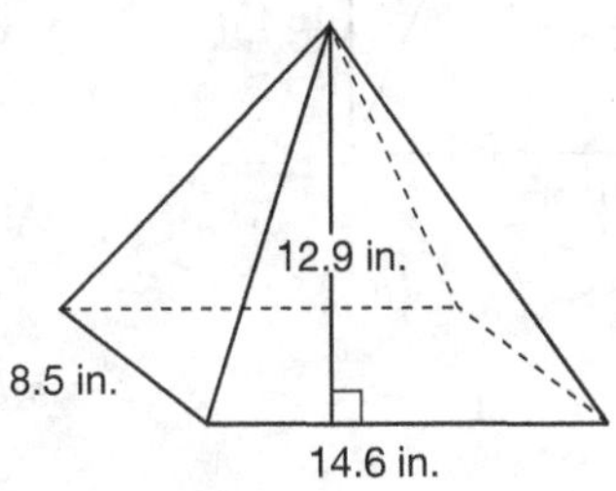

3. $L =$ ______ cm^2

   $S =$ ______ cm^2

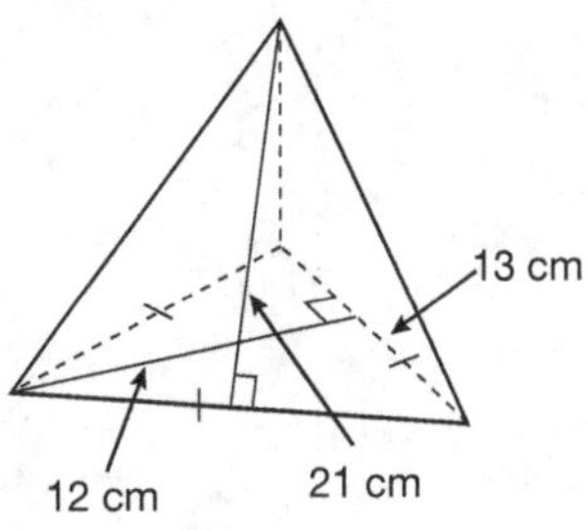

Volume of a Pyramid

The **altitude**, or **height**, of a pyramid is the perpendicular segment that measures from the vertex of the pyramid to the base of the pyramid. The **volume** of a pyramid with base area B and height h is $V = \frac{1}{3}Bh$.

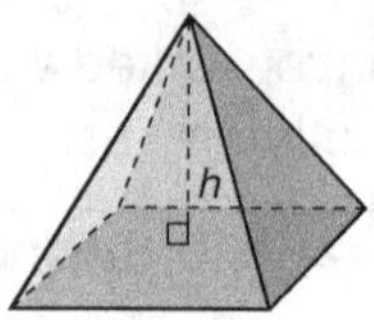

Find the volume of the pyramid. Round to the nearest tenth if necessary.

$V = \frac{1}{3}Bh$

$V = \frac{1}{3}(25)(8)$ Substitute.

$V = \frac{1}{3}(200)$ Simplify.

$V = 66.7 \text{ in}^3$ Solve.

The volume of the pyramid is 66.7 cubic inches.

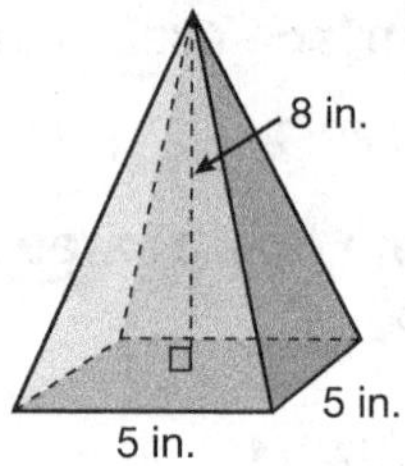

Practice

Complete the steps to find the volume of the pyramid. Round to the nearest tenth.

4. $V = \frac{1}{3}Bh$

$V = \frac{1}{3}(\_\_)\ (11)$

$V = \frac{1}{3}(\_\_)$

$V = \_\_\_\ \text{cm}^3$

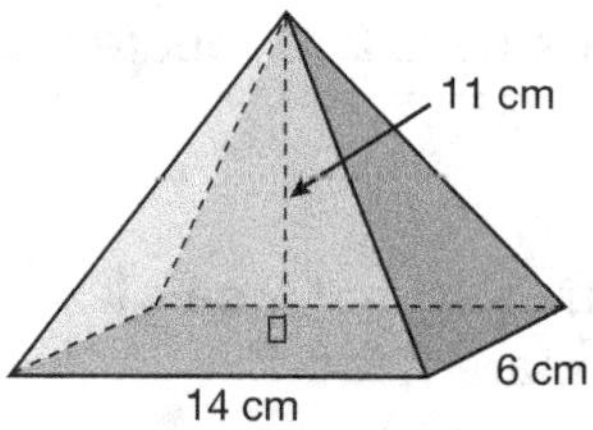

5. $V = \frac{1}{3}Bh$

$V = \frac{1}{3}(\_\_\_)\ (16.6)$

$V = \frac{1}{3}(\_\_\_)$

$V = \_\_\_\ \text{in}^3$

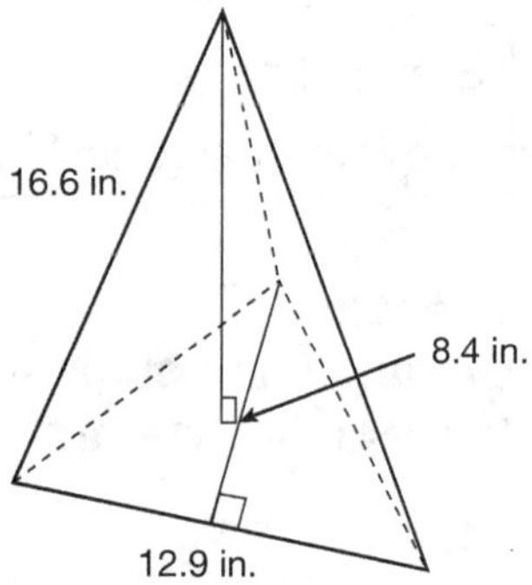

Find the volume of the pyramid. Round to the nearest tenth.

6. ______ ft^3

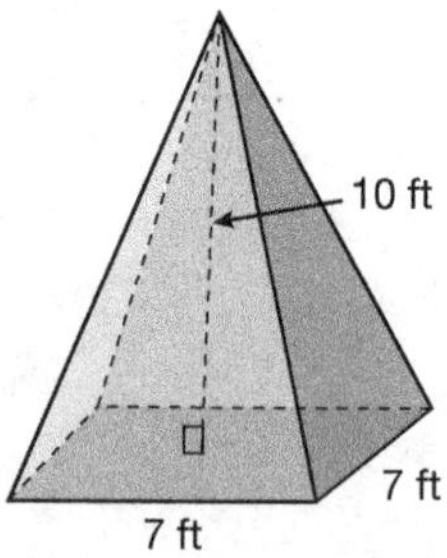

7. ______ m^3

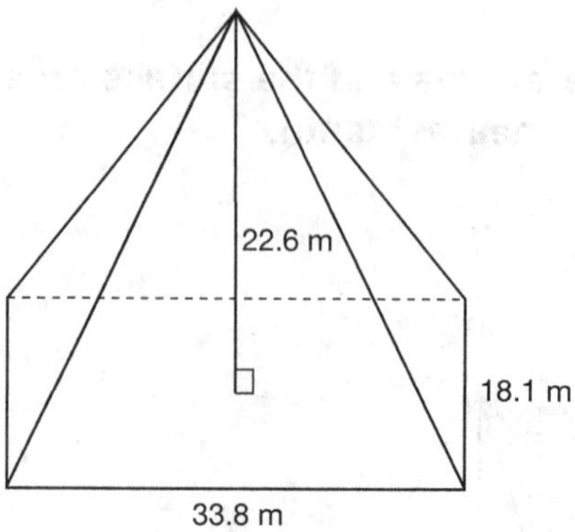

Name ______________________ Date __________ Class __________

Reteaching

INV 7

Trigonometric Functions

You have learned about right triangles and been introduced to trigonometric functions. Now you will observe and compare some of the values of the trigonometric functions as the angle measures vary between 0° and 90°.

The sine and cosine ratios for a right triangle are as follows.

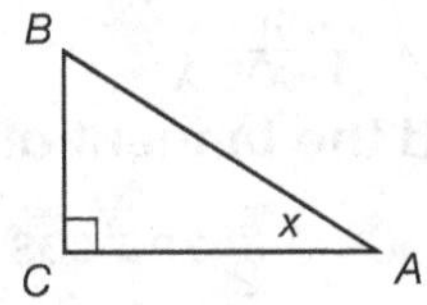

$\sin x = \dfrac{\text{opposite}}{\text{hypotenuse}} = \dfrac{BC}{AB}$

$\cos x = \dfrac{\text{adjacent}}{\text{hypotenuse}} = \dfrac{AC}{AB}$

Given a 30°–60°–90° triangle, find the sine and cosine for *x* if *x* = 30°. Then, find the sine and cosine if *x* = 60°.

For $x = 30°$:

$\sin 30° = \dfrac{1}{2}$

$\cos 30° = \dfrac{\sqrt{3}}{2}$

For $x = 60°$:

$\sin 60° = \dfrac{\sqrt{3}}{2}$

$\cos 60° = \dfrac{1}{2}$

What do you notice about the sine and cosine of these two angles? They switch values.

Practice

1. Sketch a 45°–45°–90° triangle with legs that are 1 unit long.

   Find sin 45° and cos 45°.

   ____

   What do you notice about the sine and cosine of a 45° angle?

   ____

Use your calculator for problems 2–4

2. Find the sine and cosine of 20°. $\sin 20° \approx$ _____; $\cos 20° \approx$ _____
3. Find the sine and cosine of 70°. $\sin 70° \approx$ _____; $\cos 70° \approx$ _____
4. What do you notice about the values of sine and cosine for 20° and 70°?

   ____

5. Both 20° and 70° angles and 30° and 60° angles are complementary. What is the relationship between the sine and the cosine of complementary angles?

   ____

6. In ΔXYZ, $\angle Z$ is a right angle, $m\angle X = 30°$, and $YZ = 5$. How are YZ and XZ related? How are YZ and XY related? Determine XZ and XY. Then, give exact values for sin 30° and cos 30°.

   ____

The tangent ratio for a right triangle is as follows.

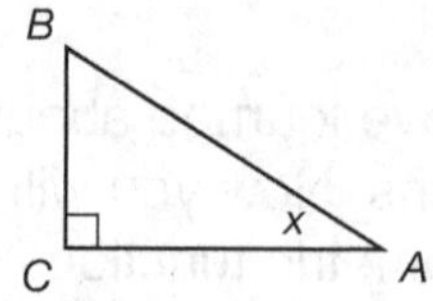

$\tan x = \frac{\text{opposite}}{\text{adjacent}} = \frac{BC}{AC}$

The tangent of an angle is also the quotient of the sine and the cosine of the angle.

$\tan x = \frac{\sin x}{\cos x}$

Find the tangent of 30° by dividing sin 30° by cos 30°.

$\sin 30° = \frac{1}{2}$ and $\cos 30° = \frac{\sqrt{3}}{2}$

$\tan 30° = \frac{\frac{1}{2}}{\frac{\sqrt{3}}{2}} = \left(\frac{1}{2}\right)\left(\frac{2}{\sqrt{3}}\right) = \frac{1}{\sqrt{3}}$

Do not forget to rationalize the denominator.

$\left(\frac{1}{\sqrt{3}}\right)\left(\frac{\sqrt{3}}{\sqrt{3}}\right) = \frac{\sqrt{3}}{3}$

Therefore, $\tan 30° = \frac{\sqrt{3}}{3}$.

Practice

Complete the steps to find the tan 60° by dividing the sin 60° by the cos 60°.

7. sin 60° = ____ and $\cos 60° = \frac{1}{2}$

tan 60° = ____ = (____)(2) = ____

Use the information you learned from problems 2 and 3 for problems 8 and 9.

8. Find the tangent of 20° by dividing sin 20° by cos 20°. ____________

9. Find the tangent of 70° by dividing sin 70° by cos 70°. ____________

10. In ΔPQR, $\angle R$ is a right angle, $m\angle P = 45°$, and $PR = 4$. Use the Converse of the Isosceles Triangle Theorem to determine how PR and RQ related. What is the length of RQ? Use the Pythagorean theorem to find PQ. Then, find and use sin 45° and cos 45° to find tan 45°.

____

____

Name ______________________ Date ___________ Class ___________

Reteaching 71
Translations

You have worked with plotting points on a coordinate grid. Now you will translate figures on a coordinate grid both in one dimension and in two dimensions.

Translation in One Dimension

A **translation,** also known as a **slide,** is an **isometry** (a transformation that does not change the size or shape of the object being translated).

A translation in one dimension changes either the *x*-values or the *y*-values, but not both. A translation left or right on the coordinate grid will change only the *x*-values of the original figure. A translation up or down on the coordinate grid will change only the *y*-values of the original figure.

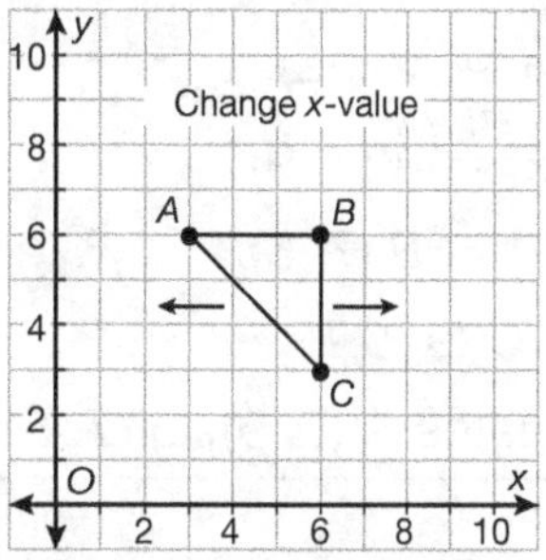

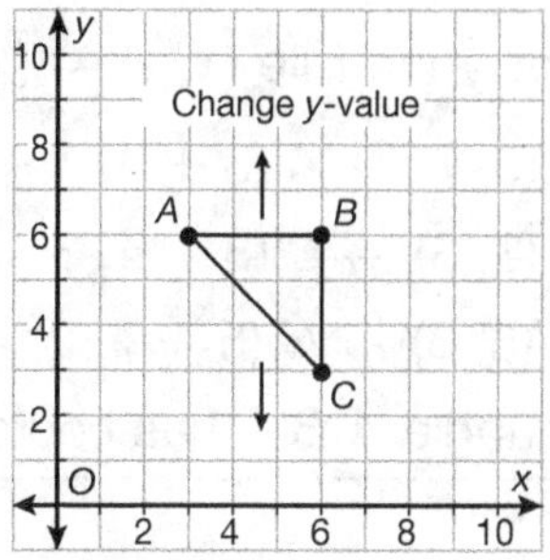

Find the coordinates of the vertices of the image of triangle *ABC* after a translation 4 units to the right. Show the preimage and image on the same coordinate grid.

Triangle *ABC* has coordinates *A* (2, 5), *B* (2, 2), and *C* (5, 2).

Apply the rule $(x, y) \rightarrow (x + 4, y)$.

$A\,(2, 5) \rightarrow A'(2 + 4, 5) = A'(6, 5)$

$B\,(2, 2) \rightarrow B'(2 + 4, 2) = B'(6, 2)$

$C\,(5, 2) \rightarrow C'(5 + 4, 2) = C'(9, 2)$

Triangle $A'B'C'$ has coordinates $A'(6, 5)$, $B'(6, 2)$, and $C'(9, 2)$.

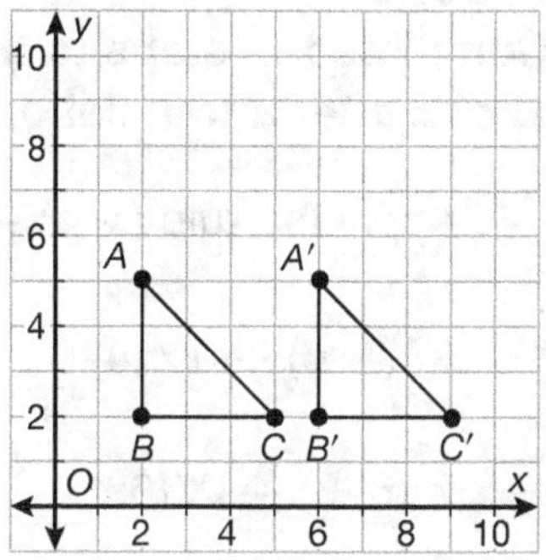

Practice

Complete the steps to find the coordinates of the image of square *JKLM* after a translation of 3 units down.

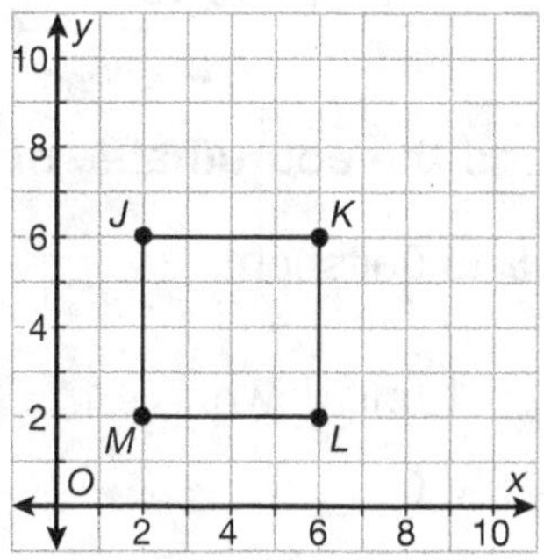

1. Apply the rule $(x, y) \rightarrow$ (___ , ______).

   $J\,(2, 6) \rightarrow J'(2, 6$ ___ $) = J'(2,$ ___ $)$

   $K\,(6, 6) \rightarrow K'(6, 6$ ___ $) = K'(6,$ _ $)$

   $L\,(6, 2) \rightarrow L'(6, 2$ ___ $) = L'(6,$ ___ $)$

   $M\,(2, 2) \rightarrow M'(2, 2$ ___ $) = M'(2,$ ___ $)$

Find the coordinates of each image after each indicated translation.

2. 5 units left

   F'(____ , _)

   G'(____ , _)

   H'(____ , ____)

   J'(____ , ____)

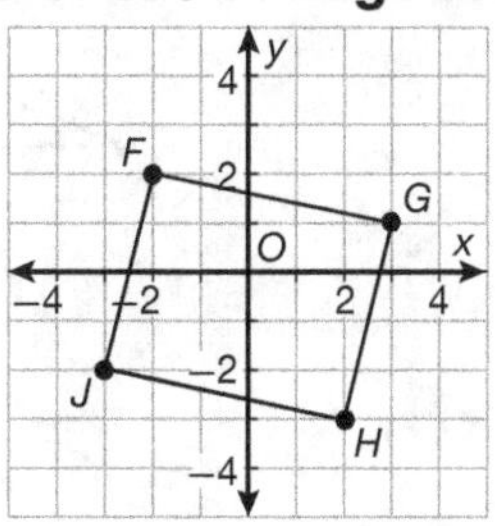

3. 6 units up

   X'(_ , ___)

   Y'(_ , ___)

   Z'(____ , _)

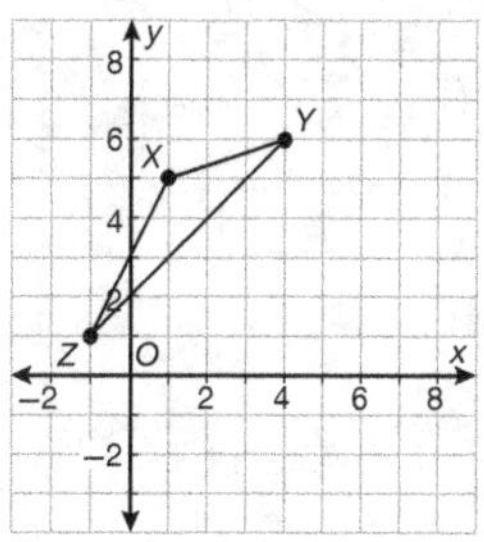

Translation in Two Dimensions

A translation in two dimensions causes both the x-values and the y-values to change. For example, a translation of 3 units right and 2 units down can be written using the rule $(x, y) \rightarrow (x + 3, y - 2)$.

Find the coordinates of the vertices of the image of triangle *ABC* after a translation of 4 units right and 2 units up. Show both the preimage and the image on the same coordinate grid.

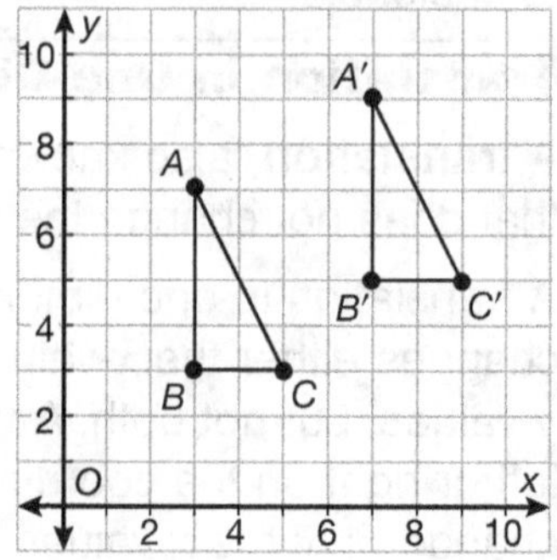

Triangle ABC has coordinates $A\,(3, 7)$, $B\,(3, 3)$, and $C\,(5, 3)$.

Apply the rule $(x, y) \rightarrow (x + 4, y + 2)$.

$A\,(3, 7) \rightarrow A'(3 + 4, 7 + 2) = A'(7, 9)$

$B\,(3, 3) \rightarrow B'(3 + 4, 3 + 2) = B'(7, 5)$

$C\,(5, 3) \rightarrow C'(5 + 4, 2) = C'(9, 5)$

Triangle $A'B'C'$ has coordinates $A'(7, 9)$, $B'(7, 5)$, and $C'(9, 5)$

Practice

Complete the steps to find the coordinates of the image of rectangle *WXYZ* after a translation of 2 units left and 3 units up.

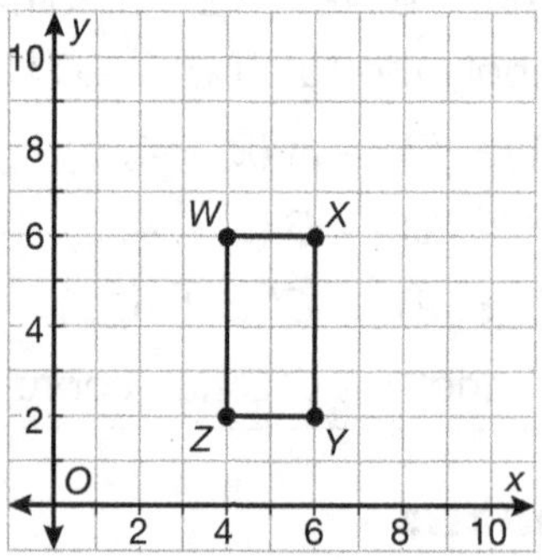

4. Apply the rule $(x, y) \rightarrow$ (______ , ______).

$W\,(4, 6) \rightarrow W'(4,$ ____ , $6 +$ ____) $= W'($_ , _$)$

$X\,(6, 6) \rightarrow X'(6,$ ____ , $6 +$ ____) $= X'($_ , _$)$

$Y\,(6, 2) \rightarrow Y'(6,$ ____ , $2 +$ ____) $= Y'($_ , _$)$

$Z\,(4, 2) \rightarrow Z'(4,$ ____ , $2 +$ ____) $= Z'($_ , _$)$

Find the coordinates of each image after each indicated translation.

5. 3 unitsright,

1 unit down

$J'($ _ , _ $)$

$K'($ _ , _ $)$

$L'($ _ , _ $)$

$M'($ _ , ____ $)$

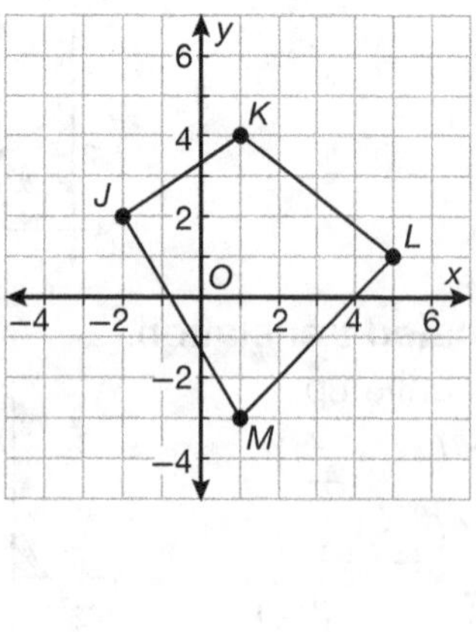

6. 5 units left,

2 units up

$D'($ ______ , ______ $)$

$E'($ ______ , ______ $)$

$F'($ ______ , ______ $)$

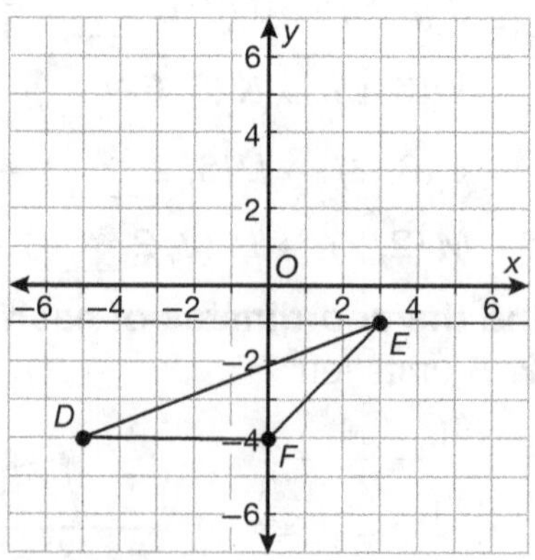

Name ______________________ Date __________ Class __________

Reteaching 72

Tangents and Circles, Part 2

You have worked with tangents and circles. Now you will solve problems involving common tangents.

Common Tangents: Internal

A **common tangent** is a tangent that is common to two circles. A common tangent is **internally tangent** if it intersects the segment that joins the centers of two circles. Two segments tangent from the same exterior point are congruent.

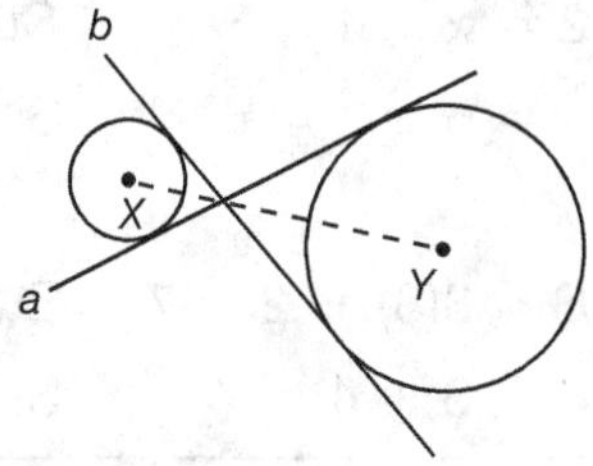

Line segments $\overline{KL}$ and $\overline{PQ}$ are common tangents to the circles. Find the lengths of $\overline{KL}$ and $\overline{PQ}$.

Step 1: Solve for *x*.

$KO = PO = 8$

$x + 3 = 8$

$x = 5$

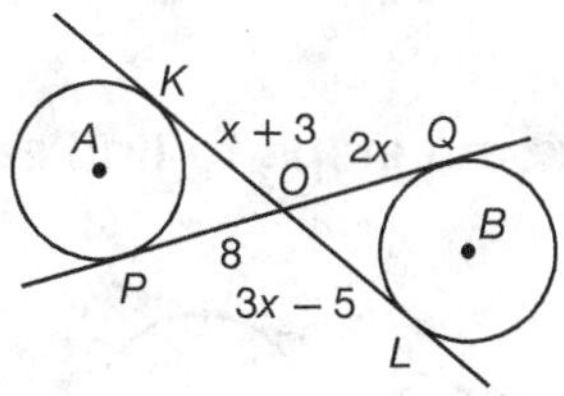

Step 2: Substitute $x = 5$ to find OQ or OL.

$OQ = 2x = 2(5) = 10$

$OL = 3x - 5 = 3(5) - 5 = 10$

Step 3: Find KL and PQ: $KL = PQ = PO + OQ = 8 + 10 = 18.$

Practice

Complete the steps to find each length.

1.

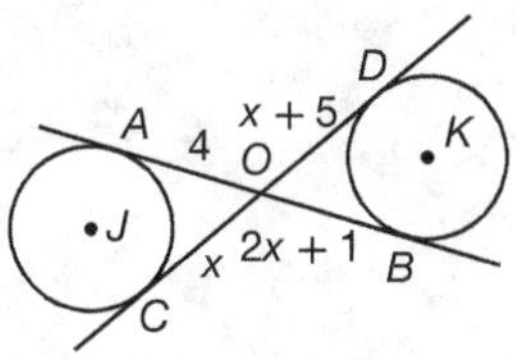

$AO = CO =$ ___

$x =$ ___

$OD =$ ___ $+ 5 =$ ___

$CD = CO + OD$

$CD =$ ___ $+$ ___ $=$ ___

$AB = CD =$ ___

2.

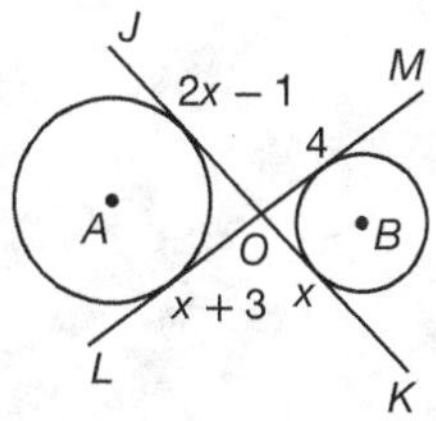

$OM = OK = 4$

$x =$ ___

$JO = 2($ ___ $) - 1 =$ ___

$JK = JO + OK$

$JK =$ ___ $+ 4 =$ ___

$LM = JK =$ ___

Find each length.

3.

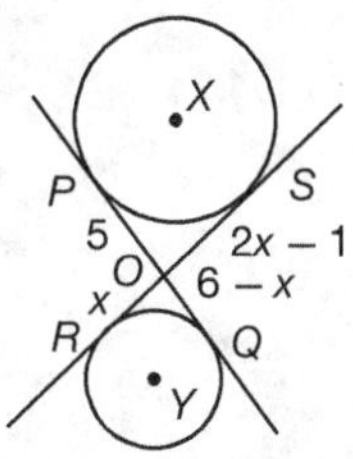

$PQ = RS =$ ___

4.

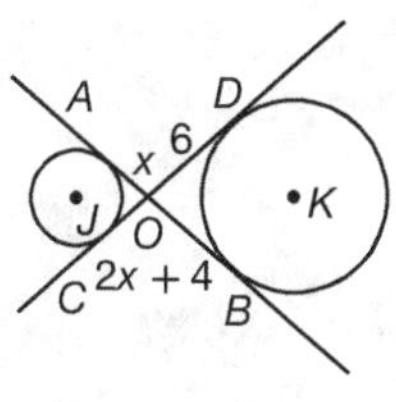

$AB = CD =$ ___

Common Tangents: External

A **common tangent** is externally tangent if the tangent does not intersect the segment joining the centers of two circles. Again, two segments tangent from the same exterior point are congruent.

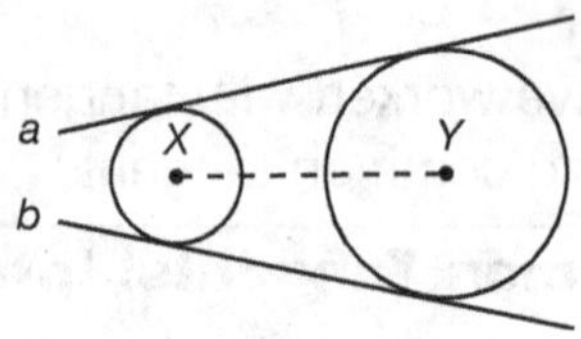

Line segments $\overline{AB}$ and $\overline{CD}$ are common tangents to the circles. Find the lengths of $\overline{AB}$ and $\overline{CD}$.

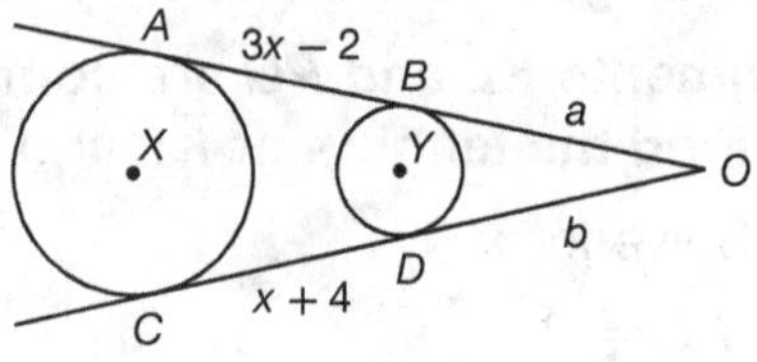

$AB = CD$	Given
$3x - 2 = x + 4$	Substitute.
$2x = 6$	Solve.
$x = 3$	
$AB = 3(3) - 2 = 7$	Substitute.
$CD = 3 + 4 = 7$	Substitute.

Practice

Complete the steps to find each length.

5.

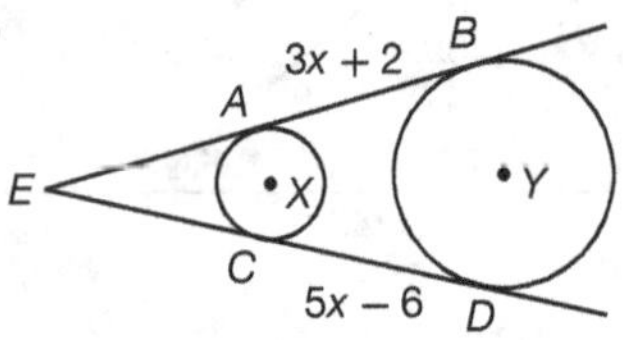

$AB = CD$

$3x + 2 = \_\_ - 6$

$8 = \_\_$

$x = \_$

$AB = 3(\_) + 2 = \_\_$

$CD = 5(\_) - 6 = \_\_$

6.

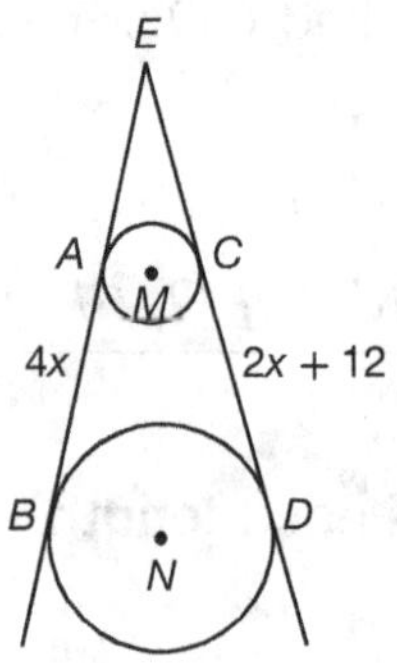

$AB = CD$

$4x = \_\_ + 12$

$\_\_ = 12$

$x = \_$

$AB = 4(\_) = \_\_$

$CD = 2(\_) + 12 = \_\_$

Find each length.

7.

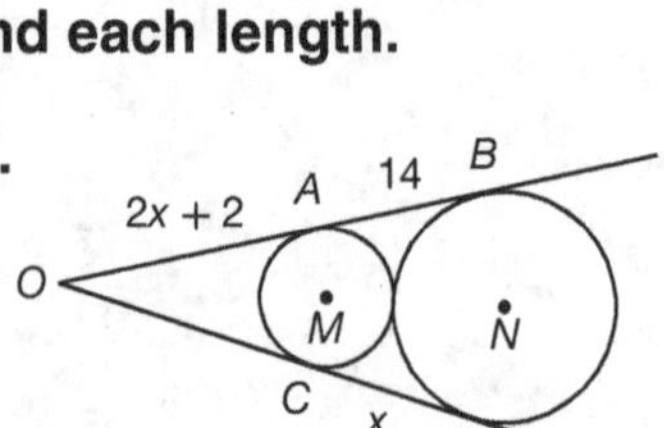

$OB = \_\_$

$OD = \_\_$

8.

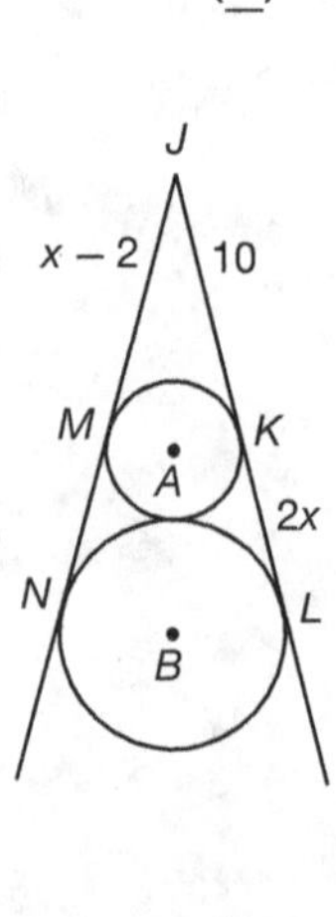

$JL = \_\_$

$JN = \_\_$

Name ______________________________ Date ___________ Class ____________

Reteaching 73

Applying Trigonometry: Angles of Elevation and Depression

You have worked with angles and trigonometry. Now you will use angles of elevation and depression to solve problems.

Angle of Elevation

An **angle of elevation** is the angle formed by a horizontal line and a line of sight to a point above. You can use trigonometry to solve problems involving angle of elevation.

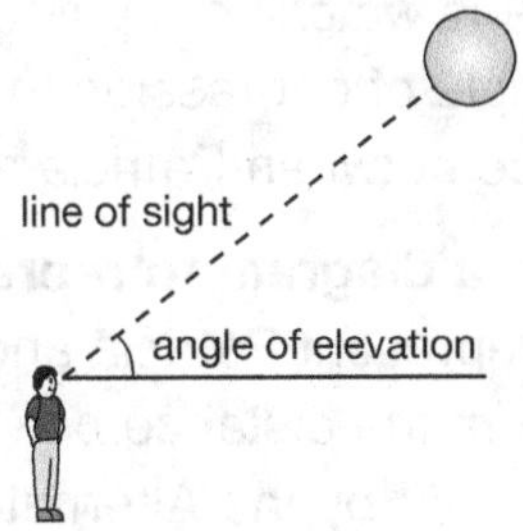

When the angle of elevation to the sun is 48°, a tree casts a shadow that is 10 meters long. What is the height of the tree? Round to the nearest tenth of a meter.

Sketch a diagram to represent the given information.

The height of the tree is represented by *x*. The angle of elevation is 48°, and the length of the shadow on the ground is 10 meters. To find *x*, use the tangent function.

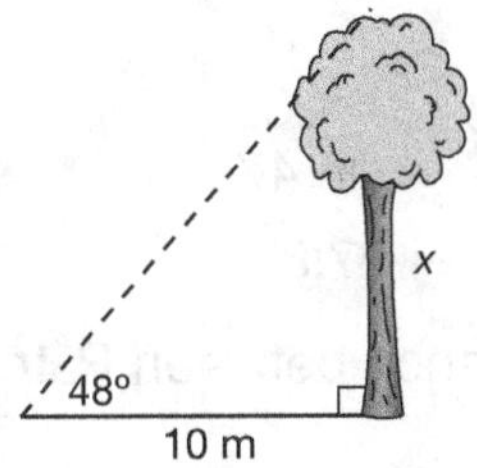

$\tan 48° = \dfrac{\text{opp}}{\text{abj}}$	Tangent Function
$\tan 48° = \dfrac{x}{10}$	Substitute.
$10 \tan 48° = x$	Simplify.
$11.1 \approx x$	Solve.

The tree is about 11.1 meters tall.

Practice

Complete the steps to solve the problem.

1. A building casts a shadow 45 feet long. The building is 120 feet tall. What is the angle of elevation from the point on the ground where the shadow of the building ends? Round to the nearest tenth of a degree.

   $\tan x = \dfrac{\text{opp}}{\text{abj}}$

   $\tan x = \dfrac{120}{\underline{\quad}} \approx \underline{\qquad}$

   $x = \underline{\qquad}$

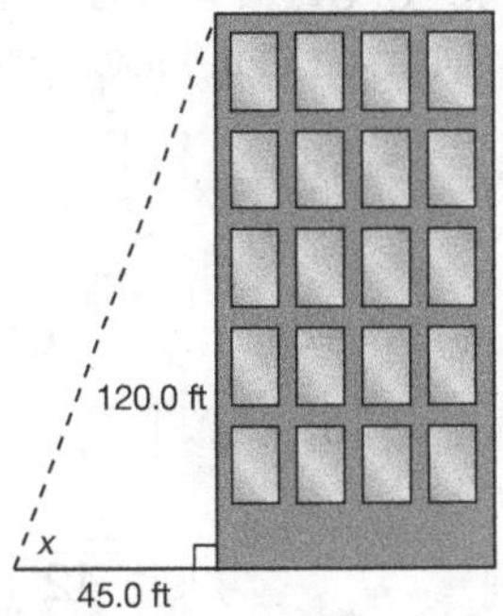

Solve the problem.

2. Andrew is 15 feet from a house his brother is painting. He measures the angle of elevation from the ground to the top of his brother's head as 52°. How high up the house is the top of his brother's head? Round to the nearest foot. ____ ft

3. Find the angle of elevation to the top of a tree for a person standing 28.6 meters from the tree, assuming the person's eye is 1.5 meters above the ground and the tree is 19.1 meters tall. Round to the nearest tenth of a degree. ______

Angle of Depression

An **angle of depression** is the angle formed by a horizontal line and a line of sight to a point below. You can use trigonometry to solve problems involving angle of depression.

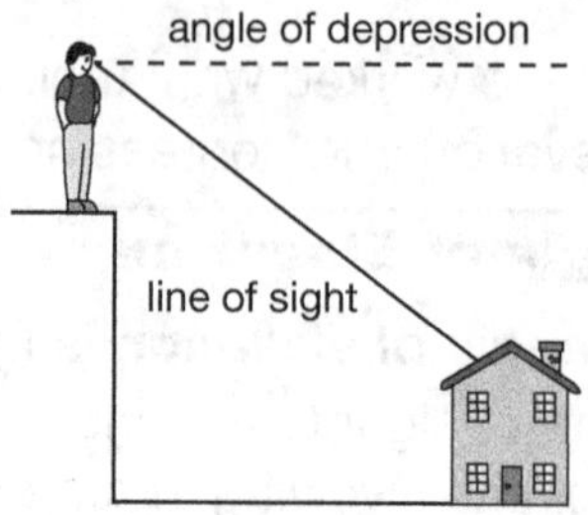

Patricia is watching a parade from a 20-foot-high balcony. The angle of depression to the parade is 47°. What is the distance between Patricia and the parade?

Sketch a diagram to represent the given information.

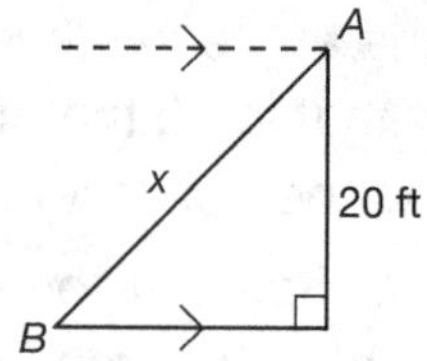

Let *A* represent Patricia and let *B* represent the parade. Let *x* represent the distance between Patricia and the parade.
$m\angle B = 47^\circ$ by the Alternate Interior Angles Theorem.
Write a sine ratio using $\angle B$

$\sin 47^\circ = \frac{20}{x}$	Sine Function
$x \sin 47^\circ = 20$	Multiply both sides by *x*.
$x = \frac{20}{\sin 47^\circ}$	Divide both sides by sin 47°.
$x \approx 27.3$	Simplify.

The distance between Patricia and the parade is about 27 feet.

Practice

Complete the steps to solve the problem.

4. A person swimming in a pool sees a quarter on the bottom of the pool at an angle of depression of 42°. She is 6 feet above the bottom of the pool. How far from the quarter is she? Round to the nearest foot.

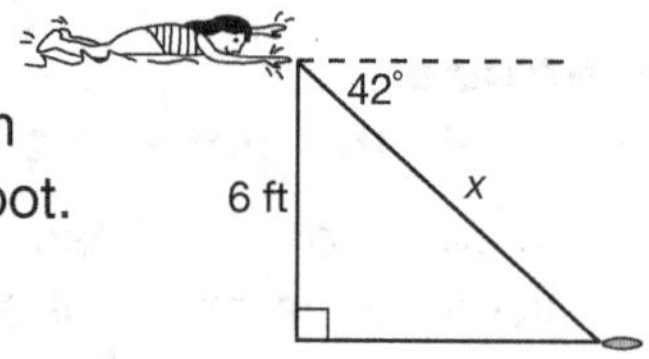

$\sin x = \frac{\text{opp}}{\text{hyp}}$

$\sin 42^\circ = \frac{\_}{x}$

$x \sin 42^\circ = \_$

$x = \frac{\_}{\sin 42^\circ}$

$x \approx \_$ ft

Solve the problem.

5. Paul is looking out a window that is 18 feet off the ground, and sees a mailbox near the street. The angle of depression to the mailbox is 28°. What is the horizontal distance from Paul to the mailbox? Round to the nearest foot. ___ ft

6. Elise is parasailing. She is attached to the boat by a rope that is 300 feet long. The angle of depression from Elise to the boat is 29°. How high up in the air is Elise? Round to the nearest foot. ___ ft

Name ______________________ Date __________ Class __________

Reteaching 74
Reflections

You have worked with translations. Now you will reflect figures across a line on a coordinate grid.

Reflecting Across an Axis

A **reflection** is an isometry (a transformation that does not change the size or shape of the object being translated). It involves a translation across a line, where each point and its image are the same distance from the line of reflection.

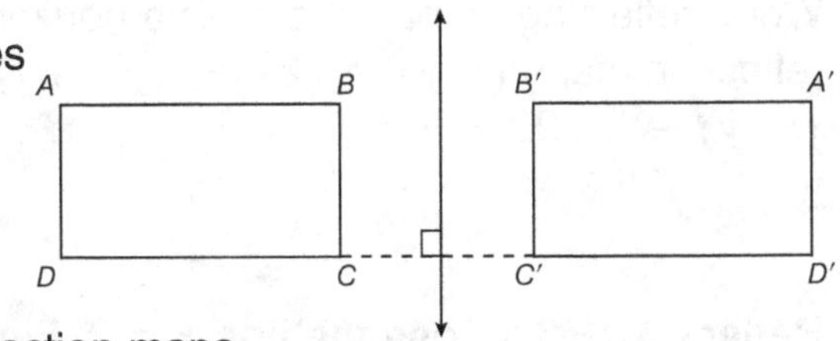

When an object is reflected across the x-axis, the reflection maps $(x, y) \rightarrow (x, -y)$. When an object is reflected across the y-axis, the reflection maps $(x, y) \rightarrow (-x, y)$.

Reflect ΔABC across the x-axis. Find the coordinates of the vertices of the reflection image.

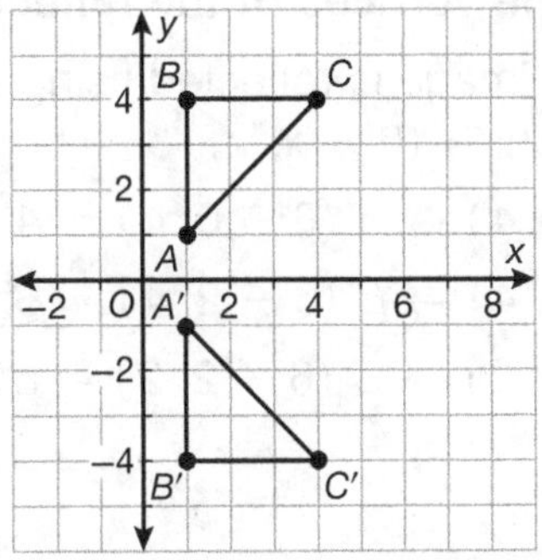

The image is reflected by using $(x, y) \rightarrow (x, -y)$.

$A(1, 1) \rightarrow A'(1, -1)$

$B(1, 4) \rightarrow B'(1, -4)$

$C(4, 4) \rightarrow C'(4, -4)$

Practice

Complete the steps to find the coordinates of each reflection.

1. Reflect ΔJKL across the y-axis.

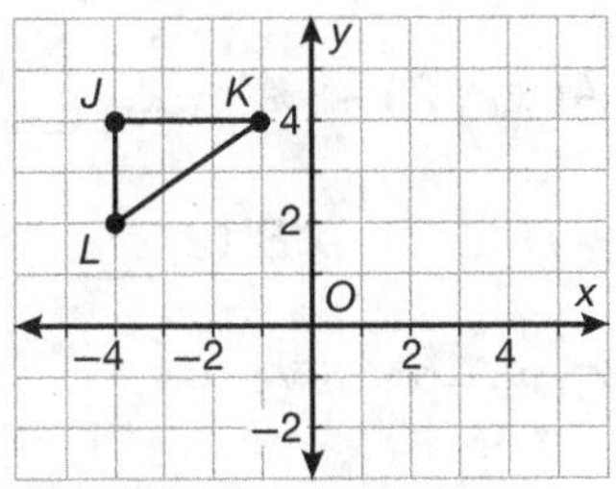

$J(-4, 4) \rightarrow J'(\_, 4)$

$K(-1, 4) \rightarrow K'(\_, \_)$

$L(\_\_\_, \_) \rightarrow L'(\_, \_)$

Find the coordinates for each reflection.

2. Reflect ▭$WXYZ$ across the y-axis.

____

____

____

____

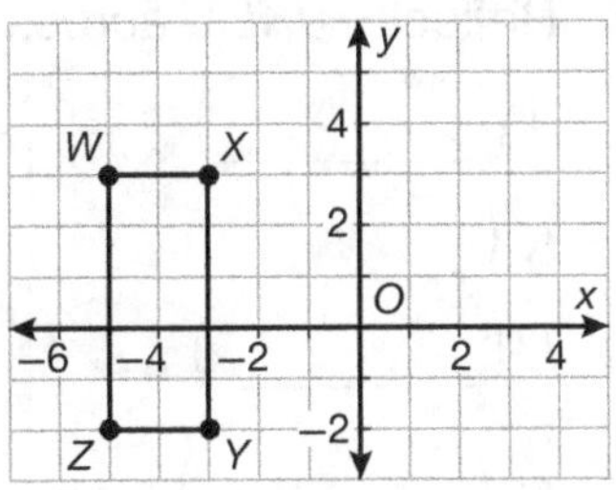

3. Reflect ΔMNO across the x-axis.

____

____

____

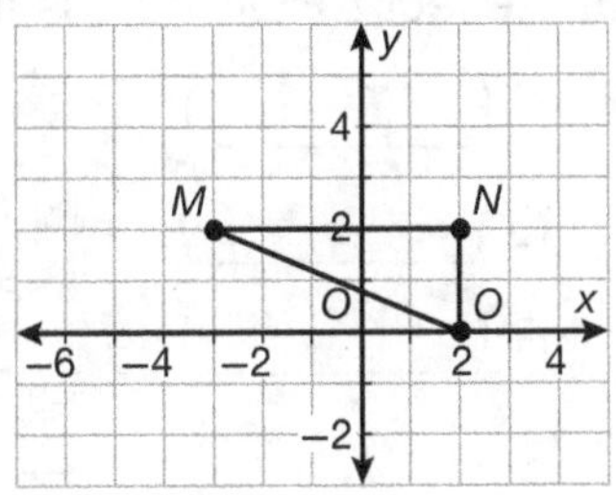

Reflecting Across a Vertical or Horizontal Line

When reflecting an object across a vertical line $x = a$, the reflection maps $(x, y) \rightarrow (a - (x - a), y)$ or $(x, y) \rightarrow (2a - x, y)$.

When reflecting an object across a horizontal line $y = b$, the reflection maps $(x, y) \rightarrow (x, b - (y - b))$ or $(x, y) \rightarrow (x, 2b - y)$.

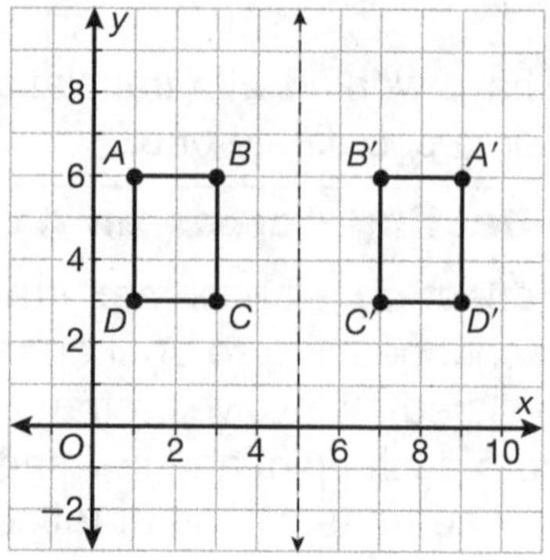

Reflect ΔABC across the line $x = 3$. Find the coordinates of the vertices of the reflection image.

The image is reflected using $(x, y) \rightarrow (3 - (x - 3), y)$ or $(x, y) \rightarrow (6 - x, y)$.

$A(0, 4) \rightarrow A'(6 - 0, 4) = A'(6, 4)$

$B(2, 4) \rightarrow B'(6 - 2, 4) = B'(4, 4)$

$C(2, 2) \rightarrow C'(6 - 2, 2) = C'(4, 2)$

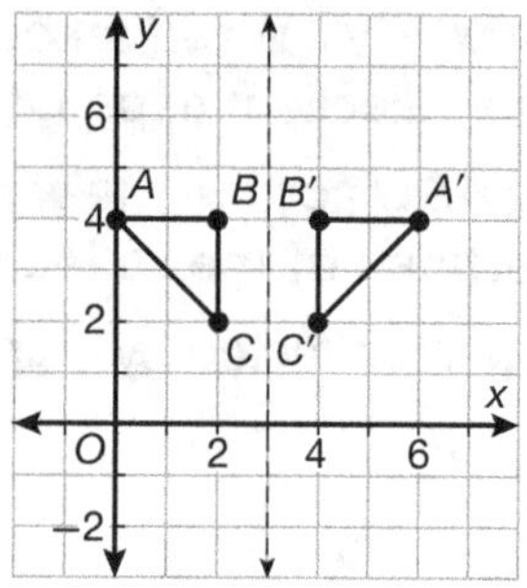

Practice

Complete the steps to find the coordinates of each reflection.

4. Reflect ▭$ABCD$ across the line $y = 3$.

$A(-2, 2) \rightarrow A'(-2,$ ______ $) = A'(-2,$ _ $)$

$B(2, 2) \rightarrow B'(2,$ ______ $) = B'(2,$ _ $)$

$C(2, -2) \rightarrow C'(2,$ ________ $) = C'(2,$ _ $)$

$D(-2, -2) \rightarrow D'(-2,$ ________ $) = D'(-2,$ _ $)$

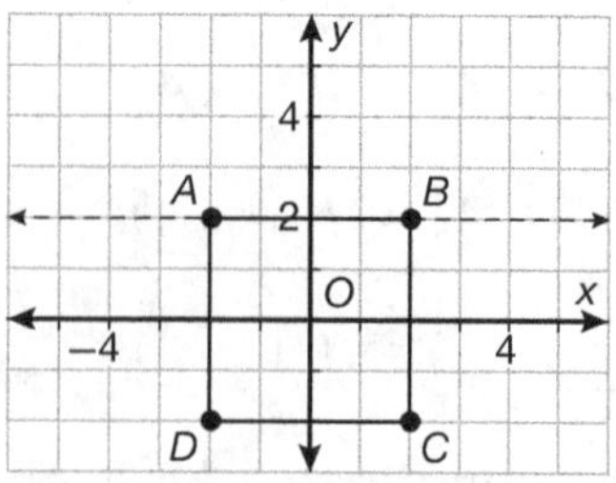

Find the coordinates of each reflection.

5. Reflect ΔPQR across the line $y = -2$.

P'(__________)

Q'(__________)

R'(__________)

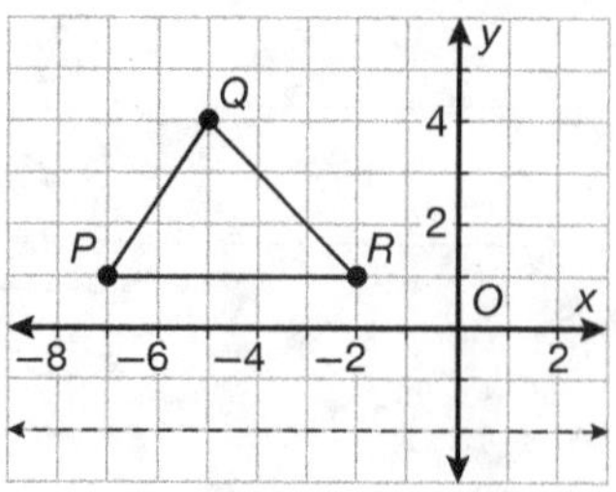

6. Reflect ▭$JKLM$ across the line $x = -1$.

J'(__________)

K'(__________)

L'(__________)

M'(__________)

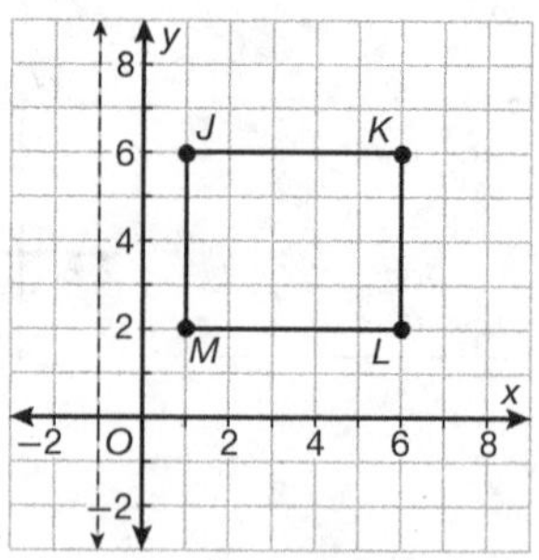

Name ______________________ Date __________ Class __________

Reteaching 75

Writing the equation of a circle

You have worked with plotting figures on a coordinate plane. Now you will find the equation of a circle.

Equation of a Circle

The equation of a circle with center (h, k) and radius r is $(x - h)^2 + (y - k)^2 = r^2$.

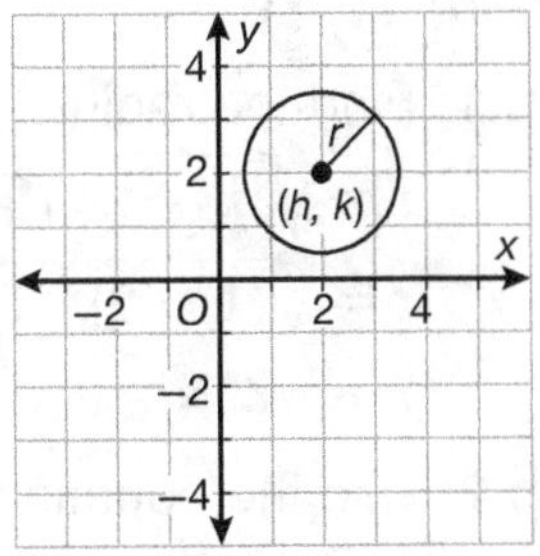

Write the equation of ⊙C with center $C(2, -1)$ and radius 6.

$(x - h)^2 + (y - k)^2 = r^2$	Equation of a Circle
$(x - 2)^2 + (y - (-1))^2 = 6^2$	Substitute.
$(x - 2)^2 + (y + 1)^2 = 36$	Simplify.

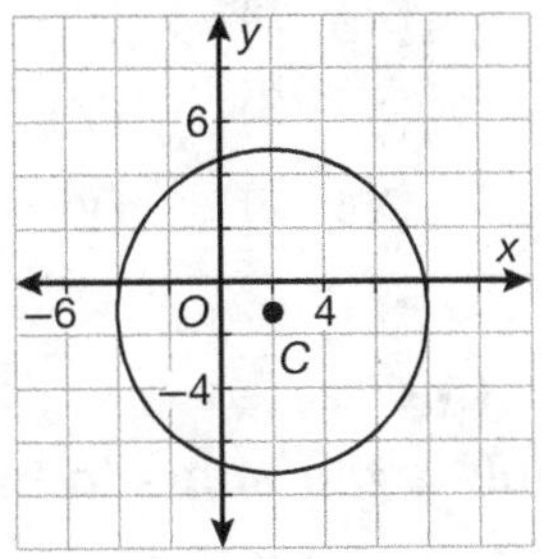

Practice

Complete the steps to write the equation of ⊙C.

1. $(x - h)^2 + (y - k)^2 = r^2$

   $(x - \underline{\quad\quad})^2 + (y - \underline{\quad\quad})^2 = \underline{\quad\quad}$

   $(\underline{\quad\quad\quad})^2 + (\underline{\quad\quad\quad})^2 = \underline{\quad}$

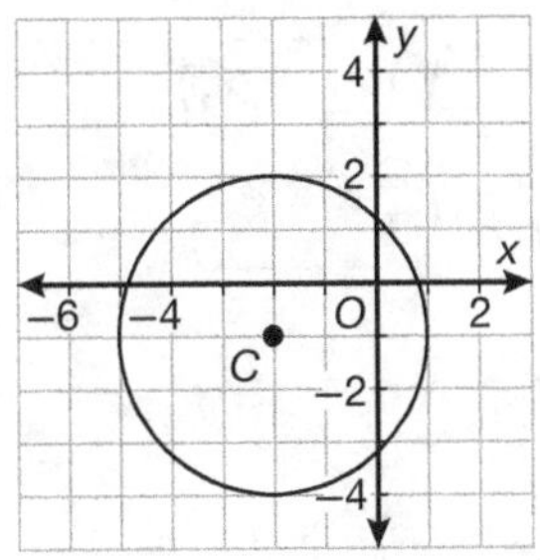

Write the equation of ⊙C.

2. ______________________

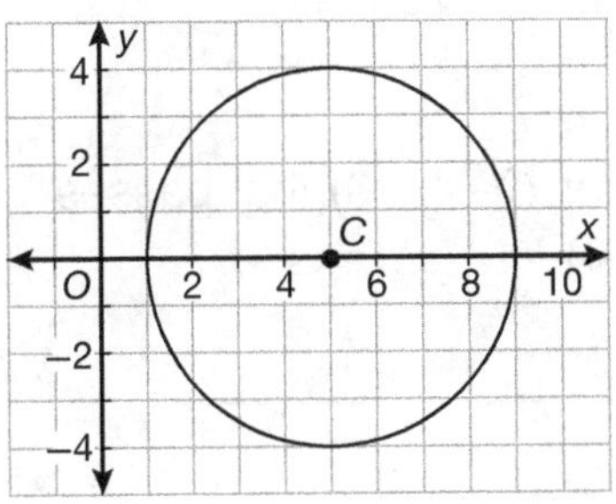

3. ______________________

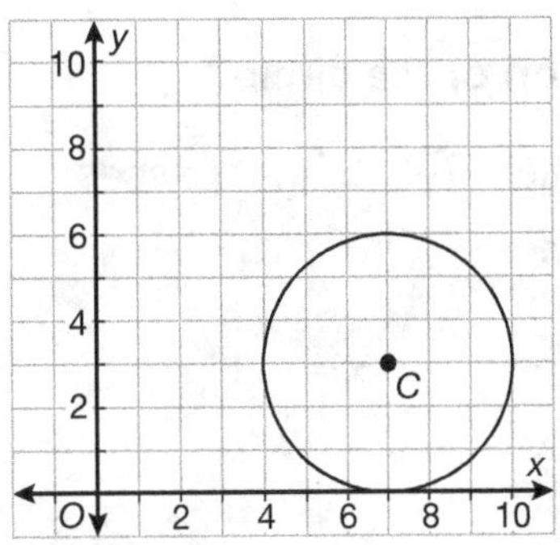

Equation of a Circle

You can also write the equation of a circle if you know the center and one point on the circle.

Write the equation of ⊙C with center $C(3, 7)$ that passes through $(1, 7)$.

Step 1: Find the radius.

$r = \sqrt{(x_2 - x_1)^2 + (y_2 - y_1)^2}$ Distance Formula

$r = \sqrt{(1 - 3)^2 + (7 - 7)^2}$ Substitute.

$r = \sqrt{4} = 2$ Simplify.

Step 2: Use the equation of a circle.

$(x - h)^2 + (y - k)^2 = r^2$ Equation of a Circle

$(x - 3)^2 + (y - 7)^2 = 2^2$ Substitute.

$(x - 3)^2 + (y - 7)^2 = 4$ Simplify.

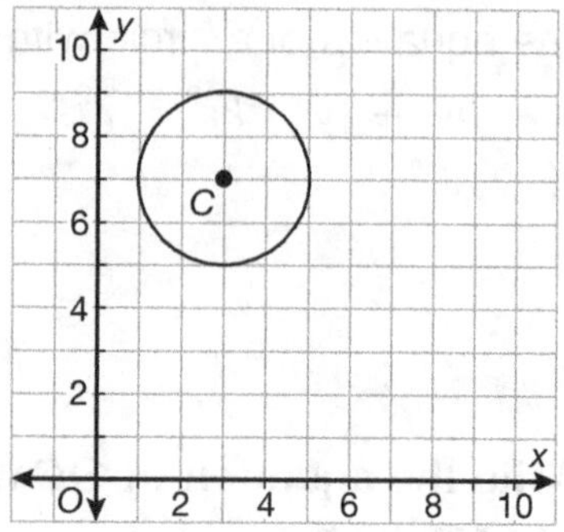

Practice

Complete the steps to find the equation of the circle.

4. ⊙C with center $C(4, 2)$ that passes through $(2, 2)$

$r = \sqrt{(x_2 - x_1)^2 + (y_2 - y_1)^2}$

$r = \sqrt{(\_\_\_\_)^2 + (\_\_\_\_)^2}$

$r = \sqrt{\_} = \_$

$(x - h)^2 + (y - k)^2 = r^2$

$(x - \_)^2 + (y - \_)^2 = \_^2$

$(x - \_)^2 + (y - \_)^2 = \_$

5. ⊙B with center $B(3, 1)$ that passes through $(7, 1)$

$r = \sqrt{(x_2 - x_1)^2 + (y_2 - y_1)^2}$

$r = \sqrt{(\_\_\_\_)^2 + (\_\_\_\_)^2}$

$r = \sqrt{\_} = \_$

$(x - h)^2 + (y - k)^2 = r^2$

$(x - \_)^2 + (y - \_)^2 = \_^2$

$(x - \_)^2 + (y - \_)^2 = \_\_$

Find the equation of the circle.

6. ⊙T with center $T(-1, 5)$ that passes through $(4, 5)$

____

7. ⊙C with center $C(-4, -3)$ that passes through $(-4, 3)$

____

Name ______________________ Date ____________ Class ____________

Reteaching 76

Symmetry

You have worked with reflections and rotations of geometric figures. Now you will work with symmetry.

Line Symmetry

A **line of symmetry** is a line that divides a plane figure into two congruent, reflected halves. Basically, you could put a mirror on the line of symmetry to see the entire object.

Line symmetry is a type of symmetry that describes a figure that can be reflected across a line so that the image coincides with the preimage. This is also called **reflection symmetry.**

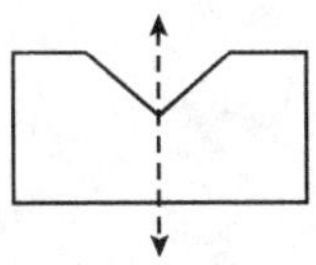
one line of symmetry

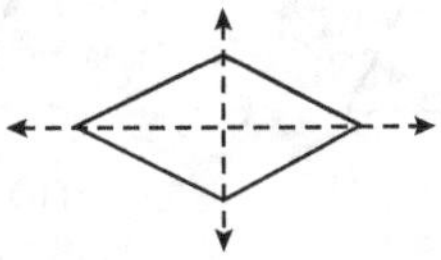
two lines of symmetry

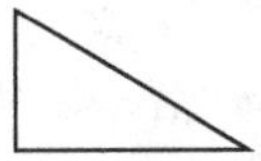
no line of symmetry

Write whether the figure below has line symmetry. If so, draw all lines of symmetry.

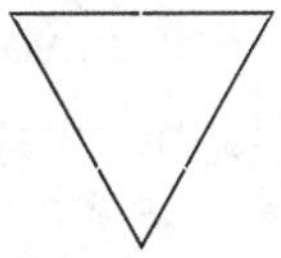

Step 1: Check for a line that divides the triangle into two congruent halves.

Step 2: If one exists, check for another line that divides the triangle into two congruent halves. Check for a third line.

Step 3: Draw the line, or lines, of symmetry.

Practice

Write whether each figure has line symmetry. If so, draw all lines of symmetry.

1.

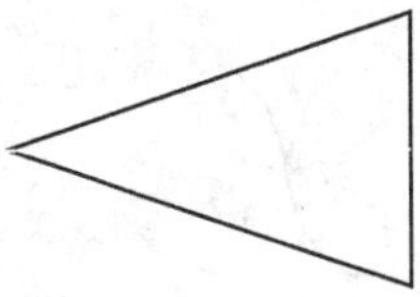

____

2.

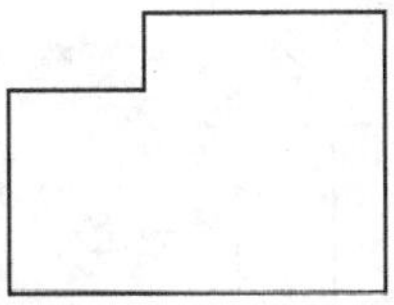

____

3.

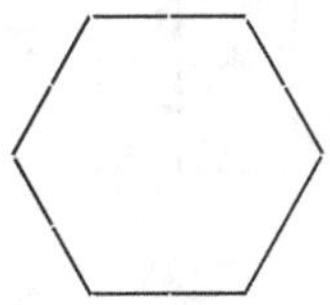

____

4.

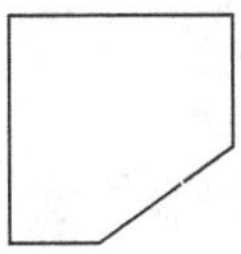

____

5.

____

6.

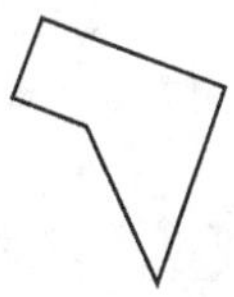

____

Rotational Symmetry

A figure has **rotational symmetry** if it can be rotated about a point by an angle between 0° and 360° so that the image coincides with the original figure.

The smallest angle through which the figure is rotated to coincide with itself is called the **angle of rotational symmetry.**

The number of times that you can get an identical figure when repeating the degree of rotation is called the **order of the rotational symmetry.**

Below are some examples of figures with and without rotationally symmetry.

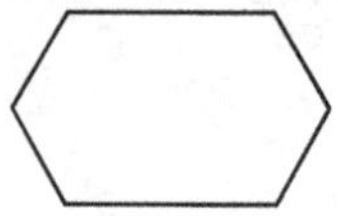
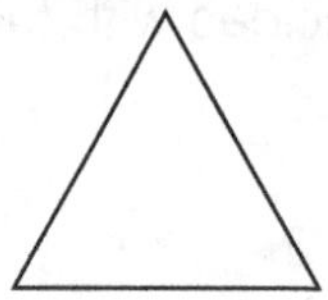
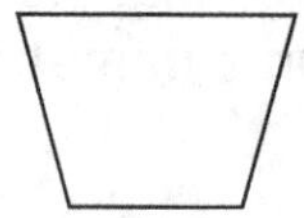

angle:	180°	120°	no rotational
order:	2	3	symmetry

Write whether the figure below has rotational symmetry. If so, give the angle of rotational symmetry and the order of symmetry.

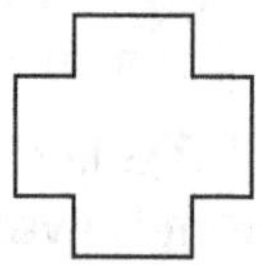

Step 1: Pick up your paper and rotate it clockwise 90°.

Step 2: Is the rotated figure congruent to the original?
Does the figure have rotational symmetry? yes

Step 3: Write the angle and order of rotational symmetry for this figure. angle: 90°; order: 4

Practice

Write whether each figure of symmetry has rotational symmetry. If so, give the angle of rotational symmetry and the order.

7.

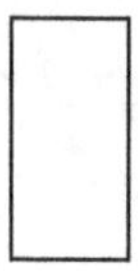

____

____

8.

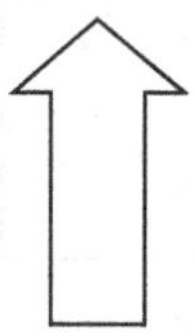

____

____

9.

____

____

10.

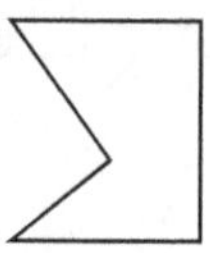

____

____

11.

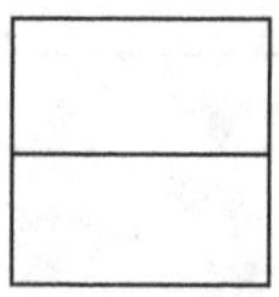

____

____

12.

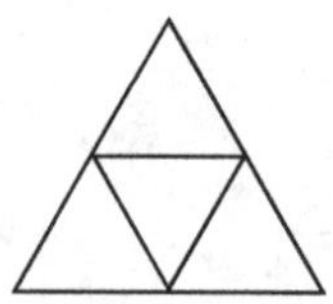

____

____

Reteaching 77

Finding Surface Areas and Volumes of Cones

You have found the lateral area, surface area, and volume of regular pyramids. Now you will find the lateral area, surface area and volume of cones.

Lateral and Surface Area of a Cone

Lateral Area: The lateral area of a cone with radius r and slant height l is $LA = \pi rl$.

Surface Area: The surface area of a cone with lateral area LA and base area B is $SA = LA + B$, where $B = \pi r^2$.

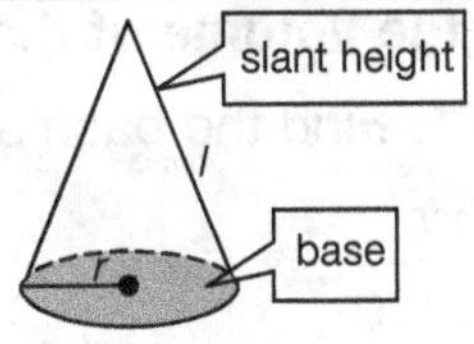

Find the lateral and surface area of the cone. Round to the nearest tenth.

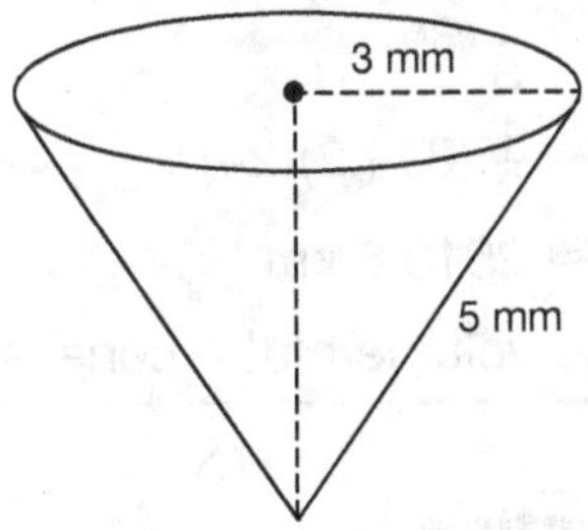

Step 1: Find LA and B using $r = 3$ and $l = 5$.

$LA = \pi rl$ $\quad\quad$ $B = \pi r^2$

$= \pi(3)(5)$ $\quad\quad$ $= \pi(3)^2$

$\approx 47.1 \text{ mm}^2$ $\quad\quad$ $= 28.3 \text{ mm}^2$

Step 2: Find SA using $LA \approx 47.1 \text{ mm}^2$ and $B \approx 28.3 \text{ mm}^2$.

$SA = LA + B$

$= (47.1) + (28.3)$

$= 75.4 \text{ mm}^2$

The surface area of the cone is approximately 75.4 mm^2.

Practice

Complete the steps to find the lateral and surface area of the cone. Round to the nearest tenth.

1. $LA = \pi rl = \pi(3)(\_) \approx$ _____ in.^{2}

   $B = \pi r^2 = \pi(\_)^2 \approx$ _____ in.^{2}

   $SA = LA + B$

   $\approx$ (____) + (____)

   $\approx$ __________

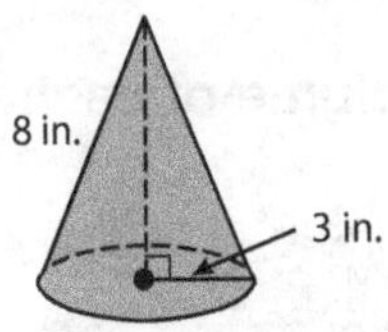

Find the lateral and surface area of each cone. Round to the nearest tenth.

2. 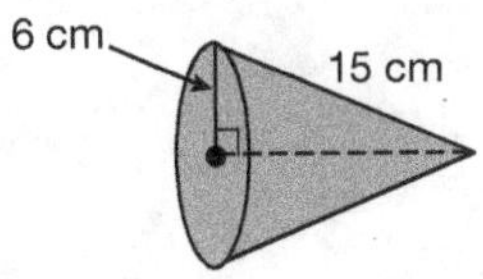

   $LA =$ __________

   $SA =$ __________

3. 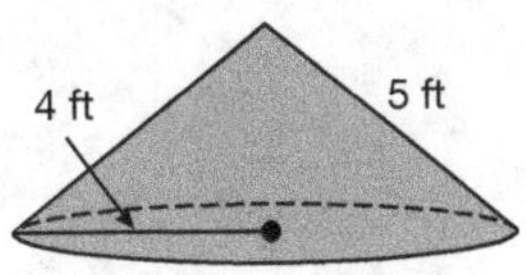

   $LA =$ __________

   $SA =$ __________

Volume of a Cone

The volume of a right cone with base area B and height h is $V = \frac{1}{3}Bh$, where $B = \pi r^2$.

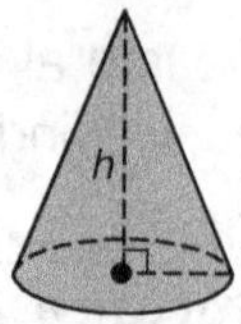

Find the volume of the cone. Round to the nearest tenth.

Step 1: Find the base area, B, using $r = 10$.

$B = \pi r^2$

$= \pi(10)^2$

$= 314.2 \text{ km}^2$

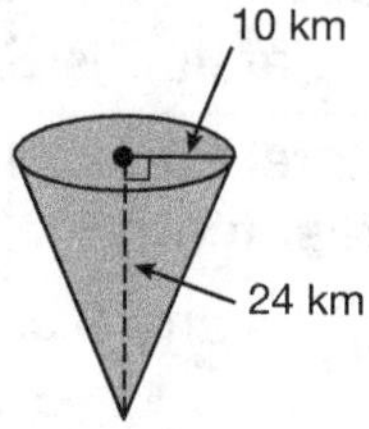

Step 2: Find the volume using $B \approx 314.2 \text{ km}^2$ and $h = 24$.

$V = \frac{1}{3} Bh$

$\approx \frac{1}{3}(314.2)(24)$

$\approx 2513.6 \text{ km}^3$

The volume of the cone is approximately 2513.6 km^3.

Practice

Complete the steps to find the volume of the cone. Round to the nearest tenth.

4. $B = \pi r^2 = \pi(\_)^2 \approx$ ____ ft^2

$V = \frac{1}{3} Bh$

$\approx \frac{1}{3}($____$)(12)$

$\approx$ ____ ft^3

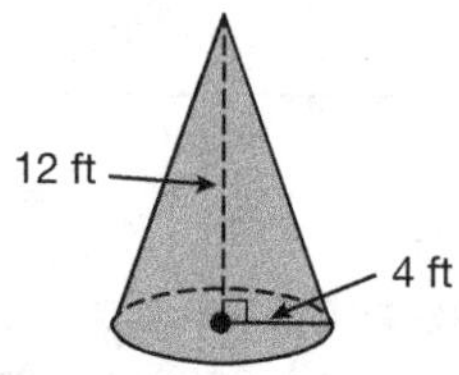

Find the volume of each cone. Round to the nearest tenth.

5.

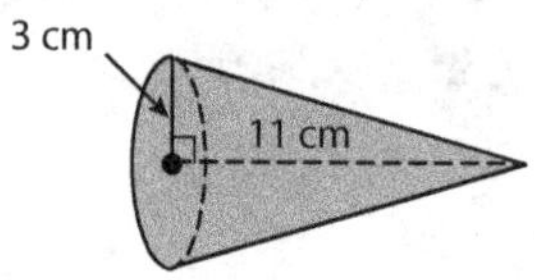

____

6.

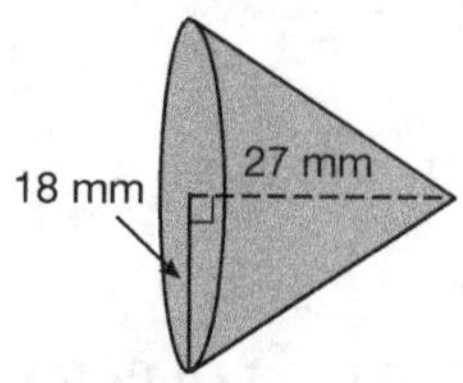

____

Name ______________________ Date __________ Class __________

Reteaching 78
Rotations

Now you will rotate figures on a coordinate plane.

Rotating About the Origin

When a figure is rotated about the origin on a coordinate plane, some simple rules apply for certain rotations.

If a point (x, y) is rotated 90°, then (x, y) becomes $(-y, x)$.

If a point (x, y) is rotated 180°, then (x, y) becomes $(-x, -y)$.

If a point (x, y) is rotated 270°, then (x, y) becomes $(y, -x)$.

In this picture, quadrilateral *LMNO* has been rotated 270° about the origin to produce quadrilateral *L′M′N′O′*.

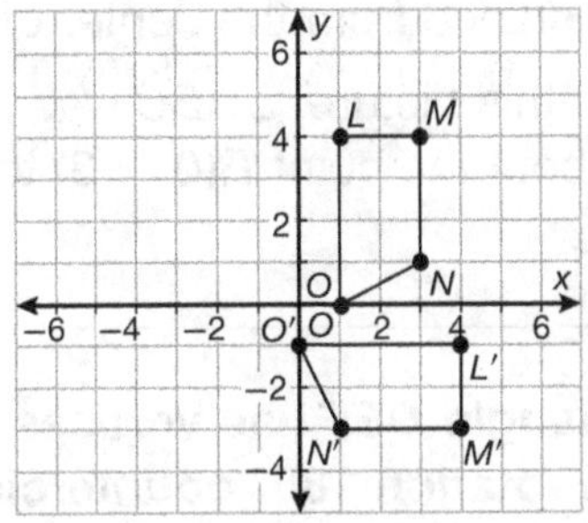

ΔABC has vertices $A(-6, 2)$, $B(-3, 5)$, and $C(-2, 2)$. Graph ΔABC and its rotation 90° counterclockwise about the origin.

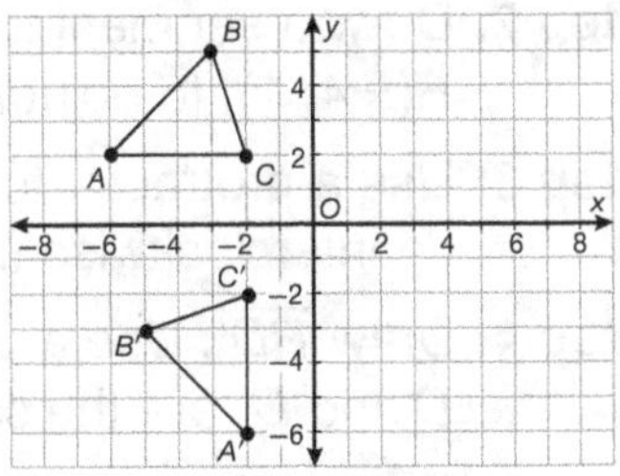

Step 1: Graph ΔABC.

Step 2: Using the rule above, (x, y) becomes $(-y, x)$.

$A(-6, 2) \rightarrow A'(-2, -6)$

$B(-3, 5) \rightarrow B'(-5, -3)$

$C(-2, 2) \rightarrow C'(-2, -2)$

Step 3: Graph $\Delta A'B'C'$.

Practice

1. ΔXYZ has vertices $X(4, -4)$, $Y(8, -4)$, and $Z(8, -7)$.

   Using the rule, (x, y). becomes $(-x, -y)$.

   $X(4, -4) \rightarrow X'(-4,$ ___ $)$

   $Y(8, -4) \rightarrow Y'($ ___ $, 4)$

   $Z(8, -7) \rightarrow Z'($ ___ , ___ $)$

   $\Delta X'Y'Z'$ has vertices $X'(4, -4)$, $Y'($ ___ , ___ $)$, and $Z'($ ___ , ___ $)$.

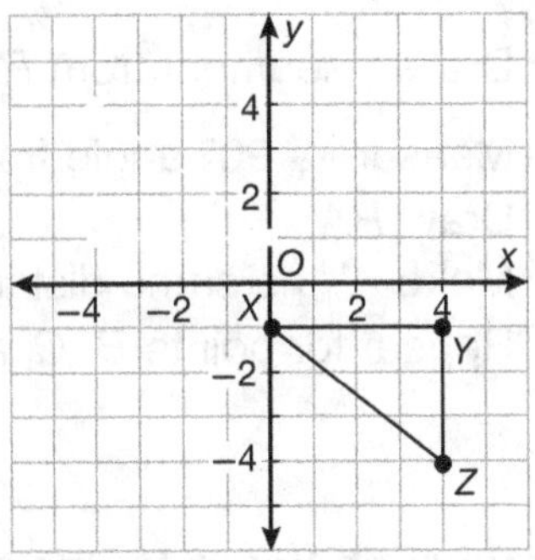

Rotate each figure by the specified angle counterclockwise about the origin. List the coordinates of the vertices of the rotated figure.

2. 270°

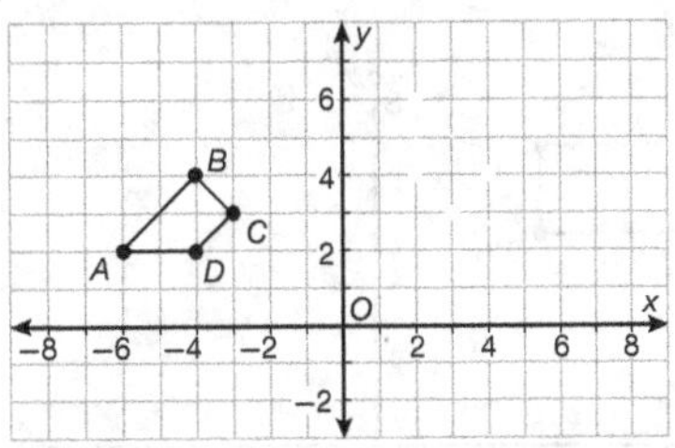

3. 90°

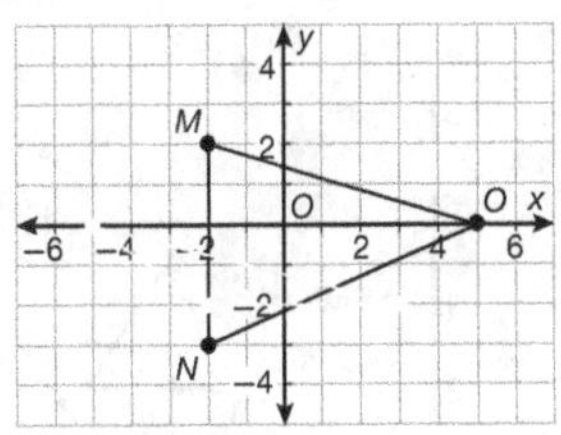

__________________________ __________________________

Rotating About a Chosen Point

When a figure is rotated about a chosen point on a coordinate plane, a protractor must be used to measure the angle of rotation. The rotated vertices will be the same distance from the center of rotation.

In this picture, ΔABC was rotated 45° counterclockwise about the point $R(0, -3)$ to produce $\Delta A'B'C'$.

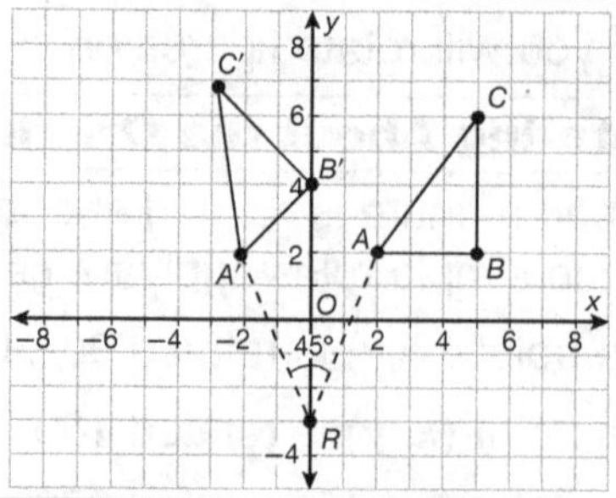

Triangle *DEF* has vertices *D*(−4, 4), *E*(−3, 2), *F*(−2, 5). Graph the triangle and its rotation 180° counterclockwise about the point *R*(−4, 1).

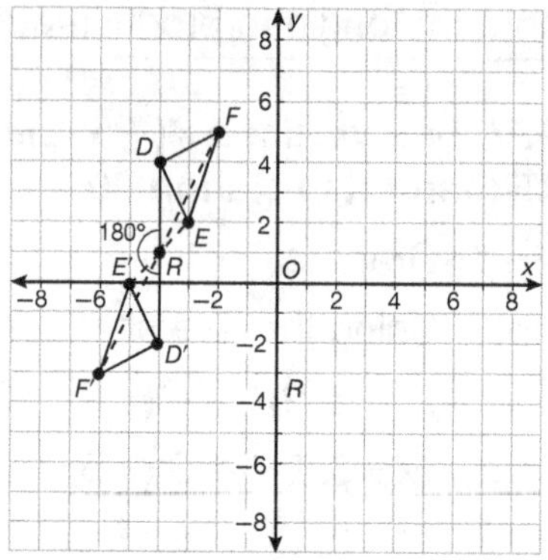

Step 1: Graph ΔDEF.

Step 2: Draw a segment from the center of rotation, $R(-4, 1)$, to point D.

Step 3: Use a protractor to measure a 180° angle counterclockwise from $\overline{RD}$.

Step 4: Draw $\overline{RD'}$. Use a compass or a ruler to make sure D and D' are the same distance from R.

Step 5: Repeat these steps for points E and F.

Practice

Complete the steps to rotate *ABCD* 90° counterclockwise about *R*(0, −1).

4. Draw a segment from $R(0, -1)$ to point A.

   Measure a 90° angle from $\overline{RA}$.
   Draw $\overline{RA'}$.
   Make A' the same distance from R as A is from R.
   Repeat for points B, C, and D.

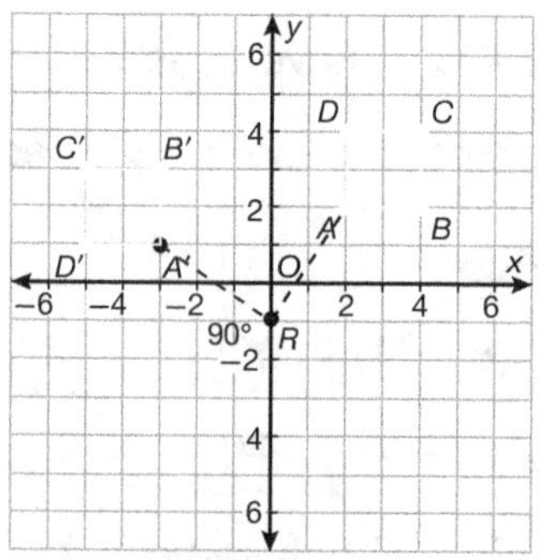

Rotate each figure by the specified angle counterclockwise about the specified point.

5. 45° about $R(3, 0)$

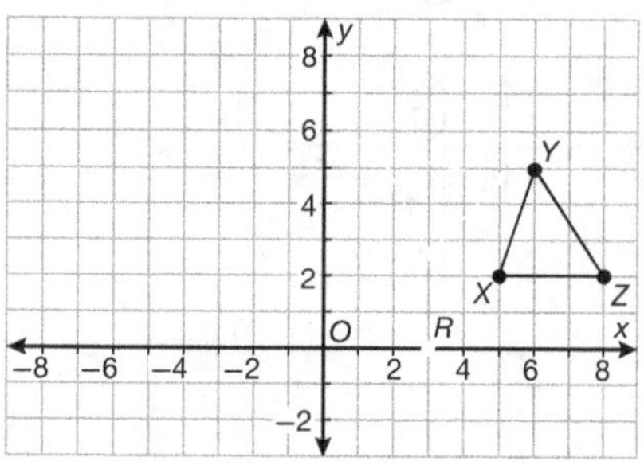

6. 180° about $R(-3, 3)$

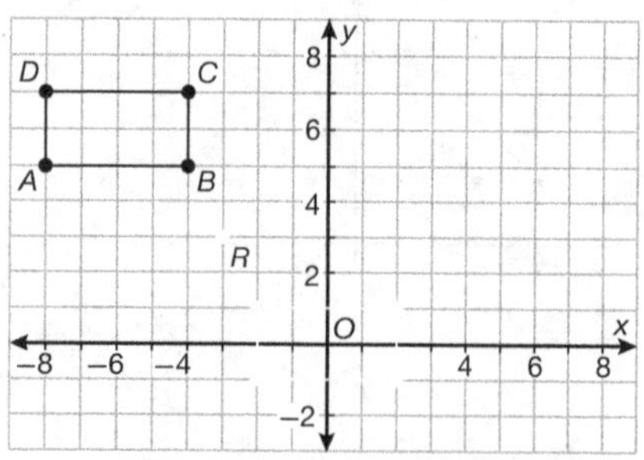

Name ______________________ Date __________ Class __________

Reteaching 79

Angles Exterior to Circles

You have worked with angles interior to circles. Now you will work with angles exterior to circles.

Intercepting by Two Secants or Two Tangents

An **exterior angle** of a circle has its vertex outside the circle with each ray intersecting the circle at least once. Below are two examples of exterior angles.

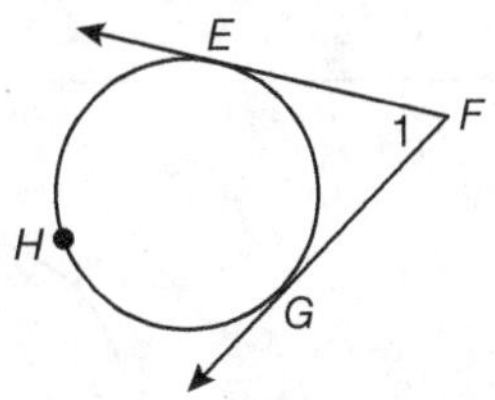

intercepting by two tangents

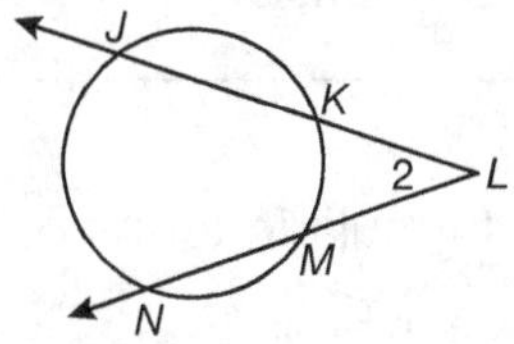

intercepting by two secants

The measure of an exterior angle is equal to half the difference of the intercepted arcs.

$m\angle 1 = \frac{1}{2}(mEHG - mEG)$ $\qquad$ $m\angle 2 = \frac{1}{2}(mJN - mKM)$

Find the value of *x*.

Step 1: Since $mPVR + mPR = 360°$, $mPVR + 142 = 360°$, and $mPVR = 218°$.

Step 2: Substitute $mPVR = 218°$ and $mPR = 142°$ into the formula and solve.

$x° = \frac{1}{2}(mPVR - mPR)$

$= \frac{1}{2}(218° - 142°)$

$x° = 38°$

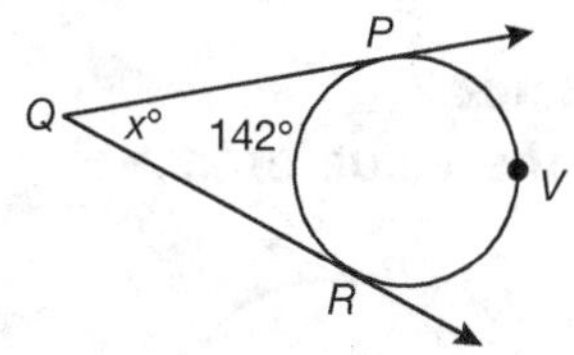

Practice

Find the value of *x*.

1.

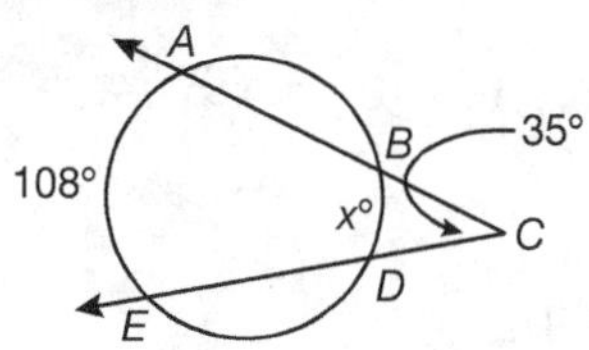

$35° = \frac{1}{2}(108° - x)$; $x =$ ___

2.

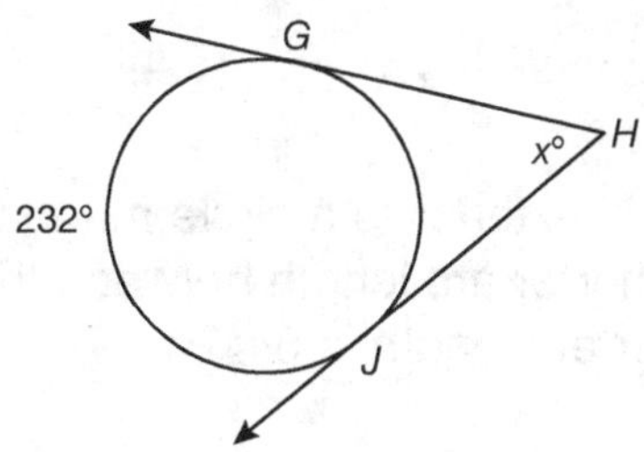

$x = \frac{1}{2}(232° - 128°)$; $x =$ ___

3.

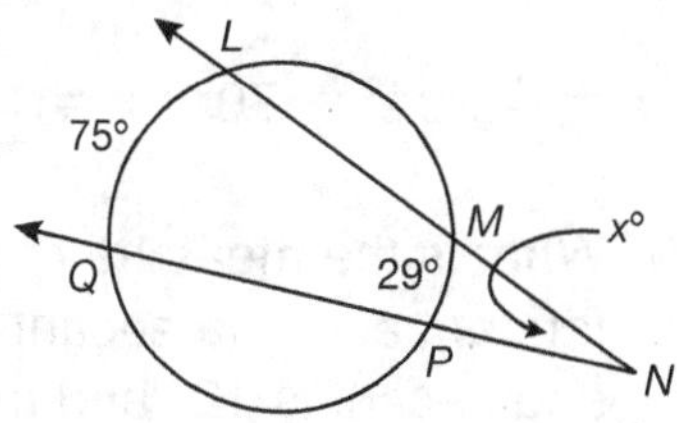

$x = \frac{1}{2}(75° - 29°)$; $x =$ ___

4. What is the measure of the angle exterior to a circle made by two secants if the shorter arc is 52° and the longer arc is 135°? ___

5. What is the measure of the angle exterior to a circle made by two tangents if there is an arc that is 267°? ___

Intercepting by a Secant and a Tangent

It is also possible for an exterior angle to intersect by a secant and a tangent. The measure of an exterior angle is still equal to half the difference of the intercepted arcs.

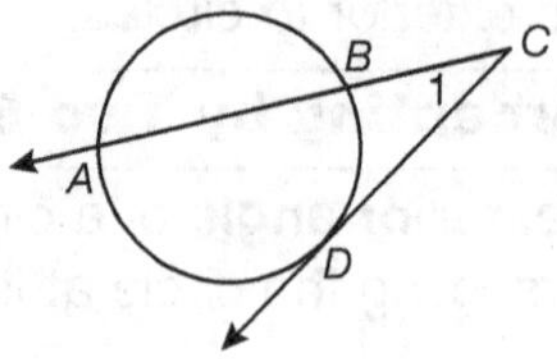

$$m\angle 1 = \frac{1}{2}(mAD - mBD)$$

Find the value of *x*.

Step 1: Write the formula in terms of the given exterior angle.

$$x^\circ = \frac{1}{2}(mTU - mSU)$$

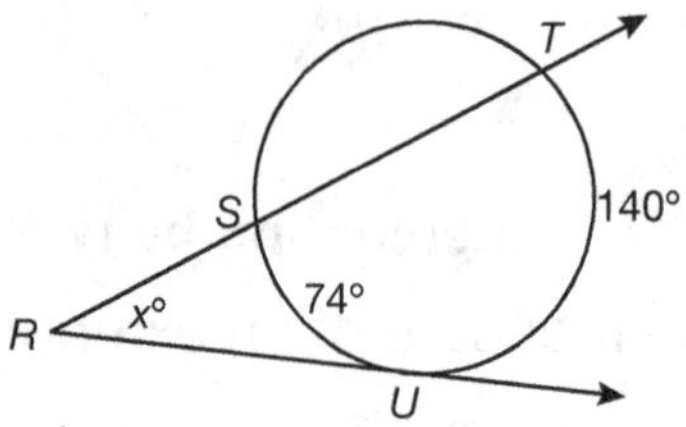

Step 2: Substitute $mSU = 74^\circ$ and $mTU = 74^\circ$ into the formula and solve.

$$x^\circ = \frac{1}{2}(mTU - mSU)$$

$$= \frac{1}{2}(140^\circ - 74^\circ)$$

$$x^\circ = 33^\circ$$

Practice

Find the value of *x*.

6.

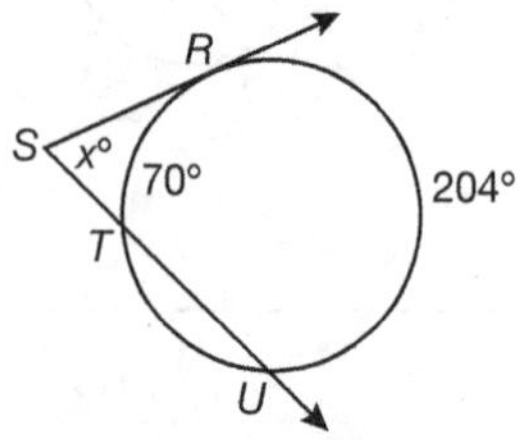

$x = \frac{1}{2}(204^\circ - 70^\circ)$; $x =$ ____

7.

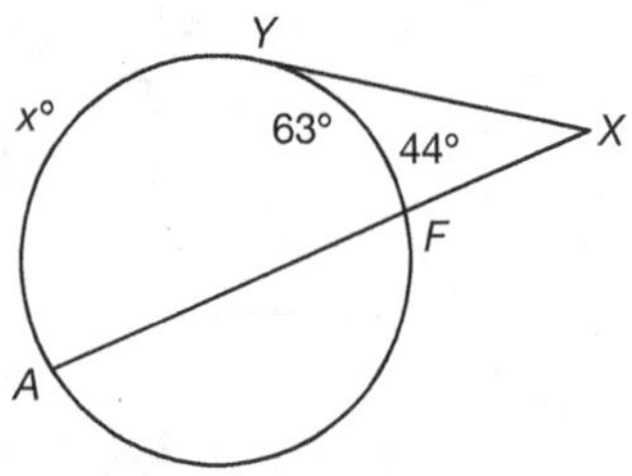

$44^\circ = \frac{1}{2}(x - 63^\circ)$; $x =$ ____

8.

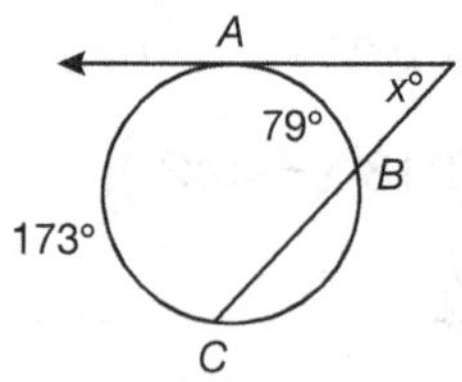

$x = \frac{1}{2}($____ $- 79^\circ)$; $x =$ ____

9. What is the measure of the angle exterior to a circle made by one tangent and one secant if the shorter arc length between the tangent and secant is 12° and the longer arc length is 64°? ____

10. If the measure of the angle exterior to a circle made by one tangent and one secant is 54° and the shorter arc length between the tangent and secant is 42°, what is the longer arc length? ____

11. What is the measure of the angle exterior to a circle made by one tangent and one secant if the shorter arc length between the tangent and secant is 67° and the longer one is 172°? ____

Name ______________________ Date __________ Class __________

Reteaching 80

Finding Surface Area and Volumes of Spheres

You have found the surface area and the volume of cylinders. Now you will find the surface area and the volume of spheres.

Surface Area of a Sphere

The surface area of a sphere with radius r is $SA = 4\pi r^2$.

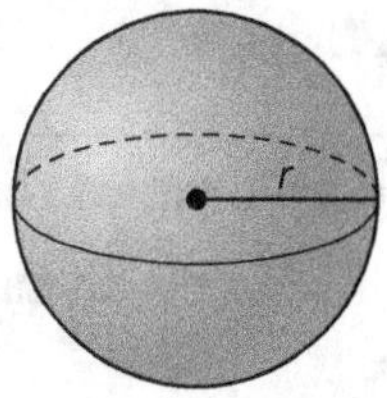

Find the surface area of the sphere below.

Round to the nearest tenth.

Step 1: Find r by dividing the diameter, $d = 20$ m, by two.

$r = \frac{20}{2}$

$= 10$ m

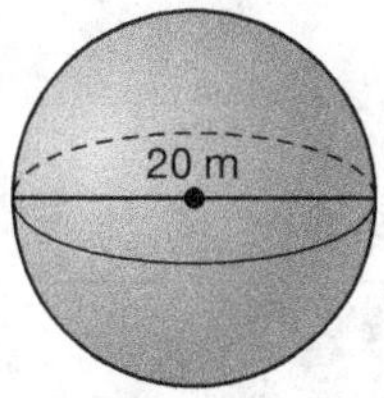

Step 2: Substitute $r = 10$ m into the formula and solve.

$SA = 4\pi r^2$

$= 4\pi(10)^2$

$\approx 1{,}256.6 \text{ m}^2$

Practice

Find the surface area of each sphere. Round each answer to the nearest tenth.

1. Find the surface area of sphere S.

The diameter of sphere S is ______________.

This means r, the radius of the sphere, is ______________.

$r^2 =$ __

$SA = 4\pi r^2 =$ __________

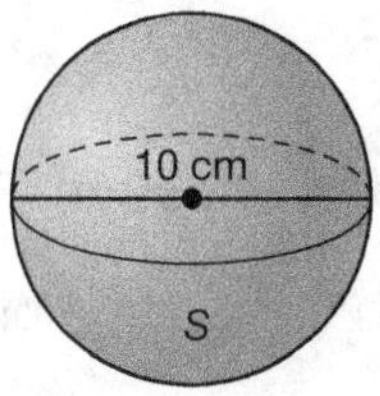

2.

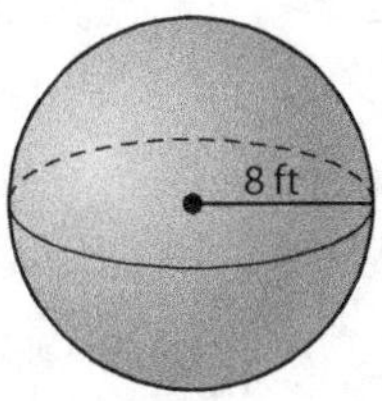

____

3.

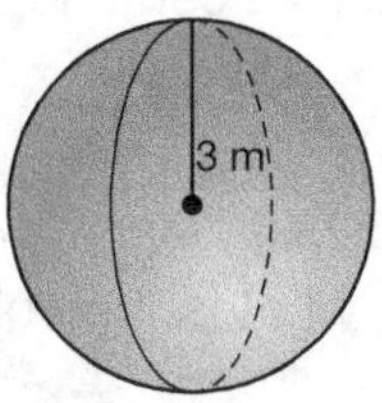

____

4.

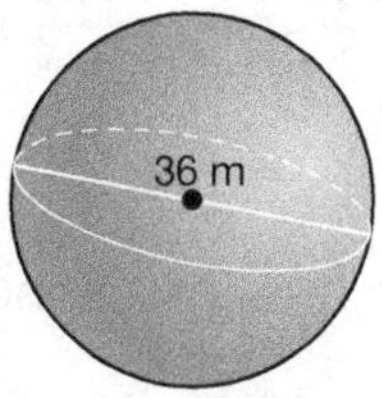

____

Volume of a Sphere

The volume of a sphere with radius r is $V = \frac{4}{3}\pi(13)^3$.

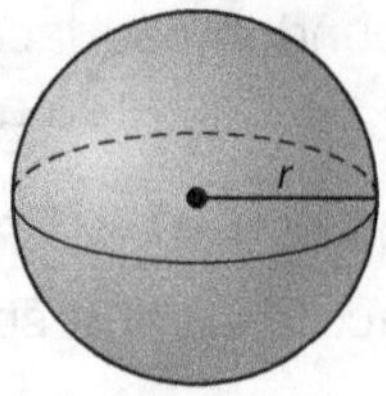

Find the volume of the sphere below. Round to the nearest tenth.

Step 1: Find r by dividing the diameter, $d = 26$ ft, by two.

$r = \frac{26}{2}$

$= 13$ ft

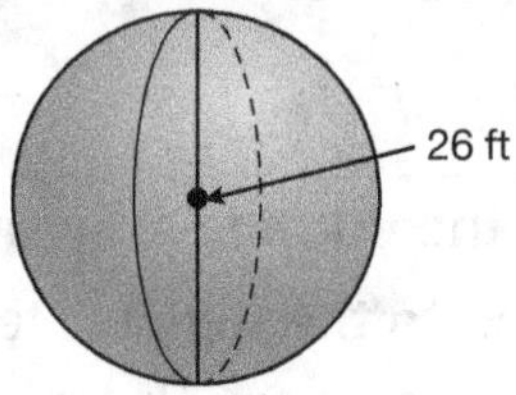

Step 2: Substitute $r = 13$ ft into the formula and solve.

$V = \frac{4}{3}\pi r^3$

$= \frac{4}{3}\pi(13)^3$

$\approx 9{,}202.8 \text{ ft}^3$

Practice

Find the volume of each sphere. Round each answer to the nearest tenth.

5. The diameter of this sphere is _______________.

   The radius, r, is ______________.

   $r^3 =$ ____

   $\frac{4}{3}\pi r^3 =$ ____________

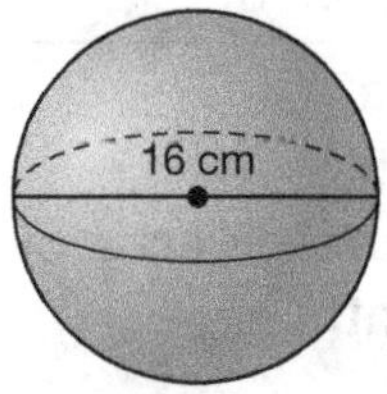

Answer each question. Round each answer to the nearest tenth.

6. What is the volume of the sphere to the right?

   ____

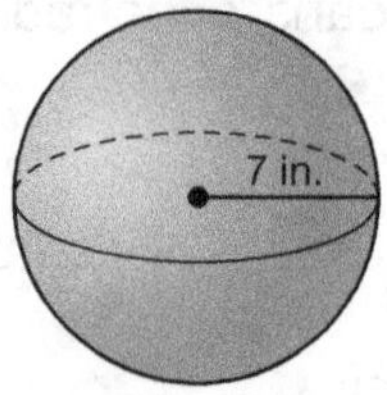

7. What is the volume of a sphere with diameter 42 centimeters?

   ____

8. If the surface area of a sphere is $64\pi \text{ mm}^2$, what is the volume of the sphere?

   ____

Name ______________________ Date __________ Class __________

Reteaching INV 8
Patterns

You have learned about transformations and symmetry. Now you will learn about geometric patterns.

Use the isosceles triangle to the right.

Step 1: Draw the triangle on a coordinate plane so that the base is on the *x*-axis with a base angle at the origin. Then, reflect the triangle over the *x*-axis and draw the result.

Step 2: Reflect the pattern from **Step 1** over the *y*-axis and draw the resulting figure.

Step 3: Does the final pattern have rotational symmetry? If so, what is the order of rotational symmetry? yes; 2

Step 4: Does the final pattern have any lines of symmetry? If so, how many? yes; 2

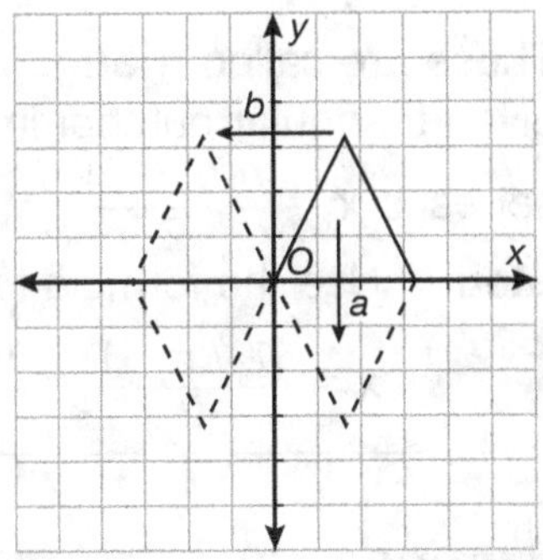

Practice

Complete the steps of the pattern on the coordinate plane.

1. Step 1: Draw an isosceles right triangle with one leg on each axis.

   Step 2: Reflect the triangle over the *y*-axis.

   Step 3: Reflect the pattern from Step 2 over the *x*-axis.

   Step 4: Does the resulting pattern have any lines of symmetry? If so, how many? ________

   Step 5: Does the final pattern have rotational symmetry? If so, what is the order of rotational symmetry? ________

   Step 6: What kind of polygon is the resulting figure? ________

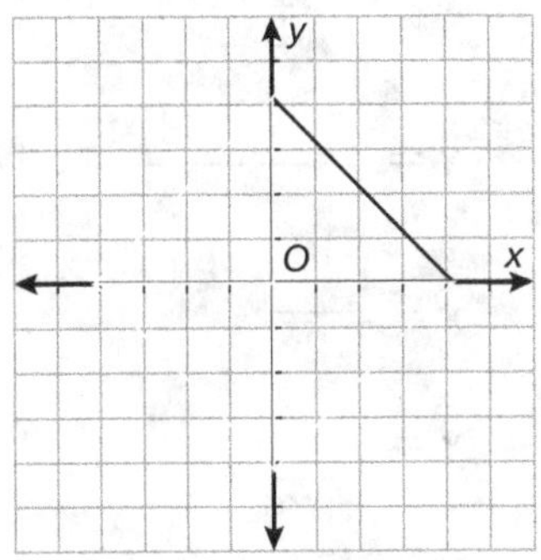

Use the rhombus to the right for problems 2–4.

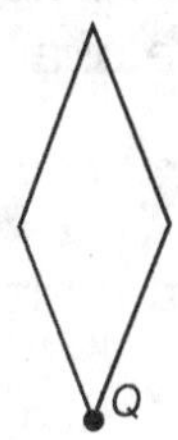

2. Sketch the resulting figure after a rotation of 120° counterclockwise around point Q.
3. If the figure is rotated 120° until it is returned to its original position, what is the order of rotational symmetry in the final pattern? ____
4. Does the resulting pattern have any lines of symmetry? If so, how many? ________

Look at the pattern on the right. The first figure has one dot; the second, three dots; the third, six dots.

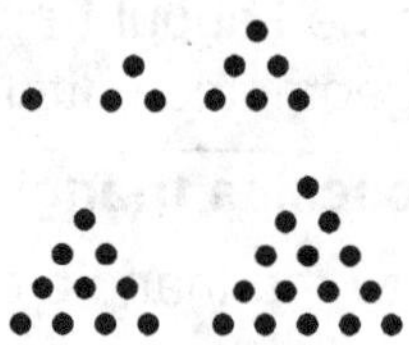

Step 1: Draw the next two figures in the pattern.

How many dots form each? 10, 15

Step 2: List the number of dots in each triangle.

1, 3, 6, 10, 15

These are called **triangular numbers.** The formula to find a given triangular number in the triangular number series is $x = \frac{n(n + 1)}{2}$.

Step 3: Use the formula to find the sixteenth triangular number.

$$x = \frac{n(n + 1)}{2} = \frac{16(16 + 1)}{2} = 136$$

Practice

Find the twenty-fifth and twenty-sixth triangular numbers.

5. Find the twenty-fifth triangular number.

$x = \frac{n(n + 1)}{2}$ Formula

$x = \frac{\_\_(\_\_ + 1)}{2}$ Substitute.

$x = \frac{\_\_}{2}$ Solve.

$x = \_\_$

The twenty-fifth triangular number is ____.

6. Find the twenty-sixth triangular number.

$x = \frac{n(n + 1)}{2}$

$x = \frac{\_\_(\_\_ + 1)}{2}$

$x = \frac{\_\_}{2}$

$x = \_\_$

The twenty-sixth triangular number is ____.

Complete the questions for the cubed numbers sequence.

7. Cubed numbers are numbers that can be used to find the volume of a cube.

The first few numbers of the sequence are 1, 8, 27, 64, . . .

What are the next three numbers in the sequence?

____

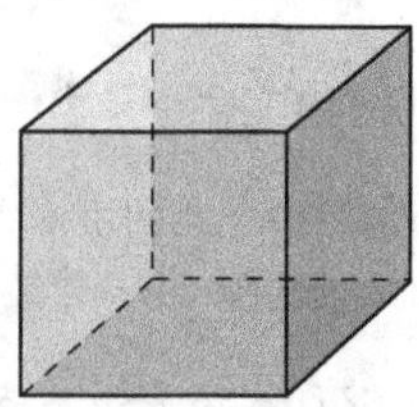

s

$V = s^3$

8. Write an equation to find the nth number in the sequence. ______

9. What is the twenty-second cubed number? ______

Name _______________________________ Date ___________ Class ___________

Reteaching 81

Graphing and Solving Linear Systems

You have found the slopes of equations and lines. Now you will solve linear systems graphically and algebraically.

Solving Linear Systems Graphically

A **system of linear equations** is a set of two or more linear equations. A system of linear equations can be solved by graphing the lines.

Solve this linear system by graphing the lines.

$y = 3x - 4$; $y = -x + 4$

Step 1: Graph each line using the equations given.

Step 2: The lines intersect at the point (2, 2).

Step 3: Check your answer by substituting $x = 2$ and $y = 2$ into both equations.

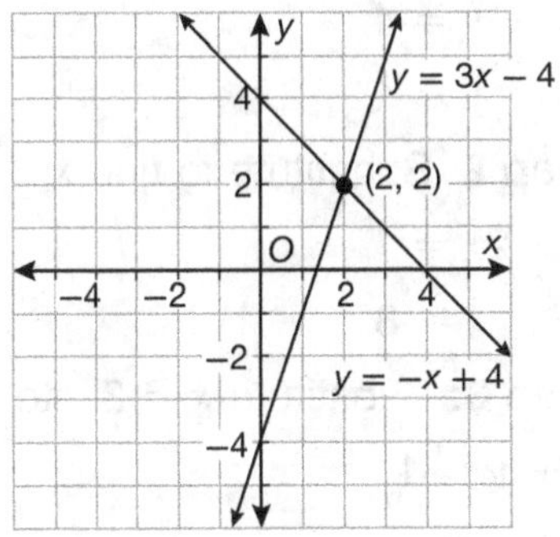

$y = 3x - 4$	$y = -x + 4$
$2 = 3(2) - 4;$	$2 = -2 + 4$
$2 = 2$	$2 = 2$

Therefore, (2, 2) is the correct solution to this system.

Practice

Complete the steps to solve the system of equations by graphing.

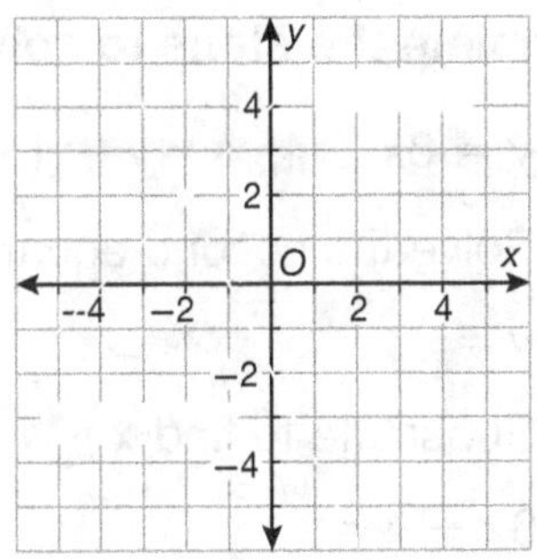

1. $y = x + 4$; $y = -4x - 6$

   Graph each line. The lines intersect at (_______).

   Check your answer.

$y = x + 4$	$y = -4x - 6$
_ $= -2 + 4;$	_ $= -4($___$) - 6$
_ $= 2$	_ = _

   The solution is ________.

Solve the system of equations by graphing. Check your answer.

2. $x + 2y = 10$; $x + y = 6$

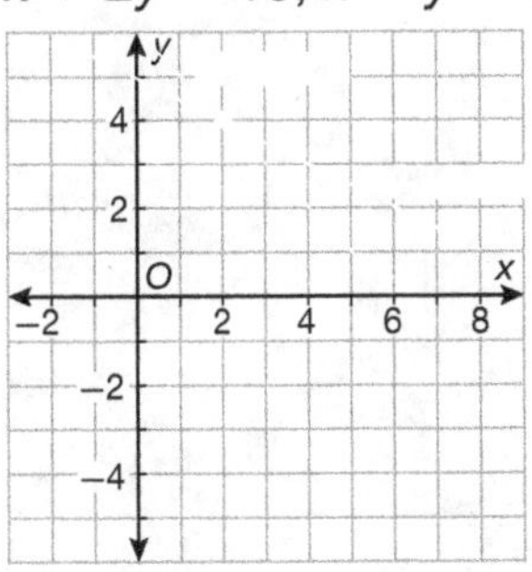

______

3. $2x + 3y = 12$; $5x - 2y = 11$

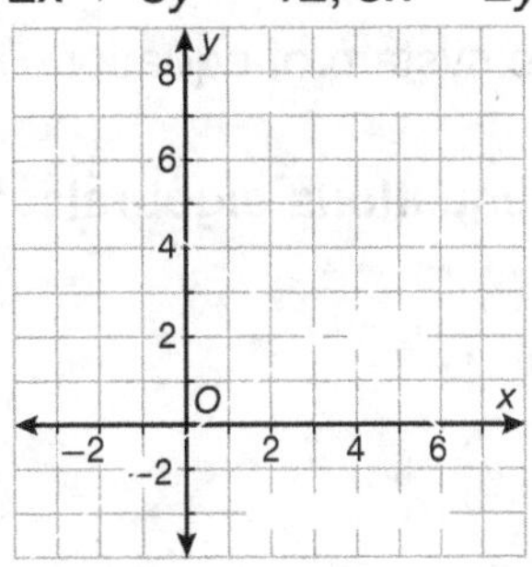

______

Solving Linear Systems Algebraically

To solve a system algebraically, solve both equations for the same variable and use substitution.

Solve this system of equations algebraically.

$2x + y = 4; x + y = 1$

Step 1: Solve both equations for y.

$2x + y = 4$ $\qquad$ $x + y = 1$

$y = 4 - 2x$ $\qquad$ $y = 1 - x$

Step 2: Substitute to find x.

$4 - 2x = 1 - x$

$x = 3$

Step 3: Substitute $x = 3$ into either of the equations.

$x + y = 1$

$3 + y = 1$

$y = -2$

Therefore, the solution to this system is the ordered pair (3, –2).

Practice

Complete the steps to solve the system of equations algebraically.

4. $y = 3x - 4; x = y - 4$

Solve the second equation for y.

$y =$ ______

Substitute to find x.

$3x - 4 =$ ______

$x =$ ______

Substitute $x = 4$ into $y = 3x - 4$.

$y = 3x - 4$

$y = 3(\_) - 4 = \_$

The solution to the system of equations is ______.

Solve the system of equations algebraically.

5. $2x + 3y = 4$
$x - 3y = 2$

______

6. $-5y = 4x - 40$
$y = x - 1$

______

Name ______________________ Date __________ Class __________

Reteaching 82

Inverse Trigonometric Functions

You have found the lengths of right triangle sides using trigonometric functions. Now you will find the measures of right triangle angles using inverse trigonometric functions.

Inverse Trigonometric Functions

With inverse trigonometric functions, you can use side lengths of a right triangle to calculate angle measures.

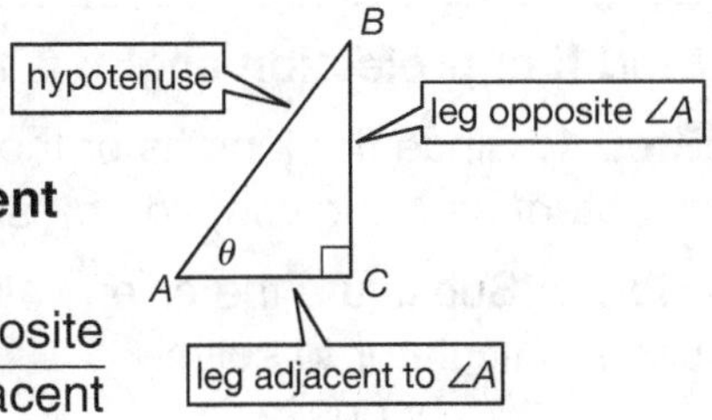

inverse sine $\theta = \sin^{-1}\dfrac{\text{opposite}}{\text{hypotenuse}}$

inverse cosine $\theta = \cos^{-1}\dfrac{\text{adjacent}}{\text{hypotenuse}}$

inverse tangent $\theta = \tan^{-1}\dfrac{\text{opposite}}{\text{adjacent}}$

Find θ to the nearest degree.

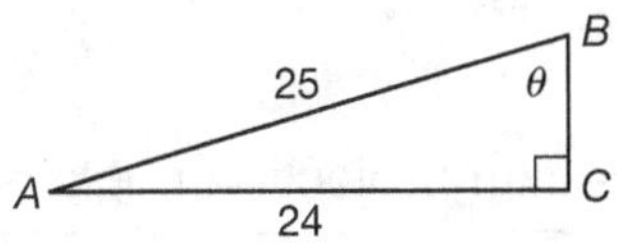

Step 1: Since the lengths of the side opposite and the hypotenuse are given, you must use sine inverse.

Step 2: Substitute the given values into the formula and use a calculator to solve.

$$\theta = \sin^{-1}\frac{\text{opposite}}{\text{hypotenuse}}$$

$$= \sin^{-1}\frac{24}{25}$$

$$\approx 74^\circ$$

Practice

Complete the steps to find θ to the nearest degree.

1. $\theta = \tan^{-1}\dfrac{\text{opposite}}{\text{adjacent}}$

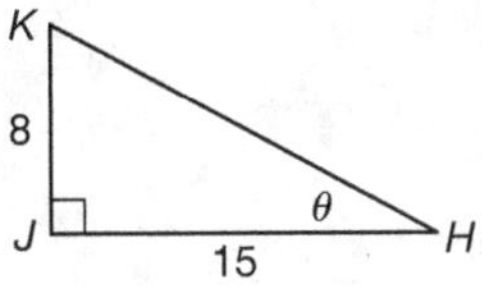

$\theta = \tan^{-1}\dfrac{\_\_}{\_\_}$

$\theta \approx$ ____

2. $\theta = \cos^{-1}\dfrac{\_\_\_\_}{\text{hypotenuse}}$

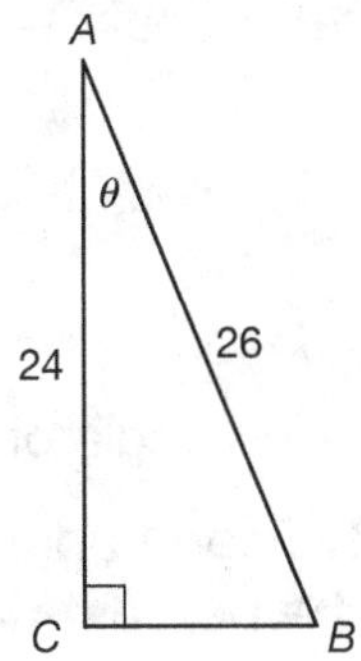

$\theta = \cos^{-1}\dfrac{\_\_}{\_\_}$

$\theta \approx$ ____

Find θ to the nearest degree.

3.

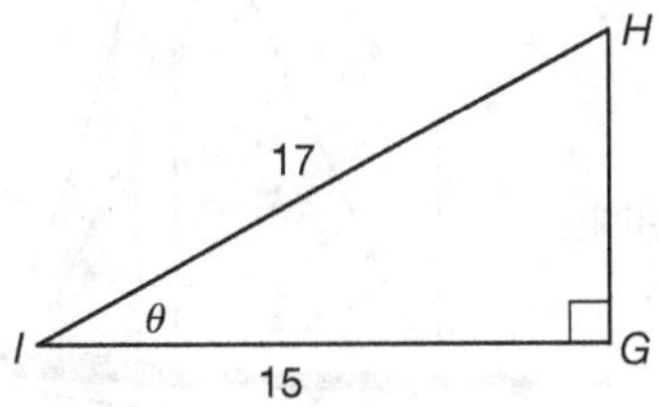

____

4.

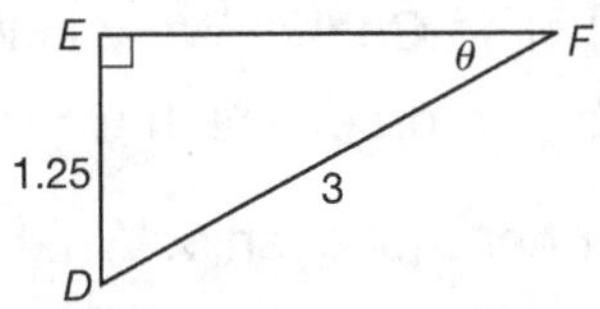

____

5.

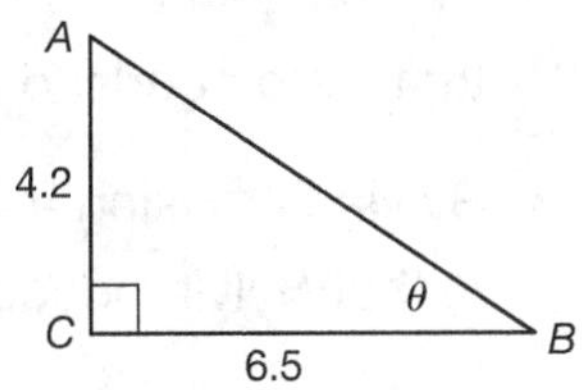

____

Applications of Inverse Trigonometric Functions

You can use inverse trigonometric functions to solve application problems. In this picture, the angle of the weight bench, θ, can be determined by using inverse tangent.

$\theta = \tan^{-1}\dfrac{3.2}{2.6}$

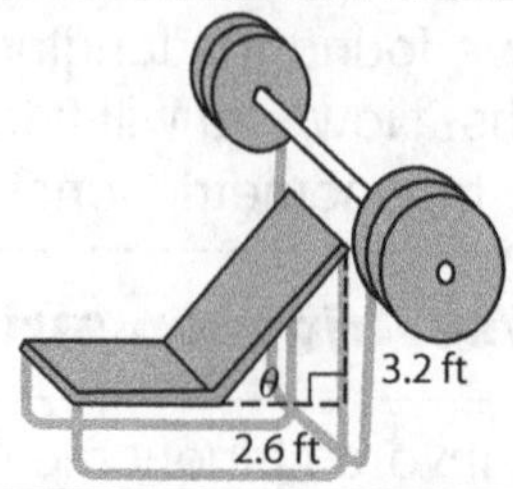

Find film projection angle, θ, to the nearest degree.

Step 1: Since the lengths of the side opposite and the hypotenuse are given, you must use sine inverse.

Step 2: Substitute the given values into the formula and use a calculator to solve.

$$\theta = \sin^{-1}\frac{\text{opposite}}{\text{hypotenuse}}$$
$$= \sin^{-1}\frac{8.2}{15}$$
$$\approx 33^\circ$$

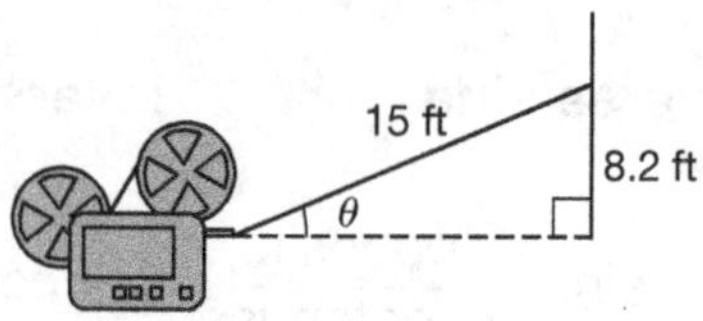

The film projection angle is approximately 33°.

Practice

Complete the steps to solve the application problem. Round to the nearest degree.

6. Yi is building a wheelchair ramp for his grandmother. What is the angle of incline, θ, that Yi needs to use for the ramp?

$$\theta = \cos^{-1}\frac{\text{adjacent}}{\text{hypotenuse}}$$

$\theta = \cos^{-1}\dfrac{\_\_}{\_\_}$

$\theta \approx$ ____

The angle of incline for the ramp is approximately ____.

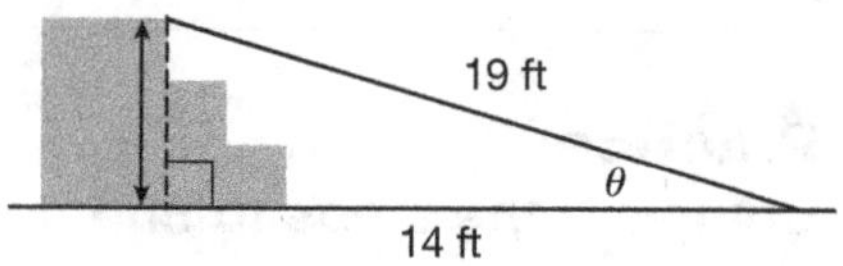

Solve each application problem. Round your answers to the nearest degree.

7. Pete is going to throw a baseball to Cedric. What is the angle, θ, that Pete needs to throw the ball in order reach Cedric? ____

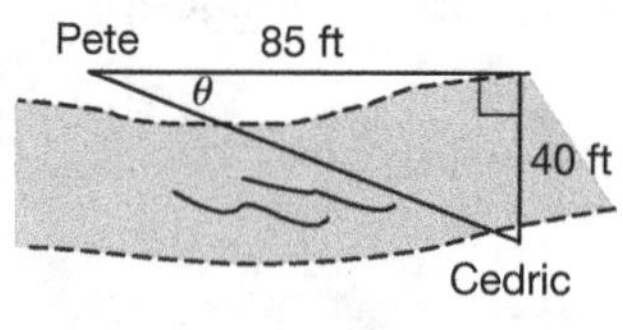

8. Sylvia is making a new sail for her model ship. What is the angle, θ, that Sylvia needs to use to cut the corner of the sail? ____

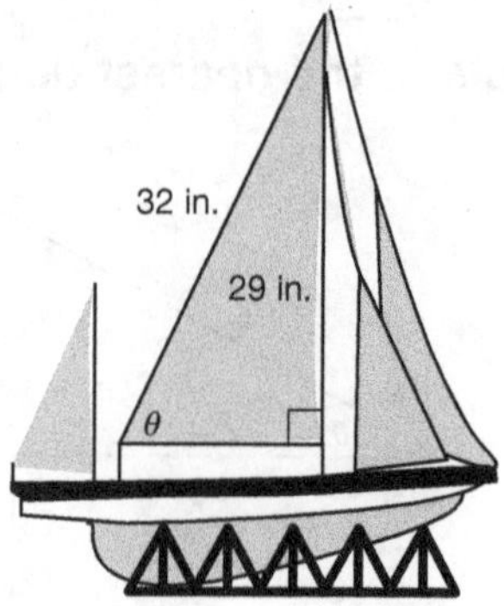

Name ________________________ Date __________ Class __________

Reteaching 83
Vector Addition

You have worked with applications of right triangles. Now you will work with adding vectors.

Adding Vectors

Example:

You can use the Parallelogram Method to add $\vec{W} = \langle 3, 4 \rangle$ and $\vec{Z} = \langle 5, -2 \rangle$.

Step 1: Draw the vectors with their tails at the same starting point on the coordinate grid

Step 2: Complete the parallelogram by sketching the other two sides.

Step 3: The resultant vector is the diagonal of the parallelogram.

Step 4: Solve for the resultant vector.

$\vec{w} + \vec{z} = \langle 3, 4 \rangle + \langle 5, -2 \rangle$

$w + \vec{z} = \langle 8, 2 \rangle$

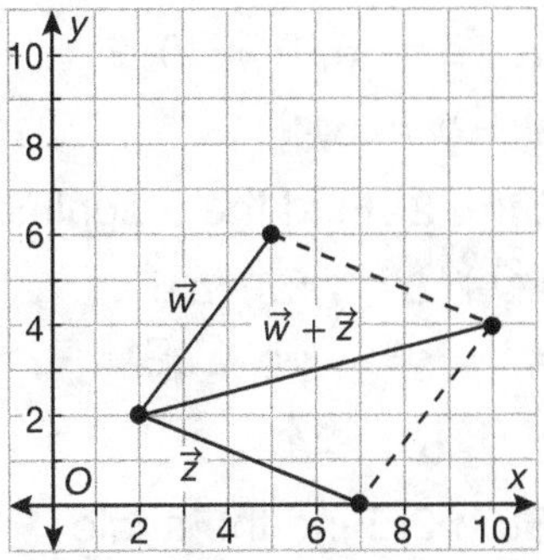

Practice

Complete the statements and diagram to add the given vectors.

1. Add $\vec{g} = \langle 2, 4 \rangle$ and $\vec{h} = \langle 4, -1 \rangle$.

Complete the diagram by finishing the parallelogram.

Sketch in the resultant vector which is the __________.

$\vec{g} + \vec{h} = \langle 2,4 \rangle +$ ________

$\vec{g} + \vec{h} =$ ______

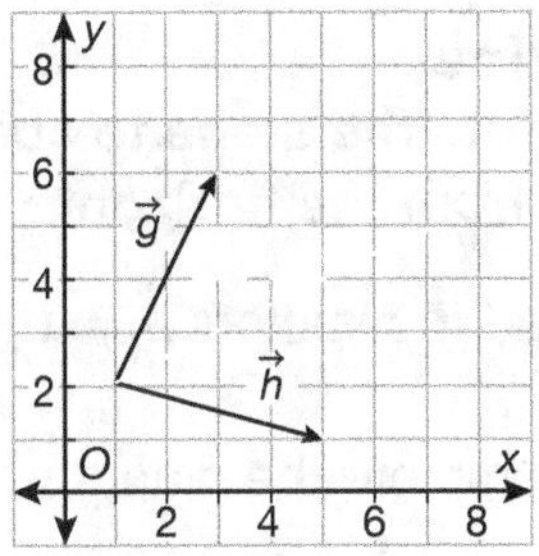

Determine the resultant vector.

2. $\vec{a} = \langle 4, 2 \rangle$ and $\vec{b} = \langle 5, -3 \rangle$

____

3. $\vec{c} = \langle 5, 0 \rangle$ and $\vec{d} = \langle 2, -1 \rangle$

____

4. $\vec{a} = \langle 3, -3 \rangle$ and $\vec{b} = \langle 6, 4 \rangle$

______

5. $\vec{c} = \langle 2, 3 \rangle$ and $\vec{d} = \langle 2, 2 \rangle$

______

6. $\vec{a} = \langle 6, 3 \rangle$ and $\vec{b} = \langle 1, -3 \rangle$

______

7. $\vec{c} = \langle 8, -2 \rangle$ and $\vec{d} = \langle 3, 4 \rangle$

____

Adding Vectors Using Trigonometry

The **magnitude** of a vector is its length. The **direction** of a vector is the angle that it makes with a horizontal line, such as the *x*-axis.

Example:

Add the vectors $\vec{a} = \langle 5, 0 \rangle$ and $\vec{b} = \langle 0, 2 \rangle$. Find the magnitude and angle from the horizontal of the resultant vector.

Step 1: Draw a diagram. Add the vectors by adding the components.

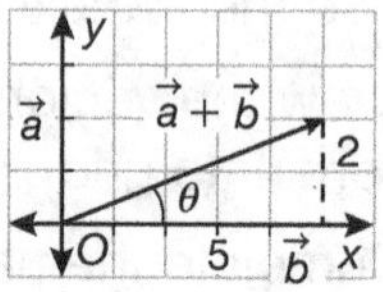

$\vec{a} + \vec{b} = \langle 5, 0 \rangle + \langle 0, 2 \rangle$

$\vec{a} + \vec{b} = \langle 5, 2 \rangle$

Step 2: Find the magnitude.

$a^2 + b^2 = (\vec{a} + \vec{b})^2$

$5^2 + 2^2 = 29 = (\vec{a} + \vec{b})^2$

$5.4 \approx \vec{a} + \vec{b}$

Step 3: Find the angle

$\tan \theta = \frac{opposite}{adjacent} = \frac{2}{5}$

$\theta = \tan^{-1}\frac{2}{5} \approx 22°$

Practice

Complete the steps to add the vectors. Find the magnitude and angle from the horizontal of the resultant vector.

8. Add the vectors $\vec{a} = \langle 0, 4 \rangle$ and $\vec{b} = \langle 4, 0 \rangle$.

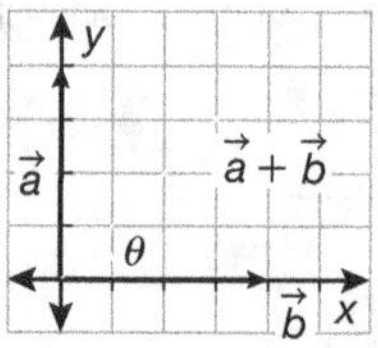

Complete the diagram. $\vec{a} + \vec{b} = \langle 0, 4 \rangle + \langle 4, 0 \rangle$

$\vec{a} + \vec{b} =$ ______

Find the magnitude. $a^2 + b^2 = (\vec{a} + \vec{b})^2$

___ $+ 4^2 = (\vec{a} + \vec{b})^2$

___ $\approx \vec{a} + \vec{b}$

Find the angle. $\tan \theta = \frac{\_\_}{4}$

$\theta = \tan^{-1}$ _ $\approx$ ___

Find the magnitude and angle from the horizontal of the resultant vector. Round to the nearest degree.

9. $\vec{a} = \langle 6, 0 \rangle$ and $\vec{b} = \langle 0, 3 \rangle$

____

10. $\vec{a} = \langle 5, 0 \rangle$ and $\vec{b} = \langle 0, 8 \rangle$

____

Name ______________________ Date __________ Class __________

Reteaching 84
Dilations

You have worked with transformations and translations. Now you will work with dilations of figures.

Dilations

A **dilation** is a transformation that changes the size of a figure but not its shape. A dilation that results in a smaller image is called a reduction or contraction. A dilation that results in a larger image is called an enlargement or expansion.

Dilation

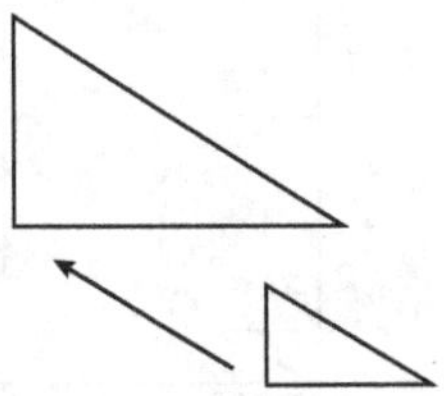

Not a Dilation

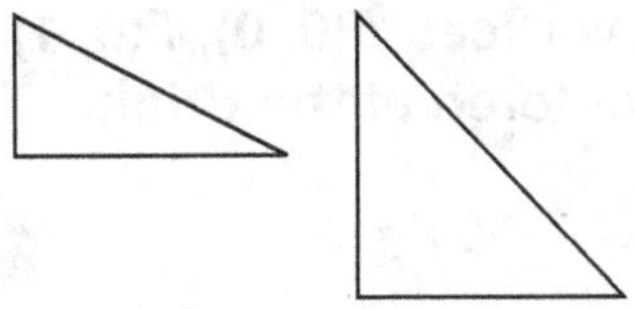

A dilation is a transformation in which the lines that connect every point *A* with its image *A*′ all intersect at point *P*, which is called the **center of dilation.**

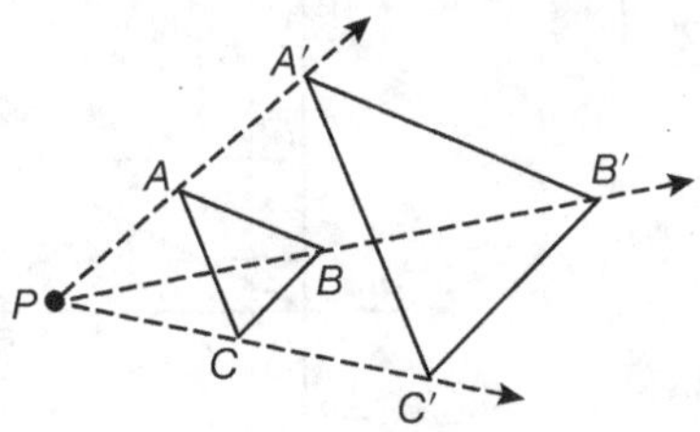

Practice

Complete the statements to determine whether the transformation appears to be a dilation.

1. Does the second figure appear to be the same shape? ___

   Does the second figure appear to be a different size? ___

   The figure appears to be a dilation.

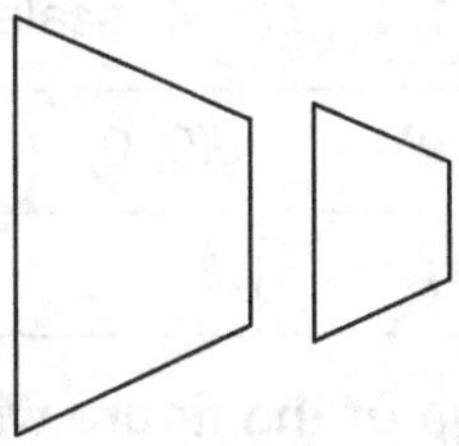

Copy each triangle and center of dilation. Draw the image of the triangle under a dilation with the given scale factor.

2. scale factor: 2

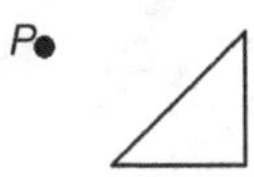

3. scale factor: 1/2

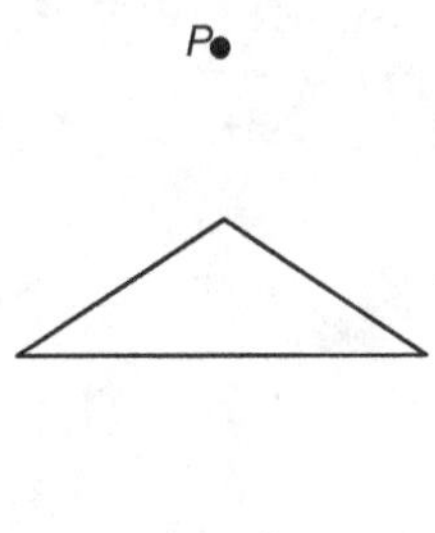

Dilations in the Coordinate Plane

For $k > 1$:

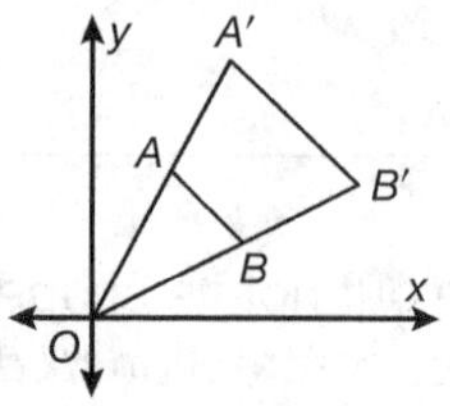

$(x, y) \rightarrow (kx, ky)$

For $0 < k < 1$:

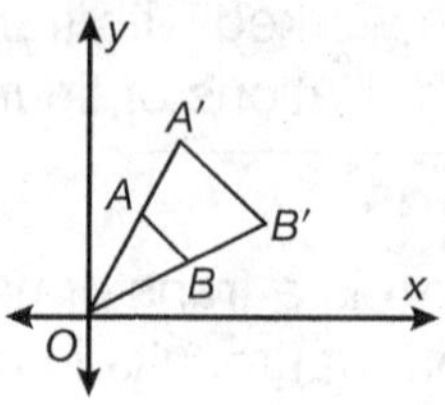

$(x, y) \rightarrow (kx, ky)$

Draw the image of ΔEFG with vertices $E(0, 0)$, $F(0, 1)$, and $G(2, 1)$. Apply a dilation centered at the origin and with a scale factor of 3.

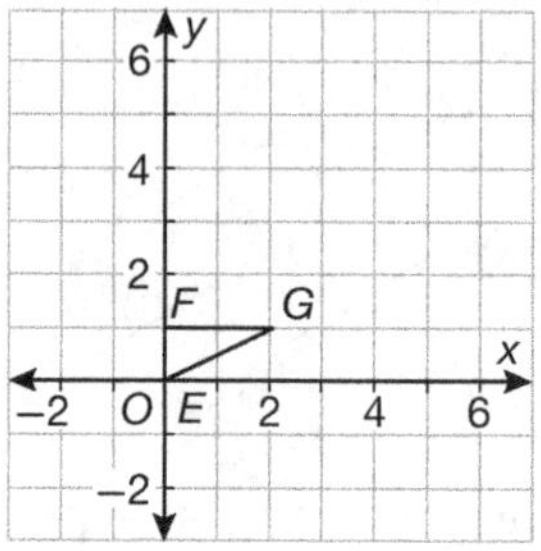

Step 1: Plot the points for ΔEFG.

Step 2: Apply the scale factor to each point. Multiply each coordinate by 3 (the scale factor).

Preimage (x, y)	E(0, 0)	F(0, 1)	G(2, 1)
Image (3x, 3y)	E′(0, 0)	F′(0, 3)	G′(6, 3)

Step 3: Plot the image points and sketch the triangle.

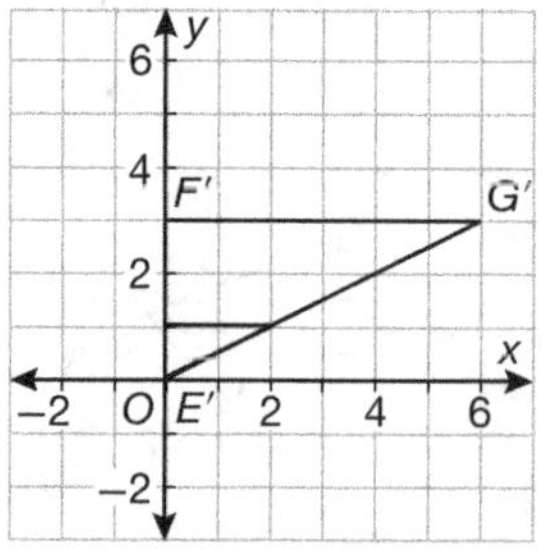

Practice

Complete the steps to dilate the figure on the coordinate plane.

4. $J(0, 0)$, $K(-1, 2)$, $L(3, 4)$ scale factor: 2

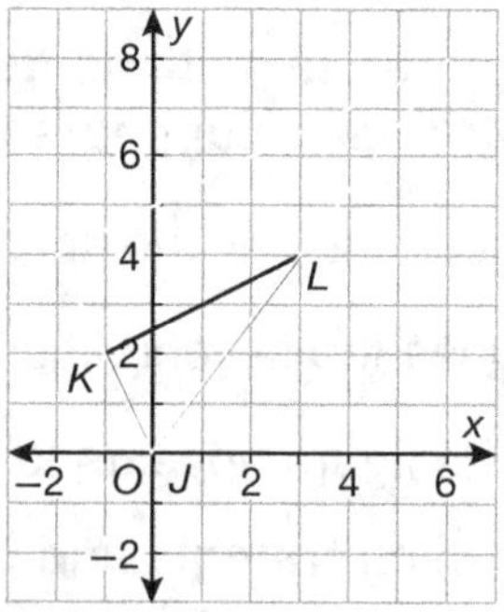

Preimage (x, y)	J(0, 0)	________	L(3, 4)
Image (2x, 2y)	J′ ______	K′(−2, 4)	L′ ______

Draw the image of the figure with the given vertices under dilation with the given scale factor and centered at the origin.

5. $A(0, 0)$, $B(0, 6)$, $C(6, 3)$; scale factor: $\frac{1}{3}$

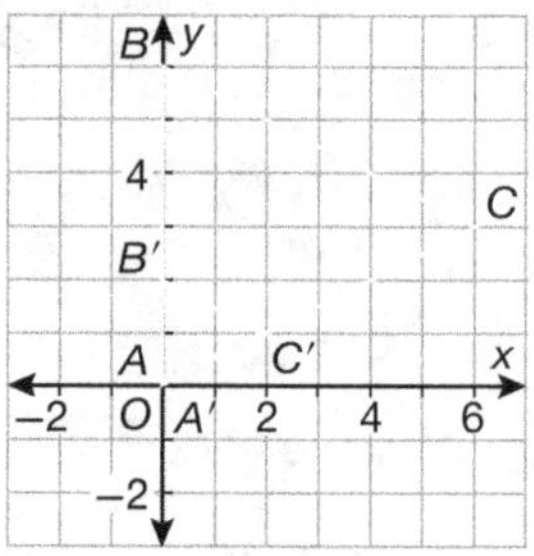

6. $R(1, 0)$, $S(1, -2)$, $T(-1, -2)$; scale factor: 2

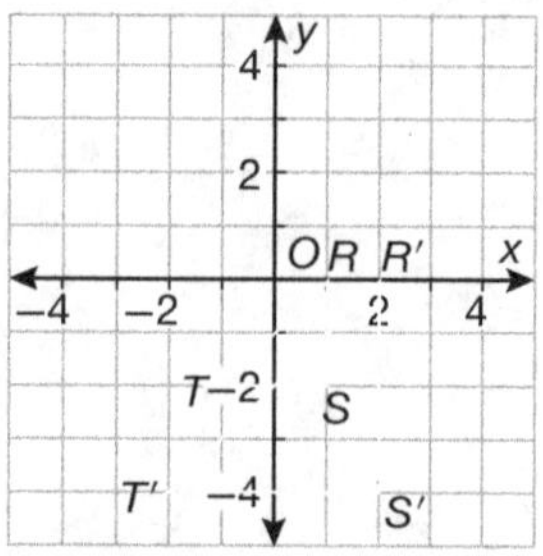

Name ______________________ Date __________ Class ____________

Reteaching 85

Cross Sections of Solids

You have worked with dilations. Now you will work with cross sections of solids.

Describing and Sketching Cross Sections of Solids

The **cross section** of a solid is the intersection of a three-dimensional figure and a plane.

Describe and draw the cross section of the solid.

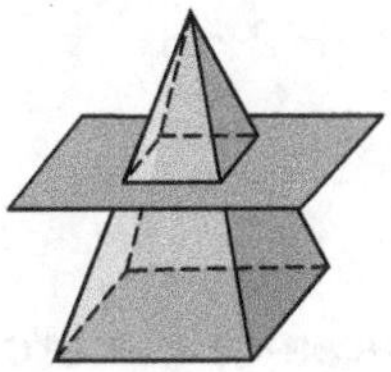

The solid is a square pyramid with a horizontal cross section.

The intersection of the plane with the pyramid is a square.

The cross section is a square.

Practice

Complete the steps to describe and draw the cross section

1. 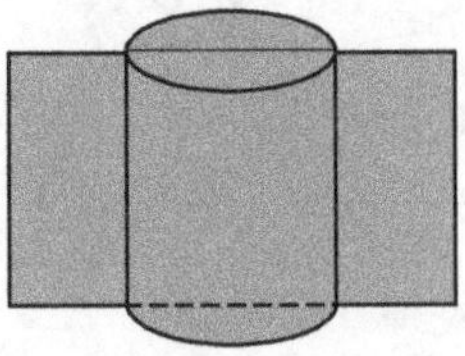

The cross section is a __________.

Draw the cross section.

2. 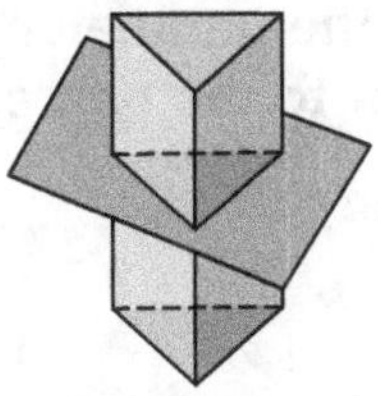

The cross section is a __________.

Draw the cross section.

Describe and draw the cross section of each solid.

3. 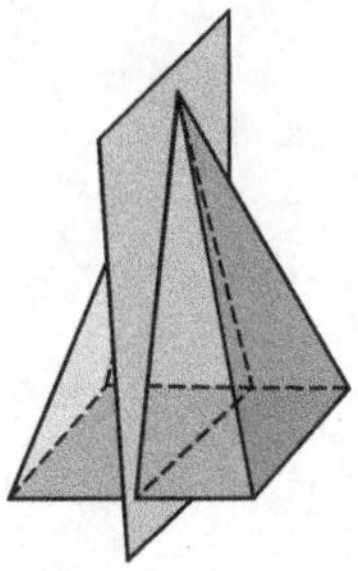

____

4. 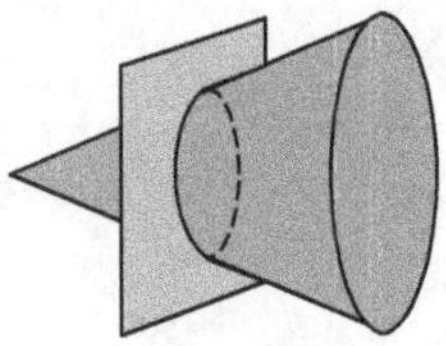

____

______

Finding Perimeter and Area of a Cross Section

Find the perimeter of the cross section of the square pyramid if the cross section is vertical.

The width of the cross section is 11 inches. The cross section is a rectangle. Its height is the same as that of the cylinder. The base of the rectangle is 11 inches.

$P = 2b + 2h$

$= 2 \cdot 11 + 2 \cdot 6$

$= 22 + 12$

$= 34$ in.

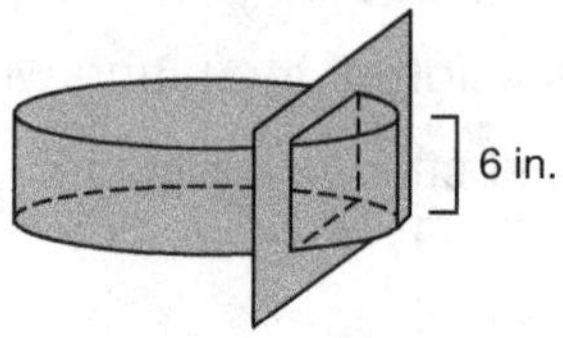

Find the area of the cross section.

$A = bh$

$= 11 \cdot 6$

$= 66 \text{ in}^2$

Practice

Complete the steps to find the perimeter or the area of each cross section. Round your answer to the nearest hundredth.

5. Perimeter of cross section

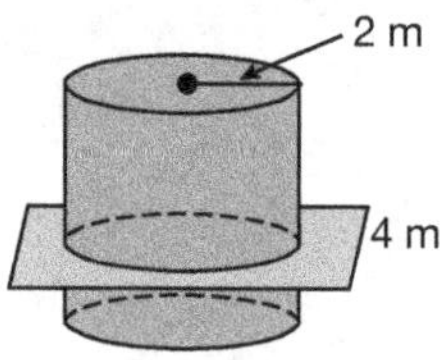

The cross section is a ______.

$P = 2\pi\underline{\quad}$

$= 2\pi\underline{\quad}$

$= \underline{\quad}\pi$

$= \underline{\qquad\quad}$

6. Area of cross section

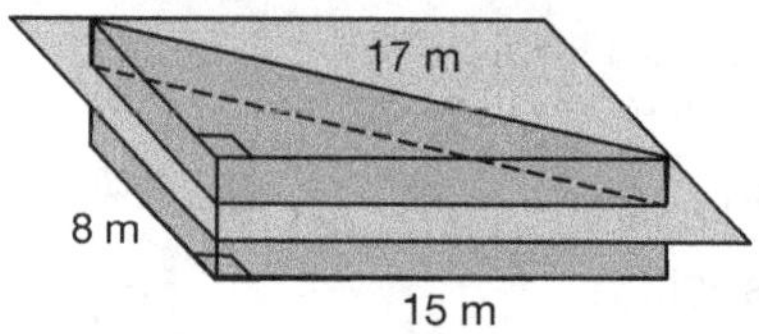

The cross section is a ____________.

$A = \frac{1}{2}b\underline{\quad}$

$= \frac{1}{2}(15)(\underline{\quad})$

$= \underline{\qquad\quad}$

Find the perimeter or area of each cross section.

7. Perimeter of cross section

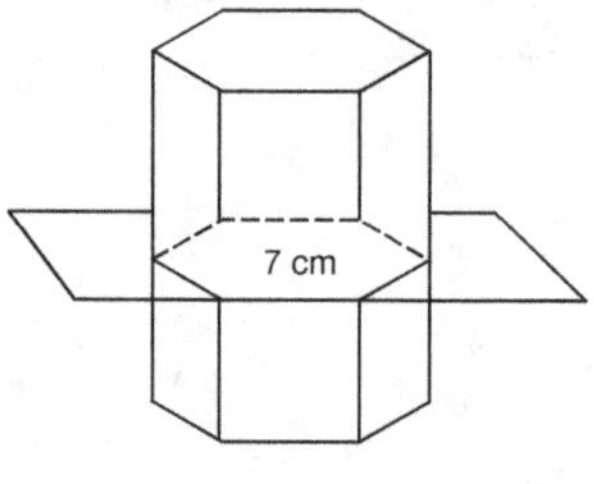

______

8. Area of cross section

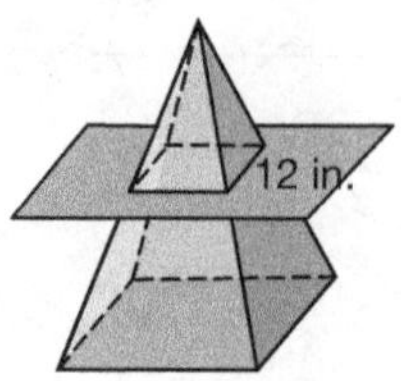

______

Name ______________________ Date __________ Class __________

Reteaching 86

Determining Chord Length

You have worked with chords and tangents in circles. Now you will learn how to determine the length of a chord.

Chords

If two chords intersect in a circle, then the products of the chord segments are equal.	$AE \cdot EB = CE \cdot ED$ 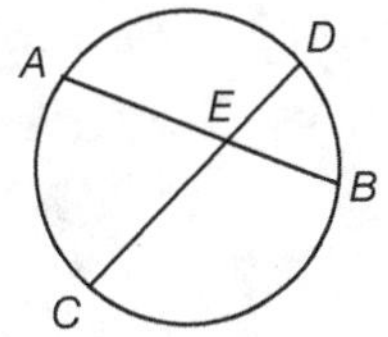

Example:

Find the value of *x* and the length of each chord.

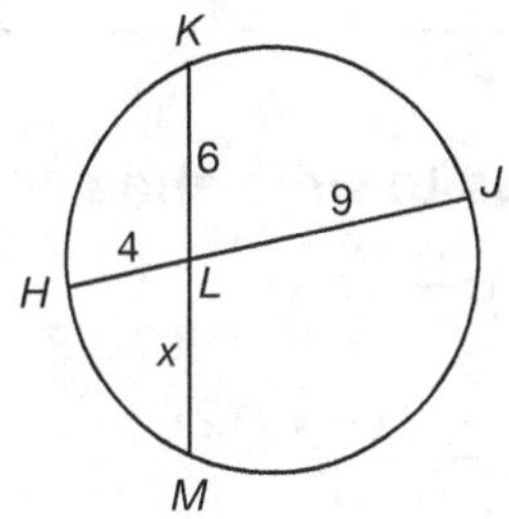

$HL \cdot LJ = KL \cdot LM$

$4 \cdot 9 = 6 \cdot x$ Substitute.

$39 = 6x$ Simplify.

$6 = x$ Divide each side by 6.

$HJ = 4 + 9 = 13$

$KM = 6 + x$

$= 6 + 6 = 12$

Practice

Complete the statements.

1. $TU \cdot UV = RU \cdot US$

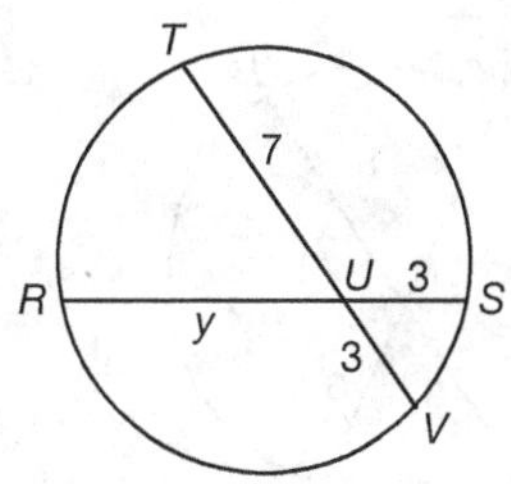

___ $\cdot 3 = y \cdot$ ___ Substitute.

___ $= 3y$ Simplify.

___ $= y$ Divide each side by 3.

$TV =$ ___ $+ 3 = 10$

$RS = 3 + y$

$= 3 +$ ___ $=$ ___

Find the value of the variable and the length of each chord.

2. ______________________

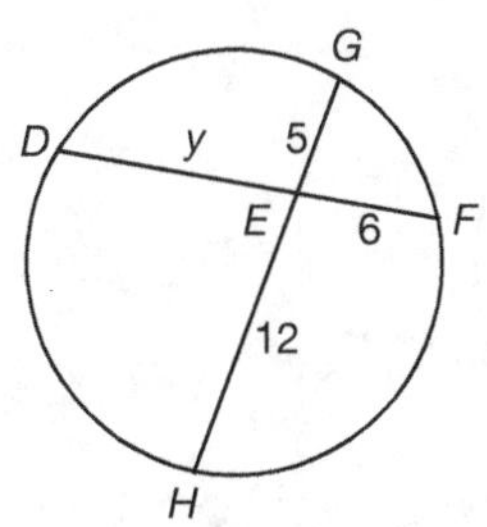

3. ______________________

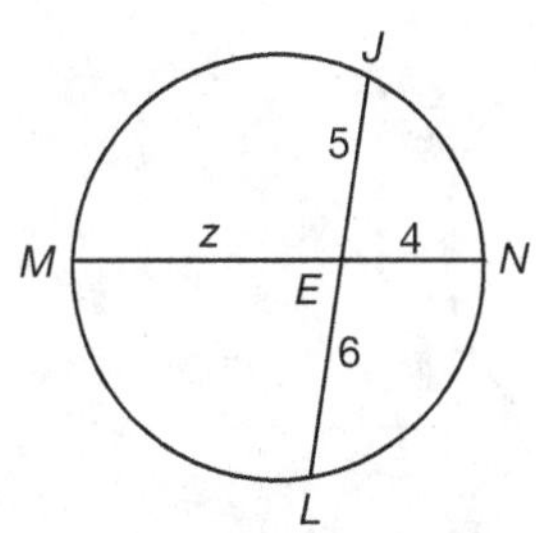

More on Chords

Example:

Use the expressions for the segment lengths to write and solve an equation for x.

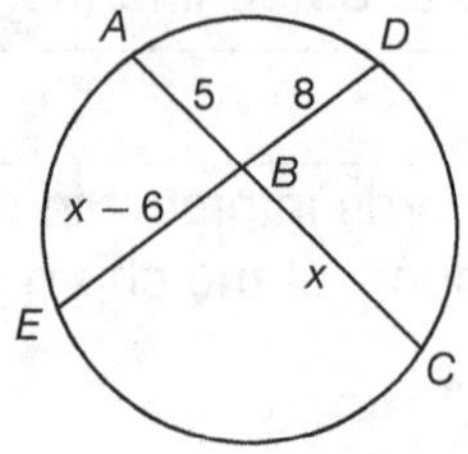

$AB \cdot BC = DB \cdot BE$

$5 \cdot x = 8(x - 6)$

$5x = 8x - 48$

$48 = 3x$

$16 = x$

Check your answer.

$BC = 16$; $BE = 16 - 6 = 10$; $(5)(16) = 8(10)$

Practice

Complete the steps to write and solve an equation for x.

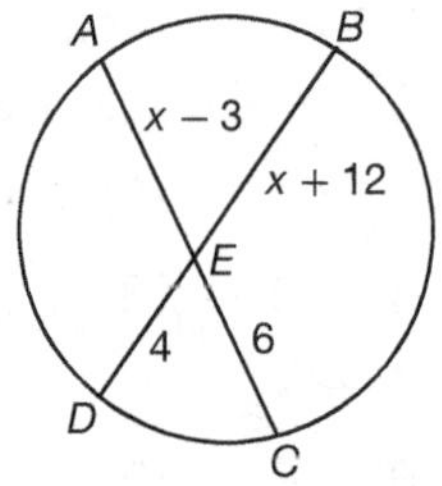

4. $AE \cdot$ ____ $= ED \cdot$ ____

____ $\cdot 6 =$ ____$(x + 12)$

____ $- 18 = 4x +$ ____

$2x =$ ____

$x =$ ____

In each circle, use the expressions for the segment lengths to write and solve an equation for x.

5. ______

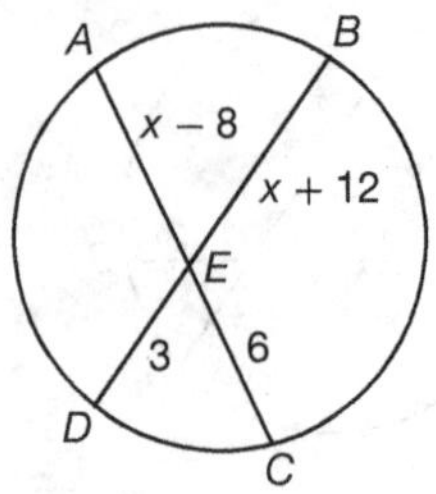

6. ______

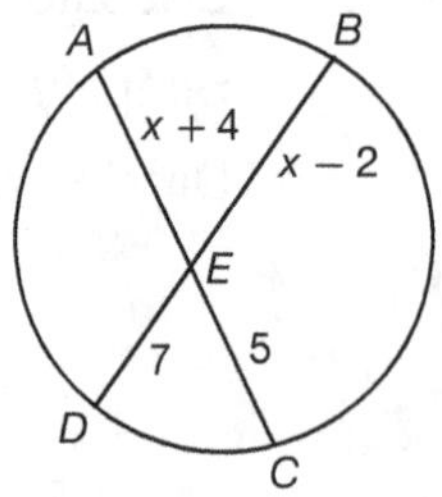

7. ______

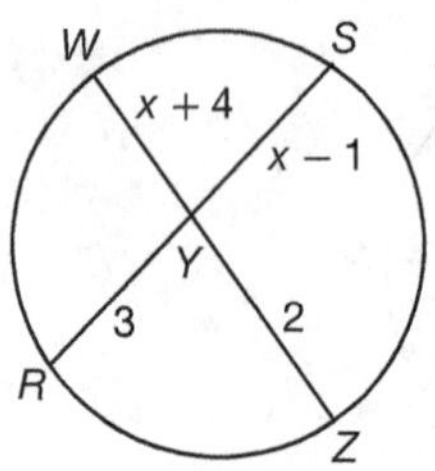

8. ______

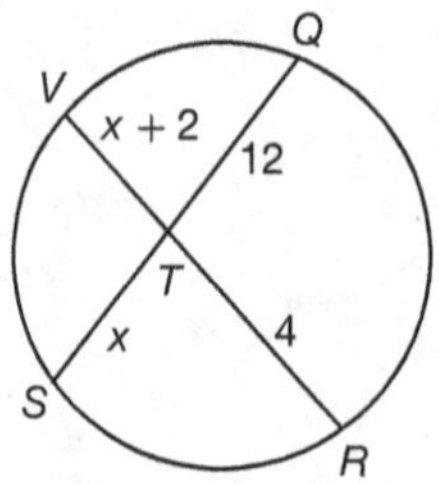

Name ______________________________ Date ___________ Class ___________

Reteaching 87

Area Ratios of Similar Figures

You have worked with similar figures and the scale ratio before. Now you will use the scale ratio to find perimeters and areas of similar figures.

Similar Figure Ratios: Perimeter

If two similar figures have a scale factor of *a*:*b*, then the ratio of their perimeters is also *a*:*b*.

The scale factor of ΔABC to ΔDEF is 3:1. The perimeter ratio is 3:1 as well.

$$\frac{\text{perimeter of } \Delta ABC}{\text{perimeter of } \Delta DEF} = \frac{36}{12} = \frac{3}{1} \rightarrow 3:1$$

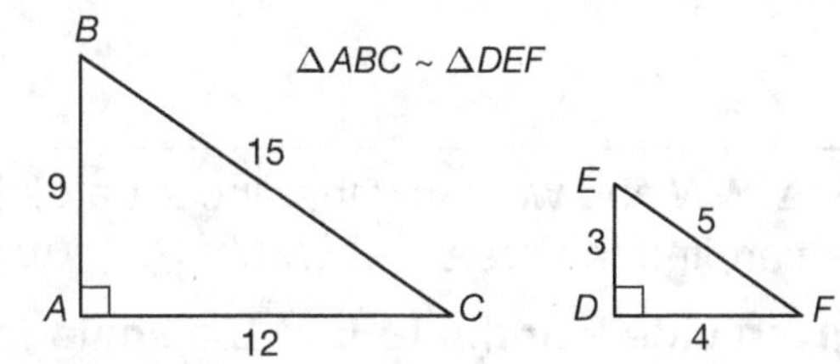

$$\frac{AB}{DE} = \frac{BC}{EF} = \frac{CA}{FD} = \frac{3}{1}$$

If $\Delta HJK \sim \Delta LMN$ and we know that the perimeter of ΔHJK is 30 inches, we can find the perimeter of ΔLMN.

Find the scale ratio using the given side lengths.

$$\frac{\Delta HJK}{\Delta LMN} = \frac{9}{12} = \frac{3}{4}$$

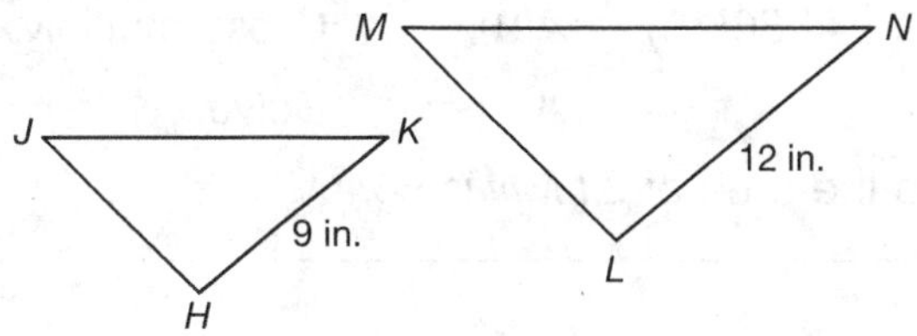

The perimeter ratio is the same, 3:4, so we can use that ratio to find the perimeter of ΔLMN.

$$\frac{\text{perimeter of } \Delta HJK}{\text{perimeter of } \Delta LMN} = \frac{3}{4}$$

$\frac{30}{P} = \frac{3}{4}$ Substitute the known values.

$30(4) = P(3)$ Cross multiply.

$40 = P$ Solve.

The perimeter of ΔLMN is 40 inches.

Practice

Complete the statements to find the perimeter of ▭*TUVW*.

1. Find the similarity ratio for $\frac{\square PQRS}{\square TUVW} = \frac{\_\_}{\_\_} = \frac{\_\_}{\_\_}$.

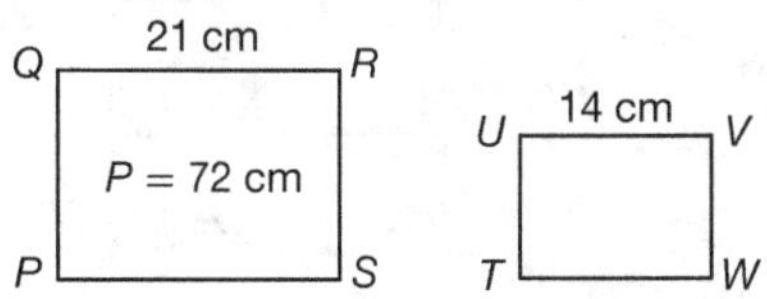

$$\frac{\text{perimeter of } \square PQRS}{\text{perimeter of } \square TUVW} = \frac{\_\_}{\_\_}$$

$\frac{\_\_}{P} = \frac{\_\_}{\_\_}$ Substitute.

$(\_\_)(\_\_) = \_\_$ Cross multiply.

$P = \_\_$ Solve.

The perimeter of ▭*TUVW* is ___ centimeters.

Similar Figure Ratios: Area

If two similar figures have a scale factor of a:b, then the ratio of their areas is $a^2 : b^2$.

$$\frac{\text{area of } \Delta ABC}{\text{area of } \Delta DEF} = \frac{54}{6} = \frac{9}{1} = \frac{3^2}{1^2}$$

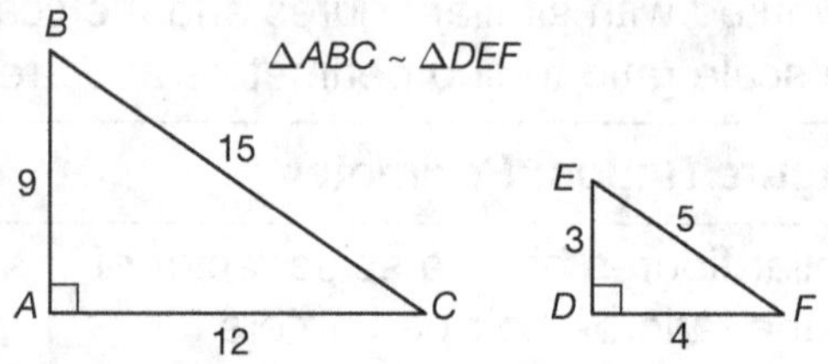

If $\Delta HJK \sim \Delta LMN$ and we know that the area of ΔHJK is 36 in.^{2}, we can find the area of ΔLMN.

Use the given side lengths to find the scale factor, 3:4. Square that scale factor to find the ratio of the areas, 9:16. Use that ratio to find the area of ΔLMN.

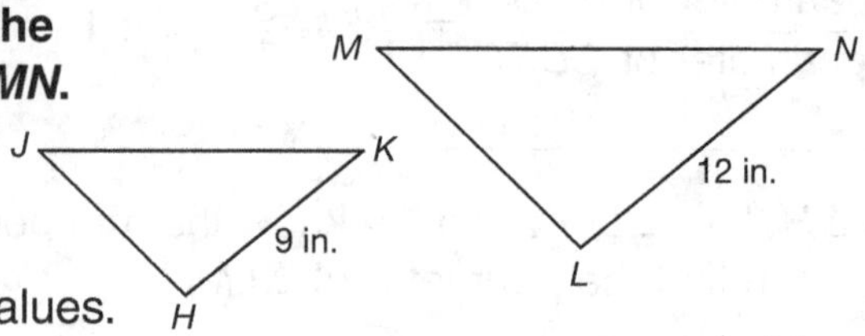

$$\frac{\text{area of } \Delta HJK}{\text{area of } \Delta LMN} = \frac{9}{16}$$

$\frac{36}{A} = \frac{9}{16}$ Substitute the known values.

$36(16) = A(9)$ Cross multiply.

$64 = A$ Solve.

So the area of ΔLMN is 64 in^2

Practice

Complete the steps to find the area of ▭TUVW.

2. Find the similarity ratio for $\frac{\square PQRS}{\square TUVW} = \frac{\_\_}{\_\_} = \left(\frac{\_}{\_}\right)^2 = \frac{\_\_}{\_\_}$

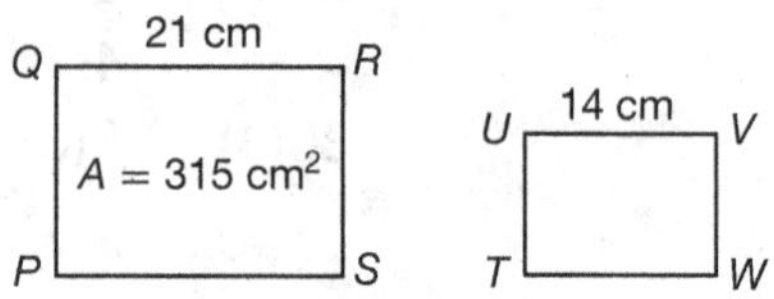

$$\frac{\text{area of } \square PQRS}{\text{area of } \square TUVW} = \frac{9}{4}$$

$\frac{\_\_\_}{A} = \frac{\_\_}{\_\_}$ Substitute the known values.

$(\_\_\_\_)(\_) = \_\_\_$ Cross multiply.

$\_\_\_\_ = \_\_\_$ Simplify.

$\_\_\_ = A$ Simplify.

The area of ▭TUVW is ____ cm^2.

3. $\Delta EFG - \Delta HJK$. Find the area of ΔHJK.

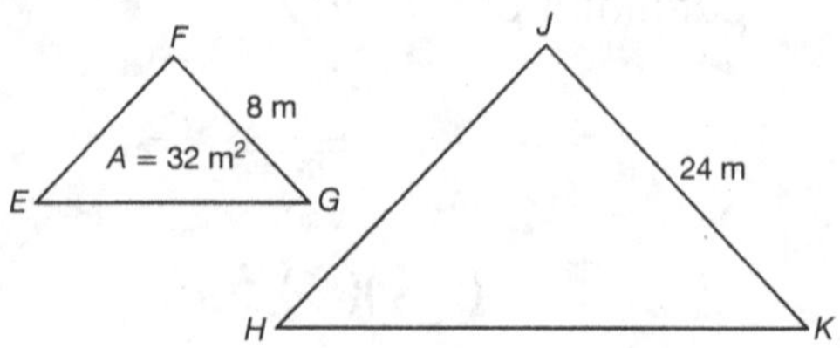

____

Name ______________________ Date ___________ Class ___________

Reteaching 88

Graphing and Solving Linear Inequalities

You have graphed linear equations before. Now you will learn how to graph linear inequalities.

Graphing Linear Inequalities

Graphing linear inequalities is similar to graphing equations. The main difference is that the solution is the entire region above or below the line. Linear inequalities can be graphed by drawing the inequality as if it were an equation and then shading the region above (for $>$ and $\geq$) or below (for $<$ and $\leq$) the line. Use a dashed line if the inequality symbol is $<$ or $>$ and a solid line if it is $\leq$ or $\geq$.

Example:

Graph the inequality $3y \geq 2x - 3$.

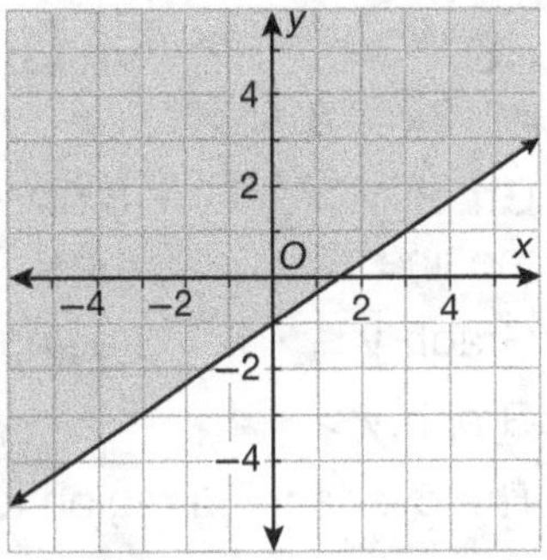

Step 1: Write the inequality in slope-intercept form.

$3y \geq 2x - 3$

$\frac{3y}{3} \geq \frac{2x}{2} - \frac{3}{3}$ Divide both sides by 3.

$y \geq \frac{2}{3}x - 1$ Finish in slope-intercept form.

Step 2: Graph the line $y = \frac{2}{3}x - 1$ using a solid line.

Step 3: The inequality is "greater than," so shade the area above the line.

Practice

Complete the statement to put the inequality in slope-intercept form, and then graph it.

1. Graph the inequality $x - y > 1$.

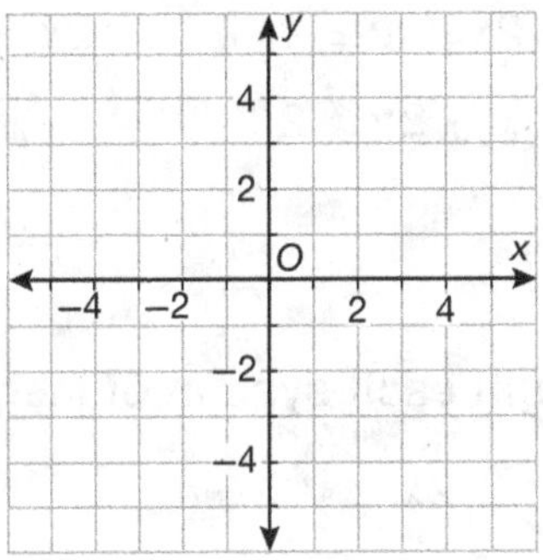

$x - y > 1$

$-y > 1 -$ ___ Subtract x from both sides.

$y < -1 +$ ___ Divide both sides by -1.

$y <$ ___ -1 Rewrite in slope-intercept form.

The slope is ___, and the y-intercept is ___. Because the symbol is $<$, the line should be made with a _______ line. Because the inequality is "less than," the solution region is ______ the line.

Graph each linear inequality.

2. $f(x) > -\frac{1}{2}x - 3$

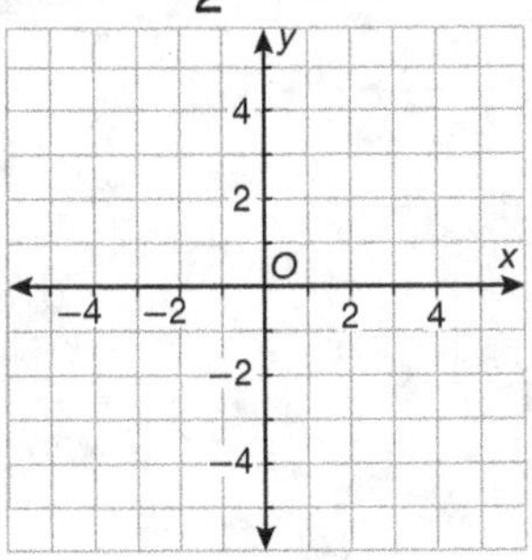

3. $2x - 3y \leq 6$

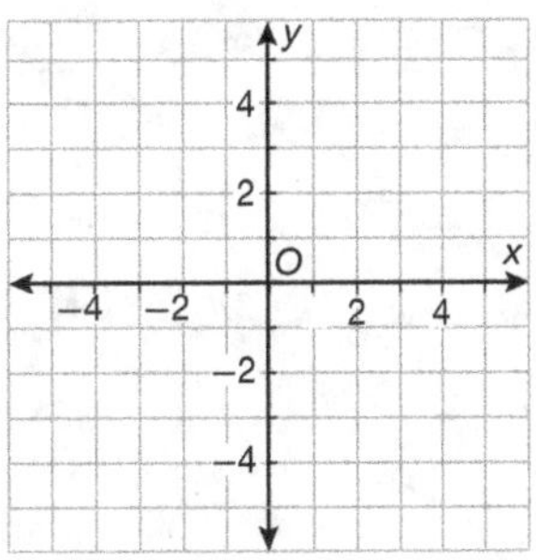

Graphing Linear Inequalities: Systems

A system of inequalities consists of two or more inequalities. The solution to a system of inequalities is the area in which the shaded regions for each inequality overlap. If there is no overlapping region, the system has no solution and is called inconsistent.

Example: Graph and solve this system of inequalities.

$3x + y > 8$

$x + y \leq 3$

$y > -4$

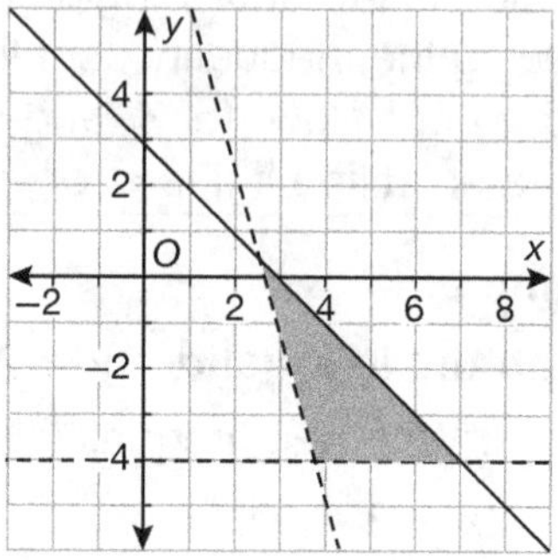

Step 1: Simplify each inequality.

$3x + y > 8$ $x + y \leq 3$ $y > -4$

$y > -3x + 8$ $y \leq -x + 3$

Step 2: Graph $y > -3x + 8$ using a dashed line. Shade above the line.

Step 3: Graph $y \leq -x + 3$ using a solid line. Shade below the line.

Step 4: Graph $y > -4$.

Step 5: The overlapping region is the solution to the system.

Practice

Complete the steps to graph the system of inequalities.

4. $2x - y \geq 2$ $2x - y \geq 2$ $x - 2y > 2$

$x - 2y > 2$ $-y \geq$ ____ $+ 2$ $-2y >$ ____ $+ 2$

$y \leq$ ___ $- 2$ $y <$ ___ $- 1$

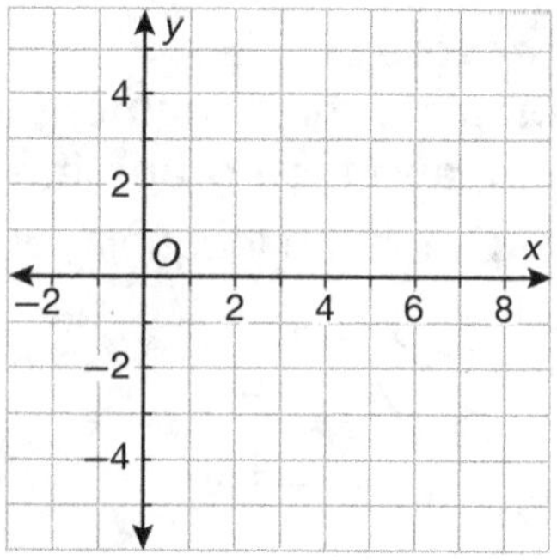

Graph each system of inequalities.

5. $2x + y > -4$

$-5x + 2y < 1$

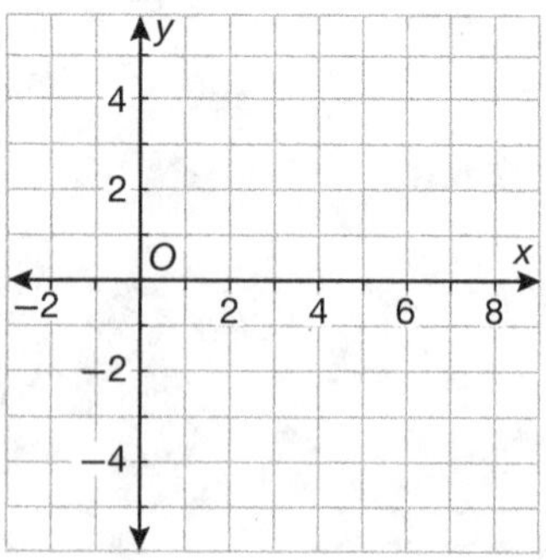

6. $y < x + 3$

$2x - 7y < 4$

$3x + 2y \leq 6$

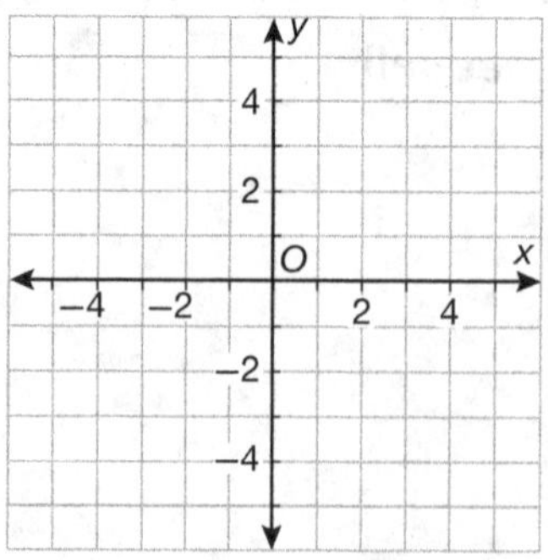

Name ______________________________ Date ____________ Class ____________

Reteaching 89

Vector Decomposition

You have worked with vectors before. Now you will learn how to decompose vectors into their horizontal and vertical components.

Decomposing Vectors

To decompose a vector is to find its horizontal and vertical components. We find the horizontal component by multiplying the vector's magnitude by cosine of the angle the vector makes with the horizontal. The vertical component is found by multiplying the magnitude by sine of the angle.

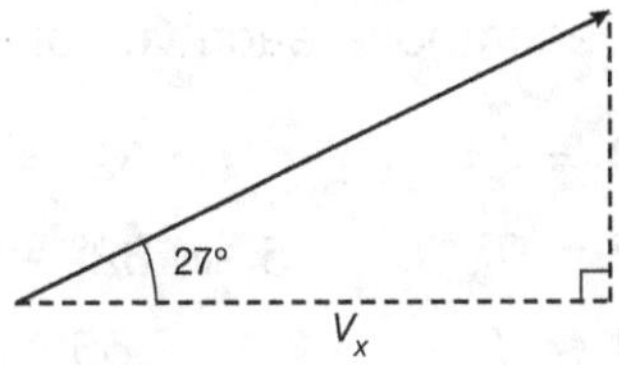

For instance, $\vec{V}$ makes a 27° angle with the horizontal and has a magnitude of 9. We can use trigonometry to find the horizontal and vertical components, V_x and V_y.

$$\cos 27° = \frac{V_x}{9} \qquad \sin 27° = \frac{V_y}{9}$$

$$9 \cos 27° = V_x \qquad 9 \sin 27° = V_y$$

$$8.01 \approx V_x \qquad 4.09 \approx V_y$$

Write the decomposition for $\vec{V}$ using brackets, $\vec{V} \approx \langle 8.01, 4.09 \rangle$.

Practice

Complete the steps to find the decomposition of the given vector.

1. $\vec{B}$ makes a 41° angle with the horizontal and has a magnitude of 25.

   cos ____ = B_x / ____ sin ____ = B_y / ____

   25 cos ____ = B_x 25 sin ____ = B_y

   ______ $\approx B_x$ ______ $\approx B_y$

   The decomposition is $\vec{B} \approx$ ______________.

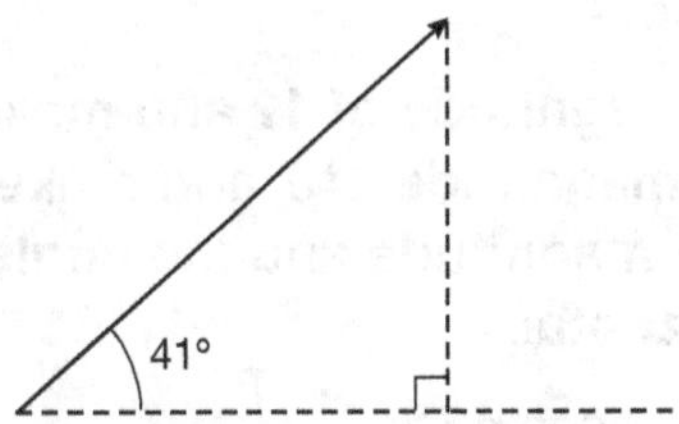

Decompose each vector.

2. $\vec{B}$ makes a 40° angle with the horizontal and has a magnitude of 4.

   ____

3. $\vec{V}$ makes a 58° angle with the horizontal and has a magnitude of 12.

   ____

4. $\vec{W}$ makes a 35° angle with the horizontal and has a magnitude of 5.

   ____

5. $\vec{Z}$ makes a 72° angle with the horizontal and has a magnitude of 9.

   ____

6. $\vec{Q}$ makes a 18° angle with the horizontal and has a magnitude of 12.

   ____

7. $\vec{R}$ makes a 25° angle with the horizontal and has a magnitude of 18.

   ____

Adding Vectors by Decomposition

When adding vectors we can find the resultant by adding the components.

$\vec{T}$ has a magnitude of 5 and makes a 32° angle with the horizontal, and $\vec{U}$ has a magnitude of 11 and makes a 64° angle with the horizontal. We'll decompose and add the components of $\vec{T}$ and $\vec{U}$ to find the magnitude and the angle of the resultant vector.

Step 1: Decompose each vector.

$$\cos 32° = \frac{T_x}{5} \qquad \sin 32° = \frac{T_y}{5} \qquad \cos 64° = \frac{U_x}{11} \qquad \sin 64° = \frac{U_y}{11}$$

$$5 \cos 32° = T_x \qquad 5 \sin 32° = T_y \qquad 11 \cos 64° = U_x \qquad 11 \sin 64° = U_y$$

$$4.24 \approx T_x \qquad 2.65 \approx T_y \qquad 4.82 \approx U_x \qquad 9.89 \approx U_y$$

Step 2: Add the components.

$T_x + U_x \approx 4.24 + 4.82 \approx 9.06$ $\qquad$ $T_y + U_y \approx 2.65 + 9.89 \approx 12.54$

Step 3: Use the Pythagorean theorem to find the magnitude of the resultant vector.

$$a^2 + b^2 = c^2$$
$$(T_x + U_x)^2 + (T_y + U_y)^2 = c^2$$
$$(9.06)^2 + (12.54)^2 \approx c^2$$
$$\sqrt{239.3352} \approx c^2$$
$$15.47 \approx c$$

Step 4: Take $\tan^{-1}$ of the vertical component divided by the horizontal component to find angle.

$$\tan\theta = \frac{12.54}{9.06}$$
$$\theta = \tan^{-1}(1.384)$$
$$\theta = 54°$$

Practice

$\vec{T}$ has a magnitude of 12 and makes a 46° angle with the horizontal. $\vec{U}$ has a magnitude of 7 and makes a 22° angle with the horizontal. Find the magnitude and the angle the resultant vector makes with the horizontal.

8. $\cos 46° = \frac{T_x}{\_\_\_\_}$ $\qquad$ $\sin 46° = \frac{T_y}{\_\_\_\_}$ $\qquad$ $\cos 22° = \frac{U_x}{\_\_\_\_}$ $\qquad$ $\sin 22° = \frac{U_y}{\_\_\_\_}$

__ $\cos 46° = T_x$ $\qquad$ __ $\sin 46° = T_y$ $\qquad$ __ $\cos 22° = U_x$ $\qquad$ __ $\sin 22 = U_y$

____ $\approx T_x$ $\qquad$ ____ $\approx T_y$ $\qquad$ ____ $\approx Ux$ $\qquad$ ____ $\approx U_y$

Add the components.

$T_x + U_x \approx 8.34 + 6.49 \approx 14.83$; $T_y + U_y \approx 8.63 + 2.62 \approx 11.25$

Determine the magnitude.

$(T_x + U_x)^2 + (T_y + U_y)^2 = c^2$

____^{2} + ____^{2} $\approx c^2$

$\sqrt{\_\_\_\_}$ $\approx c^2$

____ $\approx c$

Determine the angle.

$\tan\theta = \frac{\_\_\_\_}{\_\_\_\_}$

$\theta = \tan^{-1}$(____)

$\theta \approx$ ____

Name ______________________________ Date __________ Class __________

Reteaching 90

Composite Transformations

You have worked with translation and reflection transformations. Now you will work with composite transformations.

Performing Glide Reflection

A **composite transformation** is a combination of two or more transformations. A **glide reflection** is the composition of a translation and a reflection across a line parallel to the translation vector.

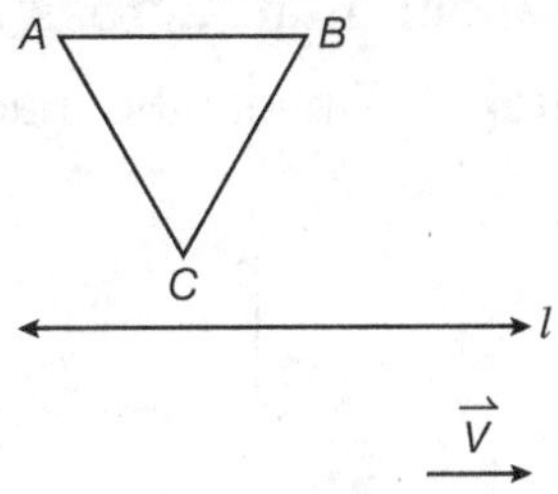

Perform a glide reflection on ΔABC by reflecting it across line ℓ and then translating it in the direction of $\vec{v}$, as indicated.

Step 1: First, reflect ΔABC across line ℓ.

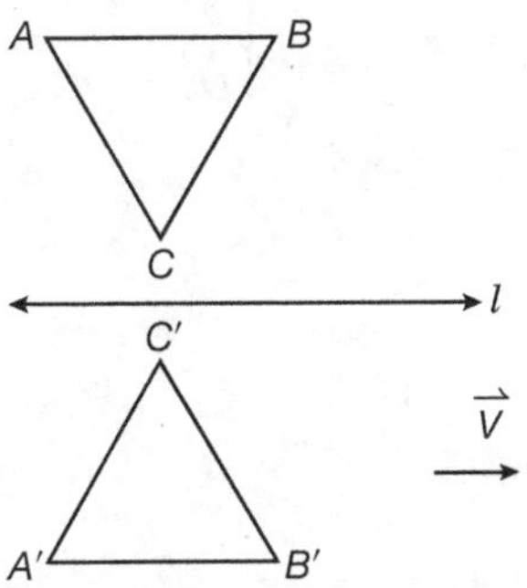

Step 2: Then, translate the triangle parallel to $\vec{v}$.

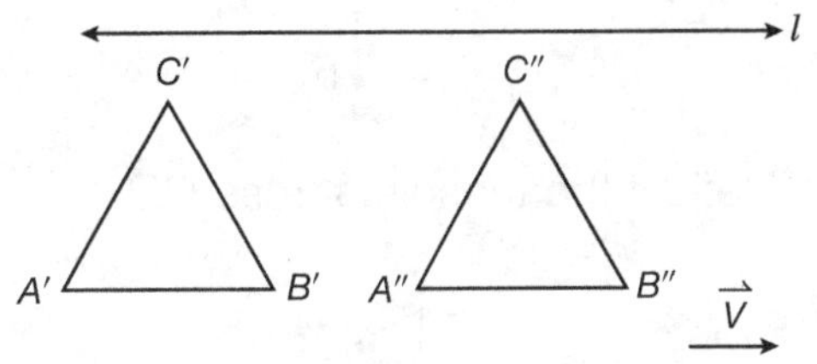

Practice

Follow the steps to perform a glide reflection on quadrilateral *GHJK*.

1. Reflect the quadrilateral across ______.
   Translate the quadrilateral in the direction of __ .

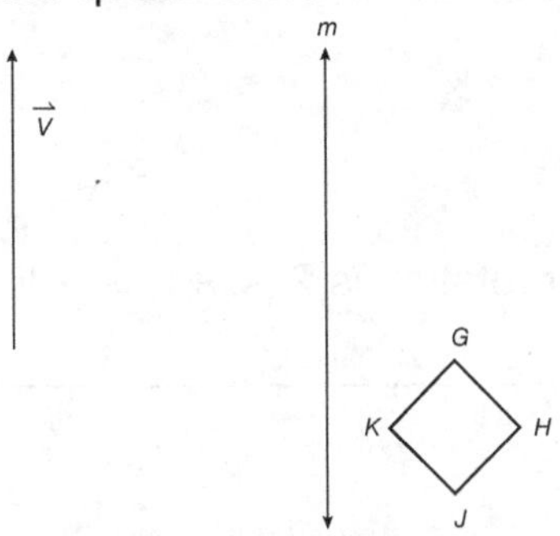

Perform the indicated glide reflections.

2.

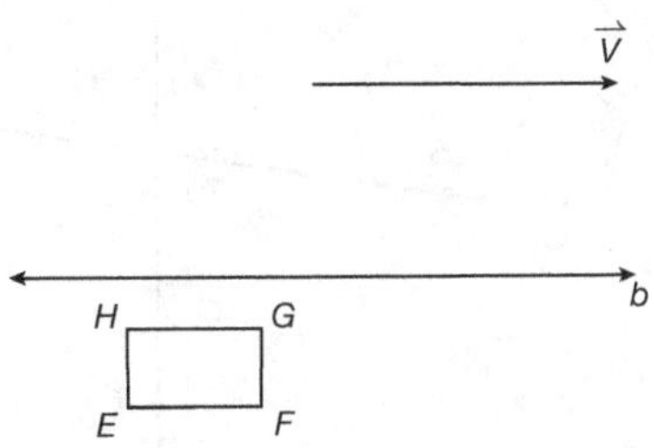

Composition of Two Reflections

The composition of two reflections across intersecting lines is a rotation about the point of intersection. The angle of rotation is twice the angle formed by the intersecting lines.

Reflect ΔXYZ across $\overleftrightarrow{AB}$ and then across $\overleftrightarrow{CD}$.

Step 1: Reflect the triangle across $\overleftrightarrow{AB}$.

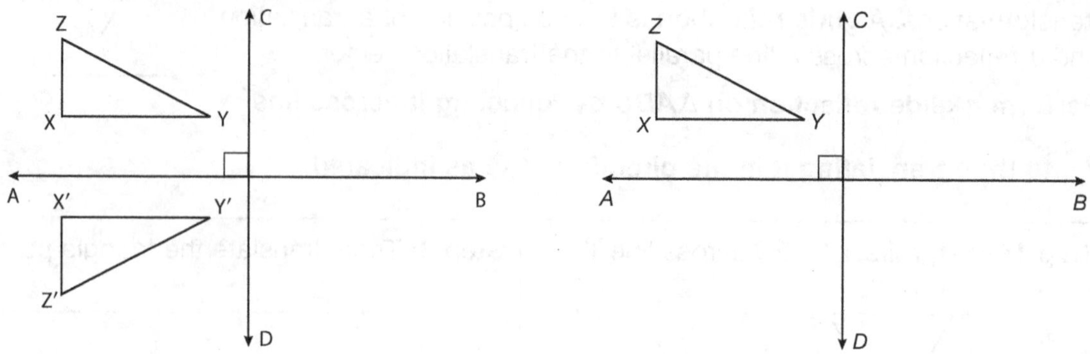

Step 2: Reflect the triangle across $\overleftrightarrow{CD}$.

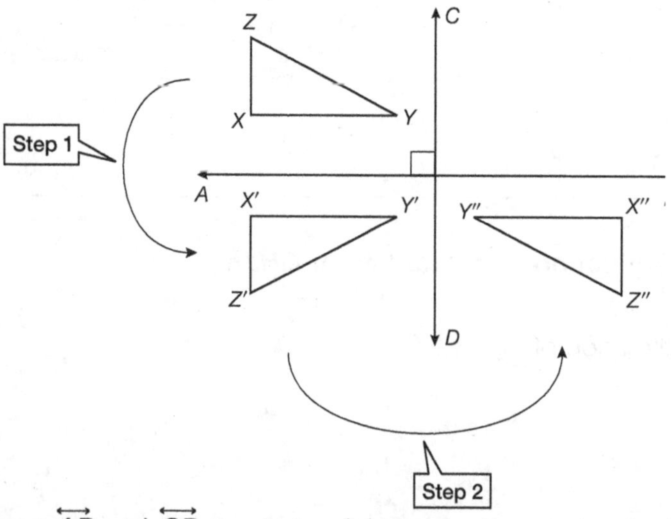

Because $\overleftrightarrow{AB}$ and $\overleftrightarrow{CD}$ meet at a right angle, the angle of rotation is $2 \times 90° = 180°$.

$V = Bh = (30)(5) = 150 \text{ ft}^3$.

Practice

Complete the steps to reflect rhombus *JKLM* across $\overleftrightarrow{CD}$ and then across $\overleftrightarrow{GH}$. Find the angle of rotation.

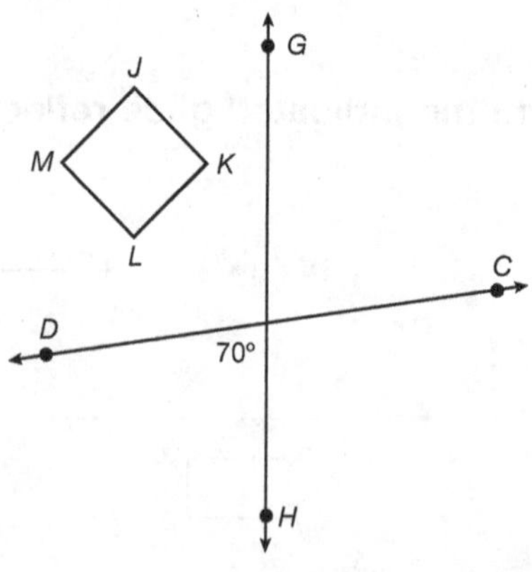

4. Reflect across $\overleftrightarrow{CD}$.

Reflect across $\overleftrightarrow{GH}$.

Because the angle of intersection is ___, the angle of rotation is ____.

Name ______________________________ Date ____________ Class ____________

Reteaching
Tessellations

INV 9

You have worked with composite transformations. Now you will investigate tessellations.

Regular Tessellations

A **tessellation** is a repeating pattern of plane figures that completely cover a plane with no gaps or overlaps. A **regular tessellation** is a tessellation with congruent regular polygons.

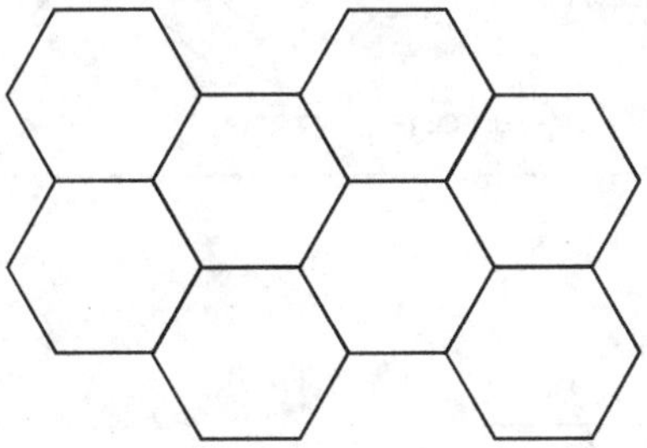

regular tessellation

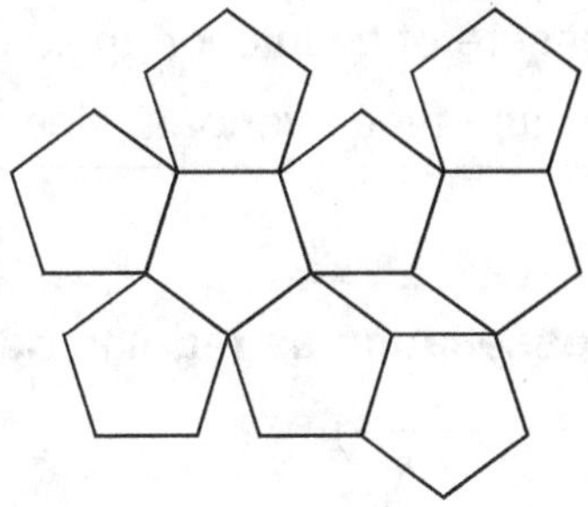

not a tessellation

In a tessellation, the angles that meet at a vertex must have a sum of 360°. In order for a regular polygon to tessellate, the measure of its interior angle must be a divisor of 360°.

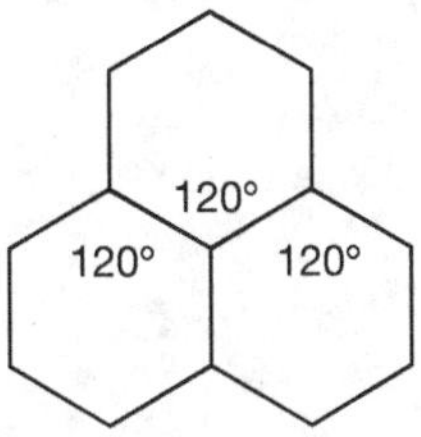

$120° + 120° + 120° = 360°$

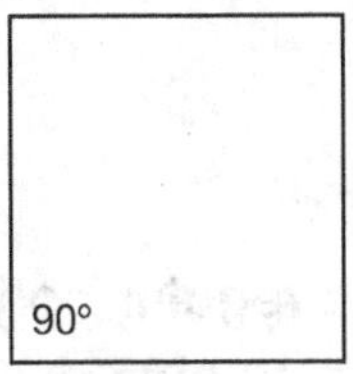

This square will tessellate.

Practice

Complete the steps to determine whether an equilateral triangle can be used to form a tessellation.

1. The measure of each interior angle is ____.

   $360° \div 60° =$ _; The measure of the interior angle is a ______ of 360°.

   The equilateral triangle will __________.

Determine whether each regular polygon can form a tessellation.

2.

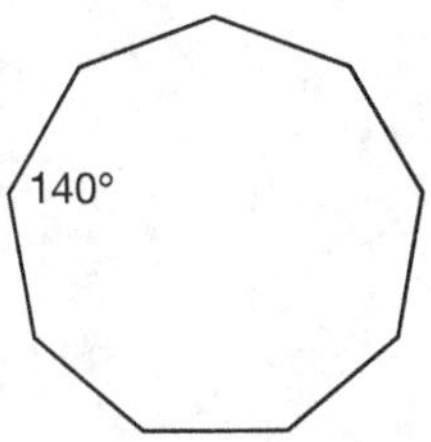

_____

3.

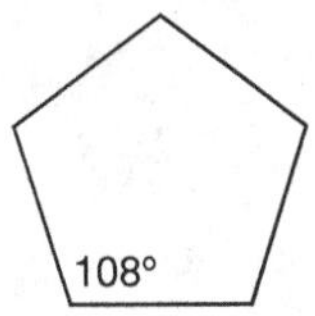

_____

Semiregular Tessellations

A **semiregular tessellation** is a repeating pattern formed by two or more polygons in which the same number of each polygon occurs in the same order at every vertex and completely covers a plane with no gaps or overlaps.

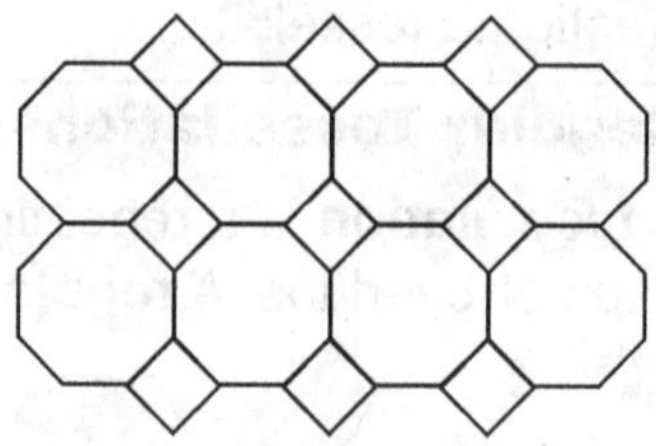

The tessellation shown is semiregular. It is made up of regular hexagons and squares. A square and two octagons meet at each vertex. The measure of an interior angle of an octagon is 135°.

Therefore, the sum of the interior angles at a vertex is $135° + 135° + 90° = 360°$.

Practice

Classify each tessellation as regular, semiregular, or neither.

4. 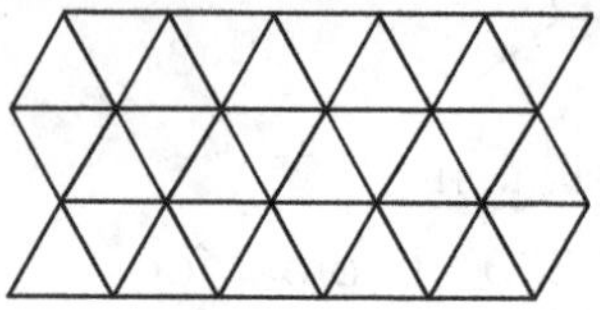

____

5. 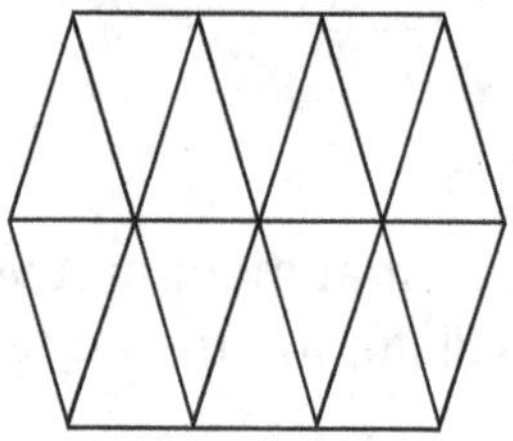

____

Determine whether the regular polygons can be used to form a tessellation. If so, draw the tessellation.

6. 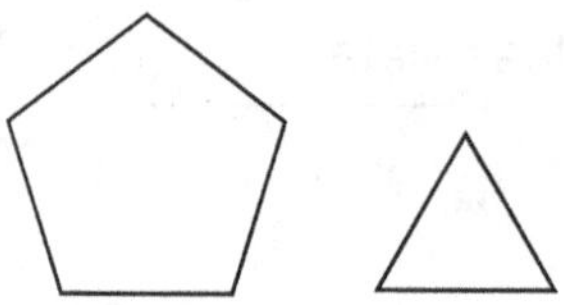

____

7. 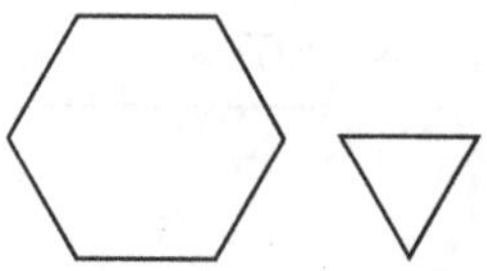

____

Name ______________________ Date __________ Class __________

Reteaching 91

Introduction to Trigonometric Functions

You have worked with composition of transformations. Now you will work with trigonometric functions.

Relating Sine and Cosine

Some important relationships between trigonometric functions are given by the following:

$\sin^2 x + \cos^2 x = 1 \quad \tan x = \dfrac{\sin x}{\cos x}$

Note that $\sin^2 x = (\sin x)^2$ and $\cos^2 x = (\cos x)^2$.

If $\sin^2 \theta = 0.9$, what is the value of $\cos \theta$?

Use the equation $\sin^2 x + \cos^2 x = 1$.

$\sin^2 \theta + \cos^2 \theta = 1$

$0.9 + \cos^2 \theta = 1$ — Substitute $\sin^2 \theta = 0.9$.

$\cos^2 \theta = 1 - 0.9$ — Substitute 0.9 from both sides.

$\cos^2 \theta = 0.1$

$\sqrt{\cos^2 \theta} = \sqrt{0.1}$ — Take the square root of both sides.

$\cos \theta \approx 0.32$

What is the value of $\tan \theta$?

Since $\sin^2 \theta = 0.9$, $\sin \theta = \sqrt{0.9} \approx 0.95$.

$\tan \theta = \dfrac{\sin \theta}{\cos \theta} \approx \dfrac{0.95}{0.32} \approx 2.97$

Practice

If $\cos \theta = 0.2$, complete the steps to find the value of $\sin \theta$ and $\tan \theta$.

1. $\cos \theta = 0.2$

   $\cos^2 \theta = (\_\_\_)^2 = \_\_\_$

   $\sin^2 \theta + \cos^2 \theta = \_$

   $\sin^2 \theta + \_\_\_ = \_$

   $\sin^2 \theta = 1 - \_\_\_$

   $\sin^2 \theta = \_\_\_$

   $\sin \theta \approx \_\_\_$

2. $\tan \theta = \dfrac{\sin \theta}{\cos \theta}$

   $\approx \dfrac{\_\_\_}{\_\_\_}$

   $\approx \_\_\_$

Find the indicated values

3. If $\sin \theta = 0.15$, find:

   $\cos \theta =$ _____

   $\tan \theta =$ _____

4. If $\cos^2 \theta = 0.25$, find:

   $\sin \theta =$ _____

   $\tan \theta =$ _____

5. If $\cos \theta = 0.7$, find:

   $\sin \theta =$ _____

   $\tan \theta =$ _____

6. If $\sin^2 \theta = 0.88$, find:

   $\cos \theta =$ _____

   $\tan \theta =$ _____

More Trigonometric Relationships

The table below outlines the relationships between sine, cosine, and tangent functions.

Function	**sin θ**	**cos θ**	**tan θ**
sin θ =	$\sin\theta$	$\sqrt{1-\cos^2\theta}$	$\dfrac{\tan\theta}{\sqrt{1+\tan\theta}}$
cos θ =	$\sqrt{1-\sin^2\theta}$	$\cos\theta$	$\dfrac{1}{\sqrt{1+\tan^2\theta}}$
tan θ =	$\dfrac{\sin\theta}{\sqrt{1-\sin^2\theta}}$	$\dfrac{\sqrt{1-\cos^2\theta}}{\cos\theta}$	$\tan\theta$

If $\tan^2\theta = 1.4$, what is the value of $\cos\theta$?

$$\cos\theta = \frac{1}{\sqrt{1+\tan^2\theta}}$$

$$= \frac{1}{\sqrt{1+1.4}} \qquad \text{Substitute } \tan^2\theta = 1.4.$$

$$= \frac{1}{\sqrt{2.4}}$$

$$\approx 0.65$$

Practice

If $\tan^2\theta = 1.8$, complete the steps to find the value of $\sin\theta$ and $\cos\theta$.

7. $\tan\theta = \sqrt{\underline{\quad}} \approx \underline{\quad\quad}$

$\sin\theta = \dfrac{\underline{\quad}}{\sqrt{1+\tan\theta}}$

$= \dfrac{\underline{\quad}}{\sqrt{1+\underline{\quad\quad}}}$

$= \dfrac{1.34}{\sqrt{\underline{\quad\quad}}}$

$\approx \underline{\quad\quad}$

8. $\cos\theta = \dfrac{\underline{\quad}}{\sqrt{1+\tan^2\theta}}$

$= \dfrac{\underline{\quad}}{\sqrt{1+\underline{\quad\quad}}}$

$= \dfrac{1}{\sqrt{\underline{\quad\quad}}}$

$\approx \underline{\quad\quad}$

Find the indicated values.

9. If $\tan^2\theta = .75$, find:

$\cos\theta = \underline{\quad\quad}$

$\sin\theta = \underline{\quad\quad}$

10. If $\tan\theta = 1.4$, find:

$\sin\theta = \underline{\quad\quad}$

$\cos\theta = \underline{\quad\quad}$

11. If $\sin\theta = 0.7$, find:

$\tan\theta = \underline{\quad\quad}$

$\cos\theta = \underline{\quad\quad}$

12. If $\cos\theta = 0.44$, find:

$\tan\theta = \underline{\quad\quad}$

$\sin\theta = \underline{\quad\quad}$

Name ______________________ Date __________ Class __________

Reteaching 92

Quadrilaterals on the Coordinate Plane

You have worked with trigonometric functions. Now you will work with quadrilaterals on the coordinate plane.

Proving Quadrilaterals are Parallelograms

If you know the coordinates of a quadrilateral, you can determine the length and the slope of each side. Therefore, you can determine which sides are congruent, parallel, or perpendicular.

Prove that quadrilateral *ABCD* is a parallelogram by determining the slope of each side.

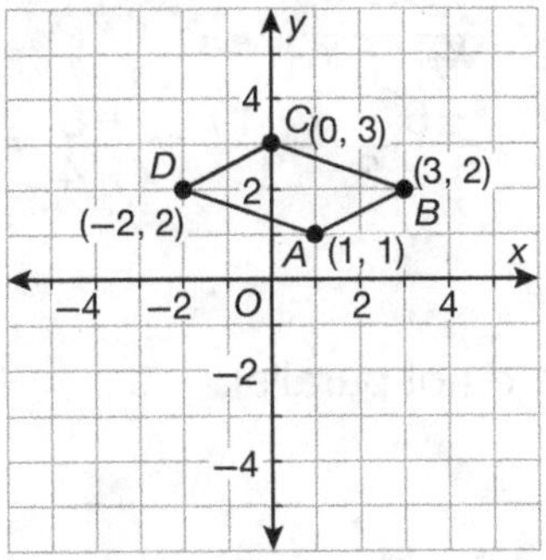

Step 1: Calculate the slope of each side.

slope of $\overline{AB} = \dfrac{y_B - y_A}{x_B - x_A} = \dfrac{2-1}{3-1} = \dfrac{1}{2}$

slope of $\overline{BC} = \dfrac{y_C - y_B}{x_C - x_B} = \dfrac{3-2}{0-3} = \dfrac{1}{-3} = -\dfrac{1}{3}$

slope of $\overline{CD} = \dfrac{y_D - y_C}{x_D - x_C} = \dfrac{2-3}{-2-0} = \dfrac{-1}{-2} = \dfrac{1}{2}$

slope of $\overline{AD} = \dfrac{y_D - y_A}{x_D - x_A} = \dfrac{2-1}{-2-1} = \dfrac{1}{-3} = -\dfrac{1}{3}$

Step 2: Compare the slopes of opposite sides.

The slopes of $\overline{AB}$ and $\overline{CD}$ are equal, so $\overline{AB} \parallel \overline{CD}$.

Similarly, $\overline{BC} \parallel \overline{AD}$.

Therefore, *ABCD* is a parallelogram.

Practice

For quadrilateral *ABCD* above, complete the steps to prove that *ABCD* is a parallelogram by determining its side lengths.

1. $AB = \sqrt{(x_B - x_A)^2 + (y_B - y_A)^2} = \sqrt{(2 - \_)^2 + (3 - \_)^2} = \sqrt{(1)^2 + (\_)^2} = \sqrt{\_}$

$BC = \sqrt{(x_C - x_B)^2 + (y_C - y_B)^2} = \sqrt{(\_ - (-3))^2 + (\_ - 2)^2} = \sqrt{(\_)^2 + (\_)^2} = \sqrt{\_\_}$

$CD = \sqrt{(\_\_ - 0)^2 + (\_ - 3)^2} = \sqrt{(\_\_)^2 + (\_\_)^2} = \sqrt{\_}$

$AD = \sqrt{(\_\_ - 1)^2 + (\_ - 1)^2} = \sqrt{(\_\_)^2 + (\_)^2} = \sqrt{\_\_}$

$AB = CD \rightarrow \overline{AB} \cong$ ___; $BC = \_AD\_ \rightarrow$ ___ $\cong$ ___

ABCD is a ______________.

Proving that Quadrilaterals Are Trapezoids

The vertices of quadrilateral *LMNP* are *L*(1, −1), *M*(4, −1), *N*(6, −6), and *P*(−3, −6).

Prove that *LMNP* is a trapezoid.

Step 1: Calculate the slope of each side.

slope of $\overline{LM} = \dfrac{y_M - y_L}{x_M - x_L} = \dfrac{-1-(-1)}{4-1} = \dfrac{0}{3} = 0$

slope of $\overline{MN} = \dfrac{y_N - y_M}{x_N - x_M} = \dfrac{-6-(-1)}{6-4} = \dfrac{-5}{2} = -\dfrac{5}{2}$

slope of $\overline{NP} = \dfrac{y_P - y_N}{x_P - x_N} = \dfrac{-6-(-6)}{-3-6} = \dfrac{0}{-9} = 0$

slope of $\overline{LP} = \dfrac{y_P - y_L}{x_P - x_L} = \dfrac{-6-(-1)}{-3-1} = \dfrac{-5}{-4} = \dfrac{5}{4}$

Step 2: Compare the slopes of opposite sides.

The slopes of $\overline{LM}$ and $\overline{NP}$ are equal, so $\overline{LM} \parallel \overline{NP}$.

However, $\overline{MN}$ and $\overline{LP}$ do not have the same slopes and are not parallel.

Therefore, *LMNP* is a trapezoid.

Practice

Complete the steps to prove that quadrilateral *ABCD* is a trapezoid.

2. slope of $\overline{AB} = \dfrac{y_B - y_A}{x_B - x_A} = \dfrac{2 - \_\_}{4 - (\_\_)} = \dfrac{\_\_}{\_\_} = -\dfrac{2}{\_\_}$

slope of $\overline{BC} = \dfrac{y_C - y_B}{x_C - x_B} = \dfrac{0 - \_\_}{1 - \_\_} = \dfrac{\_\_}{\_\_} = \dfrac{\_\_}{\_\_}$

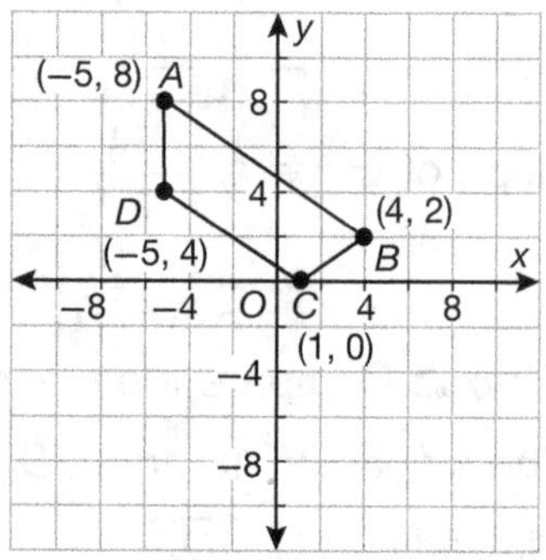

slope of $\overline{CD} = \dfrac{\_\_ - 0}{\_\_ - 1} = \dfrac{4}{\_\_} = -\dfrac{\_\_}{\_\_}$

slope of $\overline{AD} = \dfrac{\_\_ - \_\_}{\_\_ - (-5)} = \dfrac{\_\_}{0} = \_\_\_\_\_\_$

The slopes of $\overline{AB}$ and ___ are equal, so ___ $\parallel \overline{CD}$.

The slopes of $\overline{BC}$ and ___ are not equal and not parallel.

Therefore, *ABCD* is a __________.

Name ______________________________ Date __________ Class __________

Reteaching 93

Representing Solids: Orthographic Views

You have worked with quadrilaterals. Now you will draw orthographic views of solids.

Creating Orthographic Drawings

An **orthographic drawing** shows a three-dimensional object in such a way that the line of sight for each view is perpendicular to the plane of the picture.

There are usually three views—front, top, and side.

Draw the front, top, and side orthographic views of this solid.

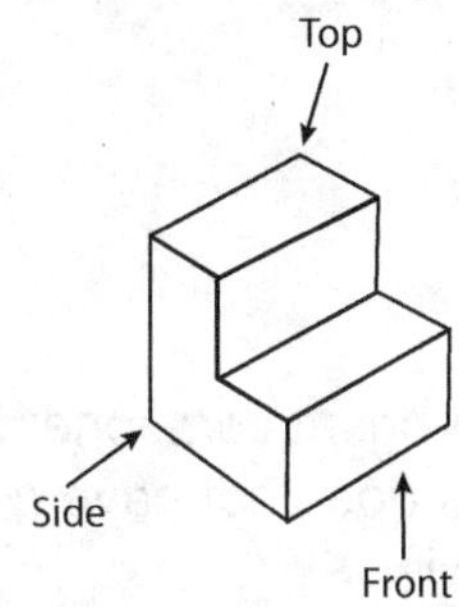

Front View: Two rectangles can be seen.	Top View: Two rectangles can be seen.	Side View:
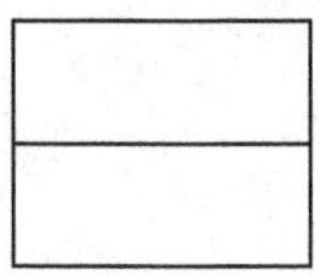	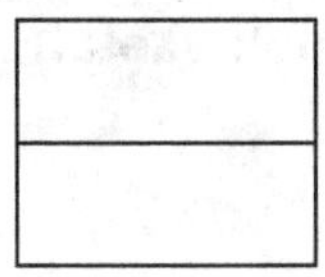	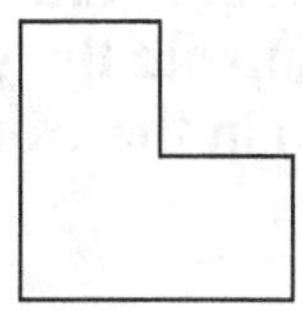

Practice

Complete the front, top, and side orthographic views of this solid.

1. Solid:

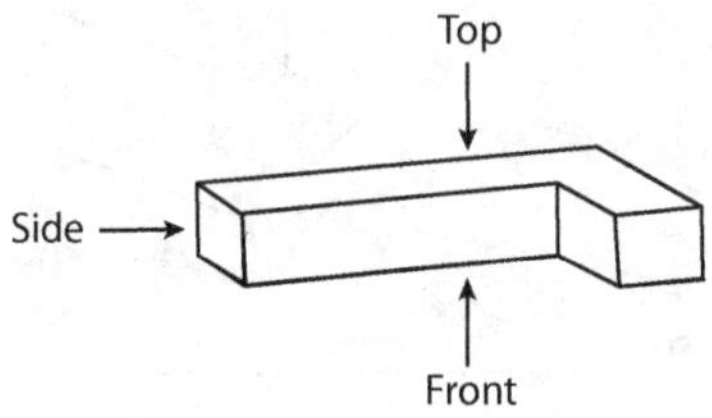

Front:

Top:

Side:

Constructing from an Orthographic Drawing

This orthographic drawing shows the front, top, and left-side views of a solid.

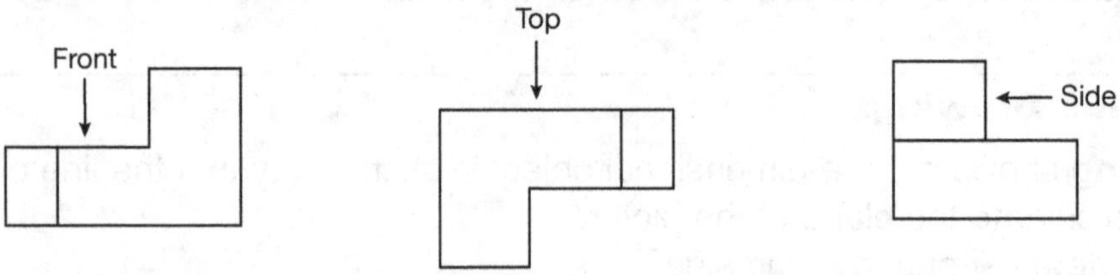

Which of these sketches represents the solid shown in the views?

A

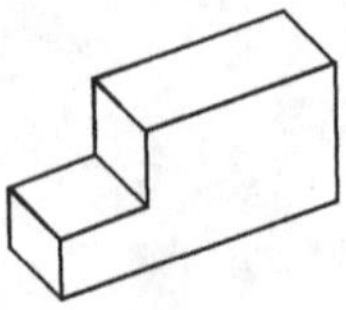

B

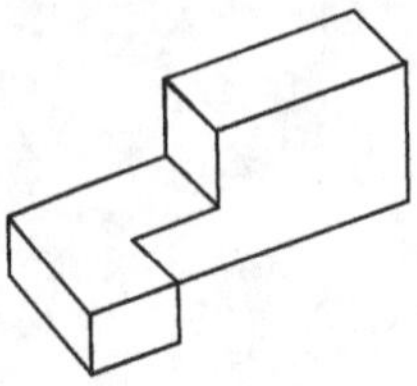

C

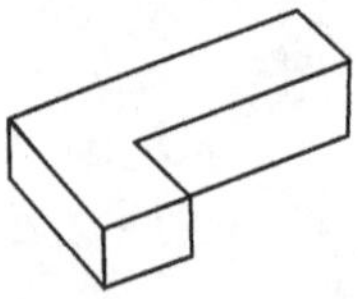

Compare sketch A to the top view. The left side of sketch A does not have a prism component shown in the top view. Compare sketch C to the top or front view. The figure does not have a top layer. Compare sketch B to all three views. Sketch B matches all three views.

Practice

The orthographic drawings show the front, top, and right-side views of a solid. Complete the steps that explain which sketch represents the solid shown in the views.

2. Front:

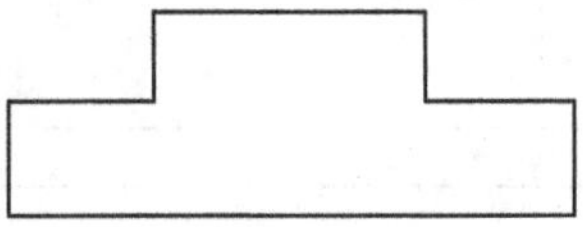

Top:

Side:

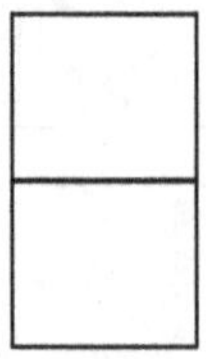

A

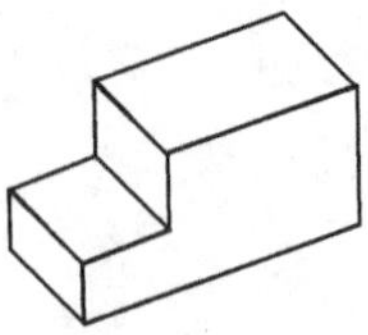

B

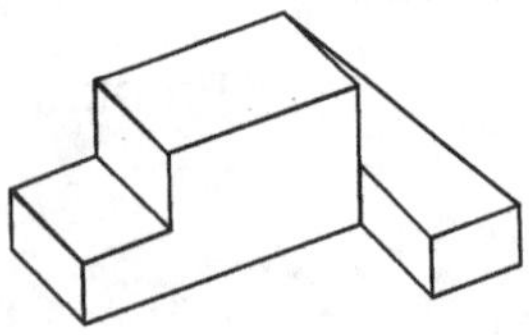

C

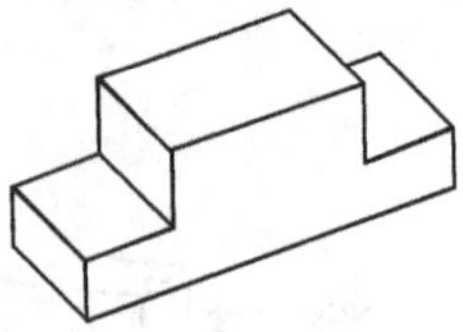

Sketch A does not represent the solid shown because ____________________

____

____

Sketch B does not represent the solid shown because ________________

____

Sketch C represents the solid shown because ______________

____

Name ______________________ Date __________ Class __________

Reteaching 94

Law of Sines

You have worked with orthographic drawings. Now you will work with the Law of Sines.

Law of Sines: Finding Distance

The **Law of Sines** is a ratio that uses given angles and sides to find other angles and sides in any triangle.

The Law of Sines states that $\frac{a}{\sin A} = \frac{b}{\sin B} = \frac{c}{\sin C}$.

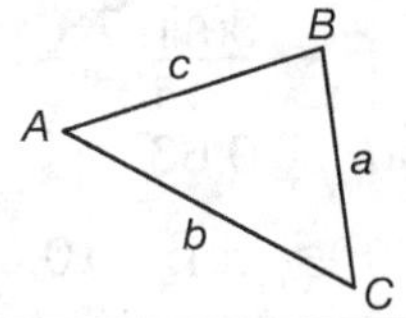

Find the length of *RT*. Round to the nearest hundredth.

Step 1: Find the measure of $\angle T$.

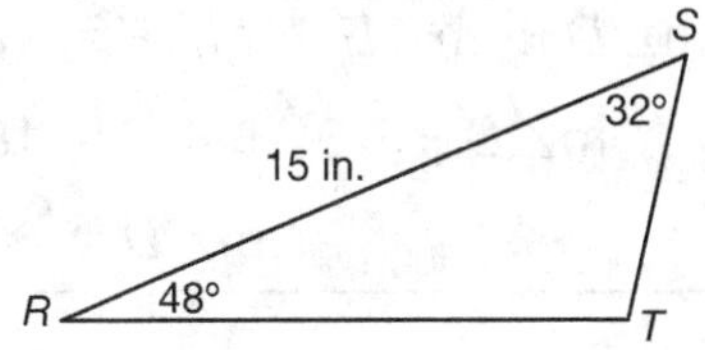

$m\angle R + m\angle S + m\angle T = 180°$	Triangle Sum Theorem
$48° + 32° + m\angle T = 180°$	Substitute values.
$m\angle T = 100°$	Simplify.

Step 2: Use $m\angle T$, *RS*, and the Law of Sines to find *RT*.

$\frac{RT}{\sin S} = \frac{RS}{\sin T}$	Law of Sines
$\frac{RT}{\sin 32°} = \frac{15}{\sin 100°}$	Substitute values.
$RT \sin 100° = 15 \sin 32°$	Cross multiply.
$RT = \frac{15 \sin 32°}{\sin 100°}$	Simplify.
$RT \approx 8.07$ in.	

Practice

Complete the steps to find the length *TU*. Round to the nearest hundredth.

1. $\frac{\_\_\_\_}{\_\_\_\_} = \frac{TU}{\sin 64°}$

$TU \sin 41° =$ ________

$TU = \frac{18 \sin 64°}{\_\_\_\_}$

$TU \approx$ ________

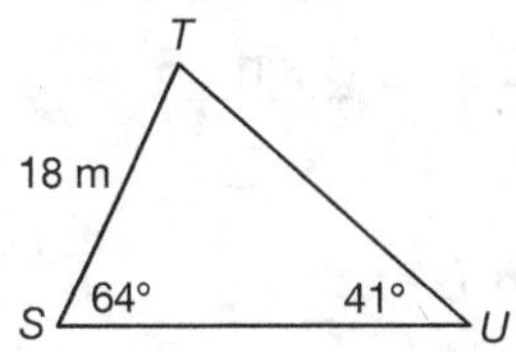

Find the given length. Round to the nearest hundredth.

2. *BC* ________

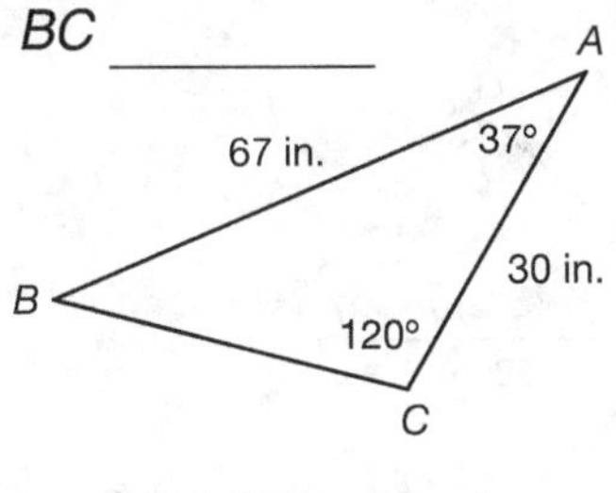

3. *DE* ________

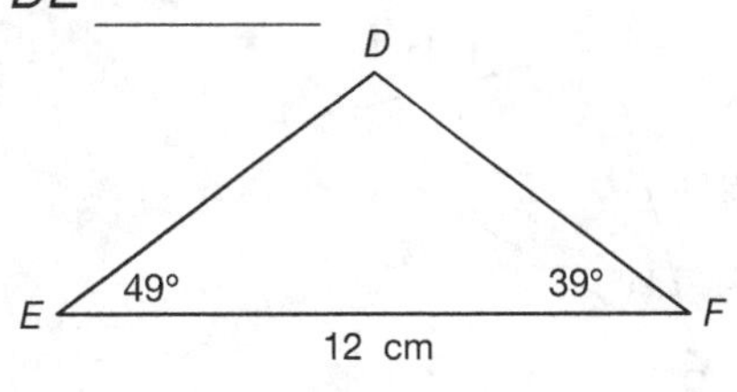

Finding a Missing Angle

Find the measure of $\angle D$. Round your answer to the nearest degree.

Step 1: The measures of $\angle D$ and its opposite side are unknown.

Use the Law of Sines to find $m\angle F$.

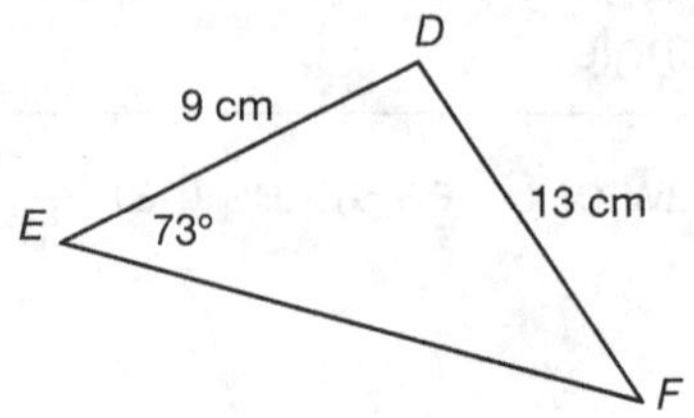

$\frac{9}{\sin F} = \frac{13}{\sin 73^\circ}$	Law of Sines
$13 \sin F = 9 \sin 73^\circ$	Cross multiply.
$\sin F = \frac{9 \sin 73^\circ}{13}$	Simplify.
$\sin F \approx 0.66$	Use calculator.
$m\angle F \approx \sin^{-1}(0.66)$	Use inverse sine function.
$m\angle F \approx 41^\circ$	Simplify.

Step 2: Use the Triangle Sum Theorem to find $m\angle D$.

$m\angle D + m\angle E + m\angle F = 180^\circ$	Triangle Sum Theorem
$m\angle D + 73^\circ + 41^\circ = 180^\circ$	Substitute values.
$m\angle D = 66^\circ$	Simplify.

Practice

Complete the steps to find the measure of $\angle K$. Round your answer to the nearest degree.

4. Find ____ first.

$\frac{KJ}{\sin L} = \frac{LK}{\_\_\_\_}$

$\frac{\_\_}{\sin L} = \frac{\_\_\_\_}{\_\_\_\_}$

$\_\_\_\_ = 3 \sin 35^\circ$

$\sin L = \frac{3 \sin 35^\circ}{\_\_}$

$\sin L \approx \_\_\_\_$

$m\angle L \approx \sin^{-1}(\_\_\_\_)$

$m\angle L \approx 14^\circ$

Find ____.

$m\angle J + m\angle K + m\angle L = \_\_\_\_$

$\_\_\_\_ + m\angle K + 14^\circ = 180^\circ$

$m\angle K = \_\_\_\_$

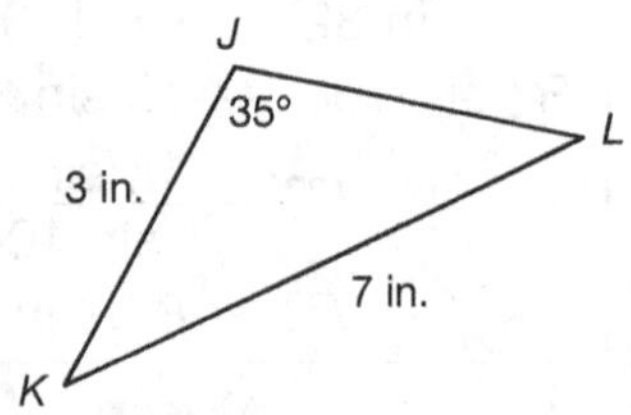

Find the given measure. Round your answer to the nearest degree.

5. $m\angle P$ ____

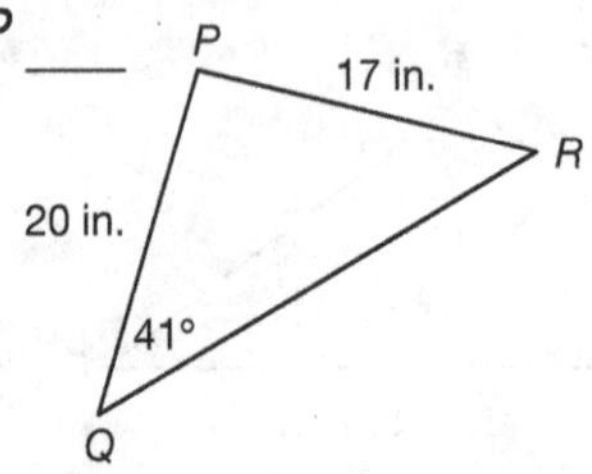

6. $m\angle G$ ____

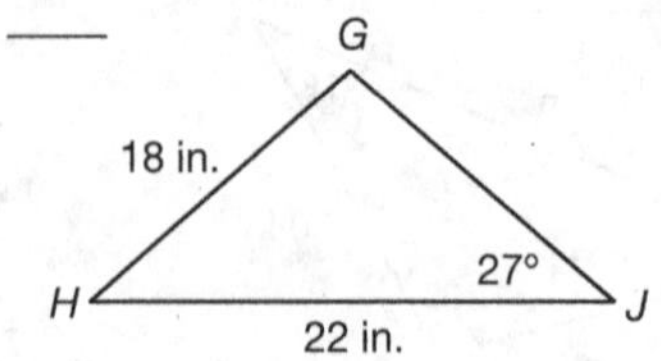

Name ______________________ Date __________ Class __________

Reteaching 95

Equations of Circles: Shifting and Dilating

You have worked with the Law of Sines. Now you will shift and dilate circles.

Shifting the Center of a Circle

The equation of a circle with radius r and center at the origin is $x^2 + y^2 = r^2$.

If the circle $x^2 + y^2 = r^2$ is shifted	the equation of the circle is
h units to the right,	$(x - h)^2 + y^2 = r^2$.
h units to the left,	$(x + h)^2 + y^2 = r^2$.
h units upward,	$x^2 + (y - h)^2 = r^2$.
h units downward,	$x^2 + (y + h)^2 = r^2$.

The equation of the circle is $x^2 + y^2 = 16$.

Shift the circle 3 units to the right. Write the equation of the shifted circle. Draw the original circle and shifted circle on the coordinate plane.

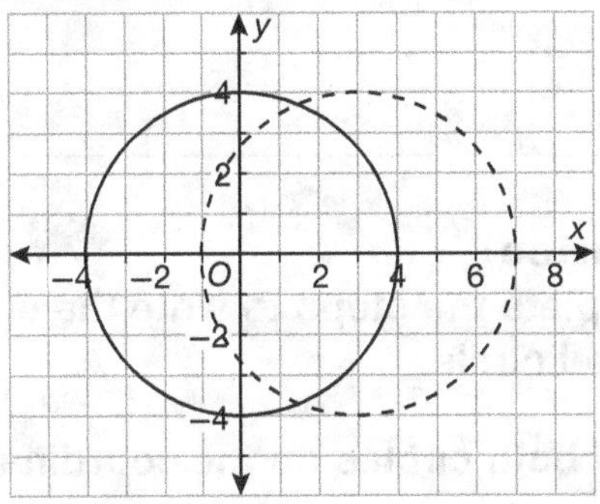

Step 1: Because the circle is shifted 3 units to the right, subtract 3 from x coordinate.

The new equation is $(x - 3)^2 + y^2 = 16$.

Step 2: The radius of the original circle is 4.

The new circle is centered at (3, 0) with radius 4.

Step 3: Draw both circles on the coordinate plane.

Practice

Complete the steps to write the equation of the shifted circle. Draw both circles on the coordinate plane.

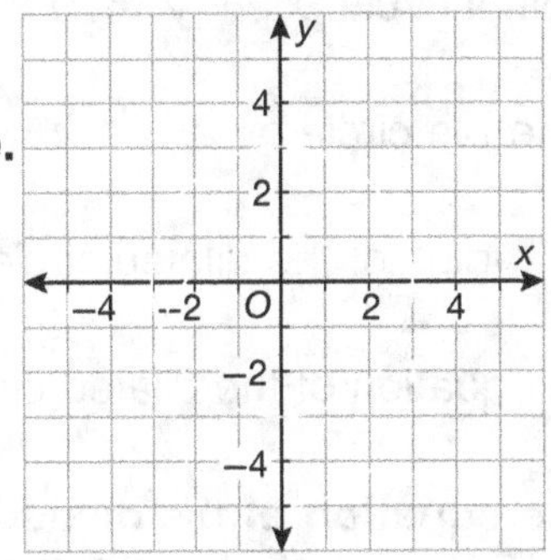

1. Original circle: $x^2 + y^2 = 25$

   Shift the circle 2 units down.

   Equation of new circle: $x^2 + (y$ ____$)^2 =$ ____

Write the equation of the new circle. Graph both circles on the coordinate plane.

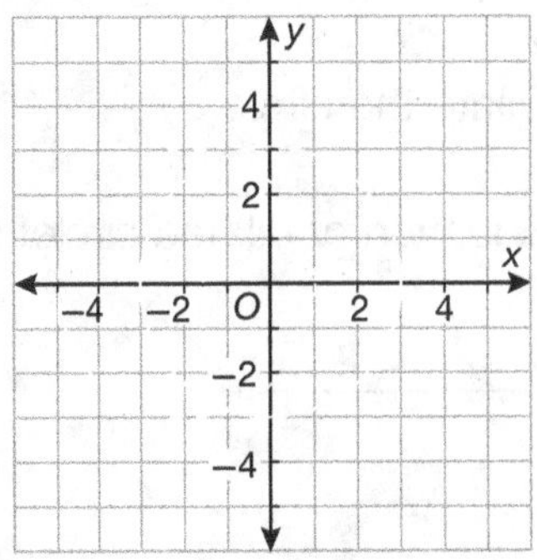

2. Original circle: $x^2 + y^2 = 9$

   Shift the circle 2 units to the left.

   Equation of new circle: ______________________

Dilating a Circle

To dilate a circle, multiply the radius by a factor *c*. If one circle has equation $x^2 + y^2 = r^2$, then a dilation of the circle has equation $x^2 + y^2 = (cr)^2$.

A dilation changes the size of a circle but does not shift the circle.

Example:

A circle has equation $x^2 + y^2 = 36$.

Find the equation of the circle after it is dilated by a factor of $\frac{1}{3}$. Draw the original circle and the dilated circle on the coordinate plane.

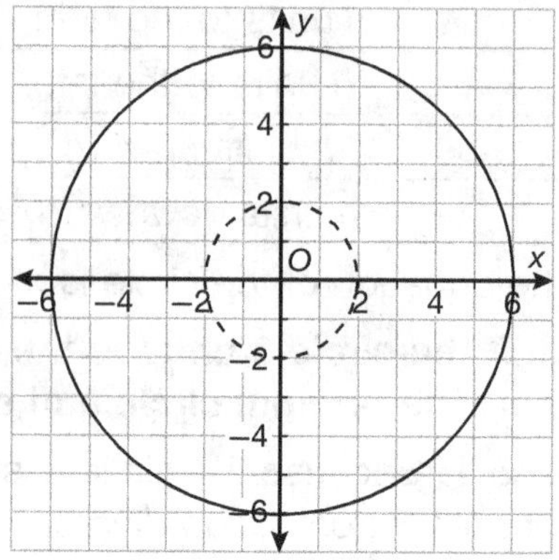

Step 1: Multiply the radius of the circle by the scale factor $\frac{1}{3}$.

$6 \times \frac{1}{3} = 2$

Step 2: Write the new equation.

$x^2 + y^2 = 2^2 = 4$

Step 3: Draw both circles on the coordinate plane.

Practice

Complete the steps to write the equation of the dilated circle.

Draw both circles on the coordinate plane.

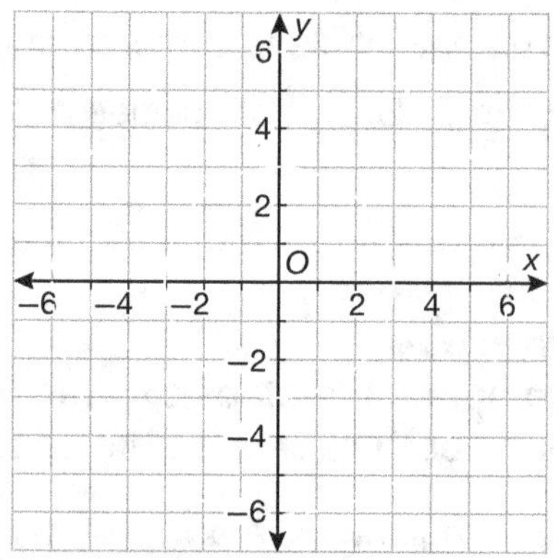

3. Original circle: $x^2 + y^2 = 9$

Dilate the circle by 2.

The radius of the dilated circle is 2 × ___ = ___.

The equation of the dilated circle is ______________.

Write the equation of the dilated circle.
Draw the original circle and dilated circle on the coordinate plane.

4. Original circle: $x^2 + y^2 = 25$

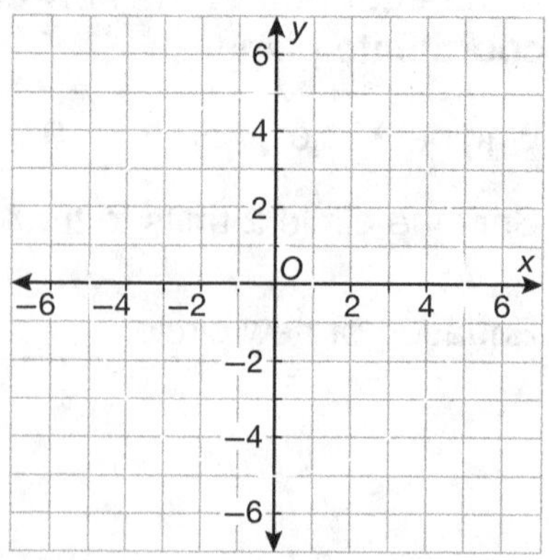

Dilate the circle by $\frac{1}{5}$.

Equation of dilated circle:

____

Name ______________________ Date __________ Class __________

Reteaching 96

Effects of Changing Dimensions on Perimeter and Area

You have shifted and dilated circles. Now you will describe changes on perimeter and area.

Effects of Changing Dimensions on Perimeter

Change in Dimensions	Change in Perimeter or in Circumference
Multiply every dimension of a polygon by a factor of *a*.	The perimeter is multiplied by a factor of *a*.
Multiply the radius of a circle by a factor of *a*.	The circumference is multiplied by a factor of *a*.

The rectangle is dilated by a factor of 2. Describe how the change affects the perimeter.

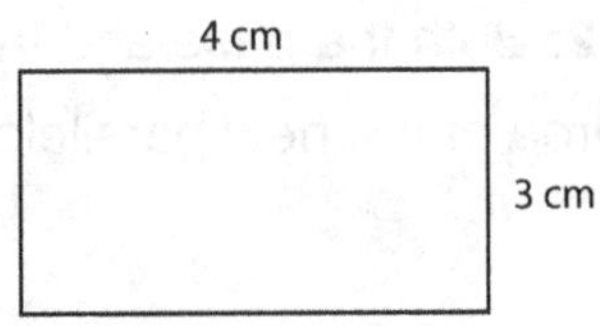

Step 1: The perimeter of the rectangle is

$P = 4 \cdot 2 + 3 \cdot 2 = 14$ cm.

Step 2: The dimensions of the rectangle are multiplied by 2.

The perimeter of the new rectangle is

$P = 8 \cdot 2 + 6 \cdot 2 = 28$ cm.

Therefore, the perimeter is multiplied by 2.

Practice

Complete the steps to describe how each change affects the perimeter or circumference.

1. The rectangle is dilated by $\frac{1}{2}$.

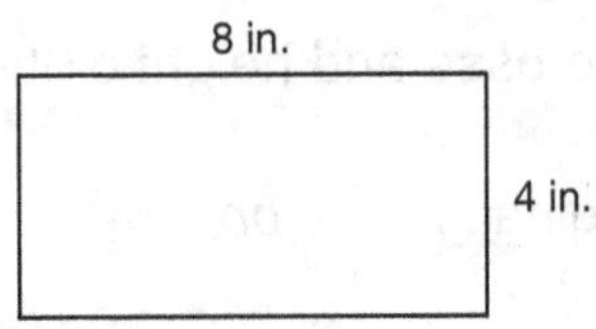

original: $P =$ ___ $\cdot 2 +$ ___ $\cdot 2 =$ ___ in.

new: $P =$ ___ $\cdot 2 +$ ___ $\cdot 2 =$ ___ in.

The perimeter is multiplied by ___.

Describe the effect of each change on the perimeter or circumference.

2. The base and the height of the triangle are multiplied by a factor of 3.

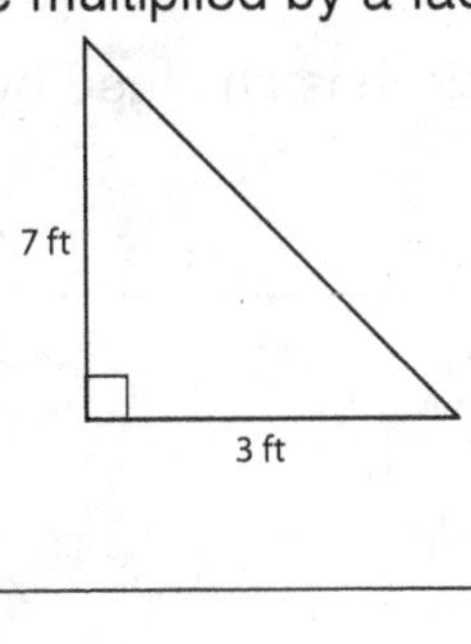

____

3. The radius of the circle is changed by a factor of $\frac{1}{5}$.

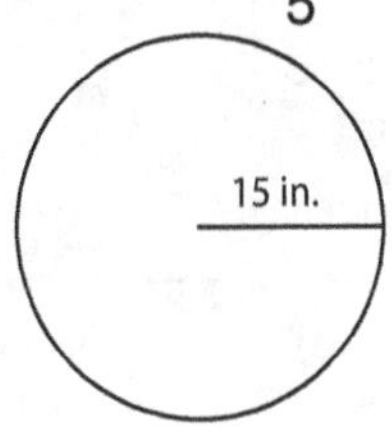

____

Effects of Changing Dimensions on Area

Change in Dimensions	Change in Area
Multiply one dimension of a polygon by a factor of a.	The area is multiplied by a.
Multiply every dimension of a polygon by a factor of a.	The area is multiplied by a^2.
Multiply the radius of a circle by a factor of a.	The area is multiplied by a^2.

The base and the height of a parallelogram are changed by a factor of 3. Describe how the change affects the area.

Step 1: The area of the parallelogram is $A = bh$.

$= 4(5)$

$= 20 \text{ cm}^2$

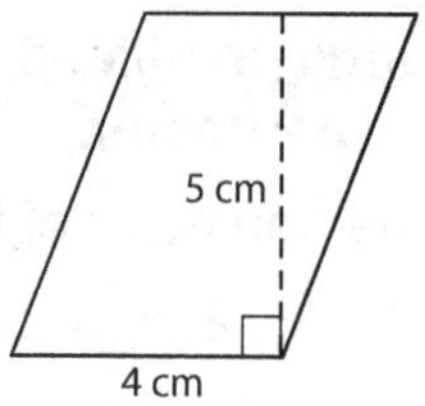

Step 2: Both the base and the height are multiplied by 3.

The area of the new parallelogram is $A = bh$.

$= (3 \cdot 4)(3 \cdot 5)$

$= 180 \text{ cm}^2$

When both the base and the height are multiplied by 3, the area is multiplied by 3^2, or 9.

Practice

Complete the steps to describe how each change affects the area of the given figure.

4. The base and height of the triangle are multiplied by 3.

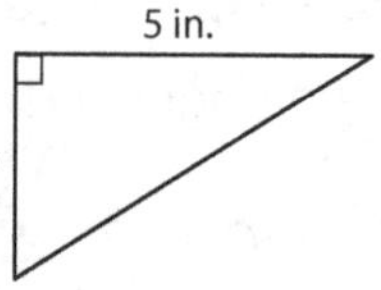

original: $A = \frac{1}{2}bh$ new: $A = \frac{1}{2}bh$

$= \frac{1}{2}(5)(\_)$ $= \frac{1}{2}(15)(\_)$

$= \_\_\_ \text{ in}^2$ $= \_\_\_ \text{ in}^2$

The area is multiplied by _.

Describe the effect of each change on the area.

5. The side length of the square is multiplied by 4.

____

6. The radius of the circle is changed by a factor of 5.

____

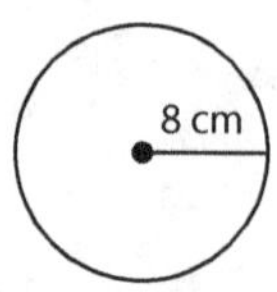

Name ______________________ Date __________ Class __________

Reteaching 97
Concentric Circles

Equations of Concentric Circles

Concentric circles are coplanar circles with the same center. There are no points of intersection between the circles.

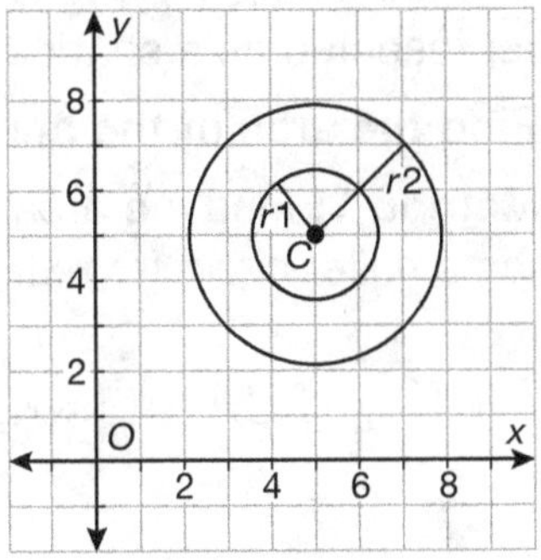

The equation of a circle is $(x - h)^2 + (y - k)^2 = r^2$. Concentric circles have the same values for h and k (because they share the same center) but the r values (the radii) differ.

Write the equations for the concentic circles shown.

$(x - h)^2 + (y - k)^2 = r^2$ Equation of a Circle

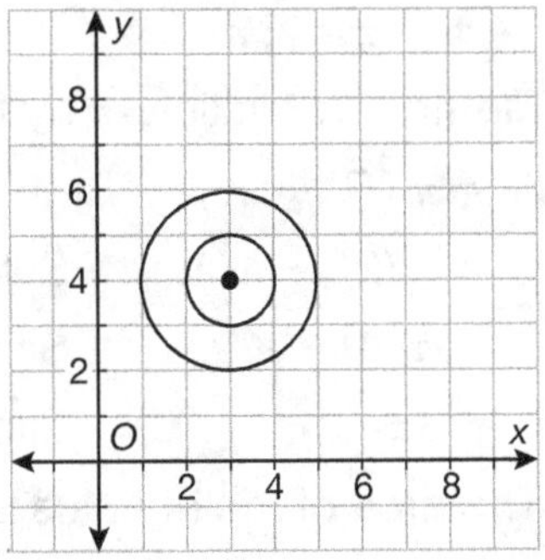

Inner circle:

$(x - 3)^2 + (y - 4)^2 = 1^2$ Substitute 3 for h, 4 for k, and 1 for r.

$(x - 3)^2 + (y - 4)^2 = 1$ Simplify.

Outer circle:

$(x - 3)^2 + (y - 4)^2 = 2^2$ Substitute 3 for h, 4 for k, and 2 for r.

$(x - 3)^2 + (y - 4)^2 = 4$ Simplify.

Practice

Complete the steps to write the equations of the concentric circles.

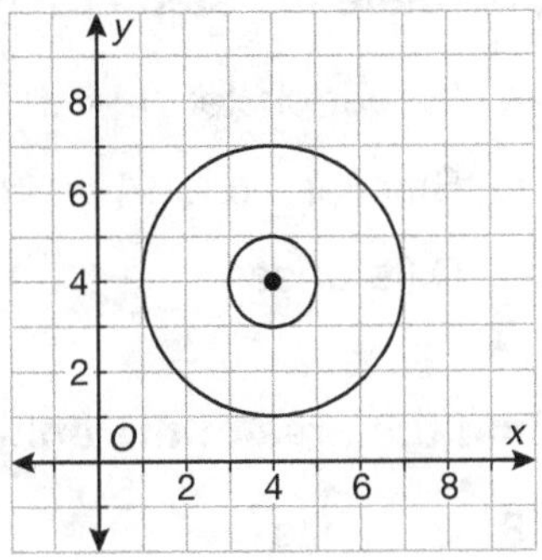

1. Inner: $(x - \_)^2 + (y - \_)^2 = \_^2$

$(x - \_)^2 + (y - \_)^2 = \_$

Outer: $(x - \_)^2 + (y - \_)^2 = \_^2$

$(x - \_)^2 + (y - \_)^2 = \_$

Write the equation of the concentric circles.

2. Inner: ______________________

Outer: ______________________

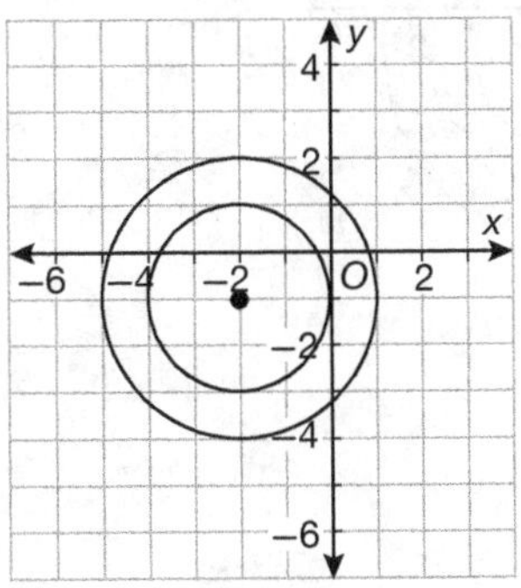

3. Inner: ______________________

Outer: ______________________

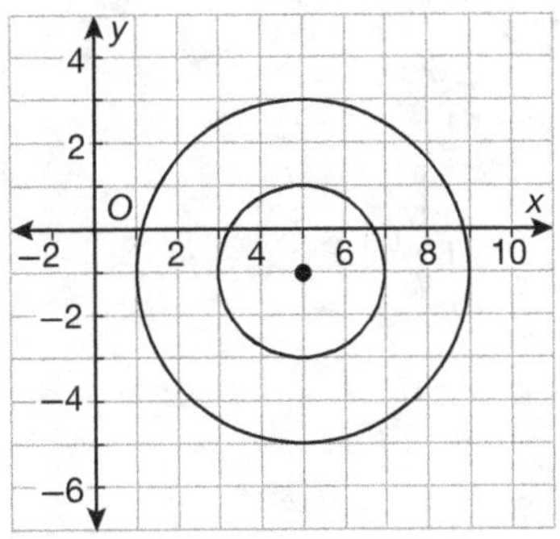

Annulus of Concentric Circles

The annulus is the region between two concentric circles, often represented as a shaded area.

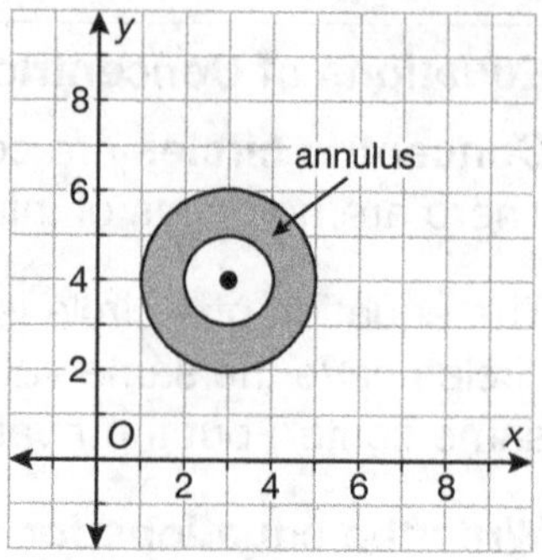

Find the area of the annulus in the concentric circles.

Method 1: Find the area of each circle. Then, subtract the area of the smaller circle from that of the larger circle.

$r = 2$	Radius of larger circle
$A = \pi r^2 = \pi(2)^2 = 4\pi$	Area of larger circle
$r = 1$	Radius of smaller circle
$A = \pi r^2 = \pi(1)^2 = \pi$	Area of smaller circle
$4\pi - \pi = 3\pi$	Difference of areas

Method 2: Multiply π by the difference of the radii squared.

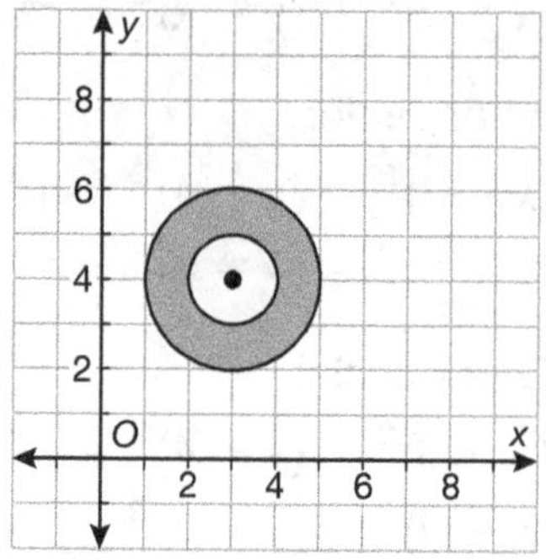

$A = \pi(R^2 - r^2)$	
$= \pi(2^2 - 1^2)$	Substitute 2 for R and 1 for r.
$= \pi(4 - 1)$	Simplify.
$= 3\pi$	Simplify.

The area of the annulus is 3π units^2.

Practice

Complete the steps to find the area of the annulus.

4. Larger circle: $A = \pi r^2 = \pi(\_)^2 = \_\,\pi$

   Smaller circle: $A = \pi r^2 = \pi(\_)^2 = \_\,\pi$

   Difference: ___ − ___ = ___ units^2

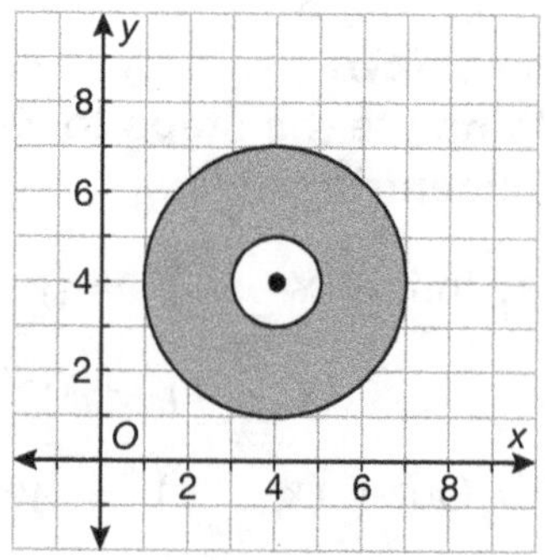

Find the area of the annulus.

5. ___ units^2

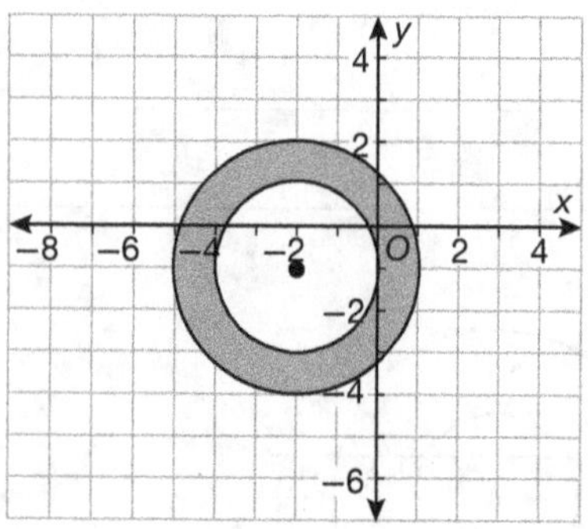

6. ___ units^2

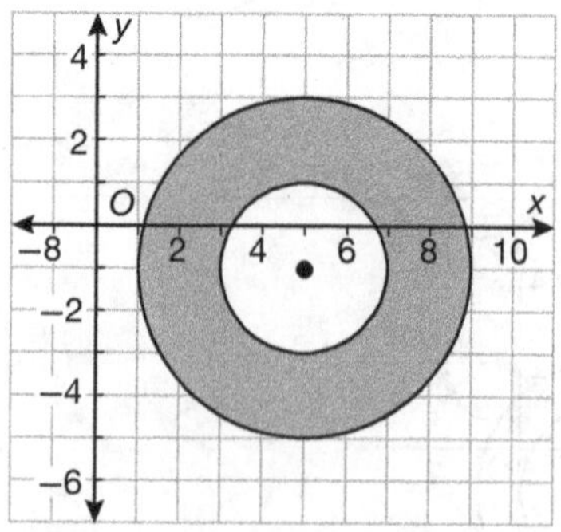

Name ______________________ Date __________ Class __________

Reteaching 98

Law of Cosines

You have worked with the Law of Sines. Now you will use the **Law of Cosines** to find side lengths and angle measures.

Law of Cosines: Side Length

For any ΔABC with side lengths a, b, and c that are opposite angles A, B, and C, the following rules apply:

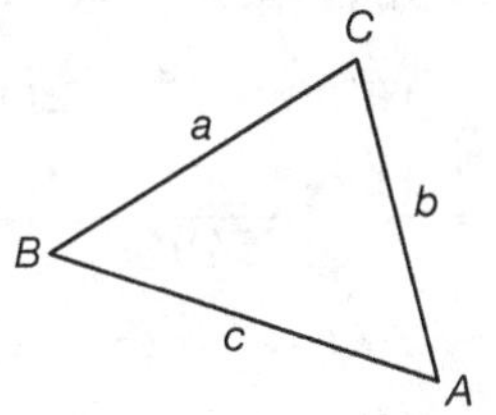

$a^2 = b^2 + c^2 - 2bc \cos A$

$b^2 = a^2 + c^2 - 2ac \cos B$

$c^2 = a^2 + b^2 - 2ab \cos C$

When you know two side lengths and the included angle measure between the two sides, you can use the Law of Cosines to find the third side.

Find AB. Round to the nearest tenth.

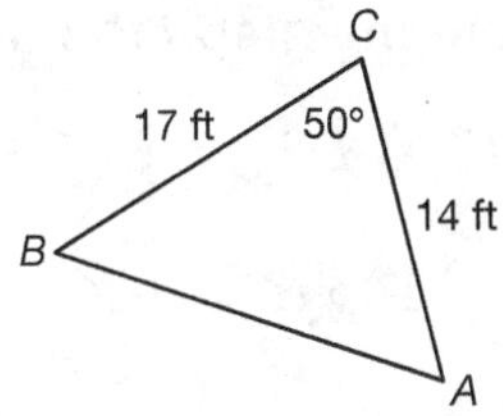

$AB^2 = BC^2 + AC^2 - 2(BC)(AC) \cos C$ — Law of Cosines

$AB^2 = 17^2 + 14^2 - 2(17)(14) \cos 50°$ — Substitute.

$AB^2 \approx 179.0331$ — Simplify.

$AB \approx 13.4$ ft — Find the square root.

Practice

Complete the steps to find the missing side length. Round to the nearest tenth.

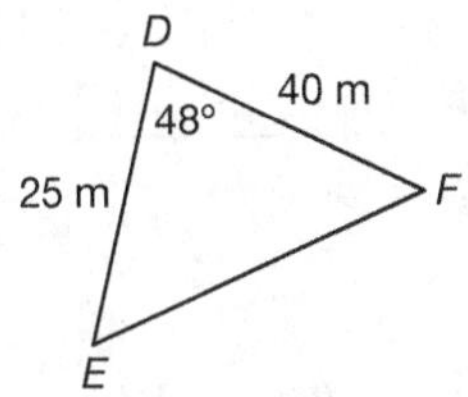

$EF^2 = DE^2 + DF^2 - 2(DE)(DF) \cos D$

1. $EF^2 =$ ___^{2} + ___^{2} − 2(__)(__) cos 48°

   $EF^2 \approx$ __________

   $EF \approx$ _____ m

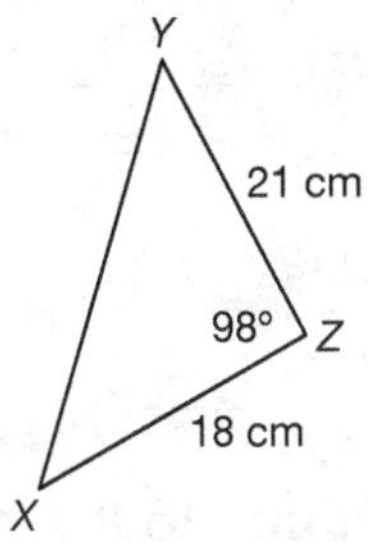

$XY^2 = XZ^2 + YZ^2 - 2(XZ)(YZ) \cos Z$

2. $XY^2 =$ ___^{2} + ___^{2} − 2(__)(__) cos ___°

   $EF^2 \approx$ __________

   $EF \approx$ _____ m

Find the missing side length. Round to the nearest tenth.

3. _____ in.

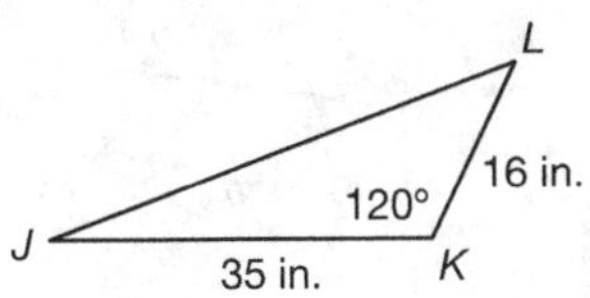

4. _____ cm

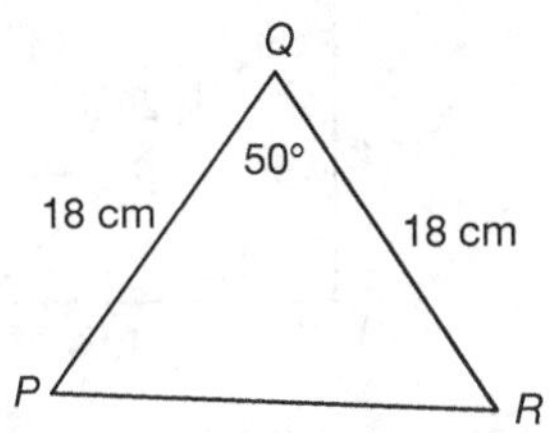

Law of Cosines: Missing Angle

You can also use the Law of Cosines to find a missing angle provided that you know all three side lengths.

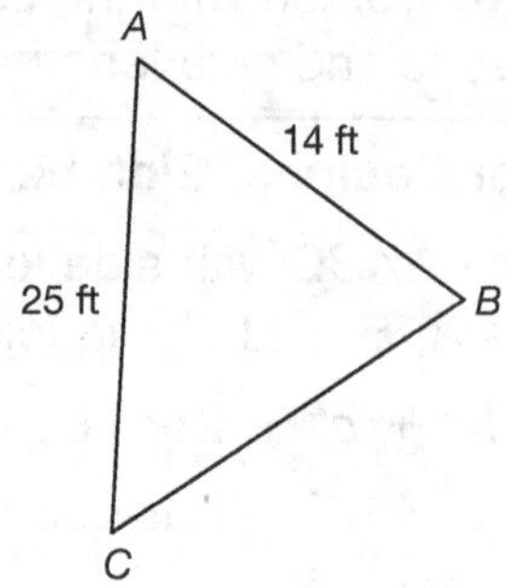

Find m∠A. Round to the nearest degree.

$BC^2 = AB^2 + AC^2 - 2(AB)(AC)\cos A$	Law of Cosines
$18^2 = 14^2 + 25^2 - 2(14)(25)\cos A$	Substitute.
$\cos A = \dfrac{14^2 + 25^2 - 18^2}{2(14)(25)}$	Isolate.
$\cos A \approx 0.71$	Simplify.
$A \approx 45°$	Solve.

Practice

Complete the steps to find θ. Round to the nearest degree.

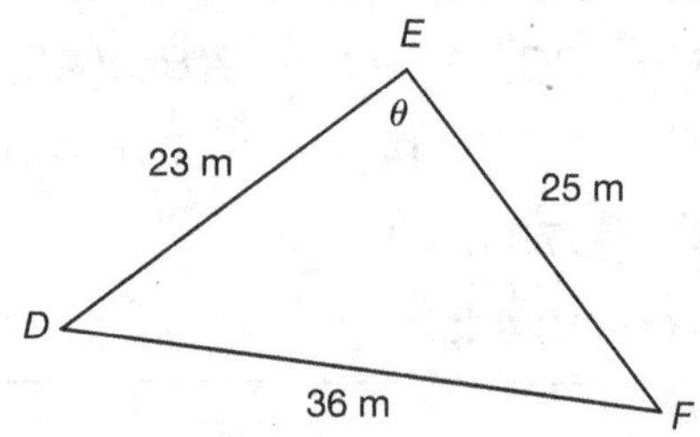

$DF^2 = DE^2 + EF^2 - 2(DE)(EF)\cos E$

$\underline{\quad}^2 = 23^2 + \underline{\quad}^2 - 2(23)(\underline{\quad})\cos\theta$

5. $\cos\theta = \dfrac{23^2 + \underline{\quad}^2 - \underline{\quad}^2}{2(23)(\underline{\quad})}$

$\cos\theta \approx$ ________

$\theta \approx \underline{\quad}°$

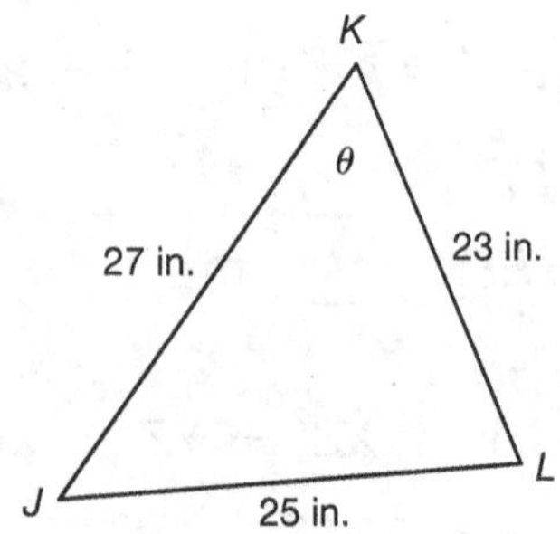

$JL^2 = JK^2 + KL^2 - 2(JK)(KL)\cos K$

$\underline{\quad}^2 = \underline{\quad}^2 + \underline{\quad}^2 - 2(\underline{\quad})(\underline{\quad})\cos\theta$

6. $\cos\theta = \dfrac{\underline{\quad}^2 + \underline{\quad}^2 - \underline{\quad}^2}{2(\underline{\quad})(\underline{\quad})}$

$\cos\theta \approx$ ________

$\theta \approx \underline{\quad}°$

Find θ. Round to the nearest degree.

7. ___°

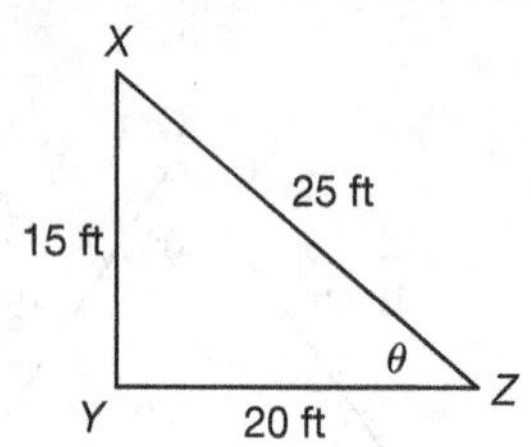

8. ___°

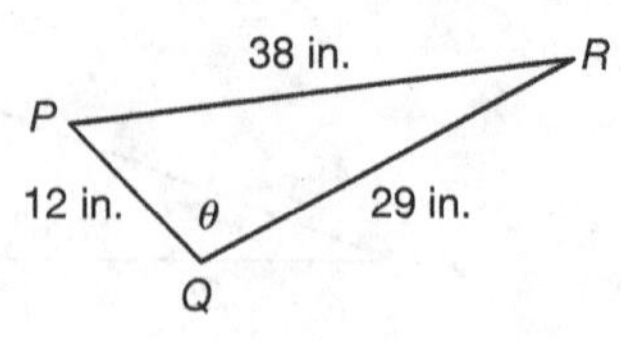

Name ______________________________ Date ____________ Class ______________

Reteaching 99

Volume Ratios of Similar Solids

You have worked with volumes of different solids. Now you will compare perimeters and areas of the faces of similar solids.

Similar Solids: Perimeters

If two similar solids have scale factor *a*:*b*, then the ratio of the perimeters of their corresponding faces is *a*:*b*.

Two similar right rectangular prisms *A* and *B* have scale factor 3:2. Determine the perimeter of *B*.

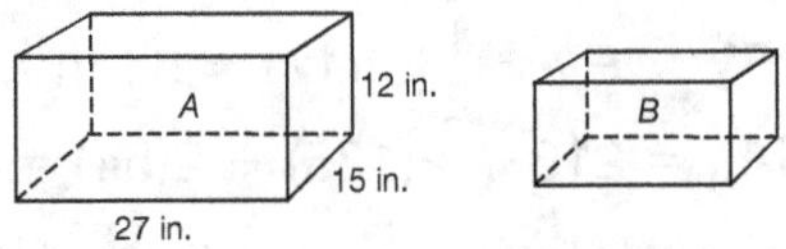

Step 1: Find the perimeter of *A*.

$P_A = 4(27) + 4(15) + 4(12)$

$P_A = 216$ in.

Step 2: Use the perimeter ratio to determine P_B.

$$\frac{P_A}{P_B} = \frac{3}{2}$$

$$2P_A = 3P_B$$

$$\frac{2P_A}{3} = P_B$$

Step 3: Find the perimeter of *B*.

$$P_B = \frac{2P_A}{3} = \frac{2(216)}{3} = 144 \text{ in.}$$

Practice

Complete the steps to find the perimeter of the other solid.

1. Scale factor is 2:1.

$P_s = 4(\_) + 4(\_\_)$

$P_s = \_\_$ ft

$\frac{P_s}{P_T} = \frac{2}{1}$

$\frac{P_s}{2} = P_T$

$P_T = \frac{\_\_}{1} = \_\_$ ft

2. Scale factor is 3:4

$P_X = 12(\_\_)$

$P_X = \_\_\_$ cm

$\frac{P_X}{P_Y} = \frac{3}{4}$

$\frac{\_P_x}{\_} = P_Y$

$P_Y = \frac{\_(\_\_\_)}{\_} = \_\_\_$ cm

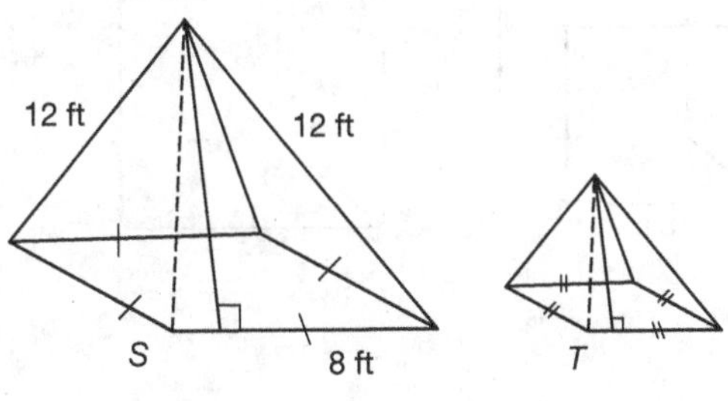

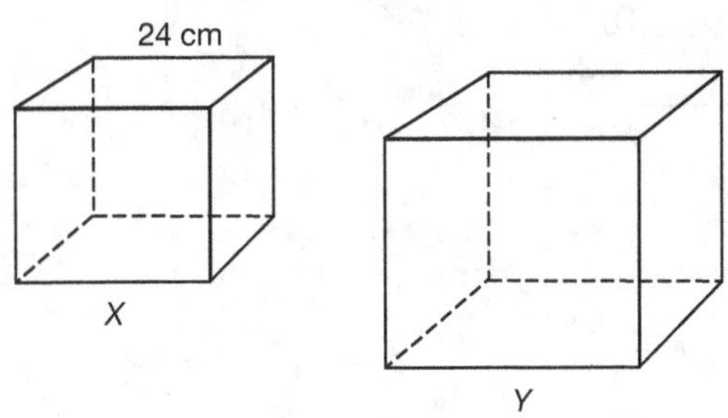

Similar Solids: Surface Areas

If two similar solids have scale factor *a*:*b*, then the ratio of the surface areas of their corresponding faces is $a^2:b^2$.

These two similar rectangular pyramids have scale factor 3:5. Determine the surface areas of both pyramids.

Step 1: Find the surface area of *A*.

$B_A = (12)(18) = 216 \text{ m}^2$ Base area

$F_A^1 = F_A^3 = \frac{1}{2}(15)(18) = 135 \text{ m}^2$ Face area

$F_A^2 = F_A^4 = \frac{1}{2}(9)(12) = 54 \text{ m}^2$ Face area

$SA_A = 216 + 2(135) + 2(54) = 594 \text{ m}^2$

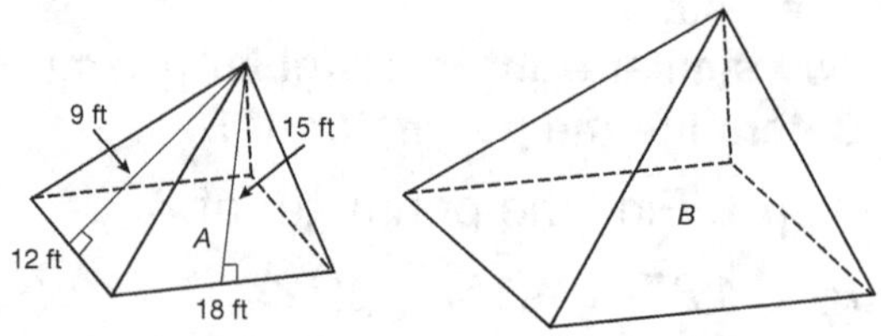

Step 2: Use the area ratio to determine SA_B.

$\frac{SA_A}{SA_B} = \left(\frac{3}{5}\right)^2$

$SA_B = \left(\frac{5}{3}\right)^2 SA_A$

Step 3: Find the surface area of *B*.

$SA_B = \left(\frac{5}{3}\right)^2 SA_A = \left(\frac{25}{9}\right)594 = 1650 \text{ m}^2$

Practice

Complete the steps to determine the surface areas of both solids.

3. Scale factor is 4:3.

$B_C = (12)(12) = 144 \text{ cm}^2$

$F_C^1 = F_C^2 = F_C^3 = F_C^4 = \frac{1}{2}(12)(\_\_\_) = \_\_\_ \text{ cm}^2$

$SA_C = 144 + 4(\_\_\_) = \_\_\_ \text{ cm}^2$

$\frac{SA_C}{SA_D} = \left(\frac{4}{3}\right)^2$

$SA_D = \frac{\_\_\_}{\_\_\_} SA_C = \frac{\quad}{\_\_\_}(\_\_\_) = \_\_\_ \text{ cm}^2$

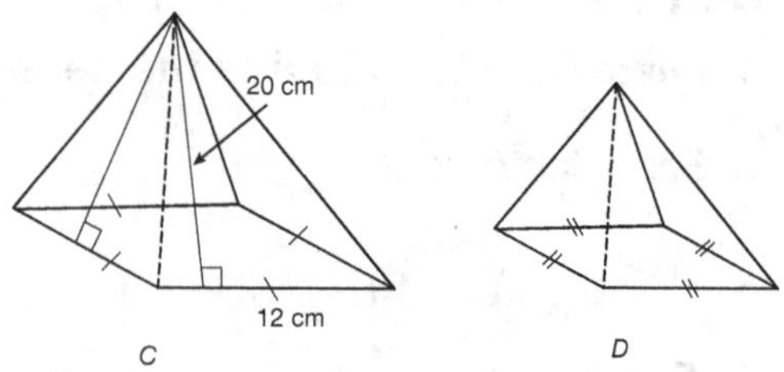

Determine the surface areas of both solids.

4. Scale factor is 1:2.

$SA_S = \_\_\_\_ \text{m}^2$

$SA_T = \_\_\_\_ \text{m}^2$

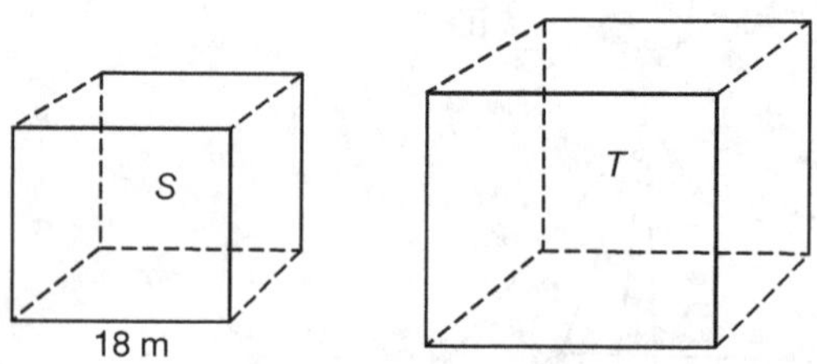

Reteaching 100

Transformation Matrices

You have worked with translations and reflections. Now you will learn to use transformation matrices.

Adding Point Matrices

A **matrix** is a rectangular array of numbers. A **point matrix** represents the coordinates of the points on a coordinate grid.

$\begin{pmatrix} 2 & 5 \\ 7 & 4 \end{pmatrix}$ ← x-coordinates (top row), ← y-coordinates (bottom row)

To add matrices, add the corresponding components.

Write a point matrix for the line segment $\overline{AB}$. Then, add the point matrix to $\begin{pmatrix} 1 & 1 \\ 1 & 1 \end{pmatrix}$. Determine the results of the addition.

$\begin{pmatrix} 3 & 5 \\ 1 & 4 \end{pmatrix}$ Point matrix for $\overline{AB}$

$\begin{pmatrix} 3 & 5 \\ 1 & 4 \end{pmatrix} + \begin{pmatrix} 1 & 1 \\ 1 & 1 \end{pmatrix}$ Add matrices.

$\begin{pmatrix} 3+1 & 5+1 \\ 1+1 & 4+1 \end{pmatrix}$ Add corresponding components.

$\begin{pmatrix} 4 & 6 \\ 2 & 5 \end{pmatrix}$ Simplify.

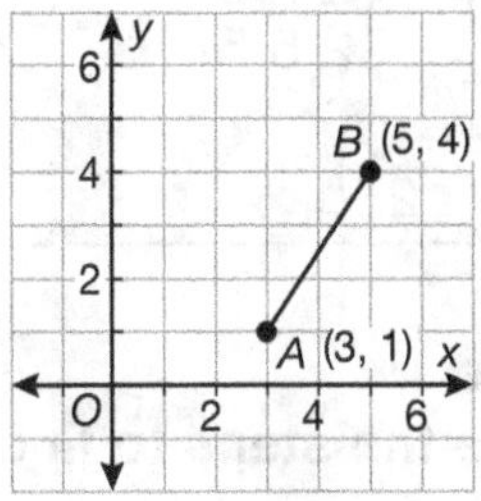

The coordinates of $\overline{A'B'}$ are $A'(4, 2)$ and $B'(6, 5)$; they have moved up 1 unit and right 1 unit.

Practice

Complete the steps to find the sum of the matrices.

1. Add the point matrix to $\begin{pmatrix} 2 & 2 & 2 \\ 1 & 1 & 1 \end{pmatrix}$.

$\Delta XYZ = \begin{pmatrix} \_ & \_ & \_ \\ \_ & \_ & \_ \end{pmatrix}$

$= \begin{pmatrix} \_ & \_ & \_ \\ \_ & \_ & \_ \end{pmatrix} + \begin{pmatrix} 2 & 2 & 2 \\ 1 & 1 & 1 \end{pmatrix}$

$= \begin{pmatrix} \_ & \_ & \_ \\ \_ & \_ & \_ \end{pmatrix}$

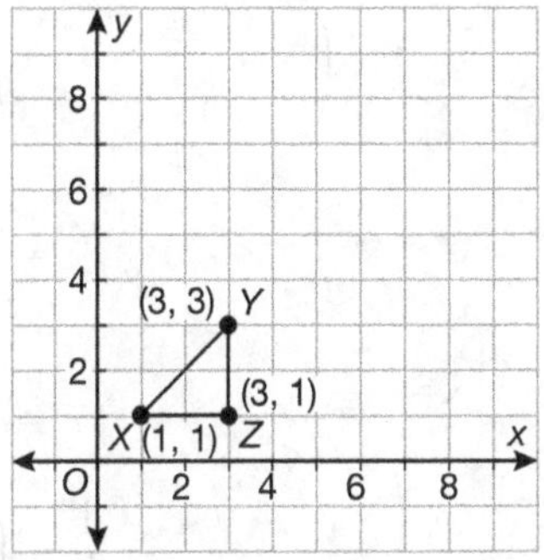

Find the sum of the matrices.

2. Add the point matrix to $\begin{pmatrix} 1 & 1 \\ 3 & 3 \end{pmatrix}$.

$\begin{pmatrix} \_ \\ \_ \end{pmatrix}$

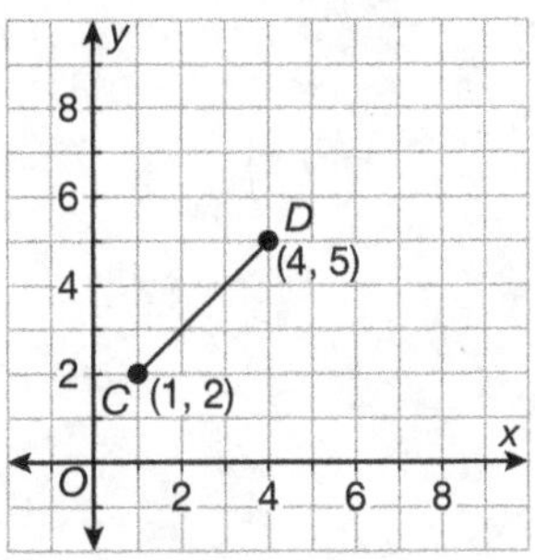

3. Add the point matrix to $\begin{pmatrix} 3 & 3 & 3 \\ -1 & -1 & -1 \end{pmatrix}$.

$\begin{pmatrix} \_ \\ \_ \end{pmatrix}$

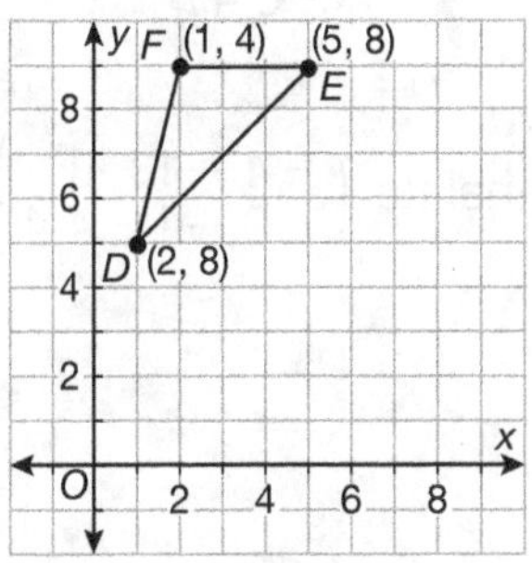

Translating with Matrices

You can use matrices to determine how one figure is translated to another figure.

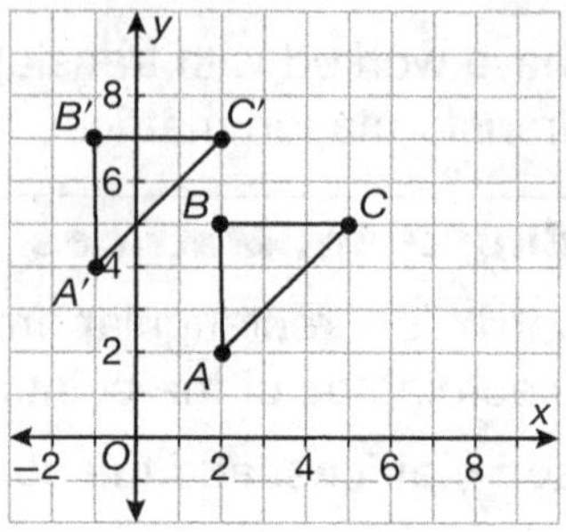

Example:

Triangle ABC is translated 3 units left and 2 units up.

Find the matrix that translates *ABC* to *A′B′C′*.

$\begin{pmatrix} 2 & 2 & 5 \\ 2 & 5 & 5 \end{pmatrix}$ ΔABC point matrix

$\begin{pmatrix} -1 & -1 & 2 \\ 4 & 7 & 7 \end{pmatrix}$ $\Delta A'B'C'$ point matrix

$\begin{pmatrix} -1 & -1 & 2 \\ 4 & 7 & 7 \end{pmatrix} - \begin{pmatrix} 2 & 2 & 5 \\ 2 & 5 & 5 \end{pmatrix}$ Subtract.

$\begin{pmatrix} -3 & -3 & -3 \\ 2 & 2 & 2 \end{pmatrix}$ Simplify.

Practice

Complete the steps to find the translation matrix.

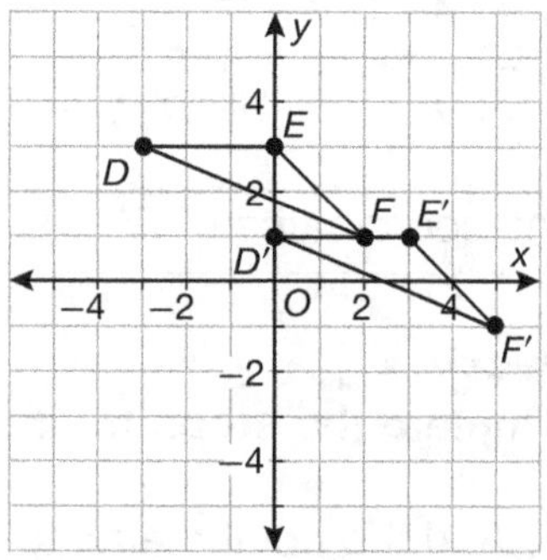

$\Delta DEF = \begin{pmatrix} -3 & 0 & 2 \\ 3 & 3 & 1 \end{pmatrix}$

4. $\Delta D'E'F' = \begin{pmatrix} \quad \end{pmatrix}$

$\begin{pmatrix} \quad \end{pmatrix} - \begin{pmatrix} -3 & 0 & 2 \\ 3 & 3 & 1 \end{pmatrix} = \begin{pmatrix} \quad \end{pmatrix}$

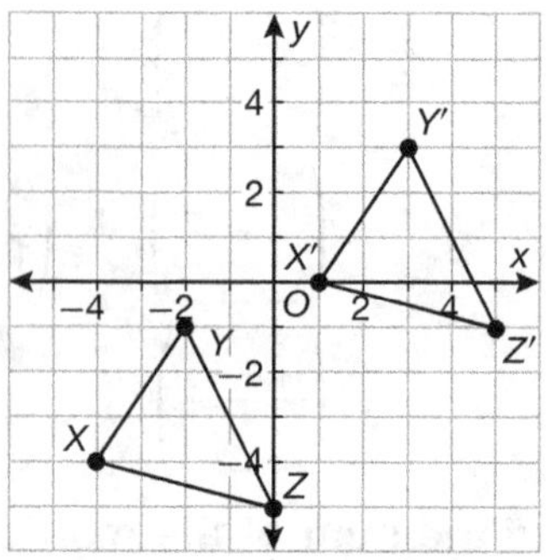

$\Delta XYZ = \begin{pmatrix} -4 & -2 & 0 \\ -4 & -1 & -5 \end{pmatrix}$

5. $\Delta X'\ Y'\ Z' = \begin{pmatrix} \quad \end{pmatrix}$

$\begin{pmatrix} \quad \end{pmatrix} - \begin{pmatrix} -4 & -2 & 0 \\ -4 & -1 & -5 \end{pmatrix} = \begin{pmatrix} \quad \end{pmatrix}$

Find the translation matrix.

6. Translate $\begin{pmatrix} 6 & 9 & 3 \\ 2 & 0 & 4 \end{pmatrix}$ 3 units right and 5 units down. $\begin{pmatrix} \_ & \_ & \_ \\ \_\_ & \_\_ & \_\_ \end{pmatrix}$

7. Translate $\begin{pmatrix} -4 & -1 & 3 \\ -3 & -1 & -4\ 4 \end{pmatrix}$ units left and 3 units up $\begin{pmatrix} \quad \end{pmatrix}$

Name ____________________ Date __________ Class __________

Reteaching INV 10
Fractals

The Sierpinski Triangle

An **iteration** is the repeated application of a rule. You can continue an iteration indefinitely. In geometry, you can generate **fractals** by iteration.

The **Sierpinski Triangle** is a famous fractal.

Start with a right isosceles triangle. The first iteration connects the midsegments of the triangle. Then, remove the triangle in the center so that there is a "hole" in the middle of the triangle.

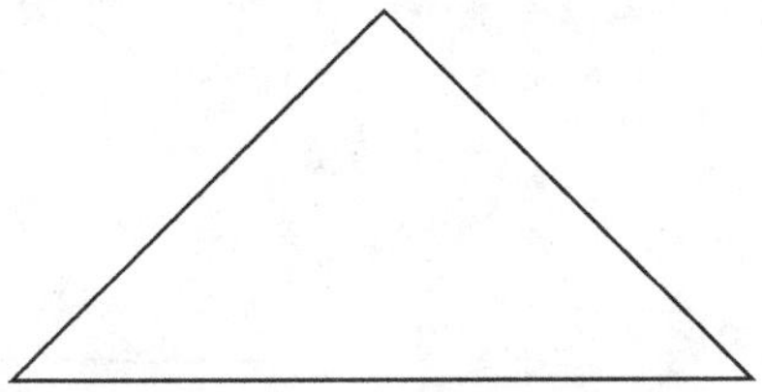

right isosceles triangle

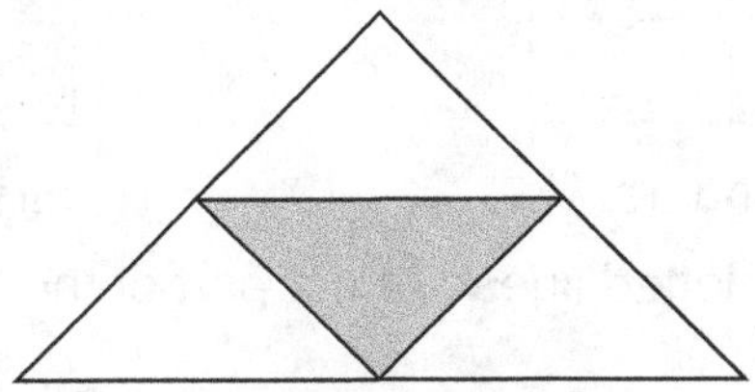

first iteration

The iteration of connecting the midsegments of the triangle and then removing the central triangle is repeated to make the Sierpinski triangle.

Practice

1. Draw a right isosceles triangle and make the first iteration for the Sierpinski triangle. Shade in the triangles you make except for the central triangle. Make a large triangle so that you can do several iterations.

   For the first iteration for the Sierpinski triangle, there are three unshaded ________.

   During the next iteration, connect the midsegments of each unshaded ________ and shade each ________ triangle.

Draw the second iteration for the Sierpinski triangle. Count the number of triangles. Remember not to count the central triangles.

2. For the second iteration, there are _____ unshaded triangles.

Draw the third iteration for the Sierpinski triangle.

3. For the third iteration, there are ____ unshaded triangles.

4. The number of unshaded triangles for each iteration of the Sierpinski triangle make a pattern. Write the number of unshaded triangles for the three iterations. ________

5. Write your numbers in the answer for problem 5 as powers of 3. ____________

6. How many unshaded triangles would the seventh iteration of the Sierpinski triangle have? ____________

The Sierpinski Carpet

The **Sierpinski carpet** is another fractal. Begin with a square. Think of the square divided into nine equal squares (look at the grid below).

The first iteration removes the middle square. Shade in the middle square as you did with the central triangle in the Sierpinski Triangle.

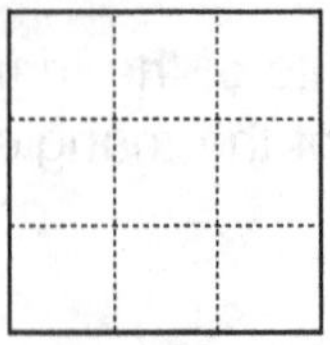

Square

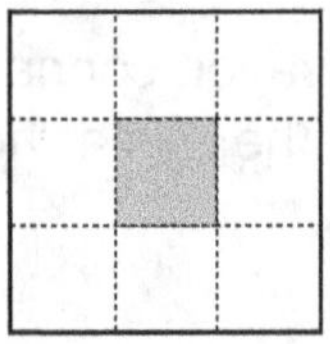

First iteration

(dotted lines are not part of the iteration)

There are 8 unshaded squares left. with each of the squares, think of the square divided into 9 equal squares. For the second iteration, remove the middle square from each of the 8 unshaded squares.

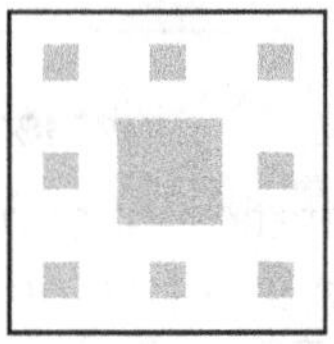

Practice

7. Draw a square and make the first iteration for the Sierpinski carpet. Shade in the central square. Be sure to make a large square so that you can do several iterations.

For the first iteration for the Sierpinski carpet, there are ______ unshaded squares.

Draw the second iteration for the Sierpinski carpet. Count the number of squares left. Remember not to count the central squares.

8. During the next iteration, remove (shade) the middle _______ from each of the eight squares.

9. There are ______ smaller squares in each of the eight squares.

There are a total of _________ unshaded squares after the second iteration.

Draw the third iteration for the Sierpinski carpet. Count the number of squares left.

10. There are ______ smaller unshaded squares in each of the sixty-four squares. There are a total of ____ unshaded squares after the second iteration.

11. How many unshaded squares would the fourth iteration of the Sierpinski carpet have? __________

Name ______________________ Date __________ Class __________

Reteaching 101

Determining Lengths of Segments Intersecting Circles

You have worked with secants and tangents of circles. Now you will learn about the relationships between secant segments and tangent segments.

According to the **Secant-Secant Product Theorem,** the product of one secant segment and its external segment equals that of the other secant segment and its external µsegment.

In the diagram to the right, the Secant-Secant Product Theorem can be stated as $AE \cdot BE = CE \cdot DE$.

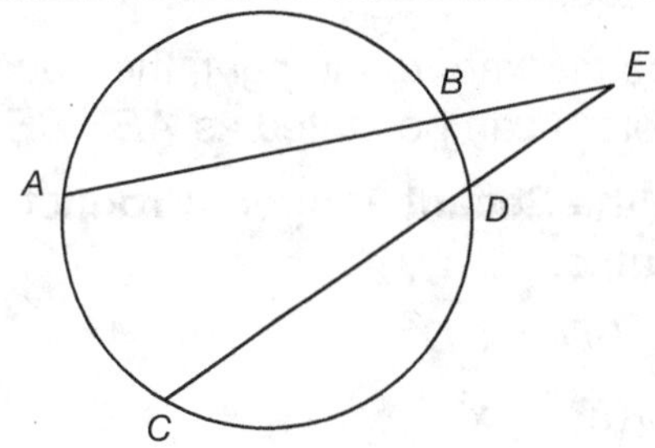

Use the Secant-Secant Product Theorem to find the value of *x* in the diagram.

$JN \cdot KN = LN \cdot MN$

$(8 + x)(8) = (20)(5)$

$64 + 8x = 100$

$8x = 36$

$x = 4.5$

$LN = 20$ and $JN = 8 + 4.5 = 12.5$

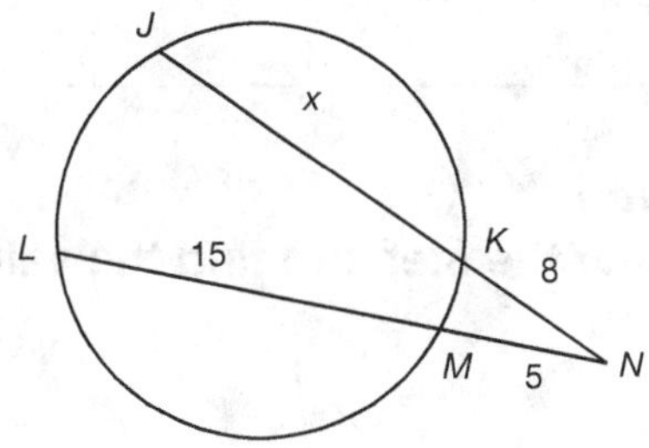

Practice

Complete the steps to find the value of *the variable*.

1.

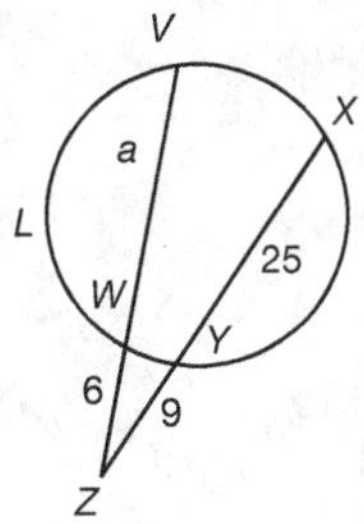

$XZ \cdot YZ = VZ \cdot WZ$

$(34)(\_) = (\_\_\_\_)(6)$

_____ = ________

_______ = _____

$a =$ _____

2.

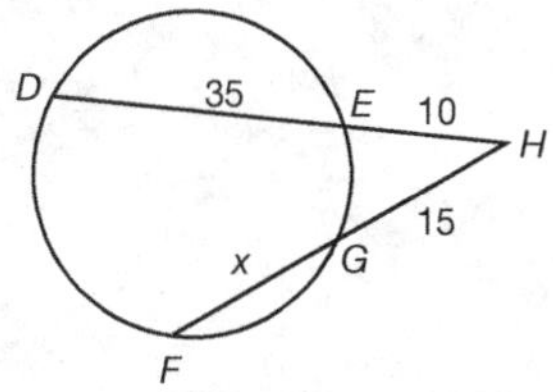

$DH \cdot EH = FH \cdot GH$

$(45)(10) = (\_\_\_\_\_\_\_\_)(\_\_\_)$

$450 =$ __________

_______ = ______

$x =$ _____

Find the value of *x*. Round your answer to the nearest tenth.

3. $x =$ _____

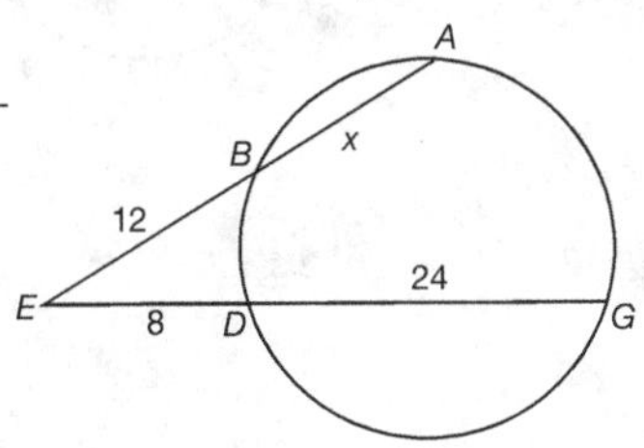

4. $x =$ _______

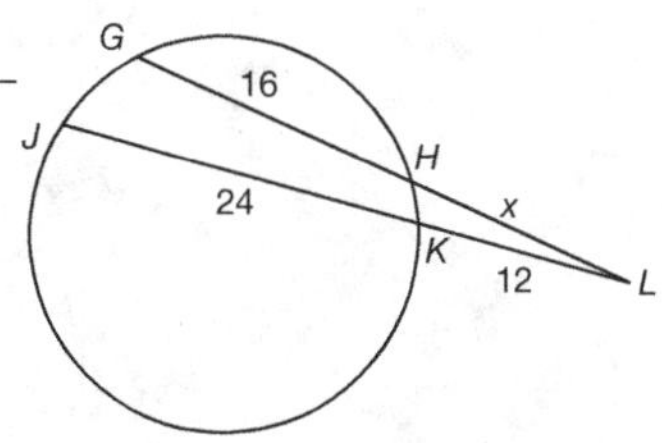

According to the **Secant-Tangent Product Theorem,** the product of the lengths of the secant segment and its external segment equals that of the tangent segment squared.

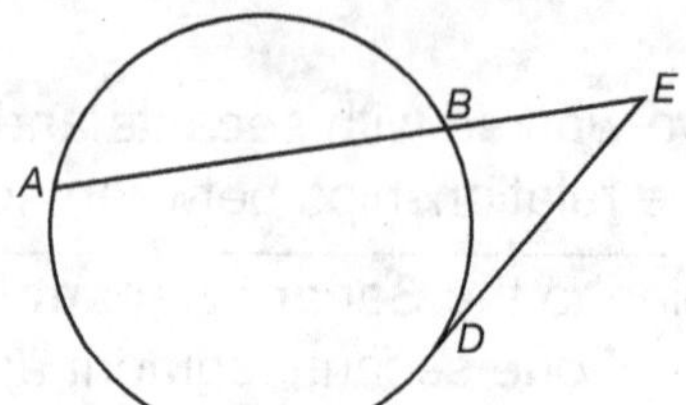

In the diagram to the right, the Secant-Tangent Product Theorem can be stated as $AE \cdot BE = DE^2$.

Use the Secant-Tangent Product Theorem to find the value of the variable.

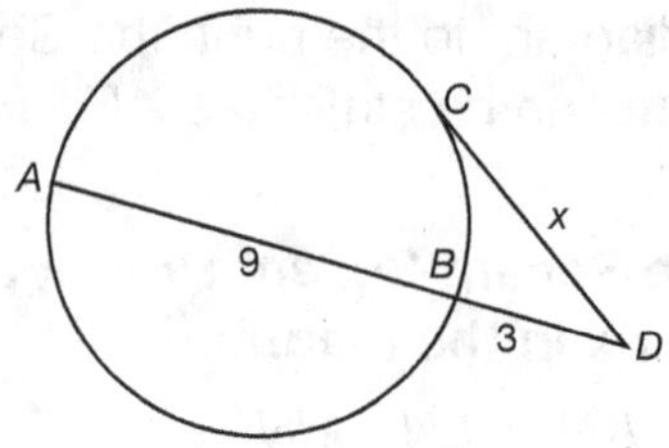

$AD \cdot BD = CD^2$

$(12)(3) = x^2$

$36 = x^2$

$x = 6$

Practice

Complete the steps to find the value *x*.

5.

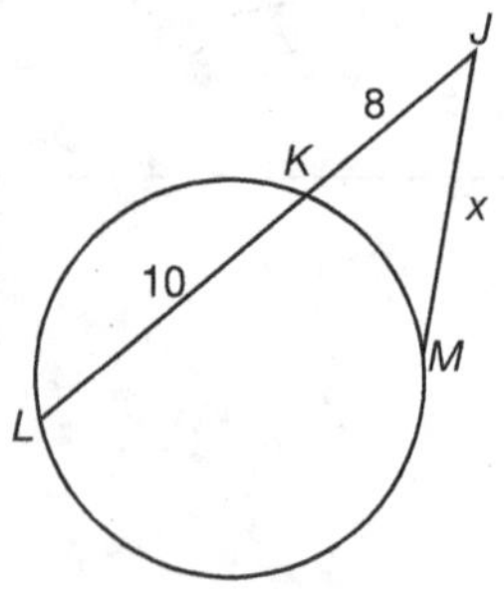

$JK \cdot JL = JM^2$

$(\_\_\_\_)(18) = x^2$

$\_\_\_\_ = x^2$

$x = \_\_\_\_$

6.

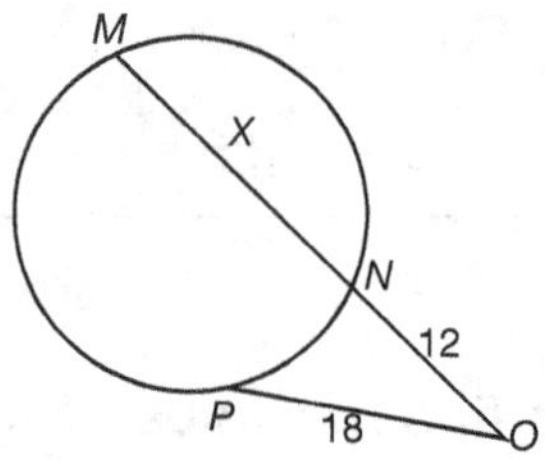

$MO \cdot NO = OP^2$

$(\_\_\_\_\_\_\_\_)(12) = 18^2$

$\_\_\_\_\_\_\_\_ = 324$

$\_\_\_\_ = \_\_\_\_$

$x = \_\_\_\_$

Find the value of *x*. Round your answer to the nearest tenth.

7. $x = \_\_\_\_\_\_$

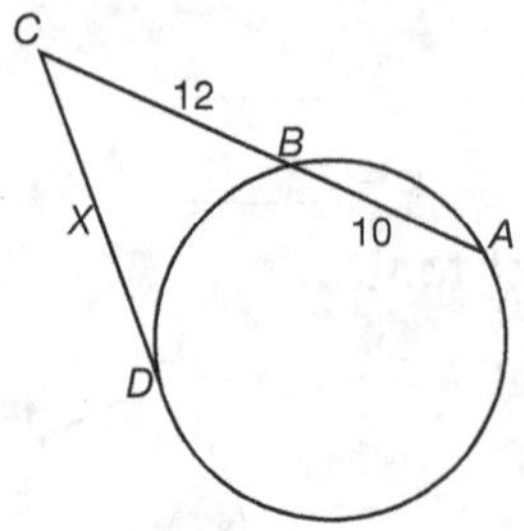

8. $x = \_\_\_\_\_\_$

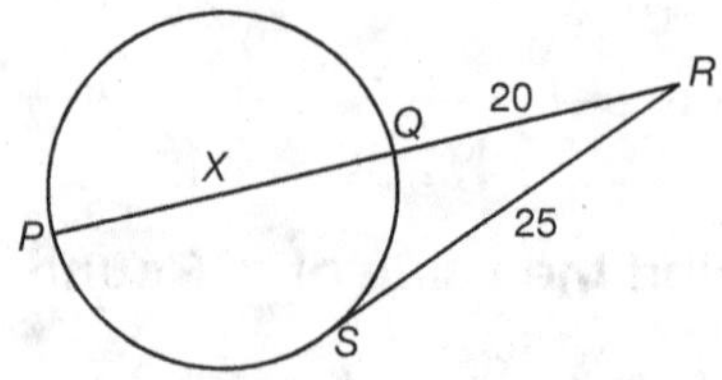

Name ______________________ Date __________ Class __________

Reteaching 102

Dilations in the Coordinate Plane

You have worked with translations and reflections. Now you will learn to make transformations using dilations in the coordinate plane.

Dilations

A **dilation** is a reduction or enlargement of a shape. Recall that the **scale factor** describes how much a figure is enlarged or reduced.

Mapping notation for a dilation with the origin at the center is written as $D_{O,\,k}(x, y) \rightarrow (kx, ky)$, where k represents the scale factor.

Example:

Determine the result of the dilation $D_{O,\,4}(x, y)$ on the points (2, 6) and (6, 4).

$D_{O,\,k}(x, y) \rightarrow (kx, ky)$

$D_{O,\,4}(2, 6) \rightarrow (4 \times 2, 4 \times 6) = (8, 24)$

$D_{O,\,4}(6, 4) \rightarrow (4 \times 6, 4 \times 4) = (24, 16)$

Practice

Complete the steps to find the result of the given dilation on the coordinates.

1. $D_{O,\,2}(x, y)$ on the points (−2, 4) and (1, 3)

   $D_{O,\,2}(-2, 4) \rightarrow (2 \times$ ___, $2 \times$ _) = (___, _)

   $D_{O,\,2}(1, 3) \rightarrow (2 \times$ _, $2 \times$ _) = (_, _)

2. $D_{O,\,3}(x, y)$ on the points (–1, 5) and (2, –4)

   $D_{O,\,3}(-1, 5) \rightarrow (3 \times$ ___, $3 \times$ _) = (___, ___)

   $D_{O,\,2}(2, -4) \rightarrow (3 \times$ _, $3 \times$ ___) = (_, ____)

Determine the result of the dilation on the given points.

3. $D_{O,\,4}(x, y)$ on the points (1, 5) and (–2, 3) ____________________
4. $D_{O,\,2.5}(x, y)$ on the points (2, 4) and (3, 6) ____________________
5. $D_{O,\,2}(x, y)$ on the points (–3, 1) and (5, 8) ____________________
6. $D_{O,\,0.5}(x, y)$ on the points (–4, 6) and (4, 8) ____________________
7. $D_{O,\,1.25}(x, y)$ on the points (1, –2) and (3, –4) ____________________
8. $D_{O,\,3}(x, y)$ on the points (0, 4) and (–4, 2) ____________________

Dilating by Point Matrix

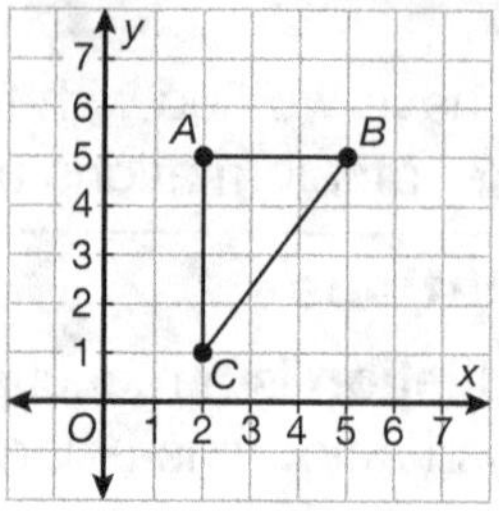

You can use a **point matrix** to determine the points of a figure on a coordinate grid. The point (a, b) would be written as $\begin{bmatrix} a \\ b \end{bmatrix}$. To apply a scale factor to a point matrix, multiply each entry in the point matrix by the scale factor.

Create a point matrix for the triangle given in the diagram.

Use this point matrix to show a dilation of the shape by a factor of 4.

Step 1: Write the point matrix. Each point is a different column in the matrix.

$$\begin{bmatrix} 5 & 5 & 1 \\ 1 & 5 & 1 \end{bmatrix}$$

Step 2: Multiply each entry in the point matrix by the scale factor.

$$4\begin{bmatrix} 5 & 5 & 1 \\ 1 & 5 & 1 \end{bmatrix} = \begin{bmatrix} 4 \cdot 5 & 4 \cdot 5 & 4 \cdot 1 \\ 4 \cdot 1 & 4 \cdot 5 & 4 \cdot 1 \end{bmatrix}$$

$$= \begin{bmatrix} 20 & 20 & 4 \\ 4 & 20 & 4 \end{bmatrix}$$

Step 3: Write the new coordinates for the vertices of the dilated triangle.
(20, 4), (20, 20), and (4, 4)

Practice

Complete the steps to create a point matrix for the given figures and scale factor.

9. Scale factor 5

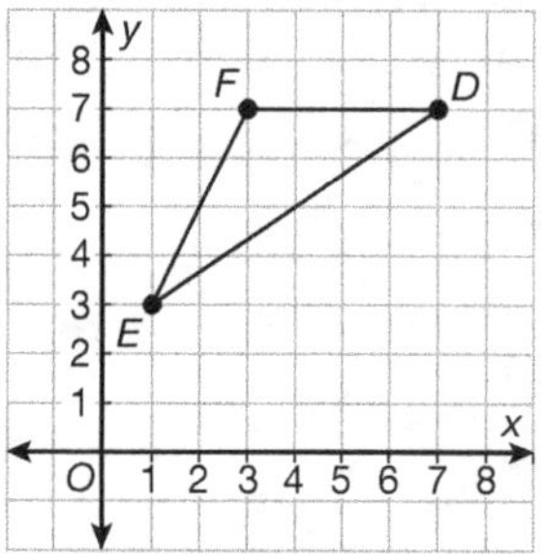

$$5\begin{bmatrix} 7 & \underline{\quad} & \underline{\quad} \\ 7 & \underline{\quad} & \underline{\quad} \end{bmatrix}$$

$$= \begin{bmatrix} 5 \cdot 7 & 5 \cdot \underline{\quad} & 5 \cdot \underline{\quad} \\ 5 \cdot 7 & 5 \cdot \underline{\quad} & 5 \cdot \underline{\quad} \end{bmatrix} \underline{\quad}$$

$$= \begin{bmatrix} 35 & \underline{\quad} & \underline{\quad} \\ 35 & \underline{\quad} & \underline{\quad} \end{bmatrix}$$

10. Scale factor 8

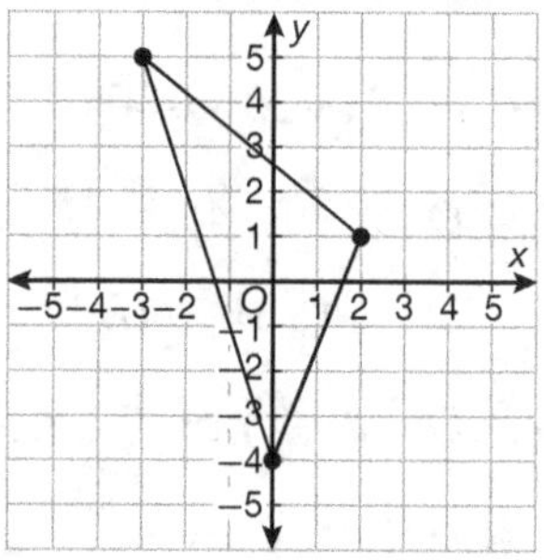

$$8\begin{bmatrix} \underline{\quad} & \underline{\quad} & \underline{\quad} \\ \underline{\quad} & \underline{\quad} & \underline{\quad} \end{bmatrix}$$

$$= \begin{bmatrix} 8 \cdot \underline{\quad} & 8 \cdot \underline{\quad} & 8 \cdot \underline{\quad} \\ 8 \cdot \underline{\quad} & 8 \cdot \underline{\quad} & 8 \cdot \underline{\quad} \end{bmatrix}$$

$$= \begin{bmatrix} \underline{\quad} & \underline{\quad} & \underline{\quad} \\ \underline{\quad} & \underline{\quad} & \underline{\quad} \end{bmatrix}$$

Show a dilation by each scale factor.

11. Scale factor 2.5

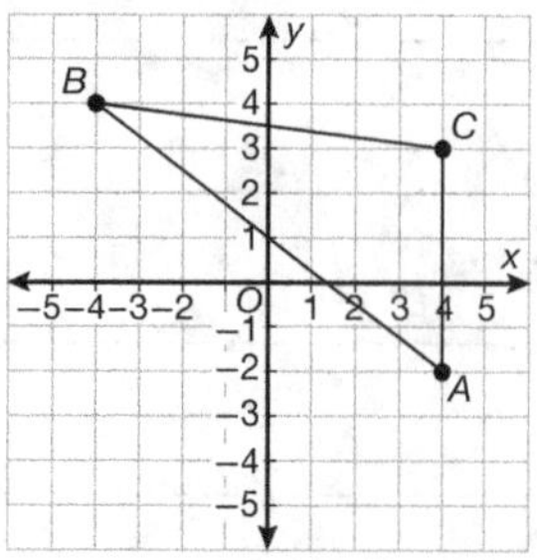

$$\begin{bmatrix} \underline{\quad} & \underline{\quad} & \underline{\quad} \\ \underline{\quad} & \underline{\quad} & \underline{\quad} \end{bmatrix}$$

12. Scale factor $\frac{1}{3}$

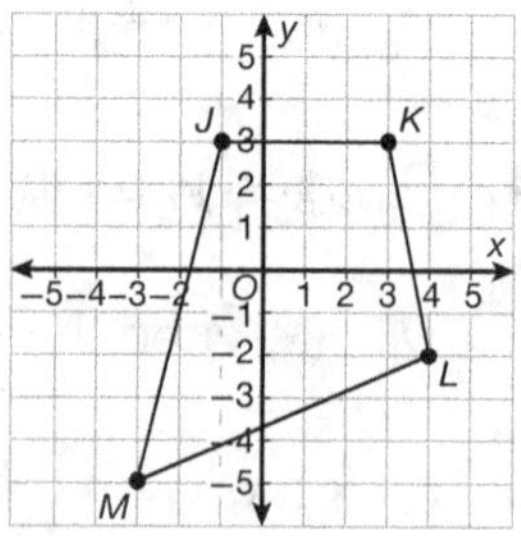

$$\begin{bmatrix} \underline{\quad} & \underline{\quad} & \underline{\quad} & \underline{\quad} \\ \underline{\quad} & \underline{\quad} & \underline{\quad} & \underline{\quad} \end{bmatrix}$$

$\underline{\quad}$ $\underline{\quad}$ $\underline{\quad}$ $\underline{\quad}$

Name ____________________ Date __________ Class __________

Reteaching 103

Frustums of Cones and Pyramids

You have worked with the volume of cones and pyramids. Now you will examine the frustums of cones and pyramids.

Frustum of a Cone

The frustum of a cone is a part of a cone with two parallel bases.

The volume of a frustum is $V = \frac{1}{3}h(B_1 + \sqrt{B_1B_2} + B_2)$, where B_1 and B_2 are the base areas and h is the height of the frustum.

Find the volume of the frustum of the cone. Round your answer to the nearest whole number.

The volume of the frustum of the cone is about 4772 cm^3.

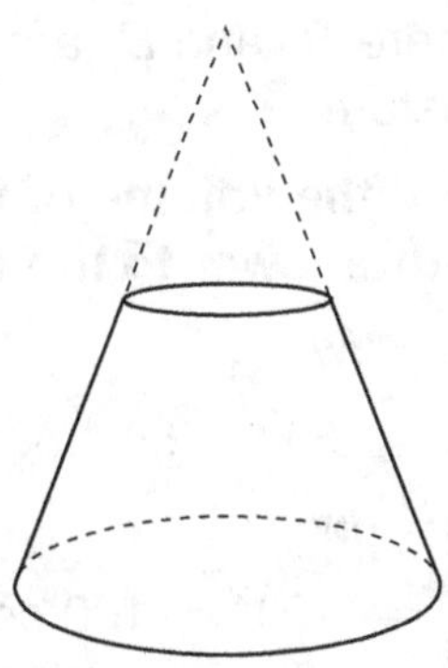

Practice

Complete the steps to find the volume of the frustum of the cone. Round your answer to the nearest whole number.

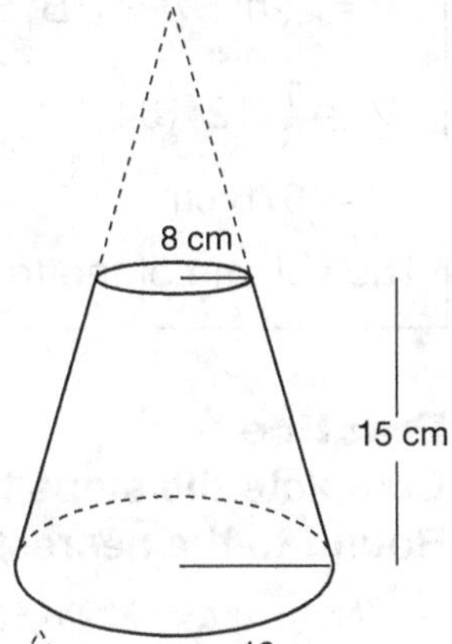

$B_1 = \pi r^2$	Base area B_1
$B_2 = \pi(8)^2 \approx 201 \text{ cm}^2$	Substitute and simplify.
$B_2 = \pi r^2$	Base area B_2
$B_2 = \pi(12)^2 \approx 452 \text{ cm}^2$	Substitute and simplify.
$V = \frac{1}{3}h(B_1 + \sqrt{B_1B_2} + B_2)$	Frustrum formula
$V \approx \frac{1}{3}(15)(201 + \sqrt{(201)(452)} + 452)$	Substitute.
$V \approx 4772 \text{ cm}^3$	Simplify.

$B_1 = \pi r^2 = \pi(10)^2 \approx$ _____ ft^2

$B_2 = \pi r^2 = \pi($_____$)^2 \approx$ _____ ft^2

1. $V = \frac{1}{3}h(B_1 + \sqrt{B_1B_2} + B_2)$

$V \approx \frac{1}{3}(16)($_____ $+ \sqrt{\text{_____}}$ $+$ _____$)$

$V \approx$ _____ ft^3

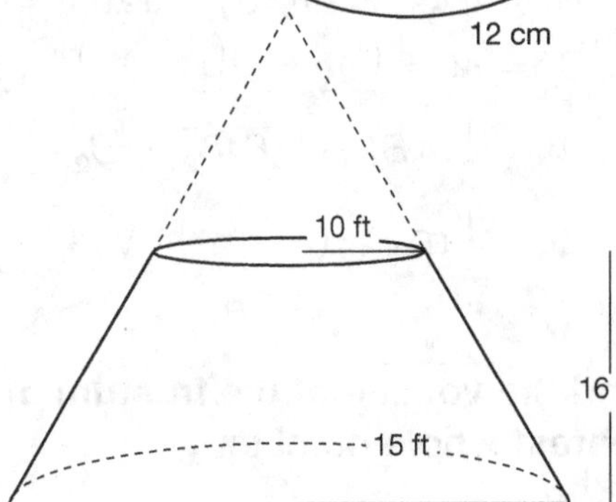

Find the volume of the frustum of the cone. Round your answer to the nearest whole number.

2. $V \approx$ _____ cm^3

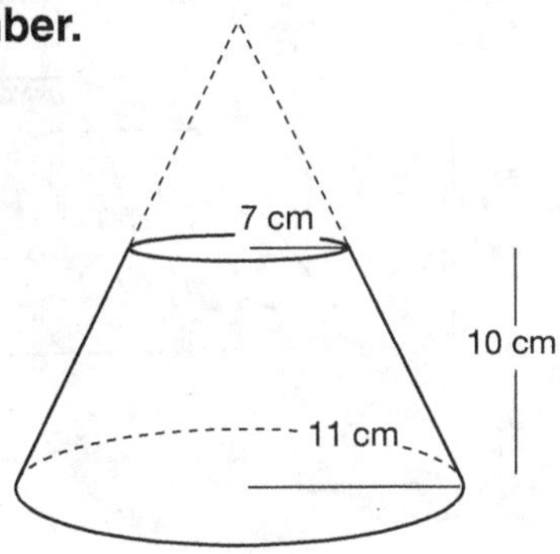

3. $V \approx$ _____ ft

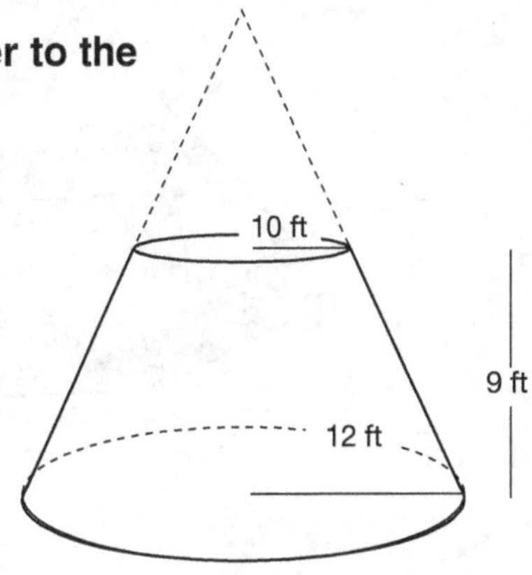

Frustum of a Pyramid

The frustum of a pyramid is a part of a pyramid with two parallel bases. The volume of a frustum is $V = \frac{1}{3}h(B_1 + \sqrt{B_1B_2} + B_2)$, where B_1 and B_2 are the base areas and h is the height of the frustum.

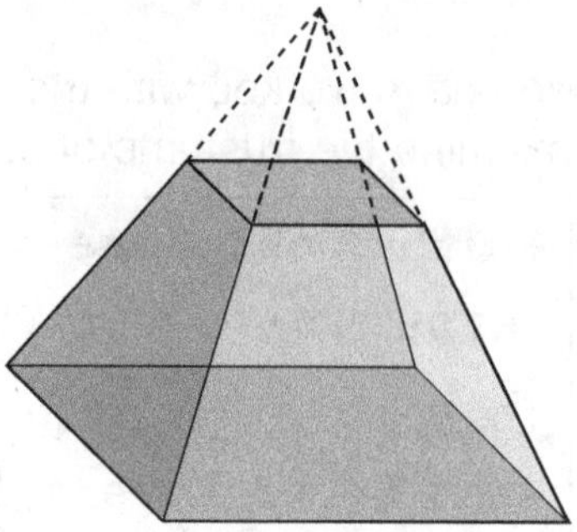

Find the volume of the frustum of the pyramid. Round your answer to the nearest whole number.

$B_1 = lw$	Base area B_1
$B_1 = (8)^2 = 64 \text{ cm}^2$	Substitute and simplify.
$B_2 = lw$	Base area B_2
$B_2 = (10)^2 = 100 \text{ cm}^2$	Substitute and simplify.
$V = \frac{1}{3}h(B_1 + \sqrt{B_1B_2} + B_2)$	Frustrum formula
$V = \frac{1}{3}(12)(64 + \sqrt{(64)(100)} + 100)$	Substitute.
$V = 976 \text{ cm}^3$	Simplify.

The volume of the frustum of the pyramid is 976 cm^3.

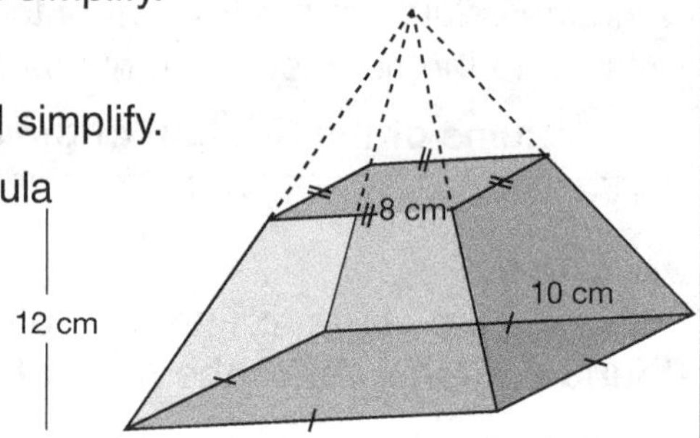

Practice

Complete the steps to find the volume of the frustum of the pyramid. Round to the nearest whole number.

$B_1 = lw = (12)(10) = 120 \text{ ft}^2$

$B_2 = lw = (\_\_\_\_)(\_\_\_\_) = \_\_\_\_ \text{ ft}^2$

4. $V = \frac{1}{3}h(B_1 + \sqrt{B_1B_2} + B_2)$

$V = \frac{1}{3}(8) + (\_\_\_\_ + \sqrt{\_\_\_\_\_\_} + \_\_\_\_)$

$V \approx \_\_\_\_ \text{ ft}^3$

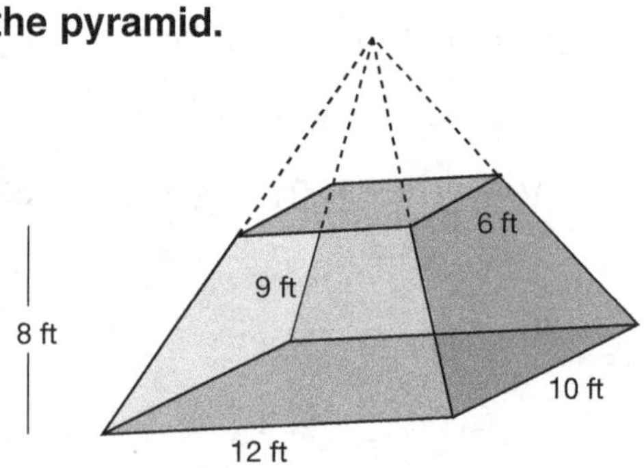

Find the volume of the frustum of the pyramid. Round to the nearest whole number.

5. $V \approx$ ______ cr

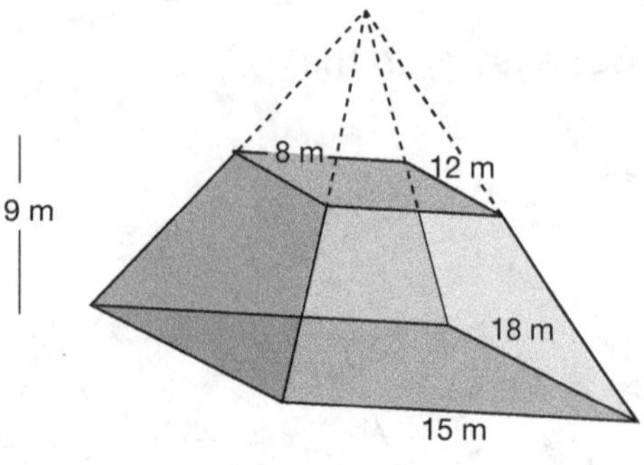

6. $V \approx$ ______

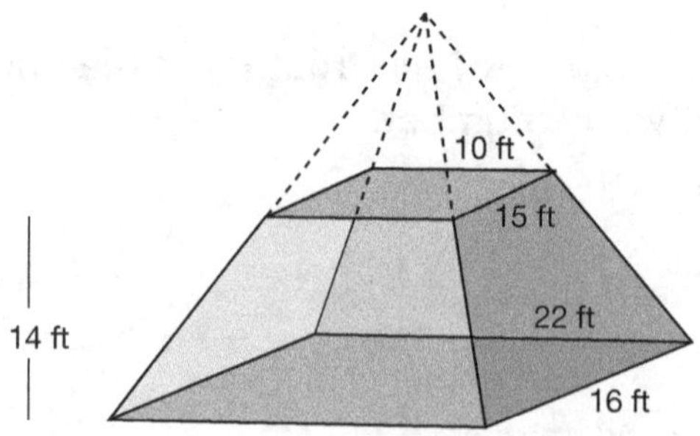

Name ______________________________ Date __________ Class __________

Reteaching 104

Relating Arc Lengths and Chords

You have worked with circles and sectors. Now you will work with arc lengths and chords.

Congruent Arcs

Congruent arcs are arcs that have the same measure.

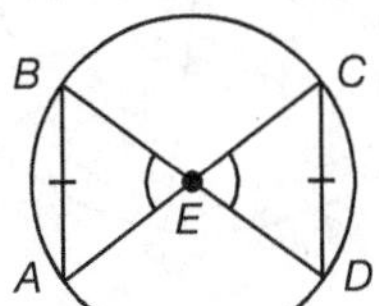 If $m\angle BEA \cong m\angle CED$, then $\overline{BA} \cong \overline{CD}$.	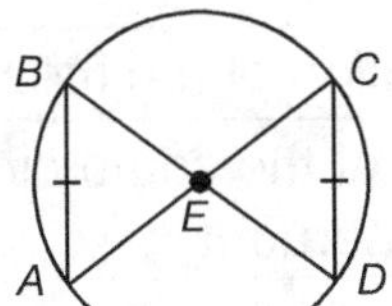 If $\overline{BA} \cong \overline{CD}$, then $\overset{\frown}{BA} \cong \overset{\frown}{CD}$.	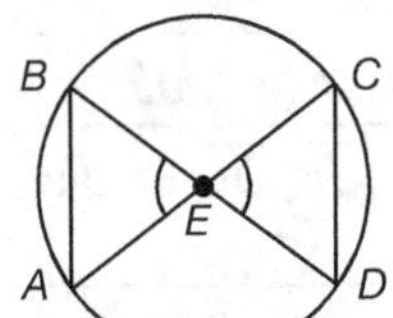 If $\overset{\frown}{BA} \cong \overset{\frown}{CD}$, then $m\angle BEA \cong m\angle CED$.
Congruent central angles have congruent chords.	Congruent chords have congruent arcs.	Congruent arcs have congruent central angles.

Example:

Given that $\angle HLG \cong \angle KLJ$, find GH.

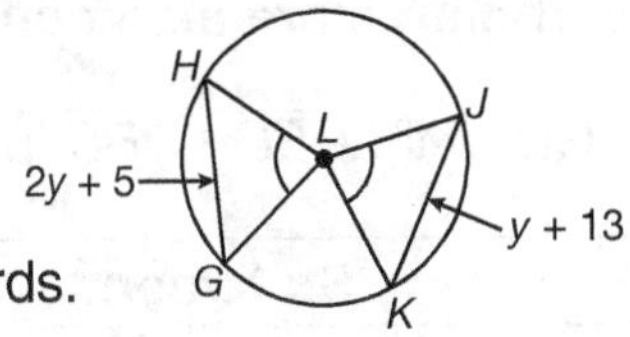

$\angle HLG \cong \angle KLJ$

$HG \cong JK$ Congruent cental angles have congruent chords.

$2y + 5 = y + 13$

$y + 5 = 13$ Subtract y from both sides.

$y = 8$ Subtract 5 from both sides.

$GH = 2y + 5 = 2(8) + 5 = 21$

Practice

Complete the statements to find m$\overset{\frown}{QR}$.

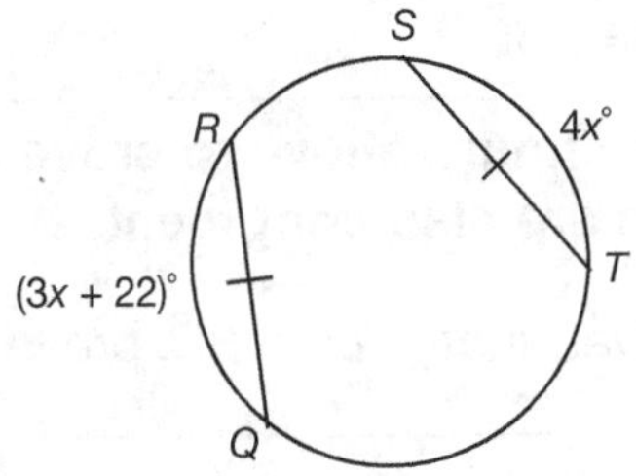

1. $\overset{\frown}{RQ} \cong \overset{\frown}{ST}$ $\qquad$ $\overset{\frown}{QR} = 3x + 22$

m$\overset{\frown}{RQ} \cong$ ____ $\qquad$ $= 3$ ____ $+ 22$

____ $+ 22 = 4x$ $\qquad$ $=$ ____ $+ 22$

____ $= x$ $\qquad$ $=$ ____

Find the each measure.

2. Given that m$\overset{\frown}{AD}$ = m$\overset{\frown}{BC}$, find m$\angle AED$. ____

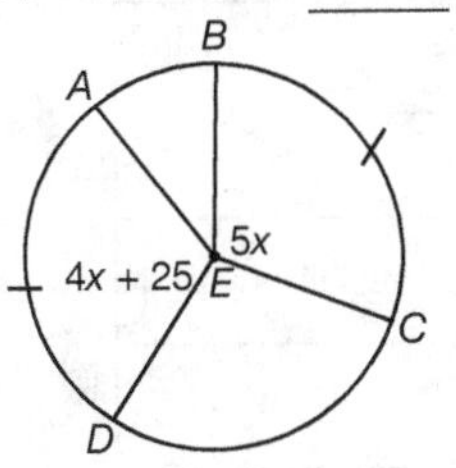

3. Given that $\angle AEB \cong \angle DEC$, find CD. ____

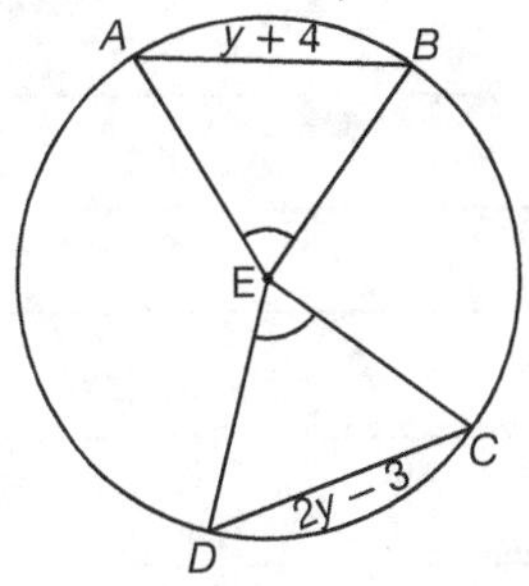

More Congruent Arcs

Use the figure shown to prove that if the arcs are congruent, then the chords are also congruent.

Given that $\overset{\frown}{KL} \cong \overset{\frown}{LM}$, prove $\overline{KL} \cong \overline{LM}$.

Statement	Reason
1. $\overset{\frown}{KL} \cong \overset{\frown}{LM}$	1. Given
2. $\angle KJL \cong \angle MJL$	2. Definition of arc measures
3. $\overline{JK} \cong \overline{JL}$; $\overline{JL} \cong \overline{JM}$	3. Radii of the same circle are all congruent.
4. $\Delta JKL \cong \Delta JLM$	4. SAS
5. $\overline{KL} \cong \overline{LM}$	5. CPCTC

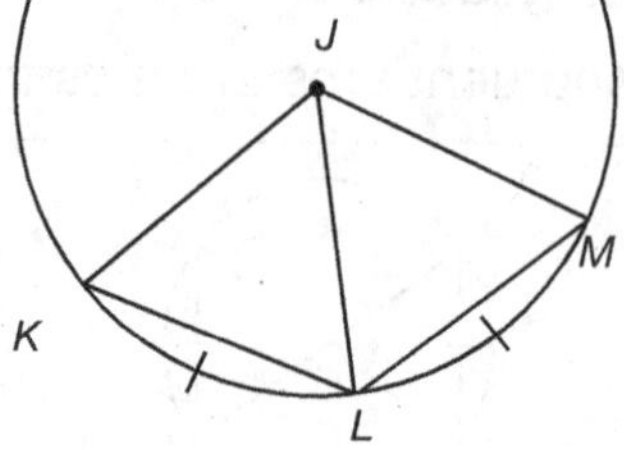

Practice

Complete the statements to prove that if the chords are congruent, then the arcs are also congruent.

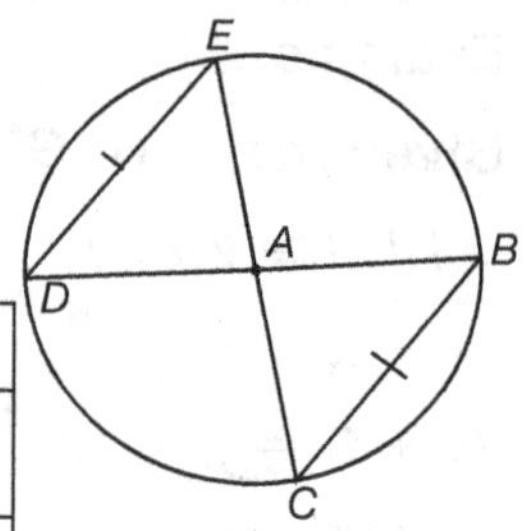

4. Given that $\overline{ED} \cong \overline{BC}$, prove $\overset{\frown}{ED} \cong \overset{\frown}{BC}$.

Statement	Reason
1. ________	1. Given
2. $\overline{AE} \cong$ ___ $\cong \overline{AB} \cong$ ___	2. Radii of the same circle are all congruent.
3. $\Delta AED \cong \Delta ABC$	3. _____
4. $\angle EAD \cong$ ______	4. CPCTC
5. $\overset{\frown}{ED} \cong \overset{\frown}{BC}$	5. ________________________

Use the figure above to prove that if the arcs are congruent, then the chords are also congruent.

5. Given that $\overset{\frown}{ED} \cong \overset{\frown}{BC}$, prove $\overline{ED} \cong \overline{BC}$.

Statement	Reason
1. $\overset{\frown}{ED} \cong \overset{\frown}{BC}$	1. _____
2. ____________	2. ________________________
3. __________________	3. ______________________ ________________
4. _____________	4. ____
5. _______	5. _______

Name ______________________ Date __________ Class __________

Reteaching 105

Rotations and Reflections on the Coordinate Plane

You have learned about geometric transformations. Now you will use matrices to perform transformations on a coordinate plane.

The vertices of ΔXYZ are $X(7, 6)$ $Y(2, 3)$, and $Z(5, -1)$. Use matrix multiplication to reflect the image in the y-axis. Draw the preimage and the image in a coordinate plane.

Multiply by the reflection matrix in the *y*-axis $\begin{pmatrix} -1 & 0 \\ 0 & 1 \end{pmatrix}$.

$$\begin{pmatrix} -1 & 0 \\ 0 & 1 \end{pmatrix}\begin{pmatrix} 7 & 2 & 5 \\ 6 & 3 & -1 \end{pmatrix} = \begin{bmatrix} (-1)(7) + (0)(6) & (-1)(2) + (0)(3) & (-1)(5) + (0)(-1) \\ (0)(7) + (1)(6) & (0)(2) + (1)(3) & (0)(5) + (1)(-1) \end{bmatrix} = \begin{pmatrix} -7 & -2 & -5 \\ 6 & 3 & -1 \end{pmatrix}$$

The image matrix is $\begin{pmatrix} -7 & -2 & -5 \\ 6 & 3 & -1 \end{pmatrix}$.

Therefore, $\Delta X'Y'Z'$ has vertices at $X'(7, 6)$, $Y'(-2, 3)$, and $Z'(-5, -1)$.

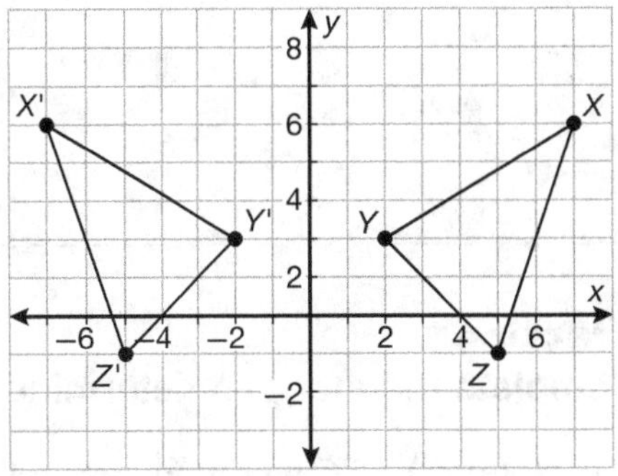

Practice

Complete the steps to find the reflected image of the triangle.

1. Reflect ΔQRS with vertices $Q(1, 6)$, $R(5, 2)$, and $S(-1, 3)$ in the x-axis.

   Draw the preimage and the image in a coordinate plane.

   Multiply by the reflection matrix in the x-axis $\begin{pmatrix} 1 & 0 \\ 0 & -1 \end{pmatrix}$.

   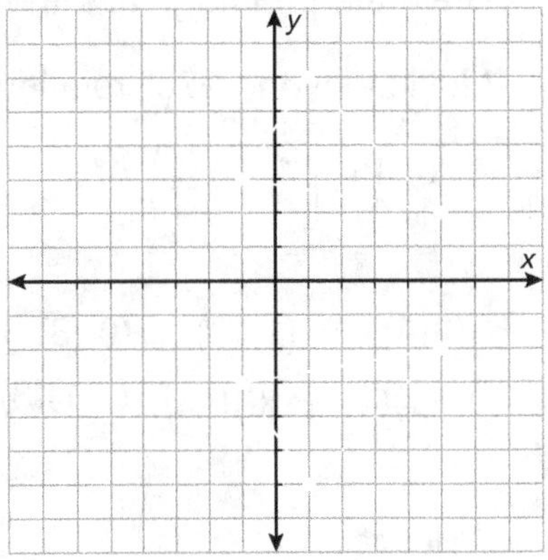

   $\begin{pmatrix} 1 & 0 \\ 0 & -1 \end{pmatrix}\begin{pmatrix} \_ & 5 & -1 \\ 6 & \_ & \_ \end{pmatrix}$

   $= \begin{bmatrix} (1)(1) + (0)(\_) & (1)(\_) + (0)(2) & (1)(-1) + (0)(\_) \\ (0)(1) + (-1)(\_) & (0)(\_) + (-1)(2) & (0)(-1) + (-)(\_) \end{bmatrix}$

   $= \begin{pmatrix} \_ & \_ & \_ \\ \_ & \_ & \_ \end{pmatrix}$

   $\Delta Q'R'S'$ has vertices at ______________________.

Use matrix multiplication to reflect the figure in the given axis. Graph the preimage and the image on a separate sheet of paper.

2. Reflect ΔABC with vertices $A(1, 1)$, $B(2, 3)$, $C(3, 2)$ in the x-axis.

____

The vertices of ΔTUV are $T(2, 1)$ $U(4, 5)$ $V(5, 2)$. Determine the image matrix for a 90° rotation of ΔTUV about the origin. Graph the preimage and image in a coordinate plane.

Multiply by the 90° rotation matrix $\begin{pmatrix} 0 & -1 \\ 1 & 0 \end{pmatrix}$**.**

$$\begin{pmatrix} 0 & -1 \\ 1 & 0 \end{pmatrix}\begin{pmatrix} 2 & 4 & 5 \\ 1 & 5 & 2 \end{pmatrix} = \begin{bmatrix} (0)(2)+(-1)(1) & (0)(4)+(-1)(5) & (0)(5)+(-1)(2) \\ (1)(2)+(0)(1) & (1)(4)+(0)(5) & (1)(5)+(0)(2) \end{bmatrix} = \begin{pmatrix} -1 & -5 & -2 \\ 2 & 4 & 5 \end{pmatrix}$$

The image matrix is $\begin{pmatrix} -1 & -5 & -2 \\ 2 & 4 & 5 \end{pmatrix}$.

So, $\Delta T'U'V'$ has vertices at $T'(-2, 1)$, $U'(-4, 5)$, $V'(-5, 2)$.

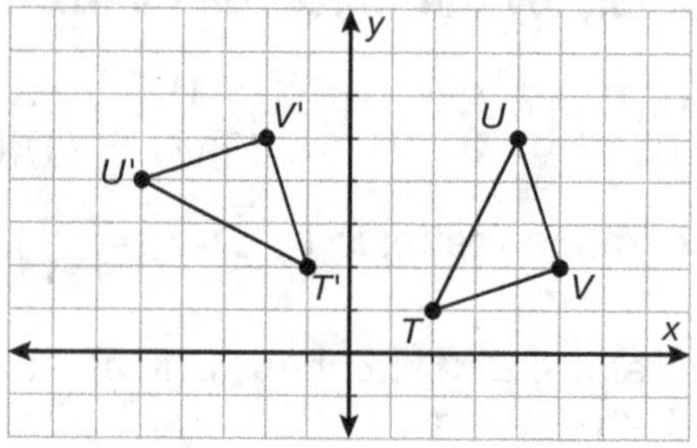

Practice

Complete the steps to determine the rotation image of the triangle.

3. Rotate ΔABC with vertices $A(4,0)$, $B(3,4)$ $C(1,1)$ 270° about the origin.

Draw the preimage and the image in a coordinate plane.

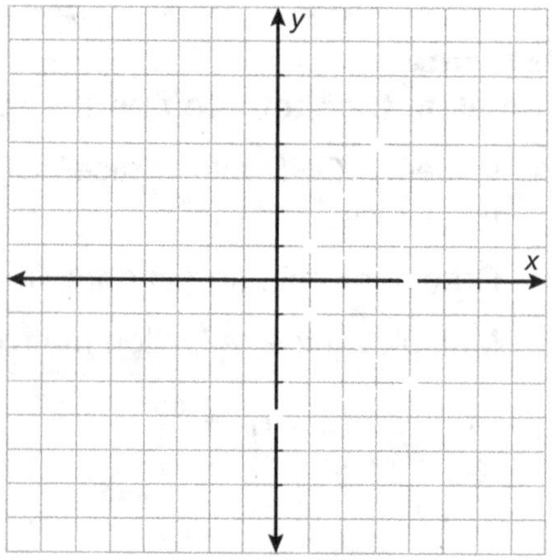

Multiply by the 270° rotation matrix $\begin{pmatrix} 0 & 1 \\ -1 & 0 \end{pmatrix}$.

$$\begin{pmatrix} 0 & 1 \\ -1 & 0 \end{pmatrix}\begin{pmatrix} 4 & \_ & \_ \\ \_ & \_ & 1 \end{pmatrix}$$

$$\begin{bmatrix} (0)(4)+(1)(\_) & (0)(3)+(1)(\_) & (0)(\_)+(1)(1) \\ (-1)(4)+(0)(\_) & (-1)(3)+(0)(\_) & (-1)(\_)+(0)(1) \end{bmatrix}$$

$$= \begin{pmatrix} \_ & \_ & \_ \\ \_ & \_ & \_ \end{pmatrix}$$

$\Delta A'B'C'$ has vertices at ______________________________.

Use matrix multiplication to rotate each figure the given degrees. Graph the preimage and the image on a separate sheet of paper.

4. Rotate rectangle $\Delta QRST$ with vertices $Q(0, 1)$, $R(4, -1)$ $S(4, 3)$, and $T(0, 3)$ 90° about the origin.

____

____

Name ______________________ Date ___________ Class ___________

Reteaching 106

Circumscribed and Inscribed Figures

You have found the perimeter and the missing side lengths of many different figures. Now you will find the perimeter of inscribed and circumscribed figures.

Find the perimeter of the largest regular octagon that can be inscribed in a circle with radius 6. Round your answer to the nearest hundredth.

Step 1: Draw a diagram. Find the center angle θ by dividing 360° by the number of sides in the figure.

$360 \div 8 = 45°$

Step 2: Then, use the law of cosines to find the length of the sides of the octagon.

$c^2 = a^2 + b^2 - 2ab \cos \theta$

$c^2 = 6^2 + 6^2 - 2(6)(6) \cos 45°$

$c^2 = 21.09$

$c = 4.59$

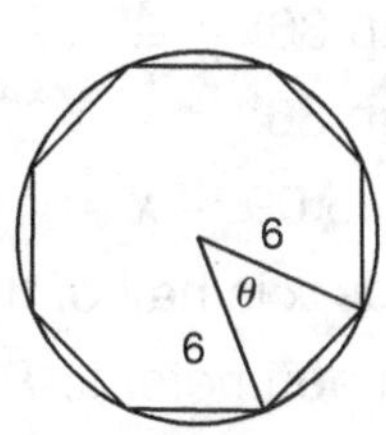

Each side of the octagon is approximately 4.59, so the perimeter of the octagon is $P = 8s = 8(4.59)$

$P = 36.72$

Practice

Complete the steps to find the perimeter of the largest regular pentagon inscribed in a circle with radius 5.

Round your answer to the nearest hundredth.

1. $360 \div 5 = 72°$

   $c^2 = a^2 + b^2 - 2ab \cos \theta$

   $c^2 = (\_)^2 + (\_)^2 - 2(\_)(\_) \cos$ ___

   $c^2 =$ ______

   $c =$ _____

   $P = 5s = 5($_____$)$

   $P =$ _____

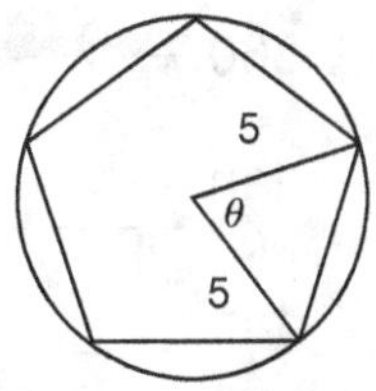

Find the perimeter of each inscribed regular polygon. Round your answer to the nearest hundredth.

2. What is the largest regular octagon that can be inscribed in a circle with radius 8? ______
3. What is the largest regular decagon that can be inscribed in a circle with radius 7? ______
4. What is the largest regular pentagon that can be inscribed in a circle with radius 9? ______
5. What is the largest regular nonagon that can be inscribed in a circle with radius 12? ______

Find the perimeter of a pentagon circumscribing a circle with radius 7. Round your answer to the nearest tenth.

Draw a diagram. Find the center angle θ by dividing 360° by the number of sides in the figure.

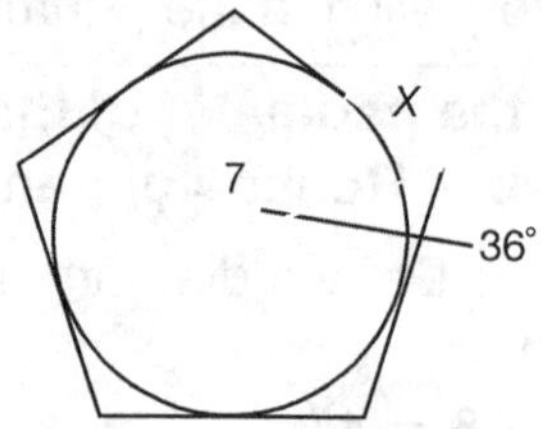

$360 \div 5 = 72°$

The radius of the circle is the apothem of the pentagon. It can be used to calculate half of the side length.

The apothem also divides the central angle in half for each side of the figure.

Half of the central angle is 36°.

$\tan 36° = \frac{x}{7}$

$7 \tan 36° = x$

$5.09 \approx x$

Since x is half of the side length, each side is $2x$.

The perimeter is $P = 5s = 5(2 \cdot 5.09)$.

$P = 50.86$

Practice

Complete the steps to find the perimeter of the polygon circumscribing a circle with the given radius. Round your answer to the nearest hundredth.

6. an octagon circumscribing a circle with radius 4

$360 \div$ ____ $=$ ____

tan ______ $= \frac{x}{\_\_\_}$

___ tan ______ $= x$

$x \approx$ _____

$P = 8s = 8(2 \cdot$ _____ $)$

$P =$ ______

7. a hexagon circumscribing a circle with radius 12

$360 \div$ ____ $=$ ____

tan ____ $= \frac{x}{\_\_\_}$

___ tan ___ $= x$

$x \approx$ _____

$P = 6s = 6(2 \cdot$ _____ $)$

$P =$ ______

Find the perimeter of each polygon circumscribing the circle with the given radius. Round your answer to the nearest hundredth.

8. a hexagon circumscribing a circle with radius 9 ______

9. a nonagon circumscribing a circle with radius 5 ______

10. a pentagon circumscribing a circle with radius 15 ____

11. a triangle circumscribing a circle with radius 10 ______

Name ______________________ Date __________ Class __________

Reteaching 107
Maximizing Area

Rectangle *VXYZ* is *l* units long and *w* units wide. It has a perimeter of 36 units.

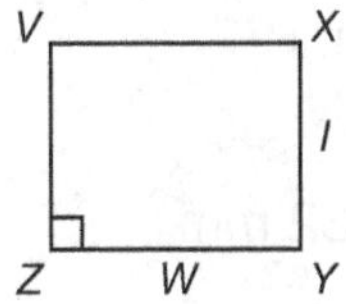

Determine the area of the rectangle for $w = 5$, $w = 7$, $w = 9$, and $w = 10$. Solve for l using the perimeter and width. Then substitute w and l into the formula for area.

$P = 2(l + w)$	$P = 2(l + w)$	$P = 2(l + w)$	$P = 2(l + w)$
$36 = 2(l + 5)$	$36 = 2(l + 7)$	$36 = 2(l + 9)$	$36 = 2(l + 10)$
$l = 13$	$l = 11$	$l = 9$	$l = 8$
$A = lw$	$A = lw$	$A = lw$	$A = lw$
$= 13 \times 5$	$= 11 \times 7$	$= 9 \times 9$	$= 8 \times 10$
$= 65$ sq. units	$= 77$ sq. units	$= 81$ sq. units	$= 80$ sq. units

Draw each rectangle on grid paper. Describe the relationship between the shape of each rectangle and the area of each rectangle.

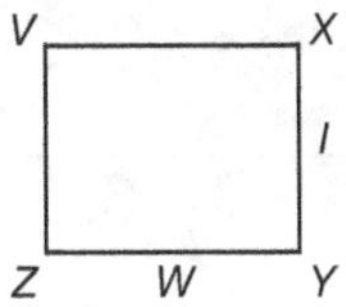

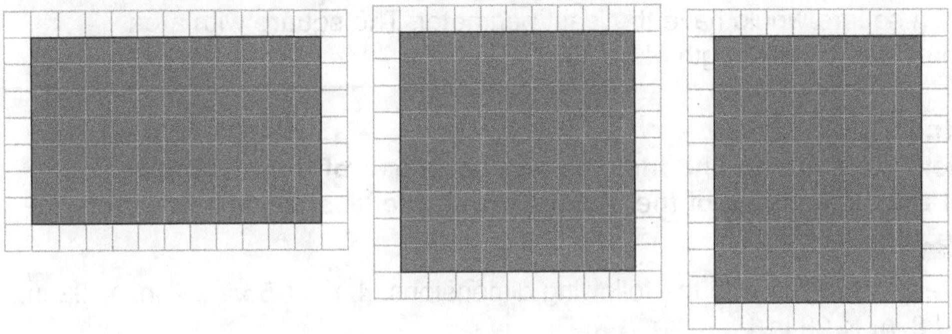

The closer to a square the rectangle is, the greater its area is. The square itself has the greatest area.

Practice

Complete the steps to find the areas of rectangle *GHJK* with the given lengths and with a perimeter of 40 inches.

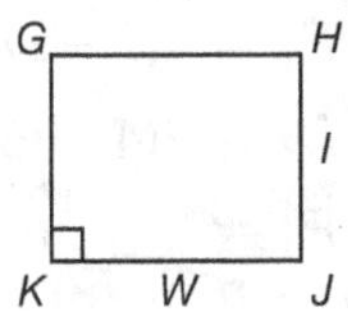

1. Determine the area of the rectangle for $l = 6$, $l = 8$, and $l = 10$.

$P = 2(l + w)$	$P = 2(l + w)$	$P = 2(l + w)$
$40 = 2(6 + w)$	$40 = 2(8 + w)$	$40 = 2(10 + w)$
$w =$ ___	$w =$ ___	$w =$ ___
$A = lw$	$A = lw$	$A = lw$
$= 6 \times$ ___	$= 8 \times$ ___	$= 10 \times$ ___
$=$ ___ in^2	$=$ ___ in^2	$=$ ___ in^2

   Which rectangle has the greatest area? ______________________

2. Determine the largest possible area of a rectangle with perimeter 48 inches. __________

3. Determine the largest possible area of a hexagon with perimeter 54 centimeters. __________

Three rectangles have the following dimensions: 4 mm $\times$ 33 mm, 6 mm $\times$ 22 mm, and 11 mm $\times$ 12 mm.

Determine the area and perimeter of each rectangle.

$A_1 = lw$	$A_2 = lw$	$A_3 = lw$
$A_1 = 4 \times 33$	$A_2 = 6 \times 22$	$A_3 = 11 \times 12$
$A_1 = 132 \text{ mm}^2$	$A_2 = 132 \text{ mm}^2$	$A_3 = 132 \text{ mm}^2$
$P_1 = 2(l + w)$	$P_2 = 2(l + w)$	$P_3 = 2(l + w)$
$P_1 = 2(4 + 33)$	$P_2 = 2(6 + 22)$	$P_3 = 2(11 + 12)$
$P_1 = 74 \text{ mm}$	$P_2 = 56 \text{ mm}$	$P_3 = 46 \text{ mm}$

Interpret the results. Predict the dimensions of the rectangle with the least possible perimeter that has the same area.

All three rectangles have the same area but different perimeters. The perimeter of each rectangle decreases as the rectangle approaches a square, which means a square would have the least perimeter. The square with area 132 mm^2 has a side length of $2\sqrt{33}$.

Practice

Complete the steps to find the area and perimeter of the rectangles. Predict the dimensions of the rectangle with the least perimeter that has the same area.

4. Three rectangles with the following dimensions: 4 in. $\times$ 56 in., 7 in. $\times$ 32 in., and 8 in. $\times$ 28 in.

$A_1 = lw$	$A_2 = lw$	$A_3 = lw$
$A_1 = 4 \times 56$	$A_2 = 7 \times 32$	$A_3 = 8 \times 28$
$A_1 =$ ____ in^2	$A_2 =$ ____ in^2	$A_3 =$ ____ in^2
$P_1 = 2(l + w)$	$P_2 = 2(l + w)$	$P_3 = 2(l + w)$
$P_1 = 2(4 + 56)$	$P_2 = 2(7 + 32)$	$P_3 = 2(8 + 28)$
$P_1 =$ ____ in.	$P_2 =$ ____ in.	$P_3 =$ ____ in.

Give the dimensions of the square with the least perimeter and the same area as the three rectangles above. ____________

Give the dimensions of each rectangle.

5. Give the dimensions of the rectangle with the least perimeter that has an area of 200 m^2. ____________

6. Give the dimensions of the reS<ctangle with the least perimeter that has an area of 135 cm^2. ____________

7. Give the dimensions of the rectangle with the least perimeter that has an area of 156 ft^2. ____________

8. Give the dimensions of the rectangle with the least perimeter that has an area of 333 in^2. ____________

Name ______________________________ Date ___________ Class ___________

Reteaching 108

Introduction to Coordinate Space

You have worked with coordinate grids. Now you will find the coordinates of points in a three-dimensional coordinate system.

Identifying Ordered Triples

A **three-dimensional coordinate system** adds a third axis, called the ***z*-axis**, to the two-dimensional coordinate system. Coordinates of points on the three-dimensional coordinate system are given by ordered triples. An **ordered triple** is a set of three numbers used to locate and plot a point (x, y, z) in a three-dimensional coordinate system.

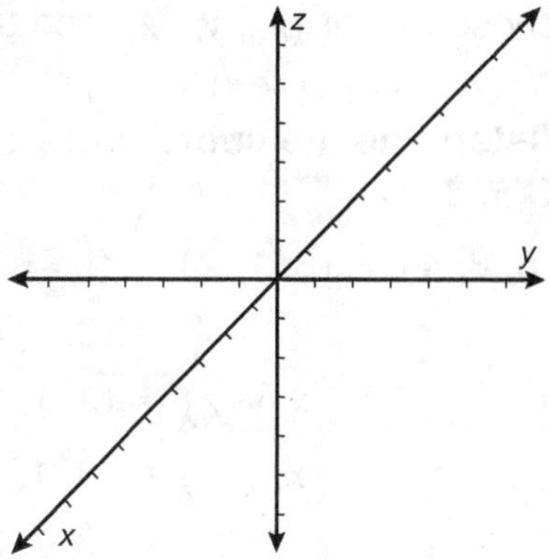

Identify the coordinates of the vertices in the given diagram.

$A(0, 3, 2) \rightarrow$ (x-coordinate, y-coordinate, z-coordinate)

$B(-2, 3, 2)$

$C(-2, 5, 2)$

$D(0, 5, 2)$

$E(0, 3, 0)$

$F(-2, 3, 0)$

$G(-2, 5, 0)$

$H(0, 5, 0)$

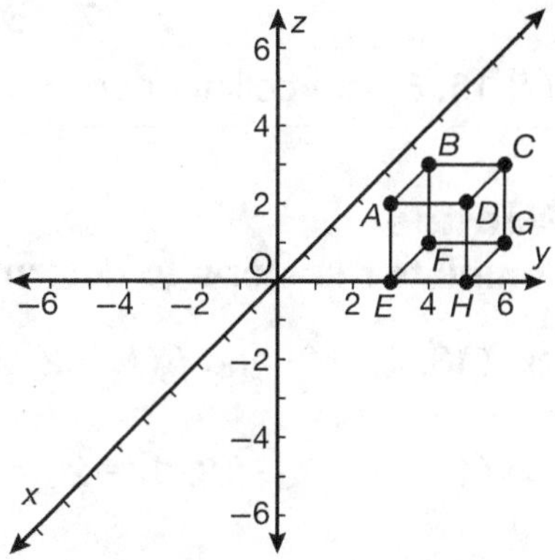

Practice

Complete the steps to identify the ordered triples.

1. $J(4, -1, 4)$, $K(2, -1, 4)$

   $L(\_, 7, 4)$, $M(4, \_, 3)$

   $N(4, \_\_\_, \_)$, $O(2, \_\_\_, \_)$

   $P(\_, \_, \_)$, $Q(\_, \_, \_)$

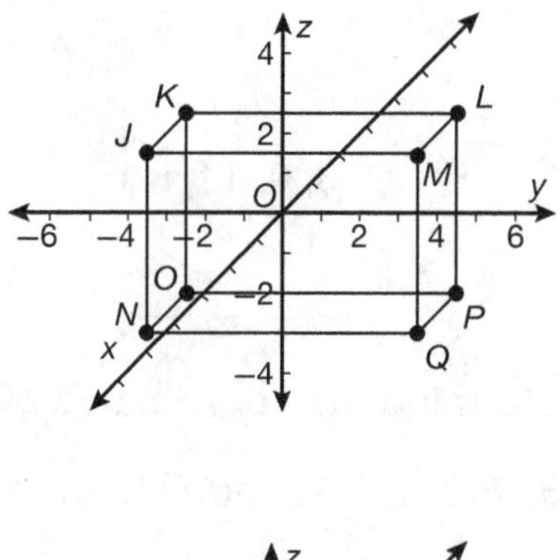

Identify the ordered triples.

2. ______________________

   ____

   ____

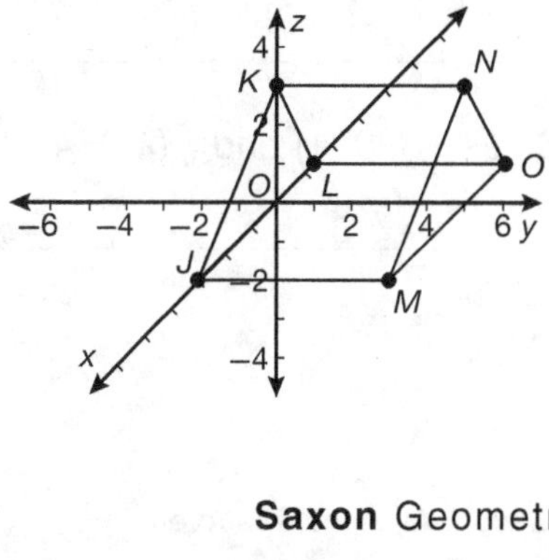

Finding Collinear Points

In two dimensions, points are **collinear** when they lie on the same line. To determine whether points are collinear in three dimensions, use a **direction vector** and one point. To find a direction vector (d_1, d_2, d_3), subtract the coordinates of any two points on the line. Use a chosen point $(x_0, y_0, z_0)$and the direction (d_1, d_2, d_3) to write the linear equation $(x, y, z) = (x_0, y_0, z_0) + t(d_1, d_2, d_3)$ for a value t, where $t \in \Re$.

Determine the coordinates of a point that is collinear with the points *A*(3, 1, 2) and *B*(5, 8, 4).

$(5, 8, 4) - (3, 1, 2) = (5 - 3, 8 - 1, 4 - 2)$	Subtract.
$= (2, 7, 2)$	Simplify.
$(x, y, z) = (3, 1, 2) + t(2, 7, 2)$	Equation of the line
$(x, y, z) = (3, 1, 2) + 2(2, 7, 2)$	Substitute $t = 2$.
$= (3, 1, 2) + (4, 14, 4)$	Multiply.
$= (7, 15, 6)$	Simplify.

(7, 15, 6) is a collinear point.

Practice

Complete the steps to determine the coordinates of the collinear point.

3. $C(3, 1, -2)$ and $D(1, -2, 2)$, $t = 3$

$(1, -2, 2) - (3, 1, -2) = (\_\_, \_\_, \_\_)$

$(x, y, z) = (3, 1, -2) + 3(\_\_, \_\_, \_\_)$

$= (3, 1, -2) + (\_\_, \_\_, \_\_)$

$= (\_\_, \_\_, \_\_)$

4. $E(-3, 0, 1)$ and $F(1, -4, 2)$, $t = -2$

$(1, -4, 2) - (\_\_, \_\_, \_\_) = (\_\_, \_\_, \_\_)$

$(x, y, z) = (-3, 0, 1) - 2(\_\_, \_\_, \_\_)$

$= (-3, 0, 1) + (\_\_, \_\_, \_\_)$

$= (\_\_, \_\_, \_\_)$

Determine the coordinates of the collinear points.

5. $E(3, 1, -2)$ and $F(1, -2, 2)$, $t = 4$, $t = -1$

__________; __________

6. $G(-3, 0, 1)$ and $H(1, -4, 2)$, $t = -5$, $t = 4$

__________; __________

Name ______________________ Date __________ Class __________

Reteaching 109

Non-Euclidean Geometry

You have worked with the Cartesian coordinate system. Now you will learn to work with the polar coordinate system.

Converting Polar Coordinates

In the polar coordinate system, a point is represented by its distance r from the origin, or pole, and the measure of angle θ. The horizontal ray from the pole along the positive horizontal axis is the polar axis.

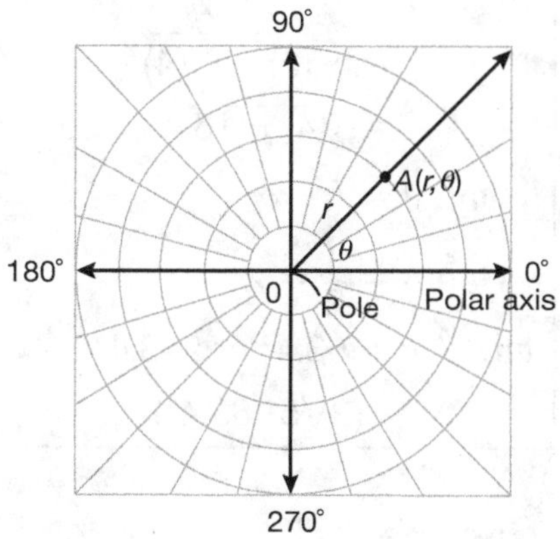

To convert polar coordinates (r, θ) into Cartesian coordinates (X, Y), use $x = r\cos\theta$ and $x = r\sin\theta$.

Convert the polar coordinates (5, 180°) into Cartesian coordinates.

$x = r\cos\theta$	Converting polar to Cartesian
$= 5\cos(180°)$	Substitute.
$= 5(-1) = (-5)$	Simplify.
$y = r\sin\theta$	Converting polar to Cartesian
$= 5\sin(180°)$	Substitute.
$= 5(0) = 0$	Simplify.

The polar coordinates (5, 180°) are converted into (−5, 0) as Cartesian coordinates.

Practice

Complete the steps to convert polar coordinates into Cartesian coordinates.

1. (−3, 90°)

$x = r\cos\theta =$ ___ cos ___

$=$ ___( ) = _

$y - r\sin\theta =$ ___ sin ___

$=$ ___( ) = ___

(_____)

2. (2, 270°)

$x = r\cos\theta =$ _ cos _____

$=$ _( ) = _

$y - r\sin\theta =$ _ sin _____

$=$ _( ) = ___

(_____)

Convert polar coordinates into Cartesian coordinates. Round to the nearest tenth.

3. (−4, 180°); (___)

4. (2, 360°); ____

5. (2, 60°); (_______)

6. (−3, 120°); (_________)

Converting Cartesian Coordinates

To convert Cartesian coordinates (x, y) into polar coordinates (r, θ), use $r^2 = x^2 + y^2$ (equation of a circle) and $\tan\theta = \frac{y}{x}$. Pay attention to the quadrant.

Convert the Cartesian coordinates (2, 4) into polar coordinates.

$r = \sqrt{x^2 + y^2}$	Equation to find r
$= \sqrt{(2)^2 + (4)^2}$	Substitute.
$= \sqrt{4 + 16} = \sqrt{20} = 2\sqrt{5}$	Simplify.
$\tan\theta = \frac{y}{x}$	Equation to find θ
$\tan\theta = \frac{4}{2} = 2$	Substitute.
$\tan^{-1}\theta \approx 63.4°$ or $243.4°$	Solve.
$(2\sqrt{5}, 63.4°)$	First quadrant, since (2, 4) is in the first quadrant

Practice

Complete the steps to convert Cartesian coordinates into polar coordinates.

7. (2, 1)

$r = \sqrt{x^2 + y^2}$

$= \sqrt{(\_)^2 + (\_)^2} = \sqrt{\_}$

$\tan\theta = \frac{y}{x} = \frac{\_}{\_}$

$\tan^{-1}\theta \approx$ _____° or _____°

(_________)

8. (−3, −4)

$r = \sqrt{x^2 + y^2}$

$= \sqrt{(\_\_)^2 + (\_\_)^2} = \_$

$\tan\theta = \frac{y}{x} = \frac{\_\_}{\_\_} = \frac{\_}{\_}$

$\tan^{-1}\theta \approx$ _____° or _____°

(_______)

Convert Cartesian coordinates into polar coordinates.

9. (−1, 3); ______________

10. (−5, 2); ______________

11. (−4, 1); (______)

12. (−4, 1); (_______)

13. (2, 2); (_________)

14. (1, 0); (_____)

15. (0, 15); (________)

16. (3, 7); (_________)

Name ______________________ Date __________ Class __________

Reteaching 110

Scale Drawings and Maps

You have worked with dilations and scale factor. Now you will use scale to make scale drawings and models

Scale Drawings

A scale **drawing** is a drawing of an object that is smaller or larger than the object's actual size. The drawing's **scale** is the ratio of any length in the drawing to the actual length of the object.

The scale for the diagram of the doghouse is 1 in. : 36 in.

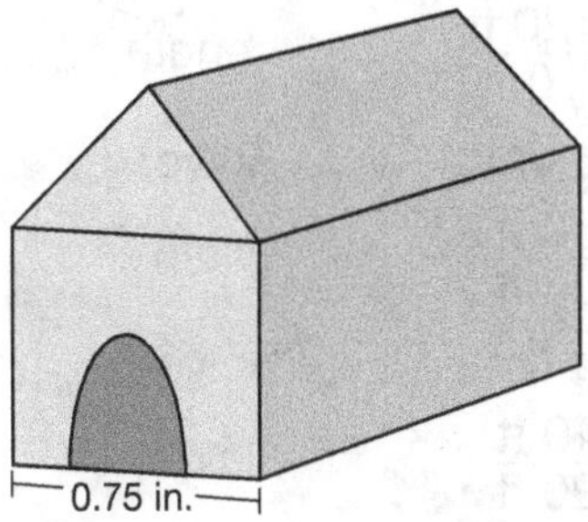

Find the width of the actual doghouse.

First convert to equivalent units.

1 in. : 36 in. (3 ft × 12 in./ft)

$\frac{1}{36} = \frac{0.75}{x}$	$\frac{\text{Diagram width}}{\text{Actual width}}$
$1x = 36(0.75)$	Cross Products Property
$x = 27$ in.	Simplify.

Practice

Complete the steps to find the actual length.

1. Scale is 1 cm : 10 m.

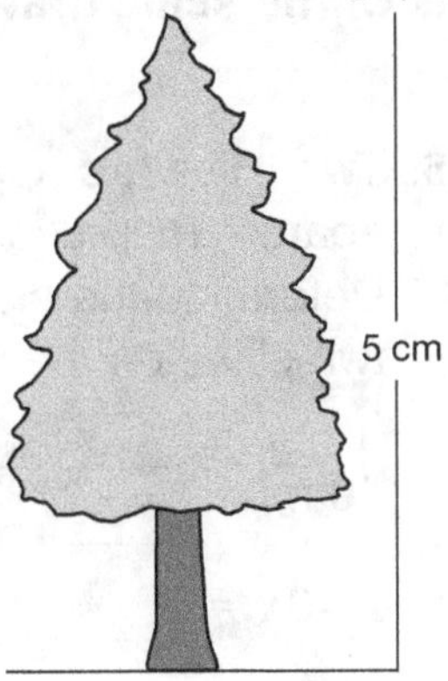

1 cm : 1000 cm

$\frac{1}{1000} = \frac{\_\_\_}{x}$

$x =$ __________

$x =$ _____ cm = ___ m

2. Scale is 1 in.: 80 ft.

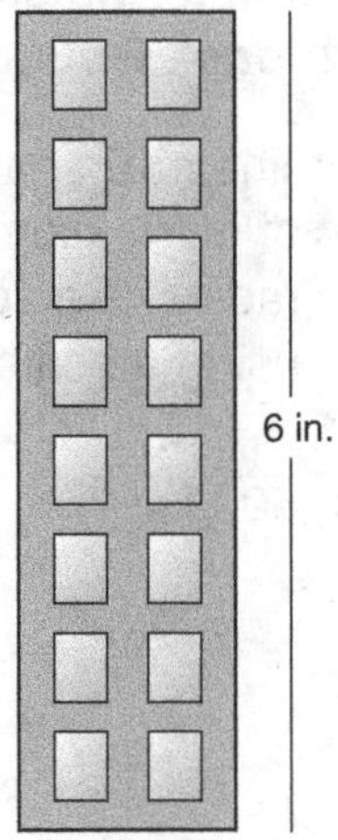

1 in. : ____ in.

$\frac{1}{\_\_\_} = \frac{\_\_\_}{x}$

$x =$ __________

$x =$ ______ in. = _____ ft

Find the actual length.

3. Scale is 1 in.: 8 ft. ______

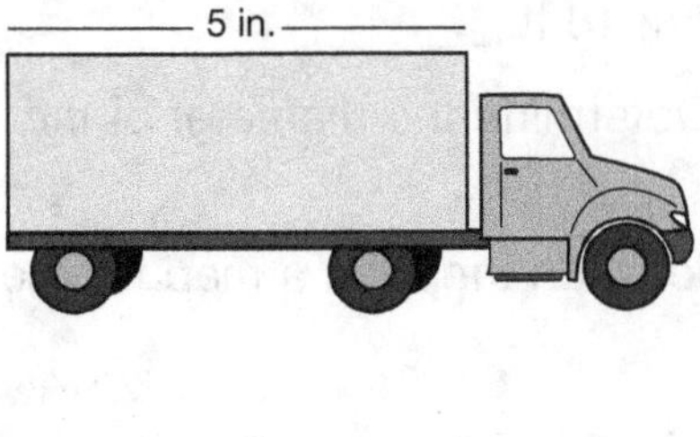

4. Scale is 1 cm: 20 m. ______

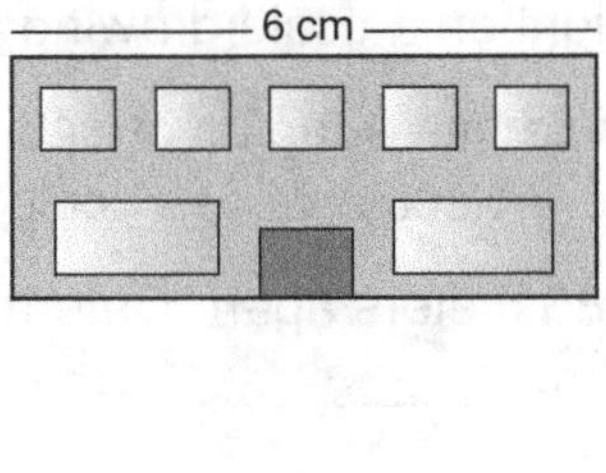

Making a Scale Drawing or Model

You can represent an object that is very large or very small on paper by creating a scale drawing or model. A **scale model** is a three-dimensional model that uses scale to represent an object as smaller than or larger than the actual object.

Example:

A school building is 120 feet long, 60 feet wide, and 40 feet high.

Using a scale of 1 cm: 20 ft, determine the dimensions of the scale model.

Draw the scale model.

$\frac{x}{1 \text{ cm}} = \frac{120 \text{ ft}}{20 \text{ ft}}$ Length

$x = 6 \text{ cm}$ Solve.

$\frac{x}{1 \text{ cm}} = \frac{60 \text{ ft}}{20 \text{ ft}}$ Width

$x = 3 \text{ cm}$ Solve.

$\frac{x}{1 \text{ cm}} = \frac{40 \text{ ft}}{20 \text{ ft}}$ Height

$x = 2 \text{ cm}$ Solve.

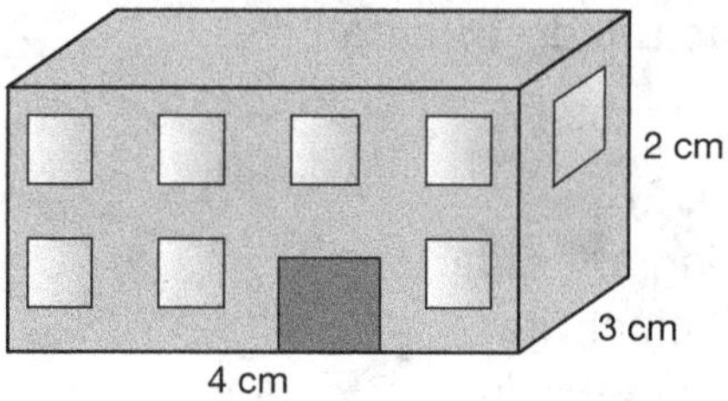

Practice

Complete the steps to determine the dimensions of the scale drawing or model. Round to the nearest tenth.

5. The earth has an approximate radius of 6378 kilometers. Determine the radius of a scale model of the earth using a scale of 1 cm : 600 km.

   $\frac{1 \text{ cm}}{600 \text{ km}} = \frac{x}{\_\_\_\_\_\_}$

   $600x =$ ______

   $x =$ ______ $\approx$ ______ cm

6. Two cities are approximately 155 miles apart. A map uses a scale of 1 cm : 40 mi Determine the distance between the two cities on the map.

   $\frac{1 \text{ cm}}{40 \text{ mi}} = \frac{x}{\_\_\_\_\_\_}$

   $40x =$ ___

   $x =$ ______ $\approx$ ______ cm

Determine the dimensions of the scale drawing or model. Round to the nearest tenth.

7. The Statue of Liberty is approximately 151 feet tall from base to torch(151 ft 1 in.). Determine the height on a scale drawing whose scale is 1 cm: 18 ft. ______

8. A spherical molecule has a radius of 140 picometers (pm). Determine the diameter of the scale model of the molecule whose scale is 1 in: 20 pm. ______

9. Two friends are 385 meters apart. Determine the distance between them on a map whose scale is 1 cm: 40 m. ______

Name ______________________________ Date __________ Class __________

Reteaching INV 11

Golden Ratio

You have investigated fractals. Now you will investigate the golden ratio.

The Golden Ratio in Line Segments

The **golden ratio** is the irrational number $\varphi = \frac{1 + \sqrt{5}}{2}$.

On the line segment to the right, since the ratio of the whole segment to the longer piece, $\frac{c}{b}$, equals the ratio of the longer piece to the shorter piece, $\frac{b}{a}$, both ratios are equal to the golden ratio, φ.

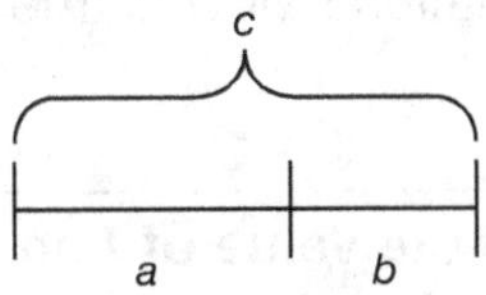

Find the value of *a* if $\frac{a}{89} = \frac{a + 89}{a}$. Round to the nearest hundredth.

Step 1: Since $\frac{a}{89} = \frac{a + 89}{a}$, $\frac{a}{89}$ and $\frac{a + 89}{a}$ are both equal to the golden ratio, φ.

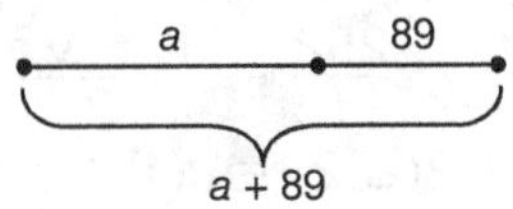

Step 2: Set $\frac{a}{89}$ equal to φ and use $\varphi = \frac{1 + \sqrt{5}}{2}$.

$\frac{a}{89} = \frac{1 + \sqrt{5}}{2}$ Substitute.

$a = 89\left(\frac{1 + \sqrt{5}}{2}\right)$ Solve.

$a \approx 144.01$

Practice

In the problem below, the ratios of the line segments equal the golden ration. Find the value of x to the nearest hundredth.

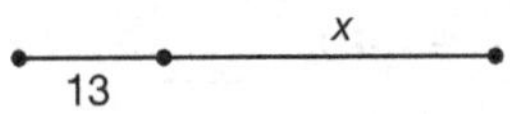

1. $\frac{x}{13} = \varphi$ Golden ratio

   $\frac{x}{13} = \frac{\qquad}{\qquad}$ Substitute.

   $2x =$ __________ Solve.

   $x =$ ______ Round to the nearest hundredth.

2. The ratios of the line segments equal the golden ratio. Find the value of x to the nearest hundredth. ______

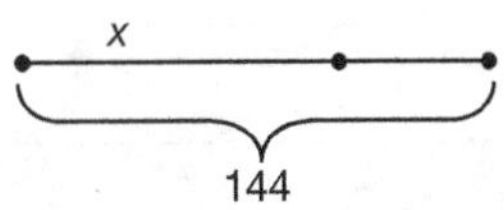

3. If the longer piece of a line segment is 45, and if the shorter piece is 18, does the line segment represent the golden ration? Explain your answer.

____

The Golden Rectangle

In a **golden rectangle,** the ratio of the length of the longer side to the length of the shorter side is the golden ratio, φ.

The rectangle pictured on the right is a golden rectangle because the equation below is true.

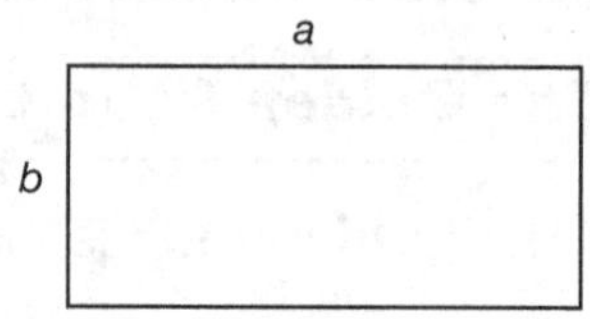

$\frac{a}{b} = \varphi$

Find the value of *t* on the golden rectangle shown below. Round to the nearest hundredth.

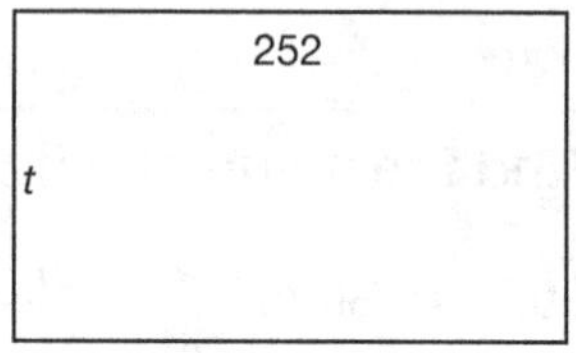

Step 1: Use the definition of a golden rectangle to get the equation $\frac{252}{t} = \varphi$.

Step 2: Solve the equation above for *t*.

$\frac{252}{t} = \frac{1 + \sqrt{5}}{2}$	Substitute.
$2(252) = t(1 + \sqrt{5})$	Solve.
$t \approx 155.74$	Round to the nearest hundredth.

Practice

Find the value of *x* on the golden rectangle shown below. Round to the nearest hundredth.

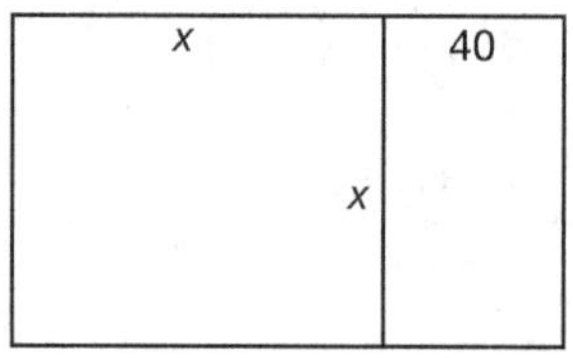

4. $\frac{x}{40} = \varphi$ Golden ratio

   $\frac{x}{40} =$ ________ Substitute.

   ___ $= 40($________$)$ Solve.

   $x \approx$ ______ Round to the nearest hundredth.

5. Use the golden rectangle to find the value of *x* to the nearest hundredth.

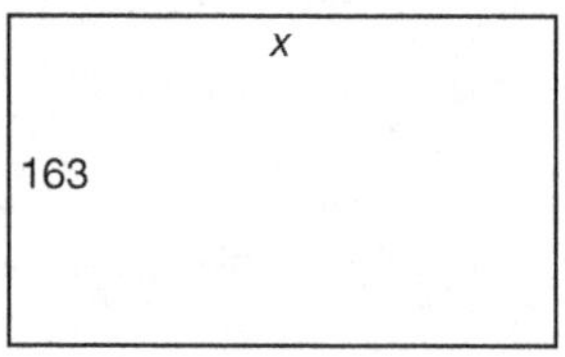

____

6. Can a square be a golden rectangle? Explain your answer.

____

____

Name ______________________ Date __________ Class __________

Reteaching 111

Finding Distance and Midpoint in Three Dimensions

You have worked with distance and midpoint. Now you will find distance and midpoint in three dimensions.

Distance Formula in Three Dimensions

Given two points, $A(x_1, y_1, z_1)$ and $B(x_2, y_2, z_2)$, and in three-dimensional space, the distance between the two points is determined by the equation

$d = \sqrt{(x_2 - x_1)^2 + (y_2 - y_1)^2 + (z_2 - z_1)^2}$.

Example:

Find the distance between $A(3,1,1)$ and $B(2, -2, -3)$.

$d = \sqrt{(x_2 - x_1)^2 + (y_2 - y_1)^2 + (z_2 - z_1)^2}$	Distance Formula for Three Dimensions
$AB = \sqrt{(2 - 3)^2 + (-2 - 1)^2 + (3 - 1)^2}$	Substitute
$= \sqrt{(-1)^2 + (-3)^2 + (-4)^2}$	
$= \sqrt{1 + 9 + 16}$	Simplify
$= \sqrt{26}$	

The distance between AB is $\sqrt{26}$.

Practice

Complete the steps to find the distance between the points.

1. $C(-3, 2, 1)$ and $D(-4, -3, 2)$

$CD = \sqrt{(x_2 - x_1)^2 + (y_2 - y_1)^2 + (z_2 - z_1)^2}$

$= \sqrt{(-4 - (-3))^2 + (\_\_ - \_)^2 + (\_ - \_)^2}$

$= \sqrt{\mathbf{1} + \_\_ + \_}$

$= \sqrt{\_\_\_}$

2. $E(0, -5, 3)$ $F(3, -1, -3)$

$EF = \sqrt{(x_2 - x_1)^2 + (y_2 - y_1)^2 + (z_2 - z_1)^2}$

$= \sqrt{(\_ - \_)^2 + (\_\_ - \_\_)^2 + (\_\_ - \_)^2}$

$= \sqrt{\_ + \_\_ + \_\_}$

$= \sqrt{\_\_}$

Find the distance between the points.

3. $G(-5, -1, -2)$ and $H(-3, -3, -3)$ _

4. $J(2, -5, -1)$ and $K(0, 3, 6)$ _____

5. $M(3, -4, -1)$ and $N(0, -5, 2)$ _____

6. $P(2, 5, 1)$ and $Q(-2, 0, 3)$ _____

Midpoint Formula in Three Dimensions

Given two points, $A(x_1, y_1, z_1)$ and $B(x_2, y_2, z_2)$, in three-dimensional space, the midpoint M is determined by the equation $M = \left(\frac{x_1 + x_2}{2}, \frac{y_1 + y_2}{2}, \frac{z_1 + z_2}{2}\right)$.

The endpoints of $\overline{AB}$ are A(5, 2, (−2) and B(1, 4, 4). Find the midpoint.

$M = \left(\frac{x_1 + x_2}{2}, \frac{y_1 + y_2}{2}, \frac{z_1 + z_2}{2}\right)$ Midpoint Formula for Three Dimensions

$= \left(\frac{5 + 1}{2}, \frac{2 + 4}{2}, \frac{-2 + 4}{2}\right)$ Substitute

$= \left(\frac{6}{2}, \frac{6}{2}, \frac{2}{2}\right)$

$= (3, 3, 1)$ Simplify

Practice

Complete the steps to find the midpoint.

7. $C(3, -3, 2)$ $D(-5, -1, 0)$

$M = \left(\frac{x_1 + x_2}{2}, \frac{y_1 + y_2}{2}, \frac{z_1 + z_2}{2}\right)$

$= \left(\frac{3 + (-5)}{2}, \frac{\_\_ + \_\_}{2}, \frac{\_ + \_}{2}\right)$

$= (-1, \_\_\_, \_)$

8. $E(-5, -3, 2)$ and $F(4, 0, 4)$

$M = \left(\frac{x_1 + x_2}{2}, \frac{y_1 + y_2}{2}, \frac{z_1 + z_2}{2}\right)$

$= \left(\frac{\_\_ + \_}{2}, \frac{\_\_ + \_}{2}, \frac{\_ + \_}{2}\right)$

$= (\_\_\_, \_\_\_, \_)$

Find the midpoint.

9. $G(-4, -6, 0)$ and $H(-3, 2, -2)$

____

10. $J(5, -5, 6)$ $K(3, -4, 8)$

____

11. $P(-3, 7, 2)$ and $Q(-5, -5, -5)$

____

12. $S(3, 10, -9)$ $T(15, -8, -11)$

____

13. $V(-18, 13, 7)$ and $W(24, 6, -12)$

____

14. $X(-11, 15, 17)$ $Q(6, -13, 22)$

____

Name ______________________ Date __________ Class __________

Reteaching 112

Finding Areas of Circle Segments

You have worked with area of a circle. Now you will find the area of circle segments.

Area of Circle Segment

A **segment of a circle** is a region bounded by an arc and its chord.

Area of segment *ABC* = area of sector *ABC* − area of ΔABC.

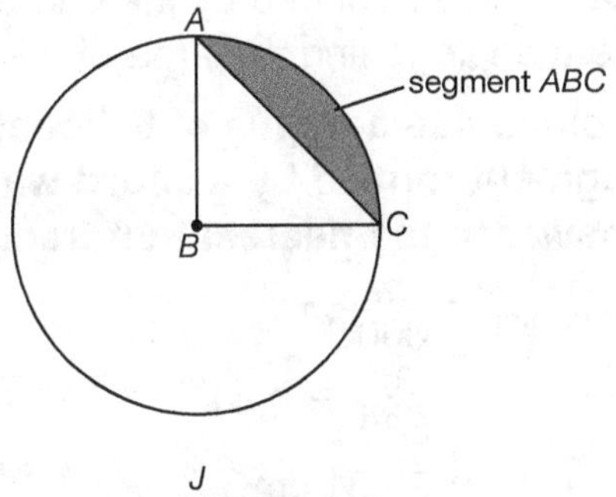

Find the area of the segment to the nearest hundredth.

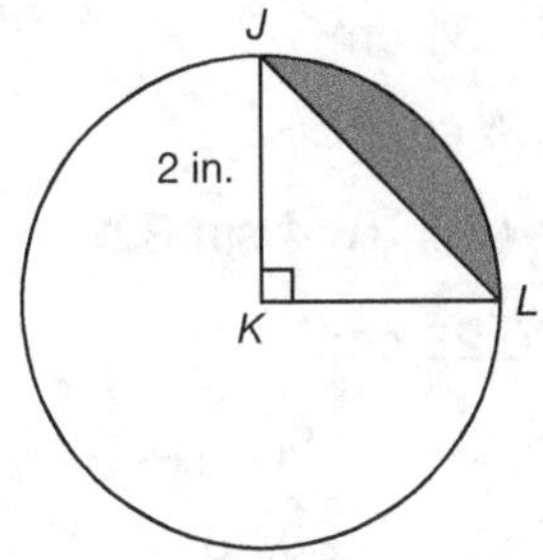

$A = \pi r^2\left(\frac{m^\circ}{360^\circ}\right)$	Area of sector
$= \pi(2)^2\left(\frac{90^\circ}{360^\circ}\right)$	Substitute.
$= \pi \text{ in}^2$	Solve.
$A = \frac{1}{2}bh$	Area of triangle
$= \frac{1}{2}(2)(2) = 2 \text{ in}^2$	Substitute.
$A_{segment} = A_{sector} - A_{triangle}$	Area of segment
$= \pi - 2$	Substitute.
$\approx 1.14 \text{ in}^2$	Solve.

Practice

Complete the steps to find the area of the segment to the nearest hundredth.

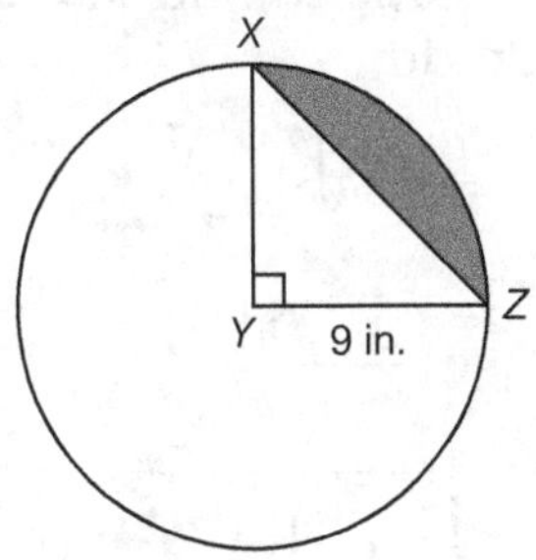

1. $A_{sector} = \pi r^2\left(\frac{m^\circ}{360^\circ}\right)$

$= \pi(9)^2\left(\frac{\_\_\_}{360}\right) = \_\_\_\_\_ \text{ in}^2$

$A_{triangle} = \frac{1}{2}(9)(9) = \_\_\_\_\_ \text{ in}^2$

$A_{segment} = \_\_\_\_\_\_\_\_ \approx \_\_\_\_\_ \text{ in}^2$

Find the area of the segment to the nearest hundredth.

2. ______ cm^2

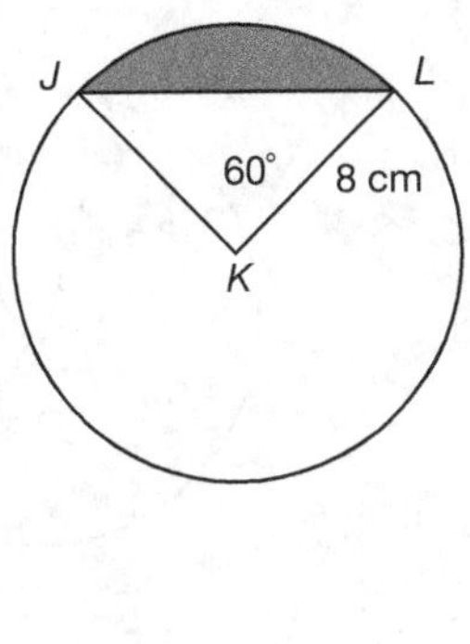

3. ________ ft^2

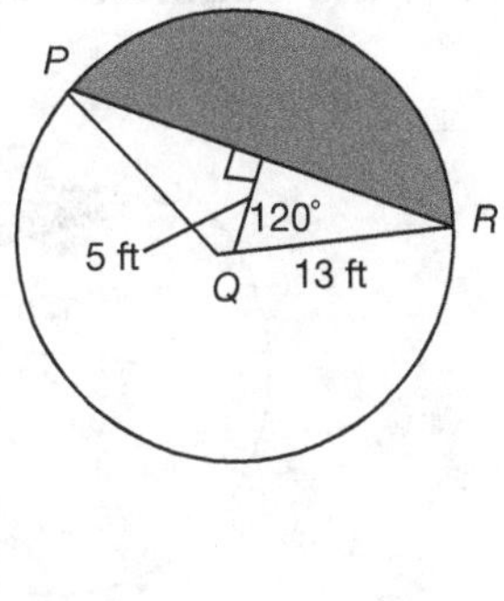

Using Trigonometry to Find Segment Area

You can use trigonometry to find the area of a circle segment. First, determine the sector area. Then, use a trigonometric function to find the height of the triangle bounded by the chord and two radii. Next, use the height of the triangle to find the area of the triangle. Finally, find the segment area.

A circle has a radius of 6.4 centimeters. Determine the area of the segment formed by a chord with central angle 82°. Give your answer to the nearest hundredth.

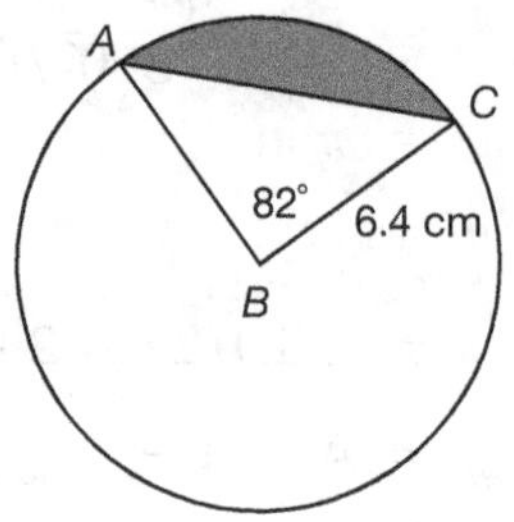

$A_{sector} = \frac{m°}{360°}\pi r^2$	Area of sector
$= \frac{82}{360}\pi(6.4)^2$	Substitute.
$\approx 29.31 \text{ cm}^2$	Solve.
$h_{triangle} = 6.4 \sin 82°$	Height of triangle
$A_{triangle} = \frac{1}{2}(6.4)(6.4 \sin 82°)$	Area of triangle
$\approx 20.28 \text{ cm}^2$	Solve.
$A_{segment} = A_{sector} - A_{triangle}$	Area of segment
$\approx 29.31 - 20.28$	Substitute.
$\approx 9.03 \text{ cm}^2$	Solve.

Practice

Complete the steps to find the area of the circle segment to the nearest hundredth.

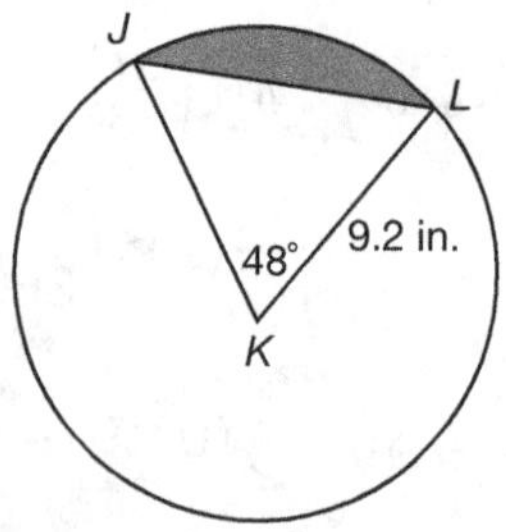

4. $A_{sector} = \frac{48°}{360°}\pi(\_\_\_)^2$

$\approx \_\_\_\_\_ \text{ in}^2$

$h_{triangle} = \_\_\_\_\_\_\_\_$

$A_{triangle} = \frac{1}{2}(\_\_\_)(\_\_\_\_\_\_\_\_)$

$\approx \_\_\_\_\_ \text{ in}^2$

$A_{segment} = \_\_\_\_\_\_\_\_\_\_ \approx \_\_\_ \text{ in}^2$

Find the area of the segment to the nearest hundredth.

5. ________ cm^2

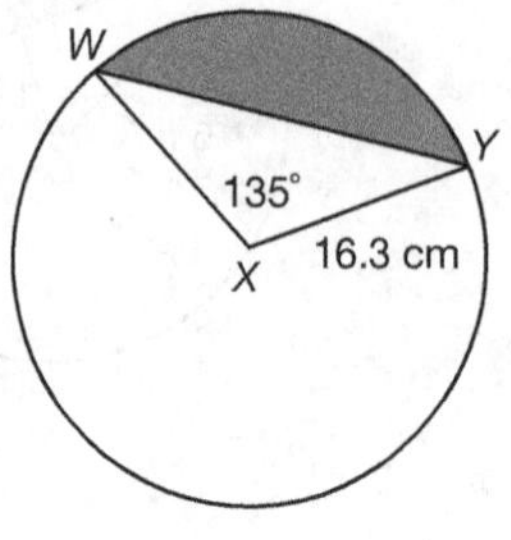

6. ______ in^2

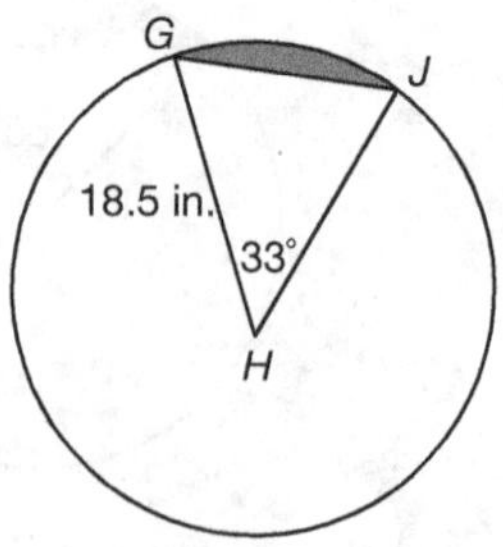

Name ______________________ Date __________ Class __________

Reteaching 113

Symmetry of Solids and Polyhedra

You have worked with symmetry of two-dimensional objects. Now you will examine plane and rotational symmetry in three-dimensional solids.

Plane Symmetry

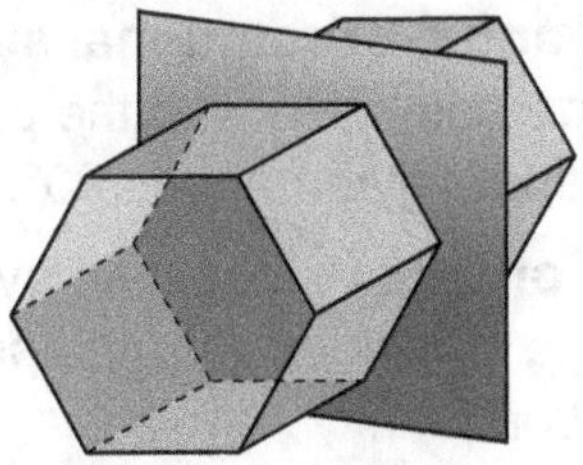

Plane symmetry describes a three-dimensional solid that can be divided into two congruent reflected halves by a plane.

Determine whether the cube has plane symmetry. If it does, show the symmetry in a drawing.

Yes, the cube has plane symmetry. A vertical plane is one of several planes that will divide the cube into two halves.

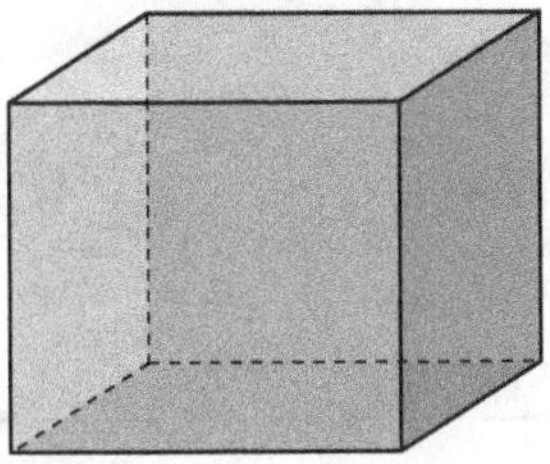

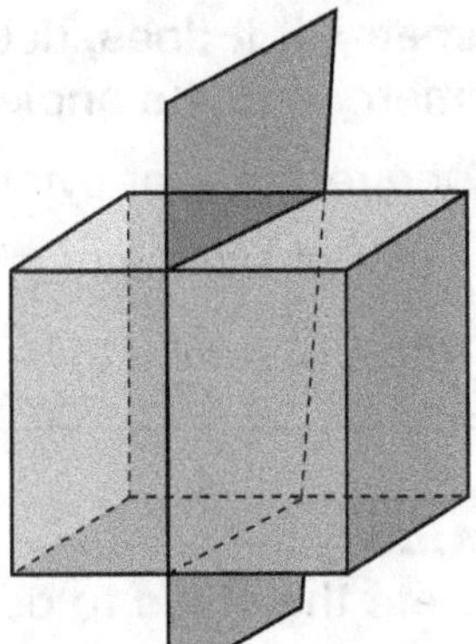

Practice

Determine whether the solid has plane symmetry. If it does, show the symmetry in a drawing.

1. ____

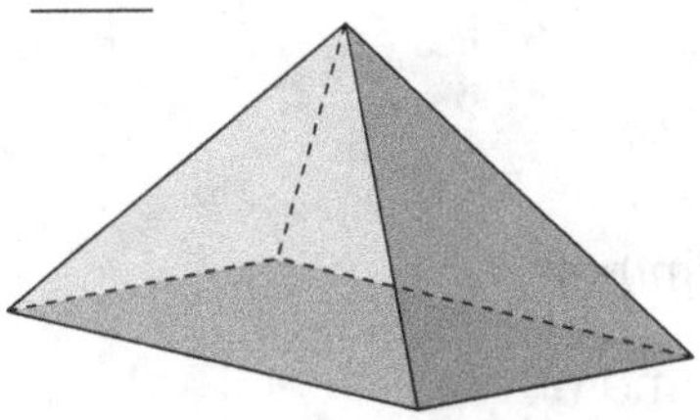

Determine whether the solid has plane symmetry.

2. ____

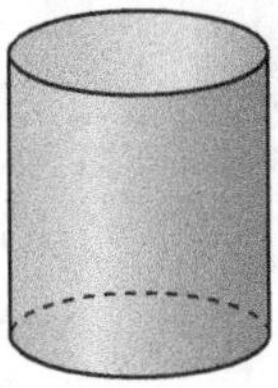

3. ____

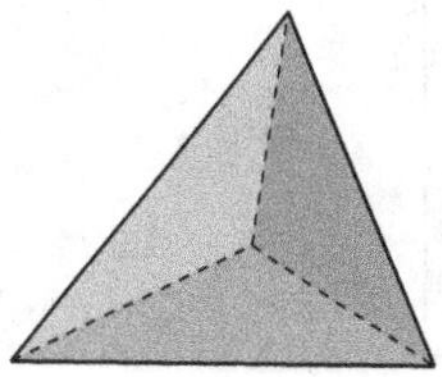

Rotational Symmetry

Rotational symmetry, or **symmetry about an axis,** describes a three-dimensional solid that can be rotated about a line so that the image coincides with the preimage.

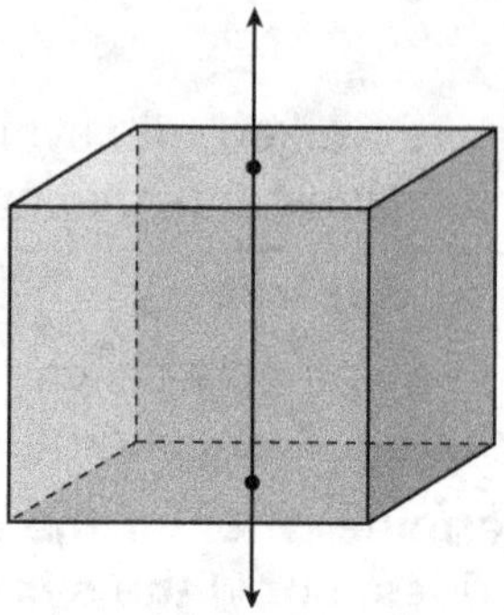

The **angle of rotational symmetry** is the angle at which the image coincides with the preimage. For a cube, the angle of rotational symmetry is 90°.

The **order of rotational symmetry** is the number of times the image coincides with the preimage. For a cube, the order of rotation is 4.

Determine whether the rectangular pyramid has rotational symmetry. If it does, determine the order of rotational symmetry and the angle of rotational symmetry.

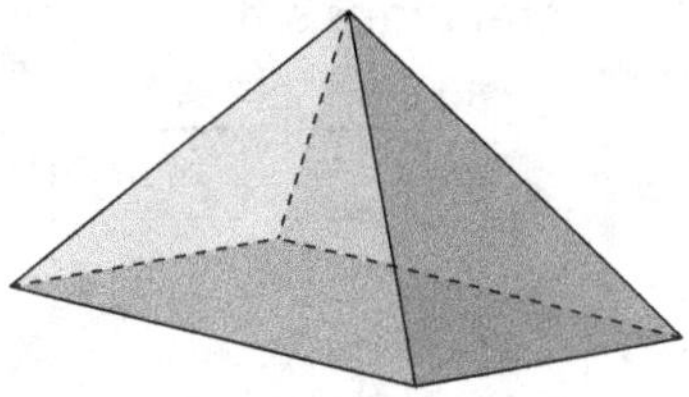

Yes, the rectangular pyramid has rotational symmetry.

The order of symmetry = 2.

The angle of symmetry $= \frac{360°}{2} = 180°$.

Practice

Complete the steps to determine the order of symmetry and the angle of symmetry (if the solid has rotational symmetry).

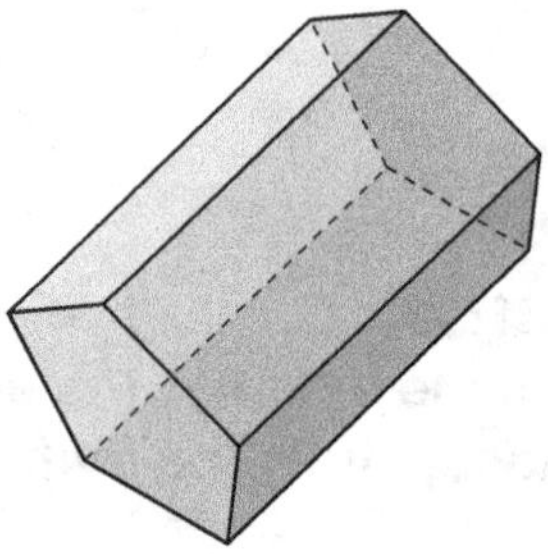

4. rotational symmetry = ____

order of symmetry = _

angle of symmetry $= \frac{360°}{\_\_\_\_} =$ ____

Determine whether the solid has rotational symmetry. If it does, determine the order of symmetry and the angle of symmetry.

5. rotational symmetry = ____

order of symmetry = _

angle of symmetry = _____

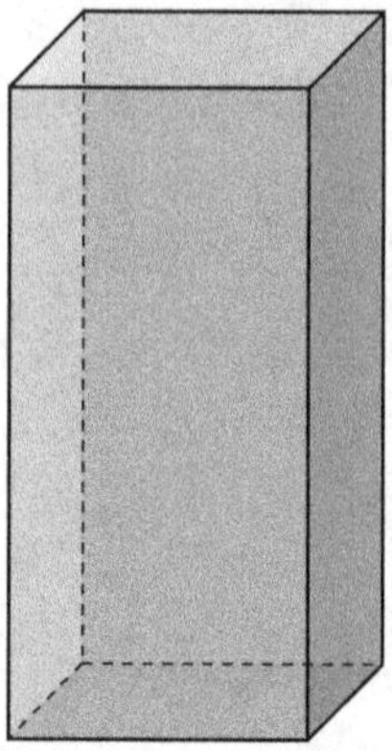

6. rotational symmetry = ____

order of symmetry = _

angle of symmetry = ____

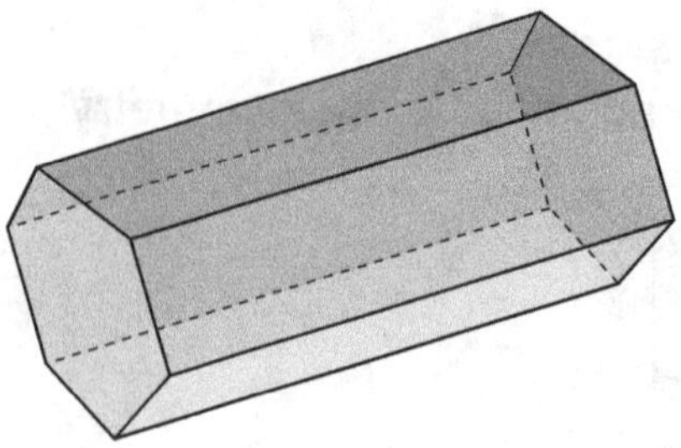

Name ______________________ Date __________ Class __________

Reteaching 114

Solving and Graphing Systems of Inequalities

You have worked with systems of inequalities on number lines. Now you will deal with systems of inequalities in two dimensions.

Systems of Inequalities: "and"

When graphing a pair of inequalities involving "and," the solution will involve an intersection of the two inequalities.

Graph the region described by the inequalities $y \leq x + 3$ and $y \geq 2x + 2$.

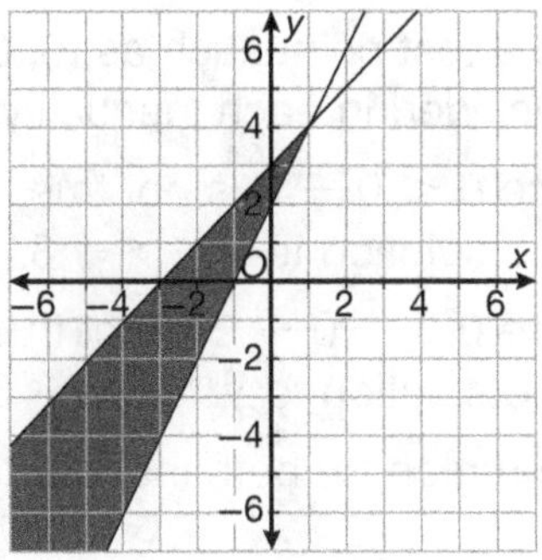

Since the inequalities are joined by "and," the solution will be the intersection of the two inequalities.

First, graph the boundary lines $y = x + 3$ and $y = 2x + 2$.

Use solid lines, since the inequalities $\leq$ and $\geq$ include "equal to."

Use a test point such as $(-2, 0)$ to determine the region that is included in each inequality.

$-2 \leq 0 + 3$ is true, so the region that includes $(-2, 0)$ is the solution to $y \leq x + 3$.

Since $0 \geq 2(-2) + 2$, the region that includes $(-2, 0)$ is the solution to $y \geq 2x + 2$.

The overlapping region is the solution to the system of inequalities.

Practice

Complete the steps to graph the region described by the inequalities $y \geq x - 4$ and $y \leq 3x + 1$.

1. Graph the _____ lines ______________________.

   Is the point (0, 0) a solution to $y \geq x - 4$? ___

   Is the point (0, 0) a solution to $y \leq 3x + 1$? ___

   Shade the region ______ the line __________ and ______ the line ___________.

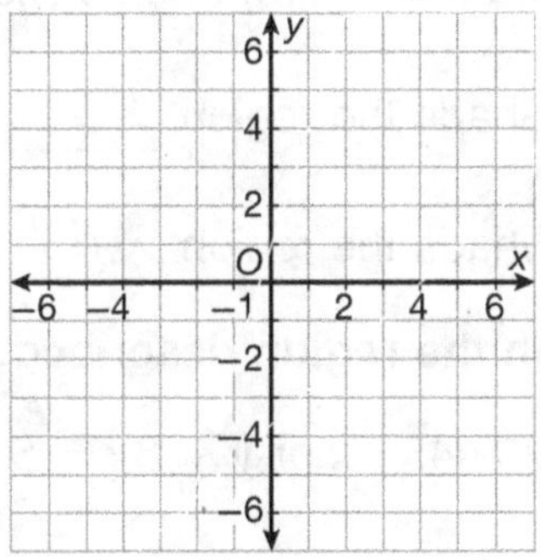

Graph the region described by the inequalities.

2. $y > 2x - 3$ and $y < 2x + 1$

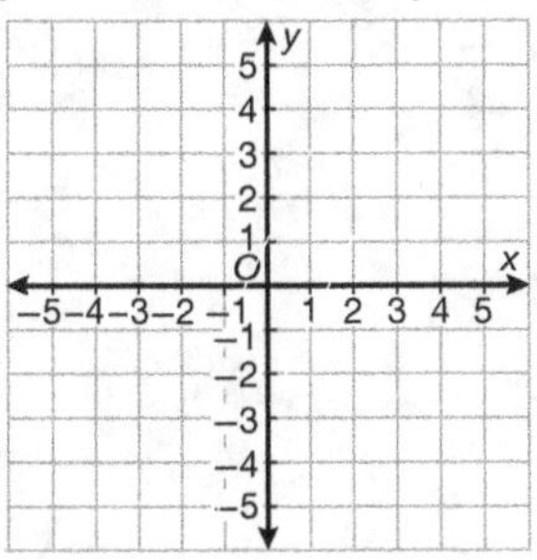

3. $y < x - 1$ and $y > 4x - 4$

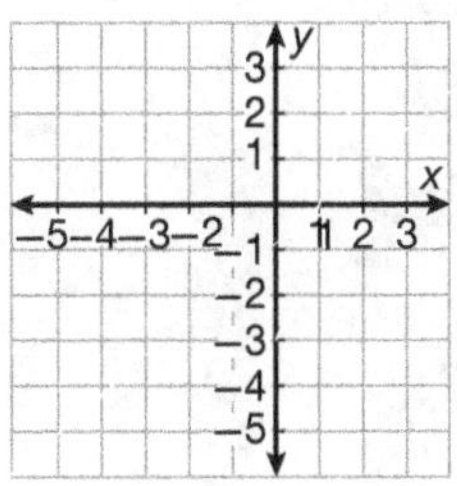

Systems of Inequalities: "or"

When graphing a pair of inequalities involving "or", the solution will involve a union of the two inequalities.

Graph the region described by the inequalities $y < x - 3$ or $y > -x + 2$.

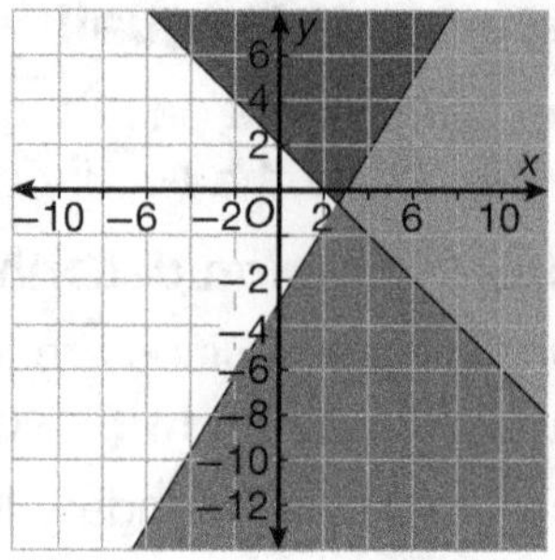

Since the inequalities are joined by "or," the solution will be the union of the two inequalities.

First, graph the boundary lines $y = x - 3$ and $y = -x + 2$.

Use dashed lines, since the inequalities $<$ and $>$ do not include "equal to."

Use a test point such as (0, 0) to determine the region that is included in each inequality.

Since $0 < 0 - 3$ is not true, the region below $y = x - 3$ is the solution to $y < x - 3$.

Since $0 > -0 + 2$ is not true, the region above $y = -x + 2$ is the solution to $y > -x + 2$.

The region for both inequalities is the solution to the system of inequalities.

Practice

Complete the steps to graph the region described by the inequalities $y \geq x - 3$ or $y \geq 2x + 3$.

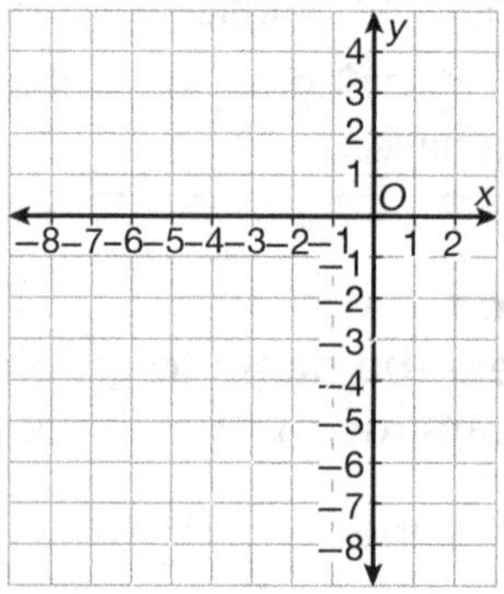

4. Graph the _____ lines ____________________________.

Is the point (0, 0) a solution to $y \geq x - 3$? ___

Is the point (0, 0) a solution to $y \geq 2x + 3$ ___

Shade the region ______ the line __________.

Shade the region ______ the line ___________.

Graph the region described by the inequalities.

5. $y > 4 - x$ or $y < 3x - 5$

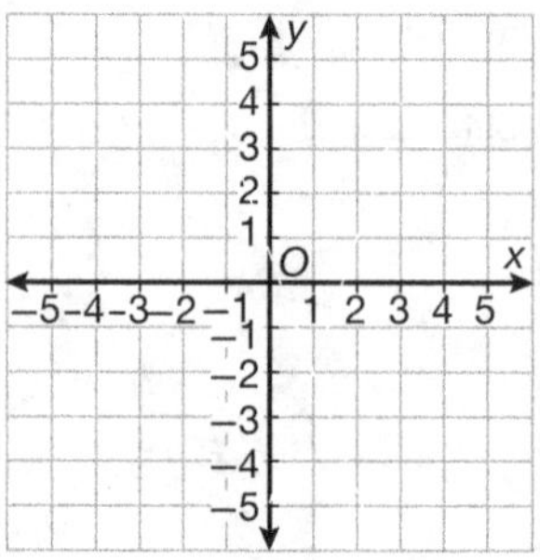

6. $y < 3 - 2x$ or $y \geq 4x - 1$

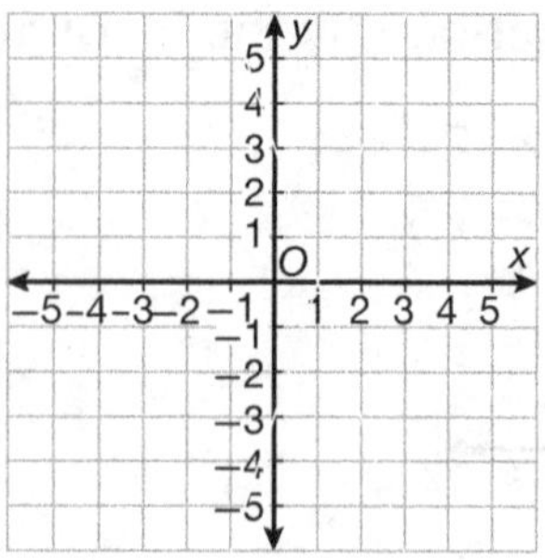

Name ______________________________ Date __________ Class __________

Reteaching 115

Finding Volumes of Composite Solids

Finding Volumes of Composite Solids: Adding

The volume of a composite solid is the sum of the volumes of the individual solids that make up the composite solid.

Find the volume of the solid. Assume that prisms are right prisms.

$V = lwh$	Volume of rectangular prism
$= (16)(12)(8)$	Substitute.
$= 1536\text{ cm}^3$	Solve.
$V = Bh$	Volume of triangular prism
$= \frac{1}{2}(16)(10)(12)$	Substitute.
$= 960\text{ cm}^3$	Solve.
$V = 1536 + 960$	Add volumes.
$= 2496\text{ cm}^3$	Solve.

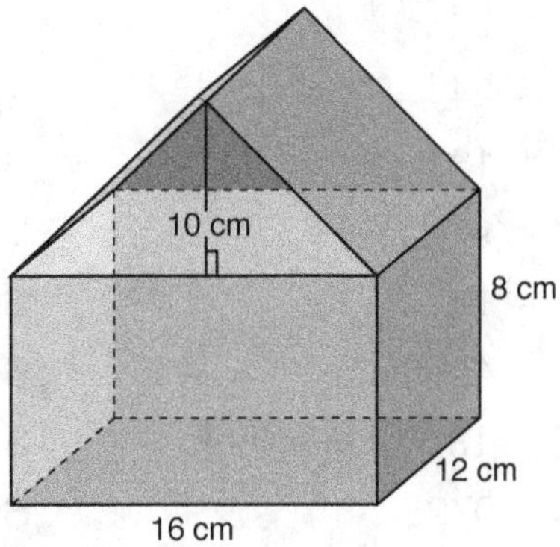

Practice

Complete the steps to find the volume of the solid by adding. Round to the nearest tenth of a cubic unit.

$V_{cube} = s^3$

$= \_\_^3$

$= \_\_\_\_\text{ in}^3$

1. $V_{cylinder} = \pi r^2 h$

$= \pi(\_)^2(\_)$

$\approx \_\_\_\_\text{ in}^3$

$V \approx \_\_\_\_ + \_\_\_\_$

$\approx \_\_\_\_\text{ in}^3$

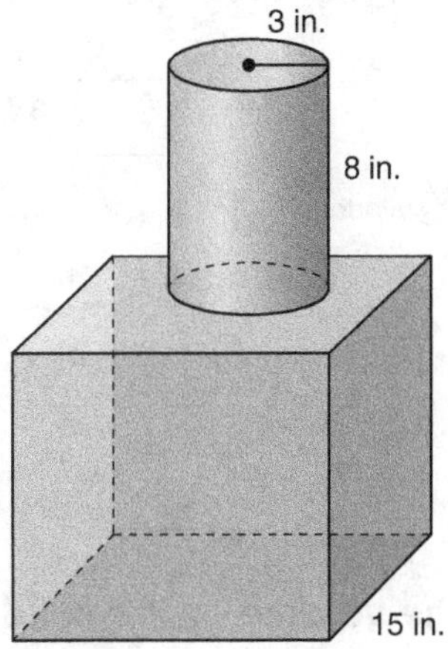

Find the volume of the solid. Round to the nearest tenth of a cubic unit.

2. ________ cm^3

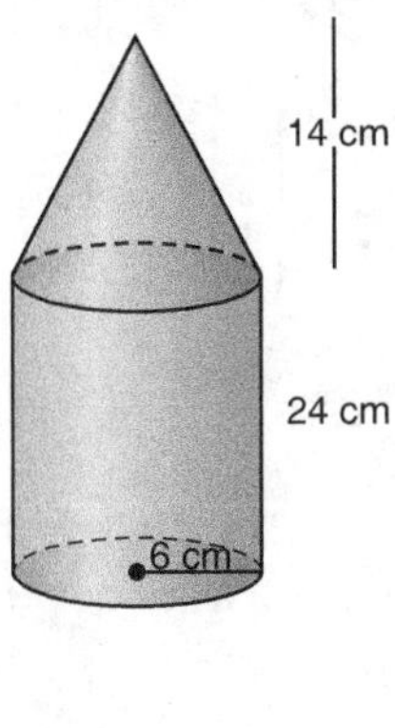

3. ______ m^3

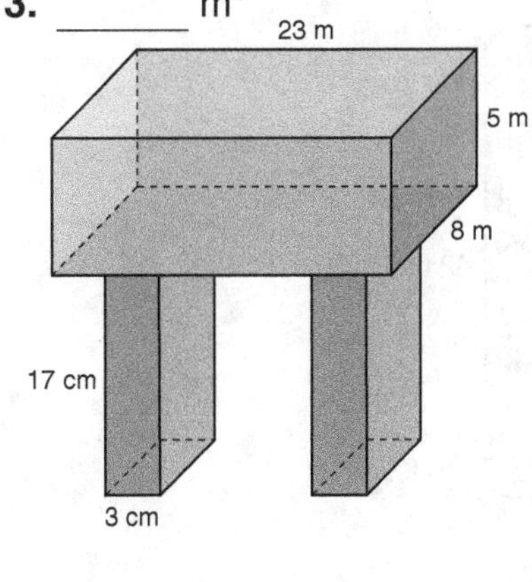

Finding Volumes of Composite Solids: Subtracting

When a larger solid is missing a portion whose volume is equivalent to that of a smaller solid, you can subtract the volume of the space from the total volume.

Find the volume of the solid. Round your answer to the nearest tenth of a cubic unit.

$V = s^3$	Volume of outer cube
$= 24^3$	Substitute.
$= 13{,}824 \text{ cm}^3$	Solve.
$V = s^3$	Volume of inner cube
$= 16^3$	Substitute.
$= 4096 \text{ cm}^3$	Solve.
$V = 13{,}824 - 4096$	Subtract volumes.
$= 9728 \text{ cm}^3$	Solve.

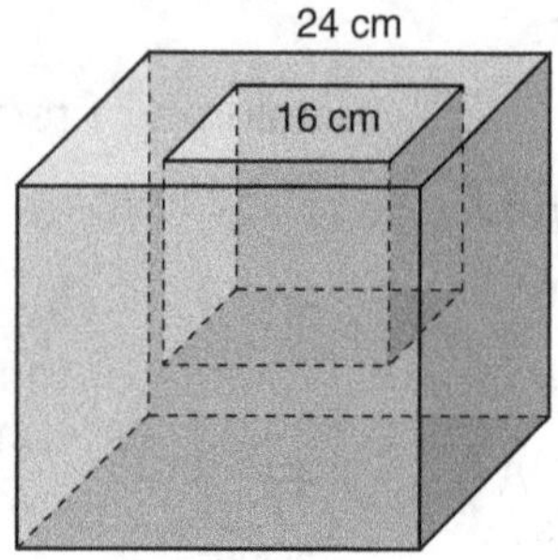

Practice

Complete the steps to find the volume of the solid. Round to the nearest tenth of a cubic unit.

$V_{prism} = lwh$

$= (40)(\_\_)(\_\_)$

$= \_\_\_\_\_\_ \text{ in}^3$

4. $V_{cylinder} = \pi r^2 h$

$= \pi(\_)^2(\_\_)$

$\approx \_\_\_\_\_\_ \text{ in}^3$

$V = \_\_\_\_\_\_ - \_\_\_\_\_\_$

$\approx \_\_\_\_\_\_ \text{ in}^3$

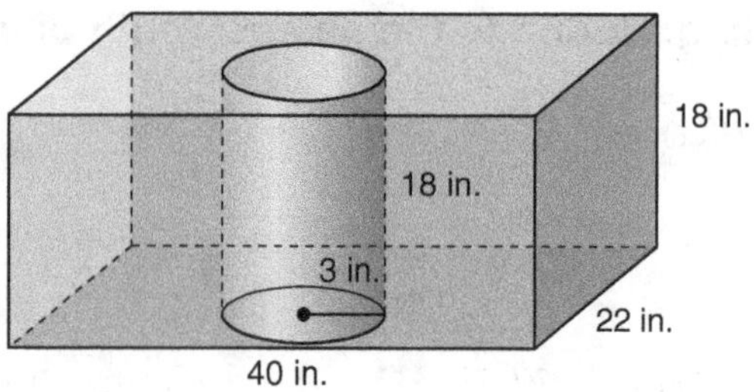

Find the volume of the solid. Round to the nearest tenth of a cubic unit.

5. ________ m^3

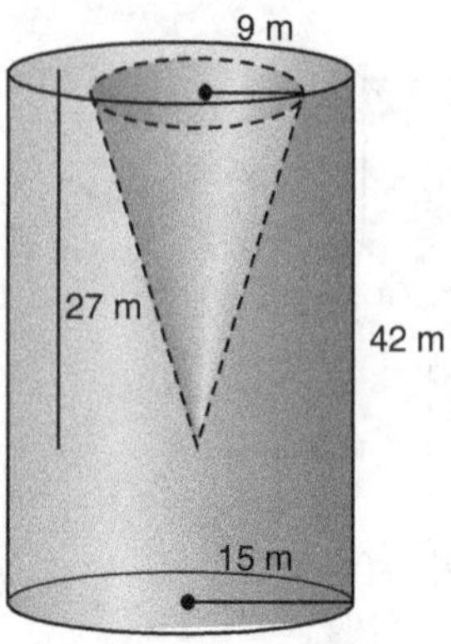

6. ________ cm^3

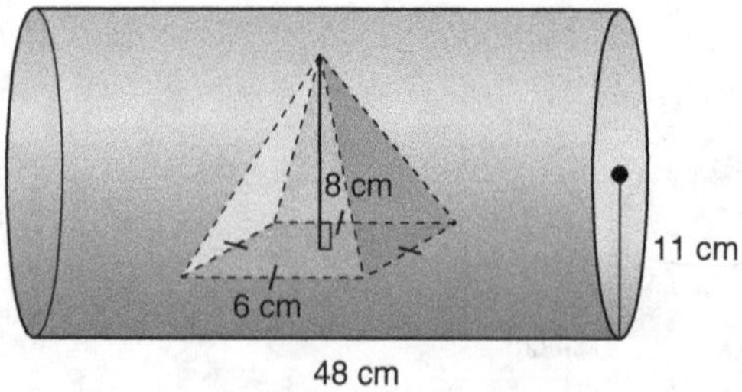

Name ______________________ Date ________ Class ________

Reteaching 116

Secant, Cosecant, and Cotangent

You have worked with three basic trigonometric functions called sine, cosine, and tangent. Now you will work with the reciprocals of those functions.

Trigonometric Ratios

cosecant: $\csc \theta = \frac{1}{\sin \theta} = \frac{\text{hypotenuse}}{\text{side opposite } \theta}$; $\sin \theta \neq 0$

secant: $\sec \theta = \frac{1}{\cos \theta} = \frac{\text{hypotenuse}}{\text{side adjacent } \theta}$; $\cos \theta \neq 0$

cotangent: $\cot \theta = \frac{1}{\tan \theta} = \frac{\text{side adjacent } \theta}{\text{side opposite } \theta}$; $\tan \theta \neq 0$

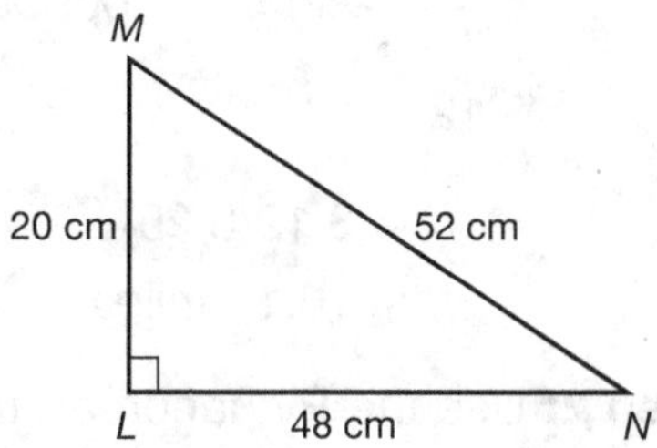

Example:

Find the ratio for cosecant, secant, and cotangent.

$$\csc N = \frac{1}{\sin N} = \frac{\text{hypotenuse}}{\text{side opposite } N} = \frac{52}{20}$$

$$\sec N = \frac{1}{\cos N} = \frac{\text{hypotenuse}}{\text{side adjacent } N} = \frac{52}{48}$$

$$\cot N = \frac{1}{\tan N} = \frac{\text{side adjacent } N}{\text{side opposite } N} = \frac{48}{20}$$

Practice

Complete the statement to determine the given trigonometric ratio.

1. $\csc K = \frac{\text{hypotenuse}}{\_\_\_\_\_\_} = \frac{\_\_}{15}$

$\sec K = \frac{\text{hypotenuse}}{\_\_\_\_\_\_} = \frac{\_\_}{\_\_}$

$\cot K = \frac{\text{side } \_\_\_ \angle K}{\text{side } \_\_\_ \angle K} = \frac{\_\_}{15}$

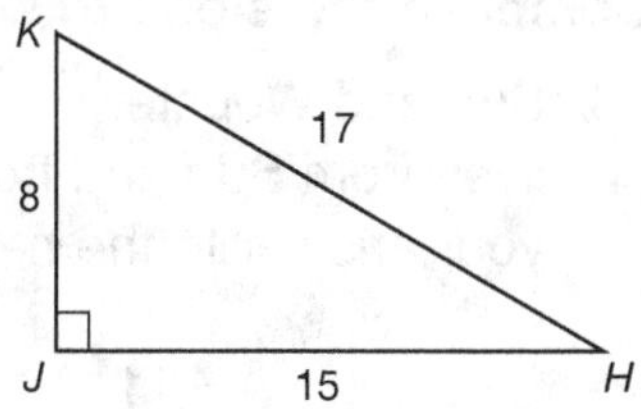

Use the diagram from Exercise 1 to write each trigonometric ratio as a fraction.

2. $\csc H =$ ______ **3.** $\cot H =$ ______ **4.** $\sec H =$ ______

Write each trigonometric ratio as a fraction.

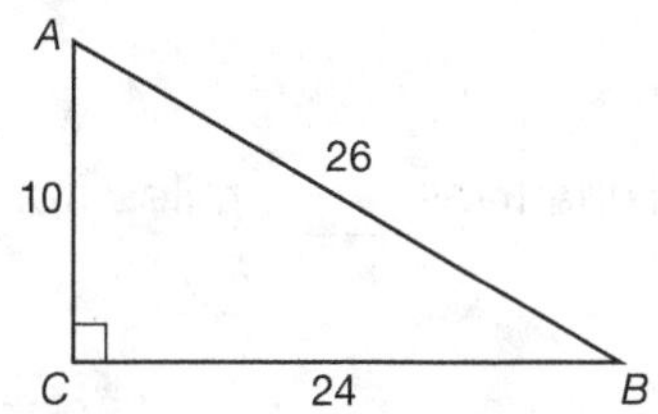

5. $\csc A =$ ______ **6.** $\cot B =$ ______

7. $\cot A =$ ______ **8.** $\csc B =$ ______

Trigonometric Ratios (continued)

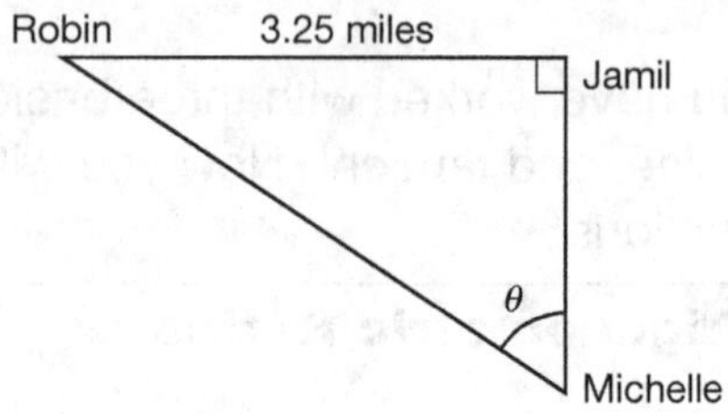

Example: The distance between Robin's home and Jamil's home is 3.25 miles. The cosecant of θ equals 3.12. How far does Michelle live from Robin and Jamil? Round your answer to the nearest tenth.

Step 1: Use the cosecant ratio to find the distance between Robin's home and Michelle's home.

$$\csc \theta = \frac{1}{\sin \theta} = \frac{\text{hypotenuse}}{\text{opposite}}$$

$$3.12 = \frac{h}{3.25} \qquad \text{Substitute.}$$

$$h = 3.12(3.25) \qquad \text{Multiply each side by 3.25.}$$

$$h \approx 10.14 \text{ miles} \qquad \text{Simplify.}$$

Step 2: Use the Pythagorean theorem to find the distance between Jamil's home and Michelle's home.

$$a^2 + b^2 = c^2$$

$$3.25^2 + x^2 = 10.14^2$$

$$x^2 = 92.2571$$

$$x \approx 9.61 \text{ miles}$$

Step 3: State the answer.

Michelle lives 10.14 miles from Robin and 9.61 miles from Jamil.

Practice

Complete the steps to solve the problem.

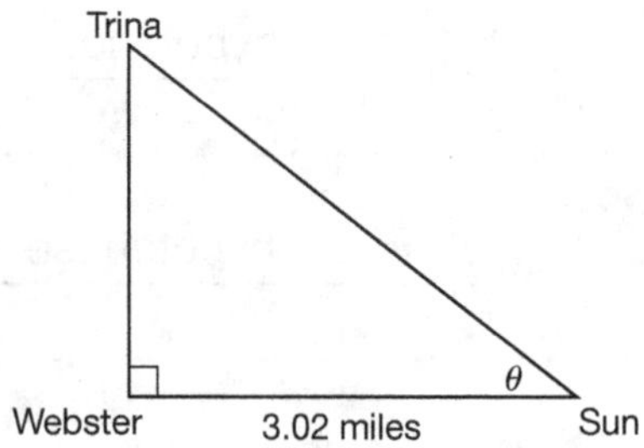

9. Sun and Webster live 3.02 miles apart. Find how far Trina lives from Sun and from Webster if the cot $\theta = 1.23$. Round your answer to the nearest hundredth.

$$\cot \theta = \frac{1}{\_\_\_\_\_\_} = \frac{\text{side } \_\_\_\_\_\_ \ \theta}{\text{side opposite } \theta}$$

$$\_\_\_\_\_\_ = \frac{3.02}{x}$$

$$\_\_\_\_\_\_ x = 3.02$$

$$x \approx \_\_\_\_\_\_ \text{ miles}$$

$$a^2 + b^2 = c^2$$

$$\_\_\_\_\_\_^2 + 2.46^2 = c^2$$

$$\_\_\_\_\_\_ = c^2$$

$$\_\_\_\_\_\_ \approx c$$

Trina lives ____ miles from Sun and ____ miles from Webster.

Name ______________________________ Date ___________ Class ___________

Reteaching 117

Determining Line of Best Fit

You have worked with equations of lines. Now you will determine the line of best fit for a set of data.

Choosing a Line of Best Fit

A **scatter plot** has points plotted to show a relationship between two sets of data.

The **line of best fit** is a line that comes closest to all the points in the data set.

Example:

Explain why the line in each graph is, or is not, a good choice for the line of best fit.

Graph A

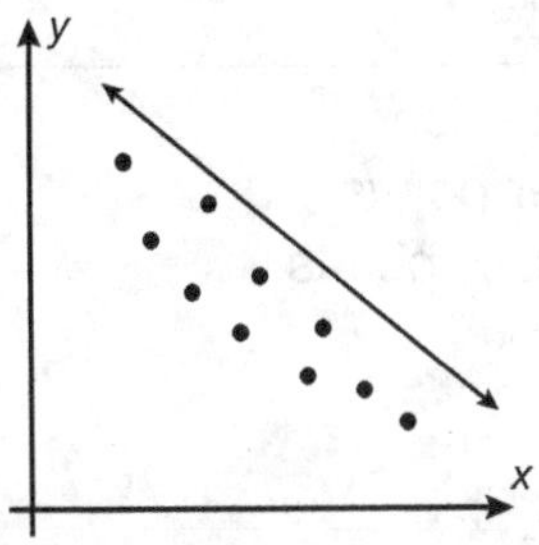

Graph B

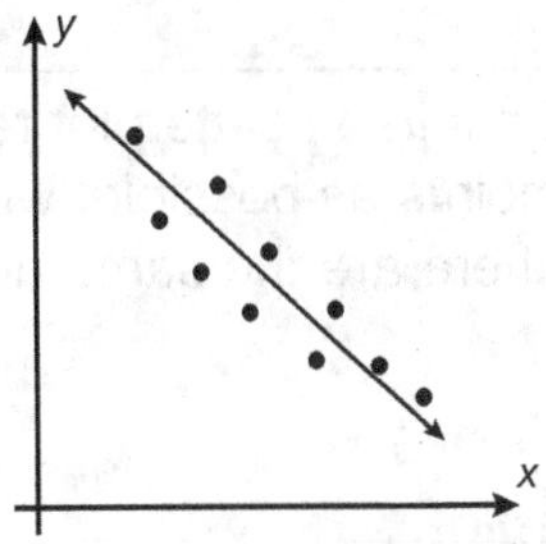

Graph C

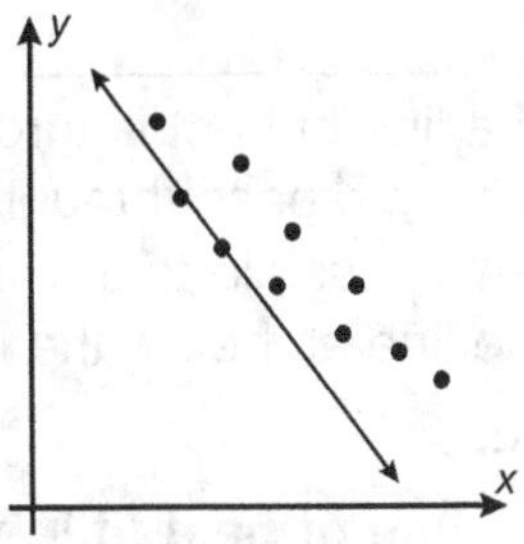

Graph A: The line has the correct slope but is not drawn as close to the data points as possible. This is not a good choice for the line of best fit.

Graph B: The line goes through the data points and has a slope that represents the trend in the data. This is a good choice for the line of best fit.

Graph C: The line goes through the data but does not have a slope that represents the trend in the data. This is not a good choice for the line of best fit.

Practice

Complete the statements to explain which line represents the line of best fit.

1. Graph A

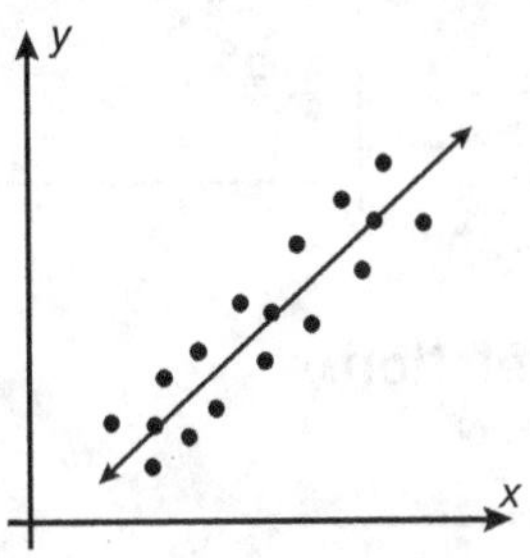

Graph B

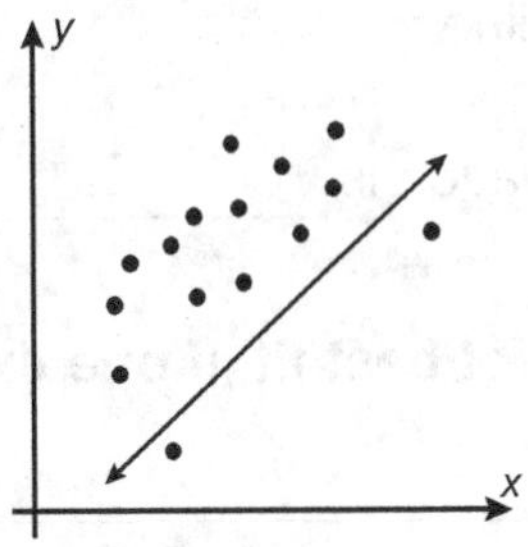

Graph A: This is a (good/bad) ______ choice for the line of best fit. The line (does/doesn't) ______ go through the middle of the data points. The line (does/doesn't) ______ have a slope that represents the trend in the data.

Graph B: This is a (good/bad) _____ choice for the line of best fit. The line (does/doesn't) _________ go through the middle of the data points. The line (does/doesn't) ______ have the correct slope.

Drawing a Line of Best Fit

Positive Correlation	Negative Correlation	No Correlation
Both quantities increase together. 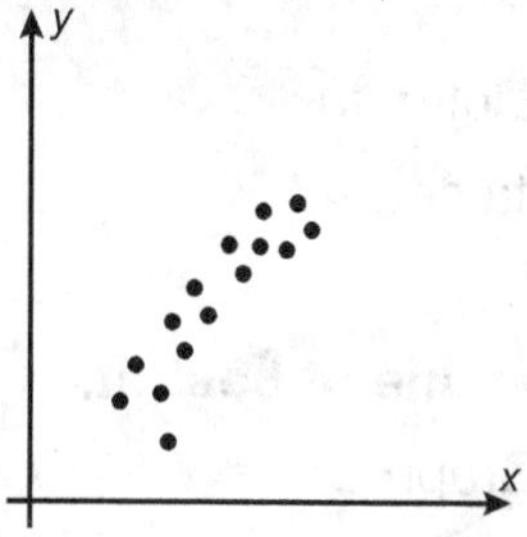	One quantity increases as the other decreases. 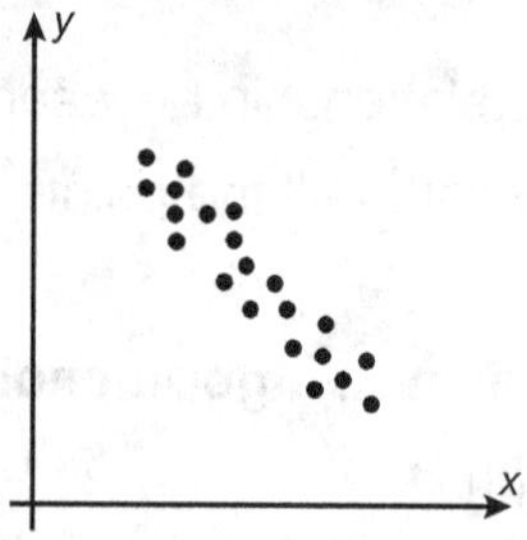	Data points are randomly scattered. 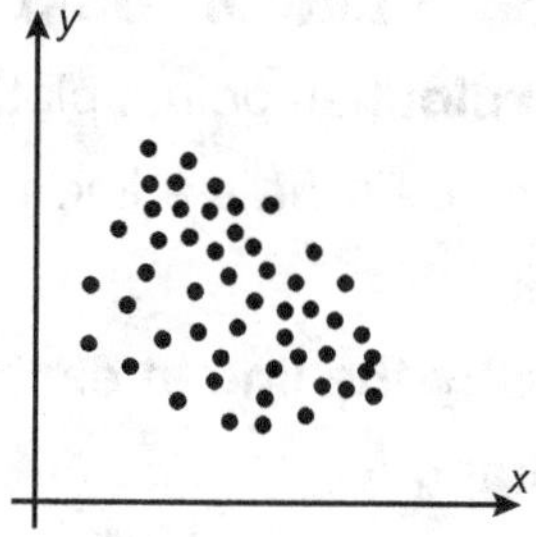

Drawing a line of best fit through the points in a scatterplot requires patience. The line should pass through as many points as possible, with the other points as close as possible to the line so that there are the same number of data points above the line as below the line.

Example:

Draw the line of best fit in the scatterplot. Then, describe the correlation in the data.

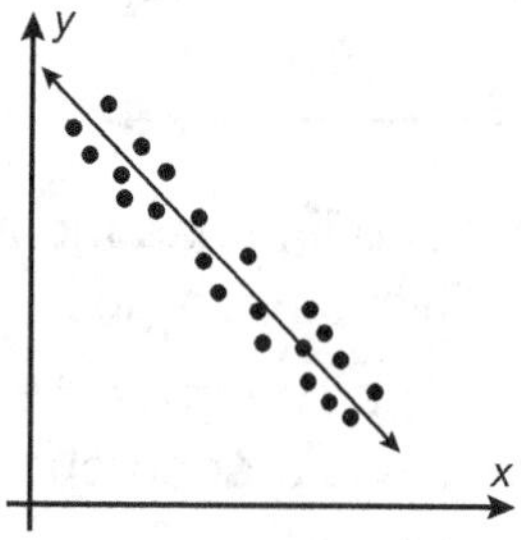

Step 1: The line is drawn with as many points above as points below.

Step 2: One quantity increases as the other decreases, so it is a strong negative correlation.

Practice

Complete the steps to draw the line of best fit. Describe the correlation.

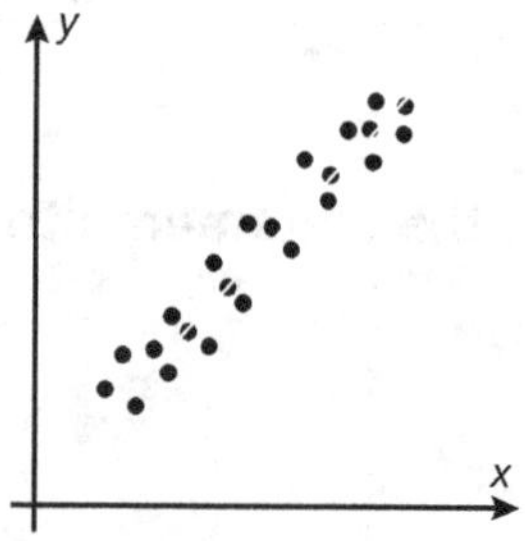

2. Draw the line.

The correlation is ______________.

Draw the line of best fit, if one exists, and describe the correlation.

3. ______________________________

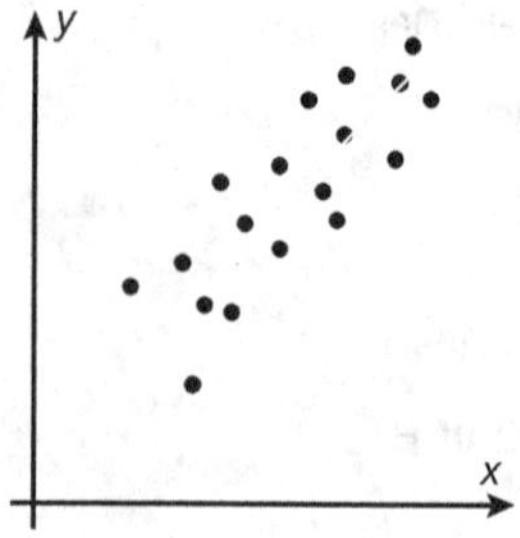

4. ______________

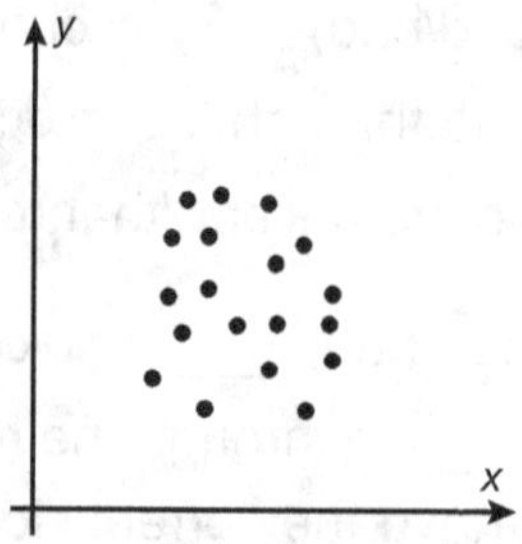

Name ______________________ Date __________ Class ____________

Reteaching 118

Finding Areas of Polygons Using Matrices

You have worked with area of polygons and of matrices. Now you will find the determinant of a matrix and use it to find area.

Determinant of a Matrix

The **determinant of a matrix** is the difference between the product of the first diagonal and that of the other diagonal. Given $A = \begin{bmatrix} x_1 & y_1 \\ x_2 & y_2 \end{bmatrix}$, where x_1, x_2, y_1, and y_2 are constants, the determinant of A is given by the formula det $A = x_1y_2 - x_2y_1$.

Calculate the determinant of $A = \begin{bmatrix} 5 & 2 \\ 8 & 4 \end{bmatrix}$.

det $A = x_1y_2 - x_2y_1$	Determinant formula
$= (5)(4) - (8)(2)$	Substitute.
$= 20 - 16$	Simplify.
$= 4$	

Practice

Complete the steps to calculate the determinant.

1. $A = \begin{bmatrix} 3 & 1 \\ 5 & 4 \end{bmatrix}$

det $A = x_1y_2 - x_2y_1$

$= (3)(\_) - (\_)(\_)$

$=$ ______

$=$ __

2. $A = \begin{bmatrix} 2 & -1 \\ -4 & 3 \end{bmatrix}$

det $A = x_1y_2 - x_2y_1$

$= (\_)(\_) - (\_)(\_\_)$

$=$ ______

$=$ __

Calculate the determinant.

3. $A = \begin{bmatrix} -2 & 0 \\ 3 & 1 \end{bmatrix}$

____

4. $A = \begin{bmatrix} -5 & -3 \\ -6 & 2 \end{bmatrix}$

____

5. $A = \begin{bmatrix} 3 & 5 \\ 0 & 0 \end{bmatrix}$

____

6. $A = \begin{bmatrix} 2 & 0 \\ 0 & -3 \end{bmatrix}$

____

7. $A = \begin{bmatrix} 4 & 2 \\ 2 & 4 \end{bmatrix}$

____

8. $A = \begin{bmatrix} -1 & 2 \\ -3 & 3 \end{bmatrix}$

____

Matrix Method of Computing Area of a Polygon

Given a complex polygon with n vertices from (x_1, y_1) to (x_n, y_n) around the polygon, the area of the polygon is equal to one-half of the determinant of the matrix listing the coordinates vertically.

The area of the polygon is given by the formula

$$A = \frac{1}{2}(x_1y_2 + x_2y_3 + \cdots + x_ny_1 - y_1x_2 - y_2x_3 - \cdots - y_nx_1).$$

Calculate the area of A rectangle with vertices (0, 2), (2, 2), (2, 8) and (0, 8).

$A = \frac{1}{2}(x_1y_2 + x_2y_3 + x_3y_4 + x_4y_1 - y_1x_2 - y_2x_3 - y_3x_4 - y_4x_1)$	Matrix formula for area.
$= \frac{1}{2}((0)(2) + (2)(8) + (2)(8) + (0)(2) - (2)(2) - (2)(2) - (8)(0) - (8)(0))$	Substitute.
$= \frac{1}{2}(0 + 16 + 16 + 0 - 4 - 4 - 0 - 0)$	Simplify.
$= \frac{1}{2}(24)$	Simplify.
$= 12$	Solve.
$\lvert 12 \rvert = 12$ sq. units	Absolute value of area

Practice

Complete the steps to calculate the area of the polygon.

9. square with vertices (3, 3), (3, 7), (7, 7), and (7, 3)

$$A = \frac{1}{2}(x_1y_2 + x_2y_3 + x_3y_4 + x_4y_1 - y_1x_2 - y_2x_3 - y_3x_4 - y_4x_1)$$

$$= \frac{1}{2}((3)(7) + (\_)(\_) + (\_)(\_) + (\_)(\_) - (3)(3) - (\_)(\_) - (\_)(\_) - (\_)(\_))$$

$$= \frac{1}{2}(21 + \_\_ + \_\_ + 21 - \_ - \_ - \_\_ - \_)$$

$$= \frac{1}{2}(\_\_\_\_)$$

$$= \_\_\_\_$$

$\lvert \_\_\_\_ \rvert = \_\_$ sq. units

Calculate the area of the polygon.

10. Pentagon with vertices (3, 3), (4, 2), (4, −1), (1, 1), and (0, 3)

___ sq. units

11. Hexagon with vertices (2, 5), (3, 4), (2, −3), (0, −3), (−1, 1), and (0, 5)

___ sq. units

Name ______________________ Date __________ Class __________

Reteaching 119
Platonic Solids

Characteristics of Platonic Solids

A **regular polyhedron** is a polyhedron whose faces are all congruent polygons, and for which the same number of faces meet at each vertex.

A **Platonic solid** is one of five regular polyhedra: a tetrahedron, a cube, an octahedron, a dodecahedron, or an icosahedron.

Regular Polyhedron	Vertices	Edges	Faces	Diagram
Tetrahedron equilateral faces	4	6	4	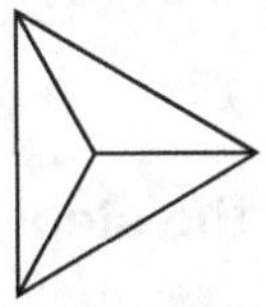
Cube square faces	8	12	6	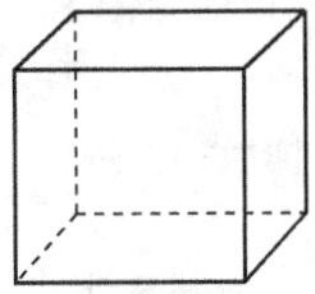
Octahedron equilateral faces	6	12	8	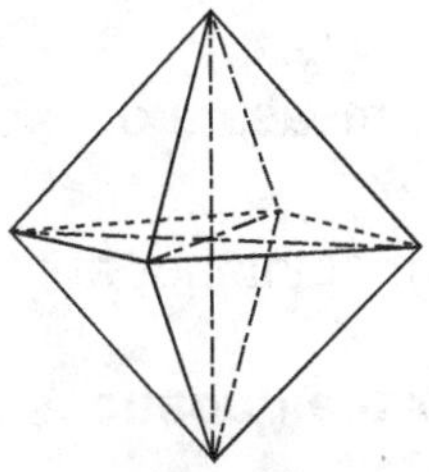
Dodecahedron pentagon faces	20	30	12	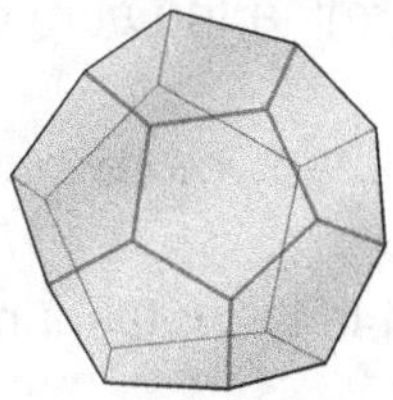
Icosahedron equilateral faces	12	30	20	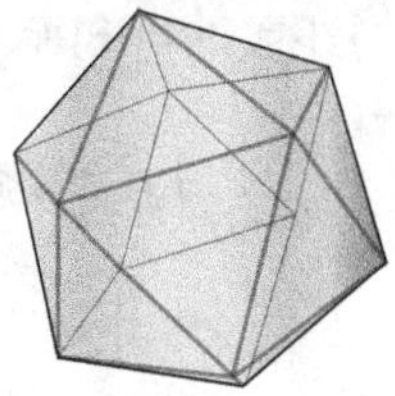

Practice

1. Each vertex is the meeting point of an ________________.

   The measure of each angle is ____. ____ × 4 = ____

Determine the sum of the measures of the angles at a vertex.

2. tetrahedron ______

3. icosahedron ______

More About Platonic Solids

Only three shapes can be used as the faces of Platonic solids: triangles, squares, and pentagons.

Example:

Explain why regular hexagons cannot be used to construct a Platonic solid.

The interior angles of a hexagon each measure 120°. If three angles of a hexagon meet at a vertex, they have a sum of $3 \times 120° = 360°$.

This would result in a tessellation, not in a polyhedron.

Practice

Complete the steps to answer the questions about Platonic solids.

4. Determine the sum of the interior angles for a vertex in a dodecahedron.

   Each face is a ___________.

   The number of faces = _____.

   The measure of each angle = ______.

   Sum of interior angles = ______ × 3 = ______.

Answer the questions about Platonic solids.

5. Determine the sum of the interior angles for a vertex in a cube. ______

6. How many more faces does an icosahedron have than an octahedron? _____

7. What is the total number of faces, vertices, and edges on a dodecahedron? _____

8. What type of shape is a tetrahedron? ____________________

9. Draw a net of a cube.

10. Draw a net of a tetrahedron.

Name ______________________ Date __________ Class __________

Reteaching 120
Topology

You have worked with different types of three-dimensional objects.
Now you will learn about topology.

Topology

Topology is the branch of geometry that deals with the properties of geometric figures that do not change under stress. Two polygons (such as a triangle and a square) are topologically the same because they both have boundaries that isolate an area within a two-dimensional space.

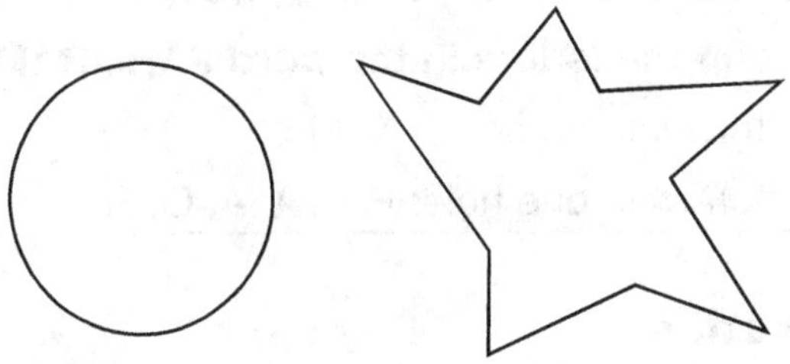

Explain why the two objects are topologically equivalent.

The circle on the left can be reshaped into a five pointed star on the right by pulling at the circle at five equally spaced points along its circumference until it forms the shape of a five pointed star.

Practice

Complete the steps to answer the following questions.

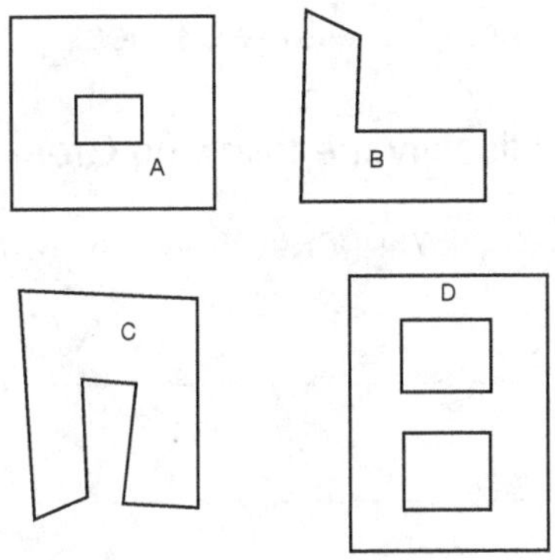

1. Which two objects are topologically equivalent?

   Objects ___ and ___ are topologically equivalent. They are the only two shapes that do not separate the two-dimensional space into different regions. Object ___ separates two-dimensional space into two regions, while Object ___ separates two-dimensional space into three regions.

Answer the following questions about topological equivalence.

2. Which numbers are topologically equivalent to the number one (1)? __________

3. Explain why a cube and a pyramid are topically equivalent.

   ____

4. Determine which two shapes are topologically equivalent. __________

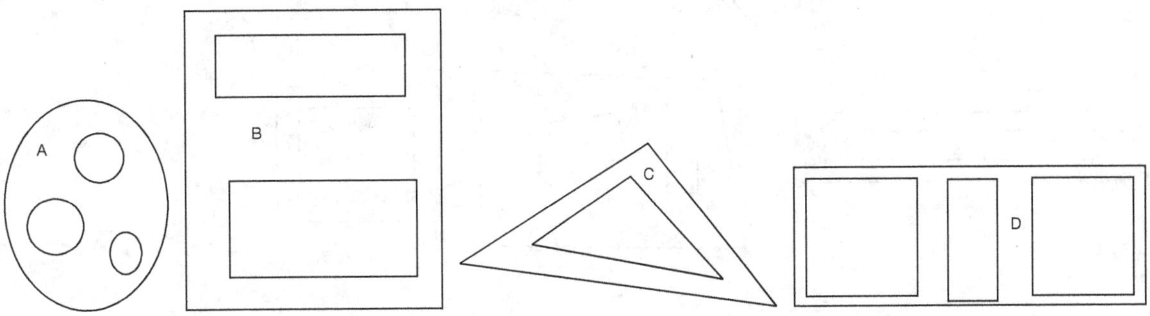

Topological Classes

Objects can be grouped into topological classes based on their characteristics. For example, the numbers 1, 2, 3, 5, and 7 can be grouped into one class (no holes); the numbers 4, 6, and 9, into another class (one hole); and the number 8, into a third class (two holes).

Group the letters in the word PANORAMA according to topological classes.

Letters with no holes: N, M

Letters with one hole: P, A, A, A, O, R

Practice

5. Classify the following symbols into topological classes.

/, >, <, @, #, $, ^, &, *, (,)

Symbols with no holes: ____________________

Symbols with one hole: ______

Symbols with two holes: ______

6. Classify the following Greek characters into topological classes,

α, β, γ, δ, ε, ζ, η, θ, ι, κ, λ, μ, ν, ξ, ο, π, ρ, ς, σ, τ, υ, φ, χ, ψ, ω

____

____

____

7. Which of the following shapes is not part of the same topological class?

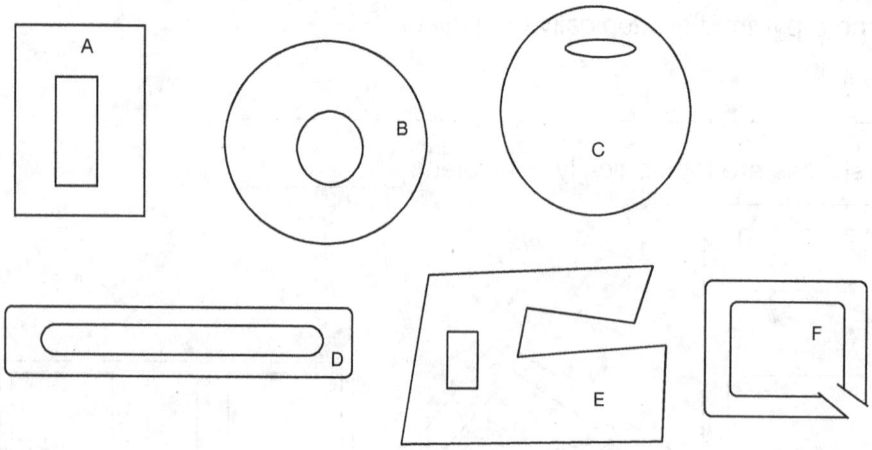

Name ______________________ Date __________ Class __________

Reteaching

Polar Coordinates

INV 12

You have plotted points and functions using Cartesian coordinates. Now you will learn how to use polar coordinates and graph polar functions.

Polar Coordinates

The polar coordinate system is based on a central point called the **pole** and a ray called the **polar axis.** Points are measured in terms of r, the distance from the point to the pole, and θ, the angle a line from the point to the pole makes with the polar axis.

Example:

Plot the point P(3, 30°).

$r = 3$; $\theta = 30°$

The angle P makes with the polar axis is 30°.

Point P is 3 units from the pole.

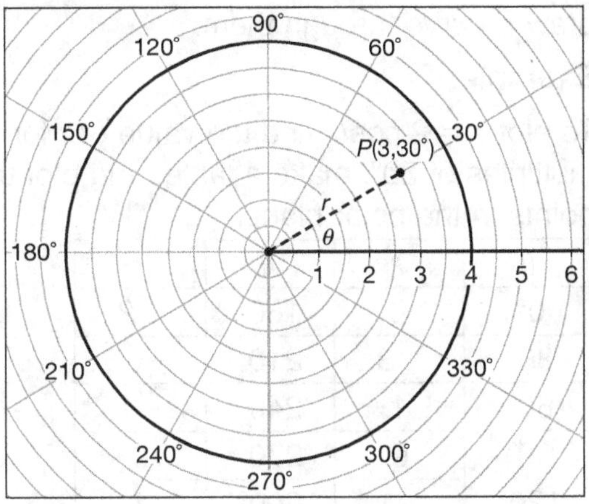

Practice

1. Plot the polar point $P(2, 45°)$.

   Because θ is ___, the angle P makes with the polar axis is ___.

   Because r is ___, P is ___ units from the pole.

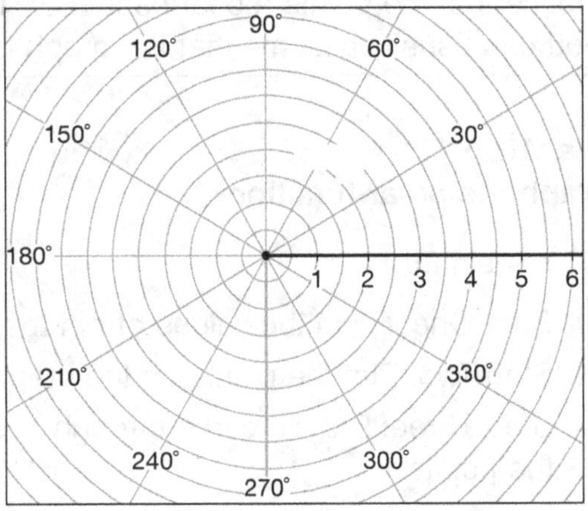

2. Plot the point $P(-2.5, 220°)$.

   If r is negative, plot the point in the direction that is 180° opposite of the angle.

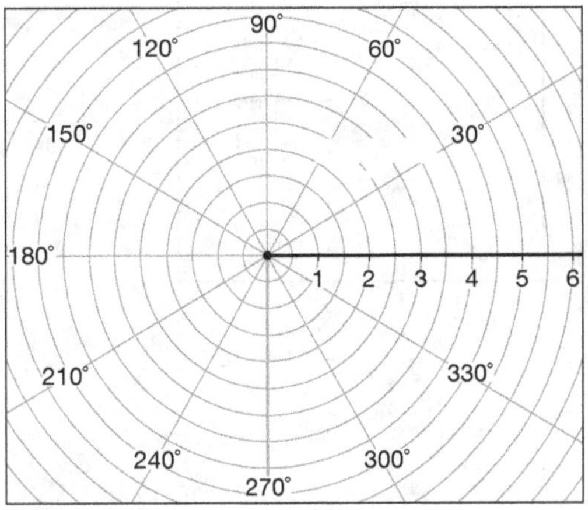

Graphing Polar Functions

When r is a function of θ, $r = f(\theta)$. This is called a **polar function.** Polar functions can be graphed in polar coordinates. Find the value of r for several values of θ, make a table with the values, plot the points, and draw a curve through them.

Example:

To plot $r = 2\cos\theta$, find the value of r for multiples of 30°, make a table, and plot the points in the polar plane.

θ	r	θ	r
0°	2	180°	−2
30°	$\sqrt{3}$	210°	$-\sqrt{3}$
60°	1	240°	−1
90°	0	270°	0
120°	−1	300°	1
120°	$-\sqrt{3}$	330°	$\sqrt{3}$

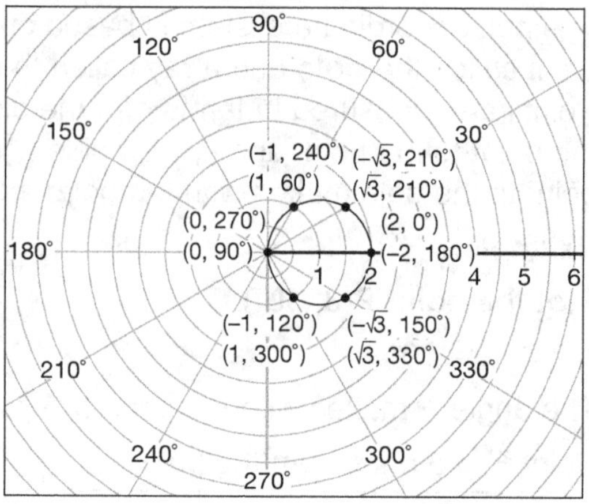

Notice that the points repeat past 180°. This is common with polar functions. When it happens you can usually assume that the graph is finished.

Practice

Graph the polar function.

3. $r = \sin 2\theta$

Step One: Fill in the values of r in the table:

Step Two: Plot the points on the graph.

Step Three: Draw a curve through the points.

θ	r	θ	r
0°	_____	180°	_____
30°	$\frac{\sqrt{3}}{2}$	210°	$\frac{\sqrt{3}}{2}$
45°	_____	225°	_____
60°	_____	240°	_____
90°	_____	270°	_____
120°	$-\frac{\sqrt{3}}{2}$	300°	_____
135°	_____	315°	□1
150°	_____	330°	$-\frac{\sqrt{3}}{2}$

Name ______________________ Date __________ Class __________

Reteaching 1
Points, Lines, and Planes

Points, lines, and planes are the building blocks for all other geometric figures. Both lines and planes extend through space forever.

Use the diagram on the right to answer the following questions.

a. Give two different names for the line.
Two possible names for the line are line *n* and $\overleftrightarrow{XY}$.

b. Name two collinear points and three noncollinear points.
Points *X* and *Y* are collinear, and points *X*, *Y*, and *Z* are noncollinear.

c. Give two different names for the plane.
Two possible names for the plane are plane *WYZ* and plane *Q*.

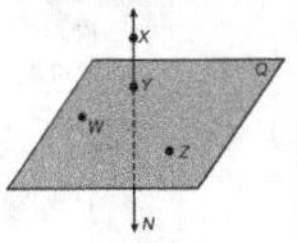

Practice

Use the diagram on the right to complete questions 1–5.

1. Label the plane as plane *W*.
2. Draw and label coplanar points *C*, *D*, and *E* on the plane.
3. Draw line *CD* on the plane.
4. Place point *K* on the plane so that it is collinear with point *C* but not on $\overleftrightarrow{CD}$.

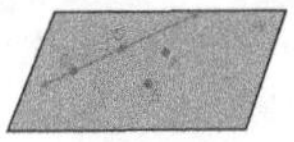

5. What is another name for plane *W*? possible answer: plane *CDE*

Use the diagram on the right to complete questions 6–10.

6. Give another name for plane *M*. possible answer: plane *HJK*
7. Name three collinear points. points *G*, *H*, and *J*
8. Give three points that are noncollinear. possible answer: points *K*, *G*, and *H*
9. Are any of the points noncoplanar? no
10. Name two coplanar lines. $\overleftrightarrow{HJ}$ and $\overleftrightarrow{GK}$

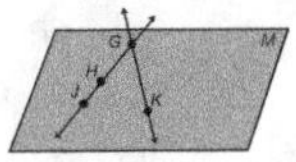

Reteaching 1
continued

Two lines intersect at exactly one point. When two planes intersect, their intersection is an infinite number of points and creates a line. When a plane and a line intersect, their intersection may be just a point or the entire line.

Use the diagram on the right to answer the following questions.

a. What is the intersection of $\overleftrightarrow{MN}$ and $\overleftrightarrow{PQ}$
Point *L* is the intersection of $\overleftrightarrow{MN}$ and $\overleftrightarrow{PQ}$.

b. What is the intersection of planes *S* and *T*?
The intersection of planes *S* and *T* is $\overleftrightarrow{MN}$.

c. What is the intersection of plane *T* and $\overleftrightarrow{PQ}$?
The intersection of plane *T* and is $\overleftrightarrow{PQ}$.

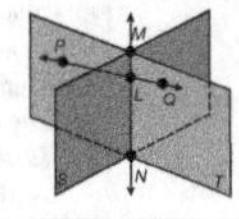

Practice

Use the diagram on the right to complete questions 15–19.

11. Label the planes *A* and *B*.
12. Draw and label the intersection of planes *A* and *B* as $\overleftrightarrow{GH}$.
13. Draw and label $\overleftrightarrow{WX}$ so that it intersects plane *A* at every point and plane *B* at point *G*.
14. Draw and label $\overleftrightarrow{YZ}$ so that it intersects $\overleftrightarrow{WX}$ at point *G*. The two lines are not on the same plane.
15. Draw and label point *T* so that it is coplanar with points *Y* and *Z*.

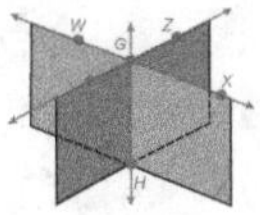

Identify each of the following from the diagram.

16. What is the intersection of plane *J* and $\overleftrightarrow{QR}$? point *Q*
17. What is the point of intersection of line *P* and $\overleftrightarrow{MN}$? point *M*
18. What is the intersection of planes *J* and *K*? line *P*
19. What is the intersection of $\overleftrightarrow{QR}$ and $\overleftrightarrow{QS}$? point *Q*
20. What is the intersection of plane *K* and $\overleftrightarrow{MN}$? point *M*
21. What is the intersection of plane *J* and $\overleftrightarrow{MN}$? $\overleftrightarrow{MN}$
22. What is the intersection of plane *K* and $\overleftrightarrow{QR}$? $\overleftrightarrow{QR}$

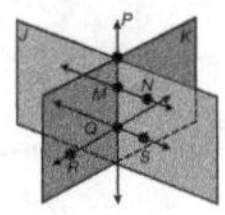

Name ______________________ Date __________ Class __________

Reteaching 2
Segments

You have learned that a line is a straight path that extends forever. Now you will work with line segments.

Find each distance.

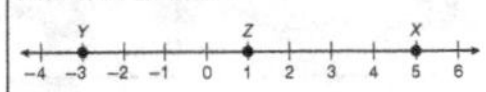

a. *XY*
Point $X = 5$ and point $Y = -3$.
So $|\text{Point } X - \text{Point } Y| = |5 - (-3)| = |8| = 8$.

b. *YZ*
Point $Y = -3$, and point $Z = 1$.
So $|\text{Point } Y - \text{Point } Z| = |(-3) - 1| = |-4| = 4$.

Practice

Complete the steps to find each distance.

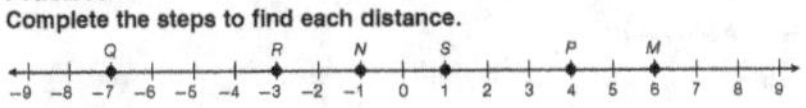

1. Find *PQ*.
$|\text{Point } P - \text{Point } Q| = |4 - \underline{(-7)}| = |\underline{11}|\ \underline{11}$

2. Find *NS*.
$|\text{Point } N - \text{Point } S| = |\underline{(-1)} - 1| = |\underline{-2}| = \underline{2}$

3. Find *MR*.
$|\text{Point } M - \text{Point } R| = |\underline{6} - (-3)| = |\underline{9}| = \underline{9}$

Find each distance.

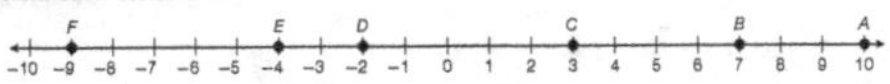

4. *AB* 3
5. *BD* 9
6. *DE* 2
7. *DC* 5
8. *FD* 7
9. *AF* 19

Reteaching 2
continued

Use the Segment Addition Postulate to find each length.

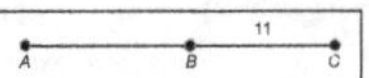

a. Find *AB* if $AC = 28$ and $BC = 11$.

$AB + BC = AC$	Segment Addition Postulate
$AB + 11 = 28$	Substitute.
$AB + 11 - 11 = 28 - 11$	Subtract 11 from both sides.
$AB = 17$	Simplify.

b. Find *PR* in terms of *x*.

$PR = PQ + QR$	Segment Addition Postulate
$PR = (3x + 4) + (4x - 1)$	Substitute.
$PR = 7x + 3$	Simplify.

Practice

Complete the steps to find each length.

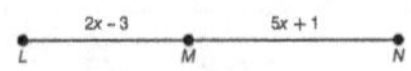

10. Find *LN* in terms of *x*.

$LN = \underline{LM} + \underline{MN}$	Segment Addition Postulate
$LN = (2x - 3) + (\underline{5x + 1})$	Substitute.
$LN = \underline{7x - 2}$	Simplify.

11. Point *G* lies on $\overline{FH}$ between *F* and *H*.
Find *GH* if $FG = 15$ and $FH = 34$.

$FH = FG + GH$	Segment Addition Postulate
$34 = \underline{15} + GH$	Substitute.
$34 - \underline{15} = 15 + GH - \underline{15}$	Subtract 15 from both sides.
$\underline{19} = GH$	Simplify.

Use the Segment Addition Postulate to find each length.

12. Point *M* lies on $\overline{LN}$ between *L* and *N*.
Find *MN* if $LN = 32$ and $LM = 14$. 18

13. Find *WY* in terms of *n*. $7n - 2$

14. Point *D* lies on $\overline{CF}$ between *C* and *F*.
Find *CD* if $DF = 22$ and $CF = 45$. 23

Name ______________________ Date __________ Class __________

Reteaching 3
Angles

You have learned about lines and line segments. Now you will learn about rays and angles.

An **acute angle** measures greater than 0° and less than 90°.
An **obtuse angle** measures greater than 90° and less than 180°.
A **right angle** measures exactly 90°.
A **straight angle** measures exactly 180°.

Classify each angle and use a protractor to find its measure.

a. ∠*MNP*

Angle *MNP* is a right angle with a measure of 90°.

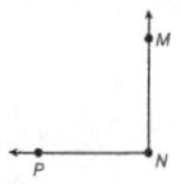

b. ∠*QRS*

Angle *QRS* is an acute angle with a measure of 30°.

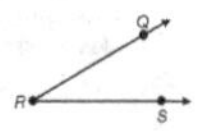

c. ∠*TUV*

Angle *TUV* is an obtuse angle with a measure of 160°.

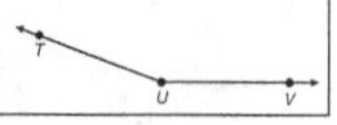

Practice

Measure each angle and complete each statement.

1. Angle *QRT* is a/an acute angle with a measure of 57°.
2. Angle *TRS* is an obtuse angle with a measure of 123°.
3. Angle *QRS* is a/an straight angle with a measure of 180°.

Give two names for each angle. Classify the angle and use a protractor to find its measure.

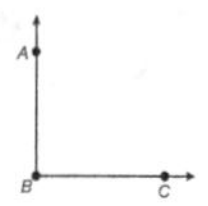

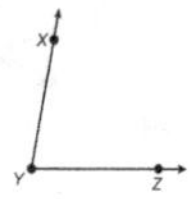

4. ∠*ABC*, ∠*B*, right, 90°
5. ∠*XYZ*, ∠*Y*, acute, 80°

Reteaching 3
continued

The measure of ∠*KLN* is 42° and the measure of ∠*NLJ* is 83°.
Find m∠*KLJ* and then classify the angle.

m∠*KLJ* = m∠*KLN* + m∠*NLJ*	Angle Addition Postulate
m∠*KLJ* = 42° + 83°	Substitute.
m∠*KLJ* = 125°	Simplify.

Angle *KLJ* is an obtuse angle.

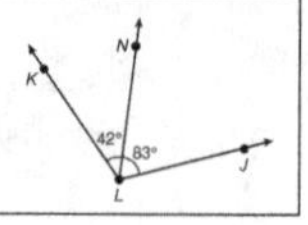

Practice

Use the diagram to complete each statement.

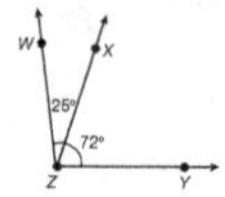

6. m∠*WZY* = m∠*WZX* + m∠*XZY* — Angle Addition Postulate
   m∠*WZY* = 25° + 72° — Substitute.
   m∠*WZY* = 97° — Simplify.
   m∠*WZY* is a/an obtuse angle.

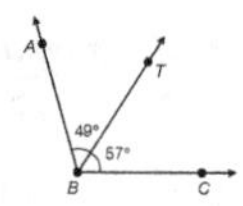

7. m∠*ABC* = m∠*ABT* + m∠*TBC* — Angle Addition Postulate
   m∠*ABC* = 49° + 57° — Substitute.
   m∠*ABC* = 106° — Simplify.
   ∠*ABC* is a/an obtuse angle.

Use the diagram to classify and find the measure of each angle.

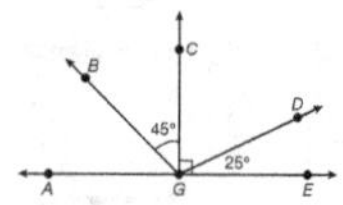

8. ∠*AGC* right, 90°
9. ∠*AGB* acute, 45°
10. ∠*CGD* acute, 65°
11. ∠*BGD* obtuse, 110°
12. ∠*AGD* obtuse, 155°

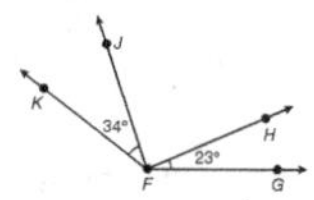

13. Determine the measure m∠*JFH* if m∠*KFG* is 125°.
    m∠*JFH* = 68°

Name ______________________ Date __________ Class __________

Reteaching 4
Postulates and Theorems About Points, Lines, and Planes

You have learned about points, lines, and planes. Now you will learn about the postulates and theorems that explain the relationships between and among points, lines, and planes.

Name the following.

a. five points
   V, W, X, Y, Z
b. two planes
   planes *G* and *H*
c. two lines
   $\overleftrightarrow{XY}$ and $\overleftrightarrow{WZ}$
d. four coplanar points
   W, *X*, *Y*, and *Z*

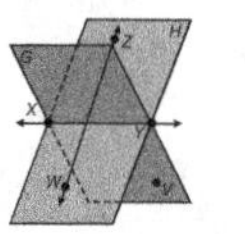

Practice

Complete the following statements.

1. Through any two points there is exactly one line.
2. Through any three noncollinear points there exists exactly one plane.
3. Give three conditions for defining a plane. Draw a figure to display each condition.

a. three noncollinear points

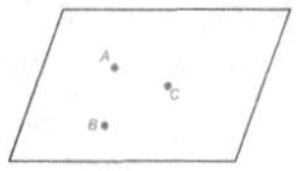

b. a line and a point not on the line

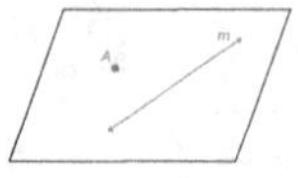

c. two intersecting line.

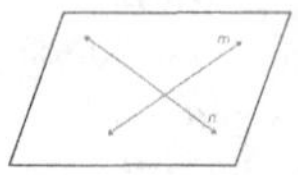

Reteaching 4
continued

a. Identify the intersection of planes *P* and *Q*.
   The intersection of planes *P* and *Q* is $\overleftrightarrow{EG}$.
b. Identify a point of intersection of plane *P* and $\overleftrightarrow{CD}$.
   Point *F*
c. Identify all points of intersection of lines on plane *Q*.
   Points *D*, *E*, and *F*

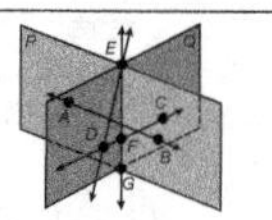

Practice

Use the figure at the right to complete problems 4–8.

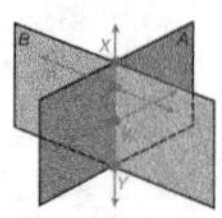

4. The intersection of plane *A* and plane *B* is $\overleftrightarrow{XY}$.
5. The intersection of line *m* and line *XY* is point *V*.
6. The intersection of line *n* and line *XY* is point *W*.
7. Plane *A* and line *m* intersect at point *V*.
8. Plane *B* and line *n* intersect at point *W*.
9. Plane *A* and line *n* intersect at line *n*.
10. Line *m* and plane *B* intersect at line *m*.

Complete the following statements.

11. If two planes intersect, then their intersection is a line.
12. Two points define a line, three points points define a plane, and four noncoplanar points define space.

Use the figure to answer problems 13–19

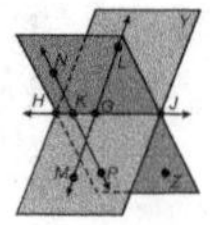

13. What is the intersection of planes *Y* and *Z*? line *HJ*
14. Identify a line on plane *Z*. $\overleftrightarrow{NP}$
15. Identify a line that intersects $\overleftrightarrow{HJ}$. What is the intersection?
    $\overleftrightarrow{NP}$, point *K* or $\overleftrightarrow{LM}$, point *G*
16. Are points P and L coplanar? no
17. Identify three points that are coplanar but not collinear.
    possible answer: points *N*, *K*, and *G*
18. What is the intersection of plane Z and $\overleftrightarrow{LM}$? point *G*
19. What is the intersection of plane Z and $\overleftrightarrow{NP}$? $\overleftrightarrow{NP}$

Name ______________________ Date __________ Class __________

Reteaching 5
More Theorems About Lines and Planes

You will learn about special relationships of lines and planes.

Both planes and lines can be **parallel** or **perpendicular**. Lines can have one additional relationship, which is called **skew**.

parallel	lines that lie in the same plane and do not intersect	$\overleftrightarrow{LK} \parallel \overleftrightarrow{MN}$
perpendicular	lines that form 90° angles	$\overleftrightarrow{LK} \perp \overleftrightarrow{JK}$
skew	lines that do not lie in the same plane and do not intersect	$\overleftrightarrow{JK}$ and $\overleftrightarrow{MN}$

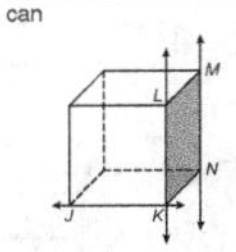

If two parallel planes are cut by a third plane, then the lines of intersection are parallel.

Planes *GHI* and *KLM* are parallel.

Name two pairs of parallel lines.

$\overleftrightarrow{GH} \parallel \overleftrightarrow{KL}$ and $\overleftrightarrow{IJ} \parallel \overleftrightarrow{MN}$

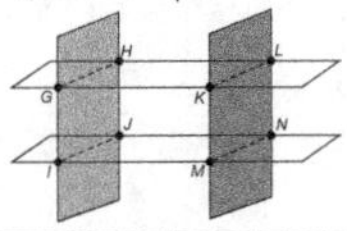

Practice

Use the diagram at the right to complete 1–7.

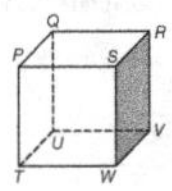

1. $\overleftrightarrow{PQ} \parallel \overleftrightarrow{TU}$ and $\overleftrightarrow{PQ} \parallel$ $\overleftrightarrow{SR}$
2. $\overleftrightarrow{PS} \parallel$ $\overleftrightarrow{QR}$ and $\overleftrightarrow{PS} \parallel$ $\overleftrightarrow{TW}$
3. $\overleftrightarrow{PT} \perp$ $\overleftrightarrow{PQ}$ and $\overleftrightarrow{PT} \perp$ $\overleftrightarrow{PS}$ on plane *PQR*
4. $\overleftrightarrow{RV} \perp$ $\overleftrightarrow{WV}$ and $\overleftrightarrow{RV} \perp$ $\overleftrightarrow{UV}$ on plane *TUV*
5. Name a line that is skew to $\overleftrightarrow{SW}$. possible answer: line *PQ*
6. Name a line that is parallel to $\overleftrightarrow{UV}$. possible answer: line *TW*
7. Name a line that is perpendicular to $\overleftrightarrow{UV}$. possible answer: line *QU*

Use the diagram at the right to answer each question.

Given: Plane *ABC* is parallel to plane *EFG*

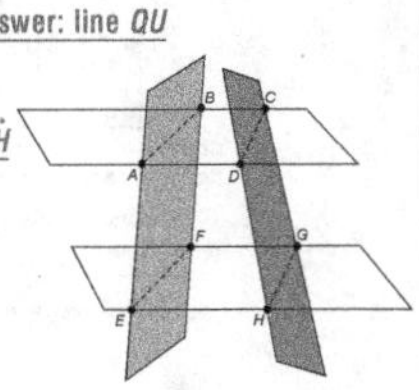

8. Name two pairs of parallel lines. $\overleftrightarrow{AB} \parallel \overleftrightarrow{EF}$, $\overleftrightarrow{CD} \parallel \overleftrightarrow{GH}$
9. Which line will intersect $\overleftrightarrow{CD}$? line $\overleftrightarrow{AB}$
10. Which line will intersect $\overleftrightarrow{EF}$? line $\overleftrightarrow{GH}$
11. Which line is skew to $\overleftrightarrow{AB}$? line $\overleftrightarrow{GH}$
12. Which line is skew to $\overleftrightarrow{EF}$? line $\overleftrightarrow{CD}$

Reteaching 5
continued

In the figure, $\angle 1 \cong \angle 2$, $\overleftrightarrow{XY} \parallel \overleftrightarrow{WZ}$, and $\overleftrightarrow{RS} \perp \overleftrightarrow{WZ}$.

If two lines form congruent adjacent angles, then they are perpendicular.

a. What is the relationship between $\overleftrightarrow{WZ}$ and $\overleftrightarrow{TU}$?

Since $\angle 1 \cong \angle 2$, $\overleftrightarrow{WZ} \perp \overleftrightarrow{TU}$.

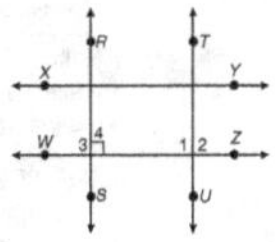

If a line is perpendicular to one of two parallel lines, then it is perpendicular to the other one.

b. What is the relationship between XY and RS?

Since $\overleftrightarrow{RS} \perp \overleftrightarrow{WZ}$ and $\overleftrightarrow{XY} \parallel \overleftrightarrow{WZ}$, $\overleftrightarrow{RS} \perp \overleftrightarrow{XY}$.

If two lines are perpendicular, then they form congruent adjacent angles.

c. What is the relationship between $\angle 3$ and $\angle 4$?

Since $\overleftrightarrow{RS} \perp \overleftrightarrow{WZ}$, $\angle 3 \cong \angle 4$.

Practice

Use the diagram at the right to answer 13–15.

Given: $\overleftrightarrow{EG} \parallel \overleftrightarrow{FH}$ and $\angle 2 \cong \angle 3$

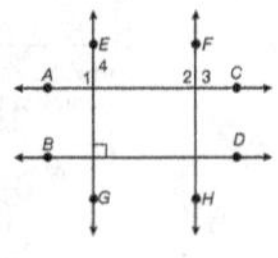

13. What is the relationship between $\overleftrightarrow{AC}$ and $\overleftrightarrow{FH}$?

Since $\angle 2 \cong \angle 3$, $\overleftrightarrow{AC} \perp \overleftrightarrow{FH}$

14. What is the relationship between $\overleftrightarrow{AC}$ and $\overleftrightarrow{EG}$?

Since $\overleftrightarrow{AC} \perp \overleftrightarrow{FH}$ and $\overleftrightarrow{EG} \parallel \overleftrightarrow{FH}$, then $\overleftrightarrow{AC} \perp \overleftrightarrow{EG}$.

15. What is the relationship between $\angle 1$ and $\angle 4$?

Since $\overleftrightarrow{EG} \perp \overleftrightarrow{AC}$, $\angle 1 \cong \angle 4$.

Use the diagram at the right to answer 16–18.

Given: $\overleftrightarrow{MN} \perp \overleftrightarrow{RV}$ and $\overleftrightarrow{WX} \parallel \overleftrightarrow{RV}$

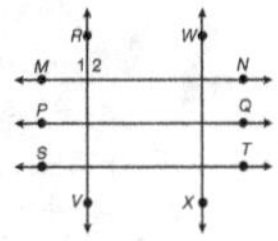

16. Name each line parallel to $\overleftrightarrow{MN}$ line $\overleftrightarrow{PQ}$ and $\overleftrightarrow{ST}$
17. What is the relationship between $\angle 1$ and $\angle 2$?

They are congruent.

18. How do you know that $\overleftrightarrow{MN}$ and $\overleftrightarrow{WX}$ are perpendicular?

If a line is perpendicular to one of two parallel lines ($\overleftrightarrow{MN} \perp \overleftrightarrow{RV}$), then it is perpendicular to the other one ($\overleftrightarrow{WX}$).

Name ______________________ Date __________ Class __________

Reteaching 6
Identifying Pairs of Angles

You have solved problems involving pairs of lines. Now you will solve problems involving pairs of angles.

Complementary and Supplementary Angles

Two angles are **complementary angles** if their combined measures total 90°.

Two angles are **supplementary angles** if their combined measures total 180°.

Find the complement of $\angle ABC$.

Step 1: $m\angle ABC = 37°$

Let x = complement of $\angle ABC$.

Step 2: $m\angle ABC$ + complement of $\angle ABC = 90°$

$37° + x = 90°$

$37° + x - 37° = 90° - 37°$

$x = 53°$

The measure of the complement of $\angle ABC = 53°$

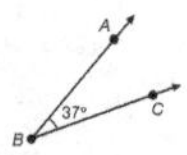

Practice

Complete the steps to find the supplement of $\angle LMN$. Let *x* represent the measure of the complement of $\angle LMN$.

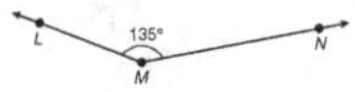

1. $135° + x = 180°$

$135° + x - 135° = 180° - 135°$

$x = 45°$

The measure of the supplement of $\angle LMN$ is 45°

Find the measure of each of the following angles.

2. complement of $\angle RST$ 25°

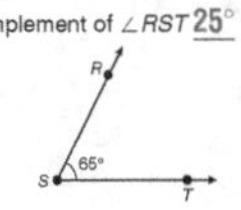

3. supplement of $\angle DEF$ 90°

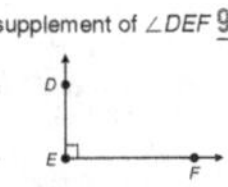

4. supplement of $\angle RST$ 115°
5. complement of $\angle DEF$ 0°

Reteaching 6
continued

Pairs of Angles

Adjacent angles have the same vertex and share a common side. In the figure, $\angle CDE$ is adjacent to $\angle EDG$.

A **linear pair** is formed by two adjacent angles whose non-common sides are opposite rays. The sum of the measures of a linear pair is 180°. $\angle FDC$ and $\angle CDG$ form a linear pair.

Vertical angles are nonadjacent angles formed by two intersecting lines. $\angle FDC$ and $\angle EDG$ are vertical angles.

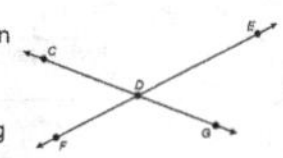

Tell whether $\angle YZW$ and $\angle WZX$ are adjacent angles, form a linear pair, or are vertical angles.

Adjacent angles: $\angle YZW$ and $\angle WZX$ have the same vertex and a common side. They are adjacent angles.

Linear pair: The two angles together do not make an angle that is 180°. The angles do not form a linear pair.

Vertical angles: The two angles are not vertical angles, because they are adjacent angles.

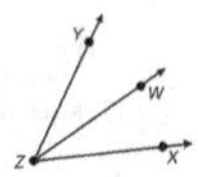

Practice

Complete the steps to show that two angles form a linear pair.

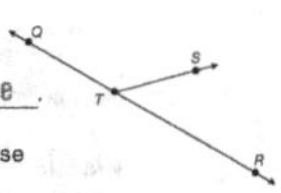

6. $\angle QTS$ and $\angle STR$ have the same vertex and a common side.

The non-common sides have an angle of measure 180° because the non-common sides form a straight line.

$\angle QTS$ and $\angle STR$ form a linear pair.

Tell whether the pair of angles are adjacent angles, form a linear pair, or are vertical angles.

7. $\angle JTY$ and $\angle YTH$ adjacent angles
8. $\angle YTZ$ and $\angle PTZ$ linear pair
9. $\angle JTP$ and $\angle YTZ$ vertical angles
10. $\angle PTZ$ and $\angle HTZ$ adjacent angles

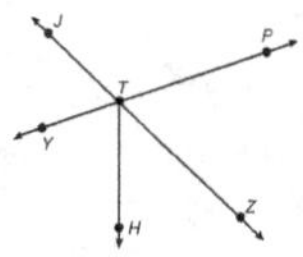

Name ______________________ Date __________ Class __________

Reteaching 7

Using Inductive Reasoning to Make Conjectures

You have solved problems involving pairs of angles. Now you will use inductive reasoning to make conjectures.

Making Conjectures

When you make a general rule or conclusion based on a pattern, you are using **inductive reasoning.** A conclusion based on a pattern is a called a **conjecture.**

Find the next two terms in the pattern.

−8, −3, 2, 7, . . .

Step 1: Study the pattern and try to find a mathematical relationship between the numbers. Test your conjecture on the given numbers.

Step 2: The correct conjecture is that each term is 5 more than the previous term.

$-8 + 5 = -3$

$-3 + 5 = 2$

$2 + 5 = 7$

Step 3: Find the next term by adding 5 to the last term: $7 + 5 = 12$.

The next term is $12 + 5 = 17$. The next two terms in the pattern are 12 and 17.

Practice

Complete the steps to find the next two items in the pattern.

1.

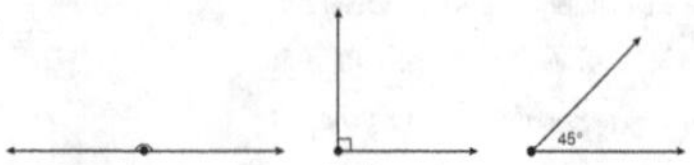

The first angle has measure __180°__.

The second angle has measure __90°__. It is __half__ of 180°.

The measure of the third angle is __45°__. It is __half__ of 90°.

The measure of the fourth angle is __half__ of 45°, or __22.5°__.

The measure of the fifth angle is __half__ of 22.5°, or __11.25°__.

Find the next two items in each pattern.

2.

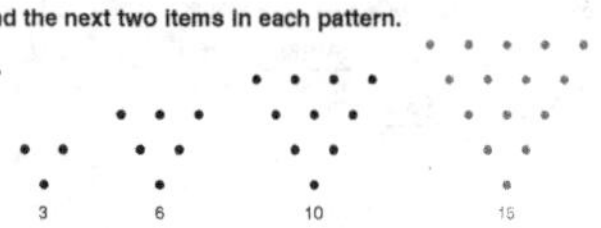

Reteaching 7 *continued*

Proving a Conjecture Is False

Since a conjecture is an educated guess, it may be true, or it may be false. It takes only one example to prove that a conjecture is false.

Show that the conjecture is false.

Conjecture: For any integer n, $n \leq 4n$.

Step 1: Make a table of sample values of n.

Step 2: Substitute each value into the inequality $n \leq 4n$ and determine whether that value makes the inequality true or false.

n	$n \leq 4n$	True or False?
3	$3 \leq 4(3)$ $3 \leq 12$	true
0	$0 \leq 4(0)$ $0 \leq 0$	true
−2	$-2 \leq 4(-2)$ $-2 \leq -8$	false

$n = -2$ makes the inequality false, so the conjecture is false.

Practice

Complete the table to show that the conjecture is false.

3. Conjecture: For any real numbers x and y, if $x > y$, then $x^2 > y^2$.

x	y	$x^2 > y^2$	True or False?
4	3	$4^2 >$ __3^2__ $16 >$ __9__	true
5	2	$5^2 >$ __2^2__ $25 >$ __4__	true
−4	−5	$(-4)^2 >$ __(-5^2)__ $16 >$ __25__	false

Show that each conjecture is false by finding one case that makes the conjecture false.

4. For any number n, $2n < n^2$.

__Possible answer: $n = 1$ or $n = 0$__

5. For any integer n, $n > -n$.

__Possible answer: any negative integer__

6. For any integer a, $a > \frac{a}{2}$.

__Possible answer: $n = -1$__

7. For any integer n, $\frac{n}{2n} = \frac{1}{2}$.

__$n = 0$__

Name ______________________ Date __________ Class __________

Reteaching 8

Using Formulas in Geometry

You have worked with conjectures. Now you will use formulas in geometry.

Perimeter and Area of Rectangles and Triangles

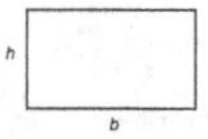

$P = 2h + 2b$

$A = bh$

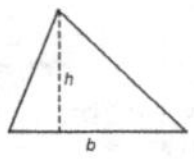

P = sum of the side lengths

$A = \frac{1}{2}bh$

Find the perimeter of the rectangle.

Step 1: Determine the base. Substitute the given information into the formula and solve for h.

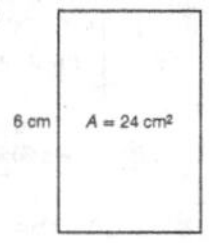

$A = bh$

$24 = (6)h$

$\frac{24}{6} = \frac{(6)h}{6}$

$4 = h$

Step 2: Substitute $b = 4$ and $h = 6$ into the perimeter formula and simplify: $P = 2b + 2h = 2(4) + 2(6) = 20$ cm.

The perimeter of the rectangle is 20 centimeters.

Practice

Complete the steps to find the area of the triangle.

1. $A = \frac{1}{2}bh$

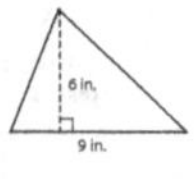

$A = \frac{1}{2}(9)$ __(6)__

$A = \frac{1}{2}$ __(54)__

$A =$ __27__ in.²

Find each measurement.

2. area of the rectangle

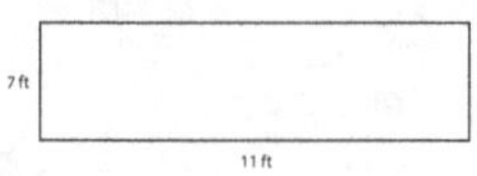

__77 ft__

3. perimeter of the triangle

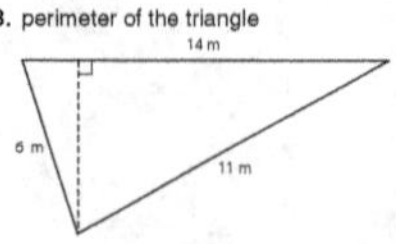

__31 m__

Reteaching 8 *continued*

The Pythagorean Theorem

In a right triangle, the sum of the square of the legs, a and b, is equal to the square of the hypotenuse, c: $a^2 + b^2 = c^2$

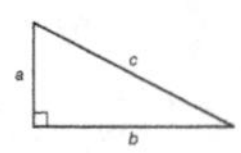

Use the Pythagorean theorem to solve for the length of b.

Step 1: Substitute $a = 12$ and $c = 20$.

$a^2 + b^2 = c^2$

$12^2 + b^2 = 20^2$

Step 2: Simplify.

$144 + b^2 = 400$

$144 + b^2 - 144 = 400 - 144$

$b^2 = 256$

$\sqrt{b^2} = \sqrt{256}$

$b = 16$

The length of leg b is 16 inches.

Practice

Complete the steps in problems 4 and 5.

4. Find the length of side b.

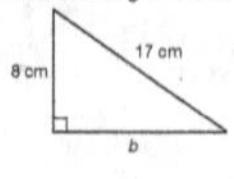

$a^2 + b^2 = c^2$

$8^2 + b^2 =$ __17^2__

__64__ $+ b^2 =$ __289__

$64 + b^2 -$ __64__ $= 289 -$ __64__

$b^2 =$ __225__

$b =$ __$\sqrt{225}$__

$b =$ __15__

5. Find the hypotenuse.

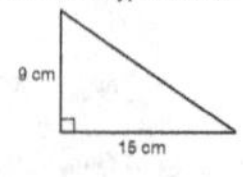

$a^2 + b^2 = c^2$

$9^2 + 15^2 =$ __c__

__81__ $+ 225 =$ __c^2__

__306__ $=$ __c^2__

__$\sqrt{306}$__ $= c^2$

__17.5__ $= c$

Name ______________________ Date __________ Class __________

Reteaching 9
Finding Length: Distance Formula

You have worked with geometric formulas. Now you will use the distance formula.

Distance Between Two Points on a Line

To find the distance between two points on a number line, take the absolute value of the difference between the points' coordinates.

$d = |a_2 - a_1|$

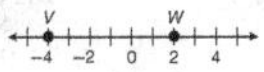

Find the distance between the points on a number line.

Step 1: Choose a point to be a_1. The other point will be a_2.

$a_1 = -4 \quad a_2 = 2$

Step 2: Substitute the values into the formula and simplify.

$d = |a_2 - a_1|$
$= |2 - (-4)|$
$= |2 + 4|$
$= |6| = 6$

The distance is 6 units.

Practice
Complete the steps to find the distance between the points on the number line.

1. $d = |a_2 - a_1|$
$= |-7 - (\underline{2})|$
$= |-7 - \underline{2}|$
$= |\underline{-9}| = \underline{9}$

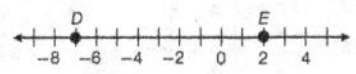

Find the distance between each pair of points.

2. 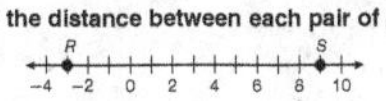

<u>12</u>

3. 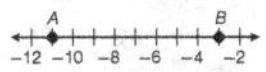

<u>9</u>

4. 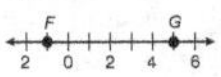

<u>6</u>

5. 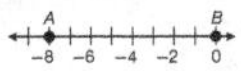

<u>8</u>

Reteaching 9
continued

Distance on a Coordinate Plane

The **Distance Formula** can be used to find the distance between two points, (x_1, y_1) and (x_2, y_2), in a coordinate plane: $d = \sqrt{(x_2 - x_1)^2 + (y_2 - y_1)^2}$.

Find the distance between the two points. Round your answer to the nearest tenth.

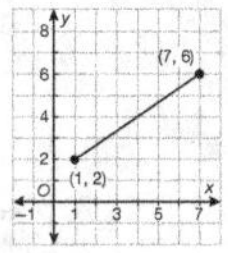

Step 1: Let $(1, 2) = (x_1, y_1)$ and $(7, 6) = (x_2, y_2)$.

Step 2: Substitute the coordinates into the distance formula and simplify. Use a calculator to find the square root.

$d = \sqrt{(x_2 - x_1)^2 + (y_2 - y_1)^2}$
$= \sqrt{(7 - 1)^2 + (6 - 2)^2}$
$= \sqrt{(6)^2 + (4)^2}$
$= \sqrt{36 + 16}$
$= \sqrt{52}$
≈ 7.2

Practice
Complete the steps to find the distance between each pair of points. Round your answer to the nearest tenth.

6. (2, 4) and (−3, 9)

$d = \sqrt{(x_2 - x_1)^2 + (y_2 - y_1)^2}$
$= \sqrt{(2 - (\underline{-3}))^2 + (4 - \underline{9})^2}$
$= \sqrt{(\underline{5})^2 + (-5)^2}$
$= \sqrt{\underline{25} + 25}$
$= \sqrt{\underline{50}}$
≈ 7.1

7. (−7, −2) and (4, 1)

$d = \sqrt{(x_2 - x_1)^2 + (y_2 - y_1)^2}$
$= \sqrt{(\underline{-7} - 4)^2 + (\underline{-2} - 1)^2}$
$= \sqrt{(\underline{-11})^2 + (\underline{-3})^2}$
$= \sqrt{\underline{121} + 9}$
$= \sqrt{\underline{130}}$
≈ 11.4

Find the distance between each pair of points. Round your answer to the nearest tenth.

8. (6, 2) and (−1, −5)

<u>9.9</u>

9. (0, 4) and (−8, 0)

<u>8.9</u>

10. (8, −3) and (5, 5)

<u>8.5</u>

11. (4, −2) (−7, 1)

<u>11.4</u>

Name ______________________ Date __________ Class __________

Reteaching 10
Using Conditional Statements

You have found the distance between two points. Now you will work with conditional statements.

Hypothesis and Conclusion

A **conditional statement** is a statement in the form "If *p*, then *q*," where *p* is the **hypothesis** and *q* is the **conclusion.**

For example:

If two lines are parallel, then they do not intersect.

The hypothesis comes after the word *if.* The conclusion comes after the word *then.*

Identify the hypothesis and conclusion of the conditional statement.

If a figure is a quadrilateral, then it has four sides.

Step 1: The hypothesis is the phrase that follows the word *if*: A figure is a quadrilateral.

Step 2: The conclusion is the phrase that follows the word *then*: It has four sides.

Practice
Complete the steps to identify the hypothesis and conclusion of the statements.

1. If *x* is an even number, then *x* is divisible by 2.

Hypothesis: <u>*x* is an even number.</u>

Conclusion: <u>*x* is divisible by 2.</u>

2. If two angles are supplementary, then they form a linear pair.

Hypothesis: <u>Two angles are supplementary.</u>

Conclusion: <u>They form a linear pair.</u>

For each conditional statement, underline the hypothesis and double-underline the conclusion.

3. If <u>two angles are not adjacent</u>, then <u>they cannot be a linear pair.</u>

4. If <u>the weather is rainy</u>, then <u>the football team will not practice after school.</u>

5. If <u>$3x - 4 = 11$</u>, then <u>$x = 5$.</u>

Reteaching 10
continued

Truth Value of a Conditional Statement

Some conditional statements are true, while others are false. This is called the **truth value** of a conditional statement. A statement is false only when the hypothesis is true and the conclusion is false.

Determine whether the conditional statement is true or false. If it is false, explain your reasoning.

If an acute angle measures 140°, then it is called a straight angle.

Step 1: Determine whether the hypothesis is true or false. The hypothesis is that an acute angle measures 140°. The hypothesis is false because the measure of an acute angle is less than 90°.

Step 2: Determine whether the conditional statement is true or false. This statement has a false hypothesis. When the hypothesis is false, the conditional statement as a whole has a truth value of "true." The statement cannot have a truth value of "false" unless a situation exists in which the hypothesis is true.

Practice
Circle the correct answers for each conditional statement.

6. If a square has a side length of 5 centimeters, then its area is 20 square centimeters.

The hypothesis is (true)/false.

The conclusion is true/(false).

The conditional situation is true/(false).

7. If two angles are right angles, then they are congruent.

The hypothesis is (true)/false.

The conclusion is (true)/false.

The conditional situation is (true)/false.

Determine whether the conditional statement is true or false. If it is false, explain your reasoning.

8. If an angle is obtuse, then it is not a right angle. <u>true</u>

9. If two right angles are complementary, then they are not congruent. <u>true</u>

Answer Key

Name ______________________ Date __________ Class __________

Reteaching INV 1

Transversals and Angle Relationships

Transversals

A transversal is a line that intersects two or more coplanar lines at different points. Line a is the transversal in the picture to the right.

When two lines are intersected by a transversal, the angle pairs are classified by type.

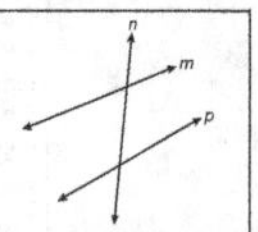

Classification	Example	Classification	Example
A pair of **corresponding angles** are two angles that lie on the same side of the transversal and on the same sides of the other two lines.	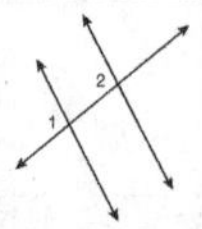	A pair of **alternate exterior angles** are two angles that lie on opposite sides of the transversal and outside the other two lines.	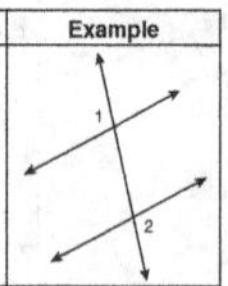
A pair of **alternate interior angles** are two nonadjacent angles that lie on opposite sides of the transversal and between the other two lines.	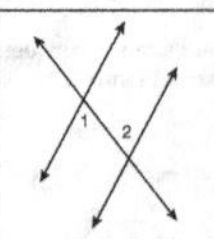	A pair of **same-side interior angles** are two angles that lie on the same side of the transversal and between the other two lines; also called **consecutive interior angles.**	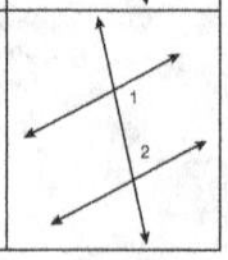

Give an example of a pair of alternate exterior angles.

One pair of alternate exterior angles is given by $\angle 2$ and $\angle 7$.

Another pair of alternate exterior angles is given by $\angle 1$ and $\angle 8$.

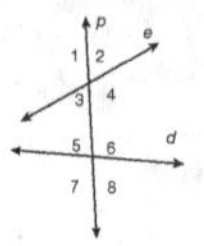

Practice

Complete each statement with the correct term.

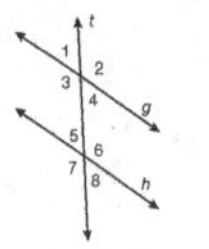

1. Line *t* is the transversal of lines *g* and *h*.
2. $\angle 3$ and $\angle 5$ are a pair of same-side interior angles.
3. $\angle 2$ and $\angle 6$ are a pair of corresponding angles.
4. $\angle 1$ and $\angle 8$ are a pair of alternate exterior angles.
5. $\angle 4$ and $\angle 5$ are a pair of alternate interior angles.

Reteaching INV 1

continued

Transversals and Parallel Lines

When a transversal intersects parallel lines, the angle pairs that are formed are either supplementary or congruent.

Corresponding Angles Postulate	Alternate Interior Angles Theorem
If two parallel lines are cut by a transversal, then the corresponding angles are congruent. 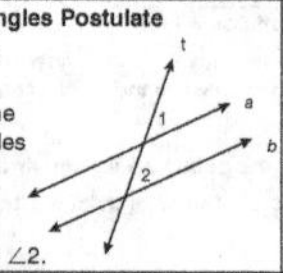 If $a \parallel b$, then $\angle 1 \cong \angle 2$.	If two parallel lines are cut by a transversal, then the alternate interior angles are congruent. 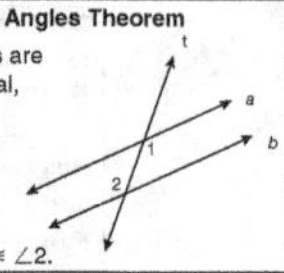 If $a \parallel b$, then $\angle 1 \cong \angle 2$.
Alternate Exterior Angles Theorem	**Same-Side Interior Angles Theorem**
If two parallel lines are cut by a transversal, then the alternate exterior angles are congruent. 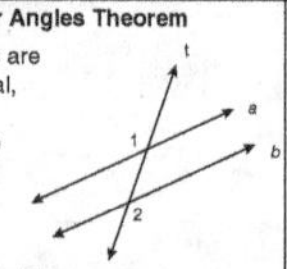 If $a \parallel b$, then $\angle 1 \cong \angle 2$.	If two parallel lines are cut by a transversal, then the same-side interior angles are supplementary. 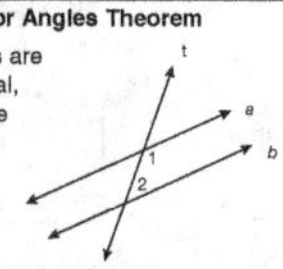 If $a \parallel b$, then $m\angle 1 + \angle 2 = 180°$.

Lines *f* and *g* are parallel lines intersected by transversal *c*.

If $m\angle 1 = 77°$, what is $m\angle 7$?

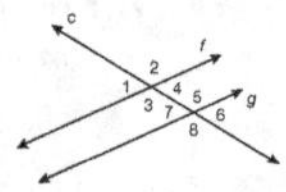

Since lines *f* and *g* are parallel and $\angle 1$ and $\angle 7$ are corresponding angles, $\angle 1$ and $\angle 7$ are congruent by the Corresponding Angles Postulate.

$m\angle 7 = 77°$

Practice

Lines *f* and *g* are parallel. Complete the steps.

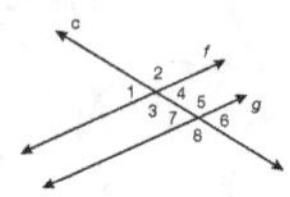

6. If $m\angle 7 = 77°$, find $m\angle 3$.

   $\angle 7$ and $\angle 3$ are same-side interior angles.

   $m\angle 7 + m\angle 3 = 180°$

   $77° + m\angle 3 = 180°$

   $m\angle 3 = 180° - 77°$

   $m\angle 3 = 103°$

Use the above picture to answer each question.

7. If you know $m\angle 8$, is it possible to know $m\angle 2$? Yes; $m\angle 2 = m\angle 8$ because the two angles are a pair of alternate exterior angles.

Name ______________________ Date __________ Class __________

Reteaching 11

Finding Midpoints

You know that a midpoint divides a segment into two congruent parts. Now you will determine the midpoint for a line segment.

The midpoint of a segment is found by taking the average of the two coordinates: $c = \frac{a+b}{2}$.

What is the midpoint of *A* and *B* on the number line?

What is the coordinate of *A*? 1

What is the coordinate of *B*? 9

Substitute the coordinates into the formula and simplify.

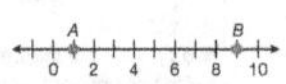

$c = \frac{a+b}{2}$

$c = \frac{1+9}{2}$

$c = \frac{10}{2}$

$c = 5$

The midpoint of *A* and *B* is 5.

Practice

Complete the steps to determine the midpoint of *C* and *D* on the number line.

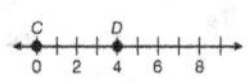

1. What is the coordinate of *C*? 0

   What is the coordinate of *D*? 4

   $c = \frac{a+b}{2}$

   $c = \frac{0+4}{2}$

   $c = \frac{4}{2}$

   $c = 2$

   The midpoint of *C* and *D* is 2.

Determine the midpoint of each segment.

2. 4
3. 3
4. 6

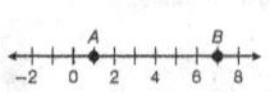

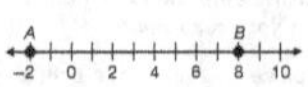

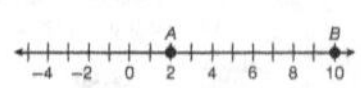

Reteaching 11

continued

The midpoint of a line segment in a coordinate plane can be found by using the Midpoint Formula.

$M = \left(\frac{x_1 + x_2}{2}, \frac{y_1 + y_2}{2}\right)$

Determine the midpoint of line segment $\overline{GH}$ connecting (1, 2) and (7, 6).

Determine the *x* and *y* coordinates for each point. Substitute the coordinates into the formula and simplify.

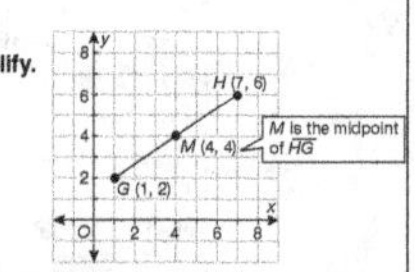

$M = \left(\frac{x_1 + x_2}{2}, \frac{y_1 + y_2}{2}\right)$

$M = \left(\frac{1+7}{2}, \frac{2+6}{2}\right)$

$M = \left(\frac{8}{2}, \frac{8}{2}\right)$

$M = (4, 4)$

The midpoint is (4, 4).

Practice

Complete the steps to determine the midpoint of the given segment.

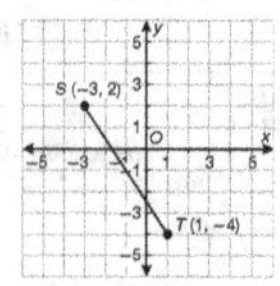

5. $M = \left(\frac{x_1 + x_2}{2}, \frac{y_1 + y_2}{2}\right)$

   $M = \left(\frac{-3+1}{2}, \frac{2+(-4)}{2}\right)$

   $M = \left(\frac{-2}{2}, \frac{-2}{2}\right)$

   $M = (-1, (-1))$

   The midpoint is $(-1, -1)$.

Determine the midpoint.

6. (1,5)
7. (3,6)
8. $\left(5, -\frac{1}{2}\right)$

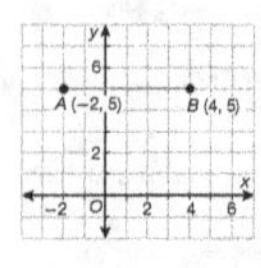

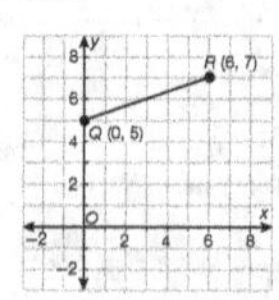

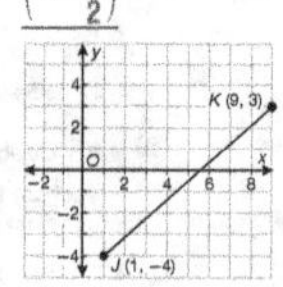

Name ______________________ Date __________ Class __________

Reteaching 12
Proving Lines Parallel

You have worked with parallel lines. Now, you will prove that lines are parallel using the converse of theorems.

Converse of the Corresponding Angles Postulate: If two lines are cut by a transversal and the corresponding angles are congruent, then the lines are parallel.

Converse of the Alternate Interior Angles Theorem: If two lines are cut by a transversal and the alternate interior angles are congruent, then the lines are parallel.

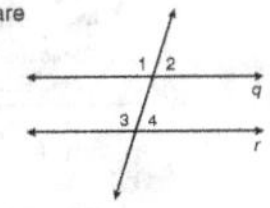

Example:

Given that ∠1 ≅ ∠3, prove that lines *q* and *r* are parallel.

Step 1: Identify the relationship between the two angles.

∠1 and ∠3 are corresponding angles.

Step 2: The lines are parallel by the Converse of the Corresponding Angles Postulate.

Practice

Complete the steps to determine whether the lines are parallel.

1. Given that ∠2 ≅ ∠3, prove that lines *a* and *b* are parallel.
   Identify the relationship between ∠2 and ∠3.
   ∠2 and ∠3 are alternate interior angles.
   The lines are parallel by the Converse of the Alternate Interior Angles Theorem.

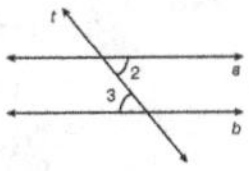

Given the information in each exercise, state the reason why lines *j* and *k* are parallel.

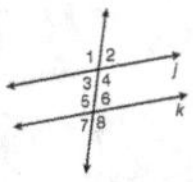

2. Given: ∠2 ≅ ∠6
   Converse of the Corresponding Angles Postulate
3. Given: ∠3 = ∠62°, ∠6 = ∠62°
   Converse of the Alternate Interior Angles Theorem
4. Given: ∠1 ≅ ∠5
   Converse of the Corresponding Angles Postulate
5. Given: ∠4 ≅ ∠5
   Converse of the Alternate Interior Angles Theorem

Reteaching 12
continued

Converse of the Alternate Exterior Angles Theorem: If two lines are cut by a transversal and the alternate exterior angles are congruent, then the lines are parallel.

Converse of the Same-Side Interior Angles Theorem: If two lines are cut by a transversal and the same-side interior angles are supplementary, then the lines are parallel.

Example:

Given that ∠1 ≅ ∠8 ⊥, prove that lines *j* and *k* are parallel.

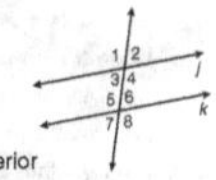

Step 1: Identify the relationship between the two angles.
∠1 and ∠8 are alternate exterior angles.

Step 2: The lines are parallel by the Converse of the Alternate Exterior Angles Theorem.

Practice

Complete the steps to determine whether the lines are parallel.

6. Given that m∠1 + m∠2 = +180° prove that lines *s* and *t* are parallel.
   Identify the relationship between ∠1 and ∠2.
   ∠1 and ∠2 are same-side interior angles whose sum is 180°.
   The lines are parallel by the Converse of the Same-Side Interior Angles Theorem.

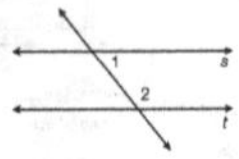

Given the information in each exercise, state the reason why lines *j* and *k* are parallel.

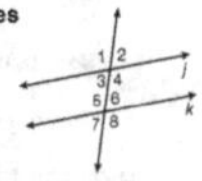

7. Given: ∠2 ≅ ∠7
   Converse of the Alternate Exterior Angles Theorem
8. Given: m∠3 = 72°, m∠5 = 108°
   Converse of the Same-Side Interior Angles Theorem
9. Given: m∠2 = 72°, m∠7 = 72°
   Converse of the Alternate Exterior Angles Theorem
10. Given: m∠4 + m∠6 = 108°
    Converse of the Same-Side Interior Angles Theorem

Name ______________________ Date __________ Class __________

Reteaching 13
Introduction to Triangles

You know that a triangle is a three-sided polygon. Now you will classify triangles by their sides and angles.

You can classify triangles by their angle measures.

Acute Triangle	Right Triangle	Obtuse Triangle
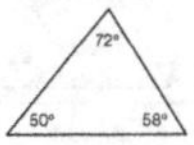	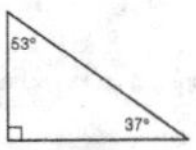	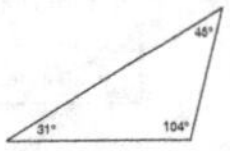
all acute angles	one right angle	one obtuse angle

Use angle measures to classify the triangle. Identify the measures of each angle.

61° = acute 70° = acute 49° = acute

All three angles are acute.

The triangle is acute.

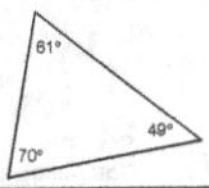

Practice

Complete the steps to classify each triangle by its angle measures.

1. 36° = acute
   54° = acute
   90° = right
   Triangle is right.

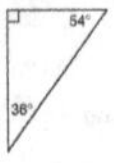

2. 47° = acute
   103° = obtuse
   30° = acute
   Triangle is obtuse.

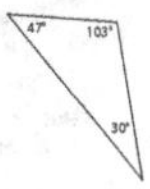

Classify each triangle by its angle measures.

3. right
4. obtuse
5. acute

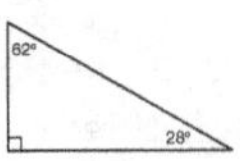

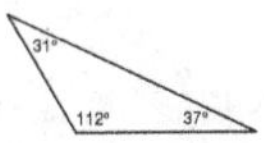

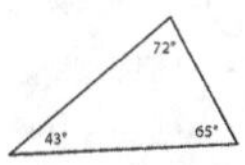

Reteaching 13
continued

You can also classify triangles by their side lengths.

Equilateral Triangle	Isosceles Triangle	Scalene Triangle
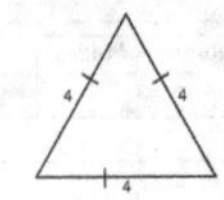	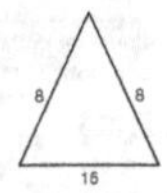	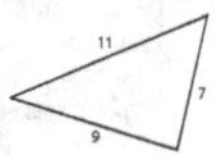
all sides congruent	at least two sides congruent	no sides congruent

Classify the triangle by its side lengths.

Three sides are the same length.

The triangle is equilateral.

The triangle is also isosceles because at least two sides are congruent.

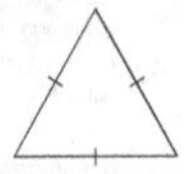

Practice

Complete the steps to classify each triangle by its side lengths.

6. one side = 3
   one side = 4
   one side = 5
   Triangle is scalene.

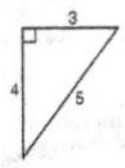

7. one side = 7
   one side = 8
   one side = 7
   Triangle is isosceles.

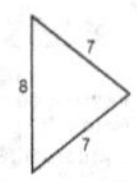

Classify each triangle by its side lengths.

8. scalene
9. equilateral; isosceles
10. isosceles

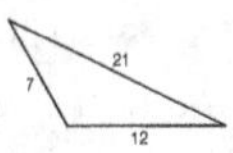

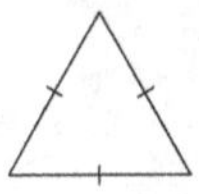

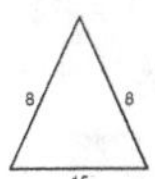

Answer Key

Name ______________________ Date __________ Class __________

Reteaching 14

Disproving Conjectures with Counterexamples

You know that a statement that is believed to be true but has not been proved is a conjecture. Now, you will disprove conjectures with counterexamples.

Geometric Conjectures

A counterexample is an example that proves a conjecture or statement is false.

Use the conjecture to answer a and b.

If $\angle A$ is an acute angle, then $\angle A = 45°$.

a. What is the hypothesis and conclusion of the conjecture?
   Hypothesis: $\angle A$ is an acute angle.
   Conclusion: $\angle A = 45°$

b. Find a counterexample to the conjecture.
   A counterexample would be an example of an angle for which the hypothesis is true but the conclusion is false.
   An acute angle has any measure between 0° and 90°.
   Counterexample: An angle of 55° is an acute angle, but it is not 45°.

Practice

Complete the steps to find a counterexample to the conjecture.

1. If two angles are congruent, then they are vertical angles.
   Hypothesis: Two angles are congruent.
   Conclusion: They are vertical angles.
   Counterexample: Two angles can be congruent in measure but not be vertical angles.

Determine the hypothesis and conclusion and find a counterexample to the conjecture.

2. If a shape is a quadrilateral, then it is a parallelogram.
   Hypothesis: A shape is a quadrilateral.
   Conclusion: It is a parallelogram.
   Counterexample: A trapezoid is a quadrilateral but not a parallelogram.

Reteaching 14

continued

Algebraic Conjectures

Find a counterexample to the conjectures.

a. Conjecture: The difference of two integers is a smaller number than either of the original numbers.
   Counterexample: The equation $-2 - (-3) = 1$ shows that the difference of two integers can be a larger number than either of the two original numbers.

b. Conjecture: If x is an integer, then $-x < 0$.
   Counterexample: If $x = -4$, then $-(-4)$ is not less than 0.

Practice

Complete the steps to find a counterexample to the algebraic conjecture.

3. If x is an even number, then $x + 2$ is divisible by 4.
   Counterexample: The expression $12 + 2$ is not divisible by 4.

4. If x and y are two different integers, then $x - y > y - x$.
   Counterexample: Possible answer: If $x = 5$ and $y = 7$, then $x = -y < y - x$.

Find a counterexample to each algebraic conjecture.

5. If $x^2 = 16$, then $x = 4$.
   Counterexample: $x = -4$

6. If $x < 0$, then $x^2 < 1$.
   Counterexample: Possible answer: $x = -2$

7. If a number is a perfect square, then its square root is even.
   Counterexample: Possible answer: 25 is a perfect square whose square root is odd; 5

8. If $(x + 7)(x - 1) = 0$, then $x = 1$.
   Counterexample: $x = -7$

9. If $x + y = 24$, then $x = 15$ and $y = 9$.
   Counterexample: Possible answer: x could be 20, and y could be 4.

Name ______________________ Date __________ Class __________

Reteaching 15

Introduction to Polygons

You have worked with congruent line segments and angles. Now you will work with polygons.

Polygons are named for the number of their sides. Some common names are given in the table.

A polygon is **equiangular** if all the angles are congruent.

A polygon is **equilateral** if all the sides are congruent.

A polygon that is both equiangular and equilateral is called a **regular** polygon.

A polygon that is not equiangular and not equilateral is called an **irregular** polygon.

Number of Sides	Polygon
3	Triangle
4	Quadrilateral
5	Pentagon
6	Hexagon
7	Heptagon
8	Octagon
9	Nonagon
10	Decagon

Name the polygon. Determine whether it is equiangular, equilateral, regular, irregular, or more than one of these.

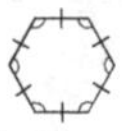

The polygon has 6 sides.

The sides and angles are all congruent.

It is a regular hexagon.

Practice

Name the polygon. Determine whether it is equiangular, equilateral, regular, irregular, or more than one of these.

1.
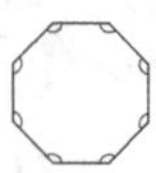

The polygon has 8 sides.

The angles are congruent, but the sides are not.

The polygon is an octagon.

It is equiangular and irregular.

2.
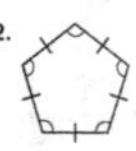

The polygon has 5 sides.

The sides and angles are all congruent.

The polygon is a regular pentagon.

Reteaching 15

continued

Interior and Exterior Angles

An **interior angle** is an angle that is inside a shape.

An **exterior angle** is any angle that is between any side of a shape and a line extended from the adjacent side.

Determine whether each angle is interior or exterior.

$\angle 1$ is formed by the side of the shape and a line extended from the adjacent side. It is an exterior angle.

$\angle 2$ is inside of the polygon. It is an interior angle.

$\angle 3$ is inside the polygon. It is an interior angle.

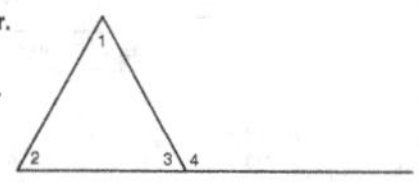

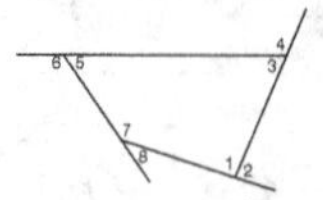

Practice

Complete each sentence.

3. $\angle 1$ is inside the polygon, so it is an interior angle.

4. $\angle 4$ is formed by the side of the shape and a line extending outside the shape, so it is an exterior angle.

Determine whether each angle is an interior angle or an exterior angle.

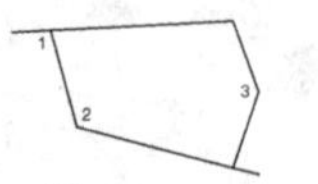

5. $\angle 1$ interior angle
6. $\angle 8$ exterior angle
7. $\angle 3$ interior angle
8. $\angle 2$ exterior angle
9. $\angle 6$ exterior angle
10. $\angle 5$ interior angle

Answer Key

Name ______________________ Date __________ Class __________

Reteaching 16
Finding Slopes and Equations of Lines

You have worked with ordered pairs. Now you will find the slope and equation of the line between two ordered pairs.

Slope

The **slope** of a line describes how steep the line is. You can find the slope by writing the ratio of the **rise** to the **run**.

slope $= \frac{\text{rise}}{\text{run}} = \frac{3}{6} = \frac{1}{2}$

You can use a formula to calculate the slope m of the line through points (x_1, y_1) and (x_2, y_2).

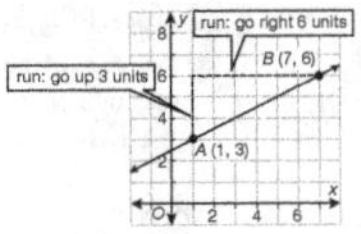

Find the slope m of $\overleftrightarrow{AB}$ using the formula.
Substitute (1, 3) for (x_1, y_1) and (7, 6) for (x_2, y_2).

$m = \frac{y_2 - y_1}{x_2 - x_1}$

$m = \frac{6 - 3}{7 - 1}$ Substitute.

$m = \frac{3}{6}$ Simplify.

$m = \frac{1}{2}$ Simplify.

Practice
Complete the steps to find the slope of each line.

1. $m = \frac{y_2 - y_1}{x_2 - x_1}$

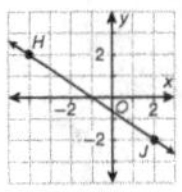

$m = \frac{-2 - 2}{2 - (-4)}$

$m = \frac{-4}{6}$

$m = -\frac{2}{3}$

2. $m = \frac{y_2 - y_1}{x_2 - x_1}$

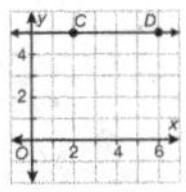

$m = \frac{5 - 5}{6 - 2}$

$m = \frac{0}{4}$

$m = 0$

Use the slope formula to determine the slope of each line.

3. (0, −2)(1, 6) $m =$ 8

4. (3, 1)(6 , −3) $m = -\frac{4}{3}$

Reteaching 16
continued

Equations of Lines

The slope-intercept form of a line is one way of writing a linear equation using the slope m and the y-intercept b of the line.

Slope-Intercept Form	Example
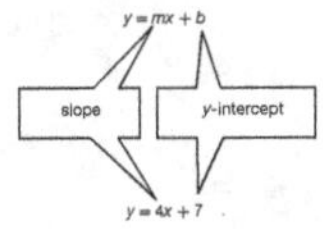	**Write the equation of the line through (0, 1) and (2, 7) in slope-intercept form.** **Step 1:** Find the slope. $m = \frac{y_2 - y_1}{x_2 - x_1}$ $\frac{7 - 1}{2 - 0} = \frac{6}{2} = 3$ **Step 2:** The y-intercept is (0, 1), so 1 is the value of b. **Step 3:** Write the equation. $y = mx + b$ $y = 3x + 1$ Substitute 3 for m and 1 for b

Practice
Complete the steps to write the equation of the line in slope-intercept form.

5. **Step 1:** Determine two points on the line to find the slope.
Use points (4, 2) and (−4, −4).

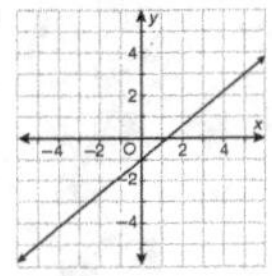

$m = \frac{y_2 - y_1}{x_2 - x_1}$ $\frac{-4 - 2}{-4 - 4} = \frac{-6}{-8} = \frac{3}{4}$

Step 2: From the graph the y-intercept is (0, −1).

Step 3: Write the equation.
$y = mx + b$

$y = \frac{3}{4}x - 1$

Write the equation of each line.

6. the line through (0, 2) and (5, 8)
$y = \frac{6}{5}x + 2$

7. the line through (0, 5) and (−2, 6)
$y = -\frac{1}{2}x + 5$

Name ______________________ Date __________ Class __________

Reteaching 17
More Conditional Statements

Now you are going to look at the converse of a statement which results from switching the hypothesis and conclusion.

Given the conditional statement below, state the converse.
If x is an even number, then x is divisible by 2.

Hypothesis	...x is an even number....
Conclusion	...x is divisible by 2....
Converse	If x is divisible by 2, then x is an even number.

Is the converse a true statement?
The converse is a true statement. We know that if a number is divisible by 2, then it is an even number.

Practice
Complete the statements for the hypothesis, conclusion, and converse.

1. If a line containing points J, K, and L lies in a plane, then J, K, and L are coplanar.
Hypothesis: A line containing points J, K, and L lies in a plane.
Conclusion: J, K, and L are coplanar.
Converse: If J, K, and L are coplanar, then A line containing points J, K, and L lies in a plane.

2. If it is Tuesday, then play practice is at 6:00.
Hypothesis: It is Tuesday.
Conclusion: Play practice is at 6:00.
Converse: If play practice is at 6:00, then it is Tuesday.

Identify the hypothesis and conclusion for each statement. Then, state the converse.

3. If you buy this cell phone, then you will receive ten free ringtones.
Hypothesis: You buy this cell phone.
Conclusion: You will receive ten free ringtones.
Converse: If you receive ten free ringtones, then you have bought this cell phone.

Reteaching 17
continued

Two other conditional statements can be formed from the hypothesis and conclusion.
Inverse: This is formed when the hypothesis and conclusion are negated.
Contrapositive: This is formed by both exchanging and negating the hypothesis and conclusion.

Statement	Example
Conditional	If a figure is a square, then it has four right angles. (Hypothesis / Conclusion)
Converse	**Switch the hypothesis and conclusion.** If a figure has four right angles, then it is a square.
Inverse	**Negate the hypothesis and conclusion.** If a figure is not a square, then it does not have four right angles.
Contrapositive	**Switch and negate the hypothesis and conclusion.** If a figure does not have four right angles, then it is not a square.

Practice
Complete the statements of the converse, inverse, and contrapositive.

4. If an animal is an armadillo, then it is nocturnal.
Converse: If an animal is nocturnal, then it is an armadillo.
Inverse: If an animal is not an armadillo, then it is not nocturnal.
Contrapositive: If an animal is not nocturnal, then it is not an armadillo.

Identify the hypothesis and conclusion of each statement. Then, state the converse, inverse, and contrapositive.

5. If an angle has a measure less than 908, then it is acute.
Converse: If an angle is acute, then it has a measure less than 90°.
Inverse: If an angle does not have a measure less than 90°, then it is not acute.
Contrapositive: If an angle is not acute, then it does not have a measure less than 90°.

6. If $y = 1$, then $y^2 = 1$.
Converse: If $y^2 = 1$, then $y = 1$.
Inverse: If $y \neq 1$, then $y^2 \neq 1$.
Contrapositive: If $y^2 \neq 1$, then $y \neq 1$.

Name ______________________ Date __________ Class __________

Reteaching 18
Triangle Theorems

You have worked with different angle measures and classified angles in triangles. Now you will work with special angle relationships in triangles.

According to the **Triangle Angle Sum Theorem,** the sum of the angle measures of a triangle is 180°.

Find the measure of ∠L.

Step 1: Write the equation. $m\angle J + m\angle K + m\angle L = 180°$

Step 2: Substitute. $62° + 73° + m\angle L = 180°$

Step 3: Solve for m∠L. $135° + m\angle L = 180°$

$m\angle L = 45°$

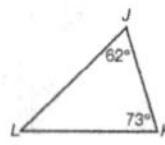

The measure of ∠L is 45°.

Practice

Complete the steps to determine the measure of the missing angle.

1. 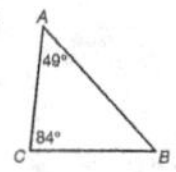

$m\angle A + m\angle B + m\angle C = $ <u>180°</u>

<u>49°</u> $+ m\angle B +$ <u>84°</u> $=$ <u>180°</u>

<u>133°</u> $+ m\angle B =$ <u>180°</u>

$m\angle B =$ <u>47°</u>

2. 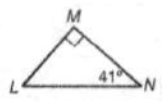

$m\angle L + m\angle M + m\angle N =$ <u>180°</u>

$m\angle L + 90° +$ <u>41°</u> $=$ <u>180°</u>

$m\angle L +$ <u>131°</u> $=$ <u>180°</u>

$m\angle L =$ <u>49°</u>

Find the measure of the missing angle.

3. 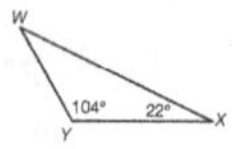

<u>54°</u>

4. 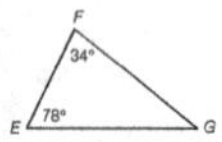

<u>68°</u>

Reteaching 18
continued

An **exterior angle** of a triangle is formed by one side of the triangle and the extension of an adjacent side.

The **Exterior Angle Theorem** states that the measure of an exterior angle of a triangle is equal to the sum of the measures of its remote interior angles.

Find the measure of ∠FHJ.

Step 1: Write the equation. $m\angle F + m\angle G = m\angle FHJ$

Step 2: Substitute. $60° + 51° = m\angle FHJ$

Step 3: Solve. $111° = m\angle FHJ$

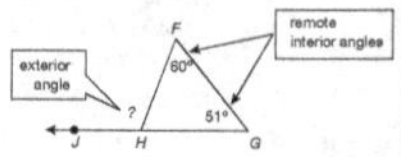

The measure of ∠FHJ is 111°.

Practice

Complete the steps to determine the measure of the angle.

5. ∠ABD

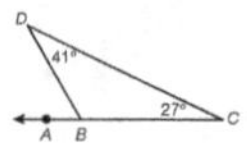

$m\angle D +$ <u>m∠C</u> $= m\angle ABD$

$41° +$ <u>27°</u> $= m\angle ABD$

<u>68°</u> $= m\angle ABD$

6. ∠HJK

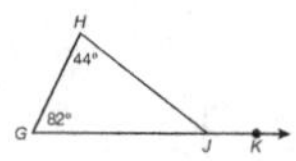

<u>m∠G</u> $+ m\angle H = m\angle HJK$

<u>82°</u> $+ 44° = m\angle HJK$

<u>126°</u> $= m\angle HJK$

Find the measure of the angle.

7. ∠MNP

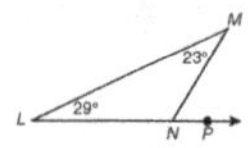

<u>52°</u>

8. ∠QRS

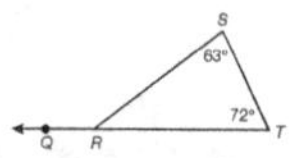

<u>135°</u>

Name ______________________ Date __________ Class __________

Reteaching 19
Introduction to Quadrilaterals

A quadrilateral is a polygon with four sides. Specific properties of figures are listed in the table below.

Figure	Properties
Parallelogram	Both pairs of opposite sides are parallel.
Kite	It has exactly two pairs of congruent consecutive sides.
Trapezoid	Exactly one pair of opposite sides is parallel.
Trapezium	No sides are parallel.
Rectangle	It is a parallelogram with four right angles.
Rhombus	It is a parallelogram with four congruent sides.
Square	It is a parallelogram with four right angles and four congruent sides.

Classify the quadrilateral. Give multiple names if possible.

Quadrilateral *EFGH*:

Sides $\overline{EF}$ and $\overline{HG}$ are parallel.

Sides $\overline{HE}$ and $\overline{GF}$ are parallel.

A figure with opposite sides parallel is a parallelogram.

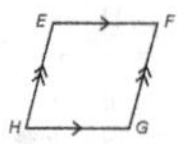

Practice

Complete the steps to classify the quadrilateral. Give multiple names if possible.

1. ∠L, ∠M, ∠N, and ∠P are <u>right</u> angles.

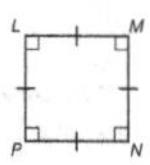

$\overline{LM}$, <u>$\overline{MN}$</u>, <u>$\overline{NP}$</u>, and <u>$\overline{PL}$</u> are congruent sides.

The figure is a <u>square</u>.

It is also a parallelogram, <u>a rhombus</u>, and a rectangle.

Classify the quadrilaterals. Give multiple names if possible.

2. 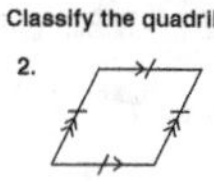

<u>rhombus; parallelogram</u>

3. 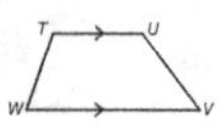

<u>trapezoid</u>

Reteaching 19
continued

Determine the perimeter, area, length and width of this rectangle.

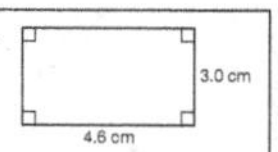

The length is 4.6 centimeters, and the width is 3.0 centimeters.

The perimeter is the sum of the side lengths.

$P = 3.0 + 4.6 + 3.0 + 4.6$

$P = 2(3.0) + 2(4.6)$

$P = 15.2$

The perimeter of the rectangle is 15.2 centimeters.

The area is the side length times the side width.

$A = lw$

$A = (4.6)(3.0)$

$A = 13.8$

The area of the rectangle is 13.8 cm².

Practice

Complete the steps to determine the perimeter and area of the figure.

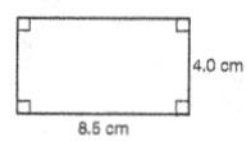

4. Perimeter

$P =$ <u>4.0</u> $+$ <u>8.5</u> $+ 4.0 + 8.5$

$P = 2$ <u>4.0</u> $+ 2$ <u>(8.5)</u>

$P =$ <u>25</u> cm

Area

$A = lw$

$A =$ <u>(8.5)</u> (4)

$A =$ <u>34</u> cm²

Find the perimeter and area of each figure.

5. 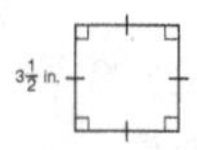

<u>$P = 14$ in.; $A = 12.25$ in^2</u>

6. 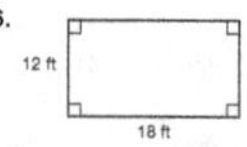

<u>$P = 60$ ft; $A = 216$ ft^2</u>

7. 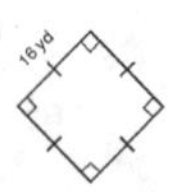

<u>$P = 64$ yd; $A = 256$ yd^2</u>

Answer Key

Name ______________________ Date __________ Class __________

Reteaching 20
Interpreting Truth Tables

You have worked with conditional statements. Now you will work with biconditional statements and truth tables.

A **biconditional statement** combines a conditional statement (if *p*, then *q*) with its converse (if *q*, then *p*).

Conditional: *p* *q*
If the sides of a triangle are congruent, then the angles are congruent.

Converse: *q* *p*
If the angles of a triangle are congruent, then the sides are congruent.

Biconditional: *p* *q*
The sides of a triangle are congruent if and only if the angles are congruent.

Practice
Complete the statements for the converse and biconditional.

1. If you can download six songs for $5.94, then each song costs $0.99.

   Converse: If each song costs $0.99, then you can download six songs for $5.94.

   Biconditional: You can download six songs for $5.94 if and only if each song costs $0.99.

2. If Lindsay works on the yearbook, then she does not play soccer.

   Converse: If Lindsay does not play soccer, then she works on the yearbook.

   Biconditional: Lindsay works on the yearbook, if and only if she does not play soccer.

For each conditional, write the converse and a biconditional statement.

3. If a figure has ten sides, then it is a decagon.

   Converse: If a figure is a decagon, then it has ten sides.

   Biconditional: A figure has ten sides if and only if it is a decagon.

4. An angle is obtuse if it measures between 90 and 180 degrees.

   Converse: If an angle measures between 90 and 180 degrees, then the angle is obtuse.

   Biconditional: An angle is obtuse if and only if it measures between 90 and 180 degrees.

Reteaching 20
continued

A compound statement combines two statements using *and* or *or*.
A compound statement that uses *and* is called a **conjunction**. A compound statement that uses *or* is called a **disjunction**.

The table below shows when a conjunction or disjunction is true or false.

p	*q*	Conjunction: *p* and *q*	Disjunction: *p* or *q*
T	T	T	T
T	F	F	T
F	T	F	T
F	F	F	F

Example: Write a conjunction using the two statements and determine whether the conjunction is true or false. All squares are rectangles.
A foot is 12 inches.

Conjunction: All squares are rectangles, and a foot is 12 inches.

The conjunction is **true** since both statements are true.

Example: Write a disjunction using the two statements and determine whether the disjunction is true or false. Pine trees are evergreens. Giraffes are blue.

Disjunction: Pine trees are evergreens, or giraffes are blue.

The disjunction is **true** since one statement is true.

Practice
Complete the statements for the conjunction and disjunction and determine whether the statement is true or false.

5. A triangle has three sides. An octagon has three sides.

   Conjunction: A triangle has three sides, and an octagon has three sides.

   Disjunction: A triangle has three sides, or an octagon has three sides.

   The conjunction is false.

   The disjunction is true since one of the statements is true.

Write the conjunction and disjunction and determine whether the statement is true or false.

6. A parallelogram has opposite parallel sides. A square has four congruent sides.

   Conjunction: A parallelogram has opposite parallel sides, and a square has four congruent sides.

   Disjunction: A parallelogram has opposite parallel sides, or a square has four congruent sides.

   The conjunction is true. The disjunction is true.

Name ______________________ Date __________ Class __________

Reteaching INV 2
Proving the Pythagorean Theorem

Right Triangles

A **right triangle** is a triangle with one right angle. The side of the right triangle that is opposite the **right angle** is called the hypotenuse. The other two sides of the triangle are called the **legs**.

The **Pythagorean theorem** states that, for a right triangle, the sum of the squares of the lengths of the legs is equal to the square of the length of the hypotenuse.

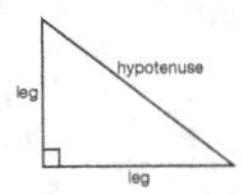

Explain how the right triangle satisfies the Pythagorean theorem.

The square of the length of one leg is 3^2, or 9. The square of the length of the other leg is 4^2, or 16. The square of the length of the hypotenuse is 5^2, or 25. The sum of the squares of the lengths of the legs is $9 + 16 = 25$, which is equal to the square of the hypotenuse's length.

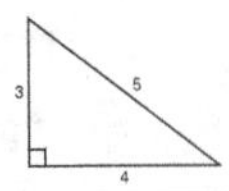

Practice

1. Draw a right triangle with side lengths of 9 centimeters, 12 centimeters, and 15 centimeters.

   Use a protractor to ensure that one angle is 90°.

   Draw and cut out three squares: one with side length of 9 centimeters, one with side length of 12 centimeters, and one with side length of 15 centimeters.

   Place each square next to the corresponding side of the triangle.

2. The length of the hypotenuse is 15 cm.
3. The lengths of the legs are 9 cm and 12 cm.
4. The area of the square on the hypotenuse is 225 cm^2.
5. The area of the squares on the legs are 81 cm^2 and 144 cm^2.
6. Calculate the area of each of the squares that you constructed.

   Show that the sum of the areas of the two smaller squares equals the area of the largest square.

   81 cm^2 + 144 cm^2 = 225 cm^2

7. How does the length of each side of the triangle relate to the area of the corresponding square?

   The length of the side of the triangle is the square root of the area of the corresponding square.

Reteaching INV 2
continued

The Pythagorean Theorem

The Pythagorean theorem states that $a^2 + b^2 = c^2$, where *a* and *b* represent the lengths of the legs and *c* represents the length of the hypotenuse.

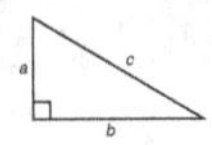

Determine *c* for this triangle.

Step 1: Determine the lengths of the legs of the right triangle.

The lengths of the legs are 5 and 12.

$a^2 + b^2 = c^2$

$5^2 + 12^2 = c^2$ Substitute values.

$25 + 144 = c^2$ Simplify.

Step 2: $169 = c^2$ Simplify.

$\sqrt{169} = \sqrt{c^2}$ Take square roots of both sides.

$13 = c$

The value of *c* is 13.

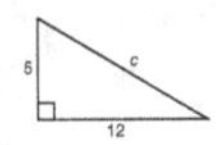

Practice
Complete the steps to find the length of the indicated side.

8. A right triangle has a hypotenuse that measures 26 inches and one leg that measures 10 inches. What is the length of the other leg?

   Represent the length of the missing side by the letter *a*.

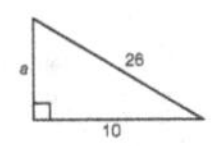

$a^2 + b^2 = c^2$

$a^2 + 10^2 = 26^2$

$a^2 + 100 = 676$

$a^2 + 100 - 100 = 676 - 100$

$a^2 = 576$

$a = 24$

The length of the leg is 24 inches.

Find the length of the indicated side.

9. A right triangle has a hypotenuse that measures 17 centimeters and one leg that measures 8 centimeters. What is the length the other leg? 15 cm
10. A right triangle has one leg that measures 20 meters and one leg that measures 15 meters. What is the length of the hypotenuse? 25 m
11. A right triangle has a hypotenuse that measures 10 inches and one leg that measures 6 inches. What is the length of the other leg? 8 in.
12. A right triangle has one leg that measures 9 yards and a hypotenuse that measures 15 yards. What is the length of the other leg? 12 yd

Answer Key

Name ______________________ Date __________ Class __________

Reteaching 21
Laws of Detachment and Syllogism

With deductive, reasoning you will use facts, definitions, and properties to draw conclusions and prove whether those conclusions are valid.

The **Law of Detachment** allows you to draw conclusions from true facts.

Given	If you have $2, then you can buy a snack. You have $2.	If you have $2, then you can buy a snack. You can buy a snack.
Therefore	You can buy a snack.	You have $2.
Valid Conclusion?	Yes, the conditional and the hypothesis are both true.	No, the hypothesis may or may not be true. For example, if you borrowed $2 from someone, you could also buy a snack.

Practice

Complete a valid concluding statement using the Law of Detachment.

1. If you ride the Titan roller coaster in Arlington, Texas, then you will drop 255 feet.

   Michael rode the Titan roller coaster.

   Therefore, Michael dropped 255 feet.

2. A segment that is a diameter of a circle has endpoints on the circle.

   $\overline{GH}$ is the diameter of a circle.

   Therefore, $\overline{GH}$ has endpoints on the circle.

For the following statements, use the Law of Detachment to write a valid concluding statement.

3. Opposite rays are two rays that have a common endpoint and form a line.

   Therefore, $\overrightarrow{YX}$ and $\overrightarrow{YZ}$ are opposite rays.

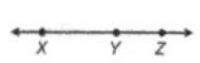

4. If a person sees penguins, then he or she is in Antarctica.

   Catherine sees penguins.

   Therefore, Catherine is in Antarctica.

5. If the sum of the digits of a number can be divided by 3, then the number can be divided by 3.

   The number 171 has a sum of 9.

   Therefore, 171 can be divided by 3.

Reteaching 21
continued

Another valid form of deductive reasoning is the **Law of Syllogism**; it is similar to the Transitive Property of Equality.

Transitive Property of Equality	**Law of Syllogism**
If $y = 10x$ and $10x = 20$, then $y = 20$.	**Given:** If you have a horse, then you have to feed it. If you have to feed a horse, then you have to get up early every morning. **Conclusion:** If you have a horse, then you have to get up early every morning.

Practice

Determine whether each conclusion is valid by the Law of Syllogism.

6. **Given:** If you buy a car, then you can drive to school. If you can drive to school, then you will not ride the bus.

   Conclusion: If you buy a car, then you will not ride the bus. valid

7. **Given:** If $\angle K$ is obtuse, then it does not have a measure of 90°. If an angle does not have measure of 90°, then it is not a right angle.

   Conclusion: If $\angle K$ is obtuse, then it is not a right angle. valid

8. **Given:** If two segments are congruent, then they have the same measure. If two segments have the same measure of 6.5 centimeters, then they are congruent.

   Conclusion: If two segments are congruent, then they each have a measure of 6.5 centimeters. invalid

Draw a conclusion from the given information.

9. If two angles are alternate interior angles, then they are congruent. If two angles are congruent, then they have the same measure. $\angle 1$ and $\angle 2$ are alternate interior angles.

   $\angle 1$ and $\angle 2$ have the same measure.

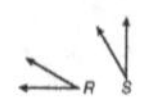

10. If $\angle R$ and $\angle S$ are complementary to the same angle, then the two angles are congruent. If two angles are congruent, then they have the same angle measure. $\angle R$ and $\angle S$ are complementary to the same angle.

    $\angle R$ and $\angle S$ have the same angle measure.

Name ______________________ Date __________ Class __________

Reteaching 22
Finding Areas of Quadrilaterals

You have found the area of a triangle. Now you will find the area of different quadrilaterals.

Area of a Trapezoid

A trapezoid with bases b^1 and b^2 and height h has area $A = \frac{1}{2}(b_1 + b_2)h$.

Find the area of the trapezoid.

Step 1: Write the formula. $A = \frac{1}{2}(b_1 + b_2)h$

Step 2: Substitute. $A = \frac{1}{2}(11\text{ m} + 15\text{ m})(7\text{ m})$

Step 3: Simplify. $A = \frac{1}{2}(26\text{ m})(7\text{ m})$

$A = 91\text{ m}^2$

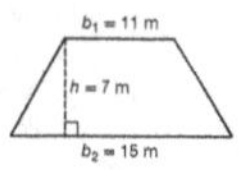

The area of the trapezoid is 91 m².

Practice

Complete the steps to find the area of each trapezoid.

1.

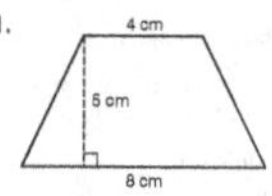

$A = \frac{1}{2}(b_1 + b_2)h$

$A = \frac{1}{2}(4\text{ cm} + \underline{8}\text{ cm})(\underline{5}\text{ cm})$

$A = \frac{1}{2}(\underline{12}\text{ cm})(\underline{5}\text{ cm})$

$A = \underline{30}\text{ cm}^2$

2.

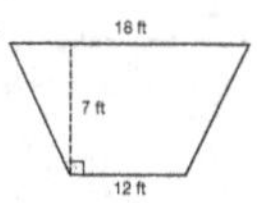

$A = \frac{1}{2}(b_1 + b_2)h$

$A = \frac{1}{2}(18\text{ ft} + \underline{12}\text{ ft})(\underline{7}\text{ ft})$

$A = \frac{1}{2}(\underline{30}\text{ ft})(\underline{7}\text{ ft})$

$A = \underline{105}\text{ ft}^2$

Find the area of each trapezoid.

3. 64 in^2

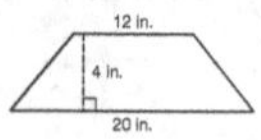

4. 96 ft^2

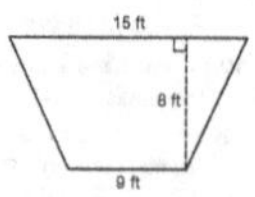

Reteaching 22
continued

Area of a Rhombus

A rhombus with diagonals d^1 and d^2 has area $A = \frac{1}{2}d_1d_2$.

Find the area of the rhombus.

Step 1: Write the formula. $A = \frac{1}{2}d_1d_2$

Step 2: Substitute. $A = \frac{1}{2}(14\text{ cm})(10\text{ cm})$

Step 3: Simplify. $A = 70\text{ cm}^2$

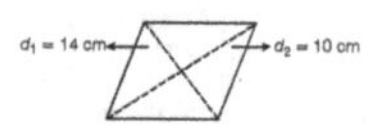

The area of the rhombus is 70 cm².

Practice

Complete the steps to find the area of each rhombus.

5.

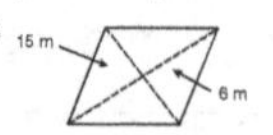

$A = \frac{1}{2}d_1d_2$

$A = \frac{1}{2}(15\text{ m})(\underline{6\text{ m}})$

$A = \underline{45}\text{ m}^2$

6.

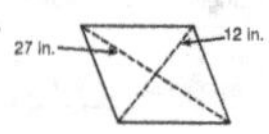

$A = \frac{1}{2}d_1d_2$

$A = \frac{1}{2}(\underline{27}\text{ in.})(\underline{12}\text{ in.})$

$A = \underline{162}\text{ in}^2$

Find the area of each rhombus.

7. 21 yd^2

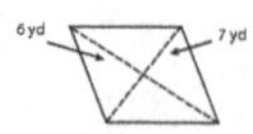

8. 12 m^2

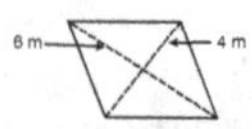

9. Find the area of a rhombus in which $d_1 = 14$ in. and $d_2 = 48$ in.

   336 in^2

10. Find the d_2 of a rhombus in which $d_1 = 15$ m and $A = 180\text{ m}^2$.

    $d_2 = 24$ m

Answer Key

Name ______________________ Date ________ Class ________

Reteaching 23
Introduction to Circles

Now you will find the circumference and area of circles

Circumference of Circles

A circle with radius r has circumference $C = 2\pi r$.

Find the circumference of the circle to the nearest hundredth inch. Use 3.14 for π.

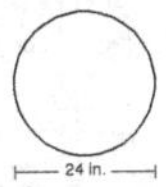

Step 1: Divide the diameter by 2 to find the radius of the circle.

24 in. ÷ 2 = 12 in.

The radius r is 12 in.

Step 2: Substitute r = 12 in. and π = 3.14 into the formula.

$C = 2\pi r$

$C \approx 2(3.14)(12 \text{ in.})$

Step 3: Multiply and round to the nearest hundredth inch.

$C \approx 75.36$ in.

The circumference is approximately 75.36 inches.

Practice

Complete the steps to find the circumference. Use 3.14 for π.

1.

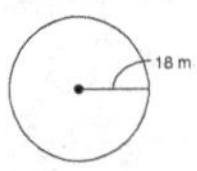

r = 18 m

$C = 2\pi r$

$C \approx 2($ 3.14 $)($ 18 m $)$

$C \approx$ 113.04 m

2.

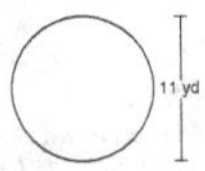

r = 5.5 yd

$C = 2\pi r$

$C \approx 2($ 3.14 $)($5.5 yd$)$

$C \approx$ 34.54 yd

Find the circumference of each circle. Use 3.14 for π.

3.

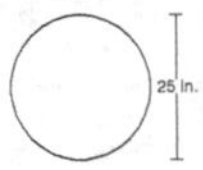

78.50 in.

4.

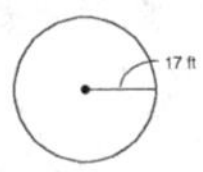

106.76 ft

Reteaching 23
continued

Area of Circles

A circle with radius r has area $A = \pi r^2$.

Find the area of the circle to the nearest hundredth square yard. Use 3.14 for π.

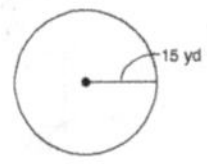

Step 1: Find the radius of the circle. The radius r is 15 yd.

Step 2: Substitute r = 15 yd and π = 3.14 into the formula for the area of a circle.

$A = \pi r^2$

$A \approx 3.14(15 \text{ yd})^2$

Step 3: Multiply and round to the nearest hundredth yard.

$A \approx 706.50 \text{ yd}^2$

The area is approximately 706.50 yd^2.

Practice

Complete the steps to find the area of the given circle to the nearest hundredth square unit. Then, complete the sentence. Use 3.14 for π.

5.

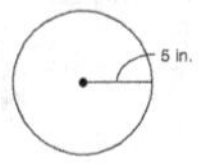

The radius of the circle is r = 5 in.

$A = \pi r^2$

$A \approx$ 3.14 (5 in. $)^2$

$A \approx$ 78.50 in^2

6.

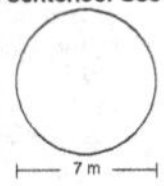

The radius of the circle is r = 3.5 m.

$A = \pi r^2$

$A \approx$ 3.14 (3.5 m$)^2$

$A \approx$ 38.47 m^2

Find the area of the given circle to the nearest hundredth. Use 3.14 for π.

7.

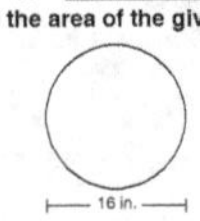

200.96 in^2

8.

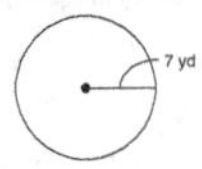

153.86 yd^2

Name ______________________ Date ________ Class ________

Reteaching 24
Algebraic Proofs

You have learned about deductive reasoning. Now you will work with algebraic proofs.

Algebraic Proofs

An **algebraic proof** uses algebraic properties—including the Distributive Property and the properties of equality—to show how a problem is solved step-by-step. The original equation is the **Given** statement.

Solve the equation $5(x - 2) = x + 6$ Provide a justification for each step.

$5(x-2) = x+6$	**Given**
$5x - 10 = x + 6$	Distributive Property
$5x - 10 + 10 = x + 6 + 10$	Addition Property of Equality
$5x = x + 16$	Simplify
$5x - x = x + 16 - x$	Subtraction Property of Equality
$4x = 16$	Simplify
$\frac{4x}{4} = \frac{16}{4}$	Division Property of Equality
$x = 4$	Simplify

Practice

Complete the proof.

	$2x + 6 = x - 1$	Given
	$2x + 6 - 6 = x - 1 - 6$	Subtraction Property of Equality
1.	$2x = x - 7$	Simplify
	$2x - x = x - 7 - x$	Subtraction Property of Equality
	$x = -7$	Simplify
2.	$\frac{x+2}{4} = 7$	Given
	$4\left(\frac{x+2}{4}\right) = 4(7)$	Multiplication Property of Equality
	$x + 2 = 28$	Simplify
	$x + 2 - 2 = 28 - 2$	Subtraction Property of Equality
	$x = 26$	Simplify

Reteaching 24
continued

The steps of the algebraic proof for solving the equation $\frac{8x - 6}{3} = 3$ are given below in the correct order. However, the justifications for each step are out of order.

Rewrite the incorrect justifications.

	Incorrect Order	Correct Order
$\frac{8x-6}{3} = 6$	Addition Property of Equality	Given
$3\left(\frac{8x-6}{3}\right) = 3(6)$	Division Property of Equality	Multiplication Property of Equality
$8x - 6 = 18$	Simplify	
$8x - 6 + 6 = 18 + 6$	Given	Addition Property of Equality
$8x = 24$	Simplify	
$\frac{8x}{8} = \frac{24}{8}$	Simplify	Division Property of Equality
$x = 3$	Simplify	

Practice

The steps for the proof below are in the correct order, but some of the justifications are not. Rewrite the incorrect justifications on the lines.

3.	$4x + 8 = 2(3x - 1)$	Given	
	$4x + 8 = 6x - 2$	Addition Property of Equality	Distributive Property
	$4x + 8 - 8 = 6x - 2 - 8$	Division Property of Equality	Subtraction Property of Equality
	$4x = 6x - 10$	Simplify	
	$4x - 6x = 6x - 10 - 6x$	Distributive Property	Subtraction Property of Equality
	$-2x = -10$	Subtraction Property of Equality	Simplify
	$\frac{-2x}{-2} = \frac{-10}{-2}$	Simplify	Division Property of Equality
	$x = 5$	Simplify	

Name ______________________ Date __________ Class __________

Reteaching 25
Congruent Triangles

The **corresponding sides** and the **corresponding angles** of a figure are the sides and angles that are in the same position.

Identify the corresponding angles and sides for $\triangle ABC$ and $\triangle MNP$.

The names of the triangles show the following:

Corresponding Angles

$\angle A$ corresponds to $\angle M$.
$\angle B$ corresponds to $\angle N$.
$\angle C$ corresponds to $\angle P$.

Corresponding Sides

$\overline{AB}$ corresponds to $\overline{MN}$.
$\overline{BC}$ corresponds to $\overline{NP}$.
$\overline{AC}$ corresponds to $\overline{MP}$.

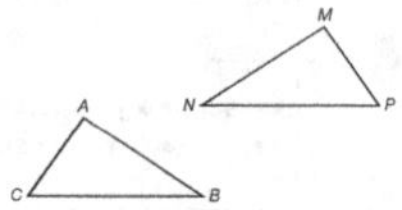

Practice

Complete each statement to identify the corresponding angles and sides for $\triangle DEF$ and $\triangle RST$.

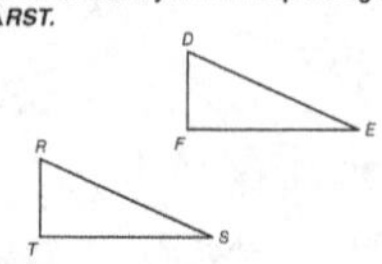

1. $\angle D$ corresponds to $\underline{\angle R}$.
   $\angle E$ corresponds to $\underline{\angle S}$.
   $\underline{\angle F}$ corresponds to $\angle T$.

2. $\overline{DE}$ corresponds to $\underline{\overline{RS}}$.
   $\overline{DF}$ corresponds to $\underline{\overline{RT}}$.
   $\underline{\overline{EF}}$ corresponds to $\overline{ST}$.

Identify the corresponding angles and sides for $\triangle XYZ$ and $\triangle KLM$.

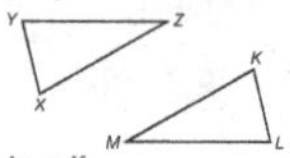

3. $\angle X$ corresponds to $\angle K$.
   $\angle Y$ corresponds to $\angle L$.
   $\angle Z$ corresponds to $\angle M$.

4. $\overline{XY}$ corresponds to $\overline{KL}$.
   $\overline{XZ}$ corresponds to $\overline{KM}$.
   $\overline{YZ}$ corresponds to $\overline{LM}$.

Reteaching 25
continued

Triangles are **congruent triangles** when all of their corresponding sides and corresponding angles are congruent.

$\triangle ABC$ and $\triangle JKL$ are congruent triangles.

Write six congruence statements for the corresponding sides and angles.

In these congruent triangles, A corresponds to J, B corresponds to K, and C corresponds to L.

Corresponding Parts	
Congruent Angles	Congruent Sides
$\angle A \cong \angle J$	$\overline{AB} \cong \overline{JK}$
$\angle B \cong \angle K$	$\overline{BC} \cong \overline{KL}$
$\angle C \cong \angle L$	$\overline{CA} \cong \overline{LJ}$

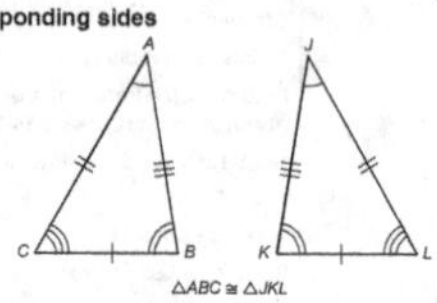

$\triangle ABC \cong \triangle JKL$

Practice

Complete the six congruence statements for the corresponding sides and angles.

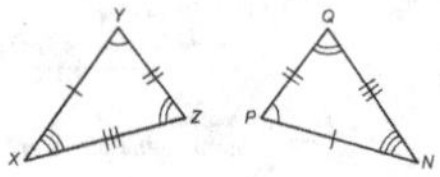

5. Given: $\triangle XYZ \cong \triangle NPQ$.

$\angle X \cong \underline{\angle N}$ $\overline{XY} \cong \underline{\overline{NP}}$
$\underline{\angle Y} \cong \angle P$ $\overline{YZ} \cong \underline{\overline{PQ}}$
$\angle Z \cong \underline{\angle Q}$ $\underline{\overline{XZ}} \cong \overline{NQ}$

Identify the congruent sides and angles for the two triangles. Write six congruence statements.

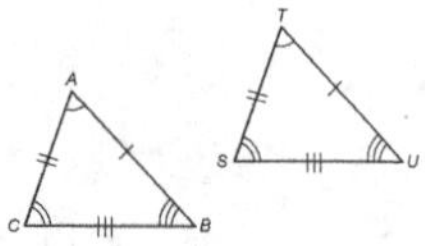

6. $\angle A \cong \underline{\angle T}$ $\overline{AB} \cong \underline{\overline{TU}}$
$\underline{\angle B} \cong \underline{\angle U}$ $\underline{\overline{BC}} \cong \underline{\overline{US}}$
$\underline{\angle C} \cong \underline{\angle S}$ $\underline{\overline{AC}} \cong \underline{\overline{TS}}$

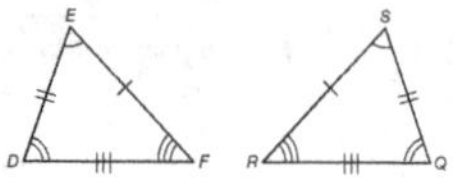

7. $\underline{\angle D} \cong \underline{\angle Q}$ $\underline{\overline{DE}} \cong \underline{\overline{QR}}$
$\underline{\angle E} \cong \underline{\angle R}$ $\underline{\overline{EF}} \cong \underline{\overline{RS}}$
$\underline{\angle F} \cong \underline{\angle S}$ $\underline{\overline{DF}} \cong \underline{\overline{QS}}$

Name ______________________ Date __________ Class __________

Reteaching 26
Central Angles and Arc Measure

You have solved problems involving congruent triangles. Now you will solve problems involving measure.

Congruent Arcs

Two arcs that are in the same circle or in congruent circles and that have the same measure are called congruent arcs.

The measure of $\overset{\frown}{AB}$ is given by the expression $4x - 12$. The measure of $\overset{\frown}{ST}$ is given by the expression $5x + 30$. It is given that $\overset{\frown}{AB} \cong \overset{\frown}{ST}$.

Determine the value of x and the measure of each arc.

Step 1: Since the arcs are congruent, set the expressions for their measures equal.

$4x - 12 = 5x + 30$

Step 2: Solve for x.

$$4x - 12 = 5x - 30$$
$$4x - 12 + 12 = 5x - 30 + 12$$
$$4x = 5x - 18$$
$$4x - 5x = 5x - 18 - 5x$$
$$-x = -18$$
$$x = 18$$

Step 3: The value of x is 18. Therefore, $m\overset{\frown}{AB} = 4(18) - 12 = 72 - 12 = 60°$.

Since the arcs are congruent, $m\overset{\frown}{ST} = 60°$.

Practice

Complete the steps to determine the value of x and the measure of each arc. Each pair of arcs is congruent.

1. $m\overset{\frown}{FG}$ is given by $5x - 14$.
   $m\overset{\frown}{JK}$ is given by $3x + 20$.

$$5x - 14 = \underline{3x + 20}$$
$$5x - \underline{14} + \underline{14} = 3x + 20 + \underline{14}$$
$$5x = 3x + \underline{34}$$
$$5x - \underline{3x} = 3x + 34 - \underline{3x}$$
$$\underline{2x} = 34$$
$$x = \underline{17}$$
$$m\overset{\frown}{FG} = 5(\underline{17}) - 14 = \underline{71°}$$
$$m\overset{\frown}{JK} = \underline{71°}$$

2. $m\overset{\frown}{AB}$ is given by $3x + 25$.
   $m\overset{\frown}{RS}$ is given by $6x - 11$.

$$\underline{3x + 25} = 6x - 11$$
$$3x + \underline{25} - \underline{25} = 6x - 11 - \underline{25}$$
$$3x = 6x - \underline{36}$$
$$3x - \underline{6x} = 6x - 36 - \underline{6x}$$
$$\underline{-6x} = -36$$
$$x = \underline{12}$$
$$m\overset{\frown}{AB} = 3(\underline{12}) + 25 = \underline{61°}$$
$$m\overset{\frown}{RS} = \underline{61°}$$

Reteaching 26
continued

The Arc Addition Postulate

The measure of an arc formed by two adjacent arcs is the sum of the measures of the two arcs.

Write an expression that represents the measure of the arc from A to C. Find the measure of the arc.

Step 1: Since measure of the arc A to C is greater than 180°, write the arc as $\overset{\frown}{ABC}$. The adjacent arcs that make up $\overset{\frown}{ABC}$ have measures less than 180°. Thus, write the arcs as $\overset{\frown}{AB}$ and $\overset{\frown}{BC}$. Therefore, $m\overset{\frown}{ABC} = m\overset{\frown}{AB} + m\overset{\frown}{BC}$.

Step 2: $m\overset{\frown}{ABC} = m\overset{\frown}{AB} + m\overset{\frown}{BC}$
$= 100° + 112°$
$= 212°$

Therefore, $m\overset{\frown}{ABC} = 212°$.

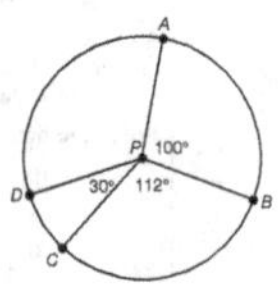

Practice

Use the Arc Addition Postulate to complete the steps.

3.

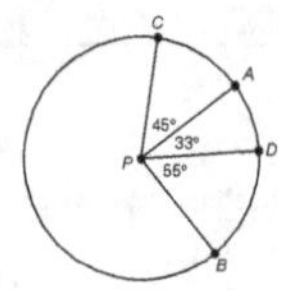

$m\overset{\frown}{CD} = \underline{m\overset{\frown}{CA}} + m\overset{\frown}{AD}$
$= 45° + \underline{33°}$
$= \underline{78°}$

4.

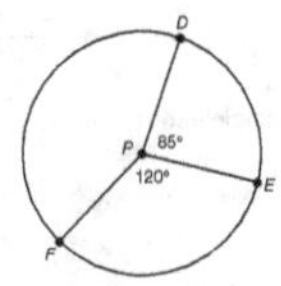

$m\overset{\frown}{DEF} = \underline{m\overset{\frown}{DE}} + m\overset{\frown}{EF}$
$= 85° + \underline{120°}$
$= \underline{205°}$

Use the Arc Addition Postulate to find the measure of the given arc.

5. $m\overset{\frown}{ABC} = \underline{195°}$

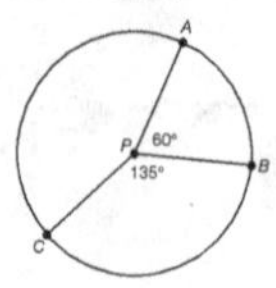

6. $m\overset{\frown}{BD} = \underline{130°}$

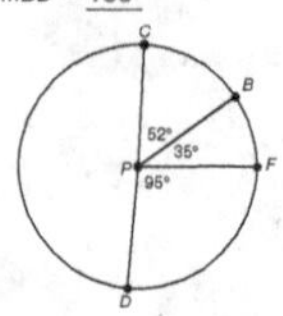

Answer Key

Name ____________________ Date __________ Class __________

Reteaching 27
Two-Column Proofs

You have worked with algebraic proofs. Now you will work with geometric two-column proofs.

Fill in the justifying statements to support the following.

Given: $\overrightarrow{BD}$ is the angle bisector of $\angle ABC$, and $\angle ABC \cong \angle 1$.

Prove: $\angle DBC \cong \angle 1$

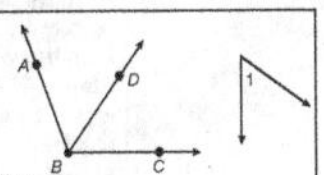

1. $\overrightarrow{BD}$ is the angle bisector of $\angle ABC$. — 1. Given
2. $\angle ABD \cong \angle DBC$ — 2. Definition of angle bisector
3. $\angle ABD \cong \angle 1$ — 3. Given
4. $\angle DBC \cong \angle 1$ — 4. Transitive Property of $\cong$

Practice

Write the letter of the correct justification next to each step. Use one justification twice.

Given: $\overrightarrow{JL}$ is the bisector of $\angle KJM$, and $\angle 1 \cong \angle 3$.

Prove: $\angle 2 \cong \angle 3$.

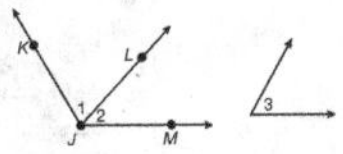

1. $\overrightarrow{JL}$ is the bisector of $\angle KJM$. B
2. $\angle 2 \cong \angle 1$. A
3. $\angle 1 \cong \angle 3$. B
4. $\angle 2 \cong \angle 3$. C

A. Definition of $\angle$bisector

B. Given

C. Transitive Property of $\cong$

Write a justification for each step to complete the proof.

Given: N is the midpoint of $\overline{MP}$, Q is the midpoint of $\overline{RP}$, and $\overline{PQ} \cong \overline{NM}$.

Prove: $\overline{PN} \cong \overline{QR}$

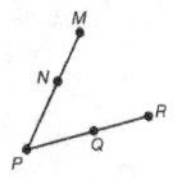

1. N is the midpoint of $\overline{MP}$. — 1. Given
2. Q is the midpoint of $\overline{RP}$. — 2. Given
3. $\overline{PN} \cong \overline{NM}$ — 3. Definition of Midpoint
4. $\overline{PQ} \cong \overline{NM}$ — 4. Given
5. $\overline{PN} \cong \overline{PQ}$ — 5. Transitive Property of Equality
6. $\overline{PQ} \cong \overline{QR}$ — 6. Definition of Midpoint
7. $\overline{PN} \cong \overline{QR}$ — 7. Transitive Property of Equality

Reteaching 27
continued

Two-column proofs have a format composed of five parts: given statements(s), prove statement, diagram, statements, and reasons.

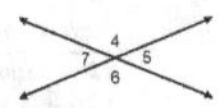

Prove the Congruent Supplements Theorem:

If two angles are supplementary to the same angle (or to two congruent angles), then the two angles are congruent.

Statements	Reasons
1. $\angle 4$ and $\angle 5$ are supplementary.	1. Given
2. $\angle 5$ and $\angle 6$ are supplementary.	2. Given
3. $m\angle 5 + m\angle 6 = 180°$	3. Definition of supplementary angles
4. $m\angle 5 + m\angle 6 = 180°$	4. Definition of supplementary angles
5. $m\angle 4 + m\angle 5 = m\angle 5 + m\angle 6$	5. Substitution Property of Equality
6. $m\angle 4 = m\angle 6$	6. Subtraction Property of Equality
7. $\angle 4 \cong \angle 6$	7. Definition of congruent angles

Practice

Fill in the blanks to complete the two-column proof of the Right Angle Congruence Theorem.

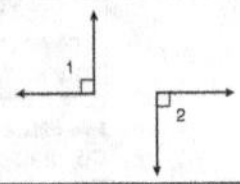

Given: $\angle 1$ and $\angle 2$ are right angles.

Prove: $\angle 1 \cong \angle 2$

Statements	Reasons
1. $\angle 1$ and $\angle 2$ are right angles.	1. Given
2. $m\angle 1 = 90°$	2. Definition of right angle
3. $m\angle 2 = 90°$	3. Definition of right angle
4. $m\angle 1 = m\angle 2$	4. Substitution Property of Equality
5. $\angle 1 \cong \angle 2$	5. Definition of congruent angles

Name ____________________ Date __________ Class __________

Reteaching 28
Triangle Congruence: SAS

You have worked with congruent figures. Now you will determine whether two triangles are congruent.

Determine whether the pair of triangles is congruent by the SAS Postulate.

If two sides and the included angle of one triangle are congruent to two sides and the included angle of another triangle, then the triangles are congruent by side-angle-side congruence.

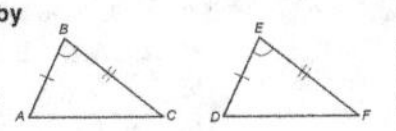

Step 1: Identify the congruent sides: $\overline{AB} \cong \overline{DE}$; $\overline{BC} \cong \overline{EF}$.

Step 2: Identify the congruent angle: $\angle B \cong \angle E$.

Step 3: Is the angle between the congruent sides in both figures? yes

Step 4: The two triangles are congruent by SAS.

Practice

Complete the steps to determine whether the pair of triangles is congruent by the SAS Postulate.

1.

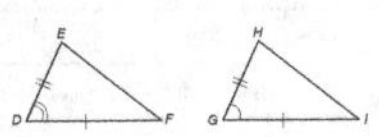

$\overline{DF} \cong \overline{GI}$; $\overline{DE} \cong \overline{GH}$

$\angle D \cong \angle G$

The angle is between the two congruent sides.

Are the two triangles congruent? yes

2.

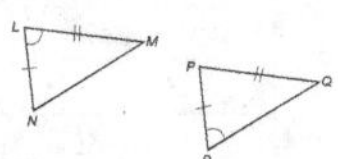

$\overline{NL} \cong \overline{RP}$; $\overline{LM} \cong \overline{PQ}$

$\angle L \cong \angle R$

The angle is not between the two congruent sides.

Are the two triangles congruent? not

Determine whether the pair of triangles is congruent by the SAS Postulate.

3. yes

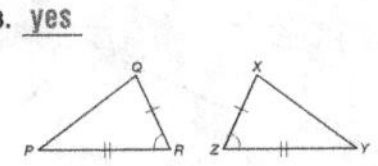

4. no

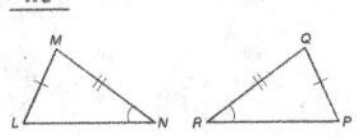

Reteaching 28
continued

Find the value of x that makes the triangles congruent.

For the two triangles to be congruent, the measures of the included angles must be equal.

Set the equations equal to each other and solve.

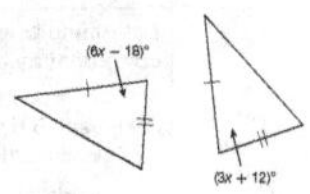

$6x - 18 = 3x + 12$

$6x - 18 + 18 = 3x + 12 + 18$

$6x = 3x + 30$

$6x - 3x = 3x + 30 - 3x$

$3x = 30$

$\frac{3x}{3} = \frac{30}{3}$

$x = 10$

Practice

Complete the steps to find the value of x that makes the triangles congruent.

5.

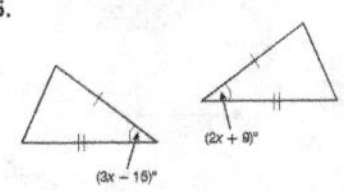

$3x - 15 = 2x + 9$

$3x - 15 + 15 = 2x + 9 + 15$

$3x = 2x + 24$

$3x - 2x = 2x + 24 - 2x$

$x = 24$

6.

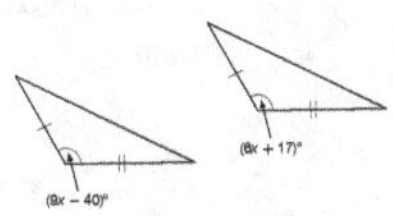

$9x - 40 = 6x + 17$

$9x - 40 + 40 = 6x + 17 + 40$

$9x = 6x + 57$

$9x - 6x = 6x + 57 - 6x$

$3x = 57$

$\frac{3x}{3} = \frac{57}{3}$

$x = 19$

Find the value of x that makes the triangles congruent.

7. $x = 7$

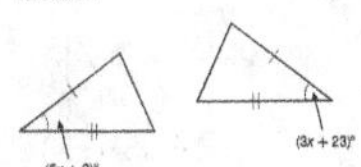

8. $x = 11$

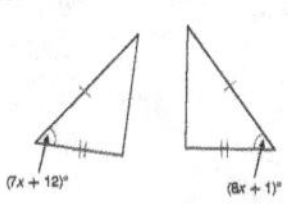

Answer Key

Name ______________________ Date __________ Class __________

Reteaching 29
Using the Pythagorean Theorem

You have solved problems involving triangle congruence with SAS.
Now you will solve problems using the Pythagorean theorem.

Pythagorean Triples

A set of three nonzero whole numbers a, b, and c such that $a^2 + b^2 = c^2$ is a **Pythagorean Triple.**

Find the unknown length in the triangle.

Do the side lengths form a Pythagorean Triple?

Step 1: Since the length of the hypotenuse is the unknown, $c = x$.

The lengths of the legs, which are represented by a and b, are 7 feet and 24 feet. It does not matter which value is a and which is b.

Step 2: Substitute $a = 7$, $b = 24$, and $c = x$ into the equation. Solve for x.

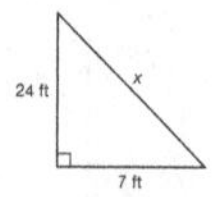

$a^2 + b^2 = c^2$
$7^2 + 24^2 = x^2$
$49 + 576 = x^2$
$625 = x^2$
$25 = x$

The length of the hypotenuse is 25 feet.

The set (7, 24, 25) is a **Pythagorean Triple** because $7^2 + 24^2 = 25^2$.

Practice

Complete the steps to find the unknown length in the triangle. Determine whether the side lengths form a Pythagorean Triple.

1. $a^2 + b^2 = c^2$
   $8^2 + x^2 = 17^2$
   $64 + x^2 = 289$
   $64 + x^2 - 64 = 289 - 64$
   $x^2 = 225$
   $x = 15$

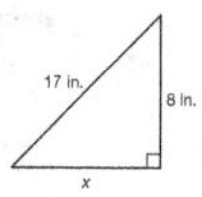

The length of the unknown leg is 15 in.

Do the side lengths form a Pythagorean Triple? yes

Find the unknown length in the triangle. Do the side lengths form a Pythagorean Triple?

2. 5 cm; yes

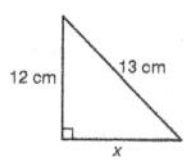

3. 20 in.; yes

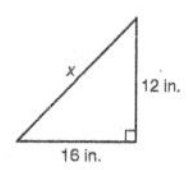

Reteaching 29
continued

Simplifying Radicals

One or more side lengths of a right triangle may be written as a **radical expression.** A radical expression is any expression that contains a root. Radical expressions are usually reduced by simplifying the root.

Find the value of x. Give your answer in simplified radical form.

Step 1: The length of the hypotenuse is 12 yards. One of the legs is 8 yards long. Since the length of other leg is unknown, it is represented by x.

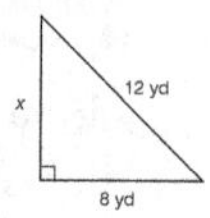

Step 2: Substitute x, 8, and 12 into the equation $a^2 + b^2 = c^2$ and solve for x.

$a^2 + b^2 = c^2$
$x^2 + 8^2 = 12^2$
$x^2 + 64 = 144$
$x^2 + 64 - 64 = 144 - 64$
$x^2 = 80$
$x = \sqrt{80}$

Step 3: Simplify the radical. Factor out all perfect square factors of the number under the radical sign. The largest perfect square that is a factor of 80 is 16, so write 80 as 16×5.

$x = \sqrt{16 \times 5} = 4\sqrt{5}$

Therefore, the length of the unknown leg is $4\sqrt{5}$ yards.

Practice

Complete the steps to find the value of x and write the answer in simplified radical form.

4. $a^2 + b^2 = c^2$
   $5^2 + 10^2 = x^2$
   $25 + 100 = x^2$
   $125 = x^2$
   $\sqrt{25 \times 5} = x$
   $5\sqrt{5} = x$

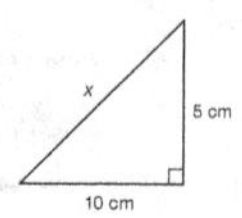

The length of the unknown side is $5\sqrt{5}$ centimeters.

Find the value of x. Give your answer in simplified radical form.

5. $x = 2\sqrt{26}$ ft

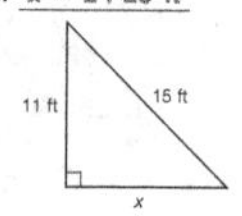

6. $x = 5\sqrt{10}$ m

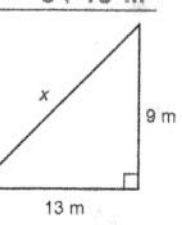

Name ______________________ Date __________ Class __________

Reteaching 30
Triangle Congruence Theorems: ASA and AAS

You have worked with triangle congruence using Side-Angle-Side.
Now you will determine triangle congruence by two other methods.

Determine whether triangles $\triangle KLM$ and $\triangle NPQ$ are congruent by the Angle-Side-Angle (ASA) Postulate.

If two angles and the included side of one triangle are congruent to two angles and the included side of another triangle, then the triangles are congruent.

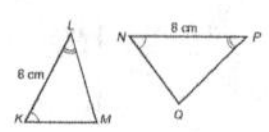

Step 1: Identify the congruent side: $\overline{KL} \cong \overline{NP}$.

Step 2: Identify the congruent angles: $\angle K \cong \angle N$, and $\angle L \cong \angle P$.

Step 3: Is the congruent side between the congruent angles in both figures? yes

Step 4: The two triangles are congruent by ASA.

Practice

Fill in the blanks to determine whether each pair of triangles is congruent by ASA.

1.

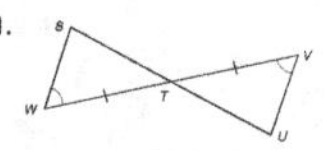

$\overline{TW} \cong \overline{TV}$, $\angle W \cong \angle V$

$\angle STW \cong \angle UTV$ by Vertical Angles Theorem.

The congruent side is between the two congruent angles.

Are the two triangles congruent by ASA? yes

2.

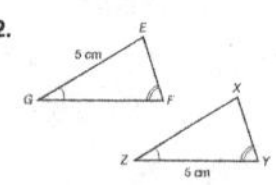

$\angle G \cong \angle Z$, $\angle F \cong \angle Y$

The congruent side is not between the two congruent angles.

Are the two triangles congruent by ASA? no

Determine whether each pair of triangles is congruent by ASA.

3. yes

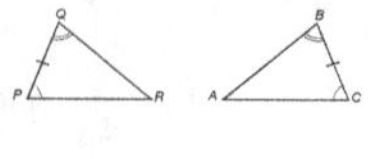

4. no

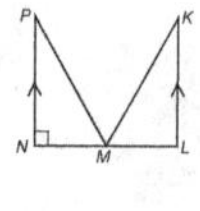

Reteaching 30
continued

Angle-Angle-Side (AAS) Congruence Postulate: If two angles and the non-included side of one triangle are congruent to two angles and the corresponding non-included side of another triangle, then the triangles are congruent.

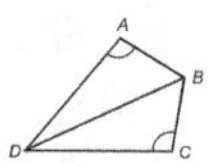

Given: $\overline{BD}$ is the angle bisector of $\angle ADC$.

Prove: $\triangle ABD \cong \triangle CBD$

1. $\angle A \cong \angle C$	1. Given
2. $\angle ADB \cong \angle CDB$	2. Definition of angle bisector
3. $\overline{BD} \cong \overline{BD}$	3. Reflexive Property of $\cong$
4. $\triangle ABD \cong \triangle CBD$	4. ASA Congruence Postulate

Practice

Complete the steps to prove the triangles are congruent by AAS.

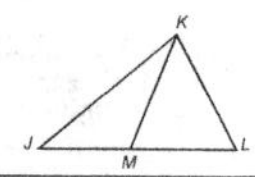

5. Given: $\overline{KM}$ bisects $\angle JKL$, $m\angle J = x$, $m\angle L = x$.

Prove: $\triangle ABD \cong \triangle CBD$

1. $\overline{KM}$ bisects $\angle JKL$.	1. Given
2. $\angle JKM \cong \angle LKM$	2. Definition of angle bisector
3. $m\angle J = x$, $m\angle L = x$	3. Given
4. $m\angle J = m\angle L$	4. Transitive Property of Equality
5. $\angle J \cong \angle L$	5. Definition of congruent angles
6. $\overline{KM} \cong \overline{KM}$	6. Reflexive Property of congruence
7. $\triangle ABD \cong \triangle CBD$	7. AAS Congruence Postulate

Determine whether you can use the AAS Congruence Postulate to prove the triangles are congruent. If not, tell what else you need to know.

7. yes

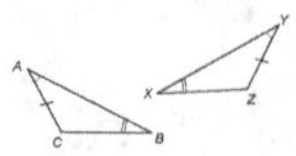

8. no; need to know $\angle L \cong \angle F$ or $\angle K \cong \angle D$

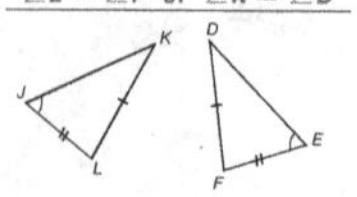

Answer Key

Name ________________________ Date __________ Class __________

Reteaching INV 3

Exploring Angles of Polygons

You have learned to identify interior and exterior angles in polygons.
Now you will determine angle measures in regular polygons.

Interior Angles

Sum of the Interior Angles of a Polygon	Interior Angle Measure of a Regular Polygon
$(n - 2)180°$, where n is the number of sides of the polygon.	$\frac{(n-2)180°}{n}$, where n is the number of sides of the polygon.
Example: **Determine the sum of the interior angles of an octagon.** **Step 1:** How many sides are in an octagon? 8 **Step 2:** Substitute and solve the formula. $(n-2)180° = (8-2)180° = 6 \cdot 180° = 1080°$	**Example:** **Determine the interior angle measure of a regular octagon.** **Step 1:** How many sides are in an octagon? 8 **Step 2:** Substitute and solve the formula. $\frac{(n-2)180°}{n} = \frac{(8-2)180°}{8} = \frac{6 \cdot 180°}{8} = 135°$

Practice

Complete the statements to solve each problem.

1. What is the sum of the interior angles of a regular 15-gon?
   How many sides are in a 15-gon? 15
   Substitute and simplify the formula.
   $(n-2)180° = (15 - 2)180°$
   $= 13 \cdot 180°$
   $= 2340°$

2. What is the interior angle measure of a regular 12-gon?
   How many sides are in a 12-gon? 12
   Substitute and simplify the formula.
   $\frac{(n-2)180°}{n} = \frac{(12-2)180°}{12}$
   $= \frac{10 \cdot 180°}{12}$
   $= 150°$

3. The sum of the measures of the interior angles of a regular 14-gon is 2160°.
4. The measure of an interior angle of a regular 30-gon is 168°.
5. The sum of the measures of the interior angles of a regular 18-gon is 2880°.

Reteaching INV 3

continued

Exterior and Central Angles

Exterior Angle Measure of a Regular Polygon	Central Angle Measure of a Regular Polygon
The measure of each exterior angle $= \frac{360°}{n}$, where n is the number of sides in the polygon.	The measure of each central angle $= \frac{360°}{n}$, where n is the number of sides in the polygon.
Example: **Determine the measure of each exterior angle of a regular 12-gon.** **Step 1:** How many sides are in a 12-gon? 12 **Step 2:** Substitute and solve the formula. $\frac{360°}{n} = \frac{360°}{12} = 30°$	**Example:** **Determine the central angle of a regular 24-gon.** **Step 1:** How many sides are in a 24-gon? 24 **Step 2:** Substitute and solve the formula. $\frac{360°}{n} = \frac{360°}{24} = 15°$

Practice

Complete the statements to solve each problem.

6. **Determine the measure of each exterior angle of a regular 15-gon. How many sides are in a 15-gon?** 15
   $\frac{360°}{n} = \frac{360°}{15}$
   $= 24°$

7. Determine the central angle of a regular 30-gon.
   How many sides are in a 30-gon? 30
   $\frac{360°}{n} = \frac{360°}{30}$
   $= 12°$

8. Determine the measure of each exterior angle for a regular 20-gon. 18°
9. Determine the central angle of a regular 36-gon. 10°
10. Determine the measure of each exterior angle of a regular 40-gon. 9°
11. Determine the central angle of a regular 72-gon. 5°

Name ________________________ Date __________ Class __________

Reteaching 31

Flowchart and Paragraph Proofs

You have worked with triangle congruence (ASA and AAS). Now you will write flowchart and paragraph proofs.

Flowchart Proofs

You can use the flowchart proof to prove conjectures are true.

Flowchart Proofs use boxes and arrows.

Steps go from top to bottom, as shown by arrows. The justification or reason for each step is written below the box.

Given: ∠1 and ∠2 are right angles.

Prove: ∠1 ≅ ∠2

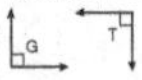

Flowchart Proof

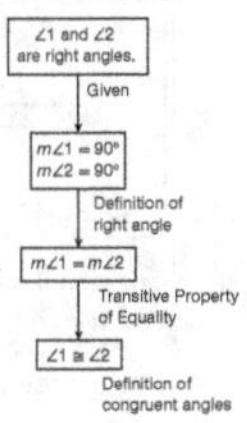

Practice

1. Complete the flowchart proof for the given two-column proof.
   Given: V is the midpoint of $\overline{SW}$, and W is the midpoint of $\overline{VT}$.
   Prove: $\overline{SV}$ is congruent to $\overline{WT}$.

Two-Column Proof

Statements	Reasons
1. V is the midpoint of $\overline{SW}$.	1. Given
2. W is the midpoint of $\overline{VT}$.	2. Given
3. $\overline{SV}$ is congruent to $\overline{VW}$. $\overline{VW}$ is congruent to $\overline{WT}$.	3. Definition of midpoint
4. $\overline{SV}$ is congruent to $\overline{WT}$.	4. Transitive Property of Equality

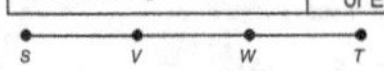

Flowchart Proof

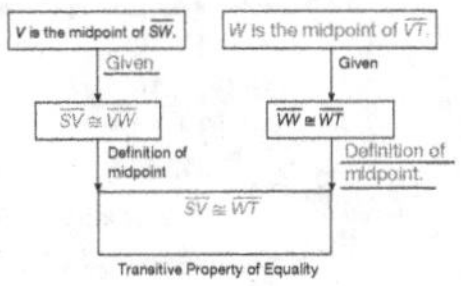

Reteaching 31

continued

Paragraph Proofs

To write a paragraph proof, use sentences to write a paragraph that presents the statements and reasons.

Use the two-column proof to write a paragraph proof.

Given: $\overline{AB} \cong \overline{BC}$ and $\overline{BC} \cong \overline{DE}$.

Prove: $\overline{AB} \cong \overline{DE}$

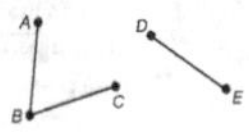

Two-Column Proof

Statements	Reasons
1. $\overline{AB} \cong \overline{BC}$	1. Given
2. $\overline{BC} \cong \overline{DE}$	2. Given
3. $AB = BC$ and $BC = DE$	3. Definition of congruent segments
4. $AB = DE$	4. Transitive Property of Equality
5. $\overline{AB} \cong \overline{DE}$	5. Definition of congruent segments

Paragraph Proof

It is given that $\overline{AB} \cong \overline{BC}$ and $\overline{BC} \cong \overline{DE}$, so $AB = BC$ and $BC = DE$ by the definition of congruent segments. $AB = DE$ by the Transitive Property of Equality. Thus, $\overline{AB} \cong \overline{DE}$ by the definition of congruent segments.

Practice

2. Use the given two-column proof to complete the paragraph proof.
   Given: ∠JKL is a right angle.
   Prove: ∠1 and ∠2 are complementary angles.

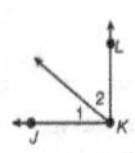

Two-Column Proof:

Statements	Reasons
1. ∠JKL is a right angle.	1. Given
2. m∠JKL = 90°	2. Definition of right angle
3. m∠JKL = m∠1 + m∠2	3. Angle Addition Postulate
4. 90° = m∠1 + m∠2	4. Substitution
5. ∠1 and ∠2 are complementary angles.	5. Definition of complementary angles

Paragraph Proof

Since it is given that ∠JKL is a right angle, m∠JKL = 90° by the definition of a right angle. By the Angle Addition Postulate, m∠JKL = m∠1 + m∠2. Using substitution, 90° = m∠1 + m∠2. Thus, by the definition of complementary angles, ∠1 and ∠2 are complementary angles.

Answer Key

Name ______________________ Date ____________ Class ____________

Reteaching 32
Altitudes and Medians of Triangles

You have written flowchart and paragraph proofs. Now you will solve problems involving altitudes and medians of triangles.

Medians of Triangles

In the picture, $\overline{AH}$, $\overline{BJ}$, and $\overline{CG}$ are **medians of a triangle.** They each join a vertex and the midpoint of the opposite side. The point of intersection of the medians is called the **centroid** of $\triangle ABC$.

Centroid Theorem

The centroid of a triangle is located $\frac{2}{3}$ of the distance from each vertex to the midpoint of the opposite side.

Example:

Given: $\overline{AH}$, $\overline{BJ}$, and $\overline{CG}$ are medians of $\triangle ABC$.

Conclusion: $AN = \frac{2}{3}AH$, $CN = \frac{2}{3}CG$, $BN = \frac{2}{3}BJ$

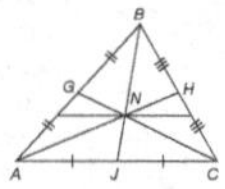

In $\triangle ABC$, suppose $AH = 18$ and $BN = 10$.

Use the Centroid Theorem to find AN and BJ.

$AN = \frac{2}{3}AH$	Centroid Theorem	$BN = \frac{2}{3}BJ$	Centroid Theorem
$AN = \frac{2}{3}(18)$	Substitute 18 for AH.	$10 = \frac{2}{3}BJ$	Substitute 10 for BN.
$AN = 12$	Simplify.	$15 = BJ$	Simplify.

Practice

In $\triangle QRS$, $RX = 48$, and $QW = 30$. Fill in the steps to find each length.

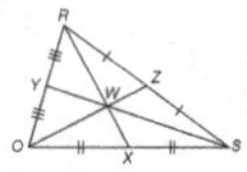

1. QZ

   $\underline{QW} = \frac{2}{3}QZ$

   $\underline{30} = \frac{2}{3}QZ$

   $\underline{45} = QZ$

2. RW

   $RW = \frac{2}{3}\,\underline{RX}$

   $RW = \frac{2}{3}\,\underline{48}$

   $RW = \underline{32}$

In $\triangle HJK$, $HD = 21$, and $BK = 18$. Find each length.

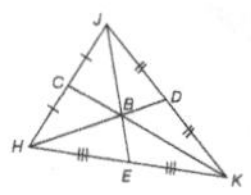

3. HB — 14
4. BD — 7
5. CK — 27
6. CB — 9

Reteaching 32
continued

Finding a Centroid on the Coordinate Plane

Find the centroid of $\triangle JKL$ with vertices at $J(0, 4)$, $K(6, 2)$ and $L(4, 6)$

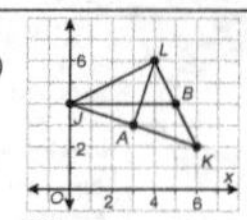

Step 1: Find the midpoint of two of the sides of the triangle. Since the three medians meet at the same point, the intersection of any two will give the location of the centroid. The midpoint of $\overline{JK}$ is (3, 3). The midpoint of is $\overline{LK}$ is (5, 4).

Step 2: Graph the two medians and find the equation of each median. $\overline{JB}$ is a horizontal line, so the equation is y = 4. To find the equation of $\overline{LA}$, use the slope and the coordinates of one endpoint. The slope of is $\overline{LA}$ is 3.

$y - y_1 = m(x - x_1)$	Point-slope Formula
$y - 6 = 3(x - 4)$	Substitute.
$y = 3x - 6$	Simplify.

Step 3: Solve the two equations as a system: $y = 4$ and $y = 3x - 6$.

The y-coordinate is 4. Solve for the x-coordinate:

$4 = 3x - 6$

$3.3 = x$

The x-coorinate is 3.3. The centroid is located at (3.3, 4).

Practice

Triangle DEF has coordinates $D(-1, 7)$, $E(0, 3)$, and $F(-5, 5)$. Complete the steps to find the centroid of the triangle.

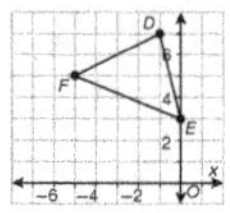

7. The midpoint of $\overline{DE}$ is $(\underline{-0.5},\ \underline{5})$, and the midpoint of $\overline{DF}$ is $(\underline{-3},\ \underline{6})$.

   The equation of the median from (−0.5, 5) to F is $y = \underline{5}$.

   The slope of the line from (−3, 6) to E is $\underline{-1}$.

   The equation of the line from (−3, 6) to E is $\underline{y = -x + 3}$.

   To find the coordinates of the centroid, solve the system of equations: $\underline{y = 5}$, and $\underline{y = -x + 3}$.

   The coordinates of the centroid are $(\underline{-2},\ \underline{5})$.

Triangle MNP has coordinates $M(-3, 3)$, $N(-1, 6)$, and $P(1, -1)$. Find the centroid of the triangle.

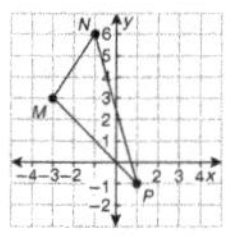

8. centroid $\underline{(-1, 2.7)}$

Name ______________________ Date ____________ Class ____________

Reteaching 33
Converse of the Pythagorean Theorem

You have worked with medians of triangles. Now you will solve problems that use the Pythagorean theorem.

The Pythagorean Theorem

The **Pythagorean theorem** states the following relationship among the lengths of the legs, a and b, and the length of the hypotenuse, c, of any right triangle.

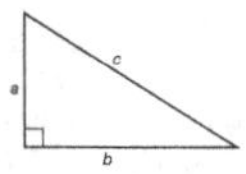

$$a^2 + b^2 = c^2$$

Use the Pythagorean theorem to find the value of x in the triangle.

Substitute 6 for a, x for b, and 12 for c in the equation $a^2 + b^2 = c^2$ and solve for x.

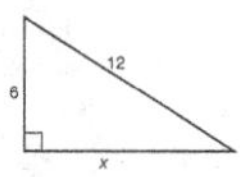

$a^2 + b^2 = c^2$	Pythagorean theorem
$6^2 + x^2 = 12^2$	Substitute
$36 + x^2 = 144$	Simplify
$x^2 = 108$	Subtract
$x = \sqrt{108}$	Square root
$x = \sqrt{36 \times 3} = 6\sqrt{3}$	Simplify radical

The length of the unknown side is $6\sqrt{3}$.

Practice

Complete the steps to find the value of x.

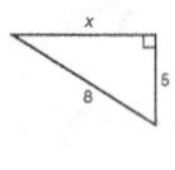

1. $\underline{5} + x^2 = \underline{8^2}$

   $\underline{25} + x^2 = \underline{64}$

   $\underline{25} + x^2 - 25 = 64 - \underline{25}$

   $x^2 = \underline{39}$

   $x = \underline{\sqrt{39}}$

Find the value of x. Give your answer in simplest radical form.

2. 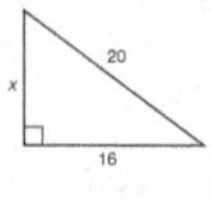

   $\underline{x = 12}$

3. 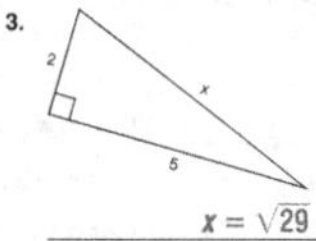

   $\underline{x = \sqrt{29}}$

Reteaching 33
continued

Pythagorean Inequality Theorem

You can use the following theorem to classify triangles by their angles if you know their side lengths. Always use the length of the longest side for c.

Pythagorean Inequality Theorem

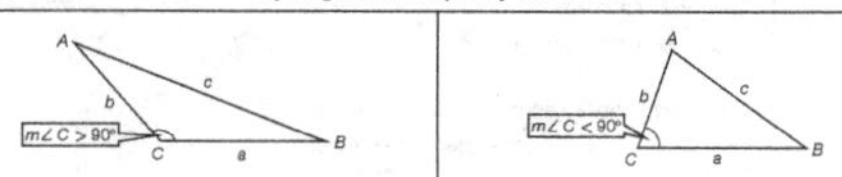

If $a^2 + b^2 < c^2$, then $\triangle ABC$ is obtuse.	If $a^2 + b^2 > c^2$, then $\triangle ABC$ is acute.

Determine if $\triangle ABC$ is an obtuse, acute, or right triangle.

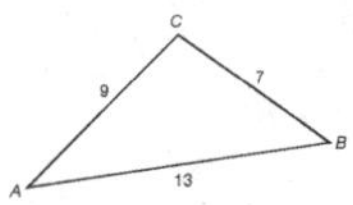

Step 1: Let $c = 13$. The other two lengths, a and b, will equal 7 and 9, respectively.

Step 2: Substitute the lengths into $a^2 + b^2 \stackrel{?}{=} c^2$.

$a^2 + b^2 \stackrel{?}{=} c^2$

$7^2 + 9^2 \stackrel{?}{=} 13^2$

$49 + 81 \stackrel{?}{=} 169$

$130 < 169$

Since $a^2 + b^2 < c^2$, the triangle is an obtuse triangle.

Practice

Complete the steps to determine if the given side lengths form an acute, obtuse, or right triangle.

4. 10, 11, 12

   $\underline{10^2} + 11^2 \stackrel{?}{=} \underline{12^2}$

   $100 + 121 \stackrel{?}{=} \underline{144}$

   $\underline{221} > 144$

   The triangle is a/an $\underline{\text{acute}}$ triangle.

5. 3, 4, 5

   $3^2 + \underline{4^2} \stackrel{?}{=} \underline{5^2}$

   $9 + 16 \stackrel{?}{=} \underline{25}$

   $\underline{25} = 25$

   The triangle is a/an $\underline{\text{right}}$ triangle.

Determine whether the triangle with the given side lengths is an obtuse, an acute, or a right triangle.

6. 8, 13, 16 — obtuse
7. 4, 7, 8 — obtuse
8. 8, 8, 11 — acute
9. 9, 12, 15 — right

Name ______________________ Date __________ Class __________

Reteaching 34
Properties of Parallelograms

All parallelograms have the following properties.

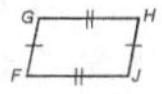 $\overline{FG} \cong \overline{JH}$, $\overline{GH} \cong \overline{FJ}$	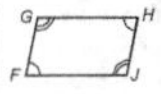 $\angle F \cong \angle H$, $\angle G \cong \angle J$
Opposite sides are congruent.	Opposite angles are congruent.

Prove opposite sides of a parallelogram are congruent.

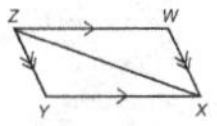

Given: *WXYZ* is a parallelogram.

Prove: $\overline{WZ} \cong \overline{XY}$, $\overline{WX} \cong \overline{ZY}$

Solution

Statements	Reasons
1. *WXYZ* is a parallelogram.	1. Given
2. $\overline{WZ} \parallel \overline{XY}$, $\overline{WZ} \parallel \overline{XY}$	2. Definition of parallelogram
3. $\overline{ZX} \cong \overline{ZX}$	3. Reflexive property
4. $\angle WXZ \cong \angle XZY$, $\angle ZXY \cong \angle WZX$	4. Corresponding Angles Postulate
5. $\Delta WZX \cong \Delta YXZ$	5. ASA triangle congruence.
6. $\overline{WZ} \cong \overline{XY}$, $\overline{WX} \cong \overline{ZY}$	6. Corresponding Parts of Congruent Triangles are Congruent.

Practice

Complete the paragraph proof to prove that opposite angles of a parallelogram are congruent.

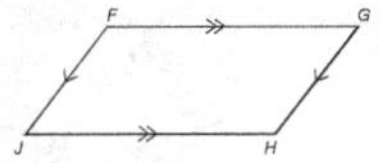

7. Given: *FGHJ* is a parallelogram.
   Prove: $\angle G \cong \angle J$, $\angle F \cong \angle H$
   It is given that FGHJ is a parallelogram. By the definition of a parallelogram, $\overline{GH} \parallel \overline{FJ}$, and $\overline{FG} \parallel \overline{HJ}$. Since same-side interior angles are supplementary, $m\angle G + m\angle H = 180°$, and $m\angle F + m\angle J = 180°$.
   For the same reason $m\angle G + m\angle F = 180°$ and $m\angle H + m\angle J = 180°$. Using substitution, $m\angle H + m\angle J = m\angle G + m\angle H$. By the Subtraction Property of Equality, $m\angle H + m\angle J - m\angle H = m\angle G + m\angle H - m\angle H$. Simplifying, $m\angle J = m\angle G$. Using substitution, $m\angle F + m\angle J = m\angle H + m\angle J$. By the Subtraction Property of Equality, $m\angle F + m\angle J - m\angle J = m\angle H + m\angle J - m\angle J$. Simplifyig, $m\angle F = m\angle H$. By the definition of congruent angles, $\angle G \cong \angle J$ and $\angle F \cong \angle H$.

Reteaching 34
continued

More Properties of Parallelograms

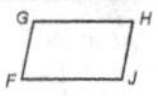 $m\angle F + m\angle G = 180°$ $m\angle G + m\angle H = 180°$ $m\angle H + m\angle J = 180°$ $m\angle J + m\angle F = 180°$	 $\overline{FP} \cong \overline{HP}$ $\overline{GP} \cong \overline{JP}$
Consecutive angles are supplementary.	The diagonals bisect each other.

You can use the properties of parallelograms to find the measures of angles and sides of parallelograms.

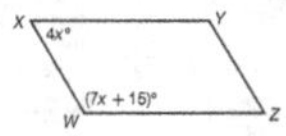

***WXYZ* is a parallelogram. Find $m\angle X$.**

Step 1: Since $\angle W$ and $\angle X$ are consecutive angles, they are supplementary. Thus, $m\angle W + m\angle X = 180°$.

Step 2: Substitute $m\angle W = (7x + 15)°$ and $m\angle X = (4x)°$ into $m\angle W + m\angle X = 180°$ and solve for *x*.

$m\angle W + m\angle X = 180°$	Consecutive angles are supplementary.
$(7x + 15) + 4x = 180$	Substitute the given values.
$11x + 15 = 180$	Combine like terms.
$11x = 165$	Subtract 15 from both sides.
$x = 15$	Divide both sides by 11.

Step 3: Substitute $x = 15$ into the expression for $m\angle X$.

$m\angle X = (4x)° = [4(15)]° = 60°$

Practice

***CDEF* is a parallelogram. Complete the steps to find the length of $\overline{DC}$.**

8.	$\overline{DC} \cong \overline{EF}$	Opposite sides are congruent
	$DC = EF$	Congruent sides have the same length.
	$4w + 8 = 5w + 1$	Substitute.
	$8 = w + 1$	Subtract 4*w* from both sides.
	$7 = w$	Subtract 1 from both sides.

$CD = 4w + 8 = 4(7) + 8 = 36$

Find each measure for the parallelogram *CDEF*.

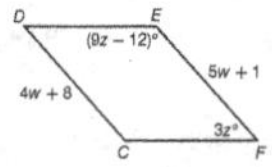

9. *EF*
   36

10. $m\angle F$
    48°

Name ______________________ Date __________ Class __________

Reteaching 35
Finding Areas of Sectors and Arc Length

You have worked with properties of parallelograms. Now you will find arc lengths and the area of sectors.

Arc Length

Arc length is the distance along a part of the circumference of a circle measured in linear units. The arc length of a circle is given by the formula $L = 2\pi r\left(\frac{m°}{360°}\right)$. The variable *m* stands for the degree measure of the arc.

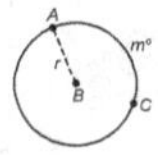

Find the arc length of $\overset{\frown}{JK}$. Give the answer in terms of π and rounded to the nearest hundredth.

From the circle, $r = 9$ cm, and $m° = 84°$. Substitute the values into the formula.

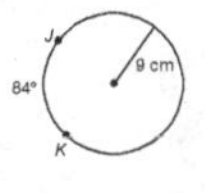

$L = 2\pi r\left(\frac{m°}{360°}\right)$	Formula for arc length.
$= 2\pi(9\text{ cm})\left(\frac{84°}{360°}\right)$	Substitute 9 cm and 84°.
$= \frac{21}{5}\pi\text{ cm}$	Solve for the answer in terms of π.
≈ 13.19 cm	Round to the nearest hundredth.

Practice

Complete the steps to find the arc length. Give your answer in terms of π and rounded to the nearest hundredth.

1. $\overset{\frown}{AB}$

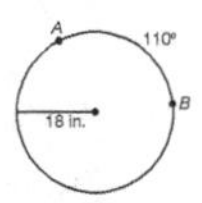

$L = 2\pi r\left(\frac{m°}{360°}\right)$

$= 2\pi(18\text{ in.})\left(\frac{110°}{360°}\right)$

$= 11\pi$ in.

≈ 34.56 in.

2. $\overset{\frown}{WX}$

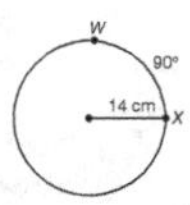

$L = 2\pi r\left(\frac{m°}{360°}\right)$

$= 2\pi(14\text{ cm})\left(\frac{90°}{360°}\right)$

$= 7\pi$ cm

≈ 21.99 cm

Reteaching 35
continued

Sector Area

A **sector of a circle** is a region bounded by two radii of the circle and their intercepted arc.

The area of a sector of a circle is given by the formula $A = \pi r^2\left(\frac{m°}{360°}\right)$.

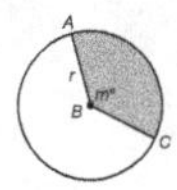

Find the area of the sector *CDE*. Give the answer in terms of π and rounded to the nearest hundredth.

From the circle, $r = 6$ cm, and $m° = 70°$. Substitute the values into the formula.

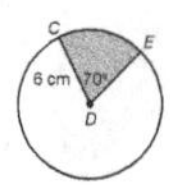

$A = \pi r^2\left(\frac{m°}{360°}\right)$	Formula for sector area.
$= \pi(6\text{ cm}^2)\left(\frac{70°}{360°}\right)$	Substitute 6 cm and 70°.
$= (36\pi\text{ cm}^2)\left(\frac{70°}{360°}\right)$	Simplify.
$= 7\pi\text{ cm}^2$	Solve for the answer in terms of π.
$\approx 21.99\text{ cm}^2$	Round to the nearest hundredth.

Practice

Complete the steps to find the area of each sector. Give your answer in terms of π and rounded to the nearest hundredth.

5. sector *QRS*

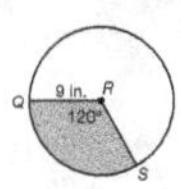

$A = \pi r^2\left(\frac{m°}{360°}\right)$

$= \pi(9\text{ in.})^2\left(\frac{120°}{360°}\right)$

$= (81\pi\text{ in}^2)\left(\frac{120°}{360°}\right)$

$= 27\pi\text{ in}^2$

$\approx 84.82\text{ in}^2$

6. sector *JHG*

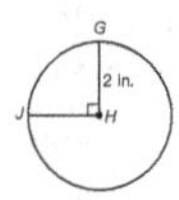

$A = \pi r^2\left(\frac{m°}{360°}\right)$

$= \pi(2\text{ in.})^2\left(\frac{90°}{360°}\right)$

$= (4\pi\text{ in}^2)\left(\frac{90°}{360°}\right)$

$= \pi\text{ in}^2$

$\approx 3.14\text{ in}^2$

Name ______________________ Date __________ Class __________

Reteaching 36

Right Triangle Congruence Theorems

Hypotenuse-Angle Congruence Theorem: If the hypotenuse and an acute angle of one right triangle are congruent to the hypotenuse and an acute angle of another right triangle, then the triangles are congruent.

Use the Hypotenuse-Angle Congruence Theorem to prove that $\Delta ABC \cong \Delta JKL$.

Statement	Reason
1. ΔABC and ΔJKL are right triangles.	1. Given
2. $\angle A \cong \angle J$	2. Given
3. $\overline{AB} \cong \overline{JK}$	3. Given
4. $\Delta ABC \cong \Delta JKL$	4. HA Congruence Theorem

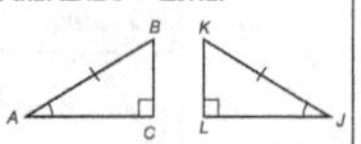

Practice

1. Use the Hypotenuse- Angle Congruence Theorem to show that $\Delta GHI \cong \Delta MNP$.

Statement	Reason
1. ΔGHI and ΔMNP are right triangles.	1. Given
2. $\overline{GI} \cong \overline{MP}$	2. Given
3. $\angle G \cong \angle M$	3. Given
4. $\Delta GHI \cong \Delta MNP$	4. HA Congruence Theorem

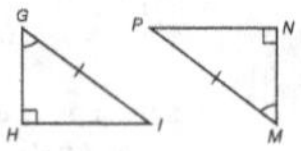

2. Use the Hypotenuse-Angle Congruence Theorem to prove that $\Delta HJK > \Delta XYZ$.

Statement	Reason
1. ΔHJK and ΔXZY are right triangles.	1. Given
2. $\overline{HJ} \cong \overline{XZ}$	2. Given
3. $\angle J \cong \angle Z$	3. Given
4. $\Delta HJK \cong \Delta XZY$	4. HA Congruence Theorem

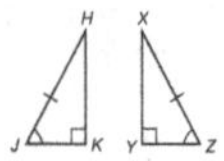

Reteaching 36

continued

Hypotenuse-Leg Congruence Theorem: If the hypotenuse and a leg of one right triangle are congruent to the hypotenuse and a leg of another right triangle, then the triangles are congruent.

Use the Hypotenuse-Leg Congruence Theorem to prove that $\Delta RST \cong \Delta UVX$.

ΔRST and ΔUVX are both given as right triangles. Leg $\overline{RT}$ is congruent to leg $\overline{UX}$ as given. Also, hypotenuse $\overline{RS}$ is congruent to hypotenuse $\overline{UV}$ as given. Therefore, $\Delta RST \cong \Delta UVX$ are congruent by the HL Congruence Theorem.

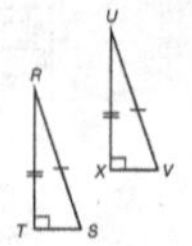

Practice

3. Use the Hypotenuse-Leg Congruence Theorem to prove that $\Delta PQR \cong \Delta CDE$.

Statement	Reason
1. ΔPQR and ΔCDE are right triangles.	1. Given
2. $\overline{PR} \cong \overline{CE}$ and $\overline{PQ} \cong \overline{CD}$	2. Given
3. $\Delta GHI \cong \Delta MNP$	4. HA Congruence Theorem

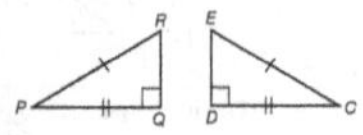

4. Use the Hypotenuse-Leg Congruence Theorem to prove that $\Delta ABC \cong \Delta VWX$.

Possible answer: ΔABC and ΔVWX are both right triangles. Leg $\overline{AC}$ is congruent to leg $\overline{VX}$ as given. Also, hypotenuse $\overline{AB}$ is congruent to hypotenuse $\overline{VW}$ as given. Therefore, $\Delta ABC \cong \Delta VWX$ by HL CongruenceTheorem.

5. Use the Hypotenuse-Leg Congruence Theorem to prove that $\Delta EFG \cong \Delta MNP$.

Statement	Reason
1. ΔEFG and ΔMNP are right triangles.	1. Given
2. $\overline{EG} \cong \overline{MP}$ and $\overline{EF} \cong \overline{MN}$	2. Given
3. $\Delta EFG \cong \Delta MNP$	3. HA Congruence Theorem

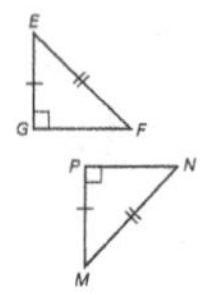

Name ______________________ Date __________ Class __________

Reteaching 37

Writing Equations of Parallel and Perpendicular Lines

Find the slopes of lines parallel and perpendicular to line *k*.

First, find the slope of line *k*.

$m = \frac{y_1 - y_2}{x_1 - x_2}$

Choose two points on line *k* to substitute into the equation.

$(x_1, y_1) = (0, 2)$ and $(x_2, y_2) = (-3, 0)$

$m = \frac{2 - 0}{0 - (-3)} = \frac{2}{3}$

Since parallel lines have the same slope, the slope of a line parallel to line k is $\frac{2}{3}$.

Since the slopes of two perpendicular lines are opposite reciprocals, the slope of a line perpendicular to line k is $-\frac{3}{2}$.

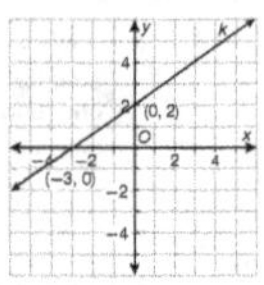

Practice

Complete the steps to find the slopes of lines parallel and perpendicular to each line.

1. line a: $m_a = \frac{(-1) - 0}{0 - 5} = \frac{1}{5}$

Slope of line parallel to line a: $\frac{1}{5}$

Slope of line perpendicular to line a: -5

2. line b: $m_b = \frac{0 - (-6)}{3 - 0} = 2$

Slope of line parallel to line b: 2

Slope of line perpendicular to line b: $\frac{1}{2}$

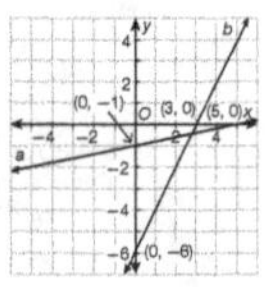

Find the slopes of the lines parallel and perpendicular to each line.

3. line g

Slope of line parallel to line g: $\frac{1}{4}$

Slope of line perpendicular to line g: -4

4. line h

Slope of line parallel to line h: $-\frac{5}{2}$

Slope of line perpendicular to line h: $-\frac{2}{5}$

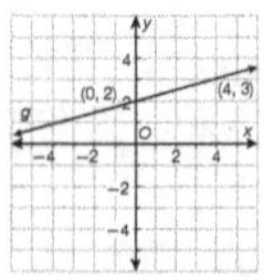

Reteaching 37

continued

Graph the line that is perpendicular to the line $y = \frac{1}{3}x + 4$ and passes through (2, 1).

The slope of the perpendicular line is the opposite reciprocal of $\frac{1}{3}$, so $m = -3$.

$y - y_1 = m(x - x_1)$ Point-slope formula

$y - 1 = -3(x - 2)$ Substitute.

$y = 3x + 7$ Solve.

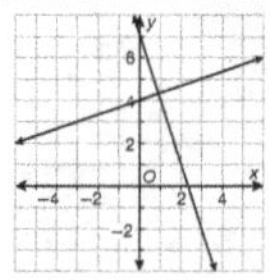

Practice

Complete the steps to graph the line that is perpendicular to $y = 2x - 1$ and passes through (2, 3).

5. $y - y_1 = m(x - x_1)$

$y - \underline{3} = \underline{-\frac{1}{2}}(x - \underline{2})$

$y = \underline{-\frac{1}{2}}(x + \underline{4})$

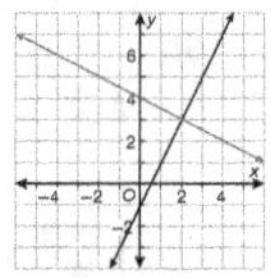

Complete the steps to graph the line that is parallel to $y = 3x + 2$ and passes through (1, -2).

6. $y - y_1 = m(x - x_1)$

$y + \underline{2} = \underline{3}(x - \underline{1})$

$y = \underline{3}x - \underline{5}$

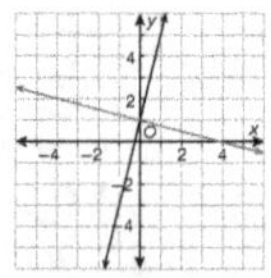

Find and graph each line.

7. Graph the line that is perpendicular to $y = 4x + 1$ and passes through $(-4, 5)$.

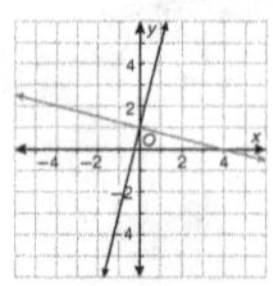

8. Graph the line that is parallel to $y = -\frac{1}{2}x - 3$ and passes through $(-4, 1)$.

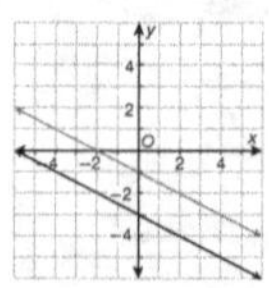

Answer Key

Name ______________________ Date __________ Class __________

Reteaching 38

Perpendicular and Angle Bisectors of Triangles

In the diagram, $\overline{XZ}$ bisects $\angle X$. According to the **Triangle Angle Bisector Theorem**, $\overline{WZ}$ and $\overline{ZY}$ are proportional to $\overline{WX}$ and $\overline{XY}$. The proportion is $\frac{WX}{WZ} = \frac{XY}{ZY}$.

Find the length of $\overline{WX}$ if $\overline{WZ} = 9$, $\overline{XY} = 24$, and $\overline{ZY} = 10$.

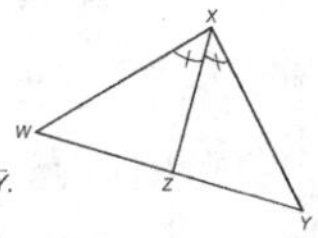

$\frac{WX}{WZ} = \frac{XY}{ZY}$ Triangle Angle Bisector Theorem

$\frac{n}{9} = \frac{24}{10}$ Substitute the values for $\overline{WZ}$, $\overline{XY}$, and $\overline{ZY}$.

$10n = 216$ Cross multiply.

$n = 21.6$ Solve.

So, $\overline{WX} = 21.6$.

Practice

1. Find the length of $\overline{TV}$ if $\overline{RS} = 18$, $\overline{ST} = 8$ and $\overline{RV} = 20$.

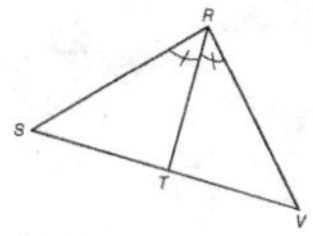

$\frac{18}{8} = \frac{20}{n}$

$18n = 160$

$n = 8.8$

2. Find the length of $\overline{RV}$ if $\overline{RS} = 27$, $\overline{ST} = 12$, and $\overline{TV} = 11$.

$\frac{27}{12} = \frac{n}{11}$

$12n = 297$

$n = 24.75$

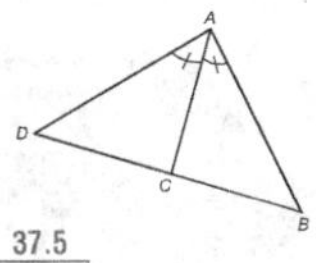

3. Find the length of $\overline{BC}$ if $\overline{AD} = 15$, $\overline{AB} = 14$, and $\overline{DC} = 9$. 8.4

4. Find the length of $\overline{AD}$ if $\overline{AB} = 30$, $\overline{BC} = 12$, and $\overline{DC} = 15$. 37.5

5. Find the length of $\overline{DC}$ if $\overline{AD} = 21$, $\overline{AB} = 25$, and $\overline{BC} = 15$. 12.6

6. Find the length of $\overline{AB}$ if $\overline{AD} = 28$, $\overline{DC} = 8$, and $\overline{BC} = 6$. 21

Reteaching 38 *continued*

The circumcenter of a triangle is the point at which all the perpendicular bisectors of the sides intersect. The circumcenter is equidistant from all the vertices of the triangle.

Find the circumcenter of the triangle with vertices at $A(3, 3)$, $B(9, 3)$, and $C(6, 7)$.

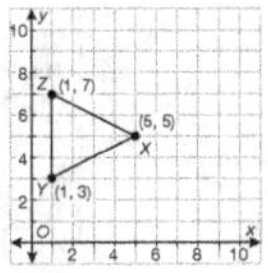

To find the circumcenter, first choose two sides to use. Then, find the midpoints of the two sides.

The midpoint of $\overline{AB}$ is (6, 3), and the midpoint of $\overline{AC}$ is (4.5, 5).

Next, find the slope of the two sides.

The slope of $\overline{AB}$ is 0, and the slope of $\overline{AC}$ is $\frac{4}{3}$.

Since the circumcenter is the point at which the perpendicular bisectors intersect, find the equations of the lines perpendicular to $\overline{AB}$ and $\overline{AC}$.

The equation of the line perpendicular to $\overline{AB}$ through (6, 3) is $x = 6$, and the equation of the line perpendicular to $\overline{AC}$ through (4.5, 5) is $y = -\frac{3}{4}x + \frac{67}{8}$.

Finally, solve the system of equations to find the point of intersection of the perpendicular bisectors: $x = 6$ and $y = -\frac{3}{4}x + \frac{67}{8}$.

The solution is (6, 3.875). This is the circumcenter of triangle ABC.

Practice

Complete the steps to find the circumcenter of the triangle.

7. ΔXYZ with vertices at $X(5, 5)$, $Y(1, 3)$, and $Z(1, 7)$.

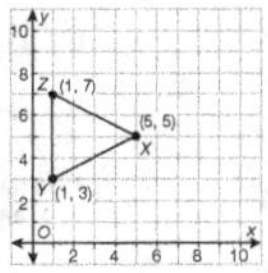

The midpoint of $\overline{XY}$ is (3, 4).

The midpoint of $\overline{YZ}$ is (1, 5).

The slope of $\overline{XY}$ is $\frac{1}{2}$.

The slope of is $\overline{YZ}$ is undefined.

The equation for the perpendicular bisector of $\overline{XY}$ is $y = -2x + 10$

The equation for the perpendicular bisector of $\overline{YZ}$ is $y = 5$.

The circumcenter for ΔXYZ is $\left(\frac{5}{2}, 5\right)$.

8. Find the circumcenter of the triangle with vertices at $A(2, 4)$, $B(8, 4)$, and $C(6, 10)$. $(5, 6.\overline{3})$

Name ______________________ Date __________ Class __________

Reteaching 39

Inequalities in a Trianglee

You learned that the sum of the angles inside a triangle is 180°. Now you will relate the angle measures to the side lengths of a triangle.

A triangle's largest side is always opposite its largest angle, and its smallest side is always opposite its smallest angle.

Order the angle measures of ΔABC from least to greatest.

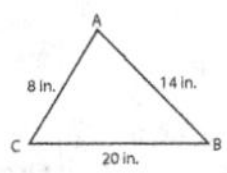

The shortest side length is $\overline{AC}$, so the smallest angle in the triangle is $\angle B$.

The longest side is $\overline{CB}$, so $\angle A$ is the largest angle.

Finally, $\overline{AB}$ is the middle side length, so $\angle C$ is the middle angle measure.

Therefore, the order of the angle measures from least to greatest is $\angle B$, $\angle C$, $\angle A$.

Practice

Fill in the blanks to put the angle measures in order from least to greatest for ΔFGH.

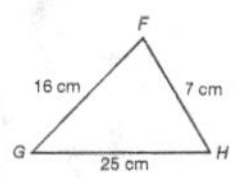

1. The shortest side is $\overline{FH}$, so the smallest angle is $\angle G$.

The longest side is $\overline{GH}$, so the largest angle is $\angle F$.

The remaining angle, $\angle H$, is the middle angle measure.

Therefore, the order of the angles from least to greatest measure is $\angle G$, $\angle H$, $\angle F$.

Fill in the blanks to put the angle measures in order from least to greatest for ΔXYZ.

2. The measure of $\angle Y$ is 100°.

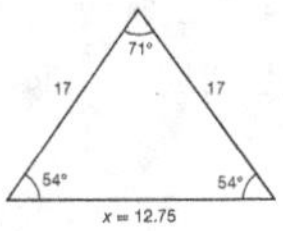

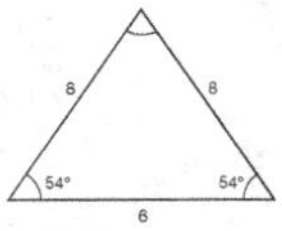

Therefore, $\overline{XZ}$ is the longest side. The angle with the smallest measure is $\angle Z$, so the shortest side length is $\overline{XY}$.

So the middle side length is $\overline{YZ}$.

Therefore, the order of the side lengths from least to greatest is $\overline{XY}$, $\overline{YZ}$, $\overline{XZ}$.

Reteaching 39 *continued*

Triangle Inequality Theorem: The sum of the lengths of any two sides of a triangle must be greater than the length of the third side.

Determine whether each set of side lengths could form a valid triangle.

(4, 5, 9) and (9, 11, 12)

For the first set of side lengths, the sum of 4 and 5 is equal to 9. This is not greater than the third side, so these side lengths cannot be a triangle.

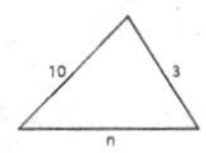

For the second set of side lengths, the sum of any pair of sides is greater than the third side, so these side lengths do form a triangle.

What is the range for the values of n in the given triangle?

Using the **Triangle Inequality Theorem**, the sum of any two sides must be greater than the third side.

First, write inequalities for the sum of each pair of sides.

$10 + 3 > n$ $10 + n > 3$ $3 + n > 10$

$13 > n$ $n > -7$ $n > 7$

Then, eliminate any side lengths that are not possible and find the solution.

Side lengths cannot be negative, so the solution is $7 < n < 13$.

Practice

Fill in the blanks to find the range for the values of x.

3. $5 + 14 > x$ $14 + x > 5$

$19 > x$ $x > -9$

$5 + x > 14$

$x > 9$

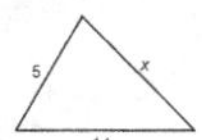

The range of values for x is $9 < x < 19$.

Determine whether each set of side lengths could form a triangle.

4. 6, 10, 5 yes

5. 4, 8, 15 no

Give the range for the values of x in the given triangles.

6. 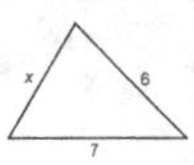

$1 < x < 13$

7. 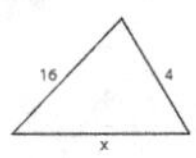

$12 < x < 20$

Name ______________________ Date __________ Class __________

Reteaching 40

Finding Perimeters and Areas of Composite Figures

You have learned how to find the perimeter and area of polygons and circles. Now you will find the perimeter and area of composite figures.

Area of Composite Figures

Find the area of this composite figure.

The figure is made up of a triangle and two rectangles.
First, find the area of each shape in the figure.

The area of the triangle is $A_1 = \frac{1}{2}bh = \frac{1}{2}(8)(2) = 8\text{ cm}^2$.

The area of the large rectangle is $A_2 = bh = (8)(12) = 96\text{ cm}^2$.

The area of the small rectangle is $A_3 = bh = (2)(5) = 10\text{ cm}^2$.

Add all the areas. By the **Angle Addition Postulate**, the area of the composite figure is $A = 8 + 96 + 10 = 114\text{ cm}^2$.

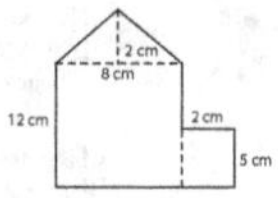

Practice

Complete the steps to find the area of each figure.

1. $A_1 = bh = (\underline{7})(\underline{7}) = \underline{49}\text{ in}^2$

   $A_2 = \frac{1}{2}bh = \frac{1}{2}(\underline{7})(\underline{6}) = \underline{21}\text{ in}^2$

   $A = \underline{49} + \underline{21} = \underline{70}\text{ in}^2$

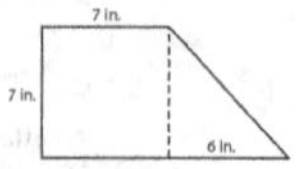

2. $A_2 = bh = (\underline{3})(\underline{9}) = \underline{27}\text{ m}^2$

   $A_2 = bh = (\underline{12})(\underline{7}) = \underline{84}\text{ m}^2$

   $A = \underline{27} + \underline{84} = \underline{111}\text{ m}^2$

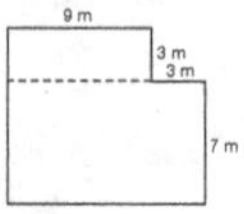

Find the area of each composite figure. When necessary, write your answer in terms of π.

3.

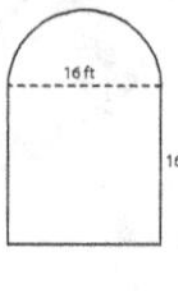

$\underline{256 + 32\pi\text{ ft}^2}$

4.

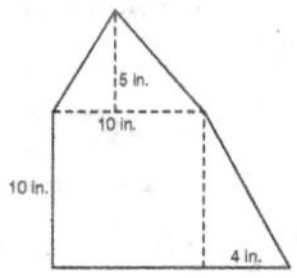

$\underline{145\text{ in}^2}$

Reteaching 40

continued

Area of Shaded Regions

Find the area of the shaded region. Leave your answer in terms of π.

The figure is made up of a semicircle inside a rectangle.

First, find the area of the whole rectangle.

$A_1 = bh = (14)(12) = 168\text{ yd}^2$

Then, find the area of the semicircle.

$A_2 = \frac{1}{2}\pi r^2 = (6)^2\pi = 18\pi\text{ yd}^2$

To find the area of the shaded region, we need to know the area of the rectangle without the area of the triangle. If A is the area of the shaded region, then $A_1 = A + A_2$.

So, $168 = A + 18\pi$, and $A = 168 - 18\pi\text{ yd}^2$.

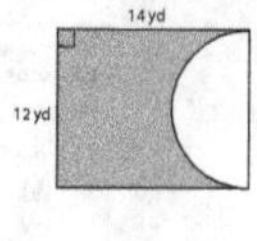

Practice

Fill in the blanks to find the area of the shaded region for each figure.

5. $A_1 = bh = (\underline{15})(\underline{12}) = \underline{180}\text{ mm}^2$

   $A_2 = \frac{1}{2}bh = \frac{1}{2}(\underline{8})(\underline{15}) = \underline{60}\text{ mm}^2$

   $\underline{180} = A + \underline{60}\text{ mm}^2$

   $A = \underline{120}\text{ mm}^2$

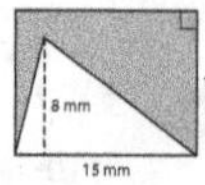

6. $A_1 = bh = (\underline{11})(\underline{11}) = \underline{121}\text{ ft}^2$

   $A_2 = \frac{1}{2}\pi r^2 = \frac{1}{2}(\underline{4})^2\pi = \underline{8}\,\pi\text{ ft}^2$

   $\underline{121} = A + \underline{8\pi}\text{ ft}^2$

   $A = \underline{121 - 8\pi}\text{ ft}^2$

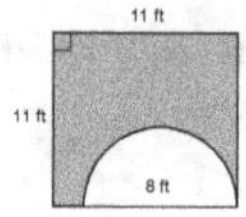

Find the area of the shaded region. When necessary, write your answer in terms of π.

7.

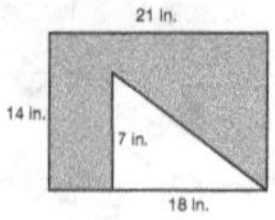

$\underline{231\text{ in}^2}$

8.

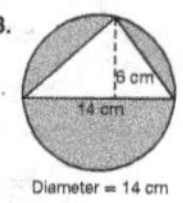

$\underline{49\pi - 42\text{ cm}^2}$

Name ______________________ Date __________ Class __________

Reteaching INV 4

Inequalities in Two Triangles

You have worked with segments and angles in triangles. Now you will explore inequalities with triangles.

Hinge Theorem

If two sides of one triangle are congruent to those of another triangle, and if the included angle of the first triangle is greater than that of the second triangle, then the third side of the first triangle is longer than that of the second triangle.

Example:

Compare the measures of *BC* and *EF*.

Step 1: Compare the angle measures of the sides opposite *BC* and *EF*.

The opposite angle of $BC = 42°$.

The opposite angle of $EF = 39°$.

Step 2: Make a comparison.

The side opposite the greater angle is longer.

$BC > EF$

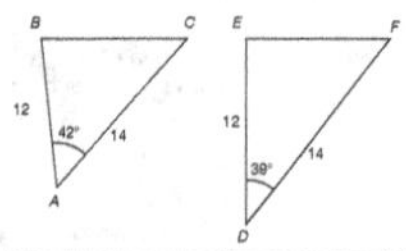

Practice

Complete the statements to solve each problem.

1. Compare the measures of *AB* and *DE*.

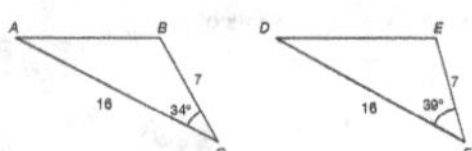

AB opposite angle = $\underline{34°}$

DE opposite angle = $\underline{39°}$

$\underline{AB} < DE$

2. Compare the lengths of *KL* and *NP*.

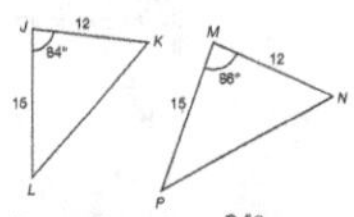

KL opposite angle = $\underline{84°}$

NP opposite angle = $\underline{86°}$

$\underline{KL} < \underline{NP}$

Compare the measures.

3. Compare the measures of *VW* and *YZ*.

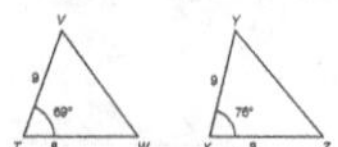

$\underline{YZ > VW}$

4. Compare the measures of *JL* and *MP*.

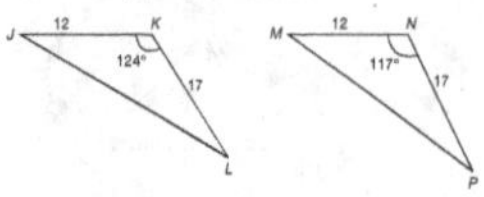

$\underline{JL > MP}$

Reteaching INV 4

continued

Converse of the Hinge Theorem

If two sides of one triangle are congruent to those of another triangle, and if the third side of the first triangle is longer than that of the second triangle, then the measure of the angle opposite the third side of the first triangle is greater than that of the second triangle.

Example:

Compare the measures of ∠*B* and ∠*E*.

Step 1: Compare the measures of the lengths of the sides opposite the angles.

The side opposite $\angle B = 4$.

The side opposite $\angle E = 4.5$.

Step 2: Make a comparison.

The angle opposite the greater side is larger.

$m\angle E > m\angle B$

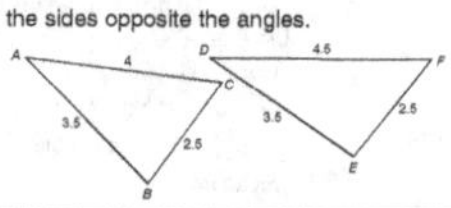

Practice

Complete the statements to solve each problem.

5. Compare the measures of ∠*B* and ∠*E*.

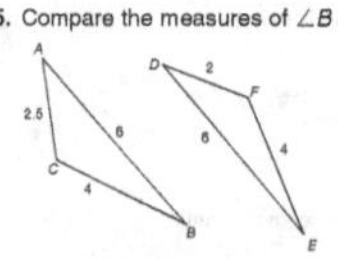

side opposite $\angle B = \underline{2.5}$

side opposite $\angle E = \underline{2}$

The angle opposite the greater side is larger.

$m\angle B \underline{>} m\angle E$

6. Compare the measures of ∠*B* and ∠*E*.

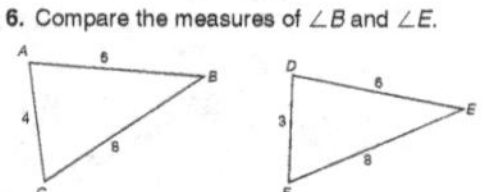

side opposite $\angle B = \underline{4}$

side opposite $\angle E = \underline{3}$

The angle opposite the greater side is larger.

$m\angle\,\underline{B} > m\angle\,\underline{E}$

Compare the measures.

7. Compare the measures of ∠*C* and ∠*F*.

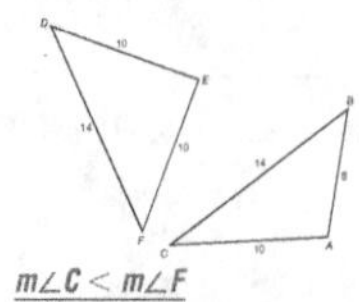

$\underline{m\angle C < m\angle F}$

8. Compare the measures of ∠*C* and ∠*F*.

$\underline{m\angle F < m\angle C}$

Name ________________ Date ________ Class ________

Reteaching 41
Ratios, Proportions, and Similarity

You have worked with corresponding sides and angles. Now you will work with ratios and proportions.

A proportion is an equation stating that two ratios are equal. In every proportion, the product of the extremes equals the product of the means.

Cross Products Property	In a proportion, if $\frac{a}{b} = \frac{c}{d}$ and b and $d \neq 0$, then $ad = bc$.

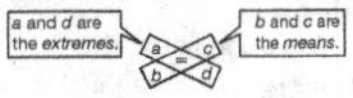

You can solve a proportion like $\frac{x}{8} = \frac{35}{56}$ by finding the cross products.

$\frac{x}{8} = \frac{35}{56}$
$x(56) = 8(35)$ Cross Products Property
$56x = 280$ Simplify.
$x = 5$ Divide both sides by 56.

Practice
Complete the statements to solve each proportion.

1. $\frac{9}{w} = \frac{36}{28}$
$w(\underline{36}) = 9(\underline{28})$ Cross Products Property
$\underline{36}\ w = \underline{252}$ Simplify.
$w = \underline{7}$ Divide both sides by 36.

2. $\frac{x}{32} = \frac{15}{16}$
$x(\underline{16}) = \underline{15}(35)$ Cross Products Property
$\underline{16x} = \underline{480}$ Simplify.
$x = \underline{30}$ Divide both sides by 16.

Solve each proportion.

3. $\frac{24}{42} = \frac{y}{7}$ $\underline{y = 4}$

4. $\frac{40}{49} = \frac{x}{98}$ $\underline{x = 80}$

5. $\frac{20}{25} = \frac{y}{45}$ $\underline{y = 36}$

6. $\frac{27}{63} = \frac{24}{x}$ $\underline{y = 56}$

Reteaching 41
continued

In **similar polygons,** the corresponding angles are congruent, and the corresponding sides are proportional.

A **similarity ratio** is the ratio of two corresponding linear measurements in a pair of similar figures.

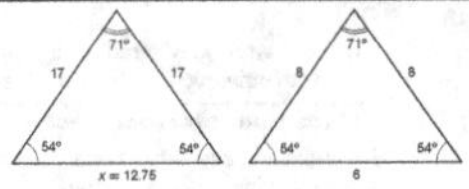

Find the unknown side length in the two similar triangles.

The triangles are similar so corresponding sides are in proportion.

Write a proportion and solve for x.

$\frac{17}{x} = \frac{8}{6}$
$(8)x = 17(6)$ Cross Products Property
$8x = 102$ Simplify.
$x = 12.75$ Divide both sides by 8.

The length of the missing side is 12.75.

Practice
Complete the statement to solve for *x*.

7. $\frac{7.0}{8.4} = \frac{2.8}{x}$
$7(x) = \underline{2.8}\ (8.4)$ Cross Products Property
$7x = \underline{23.52}$ Simplify.
$x = \underline{3.36}$ Divide

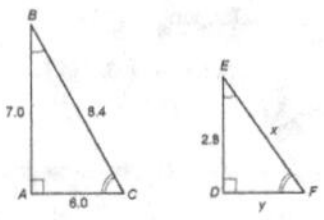

Solve for the unknown side lengths in the similar figures.

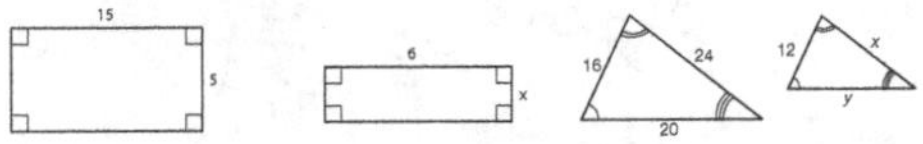

8. $\underline{x = 2}$

9. $\underline{x = 18;\ y = 15}$

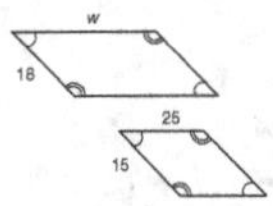

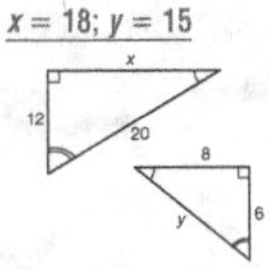

10. $\underline{w = 30}$

11. $\underline{x = 16;\ y = 10}$

Name ________________ Date ________ Class ________

Reteaching 42
Finding Distance from a Point to a Line

You have worked with finding distance on a coordinate plane. Now you will find the shortest distance between a point and a line.

Finding Distance

The shortest distance between a point and a line is always a perpendicular segment.

Find the distance from the point (1, 3) to the line $y = 1$.	Find the distance from the point (4, 2) to the line $x = 1$.
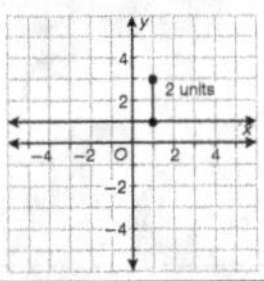	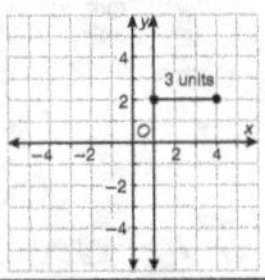
Step 1: Draw a point on the line $y = 1$, perpendicular from the given point. **Step 2:** Count the number of units between the two points. **Step 3:** The distance between the points is 2 units.	**Step 1:** Draw a point on the line $x = 1$, perpendicular from the given point. **Step 2:** Count the number of units between the two points. **Step 3:** The distance between the points is 3 units.

Practice
Complete the statements to find the distance between the point and the line.

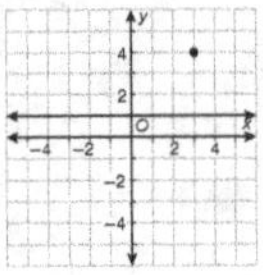

1. Find the distance between the point (3, 4) and the line $y = 1$.

The point on the line perpendicular to the given point is (3, $\underline{1}$).

Count the number of units between the two points: $\underline{3 \text{ units}}$.

The distance is $\underline{3}$ units.

2. Find the distance between the point (2, 3) and the line $y = 5$.

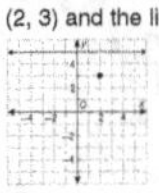

$\underline{2}$ units

3. Find the distance between the line $x = 4$ and the point (–3, 1).

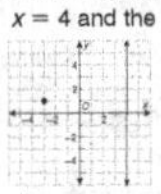

$\underline{7}$ units

Reteaching 42
continued

Finding the Closest Point

Given the equation $y = 3x + 1$ and the point $W(2, -3)$, find the point on the line that is closest to W. Find the distance from W to the line.

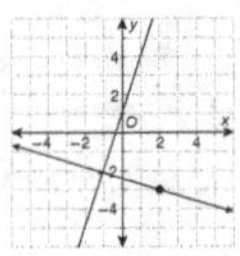

Step 1: Draw the line and point on the coordinate grid.

Step 2: Determine the slope of the line. The slope of the line is 3.

Step 3: Determine the slope of the line perpendicular to the given line. The negative reciprocal of 3 is $-\frac{1}{3}$.

Step 4: Use the slope to determine more points on the perpendicular line. Draw the line. The lines intersect at (–1, –2).

Step 5: Find the distance between (2, –3) and (–1, –2).

The distance from the point to the line is about 3.16 units.

$d = \sqrt{(x_2 - x_1)^2 + (y_2 - y_1)^2}$
$= \sqrt{(2 - (-1))^2 + (-3 - (-2))^2}$
$= \sqrt{(3)^2 + (-1)^2}$ The distance from the point to the line is about 3.16 units.
$= \sqrt{10}$
≈ 3.16

Practice
Complete the statements to find the distance.

4. Given the equation $y = 2x - 1$ and the point (2, –2), find the point on the line that is closest to the point.

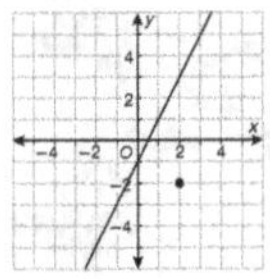

The slope of the line is $\underline{2}$. The negative reciprocal is $\underline{-\frac{1}{2}}$.

Another point on the line is (0, $\underline{-1}$).

Use the distance formula.

$d = \sqrt{(x_2 - x_1)^2 + (y_2 - y_1)^2}$
$= \sqrt{(\underline{2} - 0)^2 + (\underline{-2} - (-1))^2}$
$= \sqrt{(\underline{2})^2 + (-1)^2}$ The distance is about $\underline{2.23}$ units.
$= \sqrt{\underline{5}}$
$\approx \underline{2.23}$

Name _______________ Date _______ Class _______

Reteaching 43
Chords, Secants, and Tangents

You have working with radii and diameters of a circle. Now, you will work with other types of lines and segments in circles.

Lines that Intersect Circles

A chord is a segment whose endpoints lie on a circle. A diameter is a chord.

A **secant** is a line that intersects a circle at two points.

A **tangent** is a line in the same plane as a that of a circle that intersects the circle at exactly one point, called the **point of tangency.**

Radii and diameters also intersect circles.

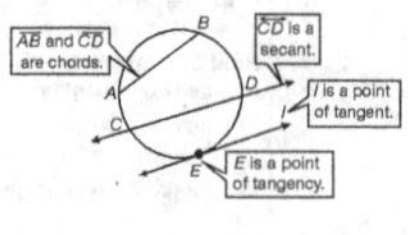

Practice

Complete the statements to identify each line or segment that intersects the circle.

1. Chord: $\overline{FG}$

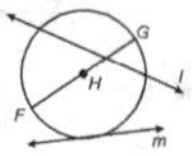

Secant: line l

Tangent: line m

Diameter: $\overline{FG}$

Radii: $\overline{HF}$ and $\overline{HG}$

Identify each line or segment that intersects each circle.

2.

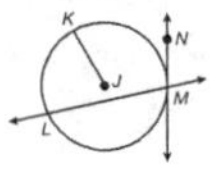

Chord: $\overline{LM}$

Secant: $\overleftrightarrow{LM}$

Tangent: $\overleftrightarrow{MN}$

Diameter: none

Radii: $\overline{JK}$

Point of tangency: M

3.

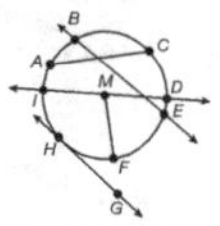

Chord: $\overline{AC}$, $\overline{ID}$, $\overline{BE}$

Secant: $\overleftrightarrow{BE}$

Tangent: $\overleftrightarrow{HG}$

Diameter: $\overline{ID}$

Radii: $\overline{MI}$, $\overline{MF}$, $\overline{MD}$

Point of tangency: H

Reteaching 43
continued

Chord–Diameter Relationships

If a diameter is perpendicular to a chord, then it bisects the chord and the chord's arcs.

If a diameter bisects a chord other than another diameter, then it is perpendicular to the chord.

The perpendicular bisector of a chord contains the center of the circle.

In a circle or congruent circle, chords equidistant from the center are congruent, and congruent chords are equidistant from the center of the circle.

Example:

Find the length of $\overline{GH}$.

The radius of circle K is 10 meters, and KJ is 6 meters.

First, draw a radius $\overline{KH}$.

Then, use the Pythagorean theorem to find $\overline{JH}$.

$KJ^2 + JH^2 = KH^2$

$6^2 + b^2 = 10^2$

$36 + b^2 = 100$ $\quad JH = 8\text{m}$

$b^2 = 64$

$b = 8$

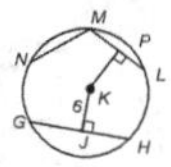

Since $\overline{KJ}$ is perpendicular to $\overline{GH}$, $\overline{JH} = \overline{JG}$.

Therefore, the length of $GH = 8 + 8 = 16$, or 16 meters.

Practice

Complete the statements to find the missing distance.

4. A circle has a diameter of 50 centimeters. Chord JK is 24 centimeters.

How far is JK from the center of the circle to the nearest hundredth?

Draw a perpendicular line from point P to chord JK.

Label the point of intersection C.

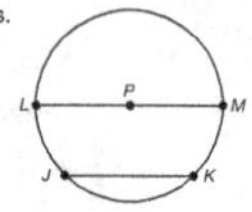

$PK =$ 25 cm; $CK =$ 12 cm

$PC^2 + CK^2 = PK^2$

$PC^2 + 12^2 = 25^2$

$PC^2 + 144 = 625$

$PC^2 = 481$

$PC \approx 21.9$

Name _______________ Date _______ Class _______

Reteaching 44
Applying Similiarity

You have worked with proportions to find unknown sides in similar triangles. Now you will work with other types of similar polygons.

Using Similarity to Find Unknown Measures

You can use properties of similar polygons to solve proportions.

The figures in the diagram are similar. Find the values of x and y.

Step 1: Write a similarity ratio. $\frac{7}{10.5}$

Step 2: Write and solve a proportion to solve for x and y.

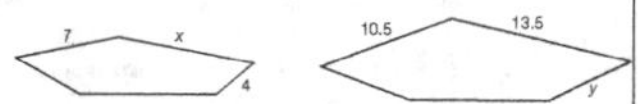

$\frac{7}{10.5} = \frac{x}{13.5}$ $\qquad$ $\frac{7}{10.5} = \frac{4}{y}$

$10.5x = 7(13.5)$ $\qquad$ $7y = 4(10.5)$

$10.5x = 94.5$ $\qquad$ $7y = 42$

$x = 9$ $\qquad$ $y = 6$

Practice

The figures in the diagram are similar. Complete the statements to find the missing values.

1. $\frac{24}{15} = \frac{x}{10}$ $\qquad$ $\frac{24}{15} = \frac{8}{y}$

$15x = 24(10)$ $\qquad$ $24y = 15(8)$

$15x = 240$ $\qquad$ $24y = 120$

$x = 16$ $\qquad$ $y = 5$

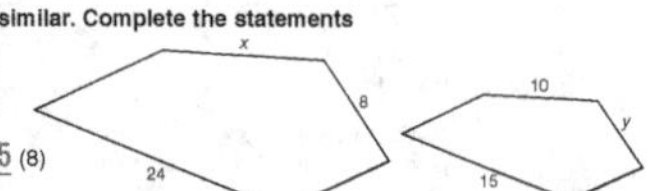

The figures in the diagram are similar. Solve for the missing lengths.

2.

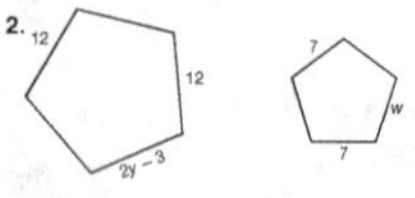

$w = 7$; $y = 7.5$

3. The similarity ratio is 3:2.

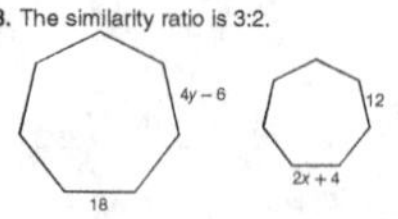

$x = 4$; $y = 6$

Reteaching 44
continued

Applying Similarity to Perimeter

Figures $ABCDE$ and $FGHIJ$ are similar polygons. Their corresponding sides have a ratio of 3:5. If the perimeter of figure $ABCDE$ is 36 inches, what is the perimeter of $FGHIJ$?

Step 1: The figures are similar polygons. The ratio of the perimeters is 3:5.

Step 2: Set up a proportion using the ratio 3:5.

$\frac{\text{ratio } ABCDE}{\text{ratio } FGHIJ} = \frac{\text{perimeter of } ABCDE}{\text{perimeter } FGHIJ}$

$\frac{3}{5} = \frac{36}{x}$

$3x = 36(5)$

$3x = 180$

$x = 60$

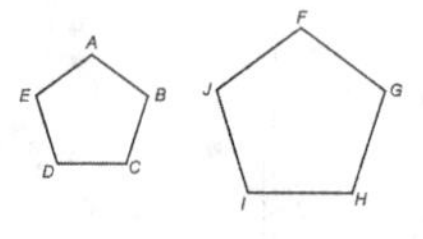

The perimeter of $FGHIJ$ is 60 inches.

Practice

Complete the statements to find the perimeter of the similar figure.

4. Figures $ABCDE$ and $FGHIJ$ are similar polygons. Their corresponding sides have a ratio of 2:7. If the perimeter of figure $ABCDE$ is 42 inches, what is the perimeter of $FGHIJ$?

$\frac{\text{ratio } ABCDE}{\text{ratio } FGHIJ} = \frac{\text{perimeter of } ABCDE}{\text{perimeter } FGHIJ}$

$\frac{2}{7} = \frac{42}{x}$

$2x = 7(42)$

$2x = 294$

$x = 147$

The perimeter is 147 inches.

5. Hexagons $ABCDEF$ and $GHIJKL$ are similar figures. The perimeter of $ABCDEF$ is 28 centimeters. The similarity ratio of $ABCDEF$ to $GHIJKL$ is 4:9. What is the perimeter of $GHIJKL$? 63 centimeters

6. Rectangle $ABCD$ is similar to $GHIJ$. The perimeter of $ABCD$ is 56 centimeters. The similarity ratio of $ABCD$ to $GHIJ$ is 1:16. What is the perimeter of $GHIJ$? 896 centimeters

7. Figures $LMNP$ and $QRST$ are similar figures. The ratio of their corresponding sides is 3:8. If the perimeter of $LMNP$ is 18 centimeters what is the perimeter of $QRST$? 48 centimeters

Name ______________________ Date __________ Class __________

Reteaching 45
Introduction to Coordinate Proofs

You have worked with a coordinate plane and proofs. Now you will work with proofs on a coordinate plane.

Positioning a Figure in the Coordinate Plane

A **coordinate proof** is a proof that uses coordinate geometry and algebra. In a coordinate proof, the first step is to position a figure in a plane. There are several ways you can do this to make your proof easier.

Keep the figure in Quadrant 1 by using the origin as a vertex.	Center a side of the figure at the origin.	Use one or both axes as sides of the figure.
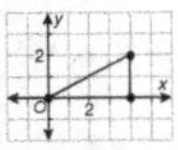	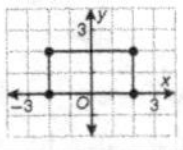	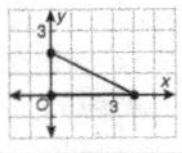

Practice

Position each figure in the coordinate plane and give the coordinates of each vertex.

1. a square with side length 4 units

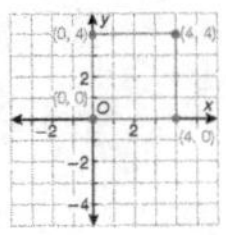

2. a right triangle with leg lengths 3 and 4 units

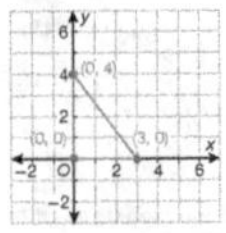

3. a triangle with base 8 units and height 2 units

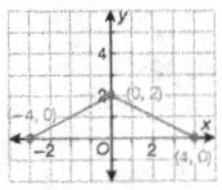

4. a rectangle with length 6 units and width 3 units

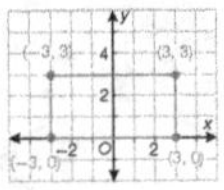

Reteaching 45
continued

You can prove that a statement about a figure is true without knowing the side lengths. To do this, assign variables as the coordinates of the vertices.

Position a right triangle with leg lengths c and d in the coordinate plane and give the coordinates of each vertex.

Step 1: Place one vertex at the origin and label (0, 0).

Step 2: The vertex on the x-axis is c units away. Label (c, 0).

Step 3: The vertex on the y-axis is d units up. Label (0, d).

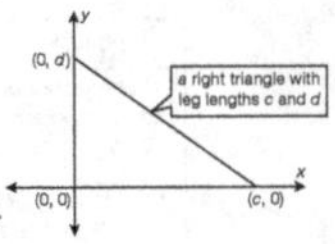

Practice

Position each figure in the coordinate plane and give the coordinates of each vertex.

5. a right triangle with leg lengths s and t

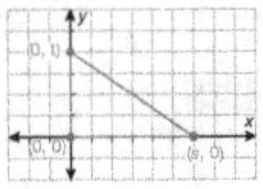

6. a square with side lengths k

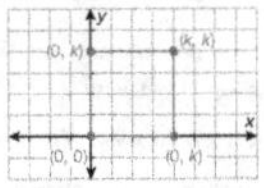

7. a rectangle with leg lengths l and w

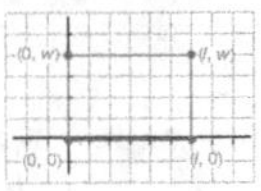

8. a triangle with base b and height h

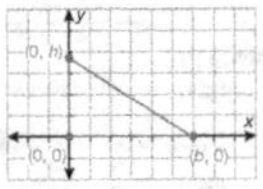

Name ______________________ Date __________ Class __________

Reteaching 46
Triangle Similarity: AA, SAS, SSS

You have worked with coordinate proofs. Now you will work with triangle similarity.

Triangle Similarity: AA, SSS

Angle-Angle (AA) Similarity	Side-Side-Side (SSS) Similarity
If two angles of one triangle are congruent to two angles of another triangle, then the triangles are similar. 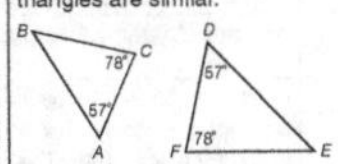 $\Delta ABC \sim \Delta DEF$	If the three sides of one triangle are proportional to the three corresponding sides of another triangle, then the triangles are similar. $\frac{TU}{XY} = \frac{UV}{YZ} = \frac{VT}{ZX} = \frac{3}{1}$ 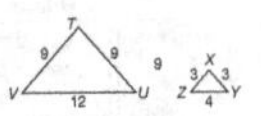 $\Delta TUV \sim \Delta XYZ$

Given the values of the two triangles as shown, show that the triangles are similar triangles.

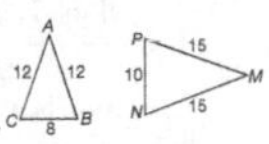

Step 1: Write similarity ratios for pairs of sides. Let the lengths of ΔABC be the top number of the proportions.

$\overline{AC} : \overline{MP}$ $\overline{AB} : \overline{MN}$ $\overline{BC} : \overline{NP}$

$\frac{AC}{MP} = \frac{12}{15} = \frac{4}{5}$ $\frac{AB}{MN} = \frac{12}{15} = \frac{4}{5}$ $\frac{BC}{NP} = \frac{8}{10} = \frac{4}{5}$

Step 2: Since the ratio of each pair of sides reduces to the same fraction, all of the side pairs of the two triangles are proportional. Therefore, $\Delta ABC \sim \Delta MNP$ by SSS Similarity.

Practice

Complete the steps to show that the two triangles are similar.

1. $\angle Q \cong$ $\underline{\angle T}$ by the definition of $\underline{\text{congruent}}$ angles.

$m\angle S =$ $\underline{39^\circ}$ by the Triangle Angle Sum Theorem.

$\angle S \cong$ $\underline{\angle V}$ by the definition of $\underline{\text{congruent}}$ angles.

Therefore, $\Delta QRS \cong \Delta TUV$ by $\underline{\text{AA}}$ Similarity.

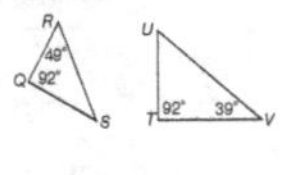

Explain how the triangles are similar and write a similarity statement.

2. $\frac{RS}{VW} = \frac{TR}{ZV} = \frac{TS}{ZW} = \frac{1}{3}$; Therefore; $\Delta RST \sim \Delta VWZ$ by SSS Similarity

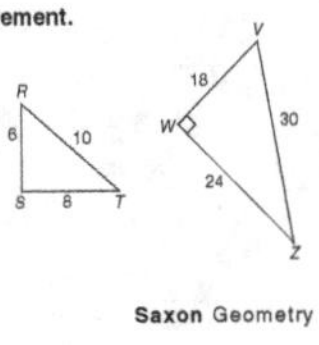

Reteaching 46
continued

Triangle Similarity: SAS

Side-Angle-Side (SAS) Similarity	
If two sides of one triangle are proportional to two sides of another triangle and their included angles are congruent, then the triangles are similar.	$\frac{AB}{DE} = \frac{AC}{DF} = \frac{5}{6}$ $\angle A = \angle D$ $\Delta ABC \sim \Delta DEF$

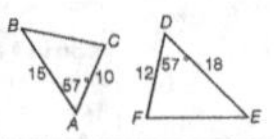

Show that $\Delta ADE \sim \Delta ABC$ Then, find BC.

Step 1: $\angle A \cong \angle A$ by the Reflexive Property.

$\frac{AD}{AB} = \frac{3}{6} = \frac{1}{2}$; $\frac{AE}{AC} = \frac{2}{4} = \frac{1}{2}$

$\Delta ADE \sim \Delta ABC$ by SAS Similarity.

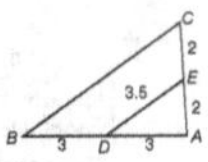

Step 2: Find BC.

$\frac{AD}{AB} = \frac{DE}{BC}$	Proportional Corresponding Sides
$\frac{3}{6} = \frac{3.5}{BC}$	Substitute.
$3(BC) = 6(3.5)$	Cross Products Property
$BC = 7$	Solve.

Practice

Complete the steps to show that the triangles are similar. Then, find GK.

3. $\overline{JK} \parallel \overline{FH}$, so $\angle J \cong$ $\underline{\angle H}$ and $\angle K \cong$ $\underline{\angle F}$ by the Alternate $\underline{\text{Interior Angles}}$ Theorem.

Δ $\underline{JGK}$ $\sim \Delta$ $\underline{FGH}$ by $\underline{\text{AA}}$ Similarity.

To find GK: $\frac{8}{12} = \frac{GK}{11}$

$8(11) = 12(GK)$

$88 = 12(GK)$

$\frac{88}{12} = GK$

$7\frac{1}{3} = GK$

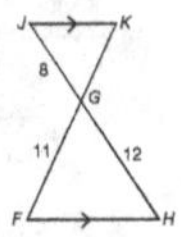

Explain why the triangles are similar and then find US.

4. It is given that $\angle TSU \cong \angle WVU$. $\angle U \cong \angle U$ by the Reflexive Property. $\Delta TSU \sim \Delta WVU$. by AA Similarity $US = 39$.

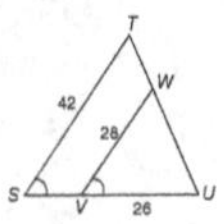

Answer Key

Name ______________________ Date __________ Class __________

Reteaching 47
Circles and Inscribed Angles

Inscribed Angles

The measure of an inscribed angle is half the measure of its intercepted arc. In the picture, $\angle ABC$ is an inscribed angle and $\overset{\frown}{AC}$ is its intercepted arc on the circle.

$m\angle ABC = \frac{1}{2}m\overset{\frown}{AC}$

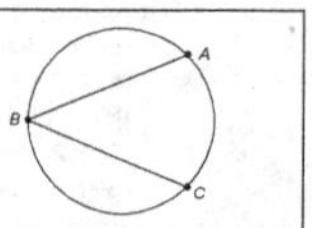

If an inscribed angle intercepts a semicircle, then it is a right angle. $\angle ACB$ intersects the semicircle, so $m\angle ACB = 90°$.

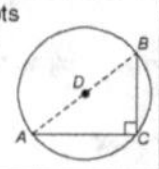

If inscribed angles intercept the same arc, then they are congruent. $\angle ABC$ and $\angle ADC$ intercept $\overset{\frown}{AC}$, so $\angle ABC \cong \angle ADC$.

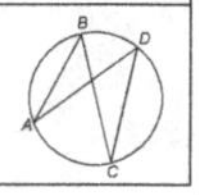

Find $m\angle GFJ$ and $m\overset{\frown}{HF}$.

Step 1: Since $m\overset{\frown}{GJ} = 110°$, $m\angle GFJ = \frac{1}{2}m\overset{\frown}{GJ} = \frac{1}{2}(110°) = 55°$.

Step 2: Since $m\angle FJH = 36°$, then $\frac{1}{2}m\overset{\frown}{HF} = m\angle FJH = 36°$.

Solve for $m\overset{\frown}{HF}$: $\frac{1}{2}m\overset{\frown}{HF} = 36°$

$m\overset{\frown}{HF} = 2(36°)$

$m\overset{\frown}{HF} = 72°$

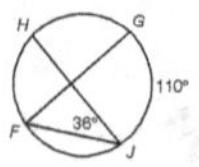

Practice

Complete the steps to solve for *x*.

1. Since $\angle RQS$ intercepts a semicircle, $m\angle RQS = 90°$.

   $(5x + 8)° = 90°$

   $5x + 8 = 90$

   $5x = 82$

   $x = 16.4$

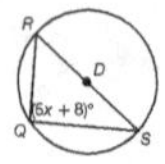

Find each value.

2. z

   9

3. $m\angle FJH$

   45°

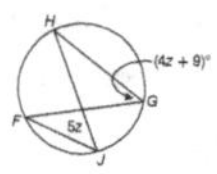

Reteaching 47
continued

Inscribed Quadrilaterals

If a quadrilateral is inscribed in a circle, then it has supplementary opposite angles.

ABCD is inscribed in circle *E*.

$\angle A + \angle C = 180°$

$\angle B + \angle D = 180°$

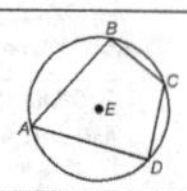

Find $m\angle G$.

Step 1: Find the value of *z*.

$m\angle E + m\angle G = 180°$ — *EFGH* is inscribed in a circle.

$8z + 14z - 18 = 180$ — Substitute.

$22z = 198$ — Simplify.

$z = 9$

Step 2: Find $m\angle G$.

$m\angle G = 14z - 18 = 14(9) - 18 = 108°$

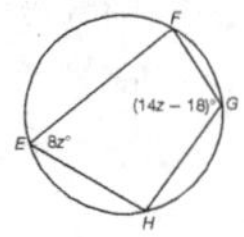

Practice

Complete the steps to find the measure of the given angles.

4. $\angle R$

   $m\angle R + m\angle T = 180°$

   $7y + 11y = 180$

   $18y = 180$

   $y = 10$

   $m\angle R = 7y = 7(10) = 70°$

5. $\angle T$

   $m\angle T = 11y$

   $m\angle T = 11(10)$

   $m\angle T = 110°$

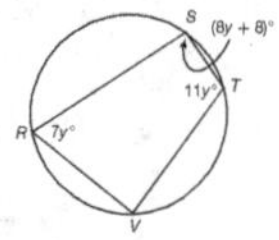

Find the measures of the given angles in RSTV.

6. $\angle S$ 88°

7. $\angle V$ 92°

Find the angle measures of each quadrilateral.

8. *ABCD*

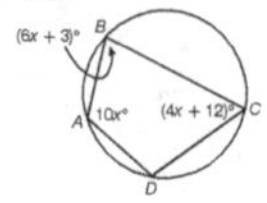

120°, 75°, 60°, 105°

9. *JKLM*

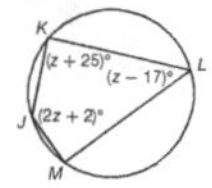

132°, 90°, 48°, 90°

Name ______________________ Date __________ Class __________

Reteaching 48
Indirect Proofs

You have worked with inscribed angles in circles. Now you will work with indirect proofs.

Indirect Proofs

In **direct reasoning**, you begin with a true hypothesis and prove that a conclusion is true. In an **indirect proof**, you begin by assuming that the conclusion is false. You then show that this assumption leads to a contradiction.

To write an indirect proof, first identify the given information and the conclusion to be proved. Next, assume either that the conclusion is false or that the opposite of the conclusion is true.

Identify the conclusion to be proved and the opposite of the conclusion to be assumed for the following statement.

If two angles are not congruent, then they are not both right angles.

Step 1: Identify the conclusion to be proved.
Prove: $\angle 1$ and $\angle 2$ are not both right angles.

Step 2: When writing an indirect proof, assume that the opposite of the conclusion is true.
Assume: $\angle 1$ and $\angle 2$ are both right angles.

Practice

Complete the steps to write the conclusion and the opposite of the conclusion for each statement.

1. If $a^2 \neq b^2$, then $a \neq b$, for any real numbers, *a* and *b*.

   Conclusion: $a \neq b$

   Assume: $a = b$

2. If the sum of the measures of two angles is not equal to 180°, then the two angles are not supplementary.

   Conclusion: $\angle 1$ and $\angle 2$ are not supplementary.

   Assume: $\angle 1$ and $\angle 2$ are supplementary.

Write the conjecture and the opposite of the conclusion for each statement.

3. If $\angle 1$ is not congruent to $\angle 1$ $\angle 2$, then line *a* intersects line *b*.

   Conclusion: Line *a* intersects line *b*.

   Assume: Line *a* is parallel to line *b*.

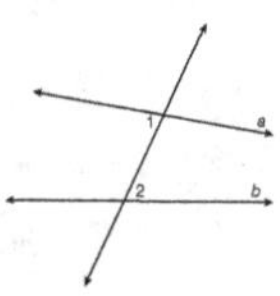

4. If ΔRST is acute, then it does not have a right angle.

   Conclusion: ΔRST does not have a right angle.

   Assume: ΔRST has a right angle.

Reteaching 48
continued

Writing an Indirect Proof

Consider the statement:

Two acute angles with a common side do not form a linear pair.

Here are the steps to write an indirect proof of the statement.

Steps	Example
1. Identify the conjecture to be proved.	Given: $\angle 1$ and $\angle 2$ are acute angles with a common side. Prove: $\angle 1$ and $\angle 2$ do not form a linear pair.
2. Assume the opposite of the conclusion is true.	Assume: $\angle 1$ and $\angle 2$ form a linear pair.
3. Use direct reasoning to show that the assumption leads to a contradiction.	$m\angle 1 + m\angle 2 = 180°$ by the definition of a linear pair. Since $m\angle 1 < 90°$ and $m\angle 2 < 90°$, $m\angle 1 + m\angle 2 < 180°$. This is a contradiction.
4. Conclude that the assumption is false. Therefore, the original conjecture must be true.	The assumption that $\angle 1$ and $\angle 2$ form a linear pair is false. Therefore, $\angle 1$ and $\angle 2$ do not form a linear pair.

Practice

Complete the indirect proof by following the directions in each step.

5. An obtuse triangle cannot have a right angle.

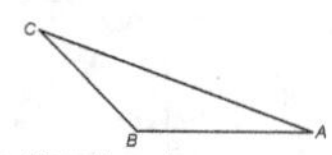

$\angle B$ is an obtuse angle.

Steps	Proof
1. Identify the conjecture to be proved.	Given: ΔABC is an obtuse triangle. $\angle B$ is an obtuse angle. Prove: ΔABC cannot have a right angle.
2. Assume the opposite of the conclusion is true. Write the assumption.	Assume: ΔABC has a right angle. Let $\angle A$ be a right angle.
3. Use direct reasoning to show that the assumption leads to a contradiction.	If $\angle A$ is a right angle, then $m\angle B + m\angle C = 90°$. But $m\angle B > 90°$ since $\angle B$ is an obtuse angle. This is a contradiction.
4. Conclude that the assumption is false. Therefore, the original conjecture must be true.	The assumption that ΔABC has a right angle is false. Therefore, ΔABC cannot have a right angle.

Name ______________________ Date __________ Class __________

Reteaching 49
Introduction to Solids

You have worked with indirect proofs. Now you will work with solid geometry.

Three-dimensional figures, or **solids**, can have flat or curved surfaces.

Prisms and pyramids are named by the shape of their **bases**.

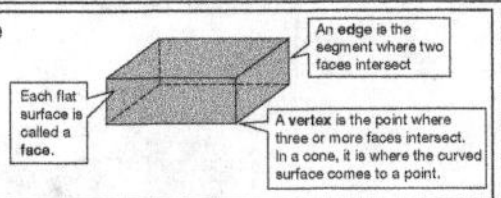

Prisms	Pyramids	Cylinders	Cones
triangular prism, rectangular prism	triangular pyramid, rectangular pyramid	Neither cylinders nor cones have edges.	

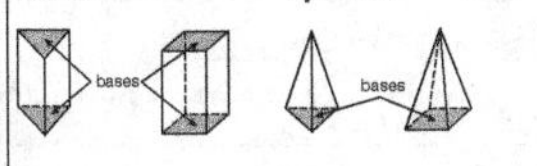

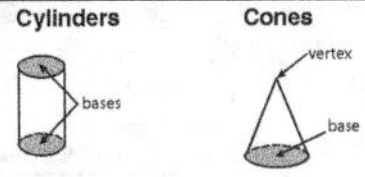

Identify the figure. Name the vertices, edges, and bases.

Step 1: The figure is a rectangular pyramid.

Step 2: The vertices are A, B, C, D, and E. The edges are $\overline{AB}$, $\overline{BD}$, $\overline{CD}$, $\overline{AC}$, $\overline{AE}$, $\overline{BE}$, $\overline{CE}$, and $\overline{DE}$. The base is quadrilateral $ABDC$.

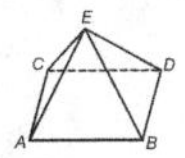

Practice

Complete the steps to identify the figure and name the vertices, edges, and bases.

1. This is a triangular pyramid.

   The vertices are Q, R, S, and T.

   The edges are $\overline{RQ}$, $\overline{SR}$, $\overline{QS}$, $\overline{QT}$, $\overline{ST}$, and $\overline{RT}$.

   The base is $\Delta\ QST$.

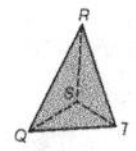

Identify each figure. Name the vertices, edges, and bases.

2. cylinder; vertices: none; edges: none; bases: circle A, circle B

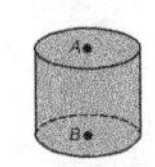

3. triangular prism: vertices: C, D, E, F, G, H; edges: $\overline{CD}, \overline{DE}, \overline{EC}, \overline{FG}, \overline{GH}, \overline{HF}, \overline{CF}, \overline{DG}, \overline{EH}$; bases: ΔCDE, ΔFGH

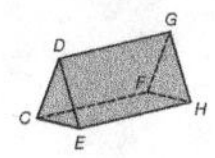

Reteaching 49
continued

Euler's Formula

A closed three-dimensional figure formed by four or more polygons that intersect only at their edges is called a **polyhedron**; for example, prisms and pyramids.

There is a unique relationship between the number of faces, vertices, and edges of any polyhedron.

Euler's formula: For any polyhedron with V vertices, E edges, and F faces, $V - E + F = 2$.

Example: Verify Euler's formula for a rectangular prism.

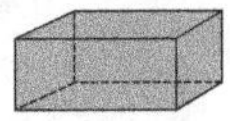

Step 1: Find the number of vertices, edges, and faces.
$V = 8$, $E = 12$, $F = 6$

Step 2: Substitute the values into Euler's formula and simplify.
$V - E + F = 8 - 12 + 6 = 2$

Practice

Complete the steps to find the missing number.

4. A triangular prism has 5 faces and 9 edges.

   How many vertices does it have?

   $V - E + F = 2$
   $V - 9 + 5 = 2$
   $V - 4 = 2$
   $V = 6$

   It has 6 vertices.

5. A rectangular pyramid has 5 faces and 5 vertices.

   How many edges does it have?

   $V - E + F = 2$
   $5 - E + 5 = 2$
   $10 - E = 2$
   $-E = -8$
   $E = 8$

   It has 8 edges

Use Euler's formula to solve each exercise.

6. A polyhedron has 6 vertices and 4 faces.

   How many edges does it have?

   8

7. A polyhedron has 12 faces and 18 vertices.

   How many edges does it have?

   28

8. A polyhedron has 16 edges and 10 vertices.

   How many faces does it have?

   8

9. A polyhedron has 9 vertices and 12 edges.

   How many faces does it have?

   5

Name ______________________ Date __________ Class __________

Reteaching 50
Geometric Mean

You have worked with solids. Now you will work with geometric mean.

Geometric Mean

The geometric mean for two positive numbers, a and b, is the positive number x such that $\frac{a}{x} = \frac{x}{b}$.

Find the geometric mean of 4 and 12.

Let x represent the geometric mean of 4 and 12. Write the algebraic expression for finding the geometric mean and solve for x:

$\frac{a}{x} = \frac{x}{b}$

$\frac{4}{x} = \frac{x}{12}$ Substitude 4 and 12.

$(x) = 4(12)$ Find the cross product.

$x^2 = 48$

$x = \sqrt{48}$ Find the square root.

$x \approx 6.9$ Round to the nearest tenth.

Practice

Complete the steps to find the geometric mean for each pair of numbers. Round your answer to the nearest tenth.

1. 9 and $\frac{2}{3}$

   $\frac{9}{x} = \frac{x}{2/3}$

   $x(\underline{x}) = 9\left(\frac{2}{3}\right)$

   $x^2 = 6$

   $x = \sqrt{6}$

   $x \approx 2.4$

2. 8 and 15

   $\frac{8}{x} = \frac{x}{15}$

   $x(\underline{x}) = 8(15)$

   $x^2 = 120$

   $x = \sqrt{120}$

   $x \approx 11$

3. 10 and 6

   $\frac{10}{x} = \frac{x}{6}$

   $x(\underline{x}) = 10(6)$

   $x^2 = 60$

   $x = \sqrt{60}$

   $x \approx 7.7$

Find the geometric mean for each pair of numbers. Round your answer to the nearest tenth.

4. 7 and $\frac{1}{3}$ 1.5
5. 9 and 4 6
6. 14 and 5 8.3
7. 25 and $\frac{3}{5}$ 3.9
8. 11 and 6 8.1
9. 4 and 16 8

Reteaching 50
continued

When an altitude is drawn from the vertex of the 90° angle of a right triangle to the hypotenuse of the right triangle, some useful relationships are formed.

If the altitude is drawn to the hypotenuse of a right triangle, then the two triangles formed are similar both to each other and to the original triangle.

$\Delta JLM \sim \Delta MLK \sim \Delta JMK$

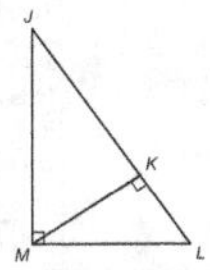

The length of the altitude is the geometric mean between the segments of the hypotenuse.	The length of a leg is the geometric mean between the hypotenuse and the segment of the hypotenuse that is closer to that leg.

Example:

Given ΔXYZ, find the missing value, t.

Step 1: Since $\overline{XW}$ is an altitude of a right triangle, t is the geometric mean of the segments of the hypotenuse. The segments measure 27 and 3.

Since t is the geometric mean of 27 and 3, write $\frac{27}{t} = \frac{t}{3}$.

Step 2: Solve for t: $\frac{27}{t} = \frac{t}{3}$
$t^2 = 81$
$t = 9$

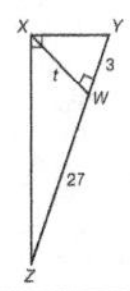

Practice

10. Find the value of a and b.

    Since a and b are the legs of the triangle, each one is a geometric mean of the hypotenuse and the segment closer to that leg.

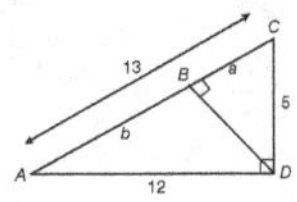

$\frac{13}{5} = \frac{5}{a}$ $\frac{13}{12} = \frac{12}{b}$

$13(a) = (5)\,5$ $13(b) = (12)\,12$

$a = \frac{25}{13}$ $b = \frac{144}{3}$

$a = 1.9$ $b = 11.1$

Name ______________________ Date __________ Class __________

Reteaching INV 5
Nets

You have worked with two and three-dimensional figures before. Now you'll work with nets, which are 2-D representations of 3-D figures.

Making a 3-D Figure from a Net

A net is a two-dimensional representation of a 3-D figure. Each shape in the net represents one face of the figure. By folding along the edges of the figures, the net can be made into the 3-D figure it represents.

Example: Figure 1 shows a net of a cube. Each square represents one face of the cube. In figure 2 the squares are being folded upwards so that it becomes the cube in figure 3.

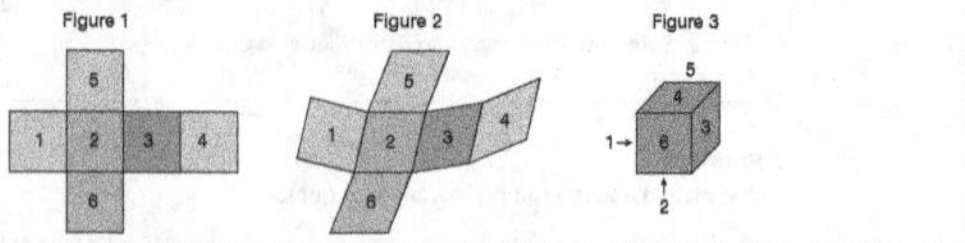

Practice

Copy the nets given in problems 1–3. Form the 3-D figures they represent. State the name of each figure.

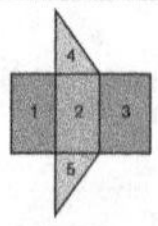

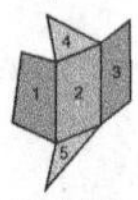

1. Step One: Fold the triangles up along the edges of the rectangle.

   Step Two: The figure has two triangular bases and three rectangular faces. The figure is a <u>prism</u>.

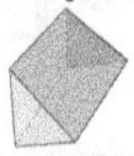

2. <u>pentagonal prism</u>

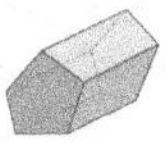

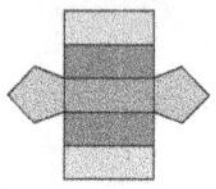

3. Can you form a rectangular prism with the net in figure 5? If so, form it. If not, explain why.
   <u>A rectangular prism cannot be formed with this net, because there is no way to give the shape a side parallel to side 6.</u>

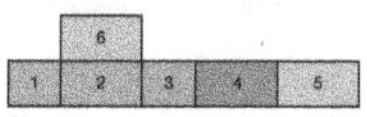

Reteaching INV 5
continued

Making Nets of a 3-D Figure

A 3-D figure with flat faces is called a polyhedron. To construct the net of a figure, imagine unfolding it along its edges so that it is a flat shape.

Example: The cube in figure 1 unfolds in figure 2 to become the net in figure 3.

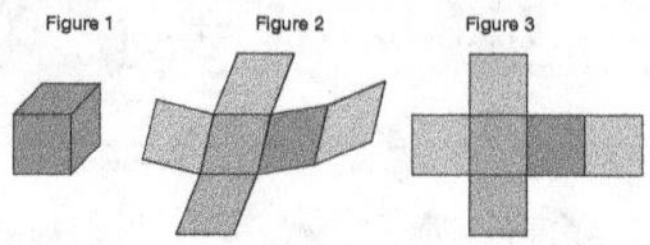

Most polyhedrons have more than one net. Can you think of a second net that represents the same cube as the one represented by figure 1?

Practice

Draw a net for the given polyhedron.

4. Draw a possible net for this triangular prism.

   Step One: Fold the triangles out.

   Step Two: Fold the rectangles out.

   Step Three: Draw the net.

   You can verify your answer by re-creating the triangular prism using the net you've made.

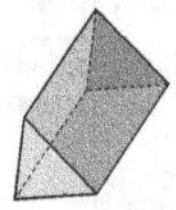

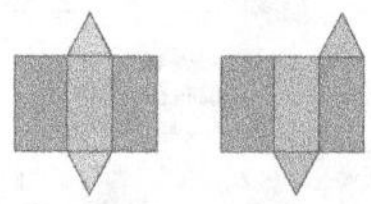

5. Draw two possible nets for this truncated pyramid.

   (Note: the sides are trapezoids and the top and bottom are squares)

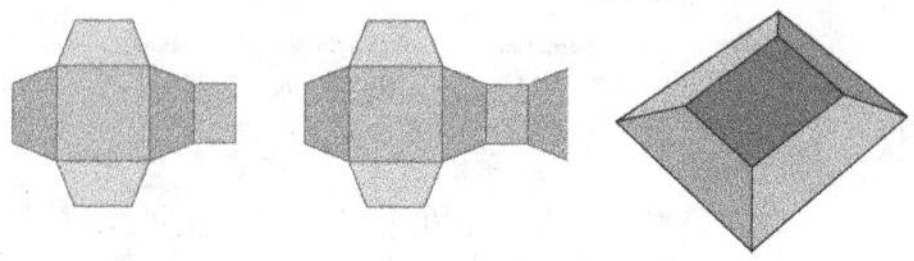

Name ______________________ Date __________ Class __________

Reteaching 51
Properties of Isosceles and Equilateral Triangles

You have worked with different types of triangles. Now you will work specifically with isosceles and equilateral triangles.

Isosceles Triangles	
Isosceles Triangle Theorem If two sides of a triangle are congruent, then the angles opposite the sides are congruent.	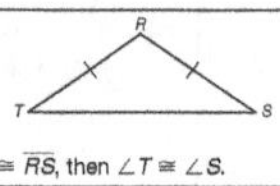 If $\overline{RT} \cong \overline{RS}$, then $\angle T \cong \angle S$.
Converse of Isosceles Triangle Theorem It two angles of a triangle are congruent, then the sides opposite those angles are congruent.	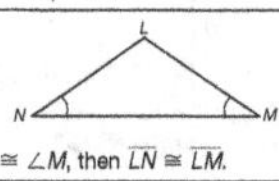 If $\angle N \cong \angle M$, then $\overline{LN} \cong \overline{LM}$.

You can use these theorems to find angle measures in isosceles triangles.

Find $m\angle E$ in ΔDEF.

$m\angle D = m\angle E$	Isosceles Triangle Theorem
$5x = 3x + 14$	Substitute.
$2x = 14$	Subtract $3x$ from both sides.
$x = 7$	Divide both sides by 2.

$m\angle E = 3(7) + 14 = 35°$

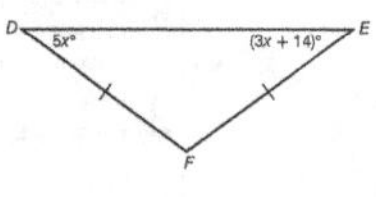

Practice

Complete the statements to find $m\angle A$ and $m\angle C$.

1. $m\angle A + m\angle B + m\angle C = 180°$

   <u>$x°$</u> + <u>$78°$</u> + $x° = 180°$

   <u>$2x$</u> + $78° = 180°$

   $2x =$ <u>$102°$</u>

   $x =$ <u>$51°$</u>

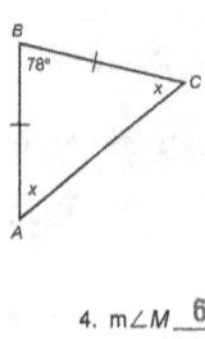

Find each angle measure.

2. $m\angle Q$ <u>$47°$</u>

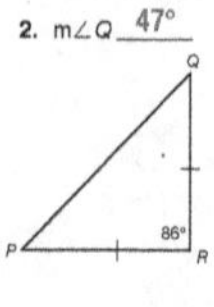

3. $m\angle H$ <u>$72°$</u>

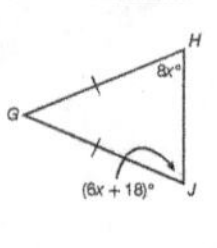

4. $m\angle M$ <u>$60°$</u>

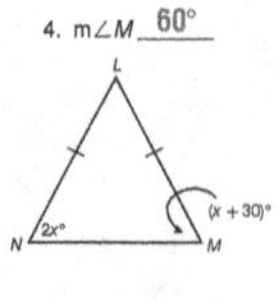

Reteaching 51
continued

Equilateral Triangles

Equilateral Triangle Corollary

If a triangle is equiangular, then it is equilateral.

If $\angle A \cong \angle B \cong \angle C$, then $\overline{AB} \cong \overline{BC} \cong \overline{CA}$.

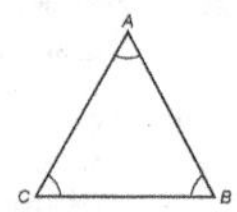

You can use this theorem to find values in equilateral triangles.

Find x in ΔSTV.

$7x + 4 = 608$

$7x = 56°$	Subtract 4 from both sides.
$x = 8$	Divide both sides by 7.

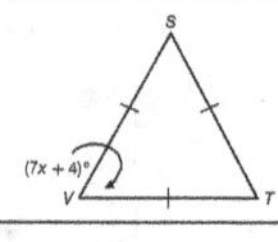

Practice

Complete the statements to find the value.

5. $2x - 6 = 60$

   <u>$2x$</u> $= 66$

   $x =$ <u>33</u>

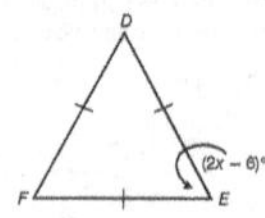

Find each value.

6. $n =$ <u>12</u>

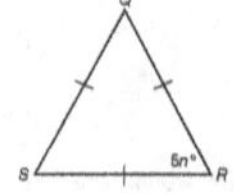

7. $VT =$ <u>18</u>

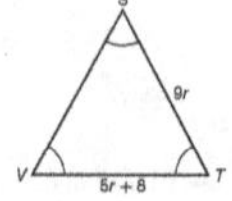

8. $MN =$ <u>13</u>

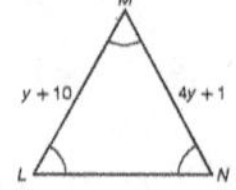

9. A triangle is equiangular and has a perimeter of 37.2 inches. Determine the lenght of each side. <u>12.4 inches</u>

Name ______________________ Date __________ Class __________

Reteaching 52
Properties of Rectangles, Rhombuses, and Squares

You have worked with special triangles. Now you will work with properties of parallelograms.

Properties of Special Parallelograms

Properties of Rectangles	Properties of a Rhombus
The diagonals of a rectangle are congruent. $\overline{GJ} \cong \overline{HK}$	The diagonals of a rhombus are perpendicular. Each diagonal of a rhombus bisects opposite angles. $\overline{QS} \perp \overline{RT}$, $\angle RQS \cong \angle SQT$

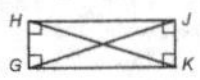

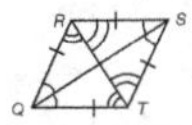

ABCD is a rectangle. Find the length of $\overline{CD}$ and of $\overline{BD}$.

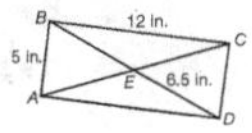

$CD = 5$ Opposite sides of a rectangle are congruent.
$\overline{BE} \cong \overline{ED}$ Diagonals of a parallelogram bisect each other.
It is given that $ED = 6.5$. Therefore, $BE = 6.5$ also.
$BD = BE + ED$ Segment Addition Postulate
$BD = 6.5 + 6.5 + 13$ inches Substitution

Practice
Complete the statements.

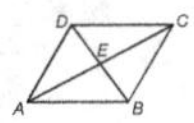

1. If $m\angle ABC = 128$, then $m\angle ABE = \underline{64}$. Diagonals of a rhombus bisect opposite angles.

   Use the sum of the angles in a quadrilateral to find and $m\angle BCD$.

   $128° + 128° + x + x = 360°$

   $\underline{256°} + 2x = \underline{360°}$

   $2x = \underline{104°}$

   $x = \underline{52°}$

   $m\angle BCD = \underline{52°}$

   $m\angle BCE = \underline{26°}$

***KLMN* is a rhombus. Find each measure.**

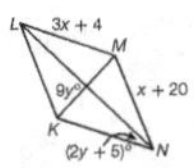

2. $KL = \underline{28}$
3. $m\angle MNK = \underline{50°}$

Reteaching 52
continued

Using Properties of Parallelograms

HJKL is a parallelogram. Decide what type of parallelogram it is by using the properties of rectangles and rhombuses.

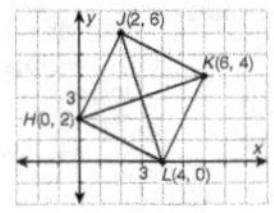

Step 1: Determine if $\overline{HK} \cong \overline{JL}$. Use the distance formula.
$HK = \sqrt{(6-0)^2 + (4-2)^2} = 2\sqrt{10}$
$JK = \sqrt{(4-2)^2 + (0-6)^2} = 2\sqrt{10}$
$HK = JL = 2\sqrt{10}$, so $\overline{HK} \cong \overline{JL}$, and *HJKL* is a rectangle.

Step 2: Determine if $\overline{HK} \perp \overline{JL}$ Find the slope.
slope of $\overline{HK} = \frac{4-2}{6-0} = \frac{1}{3}$ slope of $\overline{JL} = \frac{0-6}{4-2} = -3$
Since the product of the slopes is -1, $\overline{HK} \perp \overline{JL}$, and *HJKL* is a rhombus.

Step 3: Since *HJKL* is both a rectangle and a rhombus, it is a square.

Practice

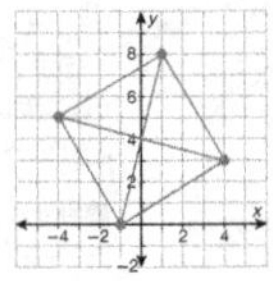

4. Determine if $\overline{AC} \cong \underline{\overline{BD}}$.

   $AC = \sqrt{(\underline{8} - 0)^2 + (\underline{1} - (-1))^2}$

   $= \sqrt{\underline{64} + 4} = \sqrt{\underline{68}}$ $AC = \underline{BD}$, so $\overline{AC} \cong \overline{BD}$.

   $\underline{BD} = \sqrt{(\underline{5} - 3)^2 + (-4 - 4)^2}$ *ABCD* is a $\underline{\text{rectangle}}$.

   $= \sqrt{\underline{4} + 64} = \sqrt{\underline{68}}$

5. Determine if $\overline{BD} \perp \underline{\overline{AC}}$. Find the slope.

   slope of $\overline{BD} = \frac{\underline{5} - 3}{\underline{-4} - 4} = \frac{\underline{2}}{-8} = -\frac{1}{4}$ slope of $\overline{AC} = \frac{\underline{8} - 0}{1 - (-1)} = \frac{8}{2} = \underline{4}$

   The product of the slopes is -1, so $\overline{BD} \perp \underline{\overline{AC}}$.

   ABCD is a $\underline{\text{rhombus}}$. Since ABCD is both a $\underline{\text{rectangle}}$ and a $\underline{\text{rhombus}}$, it is a $\underline{\text{square}}$.

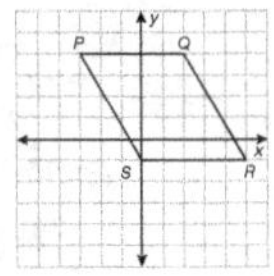

6. Find the lengths and the slopes of the diagonals. Determine what type of parallelogram *PQRS* is.

   $PR = \underline{4\sqrt{5}}$, and $QS = \underline{2\sqrt{5}}$.

   PQRS $\underline{\text{is not}}$ a rectangle.

   Slope of $PR = \underline{-\frac{1}{2}}$, and slope of $QS = \underline{2}$.

   PQRS $\underline{\text{is}}$ a rhombus.

Name ______________________ Date __________ Class __________

Reteaching 53
45°-45°-90° Right Triangles

You have worked with right triangles. Now you will work with a particular kind of right triangle.

Special Right Triangle

Side Lengths in 45°-45°-90° Triangles	Example
In a 45°-45°-90° right triangle, both legs are congruent, and the length of the hypotenuse is the length of a leg multiplied by $\sqrt{2}$.	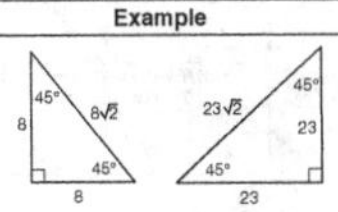

Find the value of *x* in $\triangle EFG$

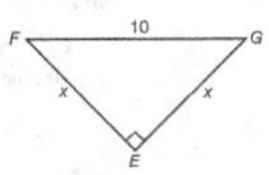

The hypotenuse has length 10.

$10 = x\sqrt{2}$ Hypotenuse is $\sqrt{2}$ times the length of a leg.
$\frac{10}{\sqrt{2}} = \frac{x\sqrt{2}}{\sqrt{2}}$ Divide both sides by $\sqrt{2}$.
$5\sqrt{2} = x$ Rationalize the denominator.

Practice
Complete the statements to find the value of *x*.

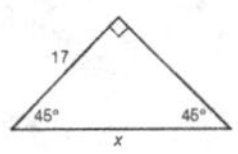

1. The leg is given. It is $\underline{17}$.

   The hypotenuse is the length of the leg times $\sqrt{2}$.

   The hypotenuse is $\underline{17\sqrt{2}}$, so $x = \underline{17\sqrt{2}}$

Find the value of *x*.

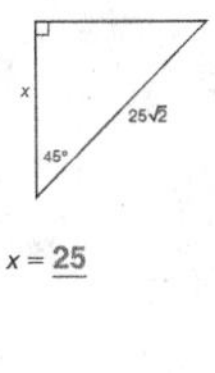

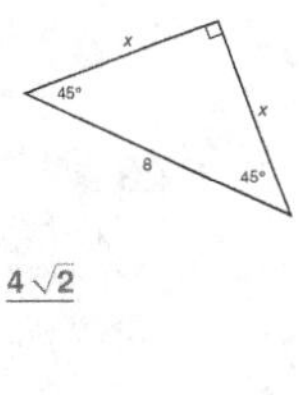

2. $x = \underline{25}$
3. $x = \underline{4\sqrt{2}}$

Reteaching 53
continued

Finding the Perimeter of a 45°-45°-90° Right Triangle

Find the perimeter of the triangle.

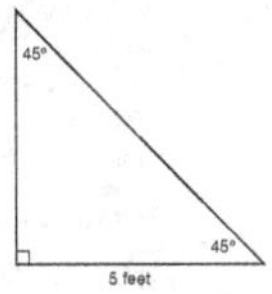

Step 1: Determine the length of each leg.
The legs are each 5 feet long.

Step 2: Determine the length of the hypotenuse.
The hypotenuse is $\sqrt{2}$ times the length of a leg, so the hypotenuse is $5\sqrt{2}$ feet.

Step 3: The perimeter is the sum of all the sides.
$P = 5 + 5 + 5\sqrt{2}$
$P = 10 + 5\sqrt{2}$
$P \approx 17.07$
The perimeter is approximately 17 feet.

Practice
Complete the statements to determine the perimeter of the triangle.

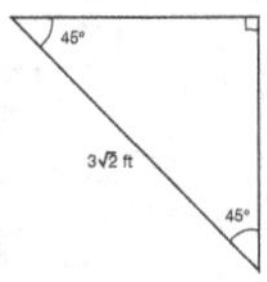

4. Determine the length of the hypotenuse.

   The hypotenuse is $\underline{3\sqrt{2}}$.

   Determine the length of each leg.
   The legs are each 3 feet long.
   The perimeter is the sum of all the sides.

   $P = \underline{3} + 3 + \underline{3\sqrt{2}}$

   $P = \underline{6} + 3\sqrt{2}$

   $P \approx \underline{10.24}$

   The perimeter is approximately $\underline{10}$ feet.

Determine the perimeter of each 45°-45°-90° right triangle.

5. 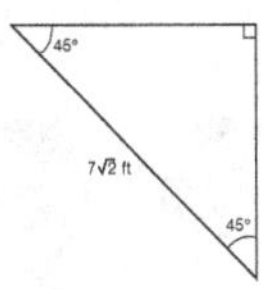

   $\underline{23.90 \approx 24\text{ ft}}$

6. 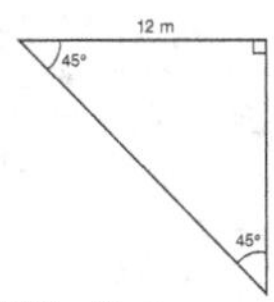

   $\underline{40.97 \approx 41\text{ m}}$

Name ______________________ Date __________ Class __________

Reteaching 54
Representing Solids

You have worked with different solids. Now you will work on drawing different perspectives of solids.

Drawing in One-Point Perspective

In a one-point perspective drawing, non-vertical lines are drawn so that they meet at a vanishing point.

Example:

Draw a triangular prism in one-point perspective.

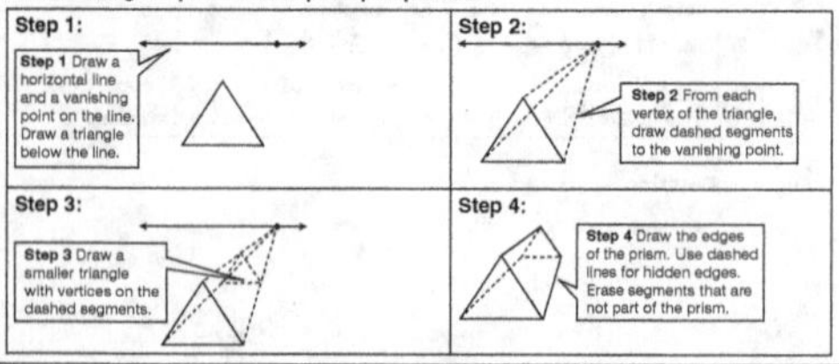

Practice

1. Draw a rectangular prism in one-point perspective that has the given rectangle as its front face and the given point as its vanishing point.

Step 1: The rectangle is the front face of the solid. The given point is the vanishing point.

Step 2: Draw a dashed line from the vanishing point to each vertex of the rectangle.

Step 3: Draw a smaller rectangle with vertices on the dashed segments.

Step 4: Draw the edges of the prism. Use dashed lines for hidden edges. Erase segments that are not part of the prism.

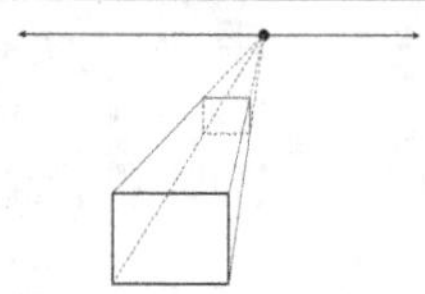

Reteaching 54
continued

Creating Isometric Drawings

An isometric drawing is a way of drawing a solid figure by using isometric dot paper. The drawings can be made by using three axes that intersect to form 120° angles.

Example:

Create an isometric drawing of a triangular prism.

Step 1: Draw the three axes on isometric dot paper.

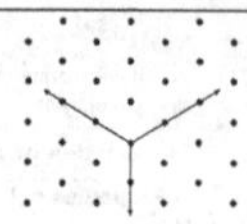

Step 2: Use the intersection of the three axes as the bottom vertex of the prism.

Step 3: Draw the prism so that the edges of the prism are parallel to the three axes.

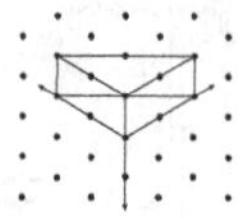

Practice

2. Follow the steps below to create an isometric drawing of a cube.

Step 1: Draw the three axes on isometric dot paper.

Step 2: Use the intersection of the three axes as the bottom vertex of the cube.

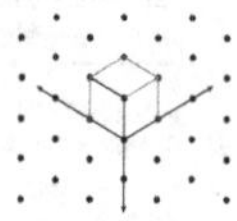

Step 3: Draw the cube so that the edges of the cube are parallel to the three axes.

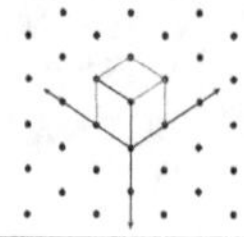

3. Create an isometric drawing of a rectangular prism.

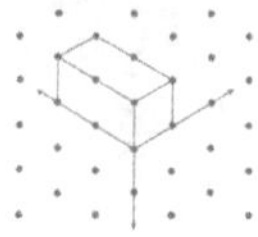

Name ______________________ Date __________ Class __________

Reteaching 55
Triangle Midsegment Theorem

You have worked with midpoints. Now you will work with midsegments of triangles.

Triangle Midsegment Theorem

The segment joining the midpoints of two sides of a triangle is parallel to, and half the length of, the third side.

$\overline{PQ}$ is the midsegment of ΔLMN.

$\overline{PQ} \parallel \overline{LN}$, $PQ = \frac{1}{2}LN$

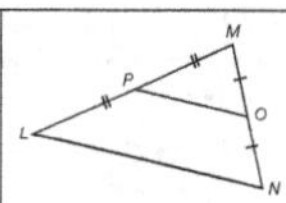

Example:

Find the following measures in ΔABC.

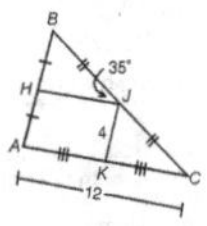

HJ

$HJ = \frac{1}{2}AC$ Δ Midsegement Theorem

$HJ = \frac{1}{2}(12)$ Substitute 12 for *AC*.

$HJ = 6$ Solve.

AB

$JK = \frac{1}{2}AB$ Δ Midsegement Theorem

$4 = \frac{1}{2}AB$ Substitute 4 for *JK*.

$8 = AB$ Solve.

m∠*BCA*

$HJ \parallel AC$ Δ Midsegment Theorem

$m\angle BCA = m\angle BJH$ Corresponding Angle Theroem

$m\angle BCA = 35°$ Substitute 35° for $m\angle BJH$.

Practice

Complete the statements to find the measure.

1. *VX*

$VX = \frac{1}{2}GH$

$VX = \frac{1}{2}(\underline{46})$

$VX = \underline{23}$

2. *HJ*

$\underline{WV} = \frac{1}{2}HJ$

$\underline{27} = \frac{1}{2}HJ$

$\underline{54} = HJ$

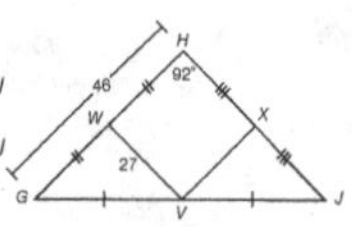

Find each measure.

3. *ST* $\underline{72}$

4. *DE* $\underline{22}$

5. m∠*DES* $\underline{48°}$

6. m∠*RCD* $\underline{48°}$

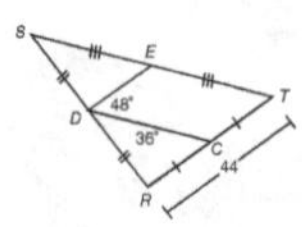

Reteaching 55
continued

Finding Midpoints of Sides of Triangles

A midsegment of a triangle joins the midpoints of two sides of the triangle. Every triangle has three midsegments. We can use the Midpoint Formula to find the coordinates of the endpoints of a midsegment.

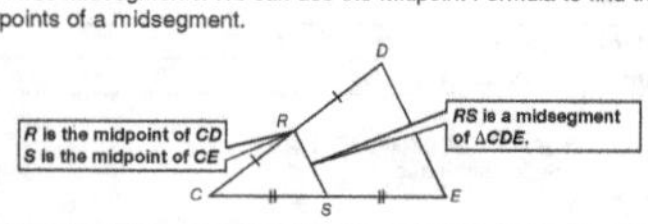

Practice

Complete the statements to find the coordinates of the midsegment.

7. Triangle *MNP* has vertices *M*(−4, 7), N(4, 5), and *P*(2, −1). $\overline{UV}$ is a midsegment of ΔMNP. Find the coordinates of *U* and *V*.

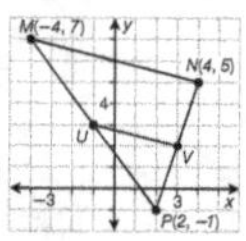

U and *V* are the midpoints of $\overline{MP}$ and $\overline{PN}$.

$U = \left(\frac{-4 + 2}{2}, \frac{7 - 1}{2}\right) = \left(\frac{-2}{2}, \frac{6}{2}\right) = (\underline{-1}, 3)$

$V = \left(\frac{4 + 2}{2}, \frac{5 - 1}{2}\right) = \left(\frac{6}{2}, \frac{4}{2}\right) = (\underline{3}, 2)$

Solve.

8. Triangle *RST* has vertices *R*(0, −3), *S*(2, 3), and *T*(6, −1). $\overline{AB}$ is the midsegment of ΔRST. Find the coordinates of *A* and *B*. *A* and *B* are the midpoints of $\overline{RS}$ and $\overline{ST}$.

$\underline{A = (1, 0); B = (4, 1)}$

9. Triangle *XYZ* has vertices *X*(3, 0), *Y*(5, 6), and *Z*(9, 2). $\overline{LM}$ is the midsegment of ΔXYZ. Find the coordinates of *L* and *M*. *L* and *M* are the midpoints of $\overline{XY}$ and $\overline{YZ}$.

$\underline{L = (4, 3); M = (7, 4)}$

10. Triangle *ABC* has vertices *A*(1, 12), *B*(9, 10), and *C*(7, 4). $\overline{DE}$ is the midsegment of ΔABC. Find the coordinates of *D* and *E*. *D* and *E* are the midpoints of $\overline{BC}$ and $\overline{AC}$.

$\underline{D = (8, 7); E = (4, 8)}$

Answer Key

Name ______________________ Date __________ Class __________

Reteaching 56

30°-60°-90° Right Triangles

You have worked with 45°-45°-90° right triangles. Now you will work with 30°-60°-90° right triangles.

Special Right Triangle

30°-60°-90° Triangle	Example
In a 30°-60°-90° right triangle, the length of the hypotenuse is twice the length of the short leg, and the length of the longer leg is the length of the shorter leg times $\sqrt{3}$.	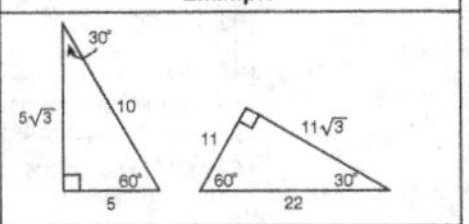

Example

Find the values of x and y in $\triangle HJK$.

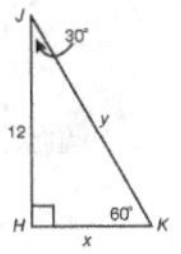

$12 = x\sqrt{3}$	Longer leg = shorter leg times $\sqrt{3}$.
$\frac{12}{\sqrt{3}} = \frac{x\sqrt{3}}{\sqrt{3}}$	Divide both sides by $\sqrt{3}$.
$4\sqrt{3} = x$	Rationalize the denominator.
$y = 2x$	Hypotenuse equals two times the shorter leg.
$y = 2(4\sqrt{3})$	Substitute $4\sqrt{3}$ for x.
$y = 8\sqrt{3}$	Simplify.

Practice

Complete the statements to find the value of x and y.

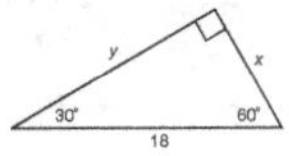

1. $18 = 2x$

   $\underline{9} = x$

2. $y = x\sqrt{3}$

   $y = \underline{9}\sqrt{3}$

Find the value of x and y.

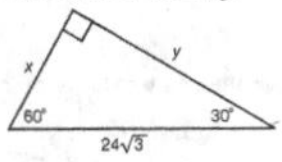

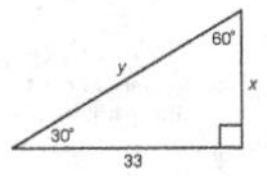

3. $\underline{x = 12\sqrt{3};\ y = 36}$

4. $\underline{x = 11\sqrt{3};\ y = 22\sqrt{3}}$

Reteaching 56

continued

Finding the Perimeter of a 30°-60°-90° Right Triangle

Example:

Find the perimeter of the triangle.

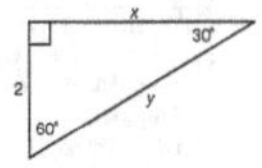

Step 1: Determine the length of each leg.

The short leg is 2 ft.

The longer leg is $2\sqrt{3}$ ft.

Step 2: Determine the length of the hypotenuse.

The hypotenuse is 2 times the shorter leg.

$2(2) = 4$ft

Step 3: The perimeter is the sum of all the sides.

$P = 2 + 4 + 2\sqrt{3}$

$P = 6 + 2\sqrt{3}$

$P \approx 9.46$

The perimeter is approximately 9 feet.

Practice

Complete the statements to determine the perimeter of the triangle.

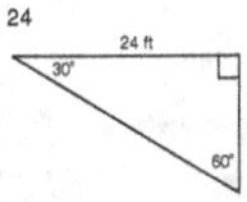

5.

Shorter Leg:	Hypotenuse:	Perimeter:
$\underline{24} = x\sqrt{3}$	$y = 2x$	$P = \underline{8\sqrt{3}} + \underline{16\sqrt{3}} + 24$
$\frac{\underline{24}}{\sqrt{3}} = \frac{x\sqrt{3}}{\sqrt{3}}$	$y = 2(\underline{8\sqrt{3}})$	$= \underline{24\sqrt{3}} + 24$
$\underline{8\sqrt{3}} = x$	$y = \underline{16\sqrt{3}}$	$\approx \underline{65.56}$

The perimeter is about 66 feet.

Determine the perimeter of each 30°-60°-90° triangle.

6. 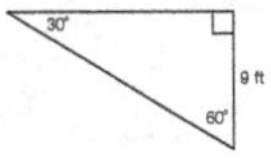

   $\underline{42.58 \approx 43 \text{ feet}}$

7. 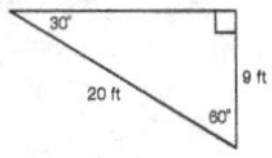

   $\underline{47.32 \approx 47 \text{ feet}}$

Name ______________________ Date __________ Class __________

Reteaching 57

Finding Perimeter and Area with Coordinates

You have worked with special triangles. Now you will find perimeter and area with coordinates.

Finding Perimeter with Coordinates

Find the perimeter of rectangle *RSTV* with coordinates $R(-5, 1)$, $S(-3, 5)$, $T(3, 2)$, and $V(1, -2)$.

Step 1: Plot the points on a coordinate plane and draw *RSTV*.

Step 2: Use the distance formula to calculate the length of each side. Find the length of $\overline{RS}$ first.

$d_{RS} = \sqrt{(x_2 - x_1)^2 + (y_2 - y_1)^2}$ Because the figure is a rectangle, $d_{TV} = 2\sqrt{5}$.

$= \sqrt{(-3-(-5))^2 + (5 - 1)^2}$

$= \sqrt{(2)^2 + (4)^2}$

$= \sqrt{20}$

$= 2\sqrt{5}$

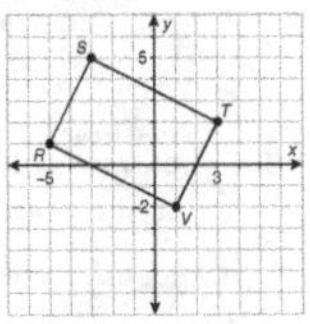

Step 3: Find the length of one of the longer sides.

$d_{RV} = \sqrt{(1 - (-5))^2 + (-2 - 1)^2} = 3\sqrt{5}$

Step 4: The length of the opposite side, *ST*, is $3\sqrt{5}$.

Step 5: Perimeter: $P = RS + ST + TV + RV$

$P = 2\sqrt{5} + 3\sqrt{5} + 2\sqrt{5} + 3\sqrt{5}$

$P = 10\sqrt{5}$ or approximately 22.36

Practice

Complete the steps to find the perimeter of the triangle.

1. $d_{CD} = \sqrt{(3 - (\underline{-1}))^2 + (2 - (\underline{-2}))^2}$

   $= \sqrt{\underline{4}^2 + \underline{4}^2}$

   $= \sqrt{\underline{32}}$

   ≈ 5.66

   $d_{BC} = \sqrt{(5 - \underline{3})^2 + (0 - \underline{2})^2}$

   $= \sqrt{\underline{4} + 4}$

   $= \sqrt{\underline{8}}$

   ≈ 2.83

   $d_{DB} = \sqrt{(5 - (\underline{-1}))^2 + (0 - (\underline{-2}))^2}$

   $= \sqrt{\underline{40}}$

   ≈ 6.32

   $P \approx 5.66 + \underline{2.83} + \underline{6.32} \approx \underline{14.81}$

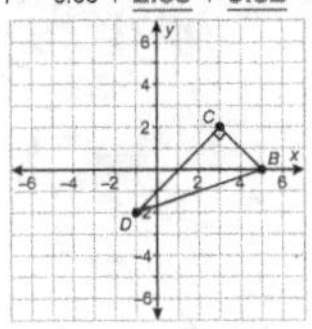

Reteaching 57

continued

Estimating Area with Coordinates

Find the area of the polygon with vertices $N(-4, -1)$, $P(-1, 3)$, $Q(4, 3)$, and $R(2, -2)$.

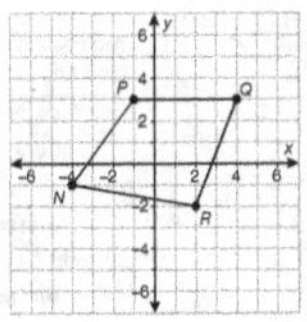

Step 1: Count all the squares completely inside the polygon: 17.

Step 2: Estimate the area of the remaining space.
Look for triangles.

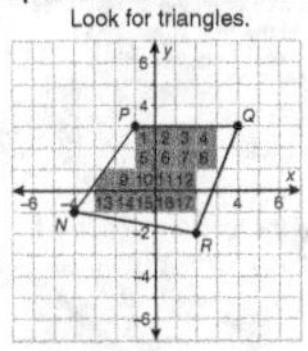

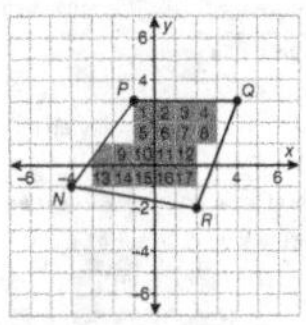

Two right triangles under point *Q* each have an area about 1.5 square units, for a total of 3 square units. Another right triangle by point *P* has an area of about 2 square units. The right triangle by point *N* has an area of about 1 square unit. The right triangle at the bottom of the polygon has an area of about 3 square units. There is one remaining area near point *N* that is almost one whole square unit.

Step 3: Add the estimates together to estimate the area of polygon *NPQR.*

$A \approx 17 + 3 + 2 + 1 + 3 + 1 \approx 27$ square units

Practice

Complete the steps to estimate the area of the polygon.

2. Find the areas of the different figures in the polygon.

   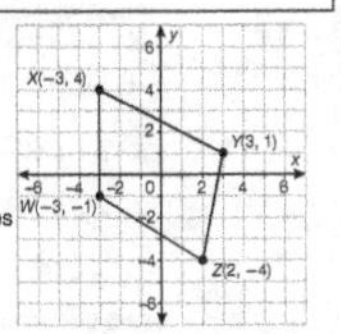

   The number of complete squares inside the polygon is $\underline{20}$.

   Estimate the remaining area by finding areas of right triangles inside the polygon: $\underline{9}$ square units.

   $A \approx 20 + \underline{9} = \underline{29}$ square units

3. Find the areas of the different figures in the polygon.

   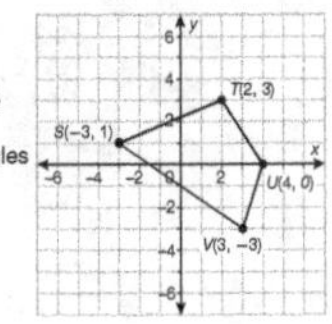

   The number of complete squares inside the polygon is $\underline{10}$.

   Estimate the remaining area by finding areas of right triangles inside the polygon: $\underline{9.5}$ square units.

   $A \approx 10 + \underline{9.5} = \underline{19.5}$ square units

Name ______________________ Date __________ Class __________

Reteaching 58
Tangents and Circles, Part 1

Tangent Lines and Angle Measures

A tangent line is a line that is in the same plane as a circle and intersects the circle at exactly one point.

Theorems		
If a line is tangent to a circle, then the line is perpendicular to a radius drawn to the point of tangency. $\overline{AB} \perp \overline{DC}$	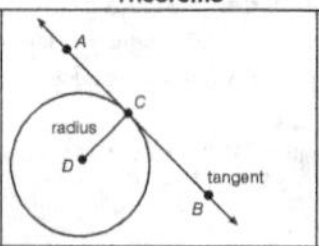	If a line in the plane of a circle is perpendicular to a radius at its endpoint on the circle, then the line is tangent to the circle. $\overline{AB}$ is tangent to $\odot D$.

Which line is tangent to $\odot A$ at point E? Line b is tangent to $\odot A$ at point E.

If $m\angle ABE = 45°$, find $m\angle BAE$.

Since $\overline{AE}$ is the radius of the circle, $\overline{AE} \perp b$ and $m\angle AEB = 90°$.

Therefore, $\triangle BAE$ is a right triangle.

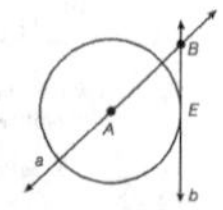

$m\angle ABE + m\angle BAE = 90°$

$45° + m\angle BAE = 90°$

$m\angle BAE = 90° - 45°$

$m\angle BAE = 45°$

Practice

If line *m* is tangent to $\odot S$ at point *R* and $m\angle STR = 28°$, find $m\angle SRT$ and $m\angle RST$.

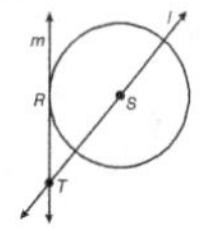

1. Since $\overline{RS}$ is the radius of the circle and is perpendicular to line m, $m\angle SRT = 90°$.

   $m\angle STR + m\angle RST = 90°$

   $28° + m\angle RST = 90°$

   $m\angle RST = 90° - 28°$

   $m\angle RST = 62°$

Line *b* is tangent to $\odot J$ at point *N*, and $m\angle NJM = 32°$.

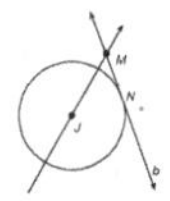

2. Find $m\angle JNM$. $90°$

3. Find $m\angle JMN$. $58°$

Reteaching 58
continued

Applying Relationships of Tangents from an Exterior Point

Theorem
If two segments are tangent to a circle from the same exterior point, then they are congruent. $\overline{EF} \cong \overline{EG}$

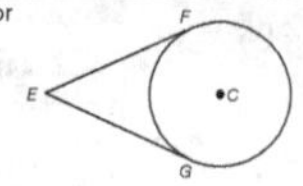

$\overline{KN}$ and $\overline{MN}$ are tangent to $\odot L$ from point *N*. Determine the perimeter of quadrilateral *KLMN*.

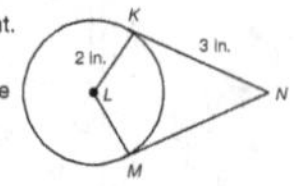

Since $\overline{LM}$ and $\overline{LK}$ are radii of the same circle, they are congruent. Therefore, $LM = 2$ in.

Since $\overline{KN}$ and $\overline{MN}$ are tangents to the same circle from the same exterior point, they are congruent.

Therefore, $MN = 3$ in.

$P = KL + LM + MN + KN$

$= 2 + 2 + 3 + 3$

$= 10$ in.

Practice

The segments are tangent to the circle from the same exterior point. Complete the steps to find the perimeter of the quadrilateral.

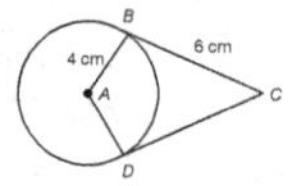

4. $AB =$ 4 cm; $AD =$ 4 cm

   $BC =$ 6 cm; $CD =$ 6 cm

   $P = AB + AD + BC + CD$

   $P = 4 + 4 + 6 + 6$

   $P = 20$ cm

The segments are tangent to the circle from the same exterior points. Find the perimeter of each quadrilateral.

5. 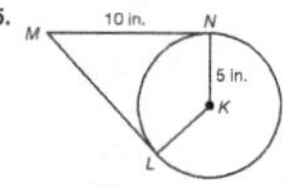

   30in.

6. 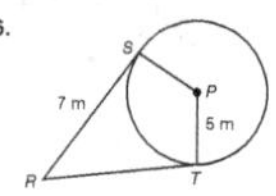

   24m

Name ______________________ Date __________ Class __________

Reteaching 59
Finding Surface Area and Volumes of Prisms

You have worked with tangents and circles. Now you will find surface area and volume of prisms.

Surface Area of Prisms

The **lateral area** of a prism is the sum of the areas of all the lateral faces. A lateral face is not a base. The **surface area** is the total area of all faces.

The **lateral area** of a prism with base perimeter p and height h is $L = ph$. The **surface area** of a prism with lateral area L and base area B is $S = L + 2B$.

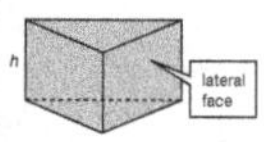

Find the lateral area and the surface area of the rectangular prism.

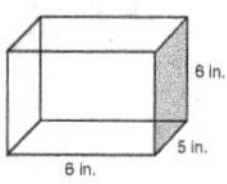

Step 1: The perimeter is $p = 2(5) + 2(6) = 22$ in.

Therefore, $L = ph = (22)(6) = 132$ in^2.

Step 2: Find the area of the base.

The area of the base is $(6)(5) = 30$ in^2.

Step 3: Find the total surface area.

$S = L + 2B = 132 + 2(30) = 192$ in^2.

The surface area of the prism is 192 square inches.

Practice

Complete the steps to find the lateral area and surface area of the prism.

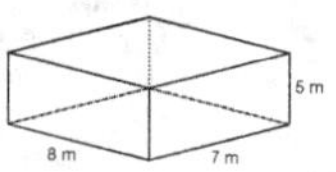

1. lateral area:

   $p = 2(8) + 2(7) =$ 30m

   surface area: area of each base $= (8)(7) =$ 56m^2

   $S = L + 2B = 150 + 2(56) =$ 262m^2

Find the lateral area and surface area of each prism.

2. 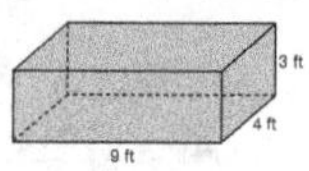

   $L = 78$ ft^2; $S = 150$ ft^2

3. 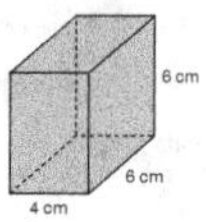

   $L = 120$ cm^2; $S = 168$ cm^2

Reteaching 59
continued

Volume of Prisms

The volume of a prism with base area B and height h is $V = Bh$.

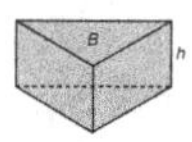

Find the volume of the right prism.

Step 1: Find the area of the base. The base is a right triangle.

The area of the base is $B = \frac{1}{2}(5)(12) = 30$ m^2.

Step 2: The height of the prism is $h = 3$ m.

The volume of the prism is $V = Bh = (30)(3) = 90$ m^3

The volume of the right prism is 90 cubic meters.

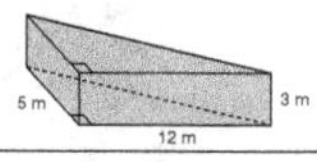

Practice

Complete the steps to find the volume of the prism.

4. $B = (2)(8) =$ 16 in^2

   The height of the prism is 4 in.

   $V = Bh = (16)(4) =$ 64 in^3

   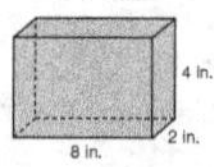

5. $B = \frac{1}{2}(4)(4) =$ 8 yd^2

   The height of the prism is 4 yd.

   $V = Bh = (8)(4) =$ 32 yd^3

   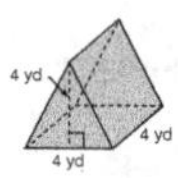

Find the volume of each prism.

6. 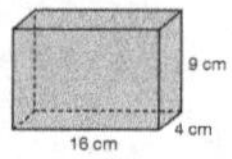

   576 cm^3

7. 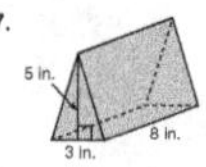

   60 in^3

Answer Key

Name ______________________ Date __________ Class __________

Reteaching 60
Proportionality Theorems

You have worked with proportional relationships that exist within triangles. Now you will work with any line that intersects a triangle and that is parallel to a base.

Triangle Proportionality Theorem

If a line parallel to one side of a triangle intersects the other two sides, then it divides those sides proportionally.

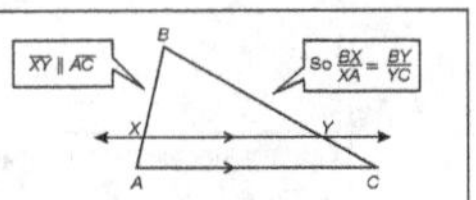

You can use the **Triangle Proportionality** Theorem to find lengths of segments in triangles.

Example:

Find the length of $\overline{EG}$.

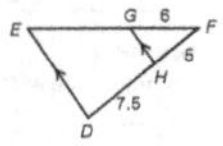

$\frac{EG}{GF} = \frac{DH}{HF}$	Triangle Proportionality Theorem
$\frac{EG}{6} = \frac{7.5}{5}$	Substitute the known values.
$EG(5) = 6(7.5)$	Cross Products Property
$5(EG) = 45$	Simplify.
$EG = 9$	Divide both sides by 5.

Practice

Complete the statements to find the length of $\overline{RQ}$.

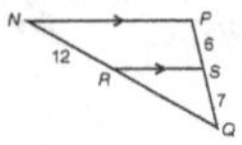

1. $\frac{NR}{RQ} = \frac{PS}{SQ}$

   $\frac{12}{RQ} = \frac{6}{7}$

   $12(7) = 6(RQ)$

   $84 = 6(RQ)$

   $14 = RQ$

Find the length of each segment.

2. $\overline{JN}$ 30.4

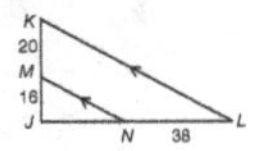

3. $\overline{BH}$ 5.4

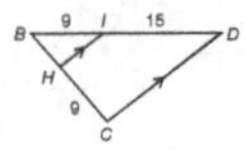

Reteaching 60
continued

Parallel Lines and Transversals

If parallel lines intersect transversals, then they divide the transversals proportionally.

$\frac{AC}{CE} = \frac{BD}{DF}$

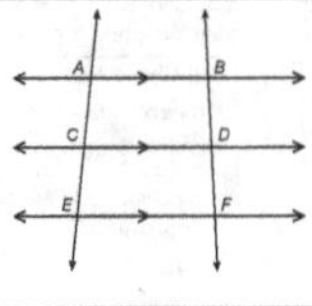

You can use the Triangle Proportionality Theorem to find lengths of segments.

Find the length of $\overline{KL}$.

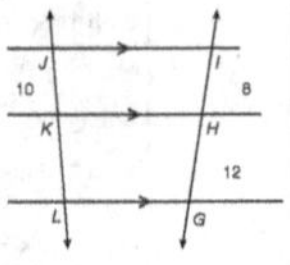

$\frac{JK}{KL} = \frac{IH}{HG}$	Triangle Proportionality Theorem
$\frac{10}{KL} = \frac{8}{12}$	Substitute the known values.
$10(12) = KL(8)$	Cross Products Property
$120 = 8(KL)$	Simplify.
$15 = KL$	Divide both sides by 8.

The length of $\overline{KL}$ is 15.

Practice

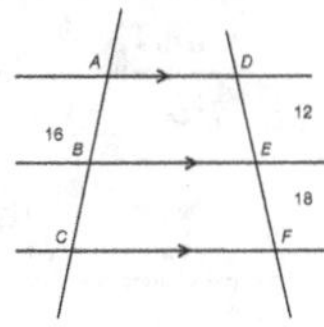

4. $\frac{AB}{BC} = \frac{DE}{EF}$

   $\frac{16}{BC} = \frac{12}{18}$

   $16(18) = BC(12)$

   $288 = 12(BC)$

   $24 = BC$

5. Find the length of segment LM 7.

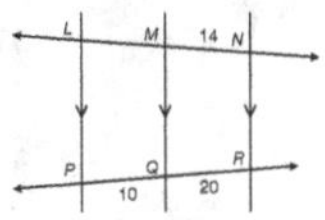

Name ______________________ Date __________ Class __________

Reteaching INV 6
Geometric Probability

You have calculated probabilities of events that occur when coins are tossed and number cubes are rolled. Now you will learn about **geometric probability.**

Geometric Probability

You can find the probability of an event occurring by dividing the number of ways the event can occur by the total number of outcomes. To know the probability of something falling into a part of a whole area or region, you can use geometric probability, which is the ratio of the size of the partial region to that of the whole region. Depending on the situation, size can be measured in degrees, length, area, or volume.

$$P(\text{partial region}) = \frac{\text{size of partial region}}{\text{size of whole region}}$$

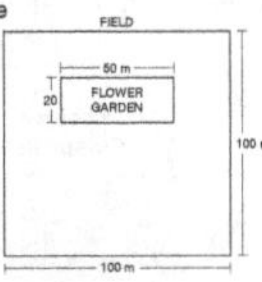

A small meteor lands in the field to the right. What is the probability that it has landed in the flower garden?

area of the flower garden = (20 m)(50 m) = 1000 m^2

area of the field = (100 m)(100 m) = 10,000 m^2

$$P(\text{meteor lands in flower garden}) = \frac{1000\ m^2}{10{,}000\ m^2} = 0.1 = 10\%$$

Practice

Compete the steps to find the probability.

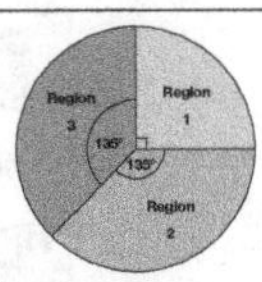

1. Find the probability that a dart thrown at the dartboard lands in region 3.

   The degree measure of region 3 is 135°.

   The dartboard is in the shape of a circle.

   The number of degrees in a circle is 360°.

   The probability of a dart hitting region 3 is $\frac{\text{region 3}}{\text{total degrees}} \frac{135°}{360°} = 0.375$, or 37.5%

Find the probabilities.

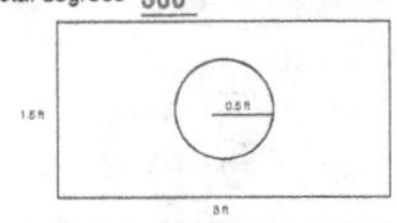

2. At the local pizza shop, if you drop a coin in the little bowl in the fish tank, you get a free slice of pizza.

   What is the probability of getting the coin in the fish tank if you have to drop it in while blindfolded?

   approximately 17.4%

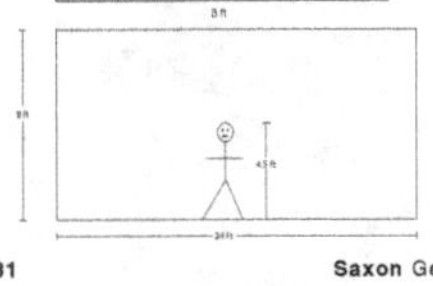

3. Find the probability that a soccer ball kicked on goal goes over the goalie's head. 43.75%

Reteaching INV 6
continued

Experimental Probability

When you say that the probability of flipping a coin and getting heads is $\frac{1}{2}$, you are talking about the **theoretical probability** of flipping heads. If we flipped a coin 50 times, the **experimental probability** of heads would be the actual percentage of flips that resulted in heads.

If a coin is tossed 50 times and heads occurs 26 times, what is the experimental probability of heads?

$$P(\text{heads}) = \frac{\text{number of heads}}{\text{number of tosses}} = \frac{26\text{ heads}}{50\text{ tosses}} = 0.52 = 52\%$$

In general, the experimental probability of an event is the number of times it occurs divided by the total number of trials:

$$\text{experimental probability} = \frac{\#\text{ of times the event occurred}}{\#\text{ of trials}}$$

As the number of trials goes up, an event's experimental probability will get closer and closer to its theoretical probability.

Practice

Find the experimental probabilities.

4. Kyle rolled a number cube 50 times and got a one 9 times.

   Find the experimental probability of a one.

   experimental probability of a one = $\frac{\text{number of ones}}{\text{number of trials}} = \frac{9}{50} = 0.18 = 18\%$

   What fraction will the experimental probability get closer to if the cube is tossed more?

   The experimental probability will get closer to the theoretical probability of $\frac{1}{6}$ because experimental probability approaches theoretical probability as the number of trials increases.

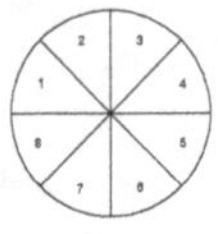

5. Make a spinner like the one to the right.

   Spin it 50 times and record the number of times it lands on three.

   Find the experimental probability of landing on three after 10, 25, and 50 spins. What number does it look like the experimental probability is getting closer to? Would it get closer to it if you performed more trials? Why?

   Answers will vary for the probabilities. The experimental probabilities should be getting closer to $\frac{1}{8}$, or 0.125 and will get closer if more trials are performed because the experimental probability should approach the theoretical probability, 0.125, as the number of trials increases.

Name ______________________ Date __________ Class __________

Reteaching 61
Quadrilateral Is a Parallelogram

You have worked with quadrilaterals. Now you will determine whether a quadrilateral is a parallelogram.

Identifying Parallelograms

A parallelogram is a quadrilateral with two pairs of parallel sides.

All parallelograms have the following properties:

Opposite sides are congruent.	Opposite angles are congruent.
$\overline{GF} \cong \overline{HJ}$ $\overline{GH} \cong \overline{FJ}$ 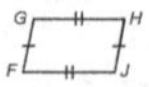	$\angle F \cong \angle H$ $\angle G \cong \angle J$ 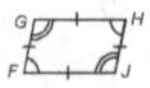

Example: In quadrilateral *JKLM*, $\overline{JK} \cong \overline{ML}$. Is *JKLM* a parallelogram?

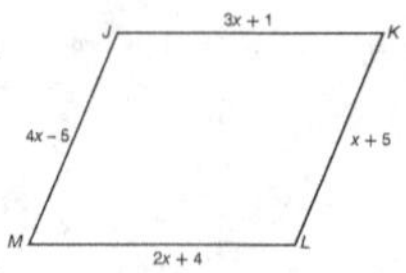

Step 1: Solve for *x*.

$JK = ML$ Given

$3x + 1 = 2x + 4$ Substitute.

$x = 3$ Solve.

Step 2: Substitute $x = 3$ to find *JM* and *KL*.

$JM = 4(3) - 5 = 7$

$KL = 3 + 4 = 7$

Since $\overline{JK} \cong \overline{ML}$ and $\overline{JM} \cong \overline{KL}$, *JKLM* is a parallelogram.

Practice

Complete the statements.

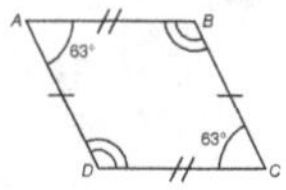

1. $\overline{AB} \cong$ $\overline{CD}$; $\overline{AD} \cong$ $\overline{BC}$ $\angle A \cong$ $\angle C$; $\angle B \cong$ $\angle D$

The figure is a parallelogram since opposite angles and sides are congruent.

Use the given information to determine whether the figures are parallelograms. State the reason.

2. Given: $\overline{AB} \cong \overline{DC}$

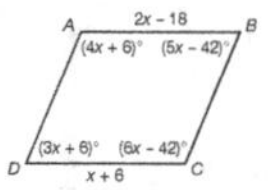

yes; $x = 24$; Opposite angles are congruent

3. Given: $\overline{PR} \cong \overline{QS}$

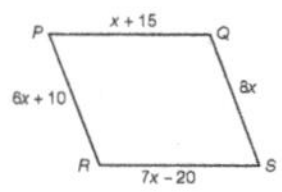

no; $x = 5$; Opposite sides are not congruent.

Reteaching 61
continued

More on Identifying Parallelograms

One pair of opposite sides of a quadrilateral are both parallel and congruent to each other.	The diagonals bisect each other.
$\overline{AB} \parallel \overline{CD}$ $\overline{AB} \cong \overline{CD}$ 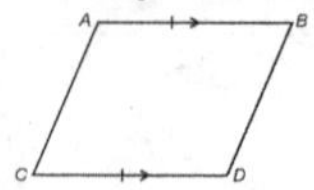	$\overline{FP} \cong \overline{PH}$ $\overline{GP} \cong \overline{PJ}$ 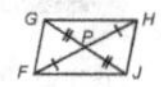

Example: Use the given information to decide whether *JLKM* is a parallelogram.

Given: $\overline{JK} \parallel \overline{LM}$ and $\overline{JK} \cong \overline{LM}$

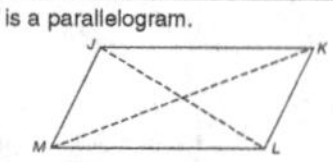

Step 1: Is there a relationship between the sides?

Opposite sides are both parallel and congruent to each other.

Step 2: If one pair of corresponding sides of a quadrilateral are both parallel and congruent to each other, the figure is a parallelogram.

Practice

Complete the statements.

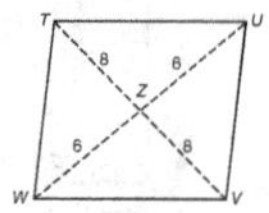

4. $\overline{TZ} \cong$ $\overline{ZV}$; $\overline{WZ} \cong$ $\overline{ZU}$

The diagonals of the quadrilateral bisect each other.

The figure is a parallelogram.

Use the given information to determine whether the figure is a parallelogram. State the reason.

5. $\overline{VZ} \cong \overline{ZX}$ and $\overline{YZ} \cong \overline{ZW}$

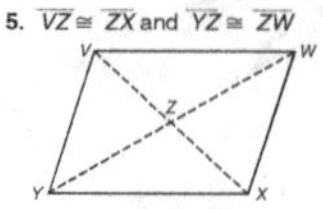

yes; Diagonals bisect each other.

6. $\overline{AB} \cong \overline{DC}$ and $\overline{AB} \parallel \overline{DC}$

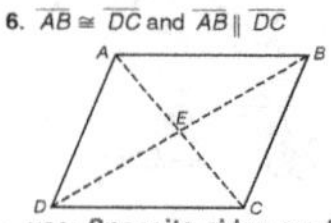

yes; Opposite sides are both parallel and congruent.

7. $\overline{PT} \cong \overline{QT}$ and $\overline{ST} \cong \overline{RT}$

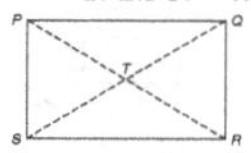

no

8. $\overline{AD} \cong \overline{AB}$ and $\overline{CD} \cong \overline{CB}$

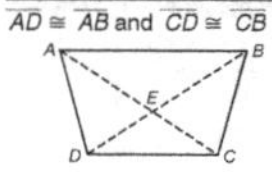

no

Name ______________________ Date __________ Class __________

Reteaching 62
Finding Surface Area and Volume of Cylinders

You have learned how to find the surface area and the volume of prisms. Now you will find the surface area and the volume of cylinders.

To find the surface area of a cylinder, use the formula $S = 2\pi r^2 + 2\pi rh$ to combine the lateral area and the area of the two bases.

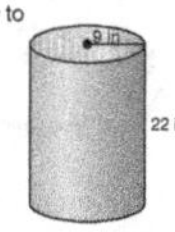

Find the area of the cylinder in terms of π.

$S = 2\pi r^2 + 2\pi rh$

$S = 2\pi(9)^2 + 2\pi(9)(22)$

$S = 162\pi + 396\pi$

$S = 558\pi \text{ in}^2$

Practice

Complete the steps to find the surface area of the cylinder in terms of π.

1. $S = 2\pi(7)^2 + 2\pi(7)(12)$

$S = 98\pi + 168\pi$

$S = 266\pi \text{ cm}^2$

Find the surface area of each cylinder in terms of π.

2.

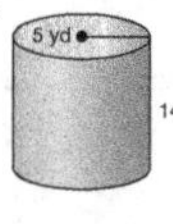

$190\ \pi \text{ yd}^2$

3.

$638\ \pi \text{ ft}^2$

4.

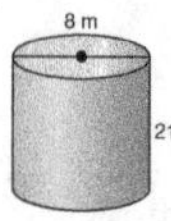

$200\ \pi \text{ m}^2$

5.

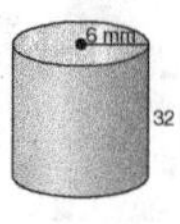

$456\ \pi \text{ mm}^2$

Reteaching 62
continued

Both right cylinders and oblique cylinders each use the same formula, $V = \pi r^2 h$, for volume.

Find the volume of the right cylinder in terms of π.

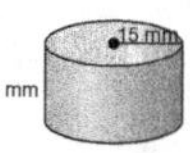

$V = \pi r^2 h$

$V = \pi(15)^2(10)$

$V = 2250\pi \text{ mm}^3$

Find the volume of the oblique cylinder in terms of π.

$V = \pi r^2 h$

$V = \pi(8)^2(5)$

$V = 320\pi \text{ ft}^3$

Practice

Complete the steps to find the volume of each cylinder in terms of π.

6. $V = \pi(5)^2(7)$

$V = 175\pi \text{ in}^3$

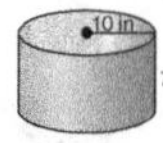

7. $V = \pi(18)^2(25)$

$V = 8100\pi \text{ yd}^3$

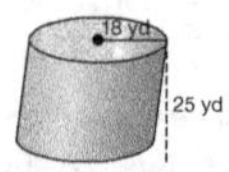

Find the volume of each cylinder in terms of π.

8.

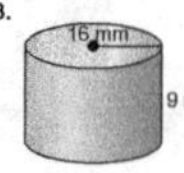

$2304\ \pi \text{ mm}^3$

9.

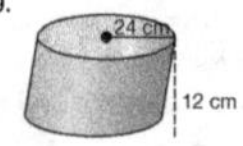

$6912\ \pi \text{ cm}^3$

10.

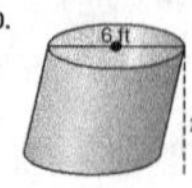

$180\ \pi \text{ ft}^3$

11.

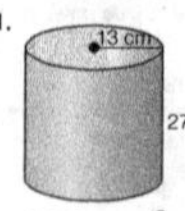

$4563\ \pi \text{ cm}^3$

Name ______________________ Date __________ Class __________

Reteaching 63
Introduction to Vectors

You have learned how to find points on a coordinate plane. Now you will determine vectors and their magnitudes.

Find $|\vec{r}|$.

Step 1: Place *M* on the origin.

Step 2: *N* will be 4 units to the right and 5 units up, so the coordinates of the terminal point *N* are (4, 5).

Step 3: Use the distance formula to find the distance between *M*(0, 0) and *N*(4, 5).

$d = \sqrt{(x_2 - x_1)^2 + (y_2 - y_1)^2}$

$d = \sqrt{(4 - 0)^2 + (5 - 0)^2}$

$d = \sqrt{4^2 + 5^2} = \sqrt{41}$

Therefore, $|\vec{r}|$ is $\sqrt{41}$.

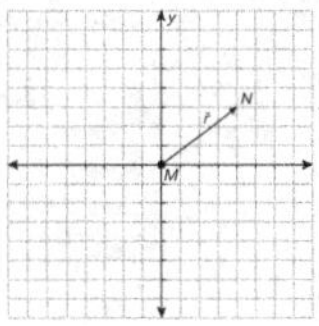

Practice

Complete the steps to find the magnitude of the vector.

1. Find $|\vec{p}|$.

Place *M* on the origin. *N* is now at (3, 4).

$d = \sqrt{(3 - 0)^2 + (4 - 0)^2}$

$d = \sqrt{(3)^2 + (4)^2} = 5$

So, $|\vec{p}| = 5$.

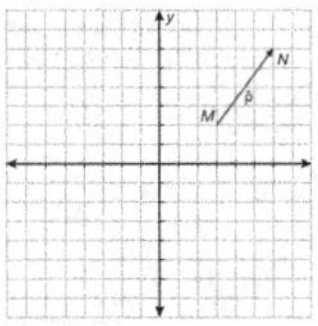

Find the magnitude of each vector in simplified radical form.

2. $|\vec{m}|$ $2\sqrt{10}$

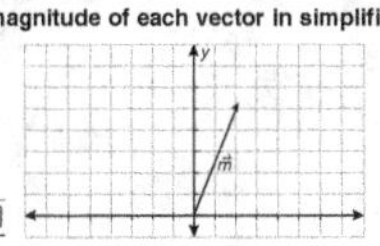

3. $|\vec{d}|$ $\sqrt{26}$

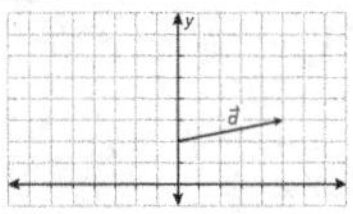

4. $|\vec{e}|$ $\sqrt{82}$

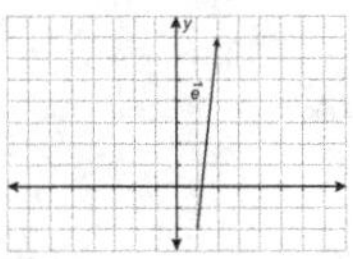

5. $|\vec{g}|$ $\sqrt{85}$

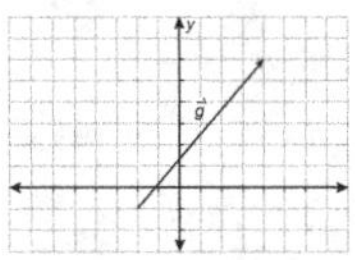

Reteaching 63
continued

Add vectors $\vec{v}$ **and** $\vec{w}$.

Step 1: Write $\vec{v}$ in component form.

The horizontal change is 7. The vertical change is 1.

The component form of $\vec{v}$ is ⟨7, 1⟩.

Step 2: Write $\vec{w}$ in component form.

The component form of $\vec{w}$ is ⟨4, 3⟩.

Step 3: Add the components: ⟨7 + 4, 1 + 3⟩ = ⟨11, 4⟩.

The resultant vector from adding $\vec{v}$ and $\vec{w}$ is ⟨11, 4⟩.

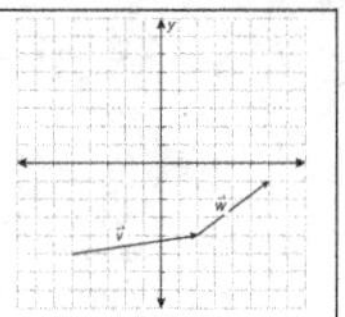

Add vectors $\vec{s}$ **and** $\vec{t}$.

The component form of $\vec{s}$ is ⟨2, 1⟩.

The component form of $\vec{t}$ is ⟨−2, −1⟩.

The components of $\vec{s}$ and $\vec{t}$ sum of 0 because they are opposite vectors.

The resultant vector is ⟨0, 0⟩.

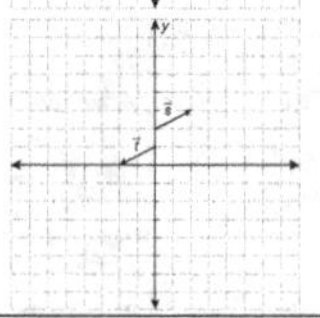

Practice

Complete the steps to find the sum of the vectors.

6. Add vectors $\vec{h}$ and $\vec{k}$.

The component form of $\vec{h}$ is ⟨5, 4⟩.

The component form of $\vec{k}$ is ⟨3, 6⟩.

Add the components: ⟨5 + 3, 4 + 6⟩.

The resultant vector is ⟨8, 10⟩.

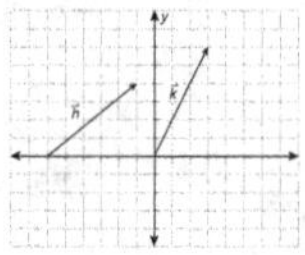

7. Add vectors $\vec{a}$ and $\vec{b}$.

The component form of $\vec{a}$ is ⟨3, −8⟩.

The component form of $\vec{b}$ is ⟨−3, 8⟩.

Add the components: ⟨3 + −3, −8 + 8⟩.

The resultant vector is ⟨0, 0⟩.

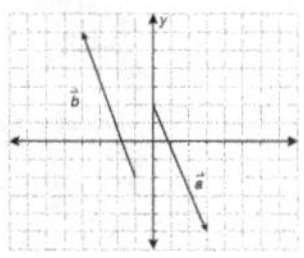

Name ______________________ Date __________ Class __________

Reteaching 64
Angles Interior to Circles

You have learned how to construct a tangent line on a circle. Now you will determine the measure of an angle formed by a chord and a line tangent to a circle.

The measure of an angle formed by a chord and a tangent is equal to half the measure of the arc that subtends it.

$m\angle ABC = \frac{1}{2}m\widehat{BEC}$

Given: $\overleftrightarrow{YZ}$ and $\overleftrightarrow{HI}$ are tangent lines.

Find the indicated measure.

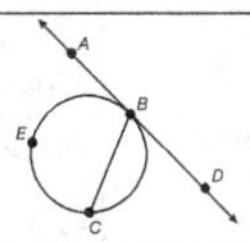

a. $m\angle XYZ$

$m\angle XYZ = \frac{1}{2}m\widehat{XY}$

$m\angle XYZ = \frac{1}{2}(170°)$

$m\angle XYZ = 85°$

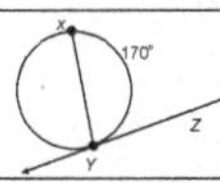

b. $m\widehat{GH}$

$m\angle GHI = \frac{1}{2}m\widehat{GH}$

$75° = \frac{1}{2}m\widehat{GH}$

$150° = m\widehat{GH}$

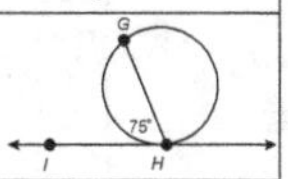

Practice

Complete the steps to find each measure.

1. $m\angle ABC$

$m\angle ABC = \frac{1}{2}\widehat{ADB}$

$m\angle ABC = \frac{1}{2}(240°)$

$m\angle ABC = 120°$

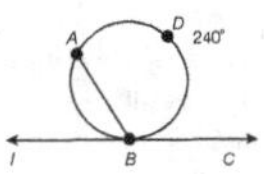

2. $m\widehat{PQR}$

$m\angle PRT = \frac{1}{2}m\widehat{PQR}$

$135° = \frac{1}{2}m\widehat{PQR}$

$270° = m\widehat{PQR}$

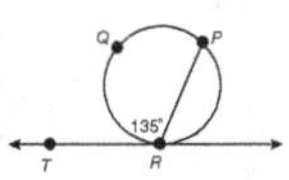

Find the value of *x* in each figure.

3. $x = 64°$

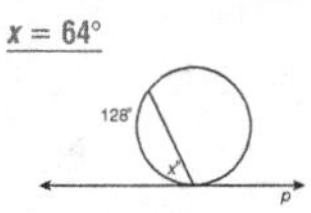

4. $x = 266°$

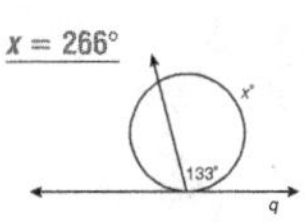

Reteaching 64
continued

The measure of an angle formed by two chords intersecting in a circle is equal to half the sum of the intersected arcs.

$m\angle 1 = \frac{1}{2}(m\widehat{AD} + m\widehat{BC})$

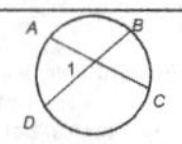

Find the measure of angle *x*.

$x = \frac{1}{2}(m\widehat{CF} + m\widehat{DE})$

$x = \frac{1}{2}(60 + 56)$

$x = 58°$

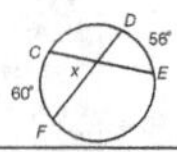

Practice

Complete the steps to find the measure of angle *x*.

5. $x = \frac{1}{2}(m\widehat{HJ} + m\widehat{LK})$

$x = \frac{1}{2}(110 + 180)$

$x = 146°$

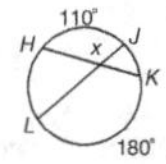

6. $x = \frac{1}{2}(m\widehat{PS} + m\widehat{QR})$

$x = \frac{1}{2}(45 + 73)$

$x = 59°$

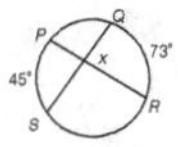

Find the measure of angle *x* for each circle.

7. $x = 53°$

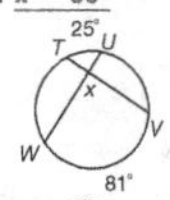

8. $x = 49.5°$

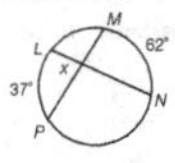

9. $x = 97°$

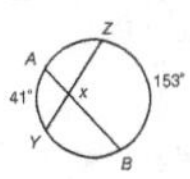

10. $x = 65.5°$

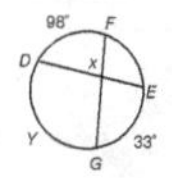

Name ______________________ Date __________ Class __________

Reteaching 65
Distinguishing Types of Parallelograms

A parallelogram is a rectangle if it has one 90° angle.

Example:

Is this parallelogram a rectangle?

Since the consecutive angles of a parallelogram are supplementary, we know that the measure of the two given expressions must add up to 180°.

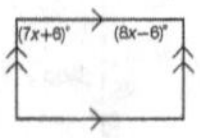

Solve for *x*.

$(8x - 6) + (7x + 6) = 180$

$15x = 180$

$x = 12$

Substitute the value of *x* into one of the expressions to find the measure of the angle.

$8(12) - 6 = 90$

One of the angle measures of the parallelogram is 90°, so the parallelogram is a rectangle.

Practice

Complete the statements to distinguish the parallelogram.

1. Is parallelogram *MNOP* a rectangle?

   For a parallelogram to be a rectangle, it must have one **90° angle**.

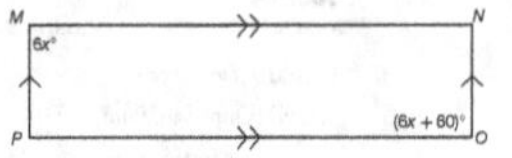

   $\underline{6x} = \underline{2x + 60}$

   $\underline{4x} = \underline{60}$

   $\underline{x} = \underline{15}$

   If $x = 15$, $\angle M = 6x = 6(\underline{15}) = \underline{90}$, and *MNOP* **is** a rectangle.

2. Is this parallelogram a rectangle? **no**

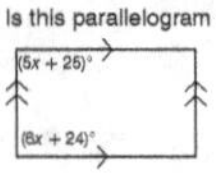

3. Is this parallelogram a rectangle? **yes**

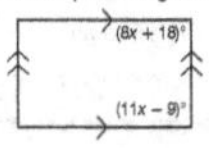

4. If $y = 10$, is this parallelogram a rectangle?

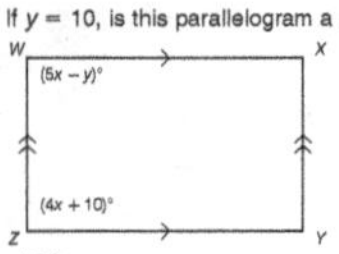

   yes

5. If $y = 3$, is this parallelogram a rectangle?

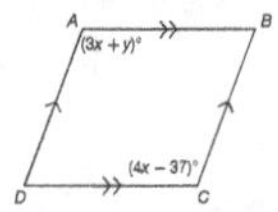

   no

Reteaching 65
continued

A parallelogram is a rhombus if it has two consecutive sides that are congruent.

Example:

If $x = 14$, is parallelogram *EFGH* a rhombus?

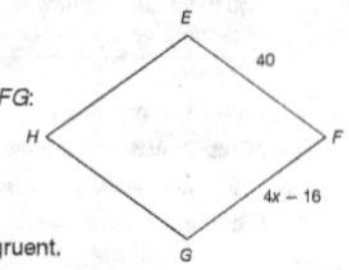

Since we know $x = 14$, we can find the length of side *FG*:

$FG = 4x - 16$

$FG = 4(16) - 16$

$FG = 40$

Since *EF* and *FG* have the same length, they are congruent.

Since *EF* and *FG* are consecutive sides of equal length, parallelogram *EFGH* is a rhombus.

Practice

6. If $x = 2$, is the parallelogram a rhombus?

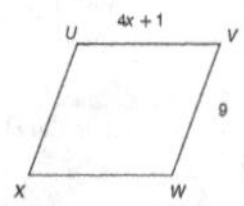

   For a parallelogram to be a rhombus, **consecutive** sides must have equal length.

   $\underline{4x + 1} = \underline{9}$

   $\underline{4(2) + 1} = \underline{9}$

   Since *UV* and *VW* are **consecutive** sides that have **equal** lengths, the parallelogram is a **rhombus**.

7. If $x = 19$, is this parallelogram a rhombus?

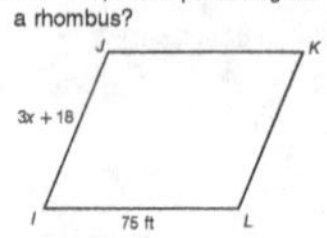

   yes

8. If $a = 3$, $b = 5$, $c = 4$, and $d = 4$, is this parallelogram a rhombus?

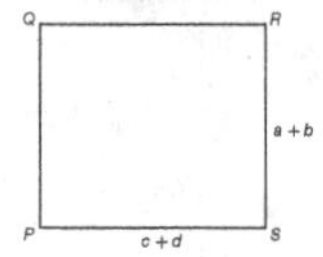

   yes

9. If $x \neq y$, is this parallelogram a rhombus? **no**

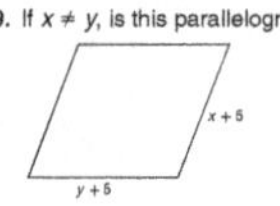

10. Is this parallelogram a rhombus? **yes**

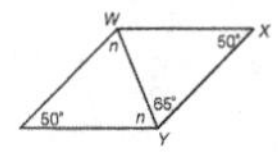

Name ______________________ Date __________ Class __________

Reteaching 66
Finding Perimeters and Areas of Regular Polygons

You have learned how to find perimeter and area of many types of figures. Now you will use formulas to find perimeters and areas of regular polygons.

The perimeter of a regular polygon can be found by multiplying the length of one of its sides by the number of sides it has.

Find the perimeter of the regular octagon.

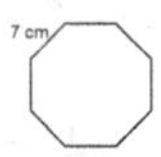

Because this is a regular octagon, the length of each side is the same as that of the labeled side, 7 centimeters. An octagon has 8 sides, so we can find the area by multiplying 8 by 7.

$P = ns$

$P = (8)(7 \text{ cm})$

$P = 56 \text{ cm}$

The perimeter of the regular octagon is 56 centimeters.

Practice

Complete the steps to find the perimeter of the regular polygon.

1. $P = (\underline{5})(\underline{9})$

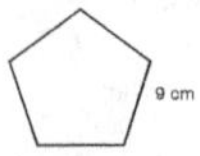

   $P = \underline{45}$ cm

2. $P = (\underline{6})(\underline{4.2})$

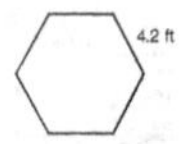

   $P = \underline{25.2}$ ft

Find the perimeter of each regular polygon.

3. **48 m**

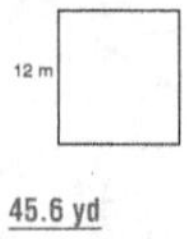

4. **80.5 mm**

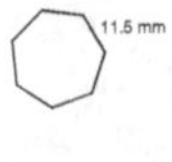

5. **45.6 yd**

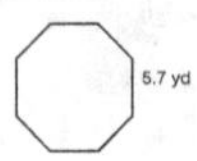

6. **108 in.**

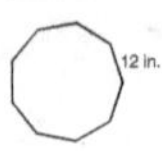

Reteaching 66
continued

The area of a regular polygon is equal to its perimeter times the length of the apothem, divided by 2 ($A = \frac{1}{2}aP$).

Here we are not given the length of the apothem, so we have to find it. When the figure is a regular hexagon, we can form an equilateral triangle with the center of the hexagon as the top vertex. The apothem forms a 30°-60°-90° with the side and hypotenuse.

Use the Pythagorean Theorem to find that the length of the apothem is $7\sqrt{3}$ millimeters.

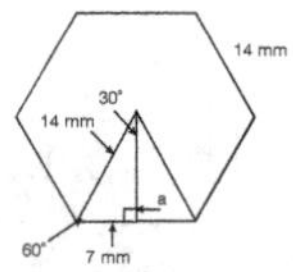

Find the perimeter:

$P = ns$

$P = (6)(14) = 84$

Find the area:

$A = \frac{1}{2}aP$

$A = \frac{1}{2}(7\sqrt{3})(84) = 294\sqrt{3}$

The area of this regular hexagon is $294\sqrt{3}$ mm².

Practice

Complete the steps to find the area of the regular polygon.

7. apothem $= 4\sqrt{3}$ ft

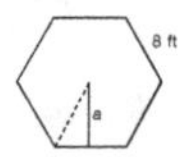

   $P = (\underline{6})(\underline{8})$

   $A = \frac{1}{2}(\underline{4\sqrt{3}})(\underline{48}) = \underline{192\sqrt{3}}$ ft²

8. apothem $= 11.5$ cm

   $P = (\underline{7})(\underline{11})$

   $A = \frac{1}{2}(\underline{11.5} \text{ cm})(\underline{77} \text{ cm}) = \underline{442.75}$ cm²

Find the area of each regular polygon.

9. **172.5 in²**

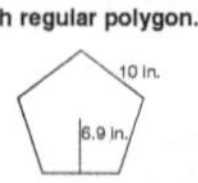

10. **672m²**

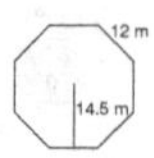

11. **$24\sqrt{3}$ yd²**

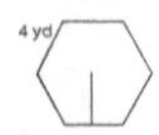

12. **315 ft²**

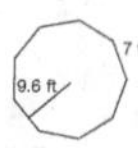

Answer Key

Name ______________________ Date __________ Class __________

Reteaching 67

Introduction to Transformations

You have worked with different figures in the coordinate plane. Now you will work with moving those figures around.

Transformations

An **isometry** is a transformation that does not change the size or the shape of a figure. There are different types of isometries.

A **translation** or a **slide** is a transformation in which all the points of a figure are moved the same distance in the same direction.

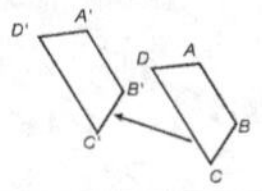

A **reflection** or a **flip** is a transformation that flips a figure across a line.

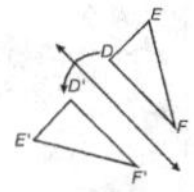

A **rotation** is a transformation that turns a figure around a fixed point, which is called the center of rotation.

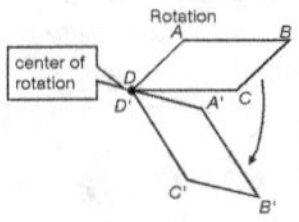

Example: Identify the type of transformation.

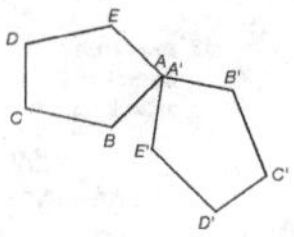

The figure is rotated about a point.

Practice

Complete the statement to determine the type of transformation.

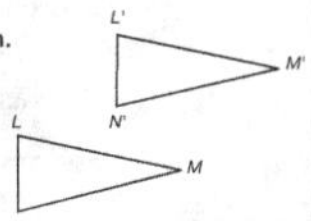

1. All the points of the figure are moved the same distance in the same direction.
   The transformation is a translation or a slide.

Determine the type of transformation.

2. reflection

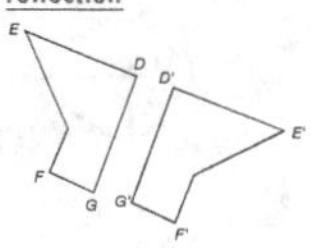

3. rotation

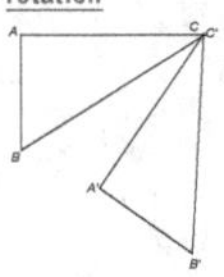

Reteaching 67

continued

Performing Transformations

For a **translation** or a **slide**, move each point of the preimage the indicated distance and direction.

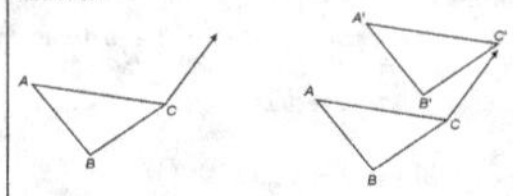

For a **reflection** or a **flip**, move each point across the line of reflection so that the point and its image are the same distance from the line.

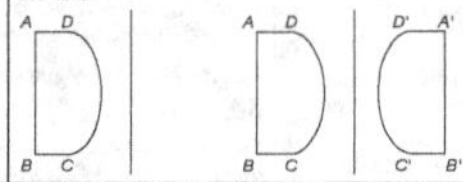

To **rotate** a figure about a point *A*, keep point *A* fixed and turn each point on a circular path around *A*, as shown.

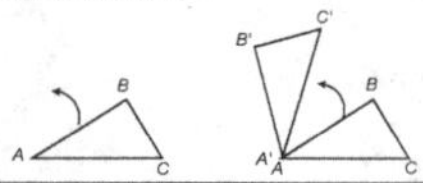

Example: Reflect the figure across the line.

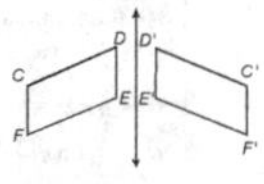

Practice

Complete each translation.

4. Draw the translation of the figure along the given path.

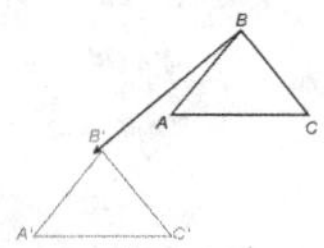

5. Draw the rotation of the figure about point *P*.

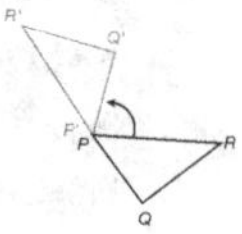

6. Draw the translation of the figure along the given path.

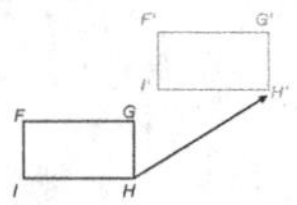

7. Draw the rotation of the figure across the line.

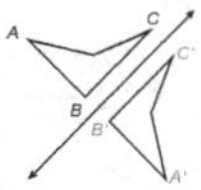

Name ______________________ Date __________ Class __________

Reteaching 68

Introduction to Trigonometric Functions

You have worked with right triangles. Now you will work with special relationships between the angles and sides in a right triangle.

Trigonometric Ratios

$\sin \angle = \dfrac{\text{leg opposite } \angle}{\text{hypotenuse}}$ $\cos \angle = \dfrac{\text{leg adjacent to } \angle}{\text{hypotenuse}}$ $\tan \angle = \dfrac{\text{leg opposite } \angle}{\text{leg adjacent to } \angle}$

Example:

Write each trigonometric ratio as a fraction.

$\sin A = \dfrac{\text{leg opposite } \angle A}{\text{hypotenuse}} = \dfrac{4}{5}$

$\cos A = \dfrac{\text{leg adjacent to } \angle A}{\text{hypotenuse}} = \dfrac{3}{5}$

$\tan A = \dfrac{\text{leg opposite } \angle A}{\text{leg adjacent to } \angle A} = \dfrac{4}{3}$

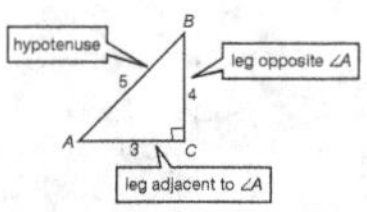

Practice

Complete the statement to determine the given trigonometric ratio.

1. $\sin K = \dfrac{\text{leg opposite } \angle K}{\text{hypotenuse}} = \dfrac{15}{17}$

$\cos K = \dfrac{\text{leg adjacent to } \angle K}{\text{hypotenuse}} = \dfrac{8}{17}$

$\tan K = \dfrac{\text{leg opposite } \angle K}{\text{leg adjacent to } \angle K} = \dfrac{15}{8}$

Use the diagram from Exercise 1 to write each trigonometric ratio as a fraction.

2. $\sin H = \dfrac{8}{17}$ 3. $\cos H = \dfrac{15}{17}$ 4. $\tan H = \dfrac{8}{15}$

Write each trigonometric ratio as a fraction.

5. $\sin A = \dfrac{24}{26}$ 6. $\cos A = \dfrac{10}{26}$

7. $\tan A = \dfrac{24}{10}$ 8. $\sin B = \dfrac{10}{26}$

9. $\tan B = \dfrac{10}{24}$ 10. $\cos B = \dfrac{24}{26}$

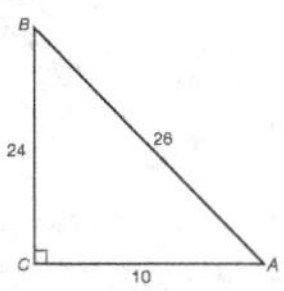

Reteaching 68

continued

Trigonometric Ratios (continued)

You can use a calculator to find the value of trigonometric ratios:

$\cos 38° \approx 0.7880107536$, or about 0.79.

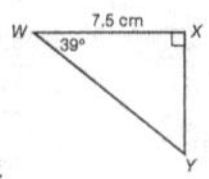

Example:

Find *WY*.

$\cos W = \dfrac{\text{adjacent leg}}{\text{hypotenuse}}$ Write a trigonometric ratio that involves *WY*.

$\cos 39° = \dfrac{7.5 \text{ cm}}{x}$ Substitute the given values.

$x \cos 39° = 7.5$ Multiply both sides by *x*.

$x = \dfrac{7.5}{\cos 39°}$ Divide both sides by os 39°

$x \approx 9.65$ cm Simplify.

The length of *WY* is approximately 9.65 centimeters.

Practice

Complete the steps to find *DE*.

11. $DE \approx 39.65$

$\sin E = \dfrac{\text{opposite leg}}{\text{hypotenuse}}$

$\sin 27° = \dfrac{18}{x}$

$x \sin 27° = 18$

$x = \dfrac{18}{\sin 27°}$

$x \approx 39.65$ m

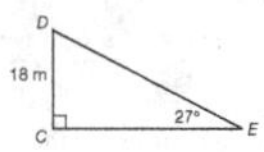

Use your calculator to find each trigonometric ratio. Round to the nearest hundredth.

12. sin 42° 0.67 13. cos 89° 0.02 14. tan 55° 1.43

Find each length. Round to the nearest hundredth.

15. *FH* 6.01 in. 16. *JK* 32.91 mm 17. *US* 55.32 cm

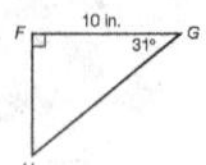

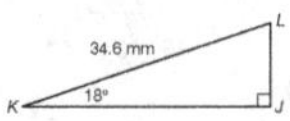

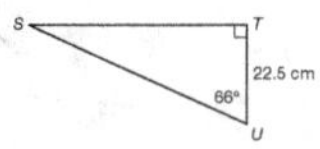

Name ______________________ Date __________ Class __________

Reteaching 69

Properties of Trapezoids and Kites

You have worked with triangles. Now you will find the lengths of sides of trapezoids and kites.

Trapezoid Midsegment Theorem

A **trapezoid** is a quadrilateral with one pair of parallel sides. The **midsegment** of a trapezoid is the segment whose endpoints are the midpoints of the legs.

The midsegment of a trapezoid is parallel to each base.

$\overline{AB} \parallel \overline{MN}$ and $\overline{AB} \parallel \overline{LP}$

The length of the midsegment is one-half of the sum of the length of the bases.

$AB = \frac{1}{2}(MN + LP)$

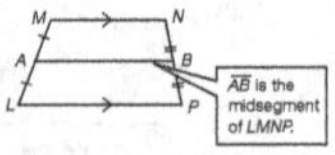

Use the formula to find the length of the midsegment $\overline{KL}$.

$KL = \frac{1}{2}(GH + FJ)$

$KL = \frac{1}{2}(18 + 26)$ Substitute.

$KL = \frac{1}{2}(44)$ Simplify.

$KL = 22$ Simplify.

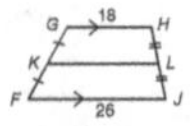

Practice

Complete the steps to find each length.

1. MN

$MN = \frac{1}{2}(AB + CD)$

$MN = \frac{1}{2}(\underline{23} + 17)$

$MN = \frac{1}{2}(\underline{40})$

$MN = \underline{20}$

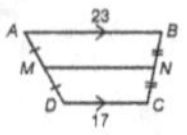

2. FJ

$XY = \frac{1}{2}(FJ + GH)$

$\underline{40} = \frac{1}{2}(FJ + 48)$

$\underline{80} = FJ + 48$

$FJ = \underline{32}$

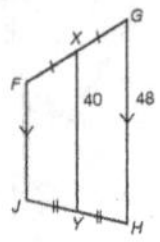

Find each length.

3. KL

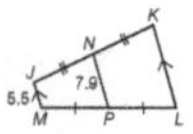

$\underline{10.3}$

4. TV

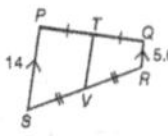

$\underline{9.8}$

Reteaching 69

continued

Diagonals of Kites

A **kite** is a quadrilateral with two pairs of congruent adjacent sides. The **diagonals** of a kite are perpendicular to each other.

$\overline{AC} \perp \overline{BD}$

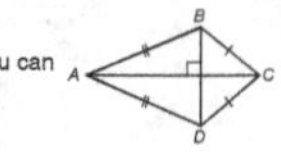

Because the diagonals of a kite are perpendicular to each other, you can use the Pythagorean theorem to find the length of each side.

Find the length of $\overline{AD}$.

Use the Pythagorean theorem to find the length of the side.

$AD^2 = AE^2 + ED^2$

$AD^2 = 12^2 + 5^2$ Substitute.

$AD^2 = 144 + 25$ Simplify.

$AD^2 = 169$ Simplify.

$AD = \sqrt{169}$ Solve.

$AD = 13$

So, the Length of $\overline{AD}$ is 13.

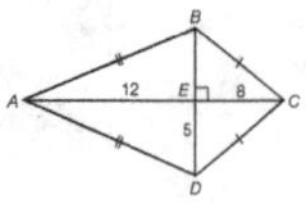

Practice

Complete the steps to find the lengths of $\overline{JK}$ and $\overline{JM}$. Round each length to the nearest tenth.

5. Use the Pythagorean theorem to find the length of each side.

$JK^2 = JN^2 + NK^2$	$JM^2 = JN^2 + NM^2$
$JK^2 = 6^2 + \underline{15}^2$	$JM^2 = 6^2 + \underline{8}^2$
$JK^2 = 36 + \underline{225}$	$JM^2 = 36 + \underline{64}$
$JK^2 = \underline{261}$	$JM^2 = \underline{100}$
$JK = \sqrt{\underline{261}}$	$JM = \sqrt{\underline{100}}$
$JK = \underline{16.2}$	$JM = \underline{10}$

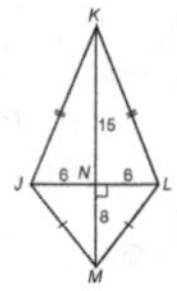

Find each length. Round to the nearest tenth.

6. Find the length of $\overline{AD}$.

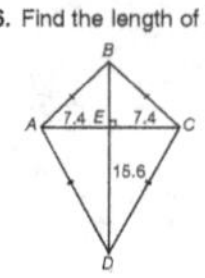

$\underline{17.3}$

7. Find the length of $\overline{ED}$.

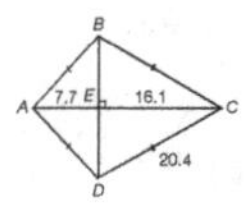

$\underline{12.5}$

Name ______________________ Date __________ Class __________

Reteaching 70

Finding Surface Areas and Volumes of Pyramids

You have worked with surface area and volume of prisms. Now you will find the lateral area, the surface area, and the volume of pyramids.

Lateral and Surface Area of a Regular Pyramid

The **lateral area** of a regular pyramid with base perimeter P and **slant height** ℓ is $L = \frac{1}{2}P\ell$.

The **surface area** of a regular pyramid with lateral area L and base area B is $S = L + B$ or $S = \frac{1}{2}P\ell + B$.

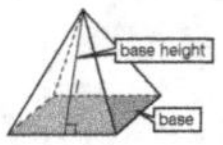

Find the lateral area and the surface area of the regular pyramid.

$L = \frac{1}{2}P\ell$ Lateral area formula

$L = \frac{1}{2}(32)(12)$ Substitute.

$L = 192\ m^2$ Simplify.

$S = L + B$ Surface area formula

$S = 192 + 6$ Substitute.

$S = 256\ m^2$ Simplify.

The lateral area is 192 square meters. The surface area is 256 square meters.

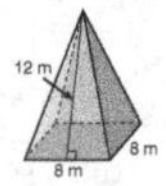

Practice

Complete the steps to find the lateral area and the surface area of the regular pyramid. Round your answer to the nearest tenth.

1. $L = \frac{1}{2}P\ell = \frac{1}{2}(\underline{48})(12) = \underline{288}\ ft^2$

$B = \frac{1}{2}\left(\frac{s\sqrt{3}}{2}\right)P = \frac{1}{2}\left(\frac{3\sqrt{3}}{2}\right)\underline{48} = \underline{62.4}\ ft^2$

$S = L + B = \underline{288} + \underline{62.4} = \underline{350.4}\ ft^2$

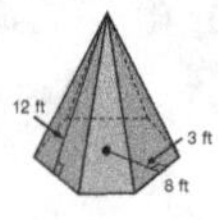

Find the lateral area and the surface area of the regular pyramid. Round to the nearest tenth.

2. $L = \underline{298.0}\ in^2$

$S = \underline{422.1}\ in^2$

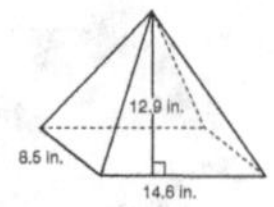

3. $L = \underline{409.5}\ cm^2$

$S = \underline{487.5}\ cm^2$

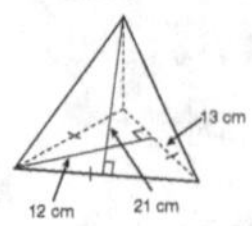

Reteaching 70

continued

Volume of a Pyramid

The **altitude**, or **height**, of a pyramid is the perpendicular segment that measures from the vertex of the pyramid to the base of the pyramid. The **volume** of a pyramid with base area B and height h is $V = \frac{1}{3}Bh$.

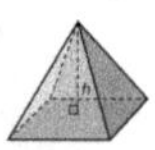

Find the volume of the pyramid. Round to the nearest tenth if necessary.

$V = \frac{1}{3}Bh$

$V = \frac{1}{3}(25)(8)$ Substitute.

$V = \frac{1}{3}(200)$ Simplify.

$V = 66.7\ in^3$ Solve.

The volume of the pyramid is 66.7 cubic inches.

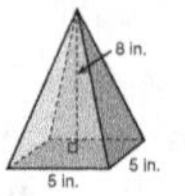

Practice

Complete the steps to find the volume of the pyramid. Round to the nearest tenth.

4. $V = \frac{1}{3}Bh$

$V = \frac{1}{3}(\underline{84})\ (11)$

$V = \frac{1}{3}(\underline{924})$

$V = \underline{308}\ cm^3$

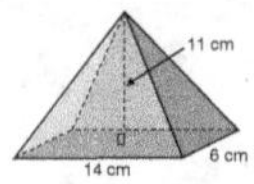

5. $V = \frac{1}{3}Bh$

$V = \frac{1}{3}\ (\underline{54.2})\ (16.6)$

$V = \frac{1}{3}\ (\underline{899.7})$

$V = \underline{299.9}\ in^3$

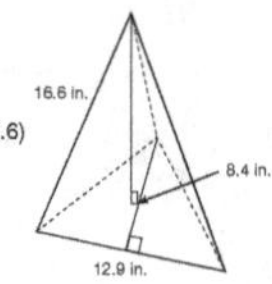

Find the volume of the pyramid. Round to the nearest tenth.

6. $\underline{163.3}\ ft^3$

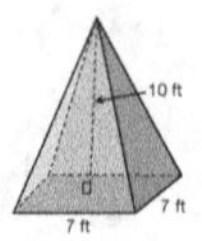

7. $\underline{4608.7}\ m^3$

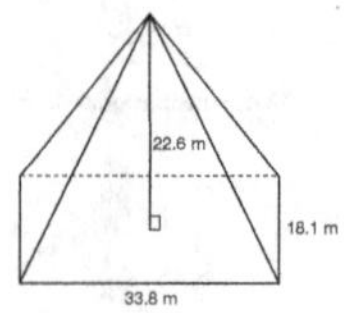

Name ______________________ Date __________ Class __________

Reteaching INV 7
Trigonometric Functions

You have learned about right triangles and been introduced to trigonometric functions. Now you will observe and compare some of the values of the trigonometric functions as the angle measures vary between 0° and 90°.

The sine and cosine ratios for a right triangle are as follows.

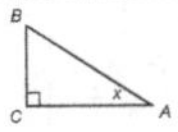

$\sin x = \frac{\text{opposite}}{\text{hypotenuse}} = \frac{BC}{AB}$

$\cos x = \frac{\text{adjacent}}{\text{hypotenuse}} = \frac{AC}{AB}$

Given a 30°–60°–90° triangle, find the sine and cosine for *x* if *x* = 30°. Then, find the sine and cosine if *x* = 60°.

For $x = 30°$: $\sin 30° = \frac{1}{2}$, $\cos 30° = \frac{\sqrt{3}}{2}$

For $x = 60°$: $\sin 60° = \frac{\sqrt{3}}{2}$, $\cos 60° = \frac{1}{2}$

What do you notice about the sine and cosine of these two angles? They switch values.

Practice

1. Sketch a 45°–45°–90° triangle with legs that are 1 unit long.
   Find sin 45° and cos 45°.
   $\sin 45° = \frac{\sqrt{2}}{2}$; $\cos 45° = \frac{\sqrt{2}}{2}$
   What do you notice about the sine and cosine of a 45° angle?
   They have the same value.

Use your calculator for problems 2–4

2. Find the sine and cosine of 20°. sin 20° ≈ 0.34; cos 20° ≈ 0.94
3. Find the sine and cosine of 70°. sin 70° ≈ 0.94; cos 70° ≈ 0.34
4. What do you notice about the values of sine and cosine for 20° and 70°?
   They switch values.
5. Both 20° and 70° angles and 30° and 60° angles are complementary. What is the relationship between the sine and the cosine of complementary angles?
   The sine of an angle equals the cosine of its complementary angle.
6. In ΔXYZ, $\angle Z$ is a right angle, $m\angle X = 30°$, and $YZ = 5$. How are *YZ* and *XZ* related? How are *YZ* and *XY* related? Determine *XZ* and *XY*. Then, give exact values for sin 30° and cos 30°.
   $XZ = YZ\sqrt{3}$, $XZ = 5\sqrt{3}$; $XY = 2YZ$, $XY = 10$; $\sin 30° = \frac{1}{2}$; $\cos 30° = \frac{\sqrt{3}}{2}$

Reteaching INV 7
continued

The tangent ratio for a right triangle is as follows.

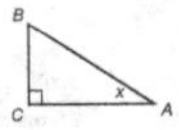

$\tan x = \frac{\text{opposite}}{\text{adjacent}} = \frac{BC}{AC}$

The tangent of an angle is also the quotient of the sine and the cosine of the angle.

$\tan x = \frac{\sin x}{\cos x}$

Find the tangent of 30° by dividing sin 30° by cos 30°.

$\sin 30° = \frac{1}{2}$ and $\cos 30° = \frac{\sqrt{3}}{2}$

$\tan 30° = \frac{\frac{1}{2}}{\frac{\sqrt{3}}{2}} = \left(\frac{1}{2}\right)\left(\frac{2}{\sqrt{3}}\right) = \frac{1}{\sqrt{3}}$

Do not forget to rationalize the denominator.

$\left(\frac{1}{\sqrt{3}}\right)\left(\frac{\sqrt{3}}{\sqrt{3}}\right) = \frac{\sqrt{3}}{3}$

Therefore, $\tan 30° = \frac{\sqrt{3}}{3}$.

Practice
Complete the steps to find the tan 60° by dividing the sin 60° by the cos 60°.

7. $\sin 60° = \frac{\sqrt{3}}{2}$ and $\cos 60° = \frac{1}{2}$
   $\tan 60° = \frac{\frac{\sqrt{3}}{2}}{\frac{1}{2}} = \left(\frac{\sqrt{3}}{2}\right)(2) = \sqrt{3}$

Use the information you learned from problems 2 and 3 for problems 8 and 9.

8. Find the tangent of 20° by dividing sin 20° by cos 20°. tan 20° ≈ 0.36
9. Find the tangent of 70° by dividing sin 70° by cos 70°. tan 70° ≈ 2.75
10. In ΔPQR, $\angle R$ is a right angle, $m\angle P = 45°$, and $PR = 4$. Use the Converse of the Isosceles Triangle Theorem to determine how *PR* and *RQ* related. What is the length of *RQ*? Use the Pythagorean theorem to find *PQ*. Then, find and use sin 45° and cos 45° to find tan 45°.
    PR and *RQ* are congruent, so $RQ = 4$; $PQ = 4\sqrt{2}$;
    $\sin 45° = \cos 45° = \frac{\sqrt{2}}{2}$, so $\tan 45° = 1$

Name ______________________ Date __________ Class __________

Reteaching 71
Translations

You have worked with plotting points on a coordinate grid. Now you will translate figures on a coordinate grid both in one dimension and in two dimensions.

Translation in One Dimension

A **translation,** also known as a **slide,** is an **isometry** (a transformation that does not change the size or shape of the object being translated).

A translation in one dimension changes either the *x*-values or the *y*-values, but not both. A translation left or right on the coordinate grid will change only the *x*-values of the original figure. A translation up or down on the coordinate grid will change only the *y*-values of the original figure.

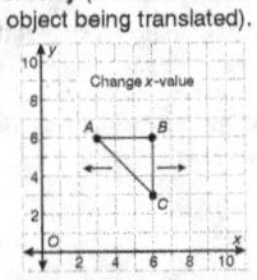

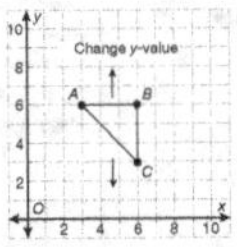

Find the coordinates of the vertices of the image of triangle *ABC* after a translation 4 units to the right. Show the preimage and image on the same coordinate grid.

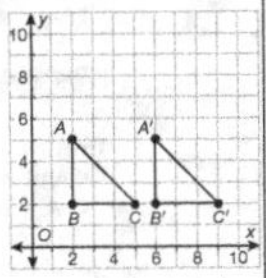

Triangle *ABC* has coordinates *A* (2, 5), *B* (2, 2), and *C* (5, 2).
Apply the rule $(x, y) \to (x + 4, y)$.
A (2, 5) → *A*′(2 + 4, 5) = *A*′(6, 5)
B (2, 2) → *B*′(2 + 4, 2) = *B*′(6, 2)
C (5, 2) → *C*′(5 + 4, 2) = *C*′(9, 2)
Triangle *A*′*B*′*C*′ has coordinates *A*′(6, 5), *B*′(6, 2), and *C*′(9, 2).

Practice
Complete the steps to find the coordinates of the image of square *JKLM* after a translation of 3 units down.

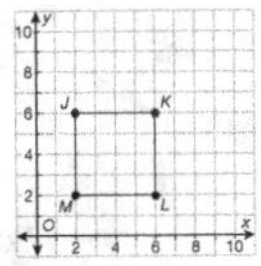

1. Apply the rule (*x*, *y*) → (x , y − 3).
   J (2, 6) → *J*′(2, 6 −3) = *J*′(2, 3)
   K (6, 6) → *K*′(6, 6 −3) = *K*′(6, 3)
   L (6, 2) → *L*′(6, 2 −3) = *L*′(6, −1)
   M (2, 2) → *M*′(2, 2 −3) = *M*′(2, −1)

Find the coordinates of each image after each indicated translation.

2. 5 units left
   F′(−7 , 2)
   G′(−2 , 1)
   H′(−3 , −3)
   J′(−8 , −2)

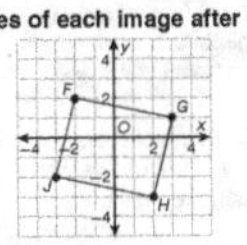

3. 6 units up
   X′(1 , 11)
   Y′(4 , 12)
   Z′(−1 , 7)

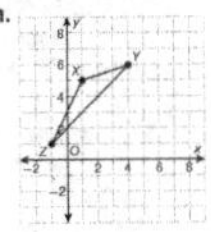

Reteaching 71
continued

Translation in Two Dimensions

A translation in two dimensions causes both the *x*-values and the *y*-values to change. For example, a translation of 3 units right and 2 units down can be written using the rule $(x, y) \to (x + 3, y - 2)$.

Find the coordinates of the vertices of the image of triangle *ABC* after a translation of 4 units right and 2 units up. Show both the preimage and the image on the same coordinate grid.

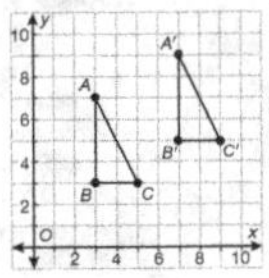

Triangle *ABC* has coordinates *A* (3, 7), *B* (3, 3), and *C* (5, 3).
Apply the rule $(x, y) \to (x + 4, y + 2)$.
A (3, 7) → *A*′(3 + 4, 7 + 2) = *A*′(7, 9)
B (3, 3) → *B*′(3 + 4, 3 + 2) = *B*′(7, 5)
C (5, 3) → *C*′(5 + 4, 2) = *C*′(9, 5)
Triangle *A*′*B*′*C*′ has coordinates *A*′(7, 9), *B*′(7, 5), and *C*′(9, 5)

Practice
Complete the steps to find the coordinates of the image of rectangle *WXYZ* after a translation of 2 units left and 3 units up.

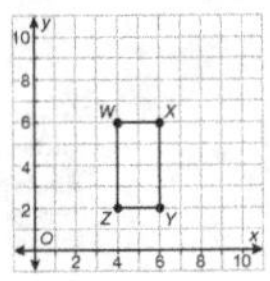

4. Apply the rule (*x*, *y*) → (x − 2 , y + 3).
   W (4, 6) → *W*′(4, −2, 6 + +3) = *W*′(2, 9)
   X (6, 6) → *X*′(6, −2, 6 + +3) = *X*′(4, 9)
   Y (6, 2) → *Y*′(6, −2, 2 + +3) = *Y*′(4, 5)
   Z (4, 2) → *Z*′(4, −2, 2 + +3) = *Z*′(2, 5)

Find the coordinates of each image after each indicated translation.

5. 3 unitsright,
   1 unit down
   J′(1, 1)
   K′(4, 3)
   L′(8, 0)
   M′(4, −4)

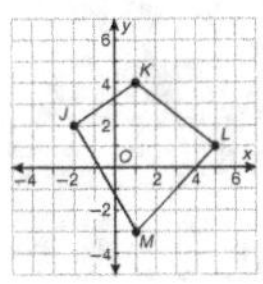

6. 5 units left,
   2 units up
   D′(−10, −2)
   E′(−2, −1)
   F′(−5, −2)

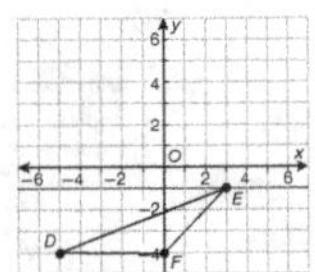

Name ______________________ Date __________ Class __________

Reteaching 72
Tangents and Circles, Part 2

You have worked with tangents and circles. Now you will solve problems involving common tangents.

Common Tangents: Internal

A **common tangent** is a tangent that is common to two circles. A common tangent is **internally tangent** if it intersects the segment that joins the centers of two circles. Two segments tangent from the same exterior point are congruent.

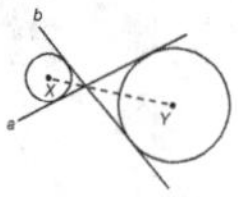

Line segments $\overline{KL}$ and $\overline{PQ}$ are common tangents to the circles. Find the lengths of $\overline{KL}$ and $\overline{PQ}$.

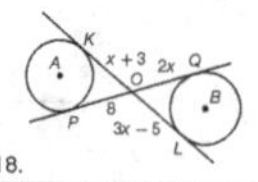

Step 1: Solve for x.

$KO = PO = 8$

$x + 3 = 8$

$x = 5$

Step 2: Substitute $x = 5$ to find OQ or OL.

$OQ = 2x = 2(5) = 10$

$OL = 3x - 5 = 3(5) - 5 = 10$

Step 3: Find KL and PQ: $KL = PQ = PO + OQ = 8 + 10 = 18$.

Practice

Complete the steps to find each length.

1.

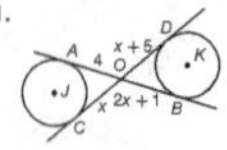

$AO = CO = \underline{4}$

$x = \underline{4}$

$OD = \underline{4} + 5 = \underline{9}$

$CD = CO + OD$

$CD = \underline{4} + \underline{9} = \underline{13}$

$AB = CD = \underline{13}$

2.

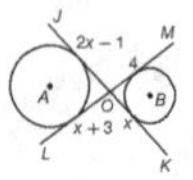

$OM = OK = 4$

$x = \underline{4}$

$JO = 2(\underline{4}) - 1 = \underline{7}$

$JK = JO + OK$

$JK = \underline{7} + 4 = \underline{11}$

$LM = JK = \underline{11}$

Find each length.

3.

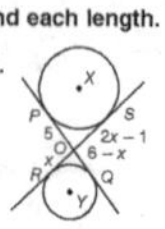

$PQ = RS = \underline{8}$

4.

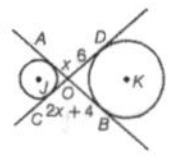

$AB = CD = \underline{7}$

Reteaching 72
continued

Common Tangents: External

A **common tangent** is externally tangent if the tangent does not intersect the segment joining the centers of two circles. Again, two segments tangent from the same exterior point are congruent.

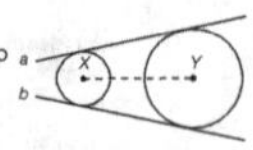

Line segments $\overline{AB}$ and $\overline{CD}$ are common tangents to the circles. Find the lengths of $\overline{AB}$ and $\overline{CD}$.

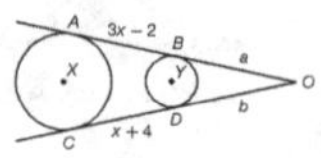

$AB = CD$	Given
$3x - 2 = x + 4$	Substitute.
$2x = 6$	Solve.
$x = 3$	
$AB = 3(3) - 2 = 7$	Substitute.
$CD = 3 + 4 = 7$	Substitute.

Practice

Complete the steps to find each length.

5.

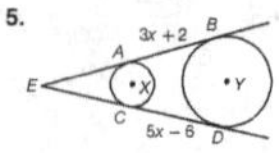

$AB = CD$

$3x + 2 = \underline{5x} - 6$

$8 = \underline{2x}$

$x = \underline{4}$

$AB = 3(\underline{4}) + 2 = \underline{14}$

$CD = 5(\underline{4}) - 6 = \underline{14}$

6.

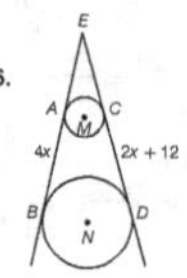

$AB = CD$

$4x = \underline{2x} + 12$

$\underline{2x} = 12$

$x = \underline{6}$

$AB = 4(\underline{6}) = \underline{24}$

$CD = 2(\underline{6}) + 12 = \underline{24}$

Find each length.

7.

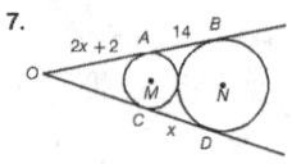

$OB = \underline{44}$

$OD = \underline{44}$

8.

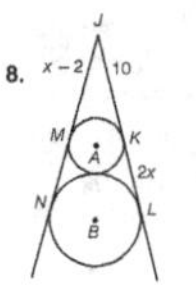

$JL = \underline{34}$

$JN = \underline{34}$

Name ______________________ Date __________ Class __________

Reteaching 73
Applying Trigonometry: Angles of Elevation and Depression

You have worked with angles and trigonometry. Now you will use angles of elevation and depression to solve problems.

Angle of Elevation

An **angle of elevation** is the angle formed by a horizontal line and a line of sight to a point above. You can use trigonometry to solve problems involving angle of elevation.

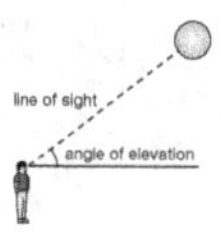

When the angle of elevation to the sun is 48°, a tree casts a shadow that is 10 meters long. What is the height of the tree? Round to the nearest tenth of a meter.

Sketch a diagram to represent the given information.

The height of the tree is represented by x. The angle of elevation is 48°, and the length of the shadow on the ground is 10 meters. To find x, use the tangent function.

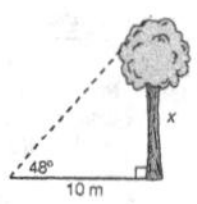

$\tan 48° = \frac{\text{opp}}{\text{abj}}$	Tangent Function
$\tan 48° = \frac{x}{10}$	Substitute.
$10 \tan 48° = x$	Simplify.
$11.1 \approx x$	Solve.

The tree is about 11.1 meters tall.

Practice

Complete the steps to solve the problem.

1. A building casts a shadow 45 feet long. The building is 120 feet tall. What is the angle of elevation from the point on the ground where the shadow of the building ends? Round to the nearest tenth of a degree.

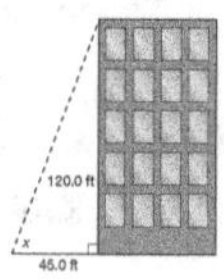

$\tan x = \frac{\text{opp}}{\text{abj}}$

$\tan x = \frac{120}{\underline{45}} \approx \underline{2.67}$

$x = \underline{69.4°}$

Solve the problem.

2. Andrew is 15 feet from a house his brother is painting. He measures the angle of elevation from the ground to the top of his brother's head as 52°. How high up the house is the top of his brother's head? Round to the nearest foot. $\underline{19}$ ft

3. Find the angle of elevation to the top of a tree for a person standing 28.6 meters from the tree, assuming the person's eye is 1.5 meters above the ground and the tree is 19.1 meters tall. Round to the nearest tenth of a degree. $\underline{31.6°}$

Reteaching 73
continued

Angle of Depression

An **angle of depression** is the angle formed by a horizontal line and a line of sight to a point below. You can use trigonometry to solve problems involving angle of depression.

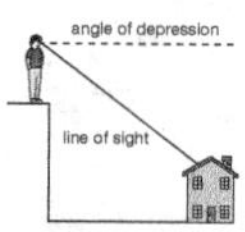

Patricia is watching a parade from a 20-foot-high balcony. The angle of depression to the parade is 47°. What is the distance between Patricia and the parade?

Sketch a diagram to represent the given information.

Let A represent Patricia and let B represent the parade. Let x represent the distance between Patricia and the parade. $m\angle B = 47°$ by the Alternate Interior Angles Theorem. Write a sine ratio using $\angle B$.

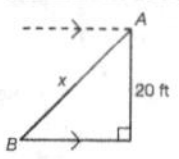

$\sin 47° = \frac{20}{x}$	Sine Function
$x \sin 47° = 20$	Multiply both sides by x.
$x = \frac{20}{\sin 47°}$	Divide both sides by sin 47°.
$x \approx 27.3$	Simplify.

The distance between Patricia and the parade is about 27 feet.

Practice

Complete the steps to solve the problem.

4. A person swimming in a pool sees a quarter on the bottom of the pool at an angle of depression of 42°. She is 6 feet above the bottom of the pool. How far from the quarter is she? Round to the nearest foot.

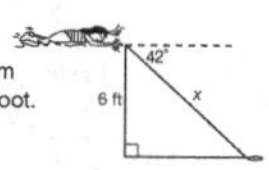

$\sin x = \frac{\text{opp}}{\text{hyp}}$

$\sin 42° = \frac{\underline{6}}{\underline{x}}$

$x \sin 42° = \underline{6}$

$x = \frac{\underline{6}}{\sin 42°}$

$x \approx \underline{9}$ ft

Solve the problem.

5. Paul is looking out a window that is 18 feet off the ground, and sees a mailbox near the street. The angle of depression to the mailbox is 28°. What is the horizontal distance from Paul to the mailbox? Round to the nearest foot. $\underline{34}$ ft

6. Elise is parasailing. She is attached to the boat by a rope that is 300 feet long. The angle of depression from Elise to the boat is 29°. How high up in the air is Elise? Round to the nearest foot. $\underline{145}$ ft

Name ______________________ Date __________ Class __________

Reteaching 74
Reflections

You have worked with translations. Now you will reflect figures across a line on a coordinate grid.

Reflecting Across an Axis

A **reflection** is an isometry (a transformation that does not change the size or shape of the object being translated). It involves a translation across a line, where each point and its image are the same distance from the line of reflection.

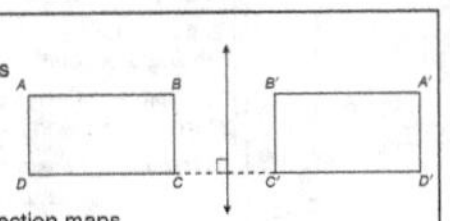

When an object is reflected across the x-axis, the reflection maps $(x, y) \rightarrow (x, -y)$. When an object is reflected across the y-axis, the reflection maps $(x, y) \rightarrow (-x, y)$.

Reflect ΔABC across the x-axis. Find the coordinates of the vertices of the reflection image.

The image is reflected by using $(x, y) \rightarrow (x, -y)$.

$A(1, 1) \rightarrow A'(1, -1)$

$B(1, 4) \rightarrow B'(1, -4)$

$C(4, 4) \rightarrow C'(4, -4)$

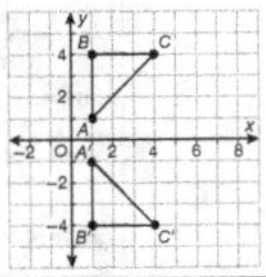

Practice

Complete the steps to find the coordinates of each reflection.

1. Reflect ΔJKL across the y-axis.

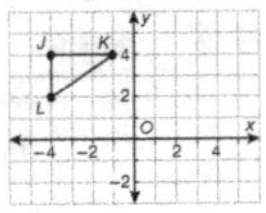

$J(-4, 4) \rightarrow J'(4, 4)$

$K(-1, 4) \rightarrow K'(1, 4)$

$L(-4, 2) \rightarrow L'(4, 2)$

Find the coordinates for each reflection.

2. Reflect ▭$WXYZ$ across the y-axis.

$W'(5, 3)$

$X'(3, 3)$

$Y'(3, -2)$

$Z'(5, -2)$

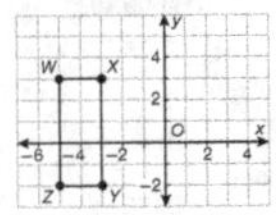

3. Reflect ΔMNO across the x-axis.

$M'(-3, -2)$

$N'(2, -2)$

$O'(2, 0)$

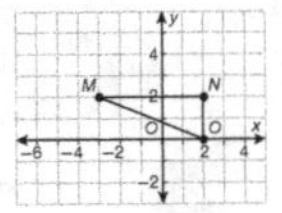

Reteaching 74
continued

Reflecting Across a Vertical or Horizontal Line

When reflecting an object across a vertical line $x = a$, the reflection maps $(x, y) \rightarrow (a - (x - a), y)$ or $(x, y) \rightarrow (2a - x, y)$.

When reflecting an object across a horizontal line $y = b$, the reflection maps $(x, y) \rightarrow (x, b - (y - b))$ or $(x, y) \rightarrow (x, 2b - y)$.

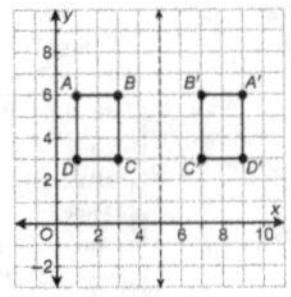

Reflect ΔABC across the line $x = 3$. Find the coordinates of the vertices of the reflection image.

The image is reflected using $(x, y) \rightarrow (3 - (x - 3), y)$ or $(x, y) \rightarrow (6 - x, y)$.

$A(0, 4) \rightarrow A'(6 - 0, 4) = A'(6, 4)$

$B(2, 4) \rightarrow B'(6 - 2, 4) = B'(4, 4)$

$C(2, 2) \rightarrow C'(6 - 2, 2) = C'(4, 2)$

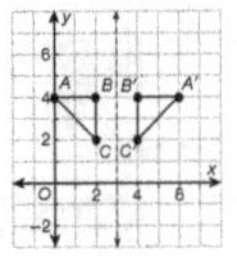

Practice

Complete the steps to find the coordinates of each reflection.

4. Reflect ▭$ABCD$ across the line $y = 3$.

$A(-2, 2) \rightarrow A'(-2, 6 - 2) = A'(-2, 4)$

$B(2, 2) \rightarrow B'(2, 6 - 2) = B'(2, 4)$

$C(2, -2) \rightarrow C'(2, 6 - (2)) = C'(2, 8)$

$D(-2, -2) \rightarrow D'(-2, 6 - (2)) = D'(-2, 8)$

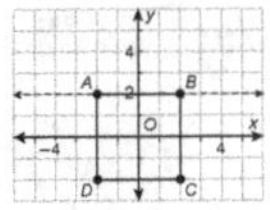

Find the coordinates of each reflection.

5. Reflect ΔPQR across the line $y = -2$.

$P'(-7, -5)$

$Q'(-5, -8)$

$R'(-2, -5)$

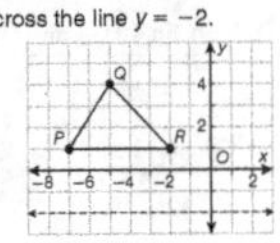

6. Reflect ▭$JKLM$ across the line $x = -1$.

$J'(-3, 6)$

$K'(-8, 6)$

$L'(-8, 2)$

$M'(-3, 2)$

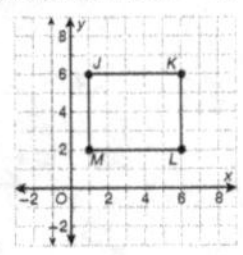

Name ______________________ Date __________ Class __________

Reteaching 75
Writing the equation of a circle

You have worked with plotting figures on a coordinate plane. Now you will find the equation of a circle.

Equation of a Circle

The equation of a circle with center (h, k) and radius r is $(x - h)^2 + (y - k)^2 = r^2$.

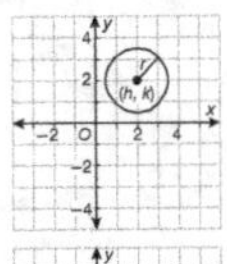

Write the equation of ⊙C with center $C(2, -1)$ and radius 6.

$(x - h)^2 + (y - k)^2 = r^2$	Equation of a Circle
$(x - 2)^2 + (y - (-1))^2 = 6^2$	Substitute.
$(x - 2)^2 + (y + 1)^2 = 36$	Simplify.

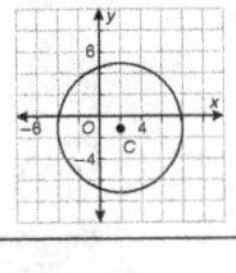

Practice

Complete the steps to write the equation of ⊙C.

1. $(x - h)^2 + (y - k)^2 = r^2$

$(x - (-2))^2 + (y - (-1))^2 = 3^2$

$(x + 2)^2 + (y + 1)^2 = 9$

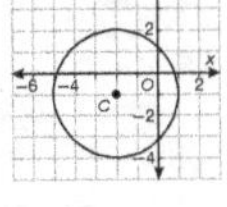

Write the equation of ⊙C.

2. $(x - 5)^2 + y^2 = 16$

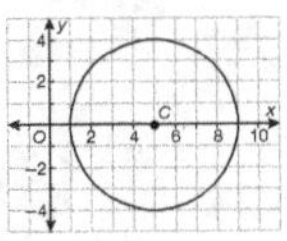

3. $(x - 7)^2 + (y - 3)^2 = 9$

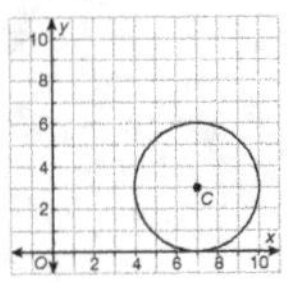

Reteaching 75
continued

Equation of a Circle

You can also write the equation of a circle if you know the center and one point on the circle.

Write the equation of ⊙C with center $C(3, 7)$ that passes through (1, 7).

Step 1: Find the radius.

$r = \sqrt{(x_2 - x_1)^2 + (y_2 - y_1)^2}$	Distance Formula
$r = \sqrt{(1 - 3)^2 + (7 - 7)^2}$	Substitute.
$r = \sqrt{4} = 2$	Simplify.

Step 2: Use the equation of a circle.

$(x - h)^2 + (y - k)^2 = r^2$	Equation of a Circle
$(x - 3)^2 + (y - 7)^2 = 2^2$	Substitute.
$(x - 3)^2 + (y - 7)^2 = 4$	Simplify.

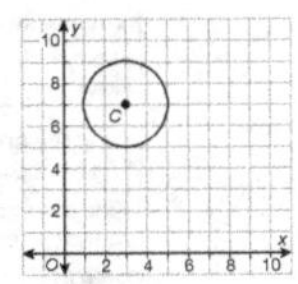

Practice

Complete the steps to find the equation of the circle.

4. ⊙C with center $C(4, 2)$ that passes through (2, 2)

$r = \sqrt{(x_2 - x_1)^2 + (y_2 - y_1)^2}$

$r = \sqrt{(4 - 2)^2 + (2 - 2)^2}$

$r = \sqrt{4} = 2$

$(x - h)^2 + (y - k)^2 = r^2$

$(x - 4)^2 + (y - 2)^2 = 2^2$

$(x - 4)^2 + (y - 2)^2 = 4$

5. ⊙B with center $B(3, 1)$ that passes through (7, 1)

$r = \sqrt{(x_2 - x_1)^2 + (y_2 - y_1)^2}$

$r = \sqrt{(7 - 3)^2 + (1 - 1)^2}$

$r = \sqrt{16} = 4$

$(x - h)^2 + (y - k)^2 = r^2$

$(x - 3)^2 + (y - 1)^2 = 4^2$

$(x - 3)^2 + (y - 1)^2 = 16$

Find the equation of the circle.

6. ⊙T with center $T(-1, 5)$ that passes through (4, 5)

$(x + 1)^2 + (y - 5)^2 = 25$

7. ⊙C with center $C(-4, -3)$ that passes through (−4, 3)

$(x + 4)^2 + (y + 3)^2 = 36$

Name ______________________ Date __________ Class __________

Reteaching 76
Symmetry

You have worked with reflections and rotations of geometric figures. Now you will work with symmetry.

Line Symmetry

A **line of symmetry** is a line that divides a plane figure into two congruent, reflected halves. Basically, you could put a mirror on the line of symmetry to see the entire object.

Line symmetry is a type of symmetry that describes a figure that can be reflected across a line so that the image coincides with the preimage. This is also called **reflection symmetry.**

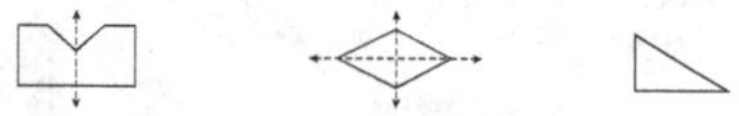

Write whether the figure below has line symmetry. If so, draw all lines of symmetry.

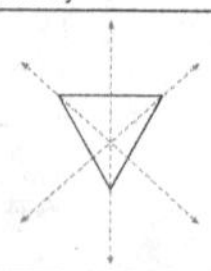

Step 1: Check for a line that divides the triangle into two congruent halves.

Step 2: If one exists, check for another line that divides the triangle into two congruent halves. Check for a third line.

Step 3: Draw the line, or lines, of symmetry.

Practice

Write whether each figure has line symmetry. If so, draw all lines of symmetry.

1. 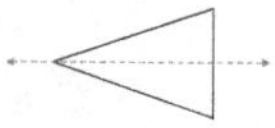 line symmetry

2. 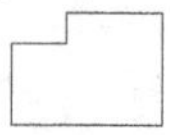 no line of symmetry

3. 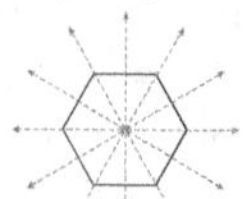 line symmetry

4. 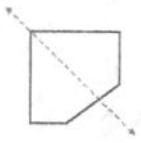 line symmetry

5.  no line of symmetry

6.  no line of symmetry

Reteaching 76
continued

Rotational Symmetry

A figure has **rotational symmetry** if it can be rotated about a point by an angle between 0° and 360° so that the image coincides with the original figure.

The smallest angle through which the figure is rotated to coincide with itself is called the **angle of rotational symmetry.**

The number of times that you can get an identical figure when repeating the degree of rotation is called the **order of the rotational symmetry.**

Below are some examples of figures with and without rotationally symmetry.

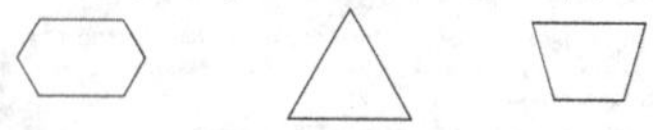

angle:	180°	120°	no rotational symmetry
order:	2	3	

Write whether the figure below has rotational symmetry. If so, give the angle of rotational symmetry and the order of symmetry.

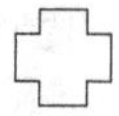

Step 1: Pick up your paper and rotate it clockwise 90°.

Step 2: Is the rotated figure congruent to the original? Does the figure have rotational symmetry? yes

Step 3: Write the angle and order of rotational symmetry for this figure. angle: 90°; order: 4

Practice

Write whether each figure of symmetry has rotational symmetry. If so, give the angle of rotational symmetry and the order.

7.  yes; angle: 180°; order: 2

8.  no rotational symmetry

9.  yes; angle: 90°; order: 4

10.  no rotational symmetry

11. 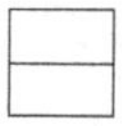 yes; angle: 180°; order: 2

12. 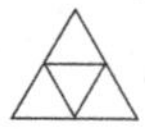 yes; angle: 120°; order: 3

Name ______________________ Date __________ Class __________

Reteaching 77
Finding Surface Areas and Volumes of Cones

You have found the lateral area, surface area, and volume of regular pyramids. Now you will find the lateral area, surface area and volume of cones.

Lateral and Surface Area of a Cone

Lateral Area: The lateral area of a cone with radius r and slant height l is $LA = \pi rl$.

Surface Area: The surface area of a cone with lateral area LA and base area B is $SA = LA + B$, where $B = \pi r^2$.

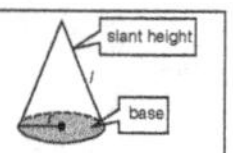

Find the lateral and surface area of the cone. Round to the nearest tenth.

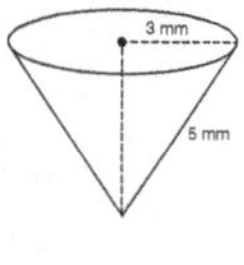

Step 1: Find LA and B using $r = 3$ and $l = 5$.

$LA = \pi rl$ $\quad B = \pi r^2$

$= \pi(3)(5)$ $\quad = \pi(3)^2$

$\approx 47.1 \text{ mm}^2$ $\quad = 28.3 \text{ mm}^2$

Step 2: Find SA using $LA \approx 47.1 \text{ mm}^2$ and $B \approx 28.3 \text{ mm}^2$.

$SA = LA + B$

$= (47.1) + (28.3)$

$= 75.4 \text{ mm}^2$

The surface area of the cone is approximately 75.4 mm^2.

Practice

Complete the steps to find the lateral and surface area of the cone. Round to the nearest tenth.

1. $LA = \pi rl = \pi(3)(\underline{8}) \approx \underline{75.4} \text{ in.}^2$

$B = \pi r^2 = \pi(\underline{3})^2 \approx \underline{28.3} \text{ in.}^2$

$SA = LA + B$

$\approx (\underline{75.4}) + (\underline{28.3})$

$\approx \underline{103.7 \text{ in.}^2}$

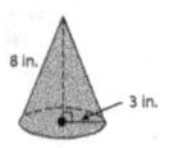

Find the lateral and surface area of each cone. Round to the nearest tenth.

2. 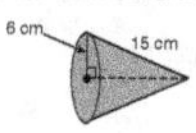

$LA = \underline{287.7 \text{ cm}^2}$

$SA = \underline{395.8 \text{ cm}^2}$

3. 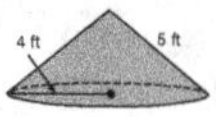

$LA = \underline{62.8 \text{ ft}^2}$

$SA = \underline{113.1 \text{ ft}^2}$

Reteaching 77
continued

Volume of a Cone

The volume of a right cone with base area B and height h is $V = \frac{1}{3}Bh$, where $B = \pi r^2$.

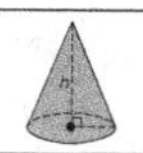

Find the volume of the cone. Round to the nearest tenth.

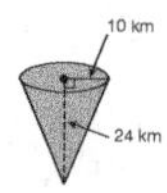

Step 1: Find the base area, B, using $r = 10$.

$B = \pi r^2$

$= \pi(10)^2$

$= 314.2 \text{ km}^2$

Step 2: Find the volume using $B \approx 314.2 \text{ km}^2$ and $h = 24$.

$V = \frac{1}{3}Bh$

$\approx \frac{1}{3}(314.2)(24)$

$\approx 2513.6 \text{ km}^3$

The volume of the cone is approximately 2513.6 km^3.

Practice

Complete the steps to find the volume of the cone. Round to the nearest tenth.

4. $B = \pi r^2 = \pi(\underline{4})^2 \approx \underline{50.3} \text{ ft}^2$

$V = \frac{1}{3}Bh$

$\approx \frac{1}{3}(\underline{50.3})(12)$

$\approx \underline{201.2} \text{ ft}^3$

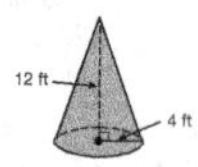

Find the volume of each cone. Round to the nearest tenth.

5. 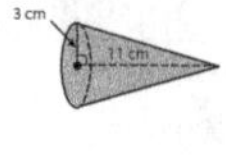

$V = 103.7 \text{ cm}^3$

6. 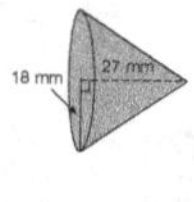

$V = 9161.1 \text{ mm}^3$

Answer Key

Name ______________________________ Date ___________ Class ___________

Reteaching 78
Rotations

Now you will rotate figures on a coordinate plane.

Rotating About the Origin

When a figure is rotated about the origin on a coordinate plane, some simple rules apply for certain rotations.

If a point (x, y) is rotated 90°, then (x, y) becomes $(-y, x)$.

If a point (x, y) is rotated 180°, then (x, y) becomes $(-x, -y)$.

If a point (x, y) is rotated 270°, then (x, y) becomes $(y, -x)$.

In this picture, quadrilateral *LMNO* has been rotated 270° about the origin to produce quadrilateral *L′M′N′O′*.

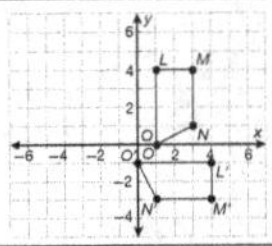

ΔABC has vertices $A(-6, 2)$, $B(-3, 5)$, and $C(-2, 2)$. Graph ΔABC and its rotation 90° counterclockwise about the origin.

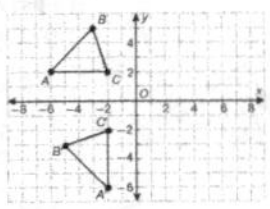

Step 1: Graph ΔABC.

Step 2: Using the rule above, (x, y) becomes $(-y, x)$.

$A(-6, 2) \rightarrow A'(-2, -6)$

$B(-3, 5) \rightarrow B'(-5, -3)$

$C(-2, 2) \rightarrow C'(-2, -2)$

Step 3: Graph $\Delta A'B'C'$.

Practice

1. ΔXYZ has vertices $X(4, -4)$, $Y(8, -4)$, and $Z(8, -7)$.

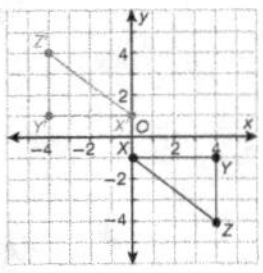

Using the rule, (x, y) becomes $(-x, -y)$.

$X(4, -4) \rightarrow X'(-4,$ <u>4</u>$)$

$Y(8, -4) \rightarrow Y'($<u>−8</u>$, 4)$

$Z(8, -7) \rightarrow Z'($<u>−8</u>, <u>7</u>$)$

$\Delta X'Y'Z'$ has vertices $X'(4, -4)$, $Y'($<u>8</u>, <u>−4</u>$)$, and $Z'($<u>−8</u>, <u>7</u>$)$.

Rotate each figure by the specified angle counterclockwise about the origin. List the coordinates of the vertices of the rotated figure.

2. 270°

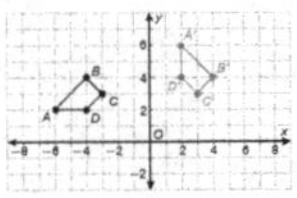

<u>$A'(2, 6)$, $B'(4, 4)$, $C'(3, 3)$, $D'(2, 4)$</u>

3. 90°

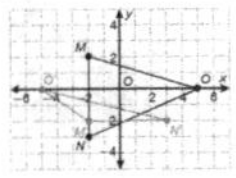

<u>$M'(-2, -2)$, $N'(3, -2)$, $O'(-5, 0)$</u>

Reteaching 78
continued

Rotating About a Chosen Point

When a figure is rotated about a chosen point on a coordinate plane, a protractor must be used to measure the angle of rotation. The rotated vertices will be the same distance from the center of rotation.

In this picture, ΔABC was rotated 45° counterclockwise about the point $R(0, -3)$ to produce $\Delta A'B'C'$.

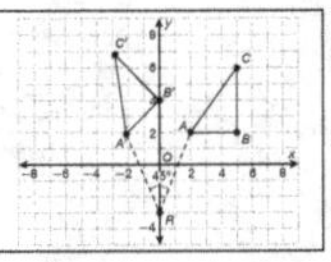

Triangle *DEF* has vertices $D(-4, 4)$, $E(-3, 2)$, $F(-2, 5)$. Graph the triangle and its rotation 180° counterclockwise about the point $R(-4, 1)$.

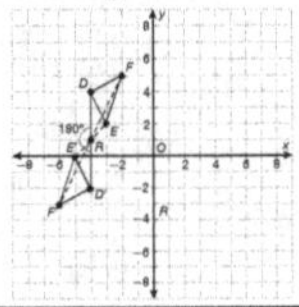

Step 1: Graph ΔDEF.

Step 2: Draw a segment from the center of rotation, $R(-4, 1)$, to point *D*.

Step 3: Use a protractor to measure a 180° angle counterclockwise from $\overline{RD}$.

Step 4: Draw $\overline{RD'}$. Use a compass or a ruler to make sure *D* and *D′* are the same distance from *R*.

Step 5: Repeat these steps for points *E* and *F*.

Practice

Complete the steps to rotate *ABCD* 90° counterclockwise about $R(0, -1)$.

4. Draw a segment from $R(0, -1)$ to point *A*.
Measure a 90° angle from $\overline{RA}$.
Draw $\overline{RA'}$.
Make *A′* the same distance from *R* as *A* is from *R*.
Repeat for points *B*, *C*, and *D*.

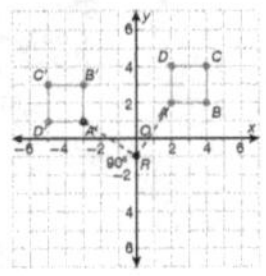

Rotate each figure by the specified angle counterclockwise about the specified point.

5. 45° about $R(3, 0)$

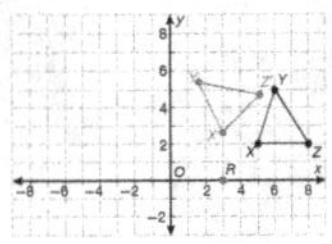

6. 180° about $R(-3, 3)$

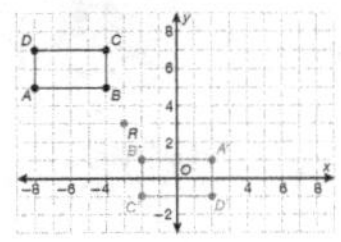

Name ______________________________ Date ___________ Class ___________

Reteaching 79
Angles Exterior to Circles

You have worked with angles interior to circles. Now you will work with angles exterior to circles.

Intercepting by Two Secants or Two Tangents

An **exterior angle** of a circle has its vertex outside the circle with each ray intersecting the circle at least once. Below are two examples of exterior angles.

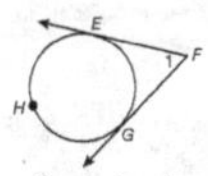

intercepting by two tangents

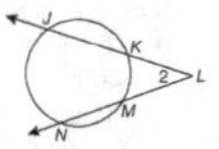

intercepting by two secants

The measure of an exterior angle is equal to half the difference of the intercepted arcs.

$m\angle 1 = \frac{1}{2}(mEHG - mEG)$ $\quad m\angle 2 = \frac{1}{2}(mJN - mKM)$

Find the value of *x*.

Step 1: Since $mPVR + mPR = 360°$, $mPVR + 142 = 360°$, and $mPVR = 218°$.

Step 2: Substitute $mPVR = 218°$ and $mPR = 142°$ into the formula and solve.

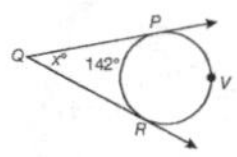

$x° = \frac{1}{2}(mPVR - mPR)$

$= \frac{1}{2}(218° - 142°)$

$x° = 38°$

Practice

Find the value of *x*.

1. 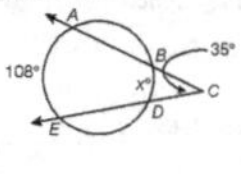

$35° = \frac{1}{2}(108° - x)$; $x =$ <u>38°</u>

2. 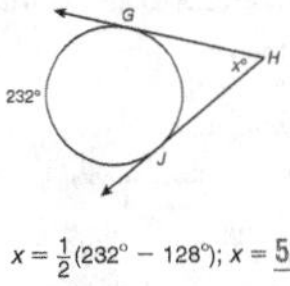

$x = \frac{1}{2}(232° - 128°)$; $x =$ <u>52°</u>

3. 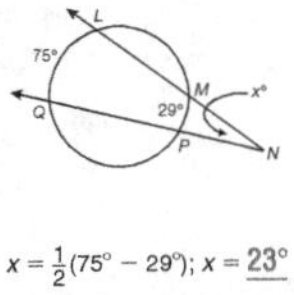

$x = \frac{1}{2}(75° - 29°)$; $x =$ <u>23°</u>

4. What is the measure of the angle exterior to a circle made by two secants if the shorter arc is 52° and the longer arc is 135°? <u>41.5°</u>

5. What is the measure of the angle exterior to a circle made by two tangents if there is an arc that is 267°? <u>87°</u>

Reteaching 79
continued

Intercepting by a Secant and a Tangent

It is also possible for an exterior angle to intersect by a secant and a tangent. The measure of an exterior angle is still equal to half the difference of the intercepted arcs.

$m\angle 1 = \frac{1}{2}(mAD - mBD)$

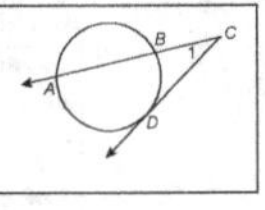

Find the value of *x*.

Step 1: Write the formula in terms of the given exterior angle.

$x° = \frac{1}{2}(mTU - mSU)$

Step 2: Substitute $mSU = 74°$ and $mTU = 74°$ into the formula and solve.

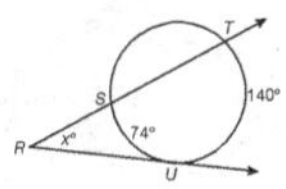

$x° = \frac{1}{2}(mTU - mSU)$

$= \frac{1}{2}(140° - 74°)$

$x° = 33°$

Practice

Find the value of *x*.

6. 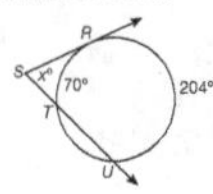

$x = \frac{1}{2}(204° - 70°)$; $x =$ <u>67°</u>

7. 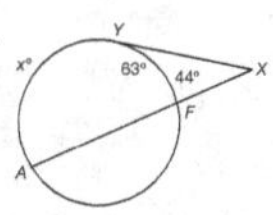

$44° = \frac{1}{2}(x - 63°)$; $x =$ <u>151°</u>

8. 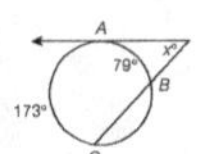

$x = \frac{1}{2}$(<u>173°</u> $- 79°)$; $x =$ <u>47°</u>

9. What is the measure of the angle exterior to a circle made by one tangent and one secant if the shorter arc length between the tangent and secant is 12° and the longer arc length is 64°? <u>26°</u>

10. If the measure of the angle exterior to a circle made by one tangent and one secant is 54° and the shorter arc length between the tangent and secant is 42°, what is the longer arc length? <u>150°</u>

11. What is the measure of the angle exterior to a circle made by one tangent and one secant if the shorter arc length between the tangent and secant is 67° and the longer one is 172°? <u>52.5°</u>

Answer Key

Name ______________________ Date __________ Class __________

Reteaching 80
Finding Surface Area and Volumes of Spheres

You have found the surface area and the volume of cylinders. Now you will find the surface area and the volume of spheres.

Surface Area of a Sphere

The surface area of a sphere with radius r is $SA = 4\pi r^2$.

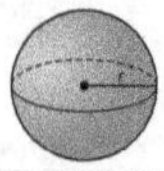

Find the surface area of the sphere below. Round to the nearest tenth.

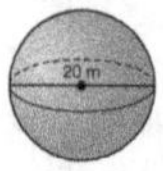

Step 1: Find r by dividing the diameter, $d = 20$ m, by two.

$r = \frac{20}{2}$

$= 10$ m

Step 2: Substitute $r = 10$ m into the formula and solve.

$SA = 4\pi r^2$

$= 4\pi(10)^2$

$\approx 1{,}256.6\ m^2$

Practice

Find the surface area of each sphere. Round each answer to the nearest tenth.

1. Find the surface area of sphere S.

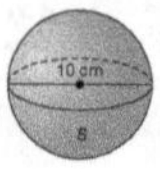

The diameter of sphere S is 10 centimeters.

This means r, the radius of the sphere, is 5 centimeters.

$r^2 = 25$

$SA = 4\pi r^2 = 314.2\ cm^2$

2. 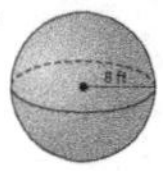

3. 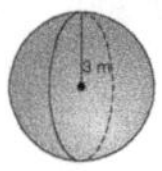

4. 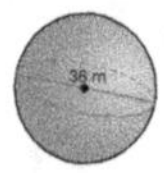

$SA = 804.2\ ft^2$ $SA = 131.1\ m^2$ $SA = 4{,}071.5\ m^2$

Reteaching 80
continued

Volume of a Sphere

The volume of a sphere with radius r is $V = \frac{4}{3}\pi(13)^3$.

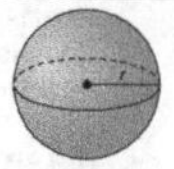

Find the volume of the sphere below. Round to the nearest tenth.

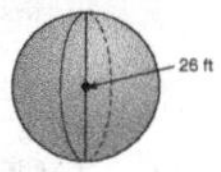

Step 1: Find r by dividing the diameter, $d = 26$ ft, by two.

$r = \frac{26}{2}$

$= 13$ ft

Step 2: Substitute $r = 13$ ft into the formula and solve.

$V = \frac{4}{3}\pi r^3$

$= \frac{4}{3}\pi(13)^3$

$\approx 9{,}202.8\ ft^3$

Practice

Find the volume of each sphere. Round each answer to the nearest tenth.

5. The diameter of this sphere is 16 centimeters.

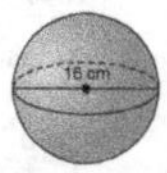

The radius, r, is 8 centimeters.

$r^3 = 512$

$\frac{4}{3}\pi r^3 = 2144.7\ cm^3$

Answer each question. Round each answer to the nearest tenth.

6. What is the volume of the sphere to the right?

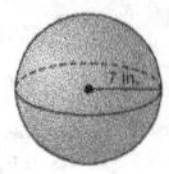

$1436.8\ in^3$

7. What is the volume of a sphere with diameter 42 centimeters?

$38{,}792.4\ cm^3$

8. If the surface area of a sphere is $64\pi\ mm^2$, what is the volume of the sphere?

$268.1\ mm^3$

Name ______________________ Date __________ Class __________

Reteaching INV 8
Patterns

You have learned about transformations and symmetry. Now you will learn about geometric patterns.

Use the isosceles triangle to the right.

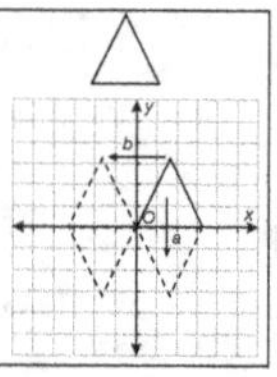

Step 1: Draw the triangle on a coordinate plane so that the base is on the x-axis with a base angle at the origin. Then, reflect the triangle over the x-axis and draw the result.

Step 2: Reflect the pattern from **Step 1** over the y-axis and draw the resulting figure.

Step 3: Does the final pattern have rotational symmetry? If so, what is the order of rotational symmetry? yes; 2

Step 4: Does the final pattern have any lines of symmetry? If so, how many? yes; 2

Practice

Complete the steps of the pattern on the coordinate plane.

1. Step 1: Draw an isosceles right triangle with one leg on each axis.

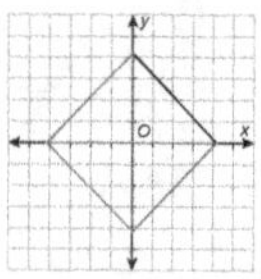

Step 2: Reflect the triangle over the y-axis.

Step 3: Reflect the pattern from Step 2 over the x-axis.

Step 4: Does the resulting pattern have any lines of symmetry? If so, how many? yes; 4

Step 5: Does the final pattern have rotational symmetry? If so, what is the order of rotational symmetry? yes; 4

Step 6: What kind of polygon is the resulting figure? a square

Use the rhombus to the right for problems 2–4.

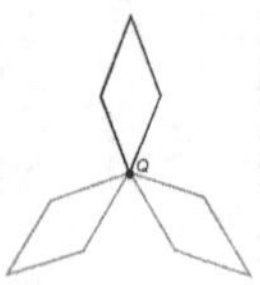

2. Sketch the resulting figure after a rotation of 120° counterclockwise around point Q.
3. If the figure is rotated 120° until it is returned to its original position, what is the order of rotational symmetry in the final pattern? 3
4. Does the resulting pattern have any lines of symmetry? If so, how many? yes; 3

Reteaching INV 8
continued

Look at the pattern on the right. The first figure has one dot; the second, three dots; the third, six dots.

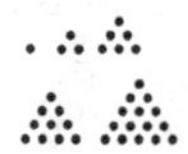

Step 1: Draw the next two figures in the pattern.

How many dots form each? 10, 15

Step 2: List the number of dots in each triangle.

1, 3, 6, 10, 15

These are called **triangular numbers.** The formula to find a given triangular number in the triangular number series is $x = \frac{n(n+1)}{2}$.

Step 3: Use the formula to find the sixteenth triangular number.

$x = \frac{n(n+1)}{2} = \frac{16(16+1)}{2} = 136$

Practice

Find the twenty-fifth and twenty-sixth triangular numbers.

5. Find the twenty-fifth triangular number.

$x = \frac{n(n+1)}{2}$ Formula

$x = \frac{25(25+1)}{2}$ Substitute.

$x = \frac{650}{2}$ Solve.

$x = 325$

The twenty-fifth triangular number is 325.

6. Find the twenty-sixth triangular number.

$x = \frac{n(n+1)}{2}$

$x = \frac{26(26+1)}{2}$

$x = \frac{702}{2}$

$x = 351$

The twenty-sixth triangular number is 351.

Complete the questions for the cubed numbers sequence.

7. Cubed numbers are numbers that can be used to find the volume of a cube.

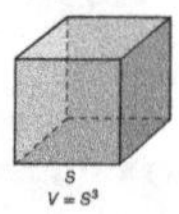

The first few numbers of the sequence are 1, 8, 27, 64, . . .

What are the next three numbers in the sequence?

125, 216, 343

8. Write an equation to find the nth number in the sequence. $x = n^3$
9. What is the twenty-second cubed number? 10,648

Name ______________________ Date __________ Class __________

Reteaching 81
Graphing and Solving Linear Systems

You have found the slopes of equations and lines. Now you will solve linear systems graphically and algebraically.

Solving Linear Systems Graphically

A **system of linear equations** is a set of two or more linear equations. A system of linear equations can be solved by graphing the lines.

Solve this linear system by graphing the lines.

$y = 3x - 4$; $y = -x + 4$

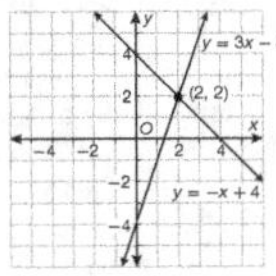

Step 1: Graph each line using the equations given.

Step 2: The lines intersect at the point (2, 2).

Step 3: Check your answer by substituting $x = 2$ and $y = 2$ into both equations.

$y = 3x - 4$ | $y = -x + 4$
$2 = 3(2) - 4$; | $2 = -2 + 4$
$2 = 2$ | $2 = 2$

Therefore, (2, 2) is the correct solution to this system.

Practice

Complete the steps to solve the system of equations by graphing.

1. $y = x + 4$; $y = -4x - 6$

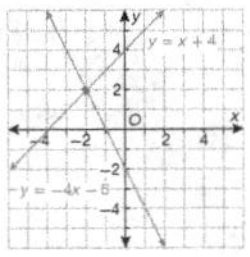

Graph each line. The lines intersect at $(\underline{-2}, \underline{2})$.

Check your answer.

$y = x + 4$ | $y = -4x - 6$
$\underline{2} = -2 + 4$; | $\underline{2} = -4(\underline{-2}) - 6$
$\underline{2} = 2$ | $\underline{2} = \underline{2}$

The solution is $\underline{(-2, 2)}$.

Solve the system of equations by graphing. Check your answer.

2. $x + 2y = 10$; $x + y = 6$

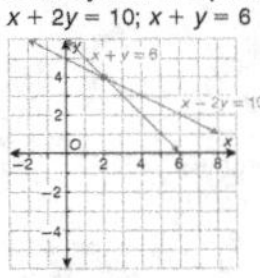

(2, 4)

3. $2x + 3y = 12$; $5x - 2y = 11$

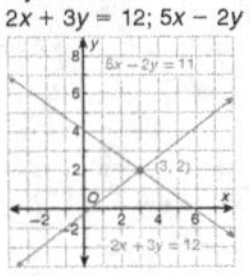

(3, 2)

Reteaching 81
continued

Solving Linear Systems Algebraically

To solve a system algebraically, solve both equations for the same variable and use substitution.

Solve this system of equations algebraically.

$2x + y = 4$; $x + y = 1$

Step 1: Solve both equations for *y*.

$2x + y = 4$ | $x + y = 1$
$y = 4 - 2x$ | $y = 1 - x$

Step 2: Substitute to find *x*.

$4 - 2x = 1 - x$
$x = 3$

Step 3: Substitute $x = 3$ into either of the equations.

$x + y = 1$
$3 + y = 1$
$y = -2$

Therefore, the solution to this system is the ordered pair (3, −2).

Practice

Complete the steps to solve the system of equations algebraically.

4. $y = 3x - 4$; $x = y - 4$

Solve the second equation for *y*.

$y = \underline{x + 4}$

Substitute to find *x*.

$3x - 4 = \underline{x + 4}$
$x = \underline{4}$

Substitute $x = 4$ into $y = 3x - 4$.

$y = 3x - 4$
$y = 3(\underline{4}) - 4 = \underline{8}$

The solution to the system of equations is $\underline{(4, 8)}$.

Solve the system of equations algebraically.

5. $2x + 3y = 4$
$x - 3y = 2$
(2, 0)

6. $-5y = 4x - 40$
$y = x - 1$
(5, 4)

Name ______________________ Date __________ Class __________

Reteaching 82
Inverse Trigonometric Functions

You have found the lengths of right triangle sides using trigonometric functions. Now you will find the measures of right triangle angles using inverse trigonometric functions.

Inverse Trigonometric Functions

With inverse trigonometric functions, you can use side lengths of a right triangle to calculate angle measures.

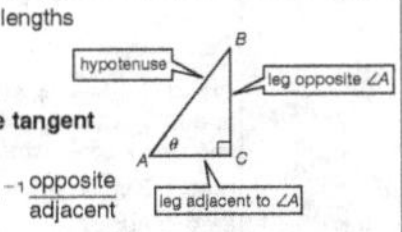

inverse sine	inverse cosine	inverse tangent
$\theta = \sin^{-1}\frac{\text{opposite}}{\text{hypotenuse}}$	$\theta = \cos^{-1}\frac{\text{adjacent}}{\text{hypotenuse}}$	$\theta = \tan^{-1}\frac{\text{opposite}}{\text{adjacent}}$

Find θ to the nearest degree.

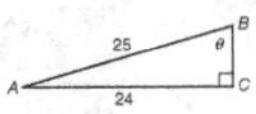

Step 1: Since the lengths of the side opposite and the hypotenuse are given, you must use sine inverse.

Step 2: Substitute the given values into the formula and use a calculator to solve.

$\theta = \sin^{-1}\frac{\text{opposite}}{\text{hypotenuse}}$
$= \sin^{-1}\frac{24}{25}$
$\approx 74°$

Practice

Complete the steps to find θ to the nearest degree.

1. $\theta = \tan^{-1}\frac{\text{opposite}}{\text{adjacent}}$

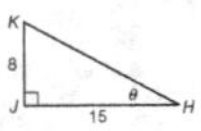

$\theta = \tan^{-1}\frac{\underline{8}}{\underline{15}}$
$\theta \approx \underline{28°}$

2. $\theta = \cos^{-1}\frac{\underline{\text{adjacent}}}{\underline{\text{hypotenuse}}}$

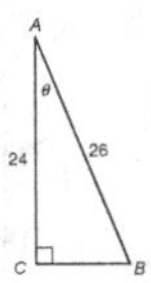

$\theta = \cos^{-1}\frac{\underline{24}}{\underline{26}}$
$\theta \approx \underline{23°}$

Find θ to the nearest degree.

3.

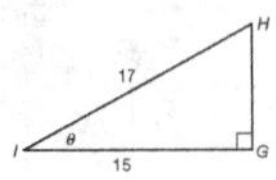

28°

4.

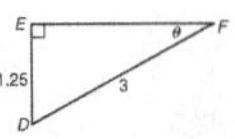

25°

5.

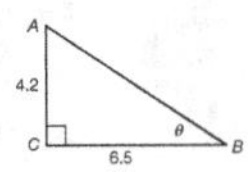

33°

Reteaching 82
continued

Applications of Inverse Trigonometric Functions

You can use inverse trigonometric functions to solve application problems. In this picture, the angle of the weight bench, θ, can be determined by using inverse tangent.

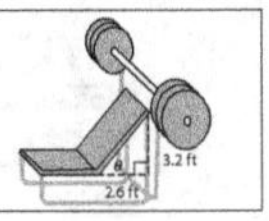

$\theta = \tan^{-1}\frac{3.2}{2.6}$

Find film projection angle, θ, to the nearest degree.

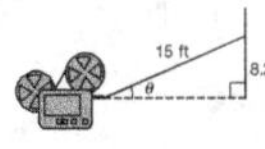

Step 1: Since the lengths of the side opposite and the hypotenuse are given, you must use sine inverse.

Step 2: Substitute the given values into the formula and use a calculator to solve.

$\theta = \sin^{-1}\frac{\text{opposite}}{\text{hypotenuse}}$
$= \sin^{-1}\frac{8.2}{15}$
$\approx 33°$

The film projection angle is approximately 33°.

Practice

Complete the steps to solve the application problem. Round to the nearest degree.

6. Yi is building a wheelchair ramp for his grandmother. What is the angle of incline, θ, that Yi needs to use for the ramp?

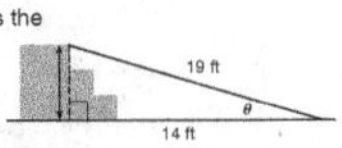

$\theta = \cos^{-1}\frac{\text{adjacent}}{\text{hypotenuse}}$
$\theta = \cos^{-1}\frac{\underline{14}}{\underline{19}}$
$\theta \approx \underline{43°}$

The angle of incline for the ramp is approximately $\underline{43°}$.

Solve each application problem. Round your answers to the nearest degree.

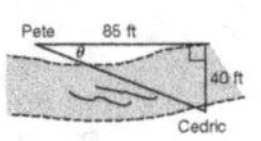

7. Pete is going to throw a baseball to Cedric. What is the angle, θ, that Pete needs to throw the ball in order reach Cedric? $\underline{33°}$

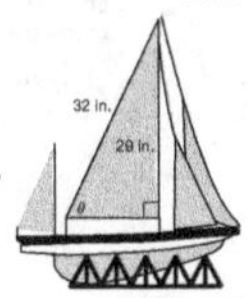

8. Sylvia is making a new sail for her model ship. What is the angle, θ, that Sylvia needs to use to cut the corner of the sail? $\underline{61°}$

Name ______________________ Date __________ Class __________

Reteaching 83
Vector Addition

You have worked with applications of right triangles. Now you will work with adding vectors.

Adding Vectors

Example:

You can use the Parallelogram Method to add $\vec{W} = \langle 3, 4\rangle$ and $\vec{Z} = \langle 5, -2\rangle$.

Step 1: Draw the vectors with their tails at the same starting point on the coordinate grid

Step 2: Complete the parallelogram by sketching the other two sides.

Step 3: The resultant vector is the diagonal of the parallelogram.

Step 4: Solve for the resultant vector.

$\vec{w} + \vec{z} = \langle 3, 4\rangle + \langle 5, -2\rangle$

$w + \vec{z} = \langle 8, 2\rangle$

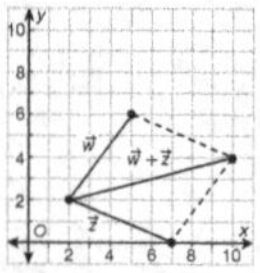

Practice

Complete the statements and diagram to add the given vectors.

1. Add $\vec{g} = \langle 2, 4\rangle$ and $\vec{h} = \langle 4, -1\rangle$.

   Complete the diagram by finishing the parallelogram.

   Sketch in the resultant vector which is the **diagonal**.

   $\vec{g} + \vec{h} = \langle 2,4\rangle + \langle 4, -1\rangle$

   $\vec{g} + \vec{h} = \langle 6, 3\rangle$

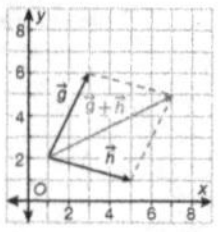

Determine the resultant vector.

2. $\vec{a} = \langle 4, 2\rangle$ and $\vec{b} = \langle 5, -3\rangle$ — $\langle 9, -1\rangle$
3. $\vec{c} = \langle 5, 0\rangle$ and $\vec{d} = \langle 2, -1\rangle$ — $\langle 7, -1\rangle$
4. $\vec{a} = \langle 3, -3\rangle$ and $\vec{b} = \langle 6, 4\rangle$ — $\langle 9, 1\rangle$
5. $\vec{c} = \langle 2, 3\rangle$ and $\vec{d} = \langle 2, 2\rangle$ — $\langle 4, 5\rangle$
6. $\vec{a} = \langle 6, 3\rangle$ and $\vec{b} = \langle 1, -3\rangle$ — $\langle 7, 0\rangle$
7. $\vec{c} = \langle 8, -2\rangle$ and $\vec{d} = \langle 3, 4\rangle$ — $\langle 11, 2\rangle$

 181 Saxon Geometry

Reteaching 83
continued

Adding Vectors Using Trigonometry

The **magnitude** of a vector is its length. The **direction** of a vector is the angle that it makes with a horizontal line, such as the *x*-axis.

Example:

Add the vectors $\vec{a} = \langle 5, 0\rangle$ and $\vec{b} = \langle 0, 2\rangle$. Find the magnitude and angle from the horizontal of the resultant vector.

Step 1: Draw a diagram. Add the vectors by adding the components.

$\vec{a} + \vec{b} = \langle 5, 0\rangle + \langle 0, 2\rangle$

$\vec{a} + \vec{b} = \langle 5, 2\rangle$

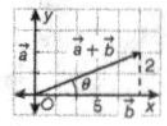

Step 2: Find the magnitude.

$a^2 + b^2 = (\vec{a} + \vec{b})^2$

$5^2 + 2^2 = 29 = (\vec{a} + \vec{b})^2$

$5.4 \approx \vec{a} + \vec{b}$

Step 3: Find the angle

$\tan\theta = \frac{opposite}{adjacent} = \frac{2}{5}$

$\theta = \tan^{-1}\frac{2}{5} \approx 22°$

Practice

Complete the steps to add the vectors. Find the magnitude and angle from the horizontal of the resultant vector.

8. Add the vectors $\vec{a} = \langle 0, 4\rangle$ and $\vec{b} = \langle 4, 0\rangle$.

   Complete the diagram. $\vec{a} + \vec{b} = \langle 0, 4\rangle + \langle 4, 0\rangle$

   $\vec{a} + \vec{b} = \langle 4, 4\rangle$

   Find the magnitude. $a^2 + b^2 = (\vec{a} + \vec{b})^2$

   $4^2 + 4^2 = (\vec{a} + \vec{b})^2$

   $5.6 \approx \vec{a} + \vec{b}$

   Find the angle. $\tan\theta = \frac{4}{4}$

   $\theta = \tan^{-1} 1 \approx 45°$

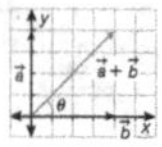

Find the magnitude and angle from the horizontal of the resultant vector. Round to the nearest degree.

9. $\vec{a} = \langle 6, 0\rangle$ and $\vec{b} = \langle 0, 3\rangle$ — 6.7; 27°
10. $\vec{a} = \langle 5, 0\rangle$ and $\vec{b} = \langle 0, 8\rangle$ — 9.4; 58°

 182 Saxon Geometry

Name ______________________ Date __________ Class __________

Reteaching 84
Dilations

You have worked with transformations and translations. Now you will work with dilations of figures.

Dilations

A **dilation** is a transformation that changes the size of a figure but not its shape. A dilation that results in a smaller image is called a reduction or contraction. A dilation that results in a larger image is called an enlargement or expansion.

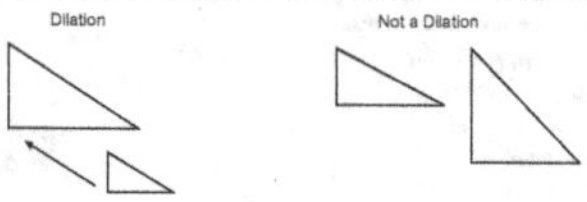

A dilation is a transformation in which the lines that connect every point *A* with its image *A′* all intersect at point *P*, which is called the **center of dilation.**

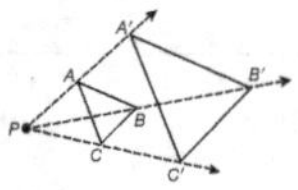

Practice

Complete the statements to determine whether the transformation appears to be a dilation.

1. Does the second figure appear to be the same shape? Yes

   Does the second figure appear to be a different size? Yes

   The figure appears to be a dilation.

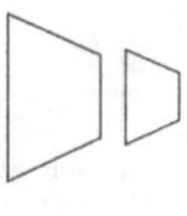

Copy each triangle and center of dilation. Draw the image of the triangle under a dilation with the given scale factor.

2. scale factor: 2

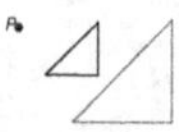

3. scale factor: 1/2

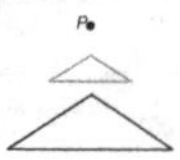

 183 Saxon Geometry

Reteaching 84
continued

Dilations in the Coordinate Plane

For $k > 1$:

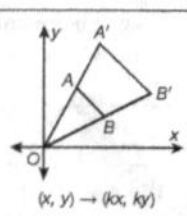

For $0 < k < 1$:

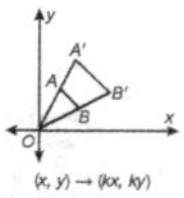

Draw the image of ΔEFG with vertices $E(0, 0)$, $F(0, 1)$, and $G(2, 1)$. Apply a dilation centered at the origin and with a scale factor of 3.

Step 1: Plot the points for ΔEFG.

Step 2: Apply the scale factor to each point. Multiply each coordinate by 3 (the scale factor).

Preimage (x, y)	E(0, 0)	F(0, 1)	G(2, 1)
Image (3x, 3y)	E′(0, 0)	F′(0, 3)	G′(6, 3)

Step 3: Plot the image points and sketch the triangle.

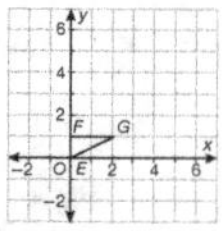

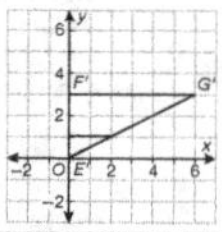

Practice

Complete the steps to dilate the figure on the coordinate plane.

4. $J(0, 0)$, $K(-1, 2)$, $L(3, 4)$ scale factor: 2

Preimage (x, y)	J(0, 0)	K(−1, 2)	L(3, 4)
Image (2x, 2y)	J′(0, 0)	K′(−2, 4)	L′(6, 8)

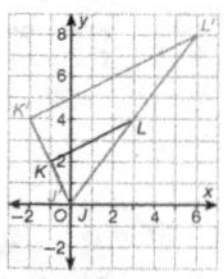

Draw the image of the figure with the given vertices under dilation with the given scale factor and centered at the origin.

5. $A(0, 0)$, $B(0, 6)$, $C(6, 3)$; scale factor: $\frac{1}{3}$

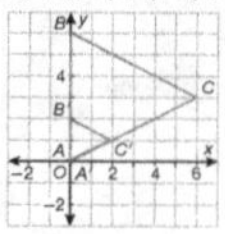

6. $R(1, 0)$, $S(1, -2)$, $T(-1, -2)$; scale factor: 2

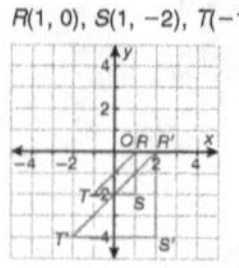

 184 Saxon Geometry

Name ______________________ Date __________ Class __________

Reteaching 85
Cross Sections of Solids

You have worked with dilations. Now you will work with cross sections of solids.

Describing and Sketching Cross Sections of Solids

The **cross section** of a solid is the intersection of a three-dimensional figure and a plane.

Describe and draw the cross section of the solid.

The solid is a square pyramid with a horizontal cross section.

The intersection of the plane with the pyramid is a square.

The cross section is a square.

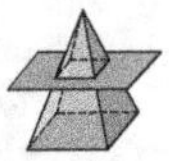

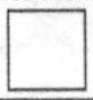

Practice

Complete the steps to describe and draw the cross section

1.

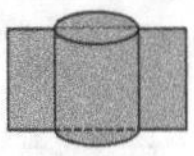

The cross section is a rectangle.

Draw the cross section.

2.

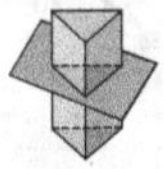

The cross section is a triangle.

Draw the cross section.

Describe and draw the cross section of each solid.

3.

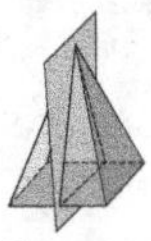

The cross section is a triangle.

4.

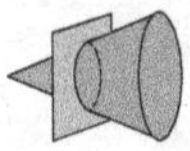

The cross section is a circle

Reteaching 85
continued

Finding Perimeter and Area of a Cross Section

Find the perimeter of the cross section of the square pyramid if the cross section is vertical.

The width of the cross section is 11 inches. The cross section is a rectangle. Its height is the same as that of the cylinder. The base of the rectangle is 11 inches.

$P = 2b + 2h$

$= 2 \cdot 11 + 2 \cdot 6$

$= 22 + 12$

$= 34$ in.

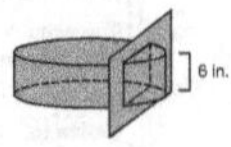

Find the area of the cross section.

$A = bh$

$= 11 \cdot 6$

$= 66$ in^2

Practice

Complete the steps to find the perimeter or the area of each cross section. Round your answer to the nearest hundredth.

5. Perimeter of cross section

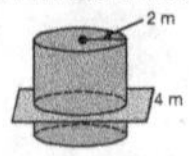

The cross section is a circle.

$P = 2\pi \underline{r}$

$= 2\pi \underline{2}$

$= \underline{4}\pi$

$= \underline{12.56 \text{ m}}$

6. Area of cross section

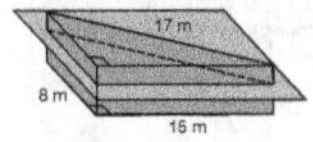

The cross section is a right triangle.

$A = \frac{1}{2} b \underline{h}$

$= \frac{1}{2}(15)(\underline{8})$

$= \underline{60 \text{ m}^2}$

Find the perimeter or area of each cross section.

7. Perimeter of cross section

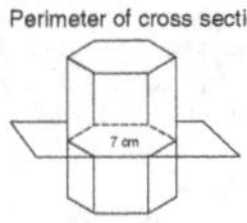

42 cm

8. Area of cross section

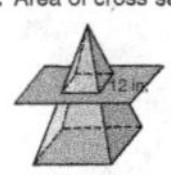

144 in^2

Name ______________________ Date __________ Class __________

Reteaching 86
Determining Chord Length

You have worked with chords and tangents in circles. Now you will learn how to determine the length of a chord.

Chords

If two chords intersect in a circle, then the products of the chord segments are equal.	$AE \cdot EB = CE \cdot ED$ 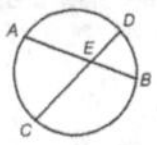

Example:

Find the value of x and the length of each chord.

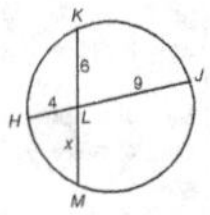

$HL \cdot LJ = KL \cdot LM$

$4 \cdot 9 = 6 \cdot x$ Substitute.

$39 = 6x$ Simplify.

$6 = x$ Divide each side by 6.

$HJ = 4 + 9 = 13$

$KM = 6 + x$

$= 6 + 6 = 12$

Practice

Complete the statements.

1. $TU \cdot UV = RU \cdot US$

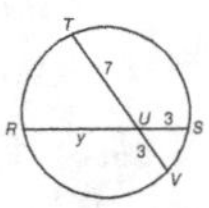

$\underline{7} \cdot 3 = y \cdot \underline{3}$ Substitute.

$\underline{21} = 3y$ Simplify.

$\underline{7} = y$ Divide each side by 3.

$TV = \underline{7} + 3 = 10$

$RS = 3 + y$

$= 3 + \underline{7} = \underline{10}$

Find the value of the variable and the length of each chord.

2. $y = 10$; $DF = 16$; $GH = 17$

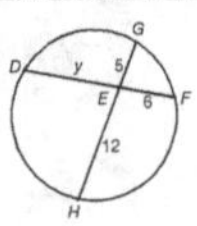

3. $z = 7.5$; $JL = 11$; $MN = 11.5$

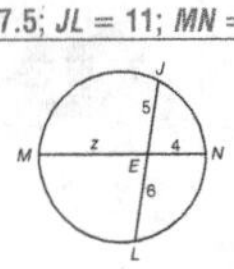

Reteaching 86
continued

More on Chords

Example:

Use the expressions for the segment lengths to write and solve an equation for x.

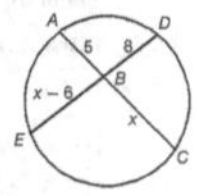

$AB \cdot BC = DB \cdot BE$

$5 \cdot x = 8(x - 6)$

$5x = 8x - 48$

$48 = 3x$

$16 = x$

Check your answer.

$BC = 16$; $BE = 16 - 6 = 10$; $(5)(16) = 8(10)$

Practice

Complete the steps to write and solve an equation for x.

4. $AE \cdot \underline{EC} = ED \cdot \underline{BE}$

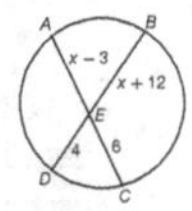

$\underline{x - 3} \cdot 6 = \underline{4}(x + 12)$

$\underline{6x} - 18 = 4x + \underline{48}$

$2x = \underline{66}$

$x = \underline{33}$

In each circle, use the expressions for the segment lengths to write and solve an equation for x.

5. $x = 28$

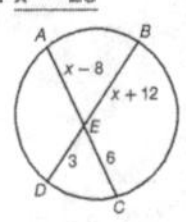

6. $x = 17$

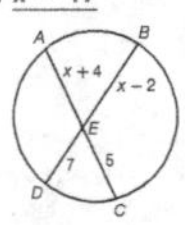

7. $x = 11$

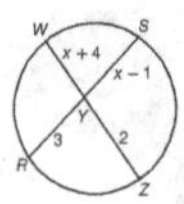

8. $x = 1$

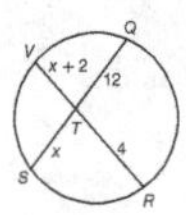

Name ______________________ Date __________ Class __________

Reteaching 87
Area Ratios of Similar Figures

You have worked with similar figures and the scale ratio before. Now you will use the scale ratio to find perimeters and areas of similar figures.

Similar Figure Ratios: Perimeter

If two similar figures have a scale factor of $a:b$, then the ratio of their perimeters is also $a:b$.

The scale factor of ΔABC to ΔDEF is 3:1. The perimeter ratio is 3:1 as well.

$\frac{\text{perimeter of } \Delta ABC}{\text{perimeter of } \Delta DEF} = \frac{36}{12} = \frac{3}{1} \rightarrow 3:1$

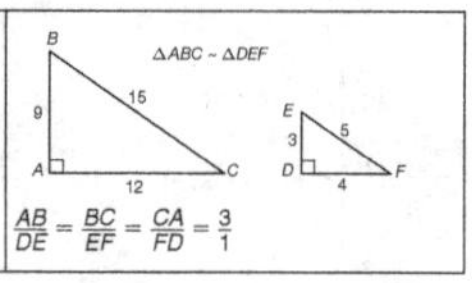

$\frac{AB}{DE} = \frac{BC}{EF} = \frac{CA}{FD} = \frac{3}{1}$

If $\Delta HJK \sim \Delta LMN$ and we know that the perimeter of ΔHJK is 30 inches, we can find the perimeter of ΔLMN.

Find the scale ratio using the given side lengths.

$\frac{\Delta HJK}{\Delta LMN} = \frac{9}{12} = \frac{3}{4}$

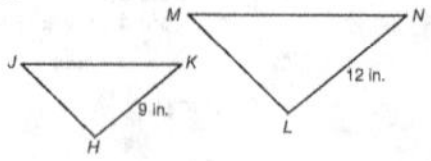

The perimeter ratio is the same, 3:4, so we can use that ratio to find the perimeter of ΔLMN.

$\frac{\text{perimeter of } \Delta HJK}{\text{perimeter of } \Delta LMN} = \frac{3}{4}$

$\frac{30}{P} = \frac{3}{4}$ Substitute the known values.

$30(4) = P(3)$ Cross multiply.

$40 = P$ Solve.

The perimeter of ΔLMN is 40 inches.

Practice

Complete the statements to find the perimeter of ▭TUVW.

1. Find the similarity ratio for $\frac{▭PQRS}{▭TUVW} = \frac{21}{14} = \frac{3}{2}$.

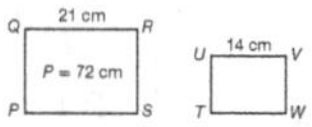

$\frac{\text{perimeter of } ▭PQRS}{\text{perimeter of } ▭TUVW} = \frac{3}{2}$

$\frac{72}{P} = \frac{3}{2}$ Substitute.

$(72)(2) = 3P$ Cross multiply.

$P = 48$ Solve.

The perimeter of ▭TUVW is 48 centimeters.

Reteaching 87 *continued*

Similar Figure Ratios: Area

If two similar figures have a scale factor of $a:b$, then the ratio of their areas is $a^2 : b^2$.

$\frac{\text{area of } \Delta ABC}{\text{area of } \Delta DEF} = \frac{54}{6} = \frac{9}{1} = \frac{3^2}{1^2}$

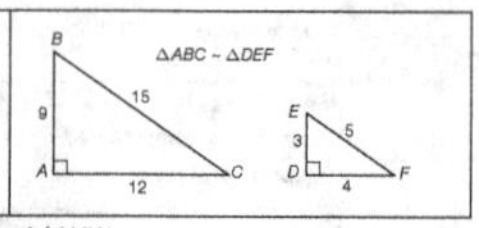

If $\Delta HJK \sim \Delta LMN$ and we know that the area of ΔHJK is 36 in.^{2}, we can find the area of ΔLMN.

Use the given side lengths to find the scale factor, 3:4. Square that scale factor to find the ratio of the areas, 9:16. Use that ratio to find the area of ΔLMN.

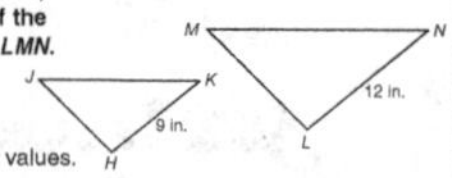

$\frac{\text{area of } \Delta HJK}{\text{area of } \Delta LMN} = \frac{9}{16}$

$\frac{36}{A} = \frac{9}{16}$ Substitute the known values.

$36(16) = A(9)$ Cross multiply.

$64 = A$ Solve.

So the area of ΔLMN is 64 in^2

Practice

Complete the steps to find the area of ▭TUVW.

2. Find the similarity ratio for $\frac{▭PQRS}{▭TUVW} = \frac{21}{14} = \left(\frac{3}{2}\right)^2 = \frac{9}{4}$

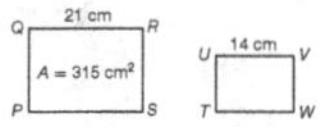

$\frac{\text{area of } ▭PQRS}{\text{area of } ▭TUVW} = \frac{9}{4}$

$\frac{315}{A} = \frac{9}{4}$ Substitute the known values.

$(315)(4) = 9A$ Cross multiply.

$1260 = 9A$ Simplify.

$140 = A$ Simplify.

The area of ▭TUVW is 140 cm^2.

3. $\Delta EFG \sim \Delta HJK$. Find the area of ΔHJK.

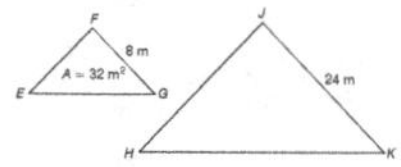

288 m^2

Name ______________________ Date __________ Class __________

Reteaching 88
Graphing and Solving Linear Inequalities

You have graphed linear equations before. Now you will learn how to graph linear inequalities.

Graphing Linear Inequalities

Graphing linear inequalities is similar to graphing equations. The main difference is that the solution is the entire region above or below the line. Linear inequalities can be graphed by drawing the inequality as if it were an equation and then shading the region above (for > and ≥) or below (for < and ≤) the line. Use a dashed line if the inequality symbol is < or > and a solid line if it is ≤ or ≥.

Example:

Graph the inequality $3y \geq 2x - 3$.

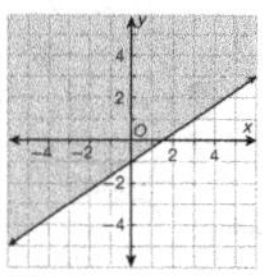

Step 1: Write the inequality in slope-intercept form.

$3y \geq 2x - 3$

$\frac{3y}{3} \geq \frac{2x}{2} - \frac{3}{3}$ Divide both sides by 3.

$y \geq \frac{2}{3}x - 1$ Finish in slope-intercept form.

Step 2: Graph the line $y = \frac{2}{3}x - 1$ using a solid line.

Step 3: The inequality is "greater than," so shade the area above the line.

Practice

Complete the statement to put the inequality in slope-intercept form, and then graph it.

1. Graph the inequality $x - y > 1$.

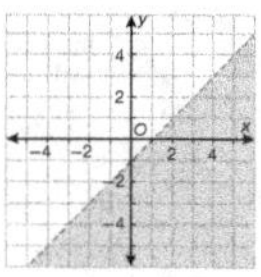

$x - y > 1$

$-y > 1 - x$ Subtract x from both sides.

$y < -1 + x$ Divide both sides by −1.

$y < x - 1$ Rewrite in slope-intercept form.

The slope is 1, and the y-intercept is −1. Because the symbol is <, the line should be made with a dashed line. Because the inequality is "less than," the solution region is below the line.

Graph each linear inequality.

2. $f(x) > -\frac{1}{2}x - 3$

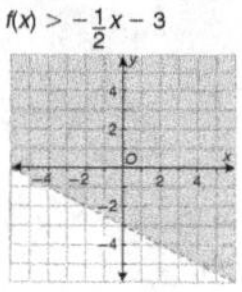

3. $2x - 3y \leq 6$

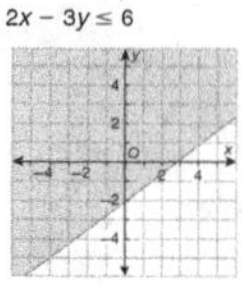

Reteaching 88 *continued*

Graphing Linear Inequalities: Systems

A system of inequalities consists of two or more inequalities. The solution to a system of inequalities is the area in which the shaded regions for each inequality overlap. If there is no overlapping region, the system has no solution and is called inconsistent.

Example: Graph and solve this system of inequalities.

$3x + y > 8$

$x + y \leq 3$

$y > -4$

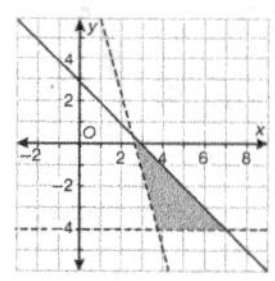

Step 1: Simplify each inequality.

$3x + y > 8$ $\quad x + y \leq 3$ $\quad y > -4$

$y > -3x + 8$ $\quad y \leq -x + 3$

Step 2: Graph $y > -3x + 8$ using a dashed line. Shade above the line.

Step 3: Graph $y \leq -x + 3$ using a solid line. Shade below the line.

Step 4: Graph $y > -4$.

Step 5: The overlapping region is the solution to the system.

Practice

Complete the steps to graph the system of inequalities.

4. $2x - y \geq 2$ $\quad 2x - y \geq 2$ $\quad x - 2y > 2$

$x - 2y > 2$ $\quad -y \geq -2x + 2$ $\quad -2y > -x + 2$

$y \leq 2x - 2$ $\quad y < \frac{x}{2} - 1$

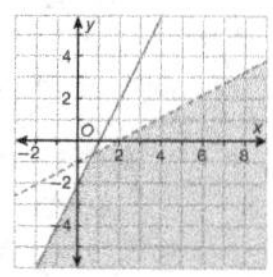

Graph each system of inequalities.

5. $2x + y > -4$

$-5x + 2y < 1$

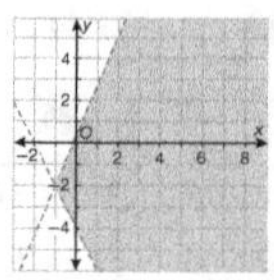

6. $y < x + 3$

$2x - 7y < 4$

$3x + 2y \leq 6$

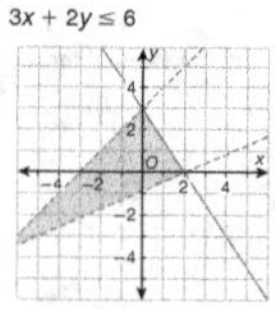

Name ______________________ Date __________ Class __________

Reteaching 89
Vector Decomposition

You have worked with vectors before. Now you will learn how to decompose vectors into their horizontal and vertical components.

Decomposing Vectors

To decompose a vector is to find its horizontal and vertical components. We find the horizontal component by multiplying the vector's magnitude by cosine of the angle the vector makes with the horizontal. The vertical component is found by multiplying the magnitude by sine of the angle.

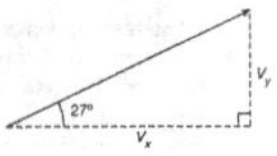

For instance, $\vec{V}$ makes a 27° angle with the horizontal and has a magnitude of 9. We can use trigonometry to find the horizontal and vertical components, V_x and V_y.

$\cos 27° = \frac{V_x}{9}$ $\quad$ $\sin 27° = \frac{V_y}{9}$

$9 \cos 27° = V_x$ $\quad$ $9 \sin 27° = V_y$

$8.01 \approx V_x$ $\quad$ $4.09 \approx V_y$

Write the decomposition for $\vec{V}$ using brackets, $\vec{V} \approx \langle 8.01, 4.09 \rangle$.

Practice

Complete the steps to find the decomposition of the given vector.

1. $\vec{B}$ makes a 41° angle with the horizontal and has a magnitude of 25.

$\cos \underline{41°} = \frac{B_x}{\underline{25}}$ $\quad$ $\sin \underline{41°} = \frac{B_y}{\underline{25}}$

$25 \cos \underline{41°} = B_x$ $\quad$ $25 \sin \underline{41°} = B_y$

$\underline{18.87} \approx B_x$ $\quad$ $\underline{16.40} \approx B_y$

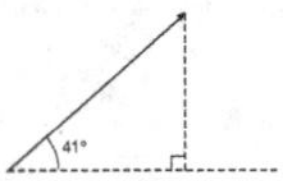

The decomposition is $\vec{B} \approx \underline{\langle 18.87, 16.40 \rangle}$.

Decompose each vector.

2. $\vec{B}$ makes a 40° angle with the horizontal and has a magnitude of 4.
$\underline{\langle 3.06, 2.57 \rangle}$

3. $\vec{V}$ makes a 58° angle with the horizontal and has a magnitude of 12.
$\underline{\langle 6.36, 10.18 \rangle}$

4. $\vec{W}$ makes a 35° angle with the horizontal and has a magnitude of 5.
$\underline{\langle 4.10, 2.87 \rangle}$

5. $\vec{Z}$ makes a 72° angle with the horizontal and has a magnitude of 9.
$\underline{\langle 2.78, 8.56 \rangle}$

6. $\vec{Q}$ makes a 18° angle with the horizontal and has a magnitude of 12.
$\underline{\langle 11.41, 3.71 \rangle}$

7. $\vec{R}$ makes a 25° angle with the horizontal and has a magnitude of 18.
$\underline{\langle 16.31, 7.61 \rangle}$

Reteaching 89
continued

Adding Vectors by Decomposition

When adding vectors we can find the resultant by adding the components.

$\vec{T}$ has a magnitude of 5 and makes a 32° angle with the horizontal, and $\vec{U}$ has a magnitude of 11 and makes a 64° angle with the horizontal. We'll decompose and add the components of $\vec{T}$ and $\vec{U}$ to find the magnitude and the angle of the resultant vector.

Step 1: Decompose each vector.

$\cos 32° = \frac{T_x}{5}$ $\quad$ $\sin 32° = \frac{T_y}{5}$ $\quad$ $\cos 64° = \frac{U_x}{11}$ $\quad$ $\sin 64° = \frac{U_y}{11}$

$5 \cos 32° = T_x$ $\quad$ $5 \sin 32° = T_y$ $\quad$ $11 \cos 64° = U_x$ $\quad$ $11 \sin 64° = U_y$

$4.24 \approx T_x$ $\quad$ $2.65 \approx T_y$ $\quad$ $4.82 \approx U_x$ $\quad$ $9.89 \approx U_y$

Step 2: Add the components.

$T_x + U_x \approx 4.24 + 4.82 \approx 9.06$ $\qquad$ $T_y + U_y \approx 2.65 + 9.89 \approx 12.54$

Step 3: Use the Pythagorean theorem to find the magnitude of the resultant vector.

$$a^2 + b^2 = c^2$$
$$(T_x + U_x)^2 + (T_y + U_y)^2 = c^2$$
$$(9.06)^2 + (12.54)^2 \approx c^2$$
$$\sqrt{239.3352} \approx c^2$$
$$15.47 \approx c$$

Step 4: Take $\tan^{-1}$ of the vertical component divided by the horizontal component to find angle.

$$\tan \theta = \frac{12.54}{9.06}$$
$$\theta = \tan^{-1}(1.384)$$
$$\theta = 54°$$

Practice

$\vec{T}$ has a magnitude of 12 and makes a 46° angle with the horizontal. $\vec{U}$ has a magnitude of 7 and makes a 22° angle with the horizontal. Find the magnitude and the angle the resultant vector makes with the horizontal.

8. $\cos 46° = \frac{T_x}{\underline{12}}$ $\quad$ $\sin 46° = \frac{T_y}{\underline{12}}$ $\quad$ $\cos 22° = \frac{U_x}{\underline{7}}$ $\quad$ $\sin 22° = \frac{U_y}{\underline{7}}$

$\underline{12} \cos 46° = T_x$ $\quad$ $\underline{12} \sin 46° = T_y$ $\quad$ $\underline{7} \cos 22° = U_x$ $\quad$ $\underline{7} \sin 22 = U_y$

$\underline{8.34} \approx T_x$ $\quad$ $\underline{8.63} \approx T_y$ $\quad$ $\underline{6.49} \approx Ux$ $\quad$ $\underline{2.62} \approx U_y$

Add the components.

$T_x + U_x \approx 8.34 + 6.49 \approx 14.83$; $T_y + U_y \approx 8.63 + 2.62 \approx 11.25$

Determine the magnitude.

$$(T_x + U_x)^2 + (T_y + U_y)^2 = c^2$$
$$\underline{(14.83)}^2 + \underline{(11.25)}^2 \approx c^2$$
$$\sqrt{\underline{346.4914}} \approx c^2$$
$$\underline{18.61} \approx c$$

Determine the angle.

$$\tan \theta = \frac{\underline{11.25}}{\underline{14.83}}$$
$$\theta = \tan^{-1}(\underline{0.7586})$$
$$\theta \approx \underline{37°}$$

Name ______________________ Date __________ Class __________

Reteaching 90
Composite Transformations

You have worked with translation and reflection transformations. Now you will work with composite transformations.

Performing Glide Reflection

A **composite transformation** is a combination of two or more transformations. A **glide reflection** is the composition of a translation and a reflection across a line parallel to the translation vector.

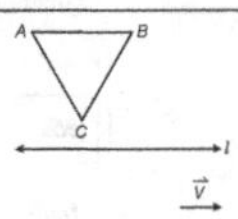

Perform a glide reflection on ΔABC by reflecting it across line ℓ and then translating it in the direction of $\vec{v}$, as indicated.

Step 1: First, reflect ΔABC across line ℓ.

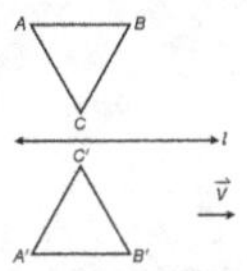

Step 2: Then, translate the triangle parallel to $\vec{v}$.

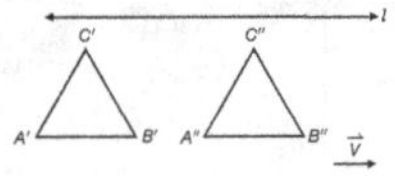

Practice

Follow the steps to perform a glide reflection on quadrilateral *GHJK*.

1. Reflect the quadrilateral across $\underline{\text{line } m}$.
Translate the quadrilateral in the direction of $\underline{\vec{V}}$.

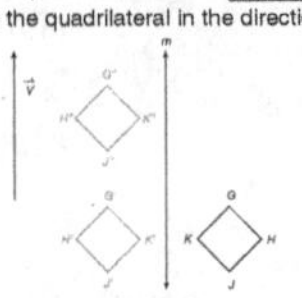

Perform the indicated glide reflections.

2.

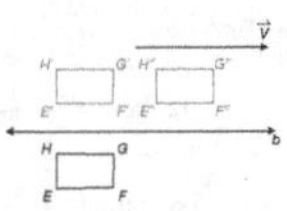

Reteaching 90
continued

Composition of Two Reflections

The composition of two reflections across intersecting lines is a rotation about the point of intersection. The angle of rotation is twice the angle formed by the intersecting lines.

Reflect ΔXYZ across $\overleftrightarrow{AB}$ and then across $\overleftrightarrow{CD}$.

Step 1: Reflect the triangle across $\overleftrightarrow{AB}$.

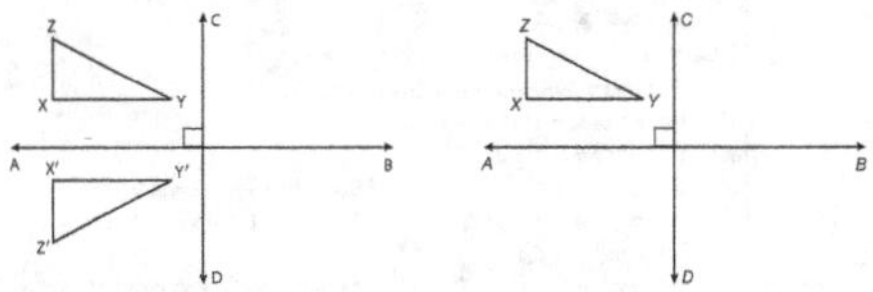

Step 2: Reflect the triangle across $\overleftrightarrow{CD}$.

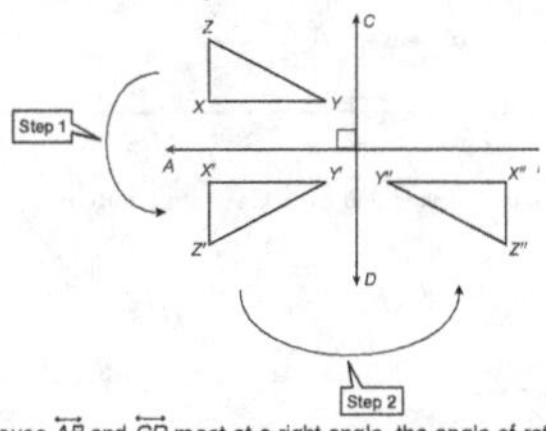

Because $\overleftrightarrow{AB}$ and $\overleftrightarrow{CD}$ meet at a right angle, the angle of rotation is $2 \times 90° = 180°$.
$V = Bh = (30)(5) = 150 \text{ ft}^3$.

Practice

Complete the steps to reflect rhombus *JKLM* across $\overleftrightarrow{CD}$ and then across $\overleftrightarrow{GH}$. Find the angle of rotation.

4. Reflect across $\overleftrightarrow{CD}$.
Reflect across $\overleftrightarrow{GH}$.
Because the angle of intersection is $\underline{70°}$, the angle of rotation is $\underline{140°}$.

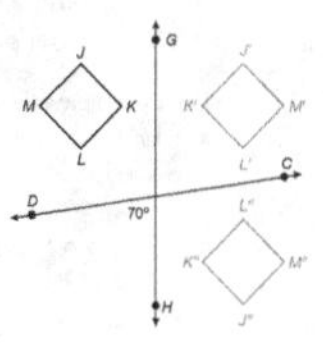

Answer Key

Name ______________________ Date __________ Class __________

Reteaching INV 9
Tessellations

You have worked with composite transformations. Now you will investigate tessellations.

Regular Tessellations

A **tessellation** is a repeating pattern of plane figures that completely cover a plane with no gaps or overlaps. A **regular tessellation** is a tessellation with congruent regular polygons.

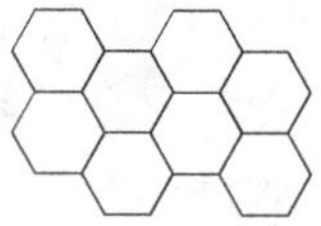

regular tessellation

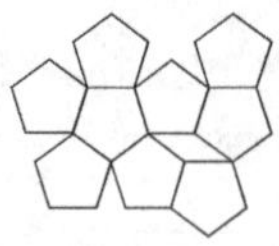

not a tessellation

In a tessellation, the angles that meet at a vertex must have a sum of 360°. In order for a regular polygon to tessellate, the measure of its interior angle must be a divisor of 360°.

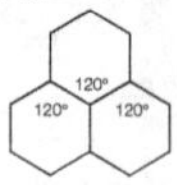

$120° + 120° + 120° = 360°$

This square will tessellate.

Practice

Complete the steps to determine whether an equilateral triangle can be used to form a tessellation.

1. The measure of each interior angle is 60°.

   $360° \div 60° =$ 6; The measure of the interior angle is a divisor of 360°.

   The equilateral triangle will tessellate.

Determine whether each regular polygon can form a tessellation.

2. 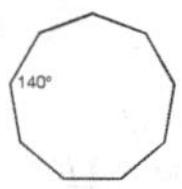

   no

3. 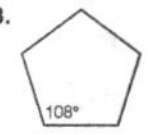

   no

Reteaching INV 9
continued

Semiregular Tessellations

A **semiregular tessellation** is a repeating pattern formed by two or more polygons in which the same number of each polygon occurs in the same order at every vertex and completely covers a plane with no gaps or overlaps.

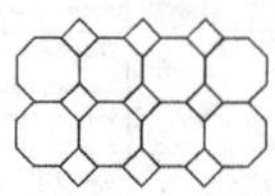

The tessellation shown is semiregular. It is made up of regular hexagons and squares. A square and two octagons meet at each vertex. The measure of an interior angle of an octagon is 135°.

Therefore, the sum of the interior angles at a vertex is $135° + 135° + 90° = 360°$.

Practice

Classify each tessellation as regular, semiregular, or neither.

4. 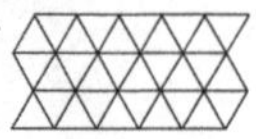

   regular

5. 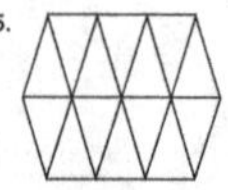

   neither

Determine whether the regular polygons can be used to form a tessellation. If so, draw the tessellation.

6. 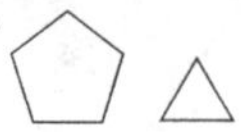

   no

7. 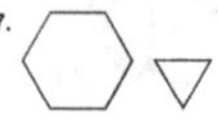

   yes

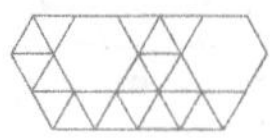

Name ______________________ Date __________ Class __________

Reteaching 91
Introduction to Trigonometric Functions

You have worked with composition of transformations. Now you will work with trigonometric functions.

Relating Sine and Cosine

Some important relationships between trigonometric functions are given by the following:

$\sin^2 x + \cos^2 x = 1 \quad \tan x = \frac{\sin x}{\cos x}$

Note that $\sin^2 x = (\sin x)^2$ and $\cos^2 x = (\cos x)^2$.

If $\sin^2 \theta = 0.9$, what is the value of $\cos \theta$?
Use the equation $\sin^2 x + \cos^2 x = 1$.

$\sin^2 \theta + \cos^2 \theta = 1$

$0.9 + \cos^2 \theta = 1$ Substitute $\sin^2 \theta = 0.9$.

$\cos^2 \theta = 1 - 0.9$ Substitute 0.9 from both sides.

$\cos^2 \theta = 0.1$

$\sqrt{\cos^2 \theta} = \sqrt{0.1}$ Take the square root of both sides.

$\cos \theta \approx 0.32$

What is the value of $\tan \theta$?

Since $\sin^2 \theta = 0.9$, $\sin \theta = \sqrt{0.9} \approx 0.95$.

$\tan \theta = \frac{\sin \theta}{\cos \theta} \approx \frac{0.95}{0.32} \approx 2.97$

Practice

If $\cos \theta = 0.2$, complete the steps to find the value of $\sin \theta$ and $\tan \theta$.

1. $\cos \theta = 0.2$

   $\cos^2 \theta = (0.2)^2 = 0.04$

   $\sin^2 \theta + \cos^2 \theta = 1$

   $\sin^2 \theta + 0.4 = 1$

   $\sin^2 \theta = 1 - 0.4$

   $\sin^2 \theta = 0.6$

   $\sin \theta \approx 0.77$

2. $\tan \theta = \frac{\sin \theta}{\cos \theta}$

   $\approx \frac{0.77}{0.2}$

   ≈ 3.85

Find the indicated values

3. If $\sin \theta = 0.15$, find:
   $\cos \theta =$ 0.99
   $\tan \theta =$ 1.74
4. If $\cos^2 \theta = 0.25$, find:
   $\sin \theta =$ 0.87
   $\tan \theta =$ 0.15
5. If $\cos \theta = 0.7$, find:
   $\sin \theta =$ 0.71
   $\tan \theta =$ 1.01
6. If $\sin^2 \theta = 0.88$, find:
   $\cos \theta =$ 0.35
   $\tan \theta =$ 2.69

Reteaching 91
continued

More Trigonometric Relationships

The table below outlines the relationships between sine, cosine, and tangent functions.

Function	$\sin \theta$	$\cos \theta$	$\tan \theta$
$\sin \theta =$	$\sin \theta$	$\sqrt{1 - \cos^2 \theta}$	$\frac{\tan \theta}{\sqrt{1 + \tan \theta}}$
$\cos \theta =$	$\sqrt{1 - \sin^2 \theta}$	$\cos \theta$	$\frac{1}{\sqrt{1 + \tan^2 \theta}}$
$\tan \theta =$	$\frac{\sin \theta}{\sqrt{1 - \sin^2 \theta}}$	$\frac{\sqrt{1 - \cos^2 \theta}}{\cos \theta}$	$\tan \theta$

If $\tan^2 \theta = 1.4$, what is the value of $\cos \theta$?

$\cos \theta = \frac{1}{\sqrt{1 + \tan^2 \theta}}$

$= \frac{1}{\sqrt{1 + 1.4}}$ Substitute $\tan^2 \theta = 1.4$.

$= \frac{1}{\sqrt{2.4}}$

≈ 0.65

Practice

If $\tan^2 \theta = 1.8$, complete the steps to find the value of $\sin \theta$ and $\cos \theta$.

7. $\tan \theta = \sqrt{1.8} \approx 1.34$

   $\sin \theta = \frac{\tan \theta}{\sqrt{1 + \tan \theta}}$

   $= \frac{1.34}{\sqrt{1 + 1.34}}$

   $= \frac{1.34}{\sqrt{2.34}}$

   ≈ 0.88

8. $\cos \theta = \frac{1}{\sqrt{1 + \tan^2 \theta}}$

   $= \frac{1}{\sqrt{1 + 1.8}}$

   $= \frac{1}{\sqrt{2.8}}$

   ≈ 0.60

Find the indicated values.

9. If $\tan^2 \theta = .75$, find:
   $\cos \theta =$ 0.76
   $\sin \theta =$ 0.42
10. If $\tan \theta = 1.4$, find:
    $\sin \theta =$ 0.81
    $\cos \theta =$ 0.58
11. If $\sin \theta = 0.7$, find:
    $\tan \theta =$ 0.98
    $\cos \theta =$ 0.71
12. If $\cos \theta = 0.44$, find:
    $\tan \theta =$ 2.05
    $\sin \theta =$ 0.90

Name ______________________ Date __________ Class __________

Reteaching 92
Quadrilaterals on the Coordinate Plane

You have worked with trigonometric functions. Now you will work with quadrilaterals on the coordinate plane.

Proving Quadrilaterals are Parallelograms

If you know the coordinates of a quadrilateral, you can determine the length and the slope of each side. Therefore, you can determine which sides are congruent, parallel, or perpendicular.

Prove that quadrilateral *ABCD* is a parallelogram by determining the slope of each side.

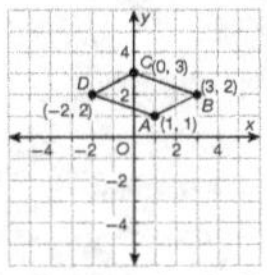

Step 1: Calculate the slope of each side.

slope of $\overline{AB} = \frac{y_B - y_A}{x_B - x_A} = \frac{2-1}{3-1} = \frac{1}{2}$ slope of $\overline{BC} = \frac{y_C - y_B}{x_C - x_B} = \frac{3-2}{0-3} = \frac{1}{-3} = -\frac{1}{3}$

slope of $\overline{CD} = \frac{y_D - y_C}{x_D - x_C} = \frac{2-3}{-2-0} = \frac{-1}{-2} = \frac{1}{2}$ slope of $\overline{AD} = \frac{y_D - y_A}{x_D - x_A} = \frac{2-1}{-2-1} = \frac{1}{-3} = -\frac{1}{3}$

Step 2: Compare the slopes of opposite sides.

The slopes of $\overline{AB}$ and $\overline{CD}$ are equal, so $\overline{AB} \parallel \overline{CD}$.

Similarly, $\overline{BC} \parallel \overline{AD}$.

Therefore, *ABCD* is a parallelogram.

Practice

For quadrilateral *ABCD* above, complete the steps to prove that *ABCD* is a parallelogram by determining its side lengths.

1. $AB = \sqrt{(x_B - x_A)^2 + (y_B - y_A)^2} = \sqrt{(2 - \underline{1})^2 + (3 - \underline{1})^2} = \sqrt{(1)^2 + (\underline{2})^2} = \sqrt{\underline{5}}$

$BC = \sqrt{(x_C - x_B)^2 + (y_C - y_B)^2} = \sqrt{(\underline{0} - (-3))^2 + (\underline{3} - 2)^2} = \sqrt{(\underline{9})^2 + (\underline{1})^2} = \sqrt{\underline{10}}$

$CD = \sqrt{(\underline{-2} - 0)^2 + (\underline{2} - 3)^2} = \sqrt{(\underline{-2})^2 + (\underline{-1})^2} = \sqrt{\underline{5}}$

$AD = \sqrt{(\underline{-2} - 1)^2 + (\underline{2} - 1)^2} = \sqrt{(\underline{-3})^2 + (\underline{1})^2} = \sqrt{\underline{10}}$

$AB = CD \rightarrow \overline{AB} \cong \underline{\overline{CD}}$; $BC =$ _AD_ $\rightarrow \underline{\overline{BC}} \cong \underline{\overline{AD}}$

ABCD is a <u>parallelogram</u>.

Reteaching 92
continued

Proving that Quadrilaterals Are Trapezoids

The vertices of quadrilateral *LMNP* are *L*(1, −1), *M*(4, −1), *N*(6, −6), and *P*(−3, −6).

Prove that *LMNP* is a trapezoid.

Step 1: Calculate the slope of each side.

slope of $\overline{LM} = \frac{y_M - y_L}{x_M - x_L} = \frac{-1-(-1)}{4-1} = \frac{0}{3} = 0$ slope of $\overline{MN} = \frac{y_N - y_M}{x_N - x_M} = \frac{-6-(-1)}{6-4} = \frac{-5}{2} = -\frac{5}{2}$

slope of $\overline{NP} = \frac{y_P - y_N}{x_P - x_N} = \frac{-6-(-6)}{-3-6} = \frac{0}{-9} = 0$ slope of $\overline{LP} = \frac{y_P - y_L}{x_P - x_L} = \frac{-6-(-1)}{-3-1} = \frac{-5}{-4} = \frac{5}{4}$

Step 2: Compare the slopes of opposite sides.

The slopes of $\overline{LM}$ and $\overline{NP}$ are equal, so $\overline{LM} \parallel \overline{NP}$.

However, $\overline{MN}$ and $\overline{LP}$ do not have the same slopes and are not parallel.

Therefore, *LMNP* is a trapezoid.

Practice

Complete the steps to prove that quadrilateral *ABCD* is a trapezoid.

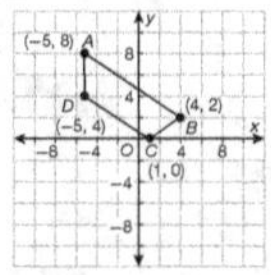

2. slope of $\overline{AB} = \frac{y_B - y_A}{x_B - x_A} = \frac{2 - \underline{8}}{4 - (\underline{-5})} = \frac{\underline{-5}}{\underline{9}} = -\frac{2}{\underline{3}}$ slope of $\overline{BC} = \frac{y_C - y_B}{x_C - x_B} = \frac{0 - \underline{2}}{1 - \underline{4}} = \frac{\underline{-2}}{\underline{-3}} = \frac{\underline{2}}{\underline{3}}$

slope of $\overline{CD} = \frac{\underline{4} - 0}{\underline{-5} - 1} = \frac{4}{\underline{-6}} = -\frac{\underline{2}}{\underline{3}}$ slope of $\overline{AD} = \frac{\underline{4} - \underline{8}}{\underline{-5} - (-5)} = \frac{\underline{-4}}{0} =$ <u>undefined</u>

The slopes of $\overline{AB}$ and $\underline{\overline{CD}}$ are equal, so $\underline{\overline{AB}} \parallel \overline{CD}$.

The slopes of $\overline{BC}$ and $\underline{\overline{AD}}$ are not equal and not parallel.

Therefore, *ABCD* is a <u>trapezoid</u>.

Name ______________________ Date __________ Class __________

Reteaching 93
Representing Solids: Orthographic Views

You have worked with quadrilaterals. Now you will draw orthographic views of solids.

Creating Orthographic Drawings

An **orthographic drawing** shows a three-dimensional object in such a way that the line of sight for each view is perpendicular to the plane of the picture.

There are usually three views—front, top, and side.

Draw the front, top, and side orthographic views of this solid.

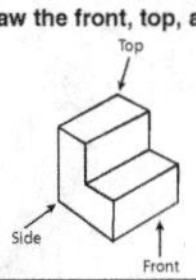

Front View: Two rectangles can be seen.	Top View: Two rectangles can be seen.	Side View:
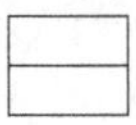		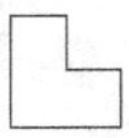

Practice

Complete the front, top, and side orthographic views of this solid.

1. Solid:

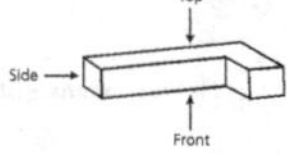

Front:

Top:

Side:

Reteaching 93
continued

Constructing from an Orthographic Drawing

This orthographic drawing shows the front, top, and left-side views of a solid.

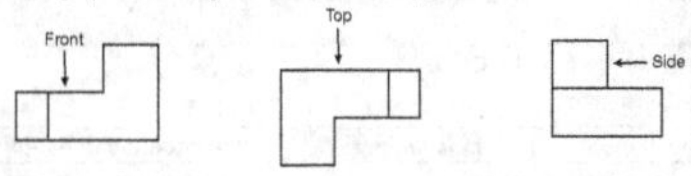

Which of these sketches represents the solid shown in the views?

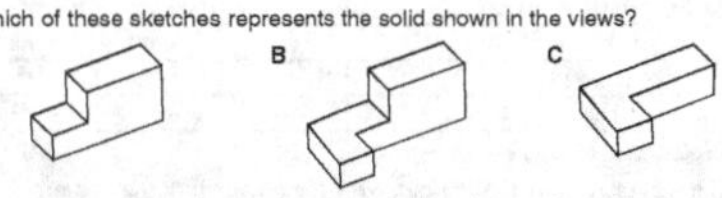

Compare sketch A to the top view. The left side of sketch A does not have a prism component shown in the top view. Compare sketch C to the top or front view. The figure does not have a top layer. Compare sketch B to all three views. Sketch B matches all three views.

Practice

The orthographic drawings show the front, top, and right-side views of a solid. Complete the steps that explain which sketch represents the solid shown in the views.

2. Front: Top: Side:

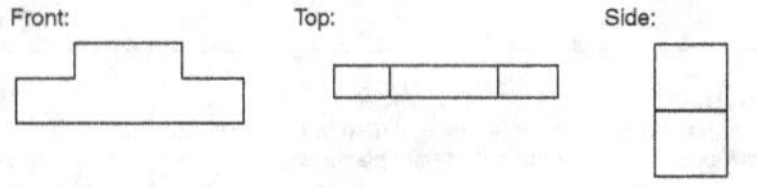

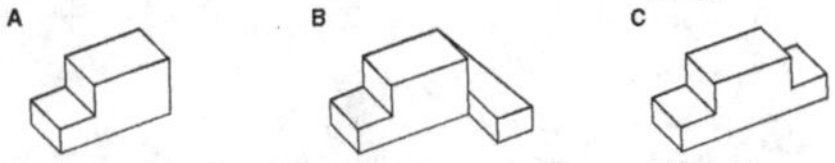

Sketch A does not represent the solid shown because <u>the front does not have a prism component shown in the front and top views.</u>

Sketch B does not represent the solid shown because <u>it has an extra prism in the front.</u>

Sketch C represents the solid shown because <u>it matches all three views.</u>

Answer Key

Name ______________________ Date __________ Class __________

Reteaching 94
Law of Sines

You have worked with orthographic drawings. Now you will work with the Law of Sines.

Law of Sines: Finding Distance

The **Law of Sines** is a ratio that uses given angles and sides to find other angles and sides in any triangle.

The Law of Sines states that $\frac{a}{\sin A} = \frac{b}{\sin B} = \frac{c}{\sin C}$.

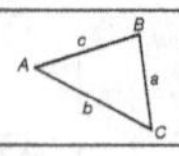

Find the length of RT. Round to the nearest hundredth.

Step 1: Find the measure of $\angle T$.

$m\angle R + m\angle S + m\angle T = 180°$	Triangle Sum Theorem
$48° + 32° + m\angle T = 180°$	Substitute values.
$m\angle T = 100°$	Simplify.

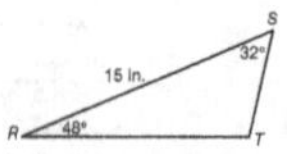

Step 2: Use $m\angle T$, RS, and the Law of Sines to find RT.

$\frac{RT}{\sin S} = \frac{RS}{\sin T}$	Law of Sines
$\frac{RT}{\sin 32°} = \frac{15}{\sin 100°}$	Substitute values.
$RT \sin 100° = 15 \sin 32°$	Cross multiply.
$RT = \frac{15 \sin 32°}{\sin 100°}$	Simplify.
$RT \approx 8.07$ in.	

Practice

Complete the steps to find the length TU. Round to the nearest hundredth.

1. $\frac{\underline{18}}{\underline{\sin 41°}} = \frac{TU}{\sin 64°}$

   $TU \sin 41° = \underline{18 \sin 64°}$

   $TU = \frac{18 \sin 64°}{\underline{\sin 41°}}$

   $TU \approx \underline{24.66 \text{ m}}$

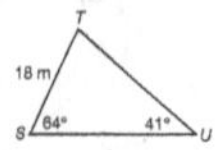

Find the given length. Round to the nearest hundredth.

2. BC $\underline{46.56 \text{ in.}}$

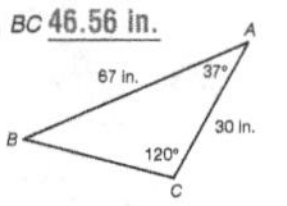

3. DE $\underline{7.56 \text{ cm}}$

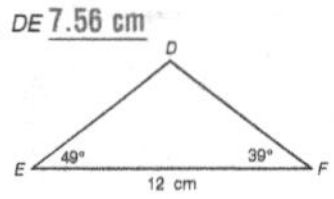

Reteaching 94
continued

Finding a Missing Angle

Find the measure of $\angle D$. Round your answer to the nearest degree.

Step 1: The measures of $\angle D$ and its opposite side are unknown.

Use the Law of Sines to find $m\angle F$.

$\frac{9}{\sin F} = \frac{13}{\sin 73°}$	Law of Sines
$13 \sin F = 9 \sin 73°$	Cross multiply.
$\sin F = \frac{9 \sin 73°}{13}$	Simplify.
$\sin F \approx 0.66$	Use calculator.
$m\angle F \approx \sin^{-1}(0.66)$	Use inverse sine function.
$m\angle F \approx 41°$	Simplify.

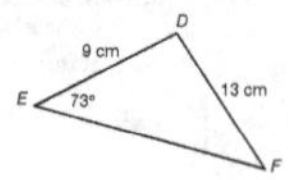

Step 2: Use the Triangle Sum Theorem to find $m\angle D$.

$m\angle D + m\angle E + m\angle F = 180°$	Triangle Sum Theorem
$m\angle D + 73° + 41° = 180°$	Substitute values.
$m\angle D = 66°$	Simplify.

Practice

Complete the steps to find the measure of $\angle K$. Round your answer to the nearest degree.

4. Find $\underline{m\angle L}$ first.

   $\frac{KJ}{\sin L} = \frac{LK}{\underline{\sin J}}$

   $\frac{\underline{3}}{\sin L} = \frac{\underline{7}}{\underline{\sin 35°}}$

   $\underline{7 \sin L} = 3 \sin 35°$

   $\sin L = \frac{3 \sin 35°}{\underline{7}}$

   $\sin L \approx \underline{0.25}$

   $m\angle L \approx \sin^{-1}(\underline{0.25})$

   $m\angle L \approx 14°$

   Find $\underline{m\angle K}$.

   $m\angle J + m\angle K + m\angle L = \underline{180°}$

   $\underline{35°} + m\angle K + 14° = 180°$

   $m\angle K = \underline{131°}$

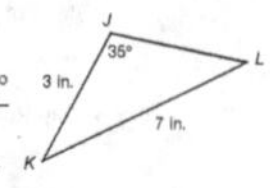

Find the given measure. Round your answer to the nearest degree.

5. $m\angle P$ $\underline{88°}$

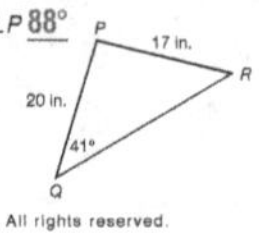

6. $m\angle G$ $\underline{33°}$

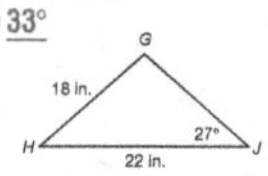

Name ______________________ Date __________ Class __________

Reteaching 95
Equations of Circles: Shifting and Dilating

You have worked with the Law of Sines. Now you will shift and dilate circles.

Shifting the Center of a Circle

The equation of a circle with radius r and center at the origin is $x^2 + y^2 = r^2$.

If the circle $x^2 + y^2 = r^2$ is shifted	the equation of the circle is
h units to the right,	$(x - h)^2 + y^2 = r^2$.
h units to the left,	$(x + h)^2 + y^2 = r^2$.
h units upward,	$x^2 + (y - h)^2 = r^2$.
h units downward,	$x^2 + (y + h)^2 = r^2$.

The equation of the circle is $x^2 + y^2 = 16$.

Shift the circle 3 units to the right. Write the equation of the shifted circle. Draw the original circle and shifted circle on the coordinate plane.

Step 1: Because the circle is shifted 3 units to the right, subtract 3 from x coordinate.

The new equation is $(x - 3)^2 + y^2 = 16$.

Step 2: The radius of the original circle is 4.

The new circle is centered at (3, 0) with radius 4.

Step 3: Draw both circles on the coordinate plane.

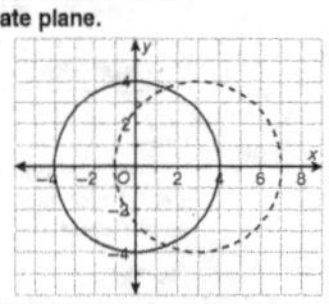

Practice

Complete the steps to write the equation of the shifted circle. Draw both circles on the coordinate plane.

1. Original circle: $x^2 + y^2 = 25$

   Shift the circle 2 units down.

   Equation of new circle: $x^2 + (y \underline{+ 2})^2 = \underline{25}$

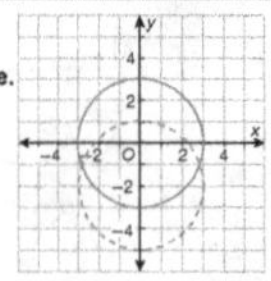

Write the equation of the new circle. Graph both circles on the coordinate plane.

2. Original circle: $x^2 + y^2 = 9$

   Shift the circle 2 units to the left.

   Equation of new circle: $\underline{(x + 2)^2 + y^2 = 9}$

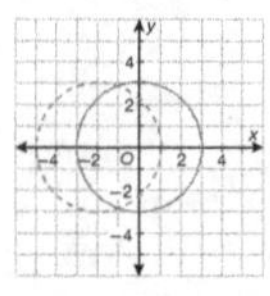

Reteaching 95
continued

Dilating a Circle

To dilate a circle, multiply the radius by a factor c. If one circle has equation $x^2 + y^2 = r^2$, then a dilation of the circle has equation $x^2 + y^2 = (cr)^2$.

A dilation changes the size of a circle but does not shift the circle.

Example:

A circle has equation $x^2 + y^2 = 36$.

Find the equation of the circle after it is dilated by a factor of $\frac{1}{3}$. Draw the original circle and the dilated circle on the coordinate plane.

Step 1: Multiply the radius of the circle by the scale factor $\frac{1}{3}$.

$6 \times \frac{1}{3} = 2$

Step 2: Write the new equation.

$x^2 + y^2 = 2^2 = 4$

Step 3: Draw both circles on the coordinate plane.

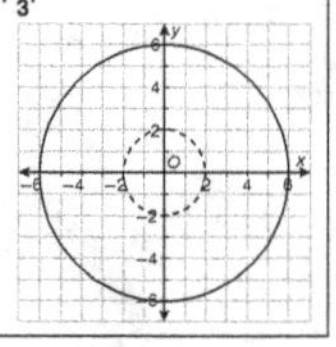

Practice

Complete the steps to write the equation of the dilated circle.

Draw both circles on the coordinate plane.

3. Original circle: $x^2 + y^2 = 9$

   Dilate the circle by 2.

   The radius of the dilated circle is $2 \times \underline{3} = \underline{6}$.

   The equation of the dilated circle is $\underline{x^2 + y^2 = 36}$.

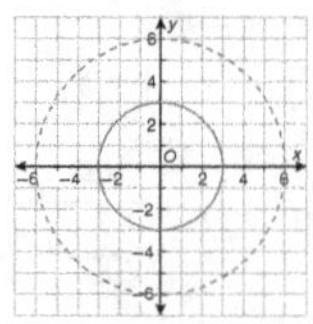

Write the equation of the dilated circle.

Draw the original circle and dilated circle on the coordinate plane.

4. Original circle: $x^2 + y^2 = 25$

   Dilate the circle by $\frac{1}{5}$.

   Equation of dilated circle:

   $\underline{x^2 + y^2 = 1}$

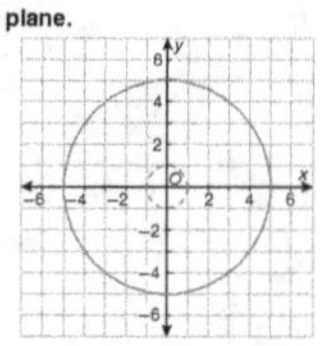

Name ______________________ Date __________ Class __________

Reteaching 96

Effects of Changing Dimensions on Perimeter and Area

You have shifted and dilated circles. Now you will describe changes on perimeter and area.

Effects of Changing Dimensions on Perimeter

Change in Dimensions	Change in Perimeter or in Circumference
Multiply every dimension of a polygon by a factor of *a*.	The perimeter is multiplied by a factor of *a*.
Multiply the radius of a circle by a factor of *a*.	The circumference is multiplied by a factor of *a*.

The rectangle is dilated by a factor of 2. Describe how the change affects the perimeter.

Step 1: The perimeter of the rectangle is

$P = 4 \cdot 2 + 3 \cdot 2 = 14$ cm.

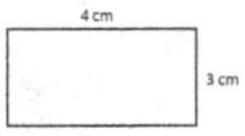

Step 2: The dimensions of the rectangle are multiplied by 2.

The perimeter of the new rectangle is

$P = 8 \cdot 2 + 6 \cdot 2 = 28$ cm.

Therefore, the perimeter is multiplied by 2.

Practice

Complete the steps to describe how each change affects the perimeter or circumference.

1. The rectangle is dilated by $\frac{1}{2}$.

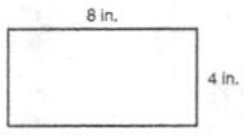

original: $P = \underline{8} \cdot 2 + \underline{4} \cdot 2 = \underline{24}$ in.

new: $P = \underline{4} \cdot 2 + \underline{2} \cdot 2 = \underline{12}$ in.

The perimeter is multiplied by $\underline{\frac{1}{2}}$.

Describe the effect of each change on the perimeter or circumference.

2. The base and the height of the triangle are multiplied by a factor of 3.

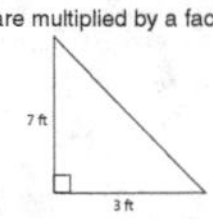

<u>*P* is multiplied by 3.</u>

3. The radius of the circle is changed by a factor of $\frac{1}{5}$.

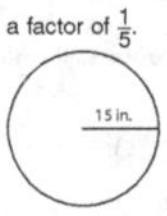

<u>*C* is multiplied by $\frac{1}{5}$.</u>

© Saxon. All rights reserved. 209 **Saxon** Geometry

Reteaching 96

continued

Effects of Changing Dimensions on Area

Change in Dimensions	Change in Area
Multiply one dimension of a polygon by a factor of *a*.	The area is multiplied by *a*.
Multiply every dimension of a polygon by a factor of *a*.	The area is multiplied by a^2.
Multiply the radius of a circle by a factor of *a*.	The area is multiplied by a^2.

The base and the height of a parallelogram are changed by a factor of 3. Describe how the change affects the area.

Step 1: The area of the parallelogram is $A = bh$.

$= 4(5)$

$= 20 \text{ cm}^2$

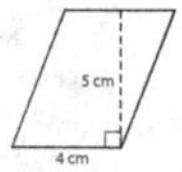

Step 2: Both the base and the height are multiplied by 3.

The area of the new parallelogram is $A = bh$.

$= (3 \cdot 4)(3 \cdot 5)$

$= 180 \text{ cm}^2$

When both the base and the height are multiplied by 3, the area is multiplied by 3^2, or 9.

Practice

Complete the steps to describe how each change affects the area of the given figure.

4. The base and height of the triangle are multiplied by 3.

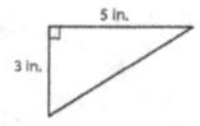

original: $A = \frac{1}{2}bh$ new: $A = \frac{1}{2}bh$

$= \frac{1}{2}(5)(\underline{3})$ $= \frac{1}{2}(15)(\underline{9})$

$= \underline{7.5} \text{ in}^2$ $= \underline{67.5} \text{ in}^2$

The area is multiplied by $\underline{9}$.

Describe the effect of each change on the area.

5. The side length of the square is multiplied by 4.

<u>The area is multiplied by 16.</u>

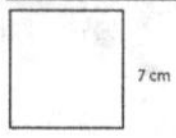

6. The radius of the circle is changed by a factor of 5.

<u>The area is multiplied by 25.</u>

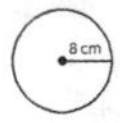

© Saxon. All rights reserved. 210 **Saxon** Geometry

Name ______________________ Date __________ Class __________

Reteaching 97

Concentric Circles

Equations of Concentric Circles

Concentric circles are coplanar circles with the same center. There are no points of intersection between the circles.

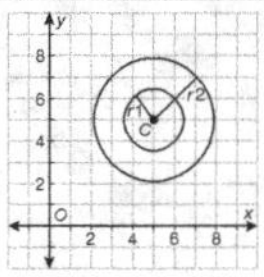

The equation of a circle is $(x - h)^2 + (y - k)^2 = r^2$. Concentric circles have the same values for *h* and *k* (because they share the same center) but the *r* values (the radii) differ.

Write the equations for the concentic circles shown.

$(x - h)^2 + (y - k)^2 = r^2$ Equation of a Circle

Inner circle:

$(x - 3)^2 + (y - 4)^2 = 1^2$ Substitute 3 for *h*, 4 for *k*, and 1 for *r*.

$(x - 3)^2 + (y - 4)^2 = 1$ Simplify.

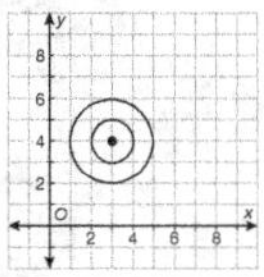

Outer circle:

$(x - 3)^2 + (y - 4)^2 = 2^2$ Substitute 3 for *h*, 4 for *k*, and 2 for *r*.

$(x - 3)^2 + (y - 4)^2 = 4$ Simplify.

Practice

Complete the steps to write the equations of the concentric circles.

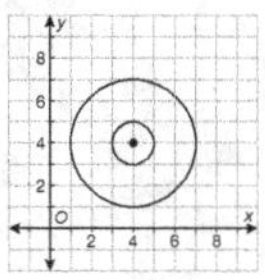

1. Inner: $(x - \underline{4})^2 + (y - \underline{4})^2 = \underline{1}^2$

$(x - \underline{4})^2 + (y - \underline{4})^2 = \underline{1}$

Outer: $(x - \underline{4})^2 + (y - \underline{4})^2 = \underline{3}^2$

$(x - \underline{4})^2 + (y - \underline{4})^2 = \underline{9}$

Write the equation of the concentric circles.

2. Inner: $\underline{(x + 2)^2 + (y + 1)^2 = 4}$

Outer: $\underline{(x + 2)^2 + (y + 1)^2 = 9}$

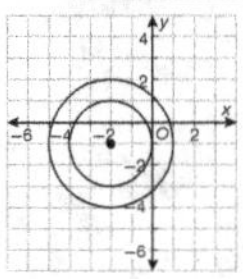

3. Inner: $\underline{(x - 5)^2 + (y + 1)^2 = 4}$

Outer: $\underline{(x - 5)^2 + (y + 1)^2 = 16}$

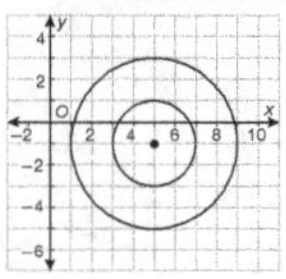

© Saxon. All rights reserved. 211 **Saxon** Geometry

Reteaching 97

continued

Annulus of Concentric Circles

The annulus is the region between two concentric circles, often represented as a shaded area.

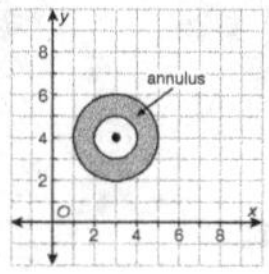

Find the area of the annulus in the concentric circles.

Method 1: Find the area of each circle. Then, subtract the area of the smaller circle from that of the larger circle.

$r = 2$ Radius of larger circle

$A = \pi r^2 = \pi(2)^2 = 4\pi$ Area of larger circle

$r = 1$ Radius of smaller circle

$A = \pi r^2 = \pi(1)^2 = \pi$ Area of smaller circle

$4\pi - \pi = 3\pi$ Difference of areas

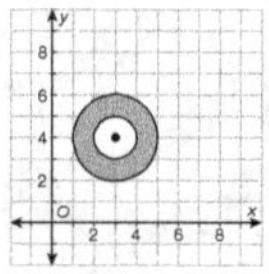

Method 2: Multiply π by the difference of the radii squared.

$A = \pi(R^2 - r^2)$

$= \pi(2^2 - 1^2)$ Substitute 2 for *R* and 1 for *r*.

$= \pi(4 - 1)$ Simplify.

$= 3\pi$ Simplify.

The area of the annulus is $3\pi \text{ units}^2$.

Practice

Complete the steps to find the area of the annulus.

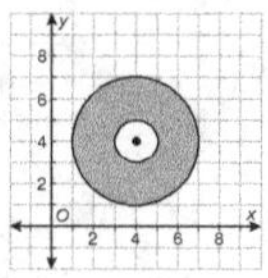

4. Larger circle: $A = \pi r^2 = \pi(\underline{3})^2 = \underline{9}\,\pi$

Smaller circle: $A = \pi r^2 = \pi(\underline{1})^2 = \underline{1}\,\pi$

Difference: $\underline{9\pi} - \underline{1\pi} = \underline{8\pi} \text{ units}^2$

Find the area of the annulus.

5. $\underline{5\pi} \text{ units}^2$

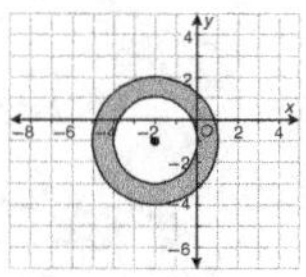

6. $\underline{12\pi} \text{ units}^2$

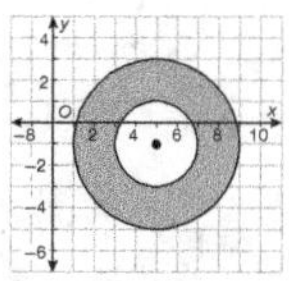

© Saxon. All rights reserved. 212 **Saxon** Geometry

Answer Key

Name ______________________ Date __________ Class __________

Reteaching 98
Law of Cosines

You have worked with the Law of Sines. Now you will use the **Law of Cosines** to find side lengths and angle measures.

Law of Cosines: Side Length

For any ΔABC with side lengths a, b, and c that are opposite angles A, B, and C, the following rules apply:

$a^2 = b^2 + c^2 - 2bc \cos A$

$b^2 = a^2 + c^2 - 2ac \cos B$

$c^2 = a^2 + b^2 - 2ab \cos C$

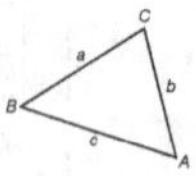

When you know two side lengths and the included angle measure between the two sides, you can use the Law of Cosines to find the third side.

Find *AB*. Round to the nearest tenth.

$AB^2 = BC^2 + AC^2 - 2(BC)(AC) \cos C$	Law of Cosines
$AB^2 = 17^2 + 14^2 - 2(17)(14) \cos 50°$	Substitute.
$AB^2 \approx 179.0331$	Simplify.
$AB \approx 13.4$ ft	Find the square root.

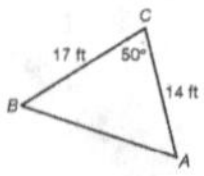

Practice

Complete the steps to find the missing side length. Round to the nearest tenth.

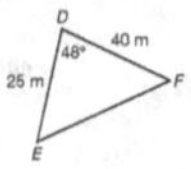

$EF^2 = DE^2 + DF^2 - 2(DE)(DF) \cos D$

1. $EF^2 = \underline{25}^2 + \underline{40}^2 - 2(\underline{25})(\underline{40}) \cos 48°$

   $EF^2 \approx \underline{886.73879}$

   $EF \approx \underline{29.8}$ m

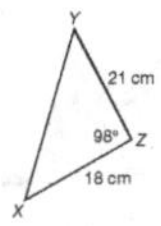

$XY^2 = XZ^2 + YZ^2 - 2(XZ)(YZ) \cos Z$

2. $XY^2 = \underline{18}^2 + \underline{21}^2 - 2(\underline{18})(\underline{21}) \cos \underline{98}°$

   $EF^2 \approx \underline{870.21486}$

   $EF \approx \underline{29.5}$ m

Find the missing side length. Round to the nearest tenth.

3. $\underline{45.2}$ in.

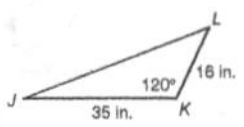

4. $\underline{15.2}$ cm

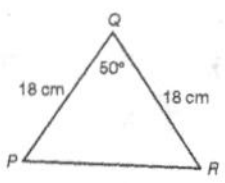

Reteaching 98
continued

Law of Cosines: Missing Angle

You can also use the Law of Cosines to find a missing angle provided that you know all three side lengths.

Find m∠*A*. Round to the nearest degree.

$BC^2 = AB^2 + AC^2 - 2(AB)(AC) \cos A$	Law of Cosines
$18^2 = 14^2 + 25^2 - 2(14)(25) \cos A$	Substitute.
$\cos A = \frac{14^2 + 25^2 - 18^2}{2(14)(25)}$	Isolate.
$\cos A \approx 0.71$	Simplify.
$A \approx 45°$	Solve.

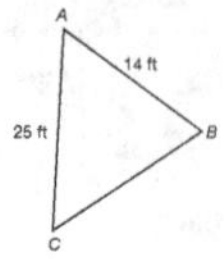

Practice

Complete the steps to find θ. Round to the nearest degree.

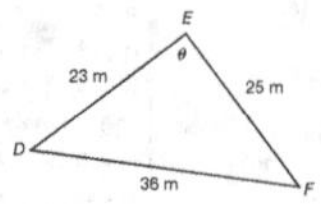

$DF^2 = DE^2 + EF^2 - 2(DE)(EF) \cos E$

$\underline{36}^2 = 23^2 + \underline{25}^2 - 2(23)(\underline{25}) \cos \theta$

5. $\cos \theta = \frac{23^2 + \underline{25}^2 - \underline{36}^2}{2(23)(\underline{25})}$

   $\cos \theta \approx \underline{-0.1235}$

   $\theta \approx \underline{97}°$

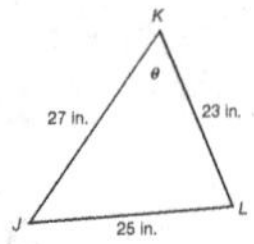

$JL^2 = JK^2 + KL^2 - 2(JK)(KL) \cos K$

$\underline{25}^2 = \underline{27}^2 + \underline{23}^2 - 2(\underline{27})(\underline{23}) \cos \theta$

6. $\cos \theta = \frac{\underline{27}^2 + \underline{23}^2 - \underline{25}^2}{2(\underline{27})(\underline{23})}$

   $\cos \theta \approx \underline{0.50966}$

   $\theta \approx \underline{59}°$

Find θ. Round to the nearest degree.

7. $\underline{37}°$

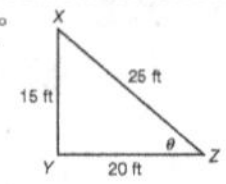

8. $\underline{131}°$

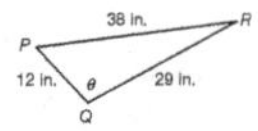

Name ______________________ Date __________ Class __________

Reteaching 99
Volume Ratios of Similar Solids

You have worked with volumes of different solids. Now you will compare perimeters and areas of the faces of similar solids.

Similar Solids: Perimeters

If two similar solids have scale factor a:b, then the ratio of the perimeters of their corresponding faces is a:b.

Two similar right rectangular prisms *A* and *B* have scale factor 3:2. Determine the perimeter of *B*.

Step 1: Find the perimeter of *A*.

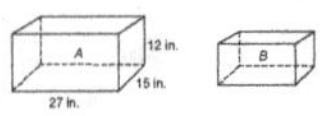

$P_A = 4(27) + 4(15) + 4(12)$

$P_A = 216$ in.

Step 2: Use the perimeter ratio to determine P_B.

$\frac{P_A}{P_B} = \frac{3}{2}$

$2P_A = 3P_B$

$\frac{2P_A}{3} = P_B$

Step 3: Find the perimeter of *B*.

$P_B = \frac{2P_A}{3} = \frac{2(216)}{3} = 144$ in.

Practice

Complete the steps to find the perimeter of the other solid.

1. Scale factor is 2:1.

   $P_s = 4(\underline{8}) + 4(\underline{12})$

   $P_s = \underline{80}$ ft

   $\frac{P_s}{P_T} = \frac{2}{1}$

   $\frac{P_s}{2} = P_T$

   $P_T = \frac{\underline{80}}{1} = \underline{40}$ ft

2. Scale factor is 3:4

   $P_X = 12(\underline{24})$

   $P_X = \underline{288}$ cm

   $\frac{P_X}{P_Y} = \frac{3}{4}$

   $\frac{\underline{4p_x}}{\underline{3}} = P_Y$

   $P_Y = \frac{\underline{4(288)}}{\underline{3}} = \underline{384}$ cm

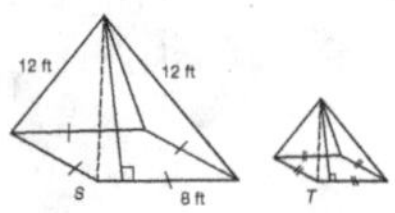

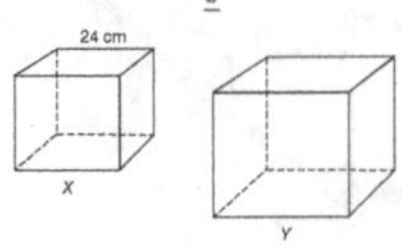

Reteaching 99
continued

Similar Solids: Surface Areas

If two similar solids have scale factor a:b, then the ratio of the surface areas of their corresponding faces is a^2:b^2.

These two similar rectangular pyramids have scale factor 3:5. Determine the surface areas of both pyramids.

Step 1: Find the surface area of *A*.

$B_A = (12)(18) = 216 \text{ m}^2$	Base area
$F_A^1 = F_A^3 = \frac{1}{2}(15)(18) = 135 \text{ m}^2$	Face area
$F_A^2 = F_A^4 = \frac{1}{2}(9)(12) = 54 \text{ m}^2$	Face area

$SA_A = 216 + 2(135) + 2(54) = 594 \text{ m}^2$

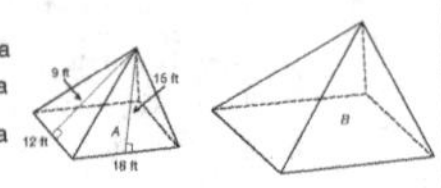

Step 2: Use the area ratio to determine SA_B.

$\frac{SA_A}{SA_B} = \left(\frac{3}{5}\right)^2$

$SA_B = \left(\frac{5}{3}\right)^2 SA_A$

Step 3: Find the surface area of *B*.

$SA_B = \left(\frac{5}{3}\right)^2 SA_A = \left(\frac{25}{9}\right)594 = 1650 \text{ m}^2$

Practice

Complete the steps to determine the surface areas of both solids.

3. Scale factor is 4:3.

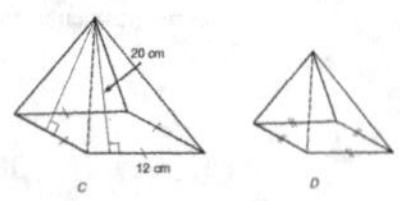

   $B_C = (12)(12) = 144 \text{ cm}^2$

   $F_C^1 = F_C^2 = F_C^3 = F_C^4 = \frac{1}{2}(12)(\underline{20}) = \underline{120} \text{ cm}^2$

   $SA_C = 144 + 4(\underline{120}) = \underline{624} \text{ cm}^2$

   $\frac{SA_C}{SA_D} = \left(\frac{4}{3}\right)^2$

   $SA_D = \underline{\left(\frac{3}{4}\right)^2} SA_C = \frac{\underline{9}}{\underline{16}}(\underline{624}) = \underline{351} \text{ cm}^2$

Determine the surface areas of both solids.

4. Scale factor is 1:2.

   $SA_S = \underline{1944} \text{ m}^2$

   $SA_T = \underline{7776} \text{ m}^2$

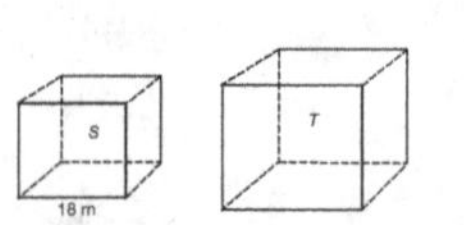

Name ______________________ Date __________ Class __________

Reteaching 100
Transformation Matrices

You have worked with translations and reflections. Now you will learn to use transformation matrices.

Adding Point Matrices

A **matrix** is a rectangular array of numbers. A **point matrix** represents the coordinates of the points on a coordinate grid.

$\begin{pmatrix} 2 & 5 \\ 7 & 4 \end{pmatrix}$ ← x-coordinates, ← y-coordinates

To add matrices, add the corresponding components.

Write a point matrix for the line segment $\overline{AB}$. Then, add the point matrix to $\begin{pmatrix} 1 & 1 \\ 1 & 1 \end{pmatrix}$. Determine the results of the addition.

$\begin{pmatrix} 3 & 5 \\ 1 & 4 \end{pmatrix}$ Point matrix for $\overline{AB}$

$\begin{pmatrix} 3 & 5 \\ 1 & 4 \end{pmatrix} + \begin{pmatrix} 1 & 1 \\ 1 & 1 \end{pmatrix}$ Add matrices.

$\begin{pmatrix} 3+1 & 5+1 \\ 1+1 & 4+1 \end{pmatrix}$ Add corresponding components.

$\begin{pmatrix} 4 & 6 \\ 2 & 5 \end{pmatrix}$ Simplify.

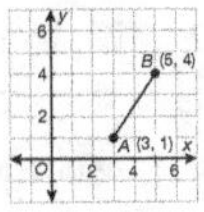

The coordinates of $\overline{A'B'}$ are $A'(4, 2)$ and $B'(6, 5)$; they have moved up 1 unit and right 1 unit.

Practice

Complete the steps to find the sum of the matrices.

1. Add the point matrix to $\begin{pmatrix} 2 & 2 & 2 \\ 1 & 1 & 1 \end{pmatrix}$.

$\Delta XYZ = \begin{pmatrix} \underline{1} & \underline{3} & \underline{3} \\ \underline{1} & \underline{3} & \underline{1} \end{pmatrix}$

$= \begin{pmatrix} \underline{1} & \underline{3} & \underline{3} \\ \underline{1} & \underline{2} & \underline{3} \end{pmatrix} + \begin{pmatrix} 2 & 2 & 2 \\ 1 & 1 & 1 \end{pmatrix}$

$= \begin{pmatrix} \underline{3} & \underline{5} & \underline{5} \\ \underline{2} & \underline{4} & \underline{2} \end{pmatrix}$

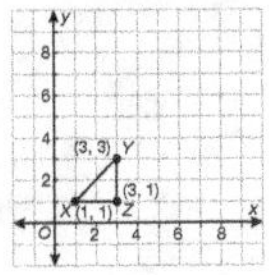

Find the sum of the matrices.

2. Add the point matrix to $\begin{pmatrix} 1 & 1 \\ 3 & 3 \end{pmatrix}$

$\begin{pmatrix} 2 & 5 \\ 5 & 8 \end{pmatrix}$

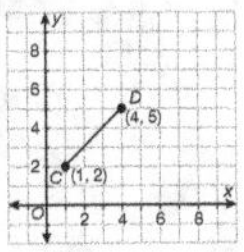

3. Add the point matrix to $\begin{pmatrix} 3 & 3 & 3 \\ -1 & -1 & -1 \end{pmatrix}$.

$\begin{pmatrix} 5 & 8 & 4 \\ 7 & 7 & 3 \end{pmatrix}$

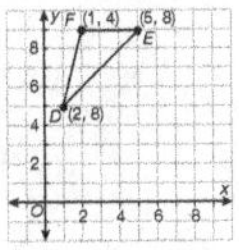

Reteaching 100
continued

Translating with Matrices

You can use matrices to determine how one figure is translated to another figure.

Example:

Triangle ABC is translated 3 units left and 2 units up.

Find the matrix that translates ABC to $A'B'C'$.

$\begin{pmatrix} 2 & 2 & 5 \\ 2 & 5 & 5 \end{pmatrix}$ ΔABC point matrix

$\begin{pmatrix} -1 & -1 & 2 \\ 4 & 7 & 7 \end{pmatrix}$ $\Delta A'B'C'$ point matrix

$\begin{pmatrix} -1 & -1 & 2 \\ 4 & 7 & 7 \end{pmatrix} - \begin{pmatrix} 2 & 2 & 5 \\ 2 & 5 & 5 \end{pmatrix}$ Subtract.

$\begin{pmatrix} -3 & -3 & -3 \\ 2 & 2 & 2 \end{pmatrix}$ Simplify.

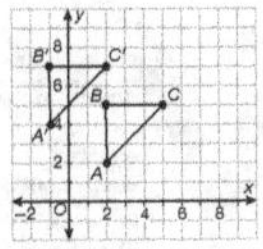

Practice

Complete the steps to find the translation matrix.

4. $\Delta DEF = \begin{pmatrix} -3 & 0 & 2 \\ 3 & 3 & 1 \end{pmatrix}$

$\Delta D'E'F' = \begin{pmatrix} \underline{0} & \underline{3} & \underline{5} \\ \underline{1} & \underline{1} & \underline{-1} \end{pmatrix}$

$\begin{pmatrix} \underline{0} & \underline{3} & \underline{5} \\ \underline{1} & \underline{1} & \underline{-1} \end{pmatrix} - \begin{pmatrix} -3 & 0 & 2 \\ 3 & 3 & 1 \end{pmatrix} = \begin{pmatrix} \underline{3} & \underline{3} & \underline{3} \\ \underline{-2} & \underline{-2} & \underline{-2} \end{pmatrix}$

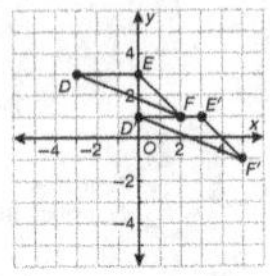

5. $\Delta XYZ = \begin{pmatrix} -4 & -2 & 0 \\ -4 & -1 & -5 \end{pmatrix}$

$\Delta X'Y'Z' = \begin{pmatrix} \underline{1} & \underline{3} & \underline{5} \\ \underline{0} & \underline{3} & \underline{-1} \end{pmatrix}$

$\begin{pmatrix} \underline{1} & \underline{3} & \underline{5} \\ \underline{0} & \underline{3} & \underline{-1} \end{pmatrix} - \begin{pmatrix} \underline{-4} & \underline{-2} & \underline{0} \\ -4 & -1 & -5 \end{pmatrix} = \begin{pmatrix} \underline{5} & \underline{5} & \underline{5} \\ \underline{4} & \underline{4} & \underline{4} \end{pmatrix}$

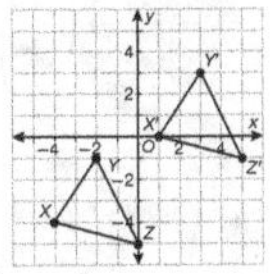

Find the translation matrix.

6. Translate $\begin{pmatrix} 6 & 9 & 3 \\ 2 & 0 & 4 \end{pmatrix}$ 3 units right and 5 units down. $\begin{pmatrix} \underline{3} & \underline{3} & \underline{3} \\ \underline{-5} & \underline{-5} & \underline{-5} \end{pmatrix}$

7. Translate $\begin{pmatrix} -4 & -1 & 3 \\ -3 & -1 & -4 & 4 \end{pmatrix}$ units left and 3 units up $\begin{pmatrix} \underline{-4} & \underline{-4} & \underline{-4} \\ \underline{3} & \underline{3} & \underline{3} \end{pmatrix}$

Name ______________________ Date __________ Class __________

Reteaching INV 10
Fractals

The Sierpinski Triangle

An **iteration** is the repeated application of a rule. You can continue an iteration indefinitely. In geometry, you can generate **fractals** by iteration.

The **Sierpinski Triangle** is a famous fractal.

Start with a right isosceles triangle. The first iteration connects the midsegments of the triangle. Then, remove the triangle in the center so that there is a "hole" in the middle of the triangle.

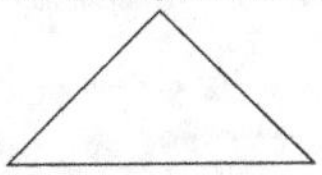

right isosceles triangle

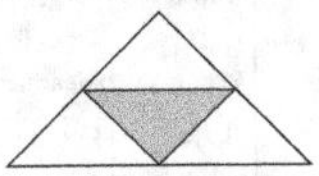

first iteration

The iteration of connecting the midsegments of the triangle and then removing the central triangle is repeated to make the Sierpinski triangle.

Practice

1. Draw a right isosceles triangle and make the first iteration for the Sierpinski triangle. Shade in the triangles you make except for the central triangle. Make a large triangle so that you can do several iterations.

   For the first iteration for the Sierpinski triangle, there are three unshaded <u>triangles</u>.

   During the next iteration, connect the midsegments of each unshaded <u>triangle</u> and shade each <u>central</u> triangle.

Draw the second iteration for the Sierpinski triangle. Count the number of triangles. Remember not to count the central triangles.

2. For the second iteration, there are <u>nine</u> unshaded triangles.

Draw the third iteration for the Sierpinski triangle.

3. For the third iteration, there are <u>27</u> unshaded triangles.
4. The number of unshaded triangles for each iteration of the Sierpinski triangle make a pattern. Write the number of unshaded triangles for the three iterations. <u>3, 9, 27</u>
5. Write your numbers in the answer for problem 5 as powers of 3. <u>$3^1, 3^2, 3^3$</u>
6. How many unshaded triangles would the seventh iteration of the Sierpinski triangle have? <u>$3^7, = 2{,}187$</u>

Reteaching INV 10
continued

The Sierpinski Carpet

The **Sierpinski carpet** is another fractal. Begin with a square. Think of the square divided into nine equal squares (look at the grid below).

The first iteration removes the middle square. Shade in the middle square as you did with the central triangle in the Sierpinski Triangle.

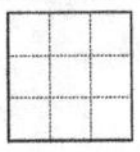

Square

(dotted lines are not part of the iteration)

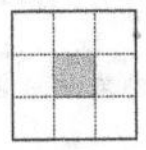

First iteration

There are 8 unshaded squares left. with each of the squares, think of the square divided into 9 equal squares. For the second iteration, remove the middle square from each of the 8 unshaded squares.

Practice

7. Draw a square and make the first iteration for the Sierpinski carpet. Shade in the central square. Be sure to make a large square so that you can do several iterations.

   For the first iteration for the Sierpinski carpet, there are <u>eight</u> unshaded squares.

Draw the second iteration for the Sierpinski carpet. Count the number of squares left. Remember not to count the central squares.

8. During the next iteration, remove (shade) the middle <u>square</u> from each of the eight squares.
9. There are <u>eight</u> smaller squares in each of the eight squares. There are a total of <u>sixty-four</u> unshaded squares after the second iteration.

Draw the third iteration for the Sierpinski carpet. Count the number of squares left.

10. There are <u>eight</u> smaller unshaded squares in each of the sixty-four squares. There are a total of <u>512</u> unshaded squares after the second iteration.
11. How many unshaded squares would the fourth iteration of the Sierpinski carpet have? <u>$8^4 = 4{,}096$</u>

Name ______________________ Date __________ Class __________

Reteaching 101
Determining Lengths of Segments Intersecting Circles

You have worked with secants and tangents of circles. Now you will learn about the relationships between secant segments and tangent segments.

According to the **Secant-Secant Product Theorem,** the product of one secant segment and its external segment equals that of the other secant segment and its external µsegment.

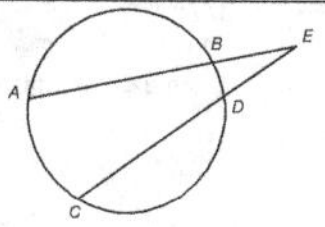

In the diagram to the right, the Secant-Secant Product Theorem can be stated as $AE \cdot BE = CE \cdot DE$.

Use the Secant-Secant Product Theorem to find the value of x in the diagram.

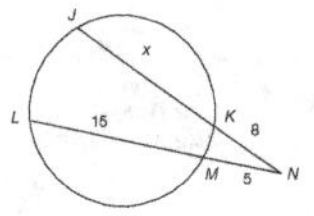

$JN \cdot KN = LN \cdot MN$
$(8 + x)(8) = (20)(5)$
$64 + 8x = 100$
$8x = 36$
$x = 4.5$
$LN = 20$ and $JN = 8 + 4.5 = 12.5$

Practice

Complete the steps to find the value of *the variable.*

1.

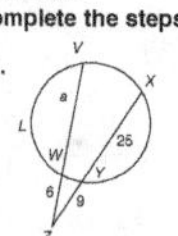

$XZ \cdot YZ = VZ \cdot WZ$
$(34)(\underline{9}) = (\underline{6 + a})(6)$
$\underline{306} = \underline{36 + 6a}$
$\underline{270} = \underline{6a}$
$a = \underline{45}$

2.

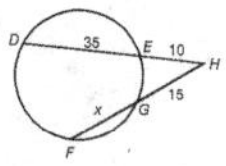

$DH \cdot EH = FH \cdot GH$
$(45)(10) = (\underline{15 + x})(\underline{15})$
$450 = \underline{225 + 15x}$
$\underline{225} = \underline{15x}$
$x = \underline{15}$

Find the value of x. Round your answer to the nearest tenth.

3. $x = \underline{9.3}$

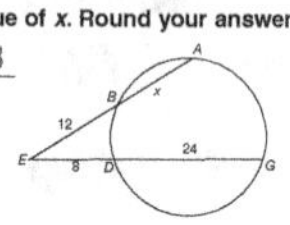

4. $x = \underline{14.3}$

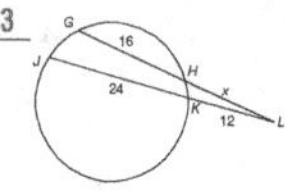

Reteaching 101
continued

According to the **Secant-Tangent Product Theorem,** the product of the lengths of the secant segment and its external segment equals that of the tangent segment squared.

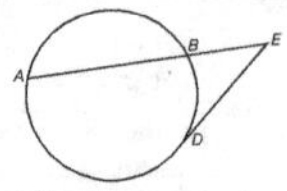

In the diagram to the right, the Secant-Tangent Product Theorem can be stated as $AE \cdot BE = DE^2$.

Use the Secant-Tangent Product Theorem to find the value of the variable.

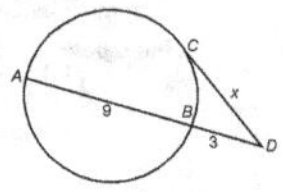

$AD \cdot BD = CD^2$
$(12)(3) = x^2$
$36 = x^2$
$x = 6$

Practice

Complete the steps to find the value x.

5.

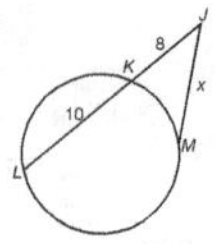

$JK \cdot JL = JM^2$
$(\underline{8})(18) = x^2$
$\underline{144} = x^2$
$x = \underline{12}$

6.

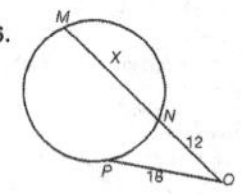

$MO \cdot NO = OP^2$
$(\underline{x + 12})(12) = 18^2$
$\underline{12x + 144} = 324$
$\underline{12x} = \underline{180}$
$x = \underline{15}$

Find the value of x. Round your answer to the nearest tenth.

7. $x = \underline{16.2}$

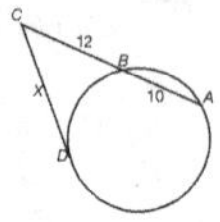

8. $x = \underline{11.3}$

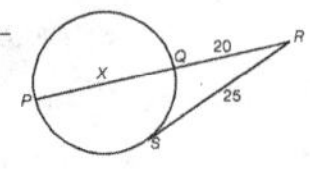

Name ______________________ Date __________ Class __________

Reteaching 102
Dilations in the Coordinate Plane

You have worked with translations and reflections. Now you will learn to make transformations using dilations in the coordinate plane.

Dilations

A **dilation** is a reduction or enlargement of a shape. Recall that the **scale factor** describes how much a figure is enlarged or reduced.

Mapping notation for a dilation with the origin at the center is written as $D_{O,k}(x, y) \rightarrow (kx, ky)$, where k represents the scale factor.

Example:

Determine the result of the dilation $D_{O,4}(x, y)$ on the points (2, 6) and (6, 4).

$D_{O,k}(x, y) \rightarrow (kx, ky)$
$D_{O,4}(2, 6) \rightarrow (4 \times 2, 4 \times 6) = (8, 24)$
$D_{O,4}(6, 4) \rightarrow (4 \times 6, 4 \times 4) = (24, 16)$

Practice

Complete the steps to find the result of the given dilation on the coordinates.

1. $D_{O,2}(x, y)$ on the points $(-2, 4)$ and $(1, 3)$

 $D_{O,2}(-2, 4) \rightarrow (2 \times \underline{-2}, 2 \times \underline{4}) = (\underline{-4}, \underline{8})$
 $D_{O,2}(1, 3) \rightarrow (2 \times \underline{1}, 2 \times \underline{3}) = (\underline{2}, \underline{6})$

2. $D_{O,3}(x, y)$ on the points $(-1, 5)$ and $(2, -4)$

 $D_{O,3}(-1, 5) \rightarrow (3 \times \underline{-1}, 3 \times \underline{5}) = (\underline{-3}, \underline{15})$
 $D_{O,2}(2, -4) \rightarrow (3 \times \underline{2}, 3 \times \underline{-4}) = (\underline{6}, \underline{-12})$

Determine the result of the dilation on the given points.

3. $D_{O,4}(x, y)$ on the points (1, 5) and (−2, 3) $\underline{(4, 20) \text{ and } (-8, 12)}$
4. $D_{O,2.5}(x, y)$ on the points (2, 4) and (3, 6) $\underline{(5, 10) \text{ and } (7.5, 15)}$
5. $D_{O,2}(x, y)$ on the points (−3, 1) and (5, 8) $\underline{(-6, 2) \text{ and } (10, 16)}$
6. $D_{O,0.5}(x, y)$ on the points (−4, 6) and (4, 8) $\underline{(-2, 3) \text{ and } (2, 4)}$
7. $D_{O,1.25}(x, y)$ on the points (1, −2) and (3, −4) $\underline{(1.25, -2.5) \text{ and } (3.75, -5)}$
8. $D_{O,3}(x, y)$ on the points (0, 4) and (−4, 2) $\underline{(0, 12) \text{ and } (-12, 6)}$

Reteaching 102
continued

Dilating by Point Matrix

You can use a **point matrix** to determine the points of a figure on a coordinate grid. The point (a, b) would be written as $\begin{bmatrix} a \\ b \end{bmatrix}$. To apply a scale factor to a point matrix, multiply each entry in the point matrix by the scale factor.

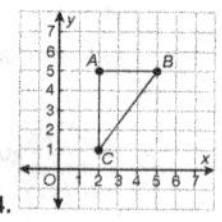

Create a point matrix for the triangle given in the diagram.

Use this point matrix to show a dilation of the shape by a factor of 4.

Step 1: Write the point matrix. Each point is a different column in the matrix.

$$\begin{bmatrix} 5 & 5 & 1 \\ 1 & 5 & 1 \end{bmatrix}$$

Step 2: Multiply each entry in the point matrix by the scale factor.

$$4\begin{bmatrix} 5 & 5 & 1 \\ 1 & 5 & 1 \end{bmatrix} = \begin{bmatrix} 4 \cdot 5 & 4 \cdot 5 & 4 \cdot 1 \\ 4 \cdot 1 & 4 \cdot 5 & 4 \cdot 1 \end{bmatrix}$$

$$= \begin{bmatrix} 20 & 20 & 4 \\ 4 & 20 & 4 \end{bmatrix}$$

Step 3: Write the new coordinates for the vertices of the dilated triangle.
(20, 4), (20, 20), and (4, 4)

Practice

Complete the steps to create a point matrix for the given figures and scale factor.

9. Scale factor 5

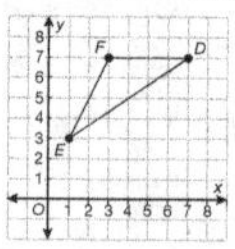

$$5\begin{bmatrix} 7 & \underline{1} & \underline{3} \\ 7 & \underline{3} & \underline{7} \end{bmatrix}$$

$$= \begin{bmatrix} 5 \cdot 7 & 5 \cdot \underline{1} & 5 \cdot \underline{3} \\ 5 \cdot 7 & 5 \cdot \underline{3} & 5 \cdot \underline{7} \end{bmatrix} \frac{1}{1}$$

$$= \begin{bmatrix} 35 & \underline{5} & \underline{15} \\ 35 & \underline{15} & \underline{35} \end{bmatrix}$$

10. Scale factor 8

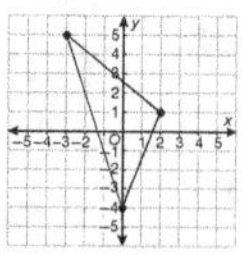

$$8\begin{bmatrix} \underline{-3} & \underline{2} & \underline{0} \\ \underline{5} & \underline{1} & \underline{4} \end{bmatrix}$$

$$= \begin{bmatrix} 8 \cdot \underline{-3} & 8 \cdot \underline{2} & 8 \cdot \underline{0} \\ 8 \cdot \underline{5} & 8 \cdot \underline{1} & 8 \cdot \underline{-4} \end{bmatrix}$$

$$= \begin{bmatrix} \underline{-24} & \underline{16} & \underline{0} \\ \underline{40} & \underline{8} & \underline{32} \end{bmatrix}$$

Show a dilation by each scale factor.

11. Scale factor 2.5

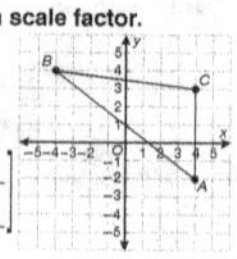

$$\begin{bmatrix} \underline{10} & \underline{-10} & \underline{10} \\ \underline{-5} & \underline{10} & \underline{7.5} \end{bmatrix}$$

12. Scale factor $\frac{1}{3}$

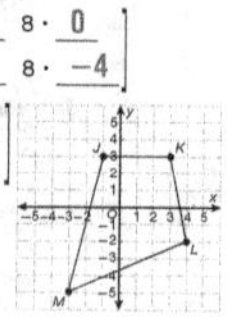

$$\begin{bmatrix} \underline{-\frac{1}{3}} & \underline{1} & \underline{\frac{4}{3}} & \underline{\frac{3}{5}} \\ \underline{1} & \underline{1} & \underline{-\frac{2}{3}} & \underline{-\frac{5}{3}} \end{bmatrix}$$

Answer Key

Name ______________________ Date __________ Class __________

Reteaching 103
Frustums of Cones and Pyramids

You have worked with the volume of cones and pyramids. Now you will examine the frustums of cones and pyramids.

Frustum of a Cone

The frustum of a cone is a part of a cone with two parallel bases.

The volume of a frustum is $V = \frac{1}{3}h(B_1 + \sqrt{B_1B_2} + B_2)$, where B_1 and B_2 are the base areas and h is the height of the frustum.

Find the volume of the frustum of the cone. Round your answer to the nearest whole number.

The volume of the frustum of the cone is about 4772 cm³.

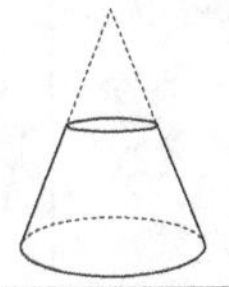

Practice

Complete the steps to find the volume of the frustum of the cone. Round your answer to the nearest whole number.

$B_1 = \pi r^2$	Base area B_1
$B_2 = \pi(8)^2 \approx 201\text{ cm}^2$	Substitute and simplify.
$B_2 = \pi r^2$	Base area B_2
$B_2 = \pi(12)^2 \approx 452\text{ cm}^2$	Substitute and simplify.
$V = \frac{1}{3}h(B_1 + \sqrt{B_1B_2} + B_2)$	Frustrum formula
$V \approx \frac{1}{3}(15)(201 + \sqrt{(201)(452)} + 452)$	Substitute.
$V \approx 4772\text{ cm}^3$	Simplify.

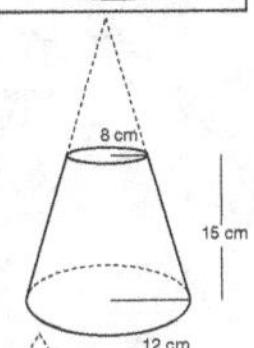

$B_1 = \pi r^2 = \pi(10)^2 \approx$ __314__ ft²

$B_2 = \pi r^2 = \pi($__15__$)^2 \approx$ __707__ ft²

1. $V = \frac{1}{3}h(B_1 + \sqrt{B_1B_2} + B_2)$

   $V \approx \frac{1}{3}(16)($__314__ $+ \sqrt{\text{314 707}}$ + __707__$)$

   $V \approx$ __7958__ ft³

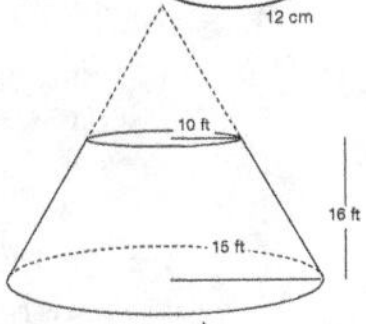

Find the volume of the frustum of the cone. Round your answer to the nearest whole number.

2. $V \approx$ __2586__ cm³

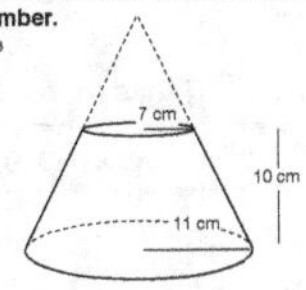

3. $V \approx$ __3428__ ft

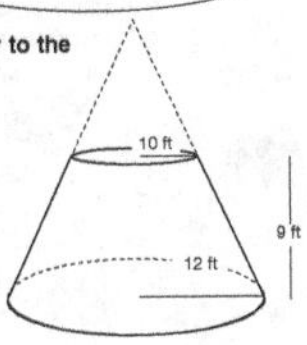

Reteaching 103
continued

Frustum of a Pyramid

The frustum of a pyramid is a part of a pyramid with two parallel bases. The volume of a frustum is $V = \frac{1}{3}h(B_1 + \sqrt{B_1B_2} + B_2)$, where B_1 and B_2 are the base areas and h is the height of the frustum.

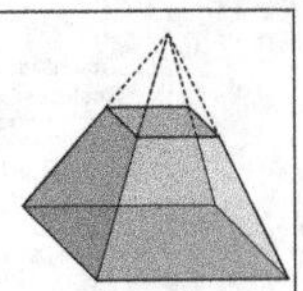

Find the volume of the frustum of the pyramid. Round your answer to the nearest whole number.

$B_1 = lw$	Base area B_1
$B_1 = (8)^2 = 64\text{ cm}^2$	Substitute and simplify.
$B_2 = lw$	Base area B_2
$B_2 = (10)^2 = 100\text{ cm}^2$	Substitute and simplify.
$V = \frac{1}{3}h(B_1 + \sqrt{B_1B_2} + B_2)$	Frustrum formula
$V = \frac{1}{3}(12)(64 + \sqrt{(64)(100)} + 100)$	Substitute.
$V = 976\text{ cm}^3$	Simplify.

The volume of the frustum of the pyramid is 976 cm³.

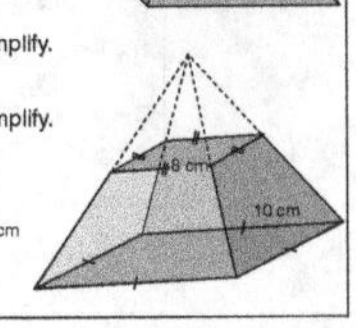

Practice

Complete the steps to find the volume of the frustum of the pyramid. Round to the nearest whole number.

$B_1 = lw = (12)(10) = 120\text{ ft}^2$

$B_2 = lw = ($__9__$)($__6__$) =$ __54__ ft²

4. $V = \frac{1}{3}h(B_1 + \sqrt{B_1B_2} + B_2)$

   $V = \frac{1}{3}(8) + ($__120__ $+ \sqrt{\text{120 54}}$ + __54__$)$

   $V \approx$ __679__ ft³

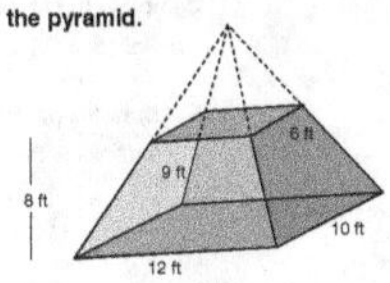

Find the volume of the frustum of the pyramid. Round to the nearest whole number.

5. $V \approx$ __1581__ cr

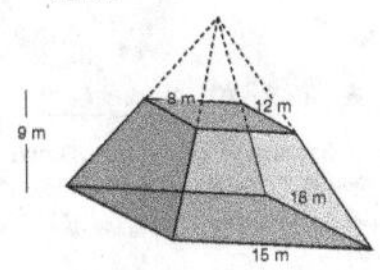

6. $V \approx$ __3415__

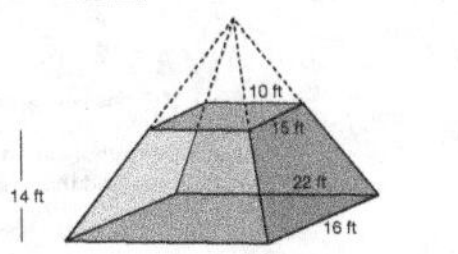

Name ______________________ Date __________ Class __________

Reteaching 104
Relating Arc Lengths and Chords

You have worked with circles and sectors. Now you will work with arc lengths and chords.

Congruent Arcs

Congruent arcs are arcs that have the same measure.

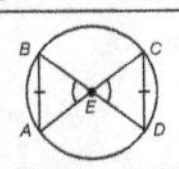 If $m\angle BEA \cong m\angle CED$, then $\overline{BA} \cong \overline{CD}$.	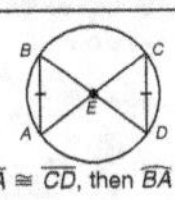 If $\overline{BA} \cong \overline{CD}$, then $\overset{\frown}{BA} \cong \overset{\frown}{CD}$.	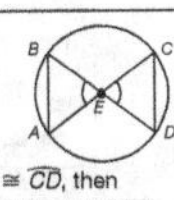 If $\overset{\frown}{BA} \cong \overset{\frown}{CD}$, then $m\angle BEA \cong m\angle CED$.
Congruent central angles have congruent chords.	Congruent chords have congruent arcs.	Congruent arcs have congruent central angles.

Example:

Given that $\angle HLG \cong \angle KLJ$, find GH.

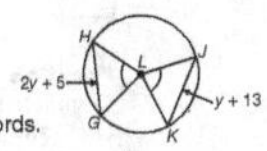

$\angle HLG \cong \angle KLJ$	
$HG \cong JK$	Congruent cental angles have congruent chords.
$2y + 5 = y + 13$	
$y + 5 = 13$	Subtract y from both sides.
$y = 8$	Subtract 5 from both sides.
$GH = 2y + 5 = 2(8) + 5 = 21$	

Practice

Complete the statements to find $m\overset{\frown}{QR}$.

1. $\overset{\frown}{RQ} \cong \overset{\frown}{ST}$ $\qquad$ $\overset{\frown}{QR} = 3x + 22$

   $m\overset{\frown}{RQ} \cong$ __$m\overset{\frown}{ST}$__ $\qquad = 3$(__22__)$ + 22$

   __3x__ $+ 22 = 4x$ $\qquad =$ __66__ $+ 22$

   __22__ $= x$ $\qquad =$ __88__

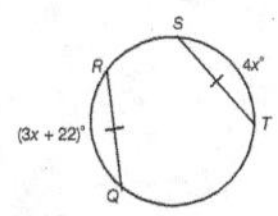

Find the each measure.

2. Given that $m\overset{\frown}{AD} = m\overset{\frown}{BC}$, find $m\angle AED$. __125°__

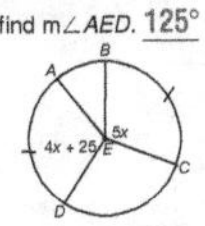

3. Given that $\angle AEB \cong \angle DEC$, find CD. __11__

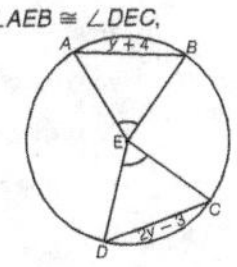

Reteaching 104
continued

More Congruent Arcs

Use the figure shown to prove that if the arcs are congruent, then the chords are also congruent.

Given that $\overset{\frown}{KL} \cong \overset{\frown}{LM}$, prove $\overline{KL} \cong \overline{LM}$.

Statement	Reason
1. $\overset{\frown}{KL} \cong \overset{\frown}{LM}$	1. Given
2. $\angle KJL \cong \angle MJL$	2. Definition of arc measures
3. $\overline{JK} \cong \overline{JL}$; $\overline{JL} \cong \overline{JM}$	3. Radii of the same circle are all congruent.
4. $\Delta JKL \cong \Delta JLM$	4. SAS
5. $\overline{KL} \cong \overline{LM}$	5. CPCTC

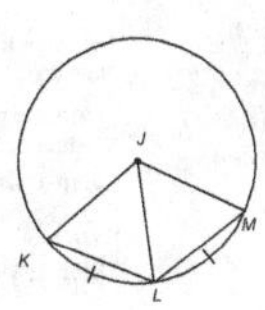

Practice

Complete the statements to prove that if the chords are congruent, then the arcs are also congruent.

4. Given that $\overline{ED} \cong \overline{BC}$, prove $\overset{\frown}{ED} \cong \overset{\frown}{BC}$.

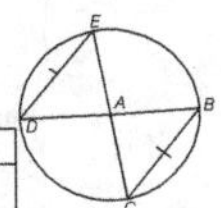

Statement	Reason
1. __$\overline{ED} \cong \overline{BC}$__	1. Given
2. $\overline{AE} \cong$ __$\overline{AD}$__ $\cong \overline{AB} \cong$ __$\overline{AC}$__	2. Radii of the same circle are all congruent.
3. $\Delta AED \cong \Delta ABC$	3. __SSS__
4. $\angle EAD \cong$ __$\angle BAC$__	4. CPCTC
5. $\overset{\frown}{ED} \cong \overset{\frown}{BC}$	5. __Definition of arc measures__

Use the figure above to prove that if the arcs are congruent, then the chords are also congruent.

5. Given that $\overset{\frown}{ED} \cong \overset{\frown}{BC}$, prove $\overline{ED} \cong \overline{BC}$.

Statement	Reason
1. $\overset{\frown}{ED} \cong \overset{\frown}{BC}$	1. __Given__
2. __$\angle EAD \cong \angle BAC$__	2. __Definition of arc measures__
3. __$\overline{AE} \cong \overline{AD} \cong \overline{AB} \cong \overline{AC}$__	3. __Radii of the same circle are all congruent.__
4. __$\Delta AED \cong \Delta ABC$__	4. __SAS__
5. __$\overline{ED} \cong \overline{BC}$__	5. __CPCTC__

Answer Key

Name ______________________ Date __________ Class __________

Reteaching 105

Rotations and Reflections on the Coordinate Plane

You have learned about geometric transformations. Now you will use matrices to perform transformations on a coordinate plane.

The vertices of ΔXYZ are $X(7, 6)$ $Y(2, 3)$, and $Z(5, -1)$. Use matrix multiplication to reflect the image in the y-axis. Draw the preimage and the image in a coordinate plane.

Multiply by the reflection matrix in the y-axis $\begin{pmatrix} -1 & 0 \\ 0 & 1 \end{pmatrix}$.

$$\begin{pmatrix} -1 & 0 \\ 0 & 1 \end{pmatrix}\begin{pmatrix} 7 & 2 & 5 \\ 6 & 3 & -1 \end{pmatrix} = \begin{bmatrix} (-1)(7) + (0)(6) & (-1)(2) + (0)(3) & (-1)(5) + (0)(-1) \\ (0)(7) + (1)(6) & (0)(2) + (1)(3) & (0)(5) + (1)(-1) \end{bmatrix} = \begin{pmatrix} -7 & -2 & -5 \\ 6 & 3 & -1 \end{pmatrix}$$

The image matrix is $\begin{pmatrix} -7 & -2 & -5 \\ 6 & 3 & -1 \end{pmatrix}$.

Therefore, $\Delta X'Y'Z'$ has vertices at $X'(7, 6)$, $Y'(-2, 3)$, and $Z'(-5, -1)$.

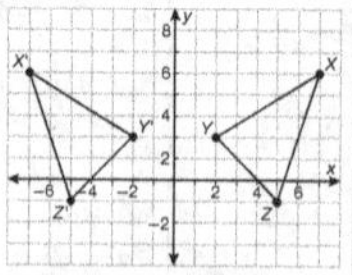

Practice

Complete the steps to find the reflected image of the triangle.

1. Reflect ΔQRS with vertices $Q(1, 6)$, $R(5, 2)$, and $S(-1, 3)$ in the x-axis.

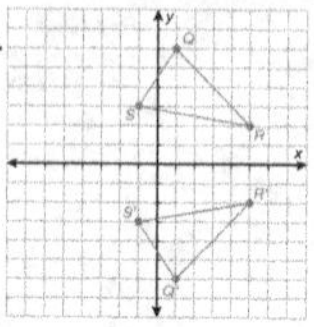

Draw the preimage and the image in a coordinate plane.

Multiply by the reflection matrix in the x-axis $\begin{pmatrix} 1 & 0 \\ 0 & -1 \end{pmatrix}$.

$$\begin{pmatrix} 1 & 0 \\ 0 & -1 \end{pmatrix}\begin{pmatrix} \underline{1} & 5 & -1 \\ 6 & \underline{2} & \underline{3} \end{pmatrix}$$

$$= \begin{bmatrix} (1)(1) + (0)(\underline{6}) & (1)(\underline{5}) + (0)(2) & (1)(-1) + (0)(\underline{3}) \\ (0)(1) + (-1)(\underline{6}) & (0)(\underline{5}) + (-1)(2) & (0)(-1) + (-)(\underline{3}) \end{bmatrix}$$

$$= \begin{pmatrix} \underline{1} & \underline{5} & \underline{-1} \\ \underline{-6} & \underline{-2} & \underline{-3} \end{pmatrix}$$

$\Delta Q'R'S'$ has vertices at <u>$Q'(1, -6)$, $R'(5, -2)$, and $S'(-1, -3)$</u>.

Use matrix multiplication to reflect the figure in the given axis. Graph the preimage and the image on a separate sheet of paper.

2. Reflect ΔABC with vertices $A(1, 1)$, $B(2, 3)$, $C(3, 2)$ in the x-axis.

<u>$\begin{pmatrix} 1 & 2 & 3 \\ -1 & -3 & -2 \end{pmatrix}$; $\Delta A'B'C'$ has vertices $A'(1, -1)$, $B'(2, -3)$, and $C'(3, -2)$.</u>

Reteaching 105

continued

The vertices of ΔTUV are $T(2, 1)$ $U(4, 5)$ $V(5, 2)$. Determine the image matrix for a 90° rotation of ΔTUV about the origin. Graph the preimage and image in a coordinate plane.

Multiply by the 90° rotation matrix $\begin{pmatrix} 0 & -1 \\ 1 & 0 \end{pmatrix}$.

$$\begin{pmatrix} 0 & -1 \\ 1 & 0 \end{pmatrix}\begin{pmatrix} 2 & 4 & 5 \\ 1 & 5 & 2 \end{pmatrix} = \begin{bmatrix} (0)(2) + (-1)(1) & (0)(4) + (-1)(5) & (0)(5) + (-1)(2) \\ (1)(2) + (0)(1) & (1)(4) + (0)(5) & (1)(5) + (0)(2) \end{bmatrix} = \begin{pmatrix} -1 & -5 & -2 \\ 2 & 4 & 5 \end{pmatrix}$$

The image matrix is $\begin{pmatrix} -1 & -5 & -2 \\ 2 & 4 & 5 \end{pmatrix}$.

So, $\Delta T'U'V'$ has vertices at $T'(-2, 1)$, $U'(-4, 5)$, $V'(-5, 2)$.

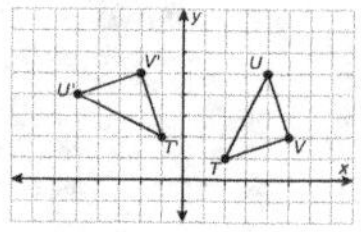

Practice

Complete the steps to determine the rotation image of the triangle.

3. Rotate ΔABC with vertices $A(4,0)$, $B(3,4)$ $C(1,1)$ 270° about the origin.

Draw the preimage and the image in a coordinate plane.

Multiply by the 270° rotation matrix $\begin{pmatrix} 0 & 1 \\ -1 & 0 \end{pmatrix}$.

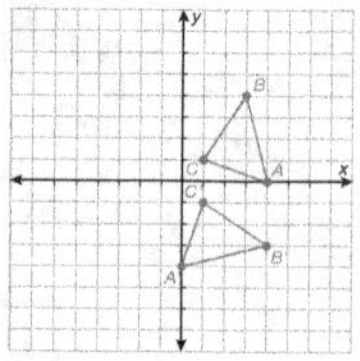

$$\begin{pmatrix} 0 & 1 \\ -1 & 0 \end{pmatrix}\begin{pmatrix} 4 & \underline{3} & \underline{1} \\ \underline{0} & \underline{4} & 1 \end{pmatrix}$$

$$\begin{bmatrix} (0)(4) + (1)(\underline{0}) & (0)(3) + (1)(\underline{4}) & (0)(\underline{1}) + (1)(1) \\ (-1)(4) + (0)(\underline{0}) & (-1)(3) + (0)(\underline{4}) & (-1)(\underline{1}) + (0)(1) \end{bmatrix}$$

$$= \begin{pmatrix} \underline{0} & \underline{4} & \underline{1} \\ \underline{-4} & \underline{-3} & \underline{-1} \end{pmatrix}$$

$\Delta A'B'C'$ has vertices at <u>$A'(0, -4)$, $B'(4, -3)$, and $C'(1, -1)$</u>.

Use matrix multiplication to rotate each figure the given degrees. Graph the preimage and the image on a separate sheet of paper.

4. Rotate rectangle $\Delta QRST$ with vertices $Q(0, 1)$, $R(4, -1)$ $S(4, 3)$, and $T(0, 3)$ 90° about the origin.

<u>$\begin{pmatrix} -1 & 1 & -3 & -3 \\ 0 & 4 & 4 & 0 \end{pmatrix}$, $\Delta Q'R'S'T'$ has vertices</u>

<u>$Q'(-1, 0)$, $R'(1, 4)$, $S'(-3, 4)$, and $T'(-3, 0)$.</u>

Name ______________________ Date __________ Class __________

Reteaching 106

Circumscribed and Inscribed Figures

You have found the perimeter and the missing side lengths of many different figures. Now you will find the perimeter of inscribed and circumscribed figures.

Find the perimeter of the largest regular octagon that can be inscribed in a circle with radius 6. Round your answer to the nearest hundredth.

Step 1: Draw a diagram. Find the center angle θ by dividing 360° by the number of sides in the figure.

$360 \div 8 = 45°$

Step 2: Then, use the law of cosines to find the length of the sides of the octagon.

$c^2 = a^2 + b^2 - 2ab\cos\theta$

$c^2 = 6^2 + 6^2 - 2(6)(6)\cos 45°$

$c^2 = 21.09$

$c = 4.59$

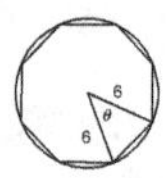

Each side of the octagon is approximately 4.59, so the perimeter of the octagon is $P = 8s = 8(4.59)$

$P = 36.72$

Practice

Complete the steps to find the perimeter of the largest regular pentagon inscribed in a circle with radius 5.

Round your answer to the nearest hundredth.

1. $360 \div 5 = 72°$

$c^2 = a^2 + b^2 - 2ab\cos\theta$

$c^2 = (\underline{5})^2 + (\underline{5})^2 - 2(\underline{5})(\underline{5})\cos \underline{72°}$

$c^2 = \underline{34.55}$

$c = \underline{5.88}$

$P = 5s = 5(\underline{5.88})$

$P = \underline{29.4}$

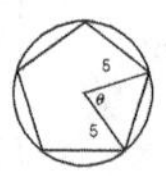

Find the perimeter of each inscribed regular polygon. Round your answer to the nearest hundredth.

2. What is the largest regular octagon that can be inscribed in a circle with radius 8? <u>48.96</u>
3. What is the largest regular decagon that can be inscribed in a circle with radius 7? <u>43.26</u>
4. What is the largest regular pentagon that can be inscribed in a circle with radius 9? <u>52.90</u>
5. What is the largest regular nonagon that can be inscribed in a circle with radius 12? <u>73.89</u>

Reteaching 106

continued

Find the perimeter of a pentagon circumscribing a circle with radius 7. Round your answer to the nearest tenth.

Draw a diagram. Find the center angle θ by dividing 360° by the number of sides in the figure.

$360 \div 5 = 72°$

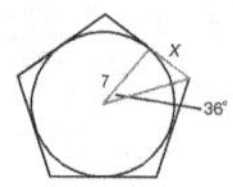

The radius of the circle is the apothem of the pentagon. It can be used to calculate half of the side length.

The apothem also divides the central angle in half for each side of the figure.

Half of the central angle is 36°.

$\tan 36° = \frac{x}{7}$

$7\tan 36° = x$

$5.09 \approx x$

Since x is half of the side length, each side is $2x$.

The perimeter is $P = 5s = 5(2 \cdot 5.09)$.

$P = 50.86$

Practice

Complete the steps to find the perimeter of the polygon circumscribing a circle with the given radius. Round your answer to the nearest hundredth.

6. an octagon circumscribing a circle with radius 4

$360 \div \underline{8} = \underline{45}$

$\tan \underline{22.5} = \frac{x}{\underline{4}}$

$\underline{4}\tan \underline{22.5} = x$

$x \approx \underline{1.66}$

$P = 8s = 8(2 \cdot \underline{1.66})$

$P = \underline{26.56}$

7. a hexagon circumscribing a circle with radius 12

$360 \div \underline{6} = \underline{60}$

$\tan \underline{30} = \frac{x}{\underline{12}}$

$\underline{12}\tan \underline{30} = x$

$x \approx \underline{6.93}$

$P = 6s = 6(2 \cdot \underline{6.93})$

$P = \underline{83.16}$

Find the perimeter of each polygon circumscribing the circle with the given radius. Round your answer to the nearest hundredth.

8. a hexagon circumscribing a circle with radius 9 <u>62.35</u>
9. a nonagon circumscribing a circle with radius 5 <u>32.76</u>
10. a pentagon circumscribing a circle with radius 15 <u>109</u>
11. a triangle circumscribing a circle with radius 10 <u>103.92</u>

Name ______________________ Date __________ Class __________

Reteaching 107
Maximizing Area

Rectangle *VXYZ* is *l* units long and *w* units wide. It has a perimeter of 36 units.

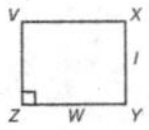

Determine the area of the rectangle for $w = 5$, $w = 7$, $w = 9$, and $w = 10$. Solve for l using the perimeter and width. Then substitute w and l into the formula for area.

$P = 2(l + w)$	$P = 2(l + w)$	$P = 2(l + w)$	$P = 2(l + w)$
$36 = 2(l + 5)$	$36 = 2(l + 7)$	$36 = 2(l + 9)$	$36 = 2(l + 10)$
$l = 13$	$l = 11$	$l = 9$	$l = 8$
$A = lw$	$A = lw$	$A = lw$	$A = lw$
$= 13 \times 5$	$= 11 \times 7$	$= 9 \times 9$	$= 8 \times 10$
= 65 sq. units	= 77 sq. units	= 81 sq. units	= 80 sq. units

Draw each rectangle on grid paper. Describe the relationship between the shape of each rectangle and the area of each rectangle.

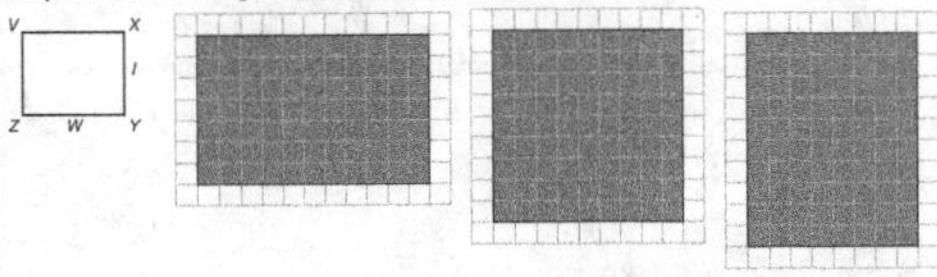

The closer to a square the rectangle is, the greater its area is. The square itself has the greatest area.

Practice

Complete the steps to find the areas of rectangle *GHJK* with the given lengths and with a perimeter of 40 inches.

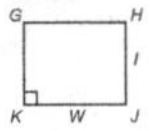

1. Determine the area of the rectangle for $l = 6$, $l = 8$, and $l = 10$.

$P = 2(l + w)$	$P = 2(l + w)$	$P = 2(l + w)$
$40 = 2(6 + w)$	$40 = 2(8 + w)$	$40 = 2(10 + w)$
$w = $ <u>14</u>	$w = $ <u>12</u>	$w = $ <u>10</u>
$A = lw$	$A = lw$	$A = lw$
$= 6 \times$ <u>14</u>	$= 8 \times$ <u>12</u>	$= 10 \times$ <u>10</u>
= <u>84</u> in²	= <u>96</u> in²	= <u>100</u> in²

Which rectangle has the greatest area? <u>The rectangle with side length 10.</u>

2. Determine the largest possible area of a rectangle with perimeter 48 inches. <u>144 in²</u>

3. Determine the largest possible area of a hexagon with perimeter 54 centimeters. <u>210.33 cm²</u>

Reteaching 107
continued

Three rectangles have the following dimensions: 4 mm × 33 mm, 6 mm × 22 mm, and 11 mm × 12 mm.

Determine the area and perimeter of each rectangle.

$A_1 = lw$	$A_2 = lw$	$A_3 = lw$
$A_1 = 4 \times 33$	$A_2 = 6 \times 22$	$A_3 = 11 \times 12$
$A_1 = 132$ mm²	$A_2 = 132$ mm²	$A_3 = 132$ mm²
$P_1 = 2(l + w)$	$P_2 = 2(l + w)$	$P_3 = 2(l + w)$
$P_1 = 2(4 + 33)$	$P_2 = 2(6 + 22)$	$P_3 = 2(11 + 12)$
$P_1 = 74$ mm	$P_2 = 56$ mm	$P_3 = 46$ mm

Interpret the results. Predict the dimensions of the rectangle with the least possible perimeter that has the same area.

All three rectangles have the same area but different perimeters. The perimeter of each rectangle decreases as the rectangle approaches a square, which means a square would have the least perimeter. The square with area 132 mm² has a side length of $2\sqrt{33}$.

Practice

Complete the steps to find the area and perimeter of the rectangles. Predict the dimensions of the rectangle with the least perimeter that has the same area.

4. Three rectangles with the following dimensions: 4 in. × 56 in., 7 in. × 32 in., and 8 in. × 28 in.

$A_1 = lw$	$A_2 = lw$	$A_3 = lw$
$A_1 = 4 \times 56$	$A_2 = 7 \times 32$	$A_3 = 8 \times 28$
$A_1 = $ <u>224</u> in²	$A_2 = $ <u>224</u> in²	$A_3 = $ <u>224</u> in²
$P_1 = 2(l + w)$	$P_2 = 2(l + w)$	$P_3 = 2(l + w)$
$P_1 = 2(4 + 56)$	$P_2 = 2(7 + 32)$	$P_3 = 2(8 + 28)$
$P_1 = $ <u>120</u> in.	$P_2 = $ <u>78</u> in.	$P_3 = $ <u>72</u> in.

Give the dimensions of the square with the least perimeter and the same area as the three rectangles above. <u>$4\sqrt{14}$ in.</u>

Give the dimensions of each rectangle.

5. Give the dimensions of the rectangle with the least perimeter that has an area of 200 m². <u>$10\sqrt{2}$ m</u>

6. Give the dimensions of the reS<ctangle with the least perimeter that has an area of 135 cm². <u>$3\sqrt{15}$ cm</u>

7. Give the dimensions of the rectangle with the least perimeter that has an area of 156 ft². <u>$2\sqrt{39}$ ft</u>

8. Give the dimensions of the rectangle with the least perimeter that has an area of 333 in². <u>$3\sqrt{37}$ in.</u>

Name ______________________ Date __________ Class __________

Reteaching 108
Introduction to Coordinate Space

You have worked with coordinate grids. Now you will find the coordinates of points in a three-dimensional coordinate system.

Identifying Ordered Triples

A **three-dimensional coordinate system** adds a third axis, called the ***z*-axis**, to the two-dimensional coordinate system. Coordinates of points on the three-dimensional coordinate system are given by ordered triples. An **ordered triple** is a set of three numbers used to locate and plot a point (*x*, *y*, *z*) in a three-dimensional coordinate system.

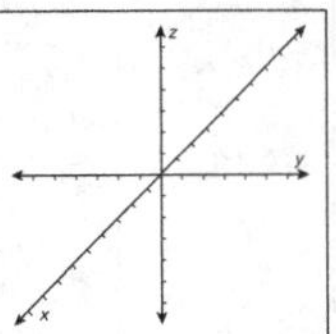

Identify the coordinates of the vertices in the given diagram.

A(0, 3, 2) → (*x*-coordinate, *y*-coordinate, *z*-coordinate)

B(−2, 3, 2)

C(−2, 5, 2)

D(0, 5, 2)

E(0, 3, 0)

F(−2, 3, 0)

G(−2, 5, 0)

H(0, 5, 0)

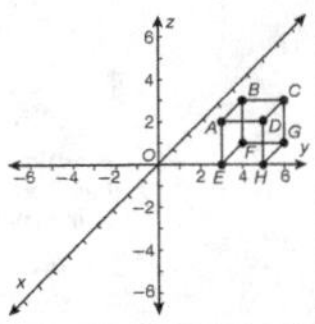

Practice

Complete the steps to identify the ordered triples.

1. *J*(4, −1, 4), *K*(2, −1, 4)

L(<u>2</u>, 7, 4), *M*(4, <u>7</u>, 3)

N(4, <u>−1</u>, <u>0</u>), *O*(2, <u>−1</u>, <u>0</u>)

P(<u>2</u>, <u>7</u>, <u>0</u>), *Q*(<u>4</u>, <u>7</u>, <u>0</u>)

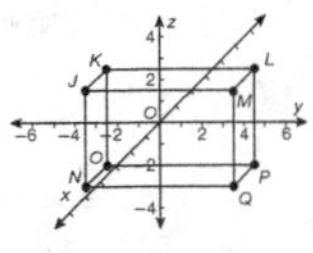

Identify the ordered triples.

2. <u>*J*(3, 0, 0), *K*(0, 0, 3), *L*(−2, 0, 0), *M*(3, 5, 0), *N*(0, 5, 3), *O*(−2, 5, 0)</u>

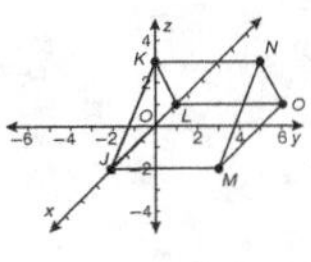

Reteaching 108
continued

Finding Collinear Points

In two dimensions, points are **collinear** when they lie on the same line. To determine whether points are collinear in three dimensions, use a **direction vector** and one point. To find a direction vector (d_1, d_2, d_3), subtract the coordinates of any two points on the line. Use a chosen point (x_0, y_0, z_0) and the direction (d_1, d_2, d_3) to write the linear equation $(x, y, z) = (x_0, y_0, z_0) + t(d_1, d_2, d_3)$ for a value *t*, where $t \in \Re$.

Determine the coordinates of a point that is collinear with the points *A*(3, 1, 2) and *B*(5, 8, 4).

$(5, 8, 4) - (3, 1, 2) = (5 - 3, 8 - 1, 4 - 2)$	Subtract.
$= (2, 7, 2)$	Simplify.
$(x, y, z) = (3, 1, 2) + t(2, 7, 2)$	Equation of the line
$(x, y, z) = (3, 1, 2) + 2(2, 7, 2)$	Substitute $t = 2$.
$= (3, 1, 2) + (4, 14, 4)$	Multiply.
$= (7, 15, 6)$	Simplify.

(7, 15, 6) is a collinear point.

Practice

Complete the steps to determine the coordinates of the collinear point.

3. *C*(3, 1, −2) and *D*(1, −2, 2), $t = 3$

(1, −2, 2) − (3, 1, −2) = (<u>−2</u>, <u>−3</u>, <u>4</u>)

(*x*, *y*, *z*) = (3, 1, −2) + 3(<u>−2</u>, <u>−3</u>, <u>4</u>)

= (3, 1, −2) + (<u>−6</u>, <u>−9</u>, <u>12</u>)

= (<u>−3</u>, <u>−8</u>, <u>10</u>)

= (−3, 0, 1) + (<u>−8</u>, <u>8</u>, <u>−2</u>)

= (<u>−11</u>, <u>8</u>, <u>−1</u>)

4. *E*(−3, 0, 1) and *F*(1, −4, 2), $t = -2$

(1, −4, 2) − (<u>−3</u>, <u>0</u>, <u>1</u>) = (<u>4</u>, <u>−4</u>, <u>1</u>)

(*x*, *y*, *z*) = (−3, 0, 1) − 2(<u>4</u>, <u>−4</u>, <u>1</u>)

Determine the coordinates of the collinear points.

5. *E*(3, 1, −2) and *F*(1, −2, 2), $t = 4$, $t = -1$

<u>(−5, −11, 14)</u>; <u>(5, 4, −6)</u>

6. *G*(−3, 0, 1) and *H*(1, −4, 2), $t = -5$, $t = 4$

<u>(−23, 20, −4)</u>; <u>(13, −16, 5)</u>

Answer Key

Name ______________________ Date __________ Class __________

Reteaching 109
Non-Euclidean Geometry

You have worked with the Cartesian coordinate system. Now you will learn to work with the polar coordinate system.

Converting Polar Coordinates

In the polar coordinate system, a point is represented by its distance r from the origin, or pole, and the measure of angle θ. The horizontal ray from the pole along the positive horizontal axis is the polar axis.

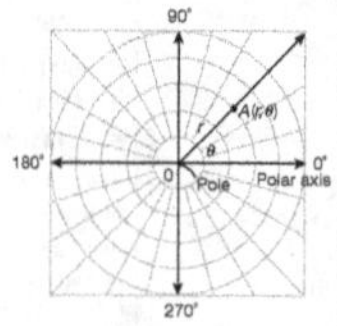

To convert polar coordinates (r, θ) into Cartesian coordinates (X, Y), use $x = r\cos\theta$ and $x = r\sin\theta$.

Convert the polar coordinates (5, 180°) into Cartesian coordinates.

$x = r\cos\theta$	Converting polar to Cartesian
$= 5\cos(180°)$	Substitute.
$= 5(-1) = (-5)$	Simplify.
$y = r\sin\theta$	Converting polar to Cartesian
$= 5\sin(180°)$	Substitute.
$= 5(0) = 0$	Simplify.

The polar coordinates (5, 180°) are converted into (−5, 0) as Cartesian coordinates.

Practice

Complete the steps to convert polar coordinates into Cartesian coordinates.

1. (−3, 90°)

$x = r\cos\theta = \underline{-3}\cos\underline{90°}$
$= \underline{-3}(\underline{0}) = \underline{0}$
$y = r\sin\theta = \underline{-3}\sin\underline{90°}$
$= \underline{-3}(\underline{1}) = \underline{-3}$
$\underline{(0, -3)}$

2. (2, 270°)

$x = r\cos\theta = \underline{2}\cos\underline{270°}$
$= \underline{2}(\underline{0}) = \underline{0}$
$y = r\sin\theta = \underline{2}\sin\underline{270°}$
$= \underline{2}(\underline{1}) = \underline{-2}$
$\underline{(0, -2)}$

Convert polar coordinates into Cartesian coordinates. Round to the nearest tenth.

3. (−4, 180°); $\underline{(4, 0)}$

4. (2, 360°); $\underline{2, 0}$

5. (2, 60°); $\underline{(1, 1.73)}$

6. (−3, 120°); $\underline{(1.5, -2.6)}$

Reteaching 109
continued

Converting Cartesian Coordinates

To convert Cartesian coordinates (x, y) into polar coordinates (r, θ), use $r^2 = x^2 + y^2$ (equation of a circle) and $\tan\theta = \frac{y}{x}$. Pay attention to the quadrant.

Convert the Cartesian coordinates (2, 4) into polar coordinates.

$r = \sqrt{x^2 + y^2}$	Equation to find r
$= \sqrt{(2)^2 + (4)^2}$	Substitute.
$= \sqrt{4 + 16} = \sqrt{20} = 2\sqrt{5}$	Simplify.
$\tan\theta = \frac{y}{x}$	Equation to find θ
$\tan\theta = \frac{4}{2} = 2$	Substitute.
$\tan^{-1}\theta \approx 63.4°$ or $243.4°$	Solve.
$(2\sqrt{5}, 63.4°)$	First quadrant, since (2, 4) is in the first quadrant

Practice

Complete the steps to convert Cartesian coordinates into polar coordinates.

7. (2, 1)

$r = \sqrt{x^2 + y^2}$
$= \sqrt{(\underline{2})^2 + (\underline{1})^2} = \sqrt{\underline{5}}$
$\tan\theta = \frac{y}{x} = \frac{\underline{1}}{\underline{2}}$
$\tan^{-1}\theta \approx \underline{26.6}°$ or $\underline{206.6}°$
$\underline{(\sqrt{5}, 26.6°)}$

8. (−3, −4)

$r = \sqrt{x^2 + y^2}$
$= \sqrt{(\underline{-3})^2 + (\underline{-4})^2} = \underline{5}$
$\tan\theta = \frac{y}{x} = \frac{\underline{-4}}{\underline{-3}} = \frac{\underline{4}}{\underline{3}}$
$\tan^{-1}\theta \approx \underline{53.1}°$ or $\underline{233.1}°$
$\underline{(5, 233.1°)}$

Convert Cartesian coordinates into polar coordinates.

9. (−1, 3); $\underline{(\sqrt{10}, 108.4°)}$

10. (−5, 2); $\underline{(\sqrt{29}, 158.2°)}$

11. (−4, 1); $\underline{(3, 90°)}$

12. (−4, 1); $\underline{(4, 180°)}$

13. (2, 2); $\underline{(2.83, 45°)}$

14. (1, 0); $\underline{(1, 0°)}$

15. (0, 15); $\underline{(15, 90°)}$

16. (3, 7); $\underline{(7.62, 67°)}$

Name ______________________ Date __________ Class __________

Reteaching 110
Scale Drawings and Maps

You have worked with dilations and scale factor. Now you will use scale to make scale drawings and models

Scale Drawings

A scale **drawing** is a drawing of an object that is smaller or larger than the object's actual size. The drawing's **scale** is the ratio of any length in the drawing to the actual length of the object.

The scale for the diagram of the doghouse is 1 in. : 36 in.

Find the width of the actual doghouse.

First convert to equivalent units.

1 in. : 36 in. (3 ft × 12 in./ft)

$\frac{1}{36} = \frac{0.75}{x}$	$\frac{\text{Diagram width}}{\text{Actual width}}$
$1x = 36(0.75)$	Cross Products Property
$x = 27$ in.	Simplify.

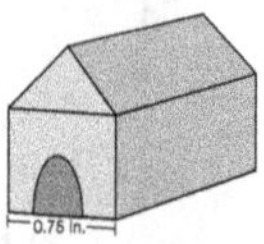

Practice

Complete the steps to find the actual length.

1. Scale is 1 cm : 10 m.

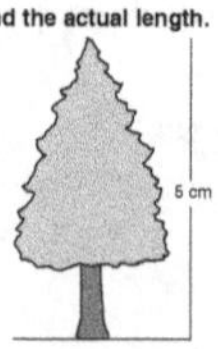

1 cm : 1000 cm
$\frac{1}{1000} = \frac{\underline{5}}{x}$
$x = \underline{(1000)(5)}$
$x = \underline{5000}$ cm $= \underline{50}$ m

2. Scale is 1 in.: 80 ft.

1 in. : $\underline{960}$ in.
$\frac{1}{\underline{960}} = \frac{\underline{6}}{x}$
$x = \underline{(960)(6)}$
$x = \underline{5760}$ in. $= \underline{480}$ ft

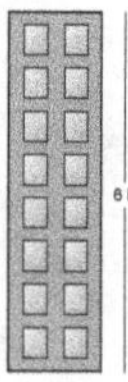

Find the actual length.

3. Scale is 1 in.: 8 ft. $\underline{40 \text{ ft}}$

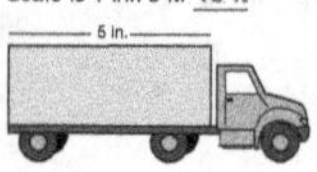

4. Scale is 1 cm: 20 m. $\underline{120 \text{ m}}$

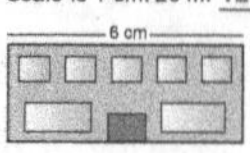

Reteaching 110
continued

Making a Scale Drawing or Model

You can represent an object that is very large or very small on paper by creating a scale drawing or model. A **scale model** is a three-dimensional model that uses scale to represent an object as smaller than or larger than the actual object.

Example:

A school building is 120 feet long, 60 feet wide, and 40 feet high.

Using a scale of 1 cm: 20 ft, determine the dimensions of the scale model.

Draw the scale model.

$\frac{x}{1\text{ cm}} = \frac{120\text{ ft}}{20\text{ ft}}$	Length
$x = 6$ cm	Solve.
$\frac{x}{1\text{ cm}} = \frac{60\text{ ft}}{20\text{ ft}}$	Width
$x = 3$ cm	Solve.
$\frac{x}{1\text{ cm}} = \frac{40\text{ ft}}{20\text{ ft}}$	Height
$x = 2$ cm	Solve.

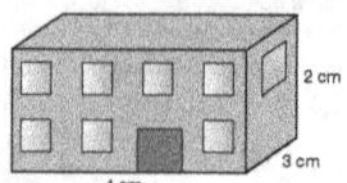

Practice

Complete the steps to determine the dimensions of the scale drawing or model. Round to the nearest tenth.

5. The earth has an approximate radius of 6378 kilometers. Determine the radius of a scale model of the earth using a scale of 1 cm : 600 km.

$\frac{1\text{ cm}}{600\text{ km}} = \frac{x}{\underline{6378\text{ km}}}$
$600x = \underline{6378}$
$x = \underline{10.63} \approx \underline{10.6}$ cm

6. Two cities are approximately 155 miles apart. A map uses a scale of 1 cm : 40 mi Determine the distance between the two cities on the map.

$\frac{1\text{ cm}}{40\text{ mi}} = \frac{x}{\underline{155\text{ mi}}}$
$40x = \underline{155}$
$x = \underline{3.875} \approx \underline{3.9}$ cm

Determine the dimensions of the scale drawing or model. Round to the nearest tenth.

7. The Statue of Liberty is approximately 151 feet tall from base to torch(151 ft 1 in.). Determine the height on a scale drawing whose scale is 1 cm: 18 ft. $\underline{8.4 \text{ cm}}$

8. A spherical molecule has a radius of 140 picometers (pm). Determine the diameter of the scale model of the molecule whose scale is 1 in: 20 pm. $\underline{14 \text{ in.}}$

9. Two friends are 385 meters apart. Determine the distance between them on a map whose scale is 1 cm: 40 m. $\underline{9.6 \text{ cm}}$

Name ______________________ Date __________ Class __________

Reteaching INV 11
Golden Ratio

You have investigated fractals. Now you will investigate the golden ratio.

The Golden Ratio in Line Segments

The **golden ratio** is the irrational number $\varphi = \frac{1+\sqrt{5}}{2}$.

On the line segment to the right, since the ratio of the whole segment to the longer piece, $\frac{c}{b}$, equals the ratio of the longer piece to the shorter piece, $\frac{b}{a}$, both ratios are equal to the golden ratio, φ.

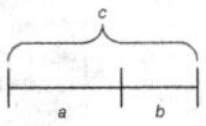

Find the value of a if $\frac{a}{89} = \frac{a+89}{a}$. Round to the nearest hundredth.

Step 1: Since $\frac{a}{89} = \frac{a+89}{a}$, $\frac{a}{89}$ and $\frac{a+89}{a}$ are both equal to the golden ratio, φ.

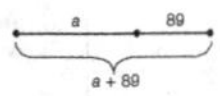

Step 2: Set $\frac{a}{89}$ equal to φ and use $\varphi = \frac{1+\sqrt{5}}{2}$.

$\frac{a}{89} = \frac{1+\sqrt{5}}{2}$ Substitute.

$a = 89\left(\frac{1+\sqrt{5}}{2}\right)$ Solve.

$a \approx 144.01$

Practice

In the problem below, the ratios of the line segments equal the golden ration. Find the value of x to the nearest hundredth.

1. $\frac{x}{13} = \varphi$ Golden ratio

   $\frac{x}{13} = \frac{1+\sqrt{5}}{2}$ Substitute.

   $2x = 13(1+\sqrt{5})$ Solve.

   $x = 21.03$ Round to the nearest hundredth.

2. The ratios of the line segments equal the golden ratio. Find the value of x to the nearest hundredth. $x = 89$

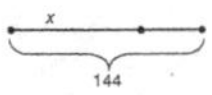

3. If the longer piece of a line segment is 45, and if the shorter piece is 18, does the line segment represent the golden ration? Explain your answer.

   No, the ratio of 45 to 18 is 2.5, which is not equal to the golden ratio.

Reteaching INV 11
continued

The Golden Rectangle

In a **golden rectangle,** the ratio of the length of the longer side to the length of the shorter side is the golden ratio, φ.

The rectangle pictured on the right is a golden rectangle because the equation below is true.

$\frac{a}{b} = \varphi$

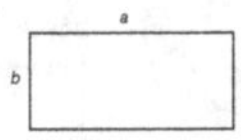

Find the value of t on the golden rectangle shown below. Round to the nearest hundredth.

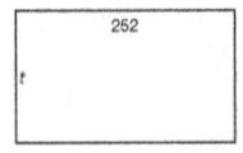

Step 1: Use the definition of a golden rectangle to get the equation $\frac{252}{t} = \varphi$.

Step 2: Solve the equation above for t.

$\frac{252}{t} = \frac{1+\sqrt{5}}{2}$ Substitute.

$2(252) = t(1+\sqrt{5})$ Solve.

$t \approx 155.74$ Round to the nearest hundredth.

Practice

Find the value of x on the golden rectangle shown below. Round to the nearest hundredth.

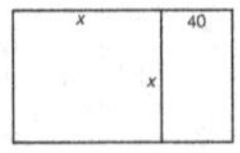

4. $\frac{x}{40} = \varphi$ Golden ratio

   $\frac{x}{40} = \frac{1+\sqrt{5}}{2}$ Substitute.

   $2x = 40(1+\sqrt{5})$ Solve.

   $x \approx 64.72$ Round to the nearest hundredth.

5. Use the golden rectangle to find the value of x to the nearest hundredth.

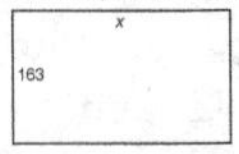

   $x = 263.74$

6. Can a square be a golden rectangle? Explain your answer.

   No, the ratio of two sides of a square is equal to 1, which is not the golden ratio.

Name ______________________ Date __________ Class __________

Reteaching 111
Finding Distance and Midpoint in Three Dimensions

You have worked with distance and midpoint. Now you will find distance and midpoint in three dimensions.

Distance Formula in Three Dimensions

Given two points, $A(x_1, y_1, z_1)$ and $B(x_2, y_2, z_2)$, and in three-dimensional space, the distance between the two points is determined by the equation

$d = \sqrt{(x_2-x_1)^2 + (y_2-y_1)^2 + (z_2-z_1)^2}$.

Example:

Find the distance between $A(3,1,1)$ and $B(2, -2, -3)$.

$d = \sqrt{(x_2-x_1)^2 + (y_2-y_1)^2 + (z_2-z_1)^2}$ Distance Formula for Three Dimensions

$AB = \sqrt{(2-3)^2 + (-2-1)^2 + (3-1)^2}$ Substitute

$= \sqrt{(-1)^2 + (-3)^2 + (-4)^2}$

$= \sqrt{1+9+16}$ Simplify

$= \sqrt{26}$

The distance between AB is $\sqrt{26}$.

Practice

Complete the steps to find the distance between the points.

1. $C(-3, 2, 1)$ and $D(-4, -3, 2)$

   $CD = \sqrt{(x_2-x_1)^2 + (y_2-y_1)^2 + (z_2-z_1)^2}$

   $= \sqrt{(-4-(-3))^2 + (-3-2)^2 + (2-0)^2}$

   $= \sqrt{1+25+4}$

   $= \sqrt{30}$

2. $E(0, -5, 3)$ $F(3, -1, -3)$

   $EF = \sqrt{(x_2-x_1)^2 + (y_2-y_1)^2 + (z_2-z_1)^2}$

   $= \sqrt{(3-0)^2 + (-1-(-5))^2 + (-3-3)^2}$

   $= \sqrt{9+16+36}$

   $= \sqrt{61}$

Find the distance between the points.

3. $G(-5, -1, -2)$ and $H(-3, -3, -3)$ 3
4. $J(2, -5, -1)$ and $K(0, 3, 6)$ $3\sqrt{13}$
5. $M(3, -4, -1)$ and $N(0, -5, 2)$ $3\sqrt{11}$
6. $P(2, 5, 1)$ and $Q(-2, 0, 3)$ $3\sqrt{5}$

Reteaching 111
continued

Midpoint Formula in Three Dimensions

Given two points, $A(x_1, y_1, z_1)$ and $B(x_2, y_2, z_2)$, in three-dimensional space, the midpoint M is determined by the equation $M = \left(\frac{x_1+x_2}{2}, \frac{y_1+y_2}{2}, \frac{z_1+z_2}{2}\right)$.

The endpoints of $\overline{AB}$ are $A(5, 2, (-2)$ and $B(1, 4, 4)$. Find the midpoint.

$M = \left(\frac{x_1+x_2}{2}, \frac{y_1+y_2}{2}, \frac{z_1+z_2}{2}\right)$ Midpoint Formula for Three Dimensions

$= \left(\frac{5+1}{2}, \frac{2+4}{2}, \frac{-2+4}{2}\right)$ Substitute

$= \left(\frac{6}{2}, \frac{6}{2}, \frac{2}{2}\right)$

$= (3, 3, 1)$ Simplify

Practice

Complete the steps to find the midpoint.

7. $C(3, -3, 2)$ $D(-5, -1, 0)$

   $M = \left(\frac{x_1+x_2}{2}, \frac{y_1+y_2}{2}, \frac{z_1+z_2}{2}\right)$

   $= \left(\frac{3+(-5)}{2}, \frac{-3+1}{2}, \frac{2+0}{2}\right)$

   $= (-1, -1, 1)$

8. $E(-5, -3, 2)$ and $F(4, 0, 4)$

   $M = \left(\frac{x_1+x_2}{2}, \frac{y_1+y_2}{2}, \frac{z_1+z_2}{2}\right)$

   $= \left(\frac{-5+4}{2}, \frac{-3+0}{2}, \frac{2+4}{2}\right)$

   $= (-0.5, -1.5, 3)$

Find the midpoint.

9. $G(-4, -6, 0)$ and $H(-3, 2, -2)$ $(-3.5, -2, -1)$
10. $J(5, -5, 6)$ $K(3, -4, 8)$ $(4, -4.5, 7)$
11. $P(-3, 7, 2)$ and $Q(-5, -5, -5)$ $(-4, 1, -1.5)$
12. $S(3, 10, -9)$ $T(15, -8, -11)$ $(9, 1, -10)$
13. $V(-18, 13, 7)$ and $W(24, 6, -12)$ $(3, 9.5, -5)$
14. $X(-11, 15, 17)$ $Q(6, -13, 22)$ $(-2.5, 1, 19.5)$

Answer Key

Name ______________________ Date __________ Class __________

Reteaching 112
Finding Areas of Circle Segments

You have worked with area of a circle. Now you will find the area of circle segments.

Area of Circle Segment

A **segment of a circle** is a region bounded by an arc and its chord.

Area of segment ABC = area of sector ABC − area of ΔABC.

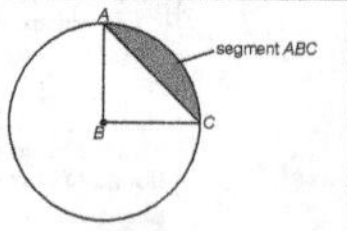

Find the area of the segment to the nearest hundredth.

$A = \pi r^2\left(\frac{m^\circ}{360^\circ}\right)$	Area of sector
$= \pi(2)^2\left(\frac{90^\circ}{360^\circ}\right)$	Substitute.
$= \pi \text{ in}^2$	Solve.
$A = \frac{1}{2}bh$	Area of triangle
$= \frac{1}{2}(2)(2) = 2 \text{ in}^2$	Substitute.
$A_{segment} = A_{sector} - A_{triangle}$	Area of segment
$= \pi - 2$	Substitute.
$\approx 1.14 \text{ in}^2$	Solve.

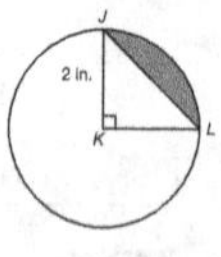

Practice

Complete the steps to find the area of the segment to the nearest hundredth.

1. $A_{sector} = \pi r^2\left(\frac{m^\circ}{360^\circ}\right)$

$= \pi(9)^2\left(\frac{\underline{90^\circ}}{\underline{360}}\right) = \underline{\frac{81}{4}}\,\pi \text{ in}^2$

$A_{triangle} = \frac{1}{2}(9)(9) = \underline{\frac{81}{2}} \text{ in}^2$

$A_{segment} = \underline{\frac{81}{4}\pi - \frac{81}{2}} \approx \underline{23.12} \text{ in}^2$

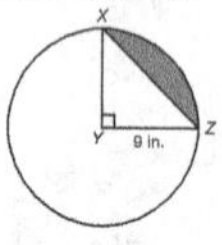

Find the area of the segment to the nearest hundredth.

2. <u>5.8</u> cm^2

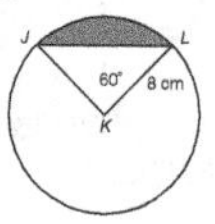

3. <u>116.98</u> ft^2

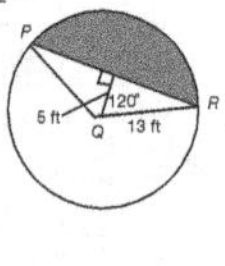

Reteaching 112
continued

Using Trigonometry to Find Segment Area

You can use trigonometry to find the area of a circle segment. First, determine the sector area. Then, use a trigonometric function to find the height of the triangle bounded by the chord and two radii. Next, use the height of the triangle to find the area of the triangle. Finally, find the segment area.

A circle has a radius of 6.4 centimeters. Determine the area of the segment formed by a chord with central angle 82°. Give your answer to the nearest hundredth.

$A_{sector} = \frac{m^\circ}{360^\circ}\pi r^2$	Area of sector
$= \frac{82}{360}\pi(6.4)^2$	Substitute.
$\approx 29.31 \text{ cm}^2$	Solve.
$h_{triangle} = 6.4 \sin 82^\circ$	Height of triangle
$A_{triangle} = \frac{1}{2}(6.4)(6.4 \sin 82^\circ)$	Area of triangle
$\approx 20.28 \text{ cm}^2$	Solve.
$A_{segment} = A_{sector} - A_{triangle}$	Area of segment
$\approx 29.31 - 20.28$	Substitute.
$\approx 9.03 \text{ cm}^2$	Solve.

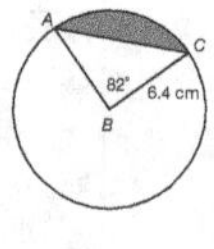

Practice

Complete the steps to find the area of the circle segment to the nearest hundredth.

4. $A_{sector} = \frac{48^\circ}{360^\circ}\pi(\underline{9.2})^2$

$\approx \underline{35.45} \text{ in}^2$

$h_{triangle} = \underline{9.2 \sin 48^\circ}$

$A_{triangle} = \frac{1}{2}(\underline{9.2})(\underline{9.2 \sin 48^\circ})$

$\approx \underline{31.45} \text{ in}^2$

$A_{segment} = \underline{35.45 - 31.45} \approx \underline{4.0} \text{ in}^2$

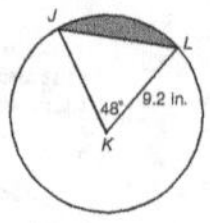

Find the area of the segment to the nearest hundredth.

5. <u>219.07</u> cm^2

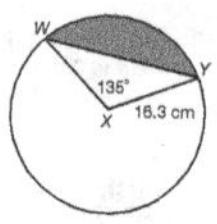

6. <u>5.36</u> in^2

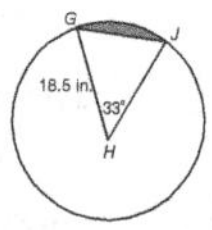

Name ______________________ Date __________ Class __________

Reteaching 113
Symmetry of Solids and Polyhedra

You have worked with symmetry of two-dimensional objects. Now you will examine plane and rotational symmetry in three-dimensional solids.

Plane Symmetry

Plane symmetry describes a three-dimensional solid that can be divided into two congruent reflected halves by a plane.

Determine whether the cube has plane symmetry. If it does, show the symmetry in a drawing.

Yes, the cube has plane symmetry. A vertical plane is one of several planes that will divide the cube into two halves.

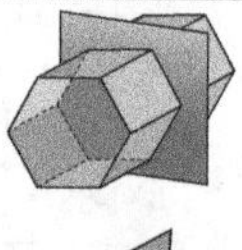

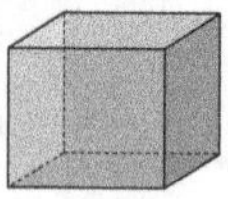

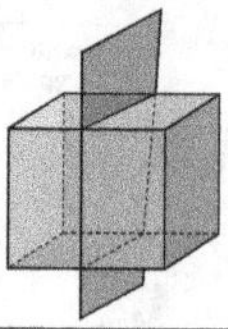

Practice

Determine whether the solid has plane symmetry. If it does, show the symmetry in a drawing.

1. <u>yes</u>

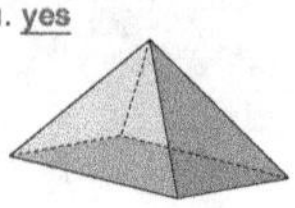

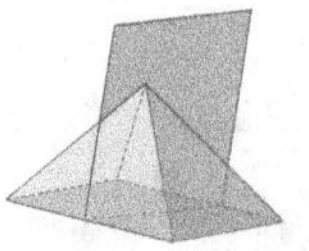

Determine whether the solid has plane symmetry.

2. <u>yes</u>

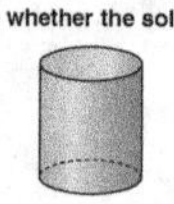

3. <u>no</u>

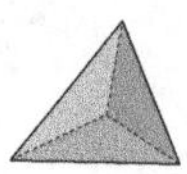

Reteaching 113
continued

Rotational Symmetry

Rotational symmetry, or **symmetry about an axis,** describes a three-dimensional solid that can be rotated about a line so that the image coincides with the preimage.

The **angle of rotational symmetry** is the angle at which the image coincides with the preimage. For a cube, the angle of rotational symmetry is 90°.

The **order of rotational symmetry** is the number of times the image coincides with the preimage. For a cube, the order of rotation is 4.

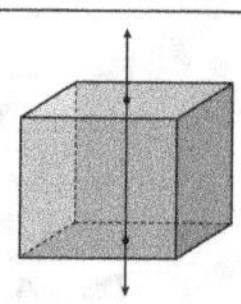

Determine whether the rectangular pyramid has rotational symmetry. If it does, determine the order of rotational symmetry and the angle of rotational symmetry.

Yes, the rectangular pyramid has rotational symmetry.

The order of symmetry = 2.

The angle of symmetry = $\frac{360^\circ}{2}$ = 180°.

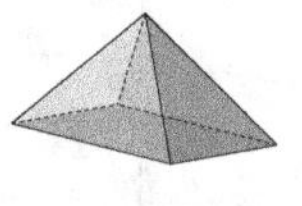

Practice

Complete the steps to determine the order of symmetry and the angle of symmetry (if the solid has rotational symmetry).

4. rotational symmetry = <u>yes</u>

order of symmetry = <u>5</u>

angle of symmetry = $\frac{360^\circ}{\underline{5}}$ = <u>72°</u>

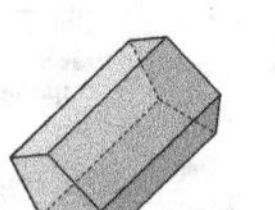

Determine whether the solid has rotational symmetry. If it does, determine the order of symmetry and the angle of symmetry.

5. rotational symmetry = <u>yes</u>

order of symmetry = <u>2</u>

angle of symmetry = <u>180°</u>

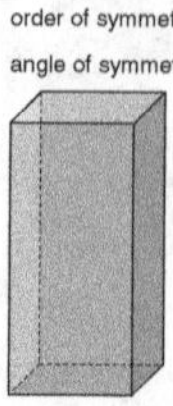

6. rotational symmetry = <u>yes</u>

order of symmetry = <u>6</u>

angle of symmetry = <u>60°</u>

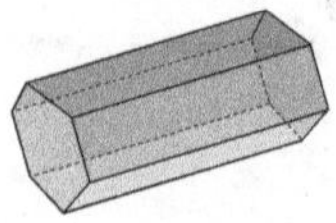

Answer Key

Name ______________________ Date __________ Class __________

Reteaching 114
Solving and Graphing Systems of Inequalities

You have worked with systems of inequalities on number lines. Now you will deal with systems of inequalities in two dimensions.

Systems of Inequalities: "and"

When graphing a pair of inequalities involving "and," the solution will involve an intersection of the two inequalities.

Graph the region described by the inequalities $y \le x + 3$ and $y \ge 2x + 2$.

Since the inequalities are joined by "and," the solution will be the intersection of the two inequalities.

First, graph the boundary lines $y = x + 3$ and $y = 2x + 2$.

Use solid lines, since the inequalities $\le$ and $\ge$ include "equal to."

Use a test point such as $(-2, 0)$ to determine the region that is included in each inequality.

$-2 \le 0 + 3$ is true, so the region that includes $(-2, 0)$ is the solution to $y \le x + 3$.

Since $0 \ge 2(-2) + 2$, the region that includes $(-2, 0)$ is the solution to $y \ge 2x + 2$.

The overlapping region is the solution to the system of inequalities.

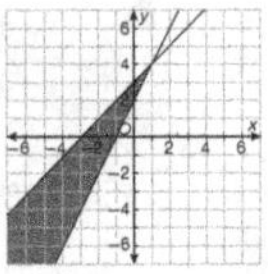

Practice

Complete the steps to graph the region described by the inequalities $y \ge x - 4$ and $y \le 3x + 1$.

1. Graph the solid lines $y = x - 4$ and $y = 3x + 1$.

   Is the point (0, 0) a solution to $y \ge x - 4$? yes

   Is the point (0, 0) a solution to $y \le 3x + 1$? yes

   Shade the region above the line $y = x - 4$ and below the line $y = 3x + 1$.

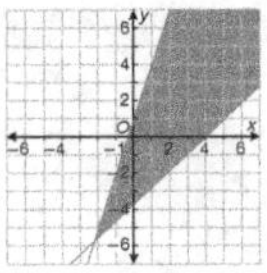

Graph the region described by the inequalities.

2. $y > 2x - 3$ and $y < 2x + 1$

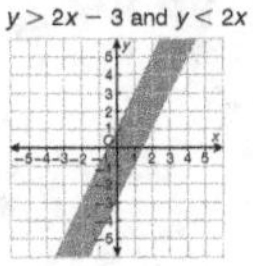

3. $y < x - 1$ and $y > 4x - 4$

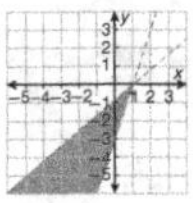

Systems of Inequalities: "or"

When graphing a pair of inequalities involving "or", the solution will involve a union of the two inequalities.

Graph the region described by the inequalities $y < x - 3$ or $y > -x + 2$.

Since the inequalities are joined by "or," the solution will be the union of the two inequalities.

First, graph the boundary lines $y = x - 3$ and $y = -x + 2$.

Use dashed lines, since the inequalities $<$ and $>$ do not include "equal to."

Use a test point such as (0, 0) to determine the region that is included in each inequality.

Since $0 < 0 - 3$ is not true, the region below $y = x - 3$ is the solution to $y < x - 3$.

Since $0 > -0 + 2$ is not true, the region above $y = -x + 2$ is the solution to $y > -x + 2$.

The region for both inequalities is the solution to the system of inequalities.

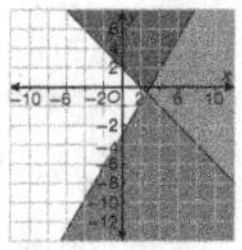

Practice

Complete the steps to graph the region described by the inequalities $y \ge x - 3$ or $y \ge 2x + 3$.

4. Graph the solid lines $y = x - 3$ and $y = 2x + 3$.

   Is the point (0, 0) a solution to $y \ge x - 3$? yes

   Is the point (0, 0) a solution to $y \ge 2x + 3$ no

   Shade the region above the line $y = x - 3$.

   Shade the region above the line $y = 2x + 3$.

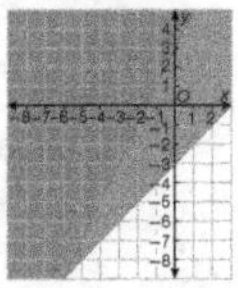

Graph the region described by the inequalities.

5. $y > 4 - x$ or $y < 3x - 5$

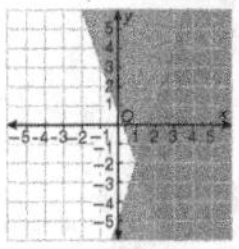

6. $y < 3 - 2x$ or $y \ge 4x - 1$

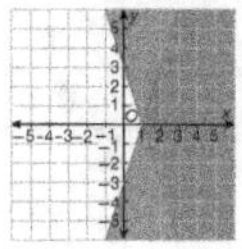

Name ______________________ Date __________ Class __________

Reteaching 115
Finding Volumes of Composite Solids

Finding Volumes of Composite Solids: Adding

The volume of a composite solid is the sum of the volumes of the individual solids that make up the composite solid.

Find the volume of the solid. Assume that prisms are right prisms.

$V = lwh$	Volume of rectangular prism
$= (16)(12)(8)$	Substitute.
$= 1536 \text{ cm}^3$	Solve.
$V = Bh$	Volume of triangular prism
$= \frac{1}{2}(16)(10)(12)$	Substitute.
$= 960 \text{ cm}^3$	Solve.
$V = 1536 + 960$	Add volumes.
$= 2496 \text{ cm}^3$	Solve.

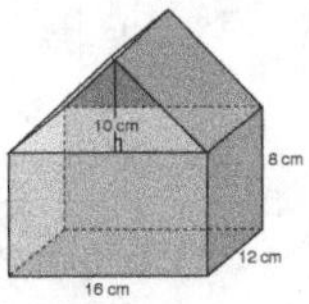

Practice

Complete the steps to find the volume of the solid by adding. Round to the nearest tenth of a cubic unit.

$V_{cube} = s^3$
$= 15^3$
$= 3375 \text{ in}^3$

1. $V_{cylinder} = \pi r^2 h$
$= \pi(3)^2(8)$
$\approx 226.2 \text{ in}^3$
$V \approx 3375 + 226.2$
$\approx 3601.2 \text{ in}^3$

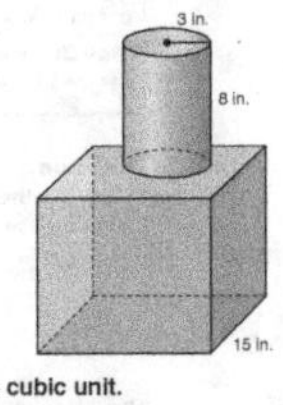

Find the volume of the solid. Round to the nearest tenth of a cubic unit.

2. 3242.1 cm^3

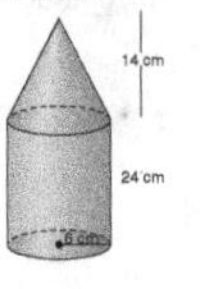

3. 1736 m^3

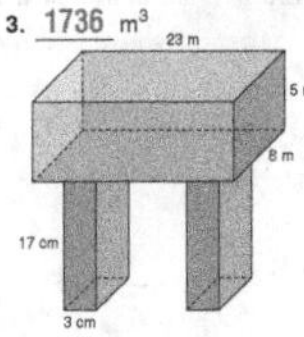

Finding Volumes of Composite Solids: Subtracting

When a larger solid is missing a portion whose volume is equivalent to that of a smaller solid, you can subtract the volume of the space from the total volume.

Find the volume of the solid. Round your answer to the nearest tenth of a cubic unit.

$V = s^3$	Volume of outer cube
$= 24^3$	Substitute.
$= 13{,}824 \text{ cm}^3$	Solve.
$V = s^3$	Volume of inner cube
$= 16^3$	Substitute.
$= 4096 \text{ cm}^3$	Solve.
$V = 13{,}824 - 4096$	Subtract volumes.
$= 9728 \text{ cm}^3$	Solve.

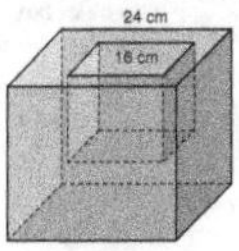

Practice

Complete the steps to find the volume of the solid. Round to the nearest tenth of a cubic unit.

$V_{prism} = lwh$
$= (40)(22)(18)$
$= 15{,}840 \text{ in}^3$

4. $V_{cylinder} = \pi r^2 h$
$= \pi(6)^2(18)$
$\approx 2035.8 \text{ in}^3$
$V = 15{,}840 - 2035.8$
$\approx 13{,}804.2 \text{ in}^3$

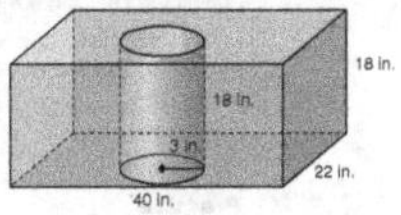

Find the volume of the solid. Round to the nearest tenth of a cubic unit.

5. 27,397.8 m^3

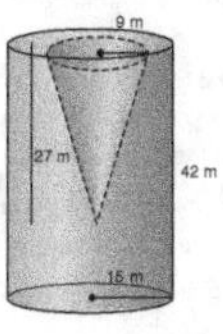

6. 18,150.4 cm^3

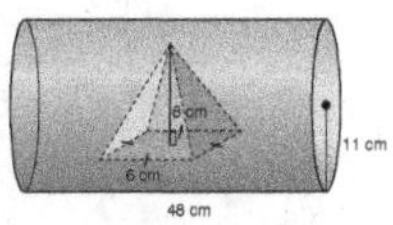

Answer Key

Name ______________________ Date __________ Class __________

Reteaching 116
Secant, Cosecant, and Cotangent

You have worked with three basic trigonometric functions called sine, cosine, and tangent. Now you will work with the reciprocals of those functions.

Trigonometric Ratios

cosecant: $\csc\theta = \frac{1}{\sin\theta} = \frac{\text{hypotenuse}}{\text{side opposite }\theta}$; $\sin\theta \neq 0$

secant: $\sec\theta = \frac{1}{\cos\theta} = \frac{\text{hypotenuse}}{\text{side adjacent }\theta}$; $\cos\theta \neq 0$

cotangent: $\cot\theta = \frac{1}{\tan\theta} = \frac{\text{side adjacent }\theta}{\text{side opposite }\theta}$; $\tan\theta \neq 0$

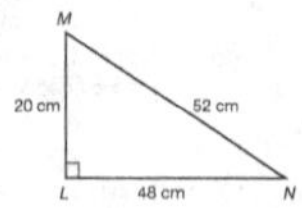

Example:

Find the ratio for cosecant, secant, and cotangent.

$\csc N = \frac{1}{\sin N} = \frac{\text{hypotenuse}}{\text{side opposite }N} = \frac{52}{20}$

$\sec N = \frac{1}{\cos N} = \frac{\text{hypotenuse}}{\text{side adjacent }N} = \frac{52}{48}$

$\cot N = \frac{1}{\tan N} = \frac{\text{side adjacent }N}{\text{side opposite }N} = \frac{48}{20}$

Practice

Complete the statement to determine the given trigonometric ratio.

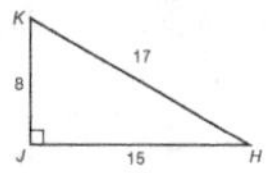

1. $\csc K = \frac{\text{hypotenuse}}{\text{side opposite }\angle K} = \frac{17}{15}$

$\sec K = \frac{\text{hypotenuse}}{\text{side adjacent }\angle K} = \frac{17}{8}$

$\cot K = \frac{\text{side adjacent }\angle K}{\text{side opposite }\angle K} = \frac{8}{15}$

Use the diagram from Exercise 1 to write each trigonometric ratio as a fraction.

2. $\csc H = \frac{17}{8}$ 3. $\cot H = \frac{15}{8}$ 4. $\sec H = \frac{17}{15}$

Write each trigonometric ratio as a fraction.

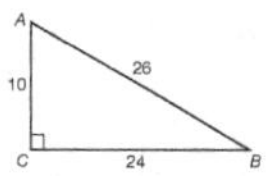

5. $\csc A = \frac{26}{24}$ 6. $\cot B = \frac{24}{10}$

7. $\cot A = \frac{10}{24}$ 8. $\csc B = \frac{26}{10}$

Reteaching 116
continued

Trigonometric Ratios (continued)

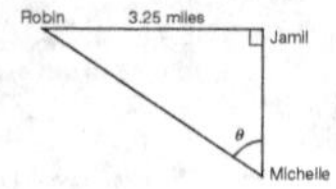

Example: The distance between Robin's home and Jamil's home is 3.25 miles. The cosecant of θ equals 3.12. How far does Michelle live from Robin and Jamil? Round your answer to the nearest tenth.

Step 1: Use the cosecant ratio to find the distance between Robin's home and Michelle's home.

$\csc\theta = \frac{1}{\sin\theta} = \frac{\text{hypotenuse}}{\text{opposite}}$

$3.12 = \frac{h}{3.25}$ Substitute.

$h = 3.12(3.25)$ Multiply each side by 3.25.

$h \approx 10.14$ miles Simplify.

Step 2: Use the Pythagorean theorem to find the distance between Jamil's home and Michelle's home.

$a^2 + b^2 = c^2$

$3.25^2 + x^2 = 10.14^2$

$x^2 = 92.2571$

$x \approx 9.61$ miles

Step 3: State the answer.

Michelle lives 10.14 miles from Robin and 9.61 miles from Jamil.

Practice

Complete the steps to solve the problem.

9. Sun and Webster live 3.02 miles apart. Find how far Trina lives from Sun and from Webster if the $\cot\theta = 1.23$. Round your answer to the nearest hundredth.

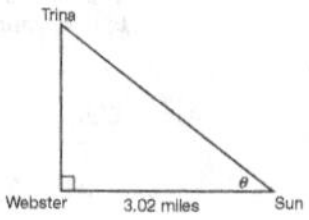

$\cot\theta = \frac{1}{\tan\theta} = \frac{\text{side adjacent }\theta}{\text{side opposite }\theta}$

$1.23 = \frac{3.02}{x}$ $a^2 + b^2 = c^2$

$1.23x = 3.02$ $3.02^2 + 2.46^2 = c^2$

$x \approx 2.46$ miles $15.172 = c^2$

$3.90 \approx c$

Trina lives 3.90 miles from Sun and 2.46 miles from Webster.

Name ______________________ Date __________ Class __________

Reteaching 117
Determining Line of Best Fit

You have worked with equations of lines. Now you will determine the line of best fit for a set of data.

Choosing a Line of Best Fit

A **scatter plot** has points plotted to show a relationship between two sets of data.

The **line of best fit** is a line that comes closest to all the points in the data set.

Example:

Explain why the line in each graph is, or is not, a good choice for the line of best fit.

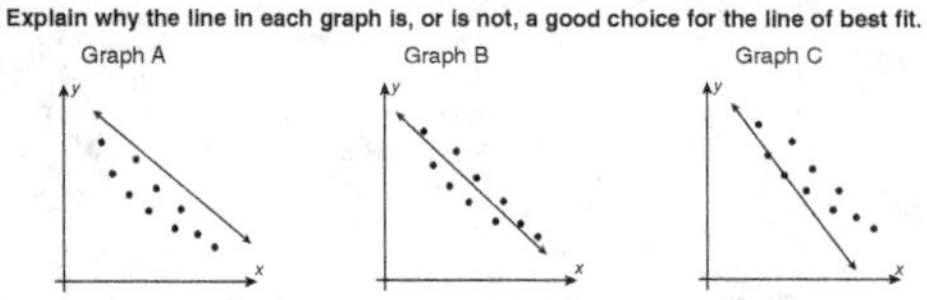

Graph A: The line has the correct slope but is not drawn as close to the data points as possible. This is not a good choice for the line of best fit.

Graph B: The line goes through the data points and has a slope that represents the trend in the data. This is a good choice for the line of best fit.

Graph C: The line goes through the data but does not have a slope that represents the trend in the data. This is not a good choice for the line of best fit.

Practice

Complete the statements to explain which line represents the line of best fit.

1. Graph A

Graph B

Graph A: This is a (good/bad) good choice for the line of best fit. The line (does/doesn't) does go through the middle of the data points. The line (does/doesn't) does have a slope that represents the trend in the data.

Graph B: This is a (good/bad) bad choice for the line of best fit. The line (does/doesn't) doesn't go through the middle of the data points. The line (does/doesn't) does have the correct slope.

Reteaching 117
continued

Drawing a Line of Best Fit

Positive Correlation	Negative Correlation	No Correlation
Both quantities increase together.	One quantity increases as the other decreases.	Data points are randomly scattered.
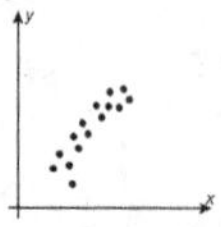	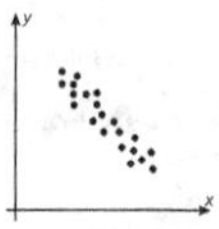	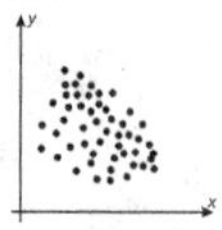

Drawing a line of best fit through the points in a scatterplot requires patience. The line should pass through as many points as possible, with the other points as close as possible to the line so that there are the same number of data points above the line as below the line.

Example:

Draw the line of best fit in the scatterplot. Then, describe the correlation in the data.

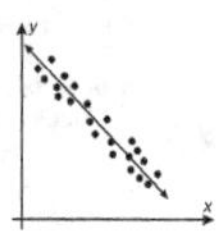

Step 1: The line is drawn with as many points above as points below.

Step 2: One quantity increases as the other decreases, so it is a strong negative correlation.

Practice

Complete the steps to draw the line of best fit. Describe the correlation.

2. Draw the line.

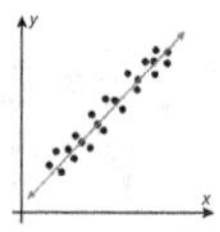

The correlation is strongly positive.

Draw the line of best fit, if one exists, and describe the correlation.

3. moderately positive correlation 4. no correlation

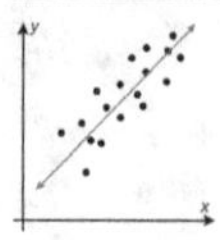

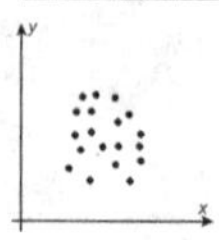

Name ______________________ Date ____________ Class ____________

Reteaching 118
Finding Areas of Polygons Using Matrices

You have worked with area of polygons and of matrices. Now you will find the determinant of a matrix and use it to find area.

Determinant of a Matrix

The **determinant of a matrix** is the difference between the product of the first diagonal and that of the other diagonal. Given $A = \begin{bmatrix} x_1 & y_1 \\ x_2 & y_2 \end{bmatrix}$, where x_1, x_2, y_1, and y_2 are constants, the determinant of A is given by the formula det $A = x_1y_2 - x_2y_1$.

Calculate the determinant of $A = \begin{bmatrix} 5 & 2 \\ 8 & 4 \end{bmatrix}$.

det $A = x_1y_2 - x_2y_1$ Determinant formula
$= (5)(4) - (8)(2)$ Substitute.
$= 20 - 16$ Simplify.
$= 4$

Practice

Complete the steps to calculate the determinant.

1. $A = \begin{bmatrix} 3 & 1 \\ 5 & 4 \end{bmatrix}$

det $A = x_1y_2 - x_2y_1$
$= (3)(4) - (5)(1)$
$= 12 - 5$
$= 7$

2. $A = \begin{bmatrix} 2 & -1 \\ -4 & 3 \end{bmatrix}$

det $A = x_1y_2 - x_2y_1$
$= (2)(3) - (4)(-1)$
$= 6 - 4$
$= 2$

Calculate the determinant.

3. $A = \begin{bmatrix} -2 & 0 \\ 3 & 1 \end{bmatrix}$ -2
4. $A = \begin{bmatrix} -5 & -3 \\ -6 & 2 \end{bmatrix}$ -28
5. $A = \begin{bmatrix} 3 & 5 \\ 0 & 0 \end{bmatrix}$ 0
6. $A = \begin{bmatrix} 2 & 0 \\ 0 & -3 \end{bmatrix}$ -6
7. $A = \begin{bmatrix} 4 & 2 \\ 2 & 4 \end{bmatrix}$ 12
8. $A = \begin{bmatrix} -1 & 2 \\ -3 & 3 \end{bmatrix}$ 3

Reteaching 118
continued

Matrix Method of Computing Area of a Polygon

Given a complex polygon with n vertices from (x_1, y_1) to (x_n, y_n) around the polygon, the area of the polygon is equal to one-half of the determinant of the matrix listing the coordinates vertically.

The area of the polygon is given by the formula

$A = \frac{1}{2}(x_1y_2 + x_2y_3 + \cdots + x_ny_1 - y_1x_2 - y_2x_3 - \cdots - y_nx_1)$.

Calculate the area of A rectangle with vertices (0, 2), (2, 2), (2, 8) and (0, 8).

$A = \frac{1}{2}(x_1y_2 + x_2y_3 + x_3y_4 + x_4y_1 - y_1x_2 - y_2x_3 - y_3x_4 - y_4x_1)$ Matrix formula for area.
$= \frac{1}{2}((0)(2) + (2)(8) + (2)(8) + (0)(2) - (2)(2) - (2)(2) - (8)(0) - (8)(0))$ Substitute.
$= \frac{1}{2}(0 + 16 + 16 + 0 - 4 - 4 - 0 - 0)$ Simplify.
$= \frac{1}{2}(24)$ Simplify.
$= 12$ Solve.
$|12| = 12$ sq. units Absolute value of area

Practice

Complete the steps to calculate the area of the polygon.

9. square with vertices (3, 3), (3, 7), (7, 7), and (7, 3)

$A = \frac{1}{2}(x_1y_2 + x_2y_3 + x_3y_4 + x_4y_1 - y_1x_2 - y_2x_3 - y_3x_4 - y_4x_1)$
$= \frac{1}{2}((3)(7) + (3)(7) + (7)(3) + (7)(3) - (3)(3) - (7)(7) - (7)(7) - (3)(3))$
$= \frac{1}{2}(21 + 21 + 21 + 21 - 9 - 9 - 49 - 9)$
$= \frac{1}{2}(-31)$
$= -16$

$|-16| = 16$ sq. units

Calculate the area of the polygon.

10. Pentagon with vertices (3, 3), (4, 2), (4, −1), (1, 1), and (0, 3)

9.5 sq. units

11. Hexagon with vertices (2, 5), (3, 4), (2, −3), (0, −3), (−1, 1), and (0, 5)

24 sq. units

Name ______________________ Date ____________ Class ____________

Reteaching 119
Platonic Solids

Characteristics of Platonic Solids

A **regular polyhedron** is a polyhedron whose faces are all congruent polygons, and for which the same number of faces meet at each vertex.

A **Platonic solid** is one of five regular polyhedra: a tetrahedron, a cube, an octahedron, a dodecahedron, or an icosahedron.

Regular Polyhedron	Vertices	Edges	Faces	Diagram
Tetrahedron equilateral faces	4	6	4	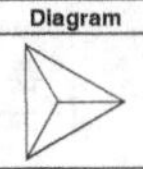
Cube square faces	8	12	6	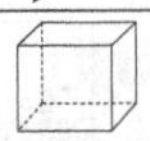
Octahedron equilateral faces	6	12	8	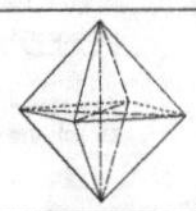
Dodecahedron pentagon faces	20	30	12	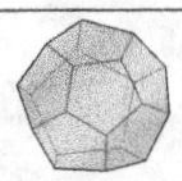
Icosahedron equilateral faces	12	30	20	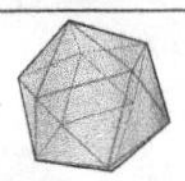

Practice

1. Each vertex is the meeting point of an equilateral triangle.

The measure of each angle is 60°. 60° × 4 = 240°

Determine the sum of the measures of the angles at a vertex.

2. tetrahedron 180°
3. icosahedron 300°

Reteaching 119
continued

More About Platonic Solids

Only three shapes can be used as the faces of Platonic solids: triangles, squares, and pentagons.

Example:

Explain why regular hexagons cannot be used to construct a Platonic solid.

The interior angles of a hexagon each measure 120°. If three angles of a hexagon meet at a vertex, they have a sum of 3 × 120° = 360°.

This would result in a tessellation, not in a polyhedron.

Practice

Complete the steps to answer the questions about Platonic solids.

4. Determine the sum of the interior angles for a vertex in a dodecahedron.

Each face is a pentagon.

The number of faces = 12.

The measure of each angle = 108°.

Sum of interior angles = 108° × 3 = 324°.

Answer the questions about Platonic solids.

5. Determine the sum of the interior angles for a vertex in a cube. 270°
6. How many more faces does an icosahedron have than an octahedron? 12
7. What is the total number of faces, vertices, and edges on a dodecahedron? 62
8. What type of shape is a tetrahedron? triangular pyramid
9. Draw a net of a cube.

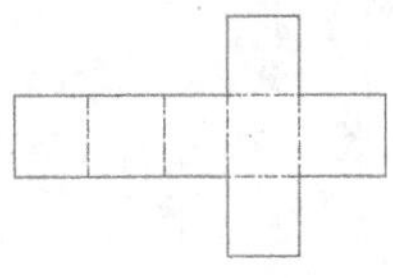

10. Draw a net of a tetrahedron.

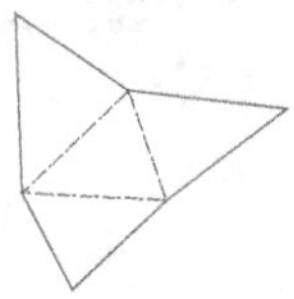

Name ______________________ Date __________ Class __________

Reteaching 120
Topology

You have worked with different types of three-dimensional objects. Now you will learn about topology.

Topology

Topology is the branch of geometry that deals with the properties of geometric figures that do not change under stress. Two polygons (such as a triangle and a square) are topologically the same because they both have boundaries that isolate an area within a two-dimensional space.

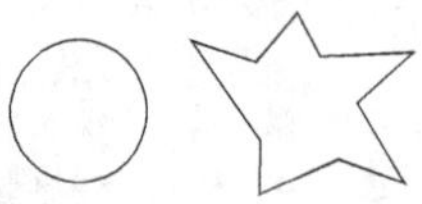

Explain why the two objects are topologically equivalent.

The circle on the left can be reshaped into a five pointed star on the right by pulling at the circle at five equally spaced points along its circumference until it forms the shape of a five pointed star.

Practice

Complete the steps to answer the following questions.

1. Which two objects are topologically equivalent?

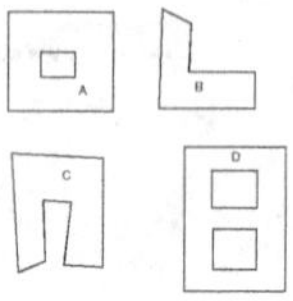

Objects B and C are topologically equivalent. They are the only two shapes that do not separate the two-dimensional space into different regions. Object A separates two-dimensional space into two regions, while Object D separates two-dimensional space into three regions.

Answer the following questions about topological equivalence.

2. Which numbers are topologically equivalent to the number one (1)? 2, 3, 5, 7
3. Explain why a cube and a pyramid are topically equivalent.

   If force is applied to a cube, it can be shaped into a pyramid.
4. Determine which two shapes are topologically equivalent. A and D

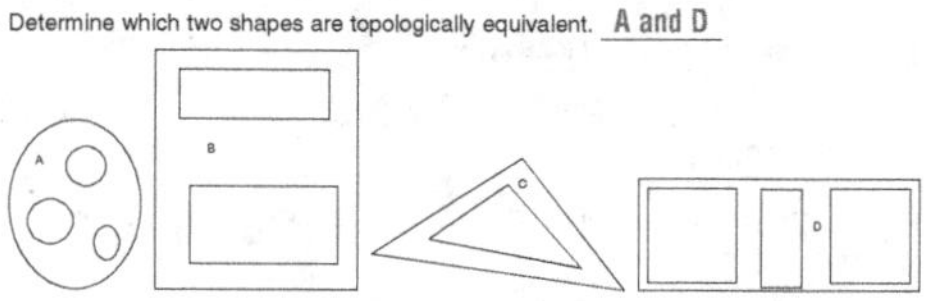

Reteaching 120
continued

Topological Classes

Objects can be grouped into topological classes based on their characteristics. For example, the numbers 1, 2, 3, 5, and 7 can be grouped into one class (no holes); the numbers 4, 6, and 9, into another class (one hole); and the number 8, into a third class (two holes).

Group the letters in the word PANORAMA according to topological classes.

Letters with no holes: N, M

Letters with one hole: P, A, A, A, O, R

Practice

5. Classify the following symbols into topological classes.

   /, >, <, @, #, $, ^, &, *, (,)

   Symbols with no holes: /, >, <, ^, *, (, and)

   Symbols with one hole: @, #

   Symbols with two holes: $, &
6. Classify the following Greek characters into topological classes,

   α, β, γ, δ, ε, ζ, η, θ, ι, κ, λ, μ, ν, ξ, ο, π, ρ, ς, σ, τ, υ, φ, χ, ψ, ω

   no holes: γ, ε, ζ, η, ι, κ, λ, μ, ν, ξ, ο, π, ς, τ, υ, χ, ψ, ω

   one hole: α, δ, ο, ρ, σ, φ

   two holes: β, θ,
7. Which of the following shapes is not part of the same topological class?

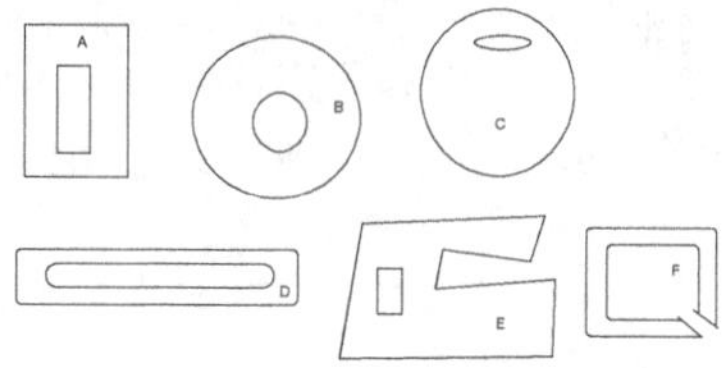

Name ______________________ Date __________ Class __________

Reteaching INV 12
Polar Coordinates

You have plotted points and functions using Cartesian coordinates. Now you will learn how to use polar coordinates and graph polar functions.

Polar Coordinates

The polar coordinate system is based on a central point called the **pole** and a ray called the **polar axis.** Points are measured in terms of r, the distance from the point to the pole, and θ, the angle a line from the pole to the point in the pole makes with the polar axis.

Example:

Plot the point P(3, 30°).

$r = 3$; $\theta = 30°$

The angle P makes with the polar axis is 30°.

Point P is 3 units from the pole.

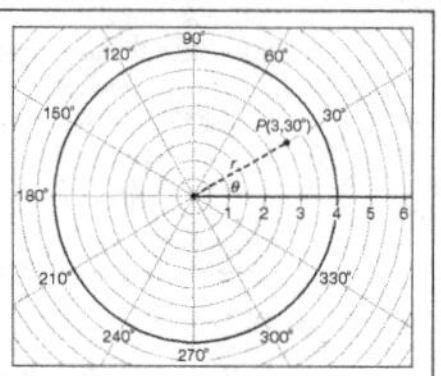

Practice

1. Plot the polar point $P(2, 45°)$.

   Because θ is 45°, the angle P makes with the polar axis is 45°.

   Because r is 2, P is 2 units from the pole.

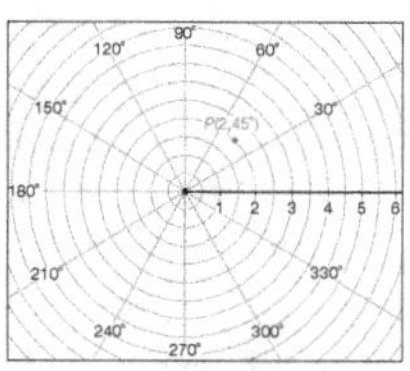

2. Plot the point $P(-2.5, 220°)$.

   If r is negative, plot the point in the direction that is 180° opposite of the angle.

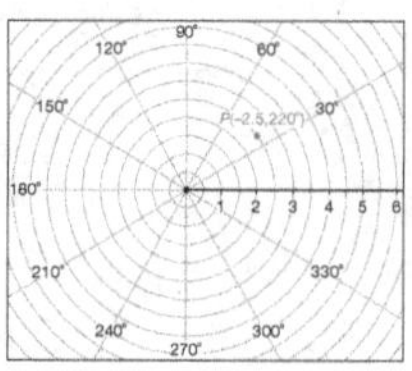

Reteaching INV 12
continued

Graphing Polar Functions

When r is a function of θ, $r = f(\theta)$. This is called a **polar function.** Polar functions can be graphed in polar coordinates. Find the value of r for several values of θ, make a table with the values, plot the points, and draw a curve through them.

Example:

To plot $r = 2\cos\theta$, find the value of r for multiples of 30°, make a table, and plot the points in the polar plane.

θ	r	θ	r
0°	2	180°	−2
30°	$\sqrt{3}$	210°	$-\sqrt{3}$
60°	1	240°	−1
90°	0	270°	0
120°	−1	300°	1
120°	$-\sqrt{3}$	330°	$\sqrt{3}$

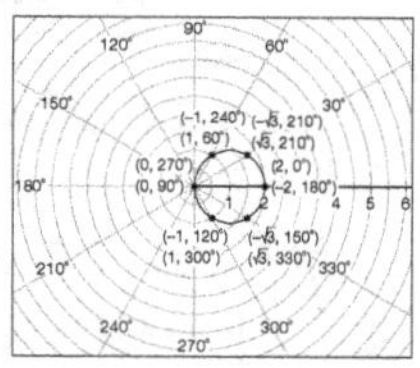

Notice that the points repeat past 180°. This is common with polar functions. When it happens you can usually assume that the graph is finished.

Practice

Graph the polar function.

3. $r = \sin 2\theta$

   Step One: Fill in the values of r in the table:

   Step Two: Plot the points on the graph.

   Step Three: Draw a curve through the points.

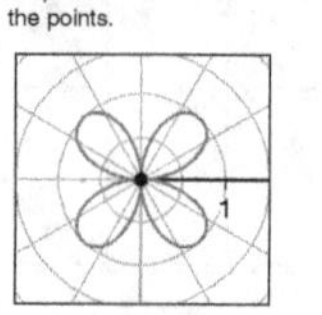

θ	r	θ	r
0°	0	180°	0
30°	$\frac{\sqrt{3}}{2}$	210°	$\frac{\sqrt{3}}{2}$
45°	1	225°	1
60°	$\frac{\sqrt{3}}{2}$	240°	$\frac{\sqrt{3}}{2}$
90°	0	270°	0
120°	$-\frac{\sqrt{3}}{2}$	300°	$-\frac{\sqrt{3}}{2}$
135°	−1	315°	□1
150°	$-\frac{\sqrt{3}}{2}$	330°	$-\frac{\sqrt{3}}{2}$

